Molecular Virology and Control of Flaviviruses

Edited by

Pei-Yong Shi

Novartis Institute for Tropical Diseases
Chromos
Singapore

Caister Academic Press

Caister Academic Press
Norfolk, UK

www.caister.com

British Library Cataloguing-in-Publication Data
A catalogue record for this book is available from the British Library

ISBN: 978-1-904455-92-9

Description or mention of instrumentation, software, or other products in this book does not imply endorsement by the author or publisher. The author and publisher do not assume responsibility for the validity of any products or procedures mentioned or described in this book or for the consequences of their use.

Cover design adapted from Figure 2.1

Printed and bound in Great Britain

Contents

Contributors

Carol D. Blair
Arthropod-borne and Infectious Diseases Laboratory
Department of Microbiology, Immunology and Pathology
Colorado State University
Fort Collins, CO
USA

cblair@colostate.edu

Yen-Liang Chen
Novartis Institute for Tropical Diseases
Chromos
Singapore

yen_liang.chen@novartis.com

Michael S. Diamond
Departments of Medicine, Molecular Microbiology, Pathology & Immunology and the Midwest Regional Center for Excellence in Biodefense and Emerging Infectious Diseseas Research
Washington University School of Medicine
St. Louis, MO
USA

diamond@borcim.wustl.edu

Gregory D. Ebel
Department of Pathology
University of New Mexico School of Medicine
Albuquerque, NM
USA

gebel@salud.unm.edu

Anneke Funk
Australian Infectious Disease Research Centre
School of Chemistry and Molecular Biosciences
The University of Queensland
Brisbane, QLD
Australia

anneke.funk@embo.org

Adolfo García-Sastre
Department of Microbiology
Mount Sinai School of Medicine
New York, NY
USA

adolfo.garcia-sastre@mssm.edu

Duane J. Gubler
Director
Signature Research Program in Emerging Infectious Diseases
Duke – NUS Graduate Medical School
Singapore

duane.gubler@duke-nus.edu.sg

Scott B. Halstead
Supportive Research and Development Program
Pediatric Dengue Vaccine Initiative
International Vaccine Institute
Seoul
Korea

halsteads@erols.com

Martin L. Hibberd
The Genome Institute of Singapore
National University of Singapore
Singapore;
Imperial College Medical School
London
UK

hibberdml@gis.a-star.edu.sg

Elizabeth Hunsperger
Serology Diagnostics and Viral Pathogenesis Laboratory Dengue Branch
Centers for Disease Control and Prevention
Division of Vector-borne Infectious Diseases
San Juan
Puerto Rico

enh4@cdc.gov

Alexander A. Khromykh
Australian Infectious Disease Research Centre
School of Chemistry and Molecular Biosciences
The University of Queensland
Brisbane, QLD
Australia

a.khromykh@uq.edu.au

Laura D. Kramer
Arbovirus Laboratory
Wadsworth Center
New York State Department of Health
Albany, NY
USA;
School of Public Health
State University of New York at Albany
New York, NY
USA

kramer@wadsworth.org

Richard J. Kuhn
Department of Biological Sciences
Markey Center for Structural Biology
Purdue University
West Lafayette, IN
USA

kuhnr@purdue.edu

Maudry Laurent-Rolle
Department of Microbiology
Mount Sinai School of Medicine
New York, NY
USA

maudry.laurent-rolle@mssm.edu

F. Joel Leong
Novartis Institute for Tropical Diseases
Chromos
Singapore

joel.leong@novartis.com

Julien Lescar
School of Biological Sciences
Nanyang Technological University
Singapore;
Architecture et Fonction des Macromolecules Biologiques
CNRS UMR 6098
Marseille
France

julien@ntu.edu.sg

Siew Pheng Lim
Novartis Institute for Tropical Diseases
Chromos
Singapore

siew_pheng.lim@novartis.com

Dahai Luo
School of Biological Sciences
Nanyang Technological University
Singapore;
Department of Molecular Biophysics and Biochemistry
Yale University
New Haven, CT
USA

dahai.luo@yale.edu

Juliet Morrison
Department of Microbiology
Mount Sinai School of Medicine
New York, NY
USA

juliet.morrison@mssm.edu

David A. Muller
Australian Infectious Diseases Research Centre
School of Chemistry and Molecular Biosciences
The University of Queensland
Brisbane, QLD
Australia

david.muller@uqconnect.edu.au

Lee-Ching Ng
Environmental Health Institute
The National Environment Agency
Singapore

NG_Lee_Ching@nea.gov.sg

Ken E. Olson
Arthropod-borne and Infectious Diseases Laboratory
Department of Microbiology, Immunology and Pathology
Colorado State University
Fort Collins, CO
USA

kolson@colostate.edu

Theodore C. Pierson
Viral Pathogenesis Section
Laboratory of Viral Diseases
National Institutes of Health
Bethesda, MD
USA

piersontc@mail.nih.gov

Justin A. Roby
Australian Infectious Disease Research Centre
School of Chemistry and Molecular Biosciences
The University of Queensland
Brisbane, QLD
Australia

justin.roby@uqconnect.edu.au

John T. Roehrig
Division of Vector-Borne Infectious Diseases
Centers for Disease Control and Prevention
Public Health Service
U.S. Department of Health and Human Services
Fort Collins, CO
USA

jtr1@cdc.gov

Mark Schreiber
Novartis Institute for Tropical Diseases
Chromos
Singapore

mark.schreiber@novartis.com

Pei-Yong Shi
Novartis Institute for Tropical Diseases
Chromos
Singapore

pei_yong.shi@novartis.com

Indra Vythilingam
Environmental Health Institute
The National Environment Agency
Singapore

indra_vythilingham@nea.gov.sg

Qing-Yin Wang
Novartis Institute for Tropical Diseases
Chromos
Singapore

qing_yin.wang@novartis.com

Paul R. Young
Australian Infectious Diseases Research Centre
School of Chemistry and Molecular Biosciences
The University of Queensland
Brisbane, QLD
Australia

p.young@uq.edu.au

Other Books of Interest

Preface

Flaviviruses are pathogens of global importance, many of which are arthropod-borne and cause hundreds of millions of human infections each year. Breakthroughs have recently been made in the understanding of the molecular details of viral replication, transmission, host responses, and pathogenesis. Such progress in basic research has provided new opportunities for development of novel vaccines and antiviral therapy.

The aim of this book was to assemble an up-to-date anthology from the leading experts in the flavivirus field. The book is divided into two major sections: Molecular Virology and Virus Prevention. I hope that this book could serve as a reference for experienced flavivirus researchers as well as open the door to new talents who are interested in joining the flavivirus research.

As an editor, I am deeply indebted to the colleagues who dedicated their research to flavivirus. I am grateful to the authors who contributed and made this book possible. I also thank Annette Griffin, Acquisition Editor at Caister Academic Press, and Emma Needs, Production Editor at Prepress Projects Ltd.

Pei-Yong Shi
Singapore

Flaviviruses: Past, Present and Future

1

Duane J. Gubler

Abstract

The flaviviruses (genus *Flavivirus*) are among the most important pathogens infecting humans and domestic animals, causing hundreds of millions of infections annually. They have a global distribution and cause a broad spectrum of illness ranging from mild viral syndrome to severe and fatal haemorrhagic and neurological disease. The genus is made up of a diverse group of 53 viral species that have evolved into three distinct groups with very different transmission cycles. The vector-borne group is transmitted among vertebrate hosts by haematophagous arthropods (mosquitoes and ticks), the no-known vector group is transmitted directly among vertebrate animals and the arthropod group is transmitted directly among arthropods. This chapter reviews the history, the present status and future trends of flaviviruses, using some of the more important species as case studies.

Introduction

The flaviviruses are among the most important human and animal pathogens causing significant morbidity and mortality worldwide (Gubler *et al.*, 2007). The taxonomy of the genus *Flavivirus* has been somewhat confused since the ICTV revised the classification in 2005 (ICTV, 2005). Previous classifications listed over 70 members of the genus based on antigenic relationships (Calisher, 1988; Calisher *et al.*, 1989; Calisher and Gould, 2003), but this resulted in classifying subtypes and serotypes as unique species. The new ICTV classification lists 53 species of flaviviruses, 40 of which can cause human infection (Table 1.1). The type species of the genus *Flavivirus*, and the family *Flaviviridae*, is yellow fever virus, which was the first filterable agent shown to cause human disease, the first virus shown to be transmitted by an arthropod vector (1900) and the first virus isolated (1927) (Gubler *et al.*, 2007).

There are three distinct groups of flaviviruses, the vector-borne (mosquito-borne and tick-borne) viruses, those viruses with no-known arthropod vector, and those that have only been isolated from mosquitoes (Table 1.1). The mosquito-borne group is the largest and most important medically with 27 species, the tick-borne group has 12 species, and 14 species make up the group with no known vector. Viruses in all these groups can cause human and animal disease, and all are zoonotic viruses.

There are two groups of mosquito-borne flaviviruses that have evolved to have unique vertebrate hosts, mosquito vectors, clinical presentation and epidemiology (Gubler *et al.*, 2007). The oldest group evolutionarily has lower primates as the principal vertebrate hosts, causes viscerotropic disease in humans and is transmitted in a sylvatic cycle by canopy-dwelling *Aedes* species mosquitoes. This group includes yellow fever (YF) and dengue (DEN) viruses, both of which have the ability to establish urban transmission cycles with domesticated *Aedes* species mosquitoes as the vectors. These two viruses are among the most important emerging viruses causing human disease. The second group of mosquito-borne flaviviruses have birds as their principal vertebrate host, cause neurological disease in humans and have bird-feeding *Culex* species mosquitoes as the principal vectors. This group includes Japanese

Table 1.1 Flaviviruses listed by mode of transmission

Mosquito-borne viruses	Tick-borne viruses	No known vector viruses
Aroa virus	**Mammalian tick-borne virus group**	**Entebbe bat virus group**
Bussuquara virus	*Gadgets Gully virus*	*Entebbe bat virus*
Iguape virus	*Kyasanur Forest disease virus*	Sokuluk virus
Naranjal virus	*Langat virus*	*Yokose virus*
Dengue virus	*Louping ill virus*	
Dengue virus 1	British subtype	**Modoc virus group**
Dengue virus 2	Irish subtype	*Apoi virus*
Dengue virus 3	Spanish subtype	*Cowbone Ridge virus*
Dengue virus 4	Turkish subtype	*Jutiapa virus*
Kedougou virus	*Omsk haemorrhagic fever virus*	*Modoc virus*
	Powassan virus	*Sal Vieja virus*
Japanese encephalitis virus serogroup	*Royal Farm virus*	*San Perlita virus*
Caci paocre virus	Karshi virus	
Japanese encephalitis virus	*Tickborne encephalitis virus*	**Rio Bravo virus group**
Koutango virus	European subtype	*Bukalasa bat virus*
Murray Valley encephalitis virus	Far Eastern subtype	*Carey Island virus*
Alfuy virus	Siberian subtype	*Dakar bat virus*
St. Louis encephalitis virus		*Montana myotis leukoencephalitis virus*
Usutu virus	**Seabird tick-borne virus group**	*Phnom Penh bat virus*
West Nile virus	*Kadam virus*	Batu cave virus
Kunjin virus	*Meaban virus*	*Rio Bravo virus*
Yaounde virus	*Saumarez Reef virus*	
Kokobera virus	*Tyuleniy virus*	**Tentative insect viruses in the genus**
Stratford virus		Cell fusing agent virus
		Tamana bat virus
Ntaya virus group		Kamiti River Virus
Bagaza virus		
Ilheus virus		
Rocio virus		
Israel turkey meningoencephalitis virus		
Ntaya virus		
Tembusu virus		
Zika virus		
Spondweni virus		
Yellow fever virus group		
Banzi virus		
Bouboui virus		
Edge Hill virus		
Jugra virus		
Saboya virus		
Potiskum virus		
Sepik virus		
Uganda S virus		
Wesselsbron virus		
Yellow fever virus		

Table adapted from Gubler *et al.* (2007).

encephalitis (JE), West Nile (WN), Murray Valley encephalitis (MVE) and St Louis encephalitis (SLE) viruses, all of which are a very closely related antigenically, being grouped in the JE serogroup of flaviviruses (Calisher, 1988); each has a unique geographic distribution (Gubler *et al.*, 2007).

There are two main groups of tick-borne flaviviruses, those with mammalian vertebrate hosts and those with sea birds as vertebrate hosts (Table 1.1). Most of the tick-borne viruses that infect humans cause neurological disease, but some also cause haemorrhagic disease.

The vector-borne flaviviruses are maintained in complex cycles in nature involving one or more vertebrate hosts and blood-sucking arthropods (Fig. 1.1). Most of these viruses have a primary vertebrate host and a primary mosquito or tick vector. Some have secondary transmission cycles involving other vertebrate and vector species. It is usually from these secondary cycles that humans are infected by 'bridge vectors' that readily feed on the feral vertebrate hosts as well as on humans and domestic animals.

For transmission of mosquito-borne flaviviruses to occur, the vertebrate hosts must have a viraemia as defined by an ID_{50} of at least 10^4 per ml in order to infect the arthropod taking a blood meal. After ingestion of the virus, there is an extrinsic incubation period of 8–14 days, depending on environmental conditions, the strain of virus and strain and species of mosquito or tick. During this period, the virus replicates in the arthropod tissues and disseminates to infect the salivary glands. Transmission can occur to another vertebrate host only after the salivary glands are infected; the virus is injected into the host with salivary fluid, while the mosquito or tick takes a blood meal. The no-known vector flaviviruses are transmitted directly among vertebrate hosts, usually in saliva, urine or faeces (Kuno and Chang, 2005).

Many of the vector-borne flaviviruses can also be transmitted vertically from infected female arthropod to her offspring through infected eggs. Mosquitoes or ticks infected in this way transmit the virus trans-stadially from egg to immature stages to adults. Male mosquitoes infected vertically, are capable of transmitting some flaviviruses venereally to the female mosquitoes via infected seminal fluid. Female arthropods infected vertically or venereally are capable of transmitting the viruses to a new vertebrate host.

Past

Historically, flaviviruses have been major public health problems for humans. Epidemics of what were later to be recognized as yellow fever and dengue fever occurred regularly in the 17th, 18th, 19th and early 20th centuries (Monath, 1991; Gubler, 2004). Dengue fever epidemics occurred on most continents whereas yellow fever was limited to Africa, Europe and the Americas. Both dengue and yellow fever were transmitted by an

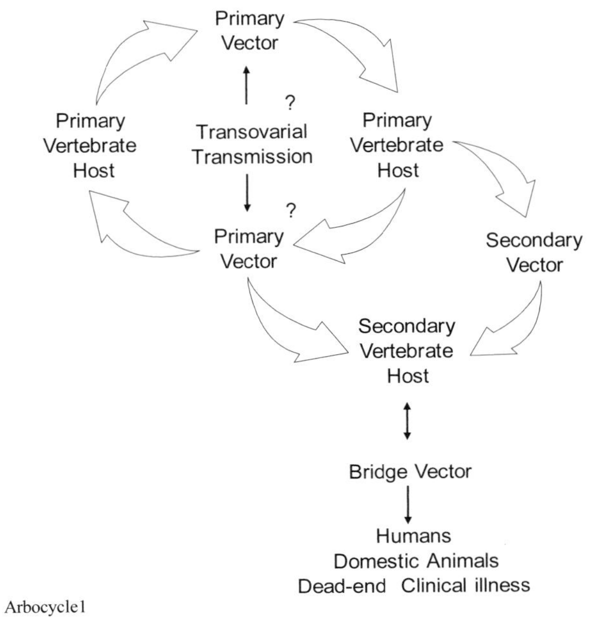

Figure 1.1 Hypothetical transmission of cycle of flaviviruses.

African mosquito, *Aedes aegypti*, which was spread to the Americas via the slave trade and from there to the Asia-Pacific region as global commerce developed in the 18th and 19th centuries. This mosquito species became highly domesticated and when introduced into a new geographic region, it first infested port cities, then moved to inland cities (Gubler, 1997). It was a highly efficient vector for both viruses and transmitted major urban epidemics.

Aedes aegypti used the water barrels on sailing ships as breeding sites and maintained a transmission cycle among the crew members and passengers of ships during the long voyages between ports. The viruses usually went ashore with the crew and passengers when they reached port. Yellow fever epidemics were limited to the Americas and Europe because the trade routes were primarily between the New World, West Africa and Europe. Transmission of this virus is not known to have occurred in Asia. If introduced to Asia, it was probably a rare event because the Panama Canal had not been constructed and trade with the Asia-pacific region was limited. Although there was trade between India and East Africa, yellow fever transmission in that part of Africa was not common. Dengue fever epidemics, on the other hand, were common in Asian cities during that time and increased in frequency as *Aedes aegypti* distribution increased (Gubler, 1997, 2004). Prior to the arrival of *Ae. aegypti* in Asia, outbreaks of dengue were small and infrequent, being transmitted by a related Asian mosquito species, *Aedes albopictus*. In 1779 and 1780, however, major epidemics of a disease compatible with dengue fever occurred on three continents almost simultaneously. This epidemiology supports the conclusion that the dengue viruses evolved in Asia, most probably from a progenitor originating in Africa (Gubler, 1997).

Both yellow fever and dengue fever were effectively controlled in the 1950s and 1960s in the Americas by the Pan American Health Organization regional *Ae. aegypti* eradication programme, and yellow fever was controlled at the same time in Francophone West Africa using a highly effective vaccine (Gubler, 1989, 2004; Monath, 1991). In Asia, dengue fever was controlled in many countries as a side benefit of the global malaria eradication programme and the use of residual insecticides, mainly DDT. Unfortunately, both the *Ae. aegypti* and malaria eradication programmes were terminated in the late 1960s and early 1970s, and both malaria and dengue fever re-emerged in the 1970s, causing major epidemics (Gubler, 1998).

Although most attention was given to the explosive epidemic diseases such as dengue and yellow fever in the early years of arboviral research, other flaviviruses were also being discovered. Disease agents associated with neurological disease were later shown to be flaviviruses, namely louping ill in Scotland (1807), Japanese encephalitis in Japan (1873), Australian X disease, later named Murray Valley encephalitis in Australia (1916), St Louis encephalitis in the USA (1932), West Nile fever in Africa (1937) and tick-borne encephalitis in Europe (1931) (Gubler *et al.*, 2007). These viruses were initially thought to be unrelated, but in the 1940s serological relationships were demonstrated by the newly developed neutralization, complement fixation and haemagglutination inhibition tests (Casals, 1957). By the 1950s, the group B arboviruses (flaviviruses) were separated antigenically from the group A arboviruses (alphaviruses), which had been linked based their transmission cycles and physicochemical characteristics, although they were still classified in the same family (*Togaviviae*) (Casals, 1957). During the 1960s and 1970s, many new arboviruses were discovered, and by the early 1980s, the flaviviruses and alphaviruses were considered to be sufficiently different to place them in separate families (ICTV, 2005).

Present

In the 1960s, the war on infectious diseases was declared won (Henderson, 1993). This declaration initiated a 30 year period of apathy and complacency, during which time effective prevention and control programmes were terminated, resources were redirected and public health infrastructure to deal with vector-borne and zoonotic diseases deteriorated in most countries of the world. As a result, field studies and laboratory research on arboviruses were grossly under-funded and career opportunities in the field disappeared (Gubler, 2001).

Coincident with this period of complacency, however, were unrecognized global trends that would ultimately drive a major re-emergence of epidemic infectious diseases in general, and vector-borne flaviviruses in particular. Unprecedented population growth in developing countries was, and still is a principal driver of environmental changes, including urbanization, and agricultural, animal husbandry and land use changes, all of which contributed to increased transmission of arboviral diseases, especially those caused by flaviviruses (Gubler, 1998, 2002). The dramatic increase in the movement of humans, animals and commodities via modern transportation, and the lack of public health infrastructure has allowed these pathogens to expand their geographic ranges and cause major epidemics.

The best example of this increased incidence and geographic expansion of flaviviruses is dengue/dengue haemorrhagic fever. The average annual number of cases reported to WHO has increased nearly 10-fold in the 40 years since 1970 (Fig. 1.2). Moreover, the viruses have spread throughout the tropical world during the intervening years (Fig. 1.3). Japanese encephalitis has also expanded its geographic distribution, moving west into Nepal and western India, east into the Pacific and south into Papua New Guinea and northern Australia (Solomon *et al.,* 2003). West Nile virus, normally limited to the Old World, jumped the Atlantic in 1999, causing an ongoing epidemic throughout the American region (Artsob *et al.,* 2009). Zika virus, not known to cause human epidemics, moved into the Western Pacific Islands, causing a large epidemic on the island of Yap (Duffy *et al.,* 2009). And yellow fever is waiting in the wings, with increasingly frequent cases of urban disease being reported in the American tropics (Gubler, 2004). These are just the most obvious examples of the re-emergence of flaviviral diseases.

At present, flaviviruses are among the most important emergent tropical diseases, causing tens of millions of cases and thousands of deaths annually. Unfortunately, little is being done to prevent or control their transmission and spread. Three flaviviruses, yellow fever, Japanese encephalitides and tick-borne encephalitis have licensed vaccines, the use of which is variable depending on the country (Gubler *et al.,* 2007). The yellow fever 17-D live virus vaccine is safe, long-lasting and inexpensive (Monath, 1999); yet it is not effectively used for prevention in many endemic countries, which still follow the failed public health policy of using the vaccine for emergency response after an epidemic has been reported

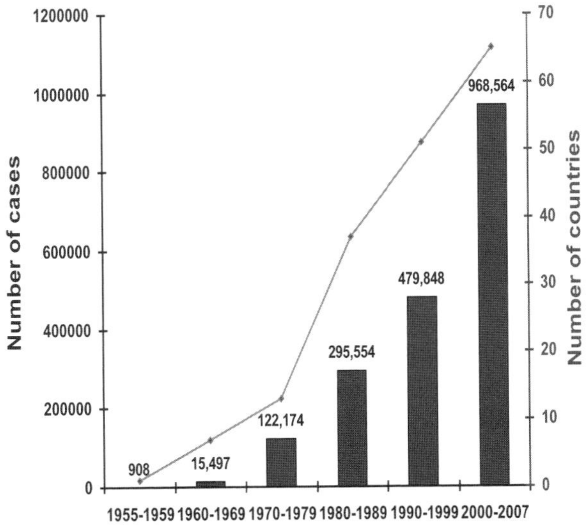

Figure 1.2 The global emergence of dengue/dengue haemorrhagic fever (DF/DHF). Average annual number of DF/DHF cases reported to World Health Organization (WHO), 1955–2007.

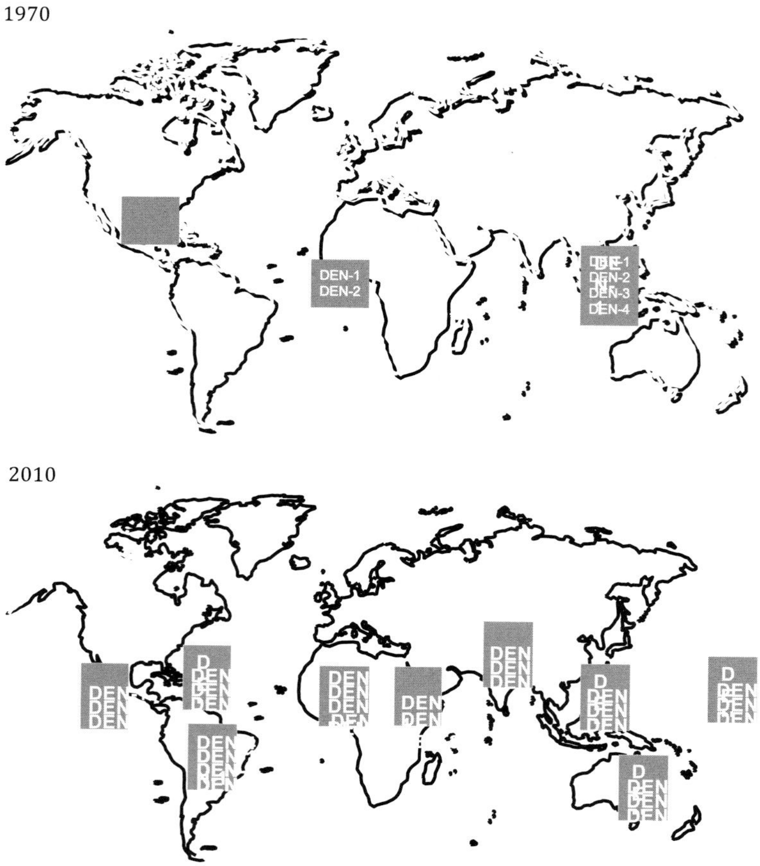

Figure 1.3 The global distribution of dengue virus serotypes, 1970 and 2010.

(Gubler, 2001). The Japanese encephalitis vaccines are used effectively for prevention in some Asian countries, but not in others, and the tick-borne encephalitis vaccine is used effectively only in Austria. There are no vaccines or antiviral drugs for the rest of the flaviviruses although there is promise for dengue (see Chapters 10 and 12). Currently, prevention and control depends on vector control, which has uniformly been a failure in preventing epidemic disease.

Challenges for the future

The near future does not look very promising for our efforts to reverse this trend of increasing flavirval disease. The global trends of population growth, urbanization, environmental change and globalization which are driving the global emergence, are projected to continue for the next 20 or more years. Unfortunately, the prospects for vaccines, antiviral drugs, and more effective vector control tools, while promising, will not be available in the near term. Thus, we can expect more of the same for the indefinite future.

To effectively reverse this trend of epidemic flaviviral diseases, major changes in public health emphasis will be required. These include:

1 The spread of viruses and vectors via modern transportation must be prevented. Unfortunately, this problem is being ignored by international health agencies because it has important political and economic implications.
2 Laboratory-based disease detection systems must be improved in those countries affected

by these diseases. This will require both laboratory and epidemiological capacity building in many resource poor countries.

3 Better international cooperation and data sharing among countries are desperately needed. Ultimately regional surveillance, prevention and control programmes must be developed.
4 The public health infrastructure in resource poor tropical countries must be developed, e.g. more and better laboratories and more and better trained personnel.
5 More and better tools to diagnose, treat, and prevent these diseases, including vaccines, drugs, diagnostic tests, and insecticides, must be developed.
6 Finally, political will and economic support are needed to develop and implement the above measures.

Given the current emphasis of international funding agencies, none of this is likely to occur in the near future. Hundreds of millions of dollars are being spent on emerging infectious disease research each year, but there is little or no co-ordination among the funding agencies, and no attempt to use the results of this extensive research to improve the public health in those countries ravaged by epidemic flaviviral diseases.

References

Artsob, H., Gubler, D.J., Enria, D.A., Morales, M.A., Pupo, M., Bunning, M.L., and Dudley, J.P. (2009). West Nile Virus in the New World: Trends in the spread and proliferation of West Nile Virus in the western hemisphere. Zoonoses Pub. Health *56*, 357–369.

Calisher, C.H. (1988). Antigenic classification and taxonomy of flaviviruses (family *Flaviviridae*) emphasizing a universal system for the taxonomy of viruses causing tick-borne encephalitis. Acta Virol. *32*, 469–478.

Calisher, C.H., and Gould, E.A. (2003). Taxonomy of the virus family *Flaviviridae*. Adv. Virus Res. *59*, 1–19.

Calisher, C.H., Karabatsos, N., Dalrymple, J.M., Shope, R.E., Porterfield, J.S., Westaway, E.G. and Brandt, W.E. (1989). Antigenic relationships between flaviviruses as determined by cross-neutralization tests with polyclonal antisera. J. Gen. Virol. *70*, 37–43.

Casals, J. (1957). Viruses: the versatile parasites; the arthropod-borne group of animal viruses. Trans. N.Y. Acad. Sci. *19*, 219–235.

Duffy, M.R., Chen, T.H., Hancock, W.T., Powers, A.M., Kool, J.L., Lanciotti, R.S., Pretrick, M., Marfel, M., Holzbauer, S., Dubray, C., *et al.* (2009). Zika virus outbreak on Yap Island, Federated States of Micronesia. N. Engl. J. Med. *360*, 2536–2543.

Gubler, D.J. (1989). *Aedes aegypti* and *Aedes aegypti*-borne disease control in the 1990s: top down or bottom up. Charles Franklin Craig Lecture. Am. J. Trop. Med. Hyg. *40*, 571–578.

Gubler D.J. (1997). Dengue and dengue hemorrhagic fever: its history and resurgence as a global public health problem. In Dengue and Dengue Hemorrhagic Fever, Gubler, D.J., and Juno, G., eds. (CAB International, Wallingford, UK), pp. 1–22.

Gubler, D.J. (1998). Resurgent vector-borne diseases as a global health problem. Emerg. Infect. Dis. *4*, 442–450.

Gubler, D.J. (2001). Prevention and control of tropical diseases in the 21st century: back to the field. Am. J. Trop. Med. Hyg. *65*, v–xi.

Gubler, D.J. (2002). The global emergence/resurgence of arboviral diseases as public health problems. Arch. Med. Res. *33*, 330–342.

Gubler, D.J. (2004). The changing epidemiology of yellow fever and dengue, 1900 to 2003: full circle? Comp. Immunol. Microbiol. Infect. Dis. *27*, 319–330.

Gubler D.J., Kuno, G., and Markoff, L. (2007). Flaviviruses. In Fields Virology, Knipe, D.M., Howley, P.M., Griffin, D., Lamb, R.A., Martin, M.A., Roizman, B., and Straus, S.E., eds. (Lippincott, Williams and Wilkins, UK), pp. 1153–1252.

Henderson, D.A. (1993). Surveillance systems and intergovernmental cooperation. In Emerging Viruses, Morse, S.S., ed. (Oxford University Press, New York), pp. 283–295.

ICTV. (2005). Virus taxonomy: classification and nomenclature of viruses. In ICTV, Fauquet, C.M., Mayo, M.A., Maniloff, J., *et al.*, eds. (Elsevier Academic Press, San Diego), pp. 982–988.

Kuno, G., and Chang, G.J. (2005). Biological transmission of arboviruses: reexamination of and new insights into components, mechanisms, and unique traits as well as their evolutionary trends. Clin. Microbiol. Rev. *18*, 608–637.

Monath, T.P. (1991). Yellow fever: Victor, Victoria? Conqueror, conquest? Epidemics and research in the last forty years and prospects for the future. Am. J. Trop. Med. Hyg. *45*, 1–43.

Monath T.P. (1999). Yellow fever vaccine. In Vaccines, Plotkin, S.A., and Orentstein, W.A., eds (WB Saunders, Philadelphia), pp. 815–879.

Solomon, T., Ni, H., Beasley, D.W., Ekkelenkamp, M., Cardosa, M.J., and Barrett, A.D. (2003). Origin and evolution of Japanese encephalitis virus in southeast Asia. J. Virol. *77*, 3091–3098.

Flavivirus Virion Structure

2

Richard J. Kuhn

Abstract

Flavivirus virions form in the endoplasmic reticulum (ER) with the recruitment of genome RNA, capsid protein, and the envelope (E) and precursor to the membrane proteins (prM). The nascent particles acquire a lipid bilayer as they bud into the ER lumen in an immature form. Glycosylation and subsequent processing of the particles occur as they proceed through the cellular secretory system. In the low pH that is encountered in the trans-Golgi network, cellular furin activates the particles by cleavage of prM into M. The particles are released from the cell in a mature and infectious form. The observations demonstrate the significant conformational and translational movements of the viral structural proteins during the virus life cycle and suggest the particles have substantial dynamic capabilities. These properties have been substantiated by analyses of antibody binding to virions and suggest novel targets for future therapeutic intervention strategies.

Introduction

Flaviviruses are a highly successful genus in the *Flaviviridae* family as demonstrated by their worldwide distribution and number of species. They are responsible for significant human disease although several effective vaccines have been developed, such as the attenuated 17D vaccine against yellow fever virus (YFV) and the inactivated tick borne-encephalitis virus (TBEV) vaccine (Lindenbach *et al.*, 2007). Despite vaccines against a select number of flaviviruses, there are no approved therapeutics against any of the flaviviruses. Recent years has seen an explosion of research into flavivirus biology, especially using structural biology as a tool. This chapter will examine the details of flavivirus particle formation, the transit of the particle through and exiting the cell, and the structure of the infectious mature virus that is released from the infected cell. The dynamic nature of the assembly and maturation process lends itself to identifying new targets for therapeutic intervention and will be highlighted in this chapter.

The RNA viral genome spans approximately 11 kb and encodes a polyprotein that is processed into individual proteins by host and virus-encoded proteases (Lindenbach *et al.*, 2007). The polyprotein is divided into an N-terminal structural protein region and a C-terminal non-structural or replication region. The structural region is so named as it contains the viral proteins that are found within the virus particle, and consists of the capsid or C protein, the precursor to the membrane protein or prM, and the envelope E protein. Structures of each of these proteins for dengue virus (DENV) have been determined and are described below.

The entry of the virus into a cell results in the release of the genome RNA and its translation into the viral polyprotein. The polyprotein is cleaved by proteases that provide the components of virus RNA replication and particle assembly. Assembly of the virus occurs in close proximity to RNA replication, and not surprisingly the two steps are coupled. The co-localization of the genome RNA, capsid protein, and the envelope proteins prM and E on the two opposite sides of the ER membrane permit formation of an immature particle. The particle will undergo processing and maturation resulting eventually in a mature virus particle that is released from the cell. The details of these events will be discussed below.

Components of the virion

A substantial effort has been expended in describing the structure of the flavivirus particle and the proteins that are used to build it. Early electron micrographs showed that mature flavivirus particles had a diameter of approximately 500 Å with an electron-dense core surrounded by a lipid bilayer (Strauss and Strauss, 1988). Extensive genetic mapping experiments, using monoclonal antibodies as probes, were performed in the late 1980s and early 1990s to deduce the antigenic structure of DENV and TBEV (Mandl *et al.*, 1989; Roehrig *et al.*, 1990, 1998a). The E protein accounted for the majority of the antigenic sites, and was predicted, based on amino acid sequence, to be the dominant surface protein found in the mature virus.

Rey and colleagues described the first atomic structure of a flavivirus protein in 1995 (Rey *et al.*, 1995a). The E protein structure of TBEV was deduced by X-ray crystallography using protein isolated from virions. Treatment of virions with trypsin released dimers of the E protein ectodomain consisting of residues 1–395 and lacking the 'stem-anchor' region (Wengler *et al.*, 1987; Heinz *et al.*, 1991). This dimeric E protein structure confirmed previous suggestions about the antigenic arrangement and provided evidence that the E protein is oriented with its long axis parallel to the viral lipid membrane. The structure of the E protein ectodomain comprises three domains: the N-terminal central domain I, the elongated dimerization domain (II), and the Ig-like domain III. The fusion peptide is located at the distal end of domain II, partially protected by the neighbouring subunit of the dimer. Structures for E protein have now been determined for several flaviviruses, including DENV types 1, 2, and 3, and WNV, and all display similar secondary structures and domain organization (Modis *et al.*, 2003, 2005; Zhang *et al.*, 2004b; Kanai *et al.*, 2006; Nybakken *et al.*, 2006). The organization of this class II fusion protein is strikingly similar to the E1 fusion protein of alphaviruses (Lescar *et al.*, 2001). However, unlike the alphaviruses, the flavivirus E protein serves as both fusion machine and receptor attachment protein for the virus. Although cellular receptors have not been widely identified for the flaviviruses, many studies have suggested that domain III serves as the receptor attachment domain. Domain III is also the target for many neutralizing monoclonal antibodies (Beasley and Barrett, 2002; Sukupolvi-Petty *et al.*, 2007).

The immature virus is converted into an infectious form by the cleavage and release of the prM protein (Guirakhoo *et al.*, 1991, 1993; Heinz *et al.*, 1994b; Elshuber *et al.*, 2003). Prior to cleavage, prM is associated with the E protein with the pr moiety capping or protecting the fusion peptide at the end of domain II of E protein. The structure of the prM–E heterodimer was solved by X-ray crystallography following the successful expression and secretion of a fused version of prM–E in *Drosophila* S2 cells (Li *et al.*, 2008). Although the density for the M residues was not visible, the pr component was shown to consist of seven β-strands that are mostly antiparallel. The arrangement of the three E protein domains is similar to the protein in its pre-fusion dimer configuration. The structure of the M protein is not known, although the polypeptide was intact in the crystal structure suggesting that it may occupy multiple positions and configurations as it parallels the length of the E protein prior to its two transmembrane domains.

The structures of the dengue and Kunjin capsid proteins have been solved to atomic resolution using NMR and X-ray crystallography, respectively (Dokland *et al.*, 2004; Ma *et al.*, 2004). In both cases the proteins lacked the C-terminal residues that constitute the signal sequence for the prM protein. The protein is a monomer in solution and is comprised of four α-helices (Jones *et al.*, 2003). Helices 1 and 2 form a hydrophobic cleft whose functional significance remains elusive, although the DENV capsid protein has been shown to bind to membranes (Markoff *et al.*, 1997). Residues in helix 2 have also been implicated in the capsid protein binding to lipid droplets that has been suggested to form a scaffold for genome encapsidation (Samsa *et al.*, 2009). Helix 4 is a long extended amphipathic helix that serves to stabilize the dimer and is important for capsid function. Attempts to assemble capsid cores *in vitro* using purified capsid protein and nucleic acid have been unsuccessful. The structure of the protein does not suggest an obvious oligomeric arrangement beyond the capsid dimer, and the physical features of the protein suggest that it may function like a cellular histone protein, acting to compact and neutralize the charge of the genome RNA.

Assembly at the ER membrane

Flaviviruses, like many viruses that replicate in the cytoplasm of infected cells, dramatically modify the intracellular environment to facilitate replication. Following infection, there is an increase in lipid biosynthesis and the formation of intracellular structures known as convoluted membranes/paracrystalline arrays (CM/PC) and vesicle packets (VPs) (Mackenzie *et al.*, 1998, 1999). The VPs appear to be sites of RNA synthesis as double strand RNA is associated with them. Tomographic analyses have shown that VPs are in close proximity to the ER and sites of virion formation (Welsch *et al.*, 2009). This spatial coupling of RNA synthesis and virus assembly is consistent with studies that demonstrate that only actively replicating viral RNA is packaged into virions (Khromykh *et al.*, 2001). However, the mechanistic details that link RNA replication with genome packaging have not been elucidated. Several non-structural proteins have been linked to genome packaging and particle assembly. Khromykh and colleagues first demonstrated that a substitution of I59N in Kunjin virus NS2A blocked production of virus but had no effect on RNA synthesis (Liu *et al.*, 2003; Leung *et al.*, 2008). Similarly, studies from the Rice laboratory showed that a mutation at residue K190S in the YFV NS2A had a similar phenotype (Kummerer and Rice, 2002). In this latter case, the mutation abrogated a cleavage of NS2A to the truncated NS2Aα form. This truncated protein is not seen in all flaviviruses, and the cleavage product has not been associated with any packaging phenotype. However, they were able to isolate a suppressor mutation that mapped to NS3 that restored infectivity. Additional studies showed that substitution in NS3 at a site in close proximity to the suppressor resulted in virus that was deficient in packaging but not RNA replication or budding of subviral particles (Patkar and Kuhn, 2008).

Virion formation occurs on ER membranes and requires the presence of the prM and E glycoproteins on the lumen side, and the capsid and genome RNA on the cytoplasmic side (Welsch *et al.*, 2009). The trigger for the coalescence of these components is not known. However, budding of just prM and E can take place without capsid and genome RNA (Mason *et al.*, 1991). This results in predominantly smaller 'subviral particles' or SVPs being released that have $T=1$ icosahedral symmetry (Ferlenghi *et al.*, 2001; Lorenz *et al.*, 2003). The expression of prM and E in the absence of capsid protein also results in the production and release of SVPs and these have been investigated for the purpose of flavivirus vaccines (Mason *et al.*, 1991; Konishi *et al.*, 1992; Pincus *et al.*, 1992). The control of the number of subunits that will produce SVPs or authentic virus particles is not understood, but appears to be restricted to 60 or 180 prM–E heterodimers, respectively. There is no evidence that capsid proteins that associate with genome RNA prior to envelopment form a spherical nucleocapsid core as seen in other enveloped viruses such as the alphaviruses.

Structure of the immature particle

Particles that bud into the ER lumen contain the prM precursor protein. The prM protein forms a heterodimer with the E protein and is necessary for proper E protein folding (Heinz *et al.*, 1994a; Lorenz *et al.*, 2002). With many flaviviruses, immature viruses are also released from infected cells. Immature particles are very inefficient in infection, presumably because they cannot undergo the required conformational and translational movements that are necessary for fusion. However, recent evidence with DENV suggests that most particles, including those that are 'mature virus' particles, contain some prM protein, and the ratio of prM to M determines the particle infectivity (Junjhon *et al.*, 2008, 2010). Thus, there is probably a continuum of infectivity that is proportional to the amount of uncleaved prM in the particles.

Treatment of infected cells with an acidotropic reagent such as ammonium chloride results in the release of exclusively immature particles. This can also be accomplished by mutation of the furin cleavage site in prM (Guirakhoo *et al.*, 1991; Heinz *et al.*, 1994b). Cell treated with acidotropic reagents have an increase in the internal pH of the trans-Golgi network (TGN) and this prevents the cleavage of prM by cellular furin protease. This technique was used to isolate a uniform population of immature DENV particles that were then examined by cryo-electron microscopy (cryo-EM) and image reconstruction techniques (Zhang *et al.*, 2003c). The 600-Å particle was found to have 60 spike-like projections (Fig. 2.1A). Similar structures were also reported for WNV and YFV, although the native structure of the 17D vaccine strain has not yet been obtained (Mukhopadhyay *et al.*, 2003; Zhang *et al.*, 2007a). The spikes were trimeric in appearance but lacked strict three-fold symmetry. Using the program EM-fit, the X-ray derived structure of the E protein ectodomain could be experimentally docked into the cryo-EM density map for an accurate placement of 180 copies of the E glycoprotein (Fig. 2.1B). This revealed that the E protein was oriented at roughly a 45° angle relative to the lipid bilayer and that the distal end of E containing the fusion protein was most distant from the virus centre. In the original structure, a difference map was generated, subtracting the density for the E protein and revealing the position of prM. This demonstrated that three prM proteins covered the top of the spike, protecting the three E proteins that come together at the centre of the projection. The prM moiety caps the fusion peptide and presumably protects the E protein from premature activation and fusion as the immature particles transit the low pH of the TGN. The determination of the X-ray structure of the prM–E heterodimer confirmed the original placement of prM into the EM density map (Li *et al.*, 2008). Furthermore, the furin site that is cleaved to release the pr moiety during maturation was found to be located on the interior of the heterotrimer and inaccessible to furin in the immature configuration of the particle. This explained the failure to *in vitro* 'mature' prM-containing particles by treatment with furin under neutral pH conditions. However, it raised the question of what conditions or state of the immature particle is required for furin cleavage and virion maturation.

As described later, the number of glycoprotein subunits in the immature virus are equivalent to those in the mature virus. However, given the placement of the E proteins in the immature particle, it is not evident how the proteins reorient themselves during conversion into the mature virion. They must transition between the trimeric arrangement that forms the spike found in the immature virion to the homodimer orientation of E found in mature particles. This requires both a translational movement of E and the collapse of the spike with the orientation of the E homodimer parallel to the viral membrane.

The infectious virion

Early physical studies of the flavivirus virion suggested that the particle was relatively smooth and approximately 500 Å in diameter. When the X-ray crystal structure for TBEV was determined, Rey and colleagues suggested the E protein, which exists as a homodimer, was oriented with its long axis parallel to the viral membrane (Rey *et al.*, 1995b). This would produce a particle with a smooth surface and with many of the antibody neutralization epitopes on the exposed surface of the E homodimer. These predictions were confirmed when the structure of the mature DENV particle was determined using cryo-EM and image reconstruction (Kuhn *et al.*, 2002). The surface-shaded image of the particle had the appearance of a golf ball, with a smooth and spherical surface characterized by gentle valleys and hilltops (Fig. 2.1E). The outer protein layer was ~25 Å in diameter and had well-defined density consistent with the position of E molecules. Immediately below this layer was a well articulated but lower density region between radii 185 and 220 Å that could be ascribed to the M ectodomain and the E protein stem region. Further down, the stem and anchor regions of the M and E proteins were shown to exist in well-defined positions with their transmembrane domains rigidly anchored into the lipid bilayer, which could clearly be identified between 140 and 185 Å from the centre of the particle (Zhang *et al.*, 2003b). Inside the bilayer,

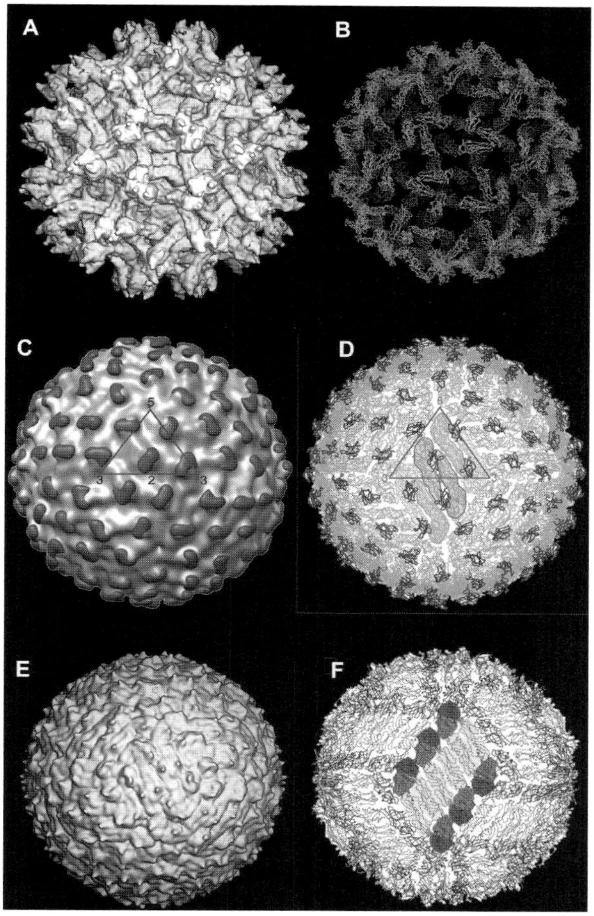

Figure 2.1 Cryo-electron microscopy (EM) reconstructions of dengue virus. Surface shaded views (A, C, E) or fitted E protein (B, D, F) of immature virus at neutral pH (A, B), immature virus at pH 6 (C, D), or mature virus at neutral pH (E, F). In B and F, the C-α backbone of the E protein is presented in standard colours with domain I in red, domain II in yellow, and domain III in blue. In D, the E protein is drawn in grey and the C-α backbone of the pr protein is shown in blue. The density corresponding to pr is also shown in blue in C, along with the symmetry axis, which is the same in all panels. The particles are not drawn to scale. A colour version of this figure is located in the plate section at the back of the book.

but separated by an ~12-Å gap of low to no density, was the density contributed by the capsid proteins and genome RNA. All reconstructions to date have been unable to identify an organized core structure, such as the one seen with alphavirus virions (Zhang *et al.*, 2007b). This suggests that the capsid protein acts as a nucleoprotein to bind to the genome RNA, neutralize its negative charge, and compact it for 'packaging' within the flavivirus particle.

Although the structure of the dengue virion provided significant insight into the particle organization it raised several questions about the assembly process. In particular, the failure to identify a defined core makes it difficult to understand the process of recruitment and packaging of genome RNA, and the engagement of the envelope glycoproteins. Furthermore, a packaging signal of the viral RNA has not been identified, and most data point towards on-going viral RNA replication as necessary for genome packaging (Khromykh and Westaway, 1996; Khromykh *et al.*, 2001). Therefore, the mechanism by which RNA is recruited to the capsid proteins and their condensation to form a nucleoprotein complex is unknown. If a core structure is not formed, what drives the association between the nucleoprotein complex found in the cytoplasm and the envelope glycoproteins prM and E in ER lumen? As was just described, there is no apparent continuous density that links the transmembrane domains of prM and E with the inner shell of RNA and capsid proteins. These proteins are coupled in the polyprotein but the capsid and prM are released from one another by the viral NS2B–NS3 protease shortly after translocation of prM into the ER lumen. Whether NS2B-NS3 is present at the site of envelopment and cleaves concomitant with budding and formation of the immature particle is not known. However, one can propose a model where replication proteins such as NS2A and NS2B–NS3 link the synthesis of nascent RNA with capsid proteins residing at the ER membrane in the polyprotein form opposed to prM and E. The condensation of protein and nucleic acid results in cleavage of the capsid protein and budding of the nascent particle into the ER. In this way, there is linkage between the capsid – RNA complex and the prM–E glycoproteins. However, the release of subviral particles in the absence of capsid and RNA remains to be resolved in the context of this model. The ability of prM and E to form SVPs and to bud in the absence of capsid/RNA suggests that the either the nucleoprotein complex is not the trigger for budding or that SVPs and regular virions bud through different mechanisms.

The flavivirus virion consists of 180 copies of the M and E proteins organized with an icosahedral arrangement (Fig. 2.1F). Three molecules of E and M are positioned within the asymmetric subunit but not in a classical T = 3 arrangement as one would predict (Kuhn *et al.*, 2002; Mukhopadhyay *et al.*, 2003; Zhang *et al.*, 2003a, 2004a). Instead a set of three E homodimers align themselves together in a raft pattern, with the central homodimer's twofold symmetry axis located on the icosahedral 2-fold axis. The 5-fold and 3-fold axes likewise contain five or three E homodimers that converge on their respective axes. Although the surface is relatively smooth, the Ig-like domain III of the E protein protrudes from the surface to give the virion a ruffled appearance analogous to a golf ball. The extension of domain III was consistent with observations that mapped many of the neutralizing antibodies to this region of the E protein. Since no icosahedral structure is observed for the nucleocapsid core, the number of capsid proteins within the virion is not known, although recent mass spectrometry results (Riley and Kuhn, unpublished results) suggests that the stoichiometry is greater than the number of E subunits.

Transit and egress of immature particles

Immature particles when treated with furin at neutral pH are resistant to cleavage (Yu *et al.*, 2008). This initially raised the possibility that the immature particles that were isolated following ammonium sulfate treatment were artefacts of the treatment and isolation process. To probe these questions, immature virus was incubated at various pH's to mimic the transit of the nascent immature virus particle as it progresses from the neutral pH of the ER lumen through the mildly acidic TGN and then the neutral pH of the extracellular medium. The particles were examined using both biochemical and structural approaches, and the results converged on an explanation. Treatment of immature particles at a pH of 6 with furin resulted in a >1000-fold increase in the level of infectivity and resulted in the cleavage of the pr moiety. In contrast, no increase in infectivity or cleavage of prM occurred under the same treatment at pH 7–8. Immature particles either treated or untreated with furin at pH 6 and examined using cryo-EM and image reconstruction showed the particles underwent a transitional and conformational change that converted the particles into a form that resembled the mature particle (Fig. 2.1C). The trimeric prM–E heterodimers moved from their upright spike configuration into one that resembled the mature virus with the E protein oriented parallel to the membrane and transformed into a dimeric organization (Fig. 2.1D). The pr moiety still covered the hydrophobic fusion peptide, but the connecting polypeptide with M now presented the furin cleavage site to an exposed region on the surface of the particle. This rendered the immature particle susceptible to cleavage. Interestingly, even after cleavage by furin at pH 6, the pr component was not released from this particle. Instead, it remained associated with the particle until exposure to neutral pH, as would be expected following transit out of the TGN and into the extracellular media.

The efficiency of cleavage of the prM protein in immature virions has received much attention. Virus harvested from infected cells invariably is a mixture of particle types depending on the virus and the cells used for propagation. Even virus within infected hosts appears to contain immature and partially mature virus, as evidenced by antibodies against the prM component of the particle (Dejnirattisai *et al.*, 2010). The mechanism by which furin acts to cleave immature particles is not known. Two potential models are that a given particle is sequentially cleaved by one or more furin proteases until the particle enters a state or site whereby furin is no longer active for cleavage. One could imagine these cleavages would occur as a wave across the particle so that one surface is initially cleaved and further cleavage would expand progressively from that site. Alternatively, the furin cleavage is carried out in the acidic TGN by a high concentration of molecules that randomly process accessible sites. In both cases, particles would be furin susceptible for a limited time and if cleavage does not occur at enough sites, the released particles would be either immature or partially mature. The presence of such immature or partially mature particles may have an influence on pathogenesis and certainly may modulate the immune response.

Partially mature flaviviruses are difficult to describe using current structural techniques, which usually rely on large numbers of homogeneous particles. The number of processed M proteins necessary to produce a particle that appears to be 'mature' is not known. Although the original assumptions were that mature particles contain 180 copies of M and E, recent biochemical analyses question whether there is complete cleavage of the prM precursor in these apparently homogeneous particles (Junjhon *et al.*, 2008, 2010). Likewise, one can ask how many prM molecules are necessary to produce an immature particle structure. Analysis of individual particles by cryo-EM demonstrates that some or even many particles have physical features that suggest a mix of immature and mature components. Currently, electron tomography, examining a single particle at a time, is the only approach to address this question and unfortunately, the resolution is probably insufficient to discern such differences. Assuming there is a continuum of maturation states of flavivirus particles, what is the infectivity of this continuum? As mentioned earlier, immature virus is not infectious, but we can expect that as the number of cleaved prM molecules increases on a particle, infectivity will be increased. The assumption is that partially mature viruses having sufficient M molecules will be able to undergo the translational and conformational movements necessary for fusion. Whether this requires a focused region of mature proteins around which the fusion pore will form, or whether the threshold of mature proteins can be distributed over the entire surface of the particle, remains to be experimentally determined.

Antibody binding and its implications

The analysis of antibody binding onto the surface of virions has been carried out in numerous virus systems with the objective of describing the molecular basis for antibody-mediated virus neutralization. Structural studies using a combination of X-ray crystallography and cryo-EM having been done with flaviviruses for this purpose; however, the initial results suggested that this analysis provided a far greater insight into the biology of the virion than simply the examination of virus–antibody binding (Kaufmann *et al.*, 2006, 2009, 2010; Lok *et al.*, 2008; Cherrier *et al.*, 2009). The intent here is not to present mechanistic insights into antibody neutralization, but rather to discuss what the structure of antibody–flavivirus complexes have contributed to virus biology.

Perhaps one of the best studying virus–antibody interactions has been the monoclonal antibody E16 developed against WNV (Nybakken *et al.*, 2005; Oliphant *et al.*, 2005; Oliphant and Diamond, 2007). This antibody has been deeply characterized and a humanized version is being developed for clinical use. The antibody neutralizes virus primarily at a post internalization step, although there is also modest reduction of virus attachment to the cell. Single particle tracking studies have demonstrated that the particles bound with E16 enter endosomes but are blocked at a stage prior to membrane fusion (Thompson *et al.*, 2009). Fab fragments derived from intact E16 provide most of these activities and thus structural studies of virus in complex with the E16 Fab should provide information into the structural correlates of neutralization. The X-ray structure of the E16 Fab–E domain III complex demonstrated binding of antibody to four discontinuous loops on the lateral ridge of DIII (Nybakken *et al.*, 2005). The subsequent structure of the Fab–WNV complex revealed 120 bound Fab molecules per virion (Kaufmann *et al.*, 2006). Binding of the Fab at the 5-fold axis was absent because of steric exclusion of incoming Fab by a neighbouring E protein DIII. That is, the binding site was blocked by the close association around the 5-fold axis of adjacent domain III molecules. It was apparent from the structure that virions with bound E16 would not be competent to undergo the structural rearrangements necessary for formation of the E homotrimers. Thus, the E16 antibody or Fab fragment neutralized virus infectivity by preventing membrane fusion. Although structures of the E protein have been determined in both the native dimer state as well as the low pH trimer state, the movement of the E proteins necessary for this transition is unclear (Bressanelli *et al.*, 2004; Modis *et al.*, 2004). Since E16 was shown to block WNV in a pre-fusion state, it was hypothesized that the E16 Fab could be used to trap the particle as it underwent the transition between neutral and low pH. Particles with Fab bound were exposed to low pH and cryo-EM was used to evaluate the trapped structure. The irreversibly trapped particles displayed a 60-Å gap between the viral membrane and the envelope protein shell suggesting that an early step in low pH induced fusion is the outward expansion of the stem region of the E. However, a detailed structural description of the virion as it undergoes fusion remains a future discovery.

The images of flavivirus virions obtained using cryo-EM and their three dimensional reconstructions showed relatively smooth spherical particles having an extensive protein layer completely covering the lipid bilayer. However, analysis of a neutralizing antibody against DENV 2 provided a very different perspective of the virion. Monoclonal antibody 1A1D2 neutralized virus infectivity primarily by blocking the entry of the particle into the cell (Roehrig *et al.*, 1998b). X-ray crystallography of the Fab complexed with E protein DIII demonstrated that the footprint of the antibody would be partially occluded in the native DENV structure that was determined by cryo-EM suggesting that there should be no binding (Fig. 2.2A) (Lok *et al.*, 2008). Consistent with this, was the observation that the 1A1D2 antibody Fab incubated in the presence of virus at room temperature did not bind. However, raising the temperature of the binding solution to 37°C did permit binding as determined by cryo-EM. The analysis of the virus–Fab complex showed that the E proteins underwent a translational movement away from the viral membrane and exposed a sufficient amount of the epitope to permit binding of 120 Fab subunits per virion (Fig. 2.2B). This temperature-dependent effect suggested that one or more Fab's captured a transient state of the particle, and prevented the E protein from collapsing back into the low temperature configuration. This dynamic motion, previously referred to as 'breathing' in flock house virus and rhinovirus, probably represents the normal state of the particle under physiological conditions (Bothner *et al.*, 1998; Che *et al.*, 1998). It is interesting to speculate how such dynamic properties of the E protein might influence the recognition of virions by antibodies (Pierson *et al.*, 2007, 2008). Cryptic epitopes may play an important role in the immune control of virus infection with rising temperatures during the febrile period facilitating greater antibody access to virus particles.

Conclusions and observations

The last several years have seen significant advances in understanding flavivirus replication and pathogenesis through the employment of structural biology approaches. To date, many features and observations from one flavivirus seems to translate well into related members of the genus. Whether structural biology will yield different observations from distinct viruses has yet to be determined, but the structural features and morphogenetic pathway among the members seem to be well conserved. The field is now on the threshold of exploiting this structural knowledge base for

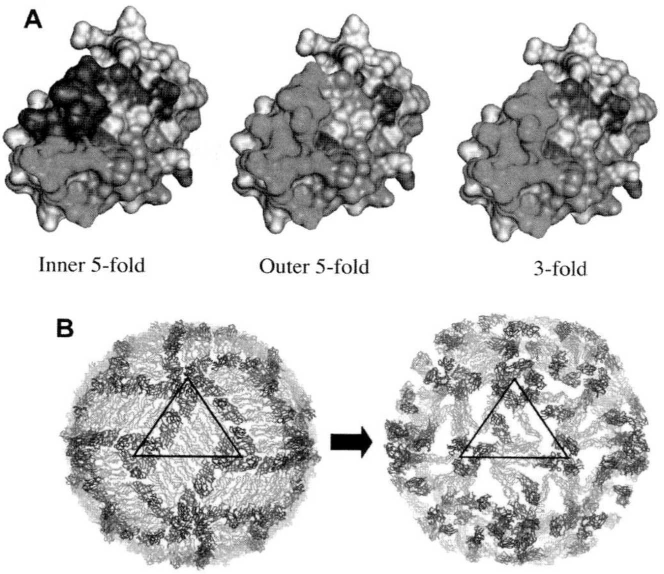

Figure 2.2 Neutralizing monoclonal binding to flaviviruses. (A) Surface shaded view of E protein domain III and the buried surface area of the epitopes recognized by Fab E16 and Fab1A1D-2 on each of the three unique E-DIIIs in the icosahedral asymmetric unit of the mature virus. Epitopes of Fab E16 and 1A1D-2 (green and pink, respectively) are partially buried under neighbouring E subunits in the mature flavivirus. Buried regions are coloured dark green and dark red. Note that 1A1D-2 has buried surface in each of the E-DIIIs of the virion, whereas E16 is only partially buried at the true 5-fold (inner 5-fold). (B) The radical change in E protein position captured by monoclonal 1A1D-2. The panel shows the Cα chains of E proteins from uncomplexed mature DENV (left), and the Fab complex structure (right). The black triangle represents the asymmetric unit of the virus. The E domains are coloured red for domain I, yellow for domain II, and blue for domain III. A colour version of this figure is located in the plate section at the back of the book.

the development of therapeutic intervention and for the design of next generation vaccines. One particular challenge will be to describe individual virus particles using high resolution techniques and to understand the molecular basis for particle infectivity. The synergy between structural biology and molecular genetics and biochemistry has reached a threshold where rapid advances are now common and our pace of discovery is accelerating.

References

Beasley, D.W., and Barrett, A.D. (2002). Identification of neutralizing epitopes within structural domain III of the West Nile virus envelope protein. J. Virol. *76*, 13097–13100.

Bothner, B., Dong, X.F., Bibbs, L., Johnson, J.E., and Siuzdak, G. (1998). Evidence of viral capsid dynamics using limited proteolysis and mass spectrometry. J. Biol. Chem. *273*, 673–676.

Bressanelli, S., Stiasny, K., Allison, S.L., Stura, E.A., Duquerroy, S., Lescar, J., Heinz, F.X., and Rey, F.A. (2004). Structure of a flavivirus envelope glycoprotein in its low-pH-induced membrane fusion conformation. EMBO J. *23*, 1–11.

Che, Z., Olson, N.H., Leippe, D., Lee, W., Mosser, A.G., Rueckert, R.R., Baker, T.S., and Smith, T.J. (1998). Antibody-mediated neutralization of human rhinovirus 14 explored by means of cryoelectron microscopy and X-ray crystallography of virus–Fab complexes. J. Virol. *72*, 4610–4622.

Cherrier, M.V., Kaufmann, B., Nybakken, G.E., Lok, S.M., Warren, J.T., Chen, B.R., Nelson, C.A., Kostyuchenko, V.A., Holdaway, H.A., Chipman, P.R., *et al.* (2009). Structural basis for the preferential recognition of immature flaviviruses by a fusion-loop antibody. EMBO J. *28*, 3269–3276.

Dejnirattisai, W., Jumnainsong, A., Onsirisakul, N., Fitton, P., Vasanawathana, S., Limpitikul, W., Puttikhunt, C., Edwards, C., Duangchinda, T., Supasa, S., *et al.* (2010). Cross-reacting antibodies enhance dengue virus infection in humans. Science *328*, 745–748.

Dokland, T., Walsh, M., Mackenzie, J.M., Khromykh, A.A., Ee, K.H., and Wang, S. (2004). West Nile virus core protein; tetramer structure and ribbon formation. Structure *12*, 1157–1163.

Elshuber, S., Allison, S.L., Heinz, F.X., and Mandl, C.W. (2003). Cleavage of protein prM is necessary for infection of BHK-21 cells by tick-borne encephalitis virus. J. Gen. Virol. *84*, 183–191.

Ferlenghi, I., Clarke, M., Ruttan, T., Allison, S.L., Schalich, J., Heinz, F.X., Harrison, S.C., Rey, F.A., and Fuller, S.D. (2001). Molecular organization of a recombinant subviral particle from tick-borne encephalitis. Mol. Cell 7, 593–602.

Guirakhoo, F., Heinz, F.X., Mandl, C.W., Holzmann, H., and Kunz, C. (1991). Fusion activity of flaviviruses: comparison of mature and immature (prM-containing) tick-borne encephalitis virions. J. Gen. Virol. *72*, 1323–1329.

Guirakhoo, F., Hunt, A.R., Lewis, J.G., and Roehrig, J.T. (1993). Selection and partial characterization of dengue 2 virus mutants that induce fusion at elevated pH. Virology *194*, 219–223.

Heinz, F.X., Mandl, C.W., Holzmann, H., Kunz, C., Harris, B.A., Rey, F., and Harrison, S.C. (1991). The flavivirus envelope protein E: isolation of a soluble form from tick-borne encephalitis virus and its crystallization. J. Virol. *65*, 5579–5583.

Heinz, F.X., Auer, G., Stiasny, K., Holzmann, H., Mandl, C., Guirakhoo, F., and Kunz, C. (1994a). The interactions of the flavivirus envelope proteins: implications for virus entry and release. Arch. Virol. Suppl. *9*, 339–348.

Heinz, F.X., Stiasny, K., Puschner-Auer, G., Holzmann, H., Allison, S.L., Mandl, C.W., and Kunz, C. (1994b). Structural changes and functional control of the tick-borne encephalitis virus glycoprotein E by the heterodimeric association with protein prM. Virology *198*, 109–117.

Jones, C.T., Ma, L., Burgner, J.W., Groesch, T.D., Post, C.B., and Kuhn, R.J. (2003). Flavivirus capsid protein is a dimeric alpha-heical protein. J. Virol. *77*, 7143–7149.

Junjhon, J., Lausumpao, M., Supasa, S., Noisakran, S., Songjaeng, A., Saraithong, P., Chaichoun, K., Utaipat, U., Keelapang, P., Kanjanahaluethai, A., *et al.* (2008). Differential modulation of prM cleavage, extracellular particle distribution, and virus infectivity by conserved residues at nonfurin consensus positions of the dengue virus pr–M junction. J. Virol. *82*, 10776–10791.

Junjhon, J., Edwards, T.J., Utaipat, U., Bowman, V.D., Holdaway, H.A., Zhang, W., Keelapang, P., Puttikhunt, C., Perera, R., Chipman, P.R., *et al.* (2010). Influence of pr–M cleavage on the heterogeneity of extracellular dengue virus particles. J. Virol. *84*, 8353–8358.

Kanai, R., Kar, K., Anthony, K., Gould, L.H., Ledizet, M., Fikrig, E., Marasco, W.A., Koski, R.A., and Modis, Y. (2006). Crystal structure of west nile virus envelope glycoprotein reveals viral surface epitopes. J. Virol. *80*, 11000–11008.

Kaufmann, B., Nybakken, G.E., Chipman, P.R., Zhang, W., Diamond, M.S., Fremont, D.H., Kuhn, R.J., and Rossmann, M.G. (2006). West Nile virus in complex with the Fab fragment of a neutralizing monoclonal antibody. Proc. Natl. Acad. Sci. U.S.A. *103*, 12400–12404.

Kaufmann, B., Chipman, P.R., Holdaway, H.A., Johnson, S., Fremont, D.H., Kuhn, R.J., Diamond, M.S., and Rossmann, M.G. (2009). Capturing a flavivirus pre-fusion intermediate. PLoS Pathog. *5*, e1000672.

Kaufmann, B., Vogt, M.R., Goudsmit, J., Holdaway, H.A., Aksyuk, A.A., Chipman, P.R., Kuhn, R.J., Diamond, M.S., and Rossmann, M.G. (2010). Neutralization of West Nile virus by cross-linking of its surface proteins with Fab fragments of the human monoclonal antibody CR4354. Proc. Natl. Acad. Sci. U.S.A. *107*, 18950–18955.

Khromykh, A.A., and Westaway, E.G. (1996). RNA binding properties of core protein of the flavivirus Kunjin. Arch. Virol. *141*, 685–699.

Khromykh, A.A., Varnavski, A.N., Sedlak, P.L., and Westaway, E.G. (2001). Coupling between replication and packaging of flavivirus RNA: evidence derived from the use of DNA-based full-length cDNA clones of Kunjin virus. J. Virol. *75*, 4633–4640.

Konishi, E., Pincus, S., Paoletti, E., Shope, R.E., Burrage, T., and Mason, P.W. (1992). Mice immunized with a subviral particle containing the Japanese encephalitis virus prM/M and E proteins are protected from lethal JEV infection. Virology *188*, 714–720.

Kuhn, R.J., Zhang, W., Rossmann, M.G., Pletnev, S.V., Corver, J., Lenches, E., Jones, C.T., Mukhopadhyay, S., Chipman, P.R., Strauss, E.G. *et al.* (2002). Structure of dengue virus: implications for flavivirus organization, maturation, and fusion. Cell *108*, 717–725.

Kummerer, B.M., and Rice, C.M. (2002). Mutations in the yellow fever virus nonstructural protein NS2A selectively block production of infectious particles. J. Virol. *76*, 4773–4784.

Lescar, J., Roussel, A., Wein, M.W., Navaza, J., Fuller, S.D., Wengler, G., Wengler, G., and Rey, F.A. (2001). The fusion glycoprotein shell of Semliki Forest virus: an icosahedral assembly primed for fusogenic activation at endosomal pH. Cell *105*, 137–148.

Leung, J.Y., Pijlman, G.P., Kondratieva, N., Hyde, J., Mackenzie, J.M., and Khromykh, A.A. (2008). Role of nonstructural protein NS2A in flavivirus assembly. J. Virol. *82*, 4731–4741.

Li, L., Lok, S.-M., Yu, I.-M., Zhang, Y., Kuhn, R., Chen, J., and Rossmann, M.G. (2008). The flavivirus precursor membrane-envelope protein complex: structure and maturation. Science *319*, 1830–1834.

Lindenbach, B.D., Thiel, H.-J., and Rice, C.M. (2007). Flaviviridae: The Viruses and Their Replication. In Fields Virology, Knipe, D.M., and Howley, P.M., eds. (Lippincott Williams & Wilkins, Philadelphia), pp. 1001–1022.

Liu, W.J., Chen, H.B., and Khromykh, A.A. (2003). Molecular and functional analyses of Kunjin virus infectious cDNA clones demonstrate the essential roles for NS2A in virus assembly and for a nonconservative residue in NS3 in RNA replication. J. Virol. *77*, 7804–7813.

Lok, S.M., Kostyuchenko, V., Nybakken, G.E., Holdaway, H.A., Battisti, A.J., Sukupolvi-Petty, S., Sedlak, D., Fremont, D.H., Chipman, P.R., Roehrig, J.T., *et al.* (2008). Binding of a neutralizing antibody to dengue virus alters the arrangement of surface glycoproteins. Nat. Struct. Mol. Biol. *15*, 312–317.

Lorenz, I.C., Allison, S.L., Heinz, F.X., and Helenius, A. (2002). Folding and dimerization of tick-borne encephalitis virus envelope proteins prM and E in the endoplasmic reticulum. J. Virol. *76*, 5480–5491.

Lorenz, I.C., Kartenbeck, J., Mezzacasa, A., Allison, S.L., Heinz, F.X., and Helenius, A. (2003). Intracellular assembly and secretion of recombinant subviral particles from tick-borne encephalitis virus. J. Virol. *77*, 4370–4382.

Ma, L., Jones, C.T., Groesch, T.D., Kuhn, R.J., and Post, C.B. (2004). Solution structure of dengue virus capsid protein reveals another fold. Proc. Natl. Acad. Sci. U.S.A. *101*, 3414–3419.

Mackenzie, J.M., Khromykh, A.A., Jones, M.K., and Westaway, E.G. (1998). Subcellular localization and some biochemical properties of the flavivirus Kunjin nonstructural proteins NS2A and NS4A. Virology *245*, 203–215.

Mackenzie, J.M., Jones, M.K., and Westaway, E.G. (1999). Markers for trans-Golgi membranes and the intermediate compartment localize to induced membranes with distinct replication functions in flavivirus-infected cells. J. Virol. *73*, 9555–9567.

Mandl, C.W., Guirakhoo, F., Holzmann, H., Heinz, F.X., and Kunz, C. (1989). Antigenic structure of the flavivirus envelope protein E at the molecular level, using tick-borne encephalitis virus as a model. J. Virol. *63*, 564–571.

Markoff, L., Falgout, B., and Chang, A. (1997). A conserved internal hydrophobic domain mediates the stable membrane integration of the dengue virus capsid protein. Virology *233*, 105–117.

Mason, P.W., Pincus, S., Fournier, M.J., Mason, T.L., Shope, R.E., and Paoletti, E. (1991). Japanese encephalitis virus-vaccinia recombinants produce particulate forms of the structural membrane proteins and induce high levels of protection against lethal JEV infection. Virology *180*, 294–305.

Modis, Y., Ogata, S., Clements, D., and Harrison, S.C. (2003). A ligand-binding pocket in the dengue virus envelope glycoprotein. Proc. Natl. Acad. Sci. U.S.A. *100*, 6986–6991.

Modis, Y., Ogata, S., Clements, D., and Harrison, S.C. (2004). Structure of the dengue virus envelope protein after membrane fusion. Nature *427*, 313–319.

Modis, Y., Ogata, S., Clements, D., and Harrison, S.C. (2005). Variable surface epitopes in the crystal structure of dengue virus type 3 envelope glycoprotein. J. Virol. *79*, 1223–1231.

Mukhopadhyay, S., Kim, B.-S., Chipman, P.R., Rossmann, M.G., and Kuhn, R.J. (2003). Structure of West Nile virus. Science *303*, 248.

Nybakken, G.E., Oliphant, T., Johnson, S., Burke, S., Diamond, M.S., and Fremont, D.H. (2005). Structural basis of West Nile virus neutralization by a therapeutic antibody. Nature *437*, 764–769.

Nybakken, G.E., Nelson, C.A., Chen, B.R., Diamond, M.S., and Fremont, D.H. (2006). Crystal structure of the West Nile virus envelope glycoprotein. J. Virol. *80*, 11467–11474.

Oliphant, T., and Diamond, M.S. (2007). The molecular basis of antibody-mediated neutralization of West Nile virus. Expert Opin. Biol. Ther. *7*, 885–892.

Oliphant, T., Engle, M., Nybakken, G.E., Doane, C., Johnson, S., Huang, L., Gorlatov, S., Mehlhop, E., Marri, A., Chung, K.M., *et al.* (2005). Development of a humanized monoclonal antibody with therapeutic potential against West Nile virus. Nat. Med. *11*, 522–530.

Patkar, C.G., and Kuhn, R.J. (2008). Yellow fever virus NS3 plays an essential role in virus assembly independent of its known enzymatic functions. J. Virol. *82*, 3342–3352.

Pierson, T.C., Xu, Q., Nelson, S., Oliphant, T., Nybakken, G.E., Fremont, D.H., and Diamond, M.S. (2007). The stoichiometry of antibody-mediated neutralization and enhancement of West Nile virus infection. Cell Host Microbe *1*, 135–145.

Pierson, T.C., Fremont, D.H., Kuhn, R.J., and Diamond, M.S. (2008). Structural insights into the mechanisms of antibody-mediated neutralization of flavivirus infection: implications for vaccine development. Cell Host Microbe *4*, 229–238.

Pincus, S., Mason, P.W., Konishi, E., Fonseca, B.A., Shope, R.E., Rice, C.M., and Paoletti, E. (1992). Recombinant vaccinia virus producing the prM and E proteins of yellow fever virus protects mice from lethal yellow fever encephalitis. Virology *187*, 290–297.

Rey, F.A., Heinz, F.X., Mandl, C., Kunz, C., and Harrison, S.C. (1995). The envelope glycoportein from tick-borne encephalitis virus at 2 Å resolution. Nature *375*, 291–298.

Roehrig, J.T., Johnson, A.J., Hunt, A.R., Bolin, R.A., and Chu, M.C. (1990). Antibodies to dengue 2 virus E-glycoprotein synthetic peptides identify antigenic conformation. Virology *177*, 668–675.

Roehrig, J.T., Bolin, R.A., and Kelly, R.G. (1998). Monoclonal antibody mapping of the envelope glycoprotein of the dengue 2 virus, Jamaica. Virology *246*, 317–328.

Samsa, M.M., Mondotte, J.A., Iglesias, N.G., Assuncao-Miranda, I., Barbosa-Lima, G., Da Poian, A.T., Bozza, P.T., and Gamarnik, A.V. (2009). Dengue virus capsid protein usurps lipid droplets for viral particle formation. PLoS Pathog. *5*, e1000632.

Strauss, J.H., and Strauss, E.G. (1988). Replication of the RNAs of alphaviruses and flaviviruses. In RNA Genetics Book *1*, Domingo, E., Holland, J.J., and Ahlquist, P., eds. (CRC Press, Inc., Boca Raton, FL), pp. 71–90.

Sukupolvi-Petty, S., Austin, S.K., Purtha, W.E., Oliphant, T., Nybakken, G.E., Schlesinger, J.J., Roehrig, J.T., Gromowski, G.D., Barrett, A.D., Fremont, D.H., *et al.* (2007). Type- and subcomplex-specific neutralizing antibodies against domain III of dengue virus type 2 envelope protein recognize adjacent epitopes. J. Virol. *81*, 12816–12826.

Thompson, B.S., Moesker, B., Smit, J.M., Wilschut, J., Diamond, M.S., and Fremont, D.H. (2009). A therapeutic antibody against west nile virus neutralizes infection by blocking fusion within endosomes. PLoS Pathog. *5*, e1000453.

Welsch, S., Miller, S., Romero-Brey, I., Merz, A., Bleck, C.K., Walther, P., Fuller, S.D., Antony, C.,

Krijnse-Locker, J., and Bartenschlager, R. (2009). Composition and three-dimensional architecture of the dengue virus replication and assembly sites. Cell Host Microbe *5*, 365–375.

Wengler, G., Wengler, G., Nowak, T., and Wahn, K. (1987). Analysis of the influence of proteolytic cleavage on the structural organization of the surface of the West Nile flavivirus leads to the isolation of a protease-resistant E protein oligomer from the viral surface. Virology *160*, 210–219.

Yu, I.-M., Zhang, W., Holdaway, H.A., Li, L., Kostyuchenko, V.A., Chipman, P.R., Kuhn, R., Rossmann, M.G., and Chen, J. (2008). Structure of immature dengue virus at low pH primes proteolytic maturation. Science *319*, 1834–1837.

Zhang, W., Chipman, P.R., Corver, J., Johnson, P.R., Zhang, Y., Mukhopadhyay, S., Baker, T.S., Strauss, J.H., Rossmann, M.G., and Kuhn, R.J. (2003a). Visualization of membrane protein domains by cryo-electron microscopy of dengue virus. Nat. Struct. Biol. *10*, 907–912.

Zhang, Y., Corver, J., Chipman, P.R., Pletnev, S.V., Sedlak, D., Baker, T.S., Strauss, J.H., Kuhn, R.J., and Rossmann, M.G. (2003b). Structures of immature flavivirus particles. J. EMBO *22*, 2604–2613.

Zhang, Y., Zhang, W., Ogata, S., Clements, D., Strauss, J.H., Baker, T.S., Kuhn, R.J., and Rossmann, M.G. (2004). Conformational changes of the flavivirus E glycoprotein. Structure *12*, 1607–1618.

Zhang, Y., Kaufmann, B., Chipman, P.R., Kuhn, R.J., and Rossmann, M.G. (2007a). Structure of immature West Nile virus. J. Virol. *81*, 6141–6145.

Zhang, Y., Kostyuchenko, V.A., and Rossmann, M.G. (2007b). Structural analysis of viral nucleocapsids by subtraction of partial projections. J. Struct. Biol. *157*, 356–364.

Flavivirus Replication and Assembly

3

Justin A. Roby, Anneke Funk and Alexander A. Khromykh

Abstract

The replication and assembly of flaviviruses are complex procedures that require the efficient coordination of a number of different steps. These stages are highly organized temporally and spatially in the infected cell and require the virus-induced establishment of host-derived membrane structures. Flavivirus RNA structures, non-structural proteins and host factors actively participate in the replication of genomic RNA within vesicle packets (VP). Progeny (+) strand RNA exits the VP pore and is incorporated into nucleocapsids by the capsid protein. Nucleocapsids are then presumably transported into the lumen of the endoplasmic reticulum at sites directly opposed to the VP pore during formation of the prM–E studded lipid envelope. These immature virions are trafficked to the Golgi network in individual vesicles for glycoprotein maturation and furin-directed prM cleavage. Mature virions (with associated, cleaved prM) are then secreted into the extracellular milieu.

Introduction

The life cycle of flaviviruses begins upon binding of the virion to an unknown receptor and subsequent uptake into the host cell by receptor-mediated endocytosis (Chu and Ng, 2004; Chu *et al.*, 2006). The low pH of the endosomal compartment triggers conformational changes in the envelope protein (E) that induce fusion of the viral membrane with the endosomal membrane (Mukhopadhyay *et al.*, 2005; van der Schaar *et al.*, 2007, 2008), which deposits the viral nucleocapsid into the host cytoplasm. The single-stranded viral RNA genome of positive polarity is then released from the nucleocapsid by an unknown mechanism and is translated at the rough endoplasmic reticulum (RER). The RNA genome thus first serves as an mRNA to produce the viral proteins required for replication. Upon translation of the viral proteins (initiated by stalling at the RNA capsid hairpin), the viral RNA replication complex (RC) assembles on the 3′-end of positive (+) polarity genomic RNA template (Westaway *et al.*, 2003) as a first step in the synthesis of minus (–) polarity genomic RNA. An alternative mechanism for initiation of (–) RNA strand synthesis has also been proposed whereby viral polymerase first binds to the 5′ end of the genome at SL-A (Filomatori *et al.*, 2006). In both models, RNA then circularizes via long-range RNA interactions between 5′ and 3′ ends (mediated by complimentary cyclization sequences, UAR and DAR elements) prior to the initiation of (–) strand synthesis. During RNA synthesis, the (–) RNA strands remain bound to the (+) strand forming a double stranded replicative form (RF). The RF then serves as the template for synthesis of more (+) RNA strands via an asymmetric semiconservative mechanism (Chu and Westaway, 1987; Cleaves *et al.*, 1981). Only a single nascent (+) strand is copied from each RF during each round of synthesis, forming the replicative intermediate (RI). The new (+) RNA strand remains bound to the RI while it displaces

This chapter is dedicated to the memory of Edwin G. Westaway, a pioneer in the flavivirus field, whose depth of knowledge, enthusiasm and generosity will be sadly missed by us and everybody in the field.

the pre-existing (+) strand (Chu and Westaway, 1987). No free (−) RNA appears to be present in cells during replication (Khromykh and Westaway, 1997). Once established, viral RNA synthesis can continue even in the absence of active protein synthesis, indicating that transient viral polyprotein precursors are not required (Cleaves *et al.*, 1981; Chu and Westaway, 1987; Westaway *et al.*, 1999). Kinetics of viral RNA synthesis indicated that on average a single nascent RNA (+) strand displaces the pre-existing RNA (+) strand in the RI every 15 min (Chu and Westaway, 1985, 1987; Westaway *et al.*, 2003). These displaced (+) RNA strands serve either as templates for synthesis of (−) strands and formation of recycled RIs, or as mRNA for translation of more viral proteins, or are directed for packaging into virus particles. In infected cells, extensive reorganization and proliferation of cytoplasmic, perinuclear endoplasmic reticulum (ER) membranes as well as redistribution of cellular cholesterol has been observed (Westaway *et al.*, 1997b; Welsch *et al.*, 2009). Flavivirus RCs are associated with perinuclear ER membranes, so called vesicle packets (VP) (Mackenzie *et al.*, 2007; Welsch *et al.*, 2009; Westaway *et al.*, 1999). Viral particles assemble at the ER membrane adjacent to the sites of RNA replication. The newly synthesized (+) strand RNA associates with the capsid protein (C) and then transported through the ER membrane to acquire a lipid envelope (Welsch *et al.*, 2009). Virus particles are then secreted via the constitutive secretion pathway through the Golgi network to infect adjacent cells (MacKenzie and Westaway, 2001).

Full-length clones and replicons

The elucidation of details of the flavivirus life cycle was greatly facilitated by the establishment of replicons (non-infectious, self-replicating RNAs) and full-length infectious clones. The construction of cDNA clones derived from the viral genomes of various positive-strand RNA viruses allowed their targeted recombinant manipulation, unveiling information about their gene expression, replication and particle assembly (reviewed in (Westaway *et al.*, 2003)), as well as targeted attenuation for new generations of live vaccines (Bukreyev *et al.*, 1997; Guirakhoo *et al.*, 2000; Kinney and Huang, 2001; Kofler *et al.*, 2002, 2003, 2004b; Hall *et al.*, 2003; Kofler *et al.*, 2004a; Mandl, 2004; Mason *et al.*, 2006; Seregin *et al.*, 2006; Shustov *et al.*, 2007; Chang *et al.*, 2008; Ishikawa *et al.*, 2008; Widman *et al.*, 2008a,b, 2009; Suzuki *et al.*, 2009). Infectious cDNA clones (either delivered as *in vitro*-transcribed RNA or as DNA under the control of a cytomegalovirus promoter) have been generated for the majority of well characterized tick- and mosquito-borne flaviviruses: tick-borne encephalitis virus (TBEV) (Mandl *et al.*, 1997; Hayasaka *et al.*, 2004a); yellow fever virus (YFV) (McElroy *et al.*, 2005); 17D YFV vaccine strain (Chambers and Nickells, 2001); Japanese encephalitis virus (JEV) (Sumiyoshi *et al.*, 1992; Zhao *et al.*, 2005); Murray Valley encephalitis virus (MVEV) (Hurrelbrink *et al.*, 1999); dengue viruses (DENV) of all four serotypes, DENV-1 (Puri *et al.*, 2000; Suzuki *et al.*, 2007); DENV-2 (Kapoor *et al.*, 1995b; Kinney *et al.*, 1997; Polo *et al.*, 1997; Gualano *et al.*, 1998; Sriburi *et al.*, 2001; Blaney *et al.*, 2004b); DENV-3 (Blaney *et al.*, 2004a, 2008); DENV-4 (Lai *et al.*, 1991); West Nile virus (WNV) (Shi *et al.*, 2002; Yamshchikov *et al.*, 2001; Rossi *et al.*, 2005); Kunjin virus (KUNV) (Khromykh and Westaway, 1994); and Langat virus (Pletnev, 2001).

Perhaps an even more useful tool for research into flavivirus replication and packaging has been the development of subgenomic replicons that incorporate reporter genes (such as luciferase, green fluorescent protein or β-galactosidase) in place of the deleted structural genes C, pre-membrane (prM) and envelope (E) (Khromykh and Westaway, 1997; Varnavski and Khromykh, 1999; Khromykh, 2000; Westaway *et al.*, 2003; Hayasaka *et al.*, 2004b; Jones *et al.*, 2005; Mosimann *et al.*, 2010; Pijlman *et al.*, 2006b; Ng *et al.*, 2007). These systems have been invaluable for the study of virus replication because mutations detrimental to genome replication result in reduced expression of a reporter gene. Furthermore, complementation of the mutated sequences in *trans* by a second helper-replicon (Khromykh *et al.*, 1998a, 1999a,b, 2000; Kummerer and Rice, 2002; Liu *et al.*, 2002, 2003; Pijlman *et al.*, 2006a; Leung *et al.*, 2008; Patkar and Kuhn, 2008) allows confirmation of the proposed functional role for the protein/

RNA sequence in question, a system aided by the construction of cell lines continuously harbouring non-cytopathic flavivirus replicons (Khromykh *et al.*, 1998a; Jones *et al.*, 2005).

Replicons may also be packaged into virus-like particles (VLPs) by providing the structural proteins in *trans*, a system first explored using KUNV (Khromykh and Westaway, 1997; Khromykh *et al.*, 1998b; Harvey *et al.*, 2004) and now routinely used for the majority of flaviviruses (Scholle *et al.*, 2004b; Gehrke *et al.*, 2005; Jones *et al.*, 2005; Yoshii *et al.*, 2005; Fayzulin *et al.*, 2006; Shustov *et al.*, 2007; Yun *et al.*, 2007; Ansarah-Sobrinho *et al.*, 2008; Lai *et al.*, 2008). Packaging of replicons into VLPs has enabled studies many facets of the flavivirus life cycle including assembly, mechanisms of virus cell entry, virus–host interactions, antibody and antiviral screening tests, and the delivery of heterologous antigens as a vaccine platform (Lindenbach and Rice, 1997; Varnavski and Khromykh, 1999; Khromykh *et al.*, 2001a; Harvey *et al.*, 2003, 2004; Westaway *et al.*, 2003; Herd *et al.*, 2004; Liu *et al.*, 2004; Scholle *et al.*, 2004a; Orlinger *et al.*, 2006; Pierson *et al.*, 2006; Yun *et al.*, 2007; Anraku *et al.*, 2008; Ansarah-Sobrinho *et al.*, 2008; Dong *et al.*, 2008; Hoenninger *et al.*, 2008; Kent *et al.*, 2008; Hoang-Le *et al.*, 2009; Puig-Basagoiti *et al.*, 2009; Qing *et al.*, 2009; Yoshii *et al.*, 2009).

Genomic RNA structures involved in replication

The ~11 kb flavivirus RNA genome has a 5′ type I cap structure ($m^7GpppAmpN_1$), but lacks a poly(A) tail at the 3′ end (Wengler and Gross, 1978; Cleaves and Dubin, 1979). The coding region is flanked by 5′ and 3′ untranslated regions (UTRs), which are approximately 100 nt and 400–700 nt in size, respectively, and upon translation a single polyprotein greater than 3000 amino acids (aa) in length is produced (Rice *et al.*, 1985;

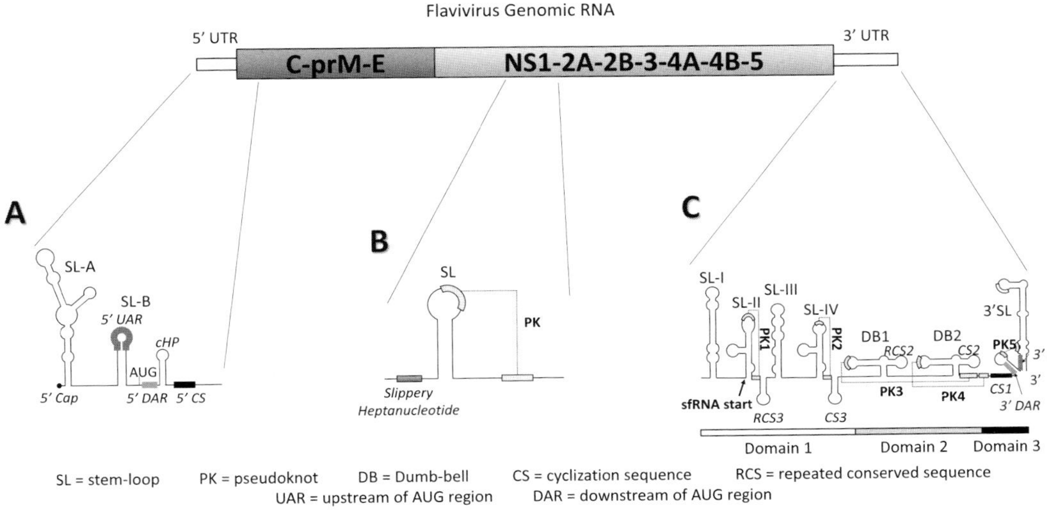

Figure 3.1 RNA secondary and tertiary structures involved in flavivirus replication. (A) Model of the 5′ UTR and N-terminal coding sequence of C showing the position of stem–loop A (SL-A) and stem–loop B (SL-B) in relation to the translation start site (AUG), cHP (c hairpin) and the 5′ CS (cyclization sequence), UAR (upstream of AUG region) and DAR (downstream of AUG region). (B) RNA sequence and structures in the beginning of NS2A gene of viruses from JEV serogroup facilitating –1 programmed ribosomal frameshift and production of unique NS1′ protein. A slippery heptanucleotide is followed by a stem–loop (SL) and pseudoknot (PK). (C) Model of highly structured 3′ UTR of mosquito-borne flaviviruses outlining the location of the three domains, variable domain 1, moderately conserved domain 2, and a highly conserved domain 3. Several SL (SL-I to SL-IV, and 3′SL) and dumb-bell (DB1 and DB2) structures incorporating five PKs (PK1 to PK5), conserved sequences CS3 and RCS3, and 3′ cyclization sequences CS1, 3′-UAR, 3′-DAR complementary to their partners at the 5′ end of the genome are shown. sfRNA start site is located immediately upstream of SL-II.

Castle *et al.*, 1986; Dalgarno *et al.*, 1986; McAda *et al.*, 1987; Coia *et al.*, 1988; Speight *et al.*, 1988) (Fig. 3.1). Co- and post-translational cleavage of the polyprotein precursor by cellular and viral proteases generates three structural (C-prM-E) proteins, which are incorporated into the virus particle, and seven non-structural (NS1-NS2A-NS2B-NS3-NS4A-NS4B-NS5) proteins, which facilitate genomic replication and virion assembly (Castle *et al.*, 1985, 1986; Rice *et al.*, 1985; Wengler *et al.*, 1985; Dalgarno *et al.*, 1986; McAda *et al.*, 1987; Coia *et al.*, 1988; Speight *et al.*, 1988; Nowak *et al.*, 1989; Speight and Westaway, 1989a,b). The various stages of the flavivirus life cycle (from polyprotein translation and RNA replication to virion assembly) appear to be coordinated and strictly regulated. Here, higher order, *cis*-acting RNA structures within the RNA genome play an important role (Liu *et al.*, 2009). These elements contribute to genome stability and participate in inter- and intramolecular interactions with RNA as well as host and viral proteins.

Flavivirus UTRs and their specific structures are involved in translation, initiation of RNA replication and probably determine genome packaging (Khromykh and Westaway, 1997; Proutski *et al.*, 1999; Westaway *et al.*, 2002; Markoff, 2003; Yun *et al.*, 2009). Though the sequences themselves differ throughout the genus, both the 5′ UTR and 3′ UTR can form secondary and tertiary structures which are highly conserved among mosquito-borne flaviviruses (Rice *et al.*, 1985; Shi *et al.*, 1996; Proutski *et al.*, 1997) and are essential for virus replication and translation events (Brinton *et al.*, 1986; Hahn *et al.*, 1987; Men *et al.*, 1996; Chiu *et al.*, 2005).

Cyclization sequences

Conserved complementary cyclization sequences of 8 nt were predicted in the 5′ end of the C coding region (5′CS) and in the 3′ UTR (3′CS or CS1) just upstream of a conserved 3′ terminal stem loop structure (3′SL) in the genomic RNA of several mosquito-borne flavivirus species and proposed to be involved in formation of panhandle structures (Hahn *et al.*, 1987) (Fig. 3.1). Genome cyclization and panhandle formation as an initiator of RNA replication has been suggested as a potential mechanism employed by flaviviruses to ensure that only undamaged, full-length genomes are replicated (Hahn *et al.*, 1987). Mutagenesis studies *in vitro* and *in vivo* showed that mutations in both the 5′ and 3′ cyclization sequences may be tolerated as long as base pairing is maintained (You and Padmanabhan, 1999; Khromykh *et al.*, 2001a). Later, the CS sequences were extended to 10 nt for DENV, 11 nt for WNV, and 18 nt for YFV (Corver *et al.*, 2003; Alvarez *et al.*, 2005; Zhang *et al.*, 2008). Importantly, the proposed long-range RNA interactions between the ends of the RNA genome were confirmed after visualization of individual RNA molecules using atomic force microscopy (Alvarez *et al.*, 2005).

The CS elements are necessary but not sufficient for genome cyclization. Using *in silico* predictions, additional complementary sequences required for viral replication were identified in the genomes of mosquito-borne flaviviruses. First, a sequence in the 5′ UTR upstream of the translation initiator AUG was found to be complementary to a sequence within the stem of the 3′SL region (Alvarez *et al.*, 2005). These 5′ and 3′UAR (upstream of AUG region) elements have also been demonstrated to base-pair (Alvarez *et al.*, 2005, 2008; Zhang *et al.*, 2008; Villordo and Gamarnik, 2009). In addition, a third RNA region has been recently identified which is required for genome cyclization (Friebe and Harris, 2010). The 5′ region is located downstream of the AUG region within the C gene (5′DAR) while the 3′ part is located upstream of the 3′SL. Thus, for mosquito-borne flaviviruses, three pairs of cyclization sequences have been identified (Fig. 3.1). In addition, it has been suggested that a complex RNA structure consisting of three 5-nt-long RNA strands is formed between the 5′ and the 3′ end of the viral genome and is required for replication (Song *et al.*, 2008). The potential contribution of viral and/or cellular proteins, which could potentially influence genome cyclization and panhandle formation, remains to be investigated.

Although the genomes of tick-borne flaviviruses such as tick-borne encephalitis virus (TBEV) have the same genomic organization as mosquito-borne flaviviruses, they do not have homologous CS elements. However, two pairs of complementary sequences (CS-A and CS-B) have been identified and are predicted to base-pair

(Mandl *et al.*, 1993). The 5′CS-A is located within the 5′ UTR while the 3′CS-A is in the stem of the 3′SL, thus these CS are located more closely to the ends of the genome than their mosquito-borne flavivirus counterparts. The CS-A regions seem to interact in a manner required for viral RNA replication (Kofler *et al.*, 2006). 5′CS-B and 3′CS-B are located similarly to their mosquito-borne flavivirus counterparts (in the C protein coding region and upstream of the 3′SL) but do not play a role in RNA replication so that the whole C coding region can be removed from the virus genome without affecting replication (Kofler *et al.*, 2006).

5′-terminal elements

The sequence of the 5′UTR (about 100 nt in length) is not well conserved between members of the genus *Flavivirus* (Liu *et al.*, 2009). In contrast, the secondary structures seem to be conserved within this region and play an important role during flavivirus replication and translation. The 5′UTR of the flavivirus genome contains two stem loops, SL-A and SL-B, which have distinct functions in viral RNA synthesis (Lodeiro *et al.*, 2009). SL-A (also called 5′SL or Y-shaped structure) is ~70 nt in length while SL-B contains the 5′UAR (Fig. 3.1). Interestingly, a short, conserved poly-U tract located immediately downstream of SL-A has been shown to be necessary for RNA synthesis (Lodeiro *et al.*, 2009). SL-A itself has been shown to be a promoter element for RNA synthesis which is recognized and directly bound by the viral RNA-dependent RNA polymerase NS5 (Filomatori *et al.*, 2006; Dong *et al.*, 2008; Zhang *et al.*, 2008) though exactly how it exerts this function remains unknown. The side loop as well as the bottom of the flavivirus SL-A are probably essential for genomic replication (Li *et al.*, 2010) and the Y-shaped structure of SL-A acts as a promoter in a manner necessary for RNA synthesis (Filomatori *et al.*, 2006). In addition, SL-A has to be located at the very 5′ end of the viral genome to exert its function (Friebe and Harris, 2010). Once NS5 is bound to SL-A, it is suggested to be delivered to the initiation site of (–) strand RNA synthesis at the 3′ end via genome cyclization (Filomatori *et al.*, 2006; Zhang *et al.*, 2008).

Finally, 14 nt into the C protein-coding region, the cHP (capsid hairpin) element directs start codon selection presumably by stalling the scanning initiation complex over the first AUG to favour its recognition (Clyde and Harris, 2006) (Fig. 3.1), and is essential for RNA replication via either stabilization of the panhandle structure after cyclization or recruitment of replication factors (Clyde *et al.*, 2008). The translational role of the cHP has most probably evolved owing to the lack of optimal Kozak initiation sequence around the start codon of the polyprotein for many flaviviruses (Clyde and Harris, 2006).

Importantly, the 5′ UTR of the (+) strand genome also forms a SL at the 3′ end of the complementary (–) strand. This 3′(–)SL binds TIAR, an RNA binding protein that regulates stress-induced translational arrest and is required for flavivirus replication (Li *et al.*, 2002; Emara *et al.*, 2008).

3′ Terminal elements

A cytosine residue is present at the penultimate and/or terminal base at the 3′ end of many non-polyadenylated viral RNAs. In flaviviruses, the conserved 5′ and 3′ terminal nucleotides end in 5′AG(U/A)-(G/A/U)CU_{OH}3′ while the negative strand ends in (A/U)CU_{OH} (Rice *et al.*, 1985; Khromykh *et al.*, 2003). Both the conserved bases in the 3′ terminal dinucleotide –CU_{OH} of (+) and (–) RNA strands are required for flavivirus replication (Khromykh *et al.*, 2003).

Generally, the 3′ UTRs of flavivirus genomes show a considerable variation in size and sequence. However, conserved secondary structures have been identified in the 3′ UTR of the genomic RNA of flaviviruses (Brinton *et al.*, 1986; Rauscher *et al.*, 1997; Markoff, 2003). The 3′ UTR of mosquito-borne flaviviruses can be divided in three domains (Fig. 3.1). The first domain is located directly downstream of the stop codon and is variable in sequence, in contrast to domain two which is moderately conserved at the sequence level containing several hairpin motifs, and domain three which is highly conserved and contains the CS1 and stable SL structures (Markoff, 2003; Liu *et al.*, 2009). The organization of the 3′ UTR of tick-borne flaviviruses differs greatly from their mosquito-borne relatives. Despite the variability, similar patterns of conserved sequences and structures have been identified (Khromykh *et al.*, 2001a) (see below).

The 5′ end of the 3′ UTR begins with an AU-rich region, which forms a stem–loop I (SL-I) upstream of SL-II, which is in turn followed by a short, repeated conserved hairpin (RCS3) and SL-III (Fig. 3.1) (Pijlman *et al.*, 2008). This combination is followed by a structure remarkably similar to the preceding structures, SL-IV and CS3 (Fig. 3.1) (Proutski *et al.*, 1997). Further downstream, one or two copies of a second conserved sequence (CS2 and RCS2) are present in the 3′ UTR of some mosquito-borne flaviviruses. These sequences fold into dumbbell structures DB 1 and 2, respectively, and are followed by a short SL and the 3′SL (Olsthoorn and Bol, 2001) (Fig. 3.1). Most of these structures were identified using *in silico* predictions, however the 3′SL of the (+) and the (−) strand were demonstrated to exist by structure probing (Brinton *et al.*, 1986; Shi *et al.*, 1996). The 3′SL of the (+) strand is about 90 to 120 nt long and is the most prominent structure at the very 3′ end of the viral genome (Brinton *et al.*, 1986; Mohan and Padmanabhan, 1991). It contains functional regions binding to viral and host factors and can, depending on the virus, enhance or inhibit translation of reporter mRNAs. The 3′SL has been shown to interact with a number of cellular proteins including phosphorylated translation elongation factor 1α (EF-1α), poly(A)-binding protein (PABP), La autoantigen and polypyrimidine tract-binding protein (PTB) (Blackwell and Brinton, 1995, 1997; De Nova-Ocampo *et al.*, 2002; Polacek *et al.*, 2009). Although the 3′SL structure in flavivirus RNA is well conserved among species, the primary nucleotide sequences involved are not completely conserved (Brinton *et al.*, 1986; Markoff, 2003). An 11 bp DENV-2 nucleotide sequence constituting the uppermost portion of the lower half of the long stem in the 3′SL is essential for DENV replication, whereas the structure rather than the sequence of the upper half of the 3′SL is a determinant for viral growth (Zeng *et al.*, 1998). Importantly, part of this domain appears to be responsible for species-specific replication. Nucleotides on the top loop of the 3′SL (the conserved pentanucleotide loop 5′CACAG(A/U)3′) are *cis*-acting and their conservation is important for replication of some but not all flaviviruses (Wengler and Castle, 1986; Westaway *et al.*, 2003; Elghonemy *et al.*, 2005; Silva *et al.*, 2007).

The described structures have been investigated in some detail for their requirement in RNA replication and translation. Generally, only a relatively short region of the flavivirus 3′ UTR consisting of the most 3′ terminal nucleotides is absolutely essential for RNA replication, although replication efficiency decreased with larger deletions into non-essential parts (Blackwell and Brinton, 1995; Men *et al.*, 1996; Khromykh and Westaway, 1997; Khromykh *et al.*, 2001a; Bredenbeek *et al.*, 2003; Yun *et al.*, 2009). Specific deletion of SL-I or SL-II did not overtly affect KUNV replication (Pijlman *et al.*, 2008) while deletion of CS2, RCS2, CS3 or RCS3 of WNV replicon RNA significantly reduced RNA replication but not viral translation (Lo *et al.*, 2003). This indicates that these elements contribute to but are not essential for RNA replication. In addition, it was shown that deletion of DB1 or DB2 resulted in a viable mutant virus restricted in growth and deletion of both DB structures resulted in a non-viable mutant (Men *et al.*, 1996). The small SL located adjacent to the 3′ terminal SL is also important for virus replication (Friebe and Harris, 2010). However, the exact functions of these regions are still unknown.

Tick-borne flaviviruses share a unique 3′ UTR organization in which a 350 nt conserved core segment in the 3′ proximal region is preceded by a variable domain (Wallner *et al.*, 1995; Gritsun *et al.*, 1997). The core region shows a high degree of sequence conservation among the tick-borne flaviviruses and is predicted to fold into defined secondary structures very similar in form and location to those of mosquito-borne flaviviruses (Proutski *et al.*, 1997; Rauscher *et al.*, 1997) and thus might fulfil similar functions in virus replication. The variable domain is dispensable for genomic replication while the conserved region is not (Mandl *et al.*, 1998b). During passaging in cell culture or mice the variable domain may even acquire extensive, spontaneous deletions (Mandl *et al.*, 1998b). In contrast, mutations in the core region strongly affect tick-borne flavivirus replication and biology (Mandl *et al.*, 1998a,b; Pletnev, 2001).

Conformational arrangements of the flavivirus 3′ UTR that facilitate and regulate genomic replication are not only limited to the secondary structures. A number of tertiary interactions (up to five pseudoknots, PK1 to PK5, Fig. 3.1) have been predicted throughout the 3′ UTR (Shi *et al.*, 1996; Olsthoorn and Bol, 2001; Pijlman *et al.*, 2008). A hairpin type (H-type) PK is formed when a single-stranded RNA region in the loop of a hairpin base-pairs with a stretch of complementary nucleotides elsewhere in the RNA (Brierley *et al.*, 2007). Typically, the final tertiary structure does not significantly alter the preformed secondary structure and can even assist in stabilizing it (Brion and Westhof, 1997). PK1 is predicted to be formed between complementary sequences in SL-II and in RCS3 (Pijlman *et al.*, 2008). PK2 has been predicted to form between complementary sequences in SL-IV and in CS3 (Pijlman *et al.*, 2008) (Fig. 3.1). Further downstream in the DENV, YFV, and JEV subgroup of flaviviruses, two PKs (PK3 and PK4) were predicted to form between DB 1 and 2 and single-stranded RNA regions further downstream (Olsthoorn and Bol, 2001). PK5 has been suggested to form between a sequence in the small 3′SL and a sequence in the 3′SL (Shi *et al.*, 1996). Although these PK interactions have been predicted using computer algorithms, their role in replication and/or translation of flavivirus RNA remain to be experimentally determined.

In addition to its canonical roles in virus replication and translation, the 3′ UTR and its conserved higher order structures in both mosquito- and tick-borne flaviviruses are important for protecting the genomic RNA from complete degradation by cellular ribonucleases resulting in accumulation of a nuclease resistant subgenomic non-coding RNA fragment designated subgenomic flavivirus RNA (sfRNA) (Pijlman *et al.*, 2008). This ~0.5 kb RNA fragment is probably generated by the 5′–3′ exoribonuclease XRN1, which is proposed to be stalled on the rigid secondary and tertiary structures located at the beginning of the 3′ UTR (Pijlman *et al.*, 2008). Recent studies have implicated a major role of PK1 interactions in providing resistance to nuclease degradation and generating sfRNA (Funk *et al.*, 2010; Silva *et al.*, 2010). Importantly, sfRNA has been shown to be required for efficient virus replication, cytopathicity, and pathogenicity (Pijlman *et al.*, 2008); however, the exact mechanism of how sfRNA exerts its function remains unclear.

Internal RNA elements

While most *cis*-acting elements are located in the 5′ or 3′ UTR of the flavivirus genome, some functional RNA structures are also present within the protein-coding region as mentioned above. So far, three of these structures were identified in the genomes of flaviviruses. The cHP is located in the capsid protein-coding region and is involved in genome cyclization as already discussed above (Clyde and Harris, 2006; Clyde *et al.*, 2008). The other two regions each responsible for ribosomal frameshifting will be discussed here.

All encephalitic flaviviruses from the JEV serogroup produce the C-terminally extended form of the NS1 protein called NS1′ (Mason, 1989; Lin *et al.*, 1998; Blitvich *et al.*, 1999) which has recently been demonstrated to contribute to efficient viral neuro-invasiveness (Melian *et al.*, 2010). Using computational analyses of RNA sequence and structure, a conserved –1 ribosomal frameshift motif located at the beginning of NS2A gene responsible for generation of NS1′ was identified (Firth and Atkins, 2009; Melian *et al.*, 2010). This motif consists of a slippery heptanucleotide followed by a short 5 nt spacer and a SL/PK structure (Fig. 3.1) and could lead to a programmed ribosomal frameshift during translation of the polyprotein with about 20–50% efficiency. This frameshift leads to generation of NS1′ protein containing the full-length NS1 protein sequence followed by the first 9 aa of NS2A and a 43 aa unique peptide (Firth and Atkins, 2009; Melian *et al.*, 2010). Thus, this newly identified internal RNA element leads to the production of an additional non-structural flavivirus protein. A direct role of the frameshift and its product, NS1′, in flavivirus replication has yet to be demonstrated. However, as the frameshift leads to a premature halt in translation of the polyprotein, it is probably to limit production of the downstream NS proteins essential for RNA replication which in turn may lead to down-regulation of replication. In addition, NS1′ itself may have yet unidentified function in virus replication and/or virus–host interactions.

Recently, bioinformatic evidence has been also provided for a similar internal RNA element in the genome of insect-specific flaviviruses (Firth *et al.*, 2009). This motif is also located at the beginning of NS2A gene and is predicted to lead to a ribosomal frameshift and the expression of a larger (~253–295 aa) protein spanning the NS2A and NS2B region (designated *fifo*) of unknown function. As discussed above, this frameshift could also indirectly affect RNA replication and/or virus–host interactions.

Viral proteins involved in replication

Co-localization studies using immunofluorescence and immune-electron microscopy assays and co-precipitation studies using antibodies raised against non-structural proteins and dsRNA suggested a role for virtually all non-structural proteins in RNA replication (Welsch *et al.*, 2009; Westaway *et al.*, 1997b). In addition to this role, some of the non-structural proteins (NS2A and NS3) have been shown to be involved in virus assembly (Kummerer and Rice, 2002; Liu *et al.*, 2002, 2003; Leung *et al.*, 2008) and most of them have been implicated in modulation of the host innate antiviral response (Munoz-Jordan *et al.*, 2003; Liu *et al.*, 2004, 2005; Best *et al.*, 2005; Munoz-Jordan *et al.*, 2005; Chung *et al.*, 2006; Lin *et al.*, 2006; Evans and Seeger, 2007; Werme *et al.*, 2008; Ashour *et al.*, 2009; Mazzon *et al.*, 2009; Laurent-Rolle *et al.*, 2010). Given the scope of this chapter, the role of non-structural proteins in RNA replication and assembly will be primarily discussed.

The ≈45 kDa NS1 is a multimeric, glycosylated protein (Winkler *et al.*, 1988; Mason, 1989; Crooks *et al.*, 1994; Pryor and Wright, 1994). NS1 exists in three discrete species in infected mammalian cells: (i) intracellular, membrane-associated, essential for viral replication (Mackenzie *et al.*, 1996b; Muylaert *et al.*, 1996; Khromykh *et al.*, 1999a), (ii) cell surface-associated, involved in signal transduction (Chung *et al.*, 2007; Wilson *et al.*, 2008), and (iii) secreted (sNS1) which inhibits complement activation (Chung *et al.*, 2006). In addition, NS1 has an important yet undefined role in early RNA replication (Mackenzie *et al.*, 1996a; Lindenbach and Rice, 1997; Clark *et al.*, 2007). In infected cells, NS1 colocalizes with the replication machinery (Mackenzie *et al.*, 1996b; Westaway *et al.*, 1997b). The strong association of NS1 with dsRNA, NS2A, NS3, NS4A and NS5 within these compartments has led to the hypothesis that, in concert with these other NS proteins, NS1 is involved in the formation of the RC (Grun and Brinton, 1986; Mackenzie *et al.*, 1996a; Westaway *et al.*, 1997b; Lindenbach and Rice, 1999). This is supported by the fact that deletion of NS1 was shown to cause a block in (−) strand synthesis, thus preventing (+) strand synthesis and virus replication (Lindenbach and Rice, 1997; Khromykh *et al.*, 1999a). In addition, it has been shown that glycosylation as well as dimer formation play an important role in NS1 function in early replication (Hori and Lai, 1990; Pryor and Wright, 1994; Muylaert *et al.*, 1996, 1997; Flamand *et al.*, 1999; Hall *et al.*, 1999). These and other functions of NS1 are discussed in more details in another chapter.

The ≈19 kDa NS2A is a highly hydrophobic, small integral membrane protein which has various functions in the virus life cycle (Rice *et al.*, 1985; Coia *et al.*, 1988; Wengler *et al.*, 1990). It has been shown that NS2A inhibits the interferon response (Liu *et al.*, 2004, 2005; Munoz-Jordan *et al.*, 2005) and plays an essential role in RNA replication, virus assembly and/or secretion (Kummerer and Rice, 2002; Liu *et al.*, 2003; Leung *et al.*, 2008). However, the function of NS2A in flavivirus replication is not well understood. It was shown that the protein strongly binds to the 3′ UTR of the flavivirus genome as well as other components of the viral RC like NS3 and NS5 (Mackenzie *et al.*, 1998). In addition, immunofluorescence analysis showed colocalization of NS2A with dsRNA, NS3 and NS5 in the perinuclear regions of cells (Mackenzie *et al.*, 1998). It was thus suggested that NS2A plays an important role in replication via facilitation of the transport of partially assembled RCs to the VP (Mackenzie *et al.*, 2001). Alternatively, it might be an attractive candidate for coordination of the switch between RNA packaging and RNA replication.

NS2B protein is another small (≈14 kDa) hydrophobic protein which is primarily involved in the proteolytic cleavage of the viral polyprotein as a co-factor for NS3 protease (Lindenbach *et al.*, 2007). It co-localizes with NS3 in flavivirus-induced specific convoluted membrane (CM) structures primarily associated with polyprotein processing (Westaway *et al.*, 1997b). KUNV NS2B was not detected in the sites of viral RNA replication (VPs), and neither co-localized with, nor co-precipitated dsRNA (Westaway *et al.*, 1997b). In contrast, DENV NS2B was detected in VPs (Welsch *et al.*, 2009). Thus, the direct involvement of NS2B in viral RNA replication as a part of the RC remains controversial.

The ≈69 kDa NS3 protein is a highly conserved, multifunctional protein, which has activities required for polyprotein processing, RNA packaging and genomic RNA replication. The N-terminal domain of the protein has high homology to chymotrypsin-like proteases and constitutes the viral protease (together with its co-factor NS2B) which mediates cleavage of the precursor polyprotein at the cytosolic face facilitating maturation of several viral proteins (Bazan and Fletterick, 1989; Gorbalenya *et al.*, 1989; Chambers *et al.*, 1990, 1991; Preugschat *et al.*, 1990; Falgout *et al.*, 1991; Wengler, 1991). Immunoelectron microscopy studies have demonstrated colocalization of NS3 and NS2B in the CM in infected cells (Westaway *et al.*, 1997b). Thus, the CM has been hypothesised to be the site of polyprotein processing, although authors of recent ultrastructural studies with DENV-infected cells speculate that the ER is the main site of viral polyprotein processing and that the CM merely represent sites of protein and lipid storage (Welsch *et al.*, 2009). In addition to protease activity, NS3 has helicase, nucleoside triphosphatase (NTPase) and RNA triphosphatase (RTPase) activities with their functional motifs residing in the C-terminal domain, which determine an essential role of NS3 in viral replication (Lindenbach *et al.*, 2007). Accordingly, mutagenesis of the helicase domain has been demonstrated to efficiently inhibit flavivirus replication (Wengler, 1991, 1993; Cui *et al.*, 1998; Li *et al.*, 1999). The RTPase activity is required for dephosphorylation of the 5′ end of the nascent viral RNA genome before capping by the NS5-encoded guanylyltransferase and methyltransferase (Wengler, 1993; Wu, 2005; Xu *et al.*, 2005; Issur *et al.*, 2009). Both NTPase and helicase activity are stimulated by interaction of NS3 with NS5, supporting the hypothesis that heterodimeric NS3-NS5 unwinds dsRNA during RNA replication (Yon *et al.*, 2005). In addition, NS3 was shown to bind to the genomic RNA at the 3′SL in association with NS5 which also increased its NTPase activity (Cui *et al.*, 1998; Yon *et al.*, 2005). Consistent with its roles in RNA replication, immunofluorescence and electron microscopy analyses revealed a perinuclear distribution of the flavivirus NS3 protein with extension into the cytoplasm of infected cells. Here NS3 colocalizes strongly with NS1 and dsRNA, presumably during formation of RC (Ng and Hong, 1989; Westaway *et al.*, 1997b). *Trans*-complementation analyses have demonstrated that mature NS3 protein associates with the N-terminus of nascent NS5 upon translation (Khromykh *et al.*, 1999b). The NS3–NS5 heterodimer then binds NS2A (which is itself complexed with the 3′SL) to begin formation of the RC (Westaway *et al.*, 2003). In addition to its enzymatic activities associated with RNA replication, NS3 is required for assembly of virions (Khromykh *et al.*, 2000; Kummerer and Rice, 2002; Liu *et al.*, 2002; Pijlman *et al.*, 2006a; Patkar and Kuhn, 2008), although how exactly NS3 exerts this function remains elusive. These and other functions of NS3 are discussed in more details in another chapter.

NS4A is a small, hydrophobic transmembrane protein of ≈16 kDa (Speight and Westaway, 1989b) that colocalizes with dsRNA in perinuclear regions with diffuse cytoplasmic staining, which implies a role for NS4A in RNA replication (Mackenzie *et al.*, 1998). This is further supported by the fact that NS4A not only forms dimers but directly interacts with NS1, NS2A, NS3, and NS5, all proteins involved in the RC (Lindenbach and Rice, 1999; Mackenzie *et al.*, 1998). More specifically, NS4A has been detected around the edges of VPs and convoluted membranes/paracrystalline arrays (CM/PC) (Mackenzie *et al.*, 1998) which led to the hypothesis that NS4A interacts with luminal NS proteins to anchor cytosolic components of the replication complex to the membrane (Khromykh *et al.*, 1999a). In addition, NS4A (including the C-terminal hydrophobic 2 kDa fragment) has been shown to induce the formation of VP and CM during flavivirus infection via dilatation and invagination of the ER membrane (Roosendaal *et al.*, 2006; Miller *et al.*, 2007), a mechanism similar to that utilized by the flock house virus protein A (Kopek *et al.*, 2007). Thus, it has been suggested that NS4A is the key protein responsible for the induction of membrane structures required for flavivirus replication (Roosendaal *et al.*, 2006). Analogous to other NS proteins, NS4A plays a role in modulating the innate immune response by blocking type I interferon signalling and STAT2 translocation to the nucleus (Munoz-Jordan *et al.*, 2003; Liu *et al.*, 2005). However, whether the ability of NS4A to modulate the innate immune response is linked to its apparent role in membrane induction requires further studies.

NS4B is the largest of the four small hydrophobic flavivirus proteins of ≈27 kDa and contains several hydrophobic regions (Coia *et al.*, 1988; Miller *et al.*, 2006; Lindenbach *et al.*, 2007). In KUNV-infected cells, it is located in the cytoplasm in perinuclear compartments as well as in the nucleus (Westaway *et al.*, 1997a) and cytoplasmic NS4B partially co-localized with dsRNA (Westaway *et al.*, 1997a). DENV NS4B has also been shown to co-localize with dsRNA as well as NS3 at the putative sites of RNA replication (Miller *et al.*, 2006). These co-localization studies indicate that NS4B is probably a part of RNA replication complex. NS4B is also required to expand and modify the ER during virus replication (Westaway *et al.*, 1997a) and to inhibit type I interferon signalling (Munoz-Jordan *et al.*, 2003, 2005; Liu *et al.*, 2005; Evans and Seeger, 2007).

NS5 is the largest (~100 kDa) and the most conserved of the flavivirus proteins (Wengler *et al.*, 1990). It is essential for RNA replication as it encodes guanyltransferase (GTase) and S-adenosyl methionine methyltransferase (MTase) in the N-terminal third of the protein and RNA-dependent RNA polymerase (RdRp) in the C-terminal two thirds of the protein (Kamer and Argos, 1984; Koonin, 1991, 1993; Issur *et al.*, 2009). The MTase and GTase activity is required for formation of the type-1 5′ RNA cap structure and deletion of the predicted active MTase centre blocks flavivirus RNA replication but not translation (Khromykh *et al.*, 1998a; Issur *et al.*, 2009). During cap formation, S-adenosyl methionine (SAM) functions as the methyl donor for methylation at both position N-7 of guanine by an RNA guanine-methyltransferase (N-7 MTase), followed by methylation of the first nucleotide at the ribose 2′-OH position by a nucleoside 2′-O MTase to form the cap 1 ($m^7GpppAmpN_1$) structure (Ray *et al.*, 2006; Zhou *et al.*, 2007). Crystal structure of the RdRp domain of NS5 revealed the classic palm, thumb and fingers domains (Malet *et al.*, 2007) with previously predicted and characterized GDD active site (Kamer and Argos, 1984; Poch *et al.*, 1989; Koonin, 1991; Khromykh *et al.*, 1998a). This motif is essential for RNA replication (Kamer and Argos, 1984; Koonin, 1991). Computer generated docking studies of the MTase and RdRp demonstrates that the MTase domain is in close proximity to where newly synthesized RNA is released from the RdRp active site. Thus, capping of the nascent RNA genome can occur after the genomic RNA leaves the RdRp domain (Malet *et al.*, 2007).

In infected cells, NS5 colocalizes with dsRNA and other NS proteins in the perinuclear region within VP as a member of the RC (Buckley *et al.*, 1992; Edward and Takegami, 1993; Kapoor *et al.*, 1995a; Westaway *et al.*, 1997b; Mackenzie *et al.*, 2007). However, in addition to its cytoplasmic localization, NS5 of some but not other flaviviruses has been shown to translocate to the nucleus (Buckley *et al.*, 1992; Kapoor *et al.*, 1995a; Malet *et al.*, 2007). The exact role this nuclear localization plays in the flavivirus life cycle is currently unclear. Interestingly, the same region of NS5 responsible for nuclear translocation has been shown to bind NS3 (Johansson *et al.*, 2001). Thus, it has been suggested that NS5 translocation to the nucleus is dependent on NS3 as a way of regulating RNA replication and switch to other viral processes (Wu *et al.*, 2005). In addition to its essential role in replication, NS5 also acts as a highly potent IFN antagonist (Best *et al.*, 2005; Lin *et al.*, 2006; Werme *et al.*, 2008; Ashour *et al.*, 2009; Mazzon *et al.*, 2009; Laurent-Rolle *et al.*, 2010).

Model of flavivirus RNA replication

Three major species of flavivirus RNA associated with RNA replication are (i) the single-stranded genomic RNA, (ii) an RNase-resistant double-stranded replicative form (RF) made up of (+) RNA and (−)RNA of the same size, and (ii) a heterogeneous population of replicative intermediate (RI) RNAs comprising both dsRNA and ssRNA which is partially resistant to RNases (Stollar *et al.*, 1967; Wengler and Gross, 1978; Cleaves *et al.*, 1981; Chu and Westaway, 1985). These RF, RI and genomic RNAs were shown to reside in close association with the flavivirus proteins NS1, NS2A, NS3, NS4A, and NS5 (Grun and Brinton, 1986; Chu and Westaway, 1992). In addition, immunofluorescence and electron microscopy studies show colocalization of dsRNA, NS1, NS2A, NS3, NS4A, and NS5 to VPs in the perinuclear region of infected cells, with labelling of NS2A and NS4A concentrated around the edges of the VP which suggests membrane association (Ng *et al.*, 1983; Mackenzie *et al.*, 1996a, 1998; Westaway *et al.*, 1997b; Welsch *et al.*, 2009). Moreover, RNA binding studies and immunoprecipitation assays showed a complex formation between dsRNA, NS1, NS2A, NS3, NS4A, and NS5 (Westaway *et al.*, 1997b; Mackenzie *et al.*, 1998). Based on the above-mentioned findings, the following model for assembly of the replication complex and subsequent RNA replication was proposed (Khromykh *et al.*, 1999a,b, 2000; Westaway *et al.*, 2003).

Replication of flaviviruses starts with translation of the incoming (+) RNA genome which acts as an mRNA at ribosomes associated with the RER (Fig. 3.2). The resulting polyprotein is then processed by a highly regulated proteolytic cascade into the individual flavivirus proteins, presumably at the CM (MacKenzie and Westaway, 2001; Westaway *et al.*, 2002; Lindenbach *et al.*, 2007). Formation of the RC has been proposed to occur during polyprotein translation (Fig. 3.2). First, the protease NS3 (and possibly NS2A) binds to conserved sequences in the N-terminal half of the RdRp NS5 (Preugschat *et al.*, 1990; Buckley *et al.*, 1992; Kapoor *et al.*, 1995a; Khromykh *et al.*, 1999b) (panel A). Then an unknown signal allows for the switch from translation to replication, and a NS2A, NS3, and NS5 complex binds to the 3′SL of the template (+) strand RNA (Chen *et al.*, 1997; Mackenzie *et al.*, 1998) (panel B). In an alternative model, viral polymerase NS5 first binds the promoter SL-A at the 5′ end of the virus genome and then reaches the initiation site on the 3′ UTR via long-range RNA–RNA interactions mediated by cyclization sequences which brings the 3′ end of the genome near the NS5–SL-A complex (Filomatori *et al.*, 2006). Upon base-pairing and genome cyclization, the bottom half of the 3′SL is predicted to open which allows the initiation of (−) strand synthesis (panel C). The partially assembled replication complex is then transported together with the attached template RNA to VP, where the hydrophobic regions of NS2A and the membrane-bound NS4A interact (Mackenzie *et al.*, 1998). Hydrophilic residues of NS4A then also associate with luminal NS1, which anchors the replication complex in the ER membrane (Chu and Westaway, 1992; Westaway *et al.*, 1997b; Mackenzie *et al.*, 1998; Lindenbach and Rice, 1999) (panel D). Interestingly, initiation of flaviviral RNA synthesis then occurs *de novo* without the use of protein or RNA primers (Ackermann, 2001). The (−) strand RNA can be detected as early as 3 h after infection (Lindenbach and Rice, 1997). During synthesis of the (−) strand RNA, the complementary RNA strands base pair to form the dsRNA RF, which is then used as template to repeatedly synthesize nascent (+) genomic RNA in a semiconservative, asymmetric fashion (Chu and Westaway, 1985) (panels E and F). A single cycle for RNA replication is completed within 15–20 minutes and yields one strand of nascent RNA per RI (Cleaves *et al.*, 1981; Chu and Westaway, 1985) (panel G). Genomic RNA synthesis is asymmetric as (+) strands accumulate about 10-fold over (−) strands (Cleaves *et al.*, 1981; Muylaert *et al.*, 1996). After completion of the progeny (+) RNA synthesis, the RF remains stable, despite being released from the replication complex (Westaway *et al.*, 1999). Both the replication complex and RF are recycled for subsequent rounds of (+) RNA synthesis (Westaway *et al.*, 1999). The newly synthesized (+) RNA can then be used either for additional rounds of RNA replication, or as mRNA for polyprotein translation, or it can be packaged with the help of structural and non-structural proteins into virus particles.

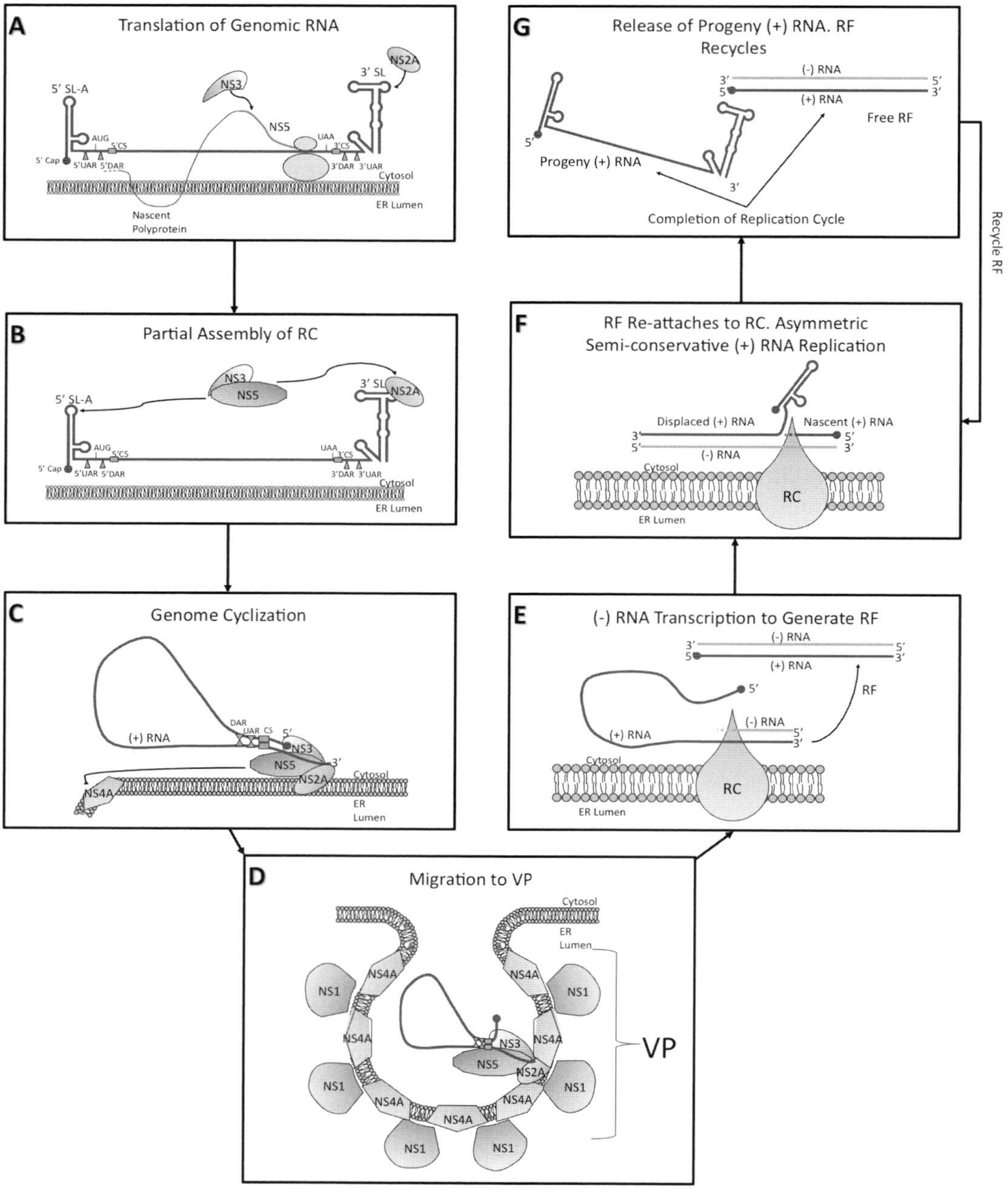
A
Translation of Genomic RNA
5′ SL-A
NS3
NS2A
3′ SL
NS5
AUG
5′CS
UAA
3′CS
5′ Cap
5′UAR 5′DAR
3′DAR 3′UAR
Cytosol
ER Lumen
Nascent
Polyprotein
B
Partial Assembly of RC
NS3
NS5
3′ SL
NS2A
5′ SL-A
AUG
5′CS
UAA
3′CS
5′ Cap
5′UAR 5′DAR
3′DAR 3′UAR
Cytosol
ER Lumen
C
Genome Cyclization
DAR
UAR CS
5′
(+) RNA
NS3
3′
NS5
NS2A
Cytosol
NS4A
ER
Lumen
D
Migration to VP
Cytosol
ER
Lumen
NS4A
NS1
NS3
NS5
NS2A
VP
E
(-) RNA Transcription to Generate RF
(-) RNA
3′
5′
5′
3′
(+) RNA
5′
RF
(-) RNA
(+) RNA
Cytosol
RC
ER Lumen
F
RF Re-attaches to RC. Asymmetric
Semi-conservative (+) RNA Replication
Displaced (+) RNA
Nascent (+) RNA
3′
5′
5′
3′
(-) RNA
Cytosol
RC
ER Lumen
G
Release of Progeny (+) RNA. RF
Recycles
(-) RNA
(+) RNA
Free RF
5′
Progeny (+) RNA
3′
Completion of Replication Cycle
Recycle RF

Figure 3.2 (opposite) Schematic model of flavivirus RNA replication. (A) The precursor polyprotein is translated from the genomic RNA at the rough endoplasmic reticulum (ER). During translation, NS3 binds to the N-terminus of nascent NS5. NS2A binds the 3′ stem–loop (SL) and NS4A intercalates with the ER membrane, dilating a necked vesicle into the lumen. (B) Following proteolytic processing the mature NS3–NS5 heterodimer binds NS2A–RNA complex forming the partially assembled replication complex (RC). Alternatively, NS5 first binds to the 5′ UTR. (C) Long-range RNA–RNA base-pairing interactions between the CSs, UARs and DARs in the 5′ and 3′ ends of the genome result in cyclization and formation of a panhandle-like RNA structure encompassed into the partially assembled RC. (D) The partially formed RC migrates to the newly formed vesicle packets (VP), associating with NS4A. Luminal NS1 and host factors join the protein-RNA aggregate thus completing the RC. (E) Cyclization of the genome brings promoter elements within the 5′ SL into contact with NS5 initiating 5′–3′ (–) strand RNA synthesis. The (+) strand RNA serves as a template for complementary (–) strand synthesis which progresses along the length of the genome generating the double stranded replicative form (RF). (F) The RF re-attaches to the RC and synthesis of a new (+) strand begins. While the nascent (+) RNA is synthesized, it displaces the pre-existing (+) strand. During this process the ssRNA–dsRNA hybrid is designated the replicative intermediate (RI). (G) Once (+) strand synthesis is complete, the RF recycles and the displaced progeny (+) RNA exits the VP for either polyprotein translation or packaging into nucleocapsids. Figure modified from Westaway *et al.* (2003).

Cellular membrane environment facilitating flavivirus replication

It is generally accepted that flavivirus RNA replication occurs exclusively in the cytoplasm although controversial reports of nuclear replication do exist (Brawner *et al.*, 1979; Uchil *et al.*, 2006). Ultrastructural changes in the perinuclear region of flavivirus-infected cells can be easily observed (Mackenzie *et al.*, 1998, 1999, 2001, 2007; MacKenzie and Westaway, 2001; Westaway *et al.*, 2002; Welsch *et al.*, 2009; Gillespie *et al.*, 2010). Thus, flaviviruses induce massive morphological changes in their host cells. The earliest event of the remodelling process induced by flaviviruses is the proliferation of ER membranes (Gillespie *et al.*, 2010). Then, smooth membrane structures are formed. These membrane structures are small clusters of 70–200 nm necked vesicles or VPs, which contain the RC as discussed above. As the infection progresses, VP accumulate and convoluted membranes (CM) are formed in close proximity. CM are continuous with the ER and can be either highly ordered (in paracrystalline arrays) or randomly folded (Murphy *et al.*, 1968; Leary and Blair, 1980; Welsch *et al.*, 2009).

As mentioned, replication occurs in close association with virus-induced, rough and smooth ER-derived membranes, which, with the help of host and viral proteins as well as host lipids, constitute an intricate factory for the efficient production of virus particles (Welsch *et al.*, 2009; Gillespie *et al.*, 2010). The RC containing the viral proteins, RNA as well as host cell factors is surrounded by the virus-induced VP membranes, which not only promote viral replication by concentrating the necessary factors but also contribute to the protection of viral factors from detection by the host (Novoa *et al.*, 2005; Hoenen *et al.*, 2007; Netherton *et al.*, 2007; Överby *et al.*, 2010) Generally, flavivirus infection induces membranes of different morphology, CM and VP, and it has been suggested that VP represent the sites of RNA replication while CM represent sites of translation and polyprotein processing (Mackenzie *et al.*, 1996b). Recently, the complex membrane architecture in infected cells has been shown to link RNA genome replication and virion assembly (Welsch *et al.*, 2009). Virus-induced membrane compartments are interconnected and constitute a single endo-membrane system in which all the steps necessary for viral replication and assembly take place (Gillespie *et al.*, 2010). Interestingly, the vesicles in which RNA replication occurs are closely linked to virus budding sites and are connected by a pore to the surrounding cytoplasm, which would allow exchange of factors between the compartments (Welsch *et al.*, 2009).

Virion assembly and maturation

Although many of the specific processes within the flavivirus life cycle remain unknown, a canonical model of virion assembly and maturation has been constructed allowing insight into the strategy of coupled replication, assembly and egress (Fig. 3.3). Following replication of the genomic flavivirus RNA in VPs, progeny strands are incorporated into the nucleocapsid as they exit VP pores (Khromykh *et al.*, 2001b; Welsch *et al.*, 2009). Nucleocapsids are then transported through an adjacent ER membrane and acquire the independently forming lipid envelope containing prM–E heterodimers (Welsch *et al.*, 2009). These newly formed virions accumulate in ER cisternae prior to transport to the Golgi apparatus in individual protuberant vesicles (MacKenzie and Westaway, 2001; Welsch *et al.*, 2009). Virus particles are then transported rapidly through the Golgi, allowing maturation of the

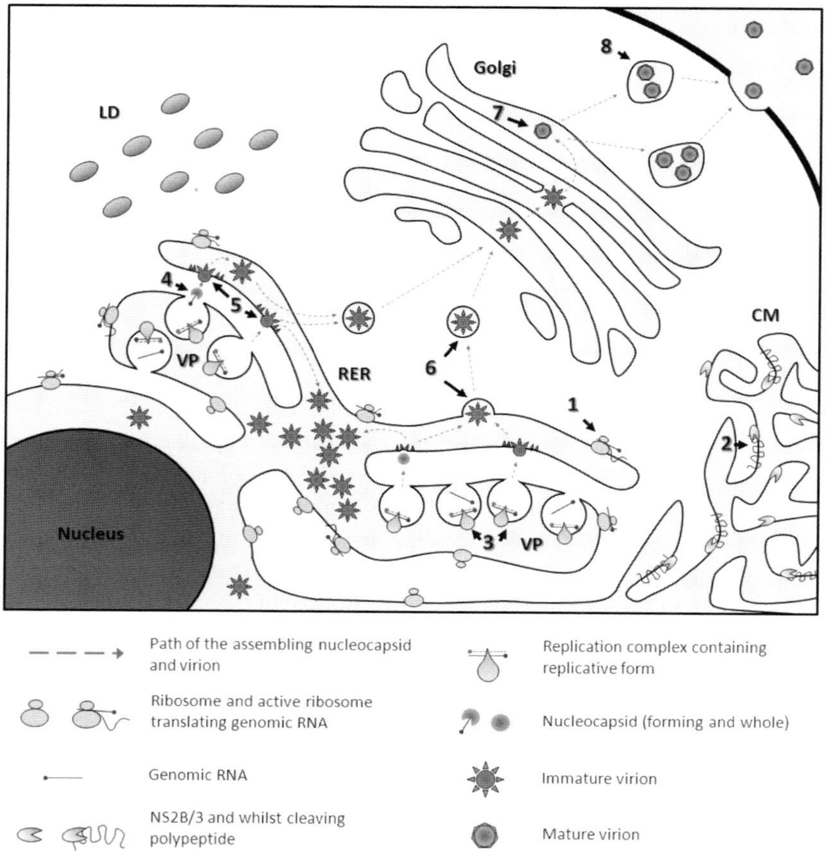

Figure 3.3 Canonical model of flavivirus replication and assembly. (1) Genomic RNA is translated at the rough endoplasmic reticulum (RER) forming a polyprotein that crosses the RER membrane multiple times. (2) The polyprotein is cleaved post-translationally by viral NS2B/3 protease and host cell signalase at either the RER or within virus-induced convoluted membranes (CM). (3) The replication complex consisting of NS1, NS2A, NS4A, NS3 and NS5, sequestered within virus-induced vesicle packets (VP), generates and processes the RNA replicative form (via firstly minus-strand synthesis, then generation of positive strand progeny RNA). (4) Progeny RNA exits the VP pore and is incorporated into the nucleocapsid by C protein. (5) Nucleocapsids collect the envelope glycoproteins prM and E and form immature virions in the ER lumen. (6) Immature virions are transported to the Golgi apparatus via individual vesicles. (7) Virions traverse the Golgi apparatus allowing maturation of carbohydrate groups on envelope glycoproteins. (8) Virions are transported in endosomes and via the *trans*-Golgi network, where acidification allows furin to cleave prM to M, thus forming mature particles. Mature virions then accumulate for secretion at the plasma membrane. Figure modified from Mackenzie and Westaway (2001) and Welsch *et al.* (2009).

carbohydrate groups on the prM and E glycoproteins (MacKenzie and Westaway, 2001; Lorenz *et al.*, 2003; Hanna *et al.*, 2005; Welsch *et al.*, 2009). Immature virions are then conveyed to vesicles of the *trans*-Golgi network (MacKenzie and Westaway, 2001). Here they are exposed to low pH allowing the host enzyme furin to proteolytically cleave prM (Stadler *et al.*, 1997; MacKenzie and Westaway, 2001; Li *et al.*, 2008; Yu *et al.*, 2008). The pr portion of prM is released upon exit of the virion from the cell, resulting in the formation of a mature virion (MacKenzie and Westaway, 2001).

Encapsidation of nascent genomic RNA

Following RNA replication progeny flavivirus genomes are incorporated into nucleocapsids with the capsid protein via an undefined association, as no packaging signal has yet been identified. NMR and X-ray crystallography experiments (Dokland *et al.*, 2004; Ma *et al.*, 2004) have revealed that the capsid protein consists of four structurally conserved alpha helices and contains a conserved internal hydrophobic domain. C forms dimers in solution mediated by association of the internal hydrophobic regions, and in particular a conserved tryptophan residue (W69 in WNV) in α-3 (Kiermayr *et al.*, 2004; Bhuvanakantham and Ng, 2005). The α-4 helices associate to form a positively charged surface that has been proposed to act alongside the flexible N-terminus in stabilizing the interaction with RNA (Khromykh and Westaway, 1996; Dokland *et al.*, 2004; Ma *et al.*, 2004). Highly charged basic residues found at the N- and C-termini (aa 1–32 and 82–105 in WNV) associate with nucleic acids, particularly with the 5′ and 3′ UTRs of flavivirus genomic RNA (Khromykh and Westaway, 1996; Khromykh *et al.*, 1998b).

The structure of flavivirus nucleocapsids packaged within mature and immature virus particles has recently been investigated by subtracting the prM–E projections from cryo-electron microscopy reconstructions (Zhang *et al.*, 2007). The results indicate that there is no detectable ordered icosahedral arrangement of C protein around the genomic RNA but do not exclude the possibility of another, unanticipated ordered arrangement being present. The underlying structure of the nucleocapsid remains elusive despite reports of *in vitro* assembly of spherical nucleocapsid-like particles when combining purified C with either *in vitro*-transcribed viral RNA or ssDNA (Kiermayr *et al.*, 2004; Lopez *et al.*, 2009). Whether this apparent spherical architecture is due to clustering of C tetramer ribbons (Dokland *et al.*, 2004), RNA structural rearrangement due to C chaperone activity (Ivanyi-Nagy *et al.*, 2008), an unforeseen association with lipids (Samsa *et al.*, 2009), or something else entirely needs to be further resolved.

A recent study suggested that C protein association with ER-derived lipid droplets (LDs) is an important step in the assembly of virus particles (Samsa *et al.*, 2009). Treatment of DENV-infected BHK cells with the drug disrupting LD formation also led to a profound inhibition of infectious particle production without interfering with RNA synthesis or polypeptide translation (Samsa *et al.*, 2009). Hepatitis C virus has also been shown to utilize LDs in virus assembly (Miyanari *et al.*, 2007). Previous study demonstrated C localization to ER membranes adjacent to VPs and to the nucleus (Westaway *et al.*, 1997a) and it is possible that interfering with C-lipid association by the LD-disrupting drug treatment may simply lead to mistargeting of C away from the sites of RNA replication (VPs), thus inhibiting nucleocapsid formation without any active involvement of LDs.

Budding of virions at the rough endoplasmic reticulum

Whilst direct visualization of the encapsidation process remains elusive, Welsch *et al.* (2009) have used electron tomography to demonstrate occasional diffuse electron density at the cytosolic junction of vesicle pores, as well as putative virion budding events at opposing ER membranes. It is probable that the nucleocapsid formation and virus budding occur simultaneously, possibly mediated by a host protein or lipid interaction in association with C. Empty prM–E particle (subviral particle; SVP) secretion has been well documented in flavivirus infection (Lindenbach *et al.*, 2007; Murray *et al.*, 2008), and SVPs are able to be assembled and secreted when the prM and E genes are expressed independently of other viral proteins (Konishi *et al.*, 1992, 2001; Konishi and Mason, 1993; Allison *et al.*, 1994, 1995, 2003; Schalich *et al.*, 1996; Corver *et al.*, 2000; Ferlenghi *et al.*, 2001; Kroeger and McMinn, 2002; Lorenz *et al.*, 2003; Wang *et al.*, 2009).

The source of the flavivirus envelope has long been believed to be the ER membrane, a logical assumption based upon the observed clustering of immature virions in the lumen of dilated ER cisternae (Westaway *et al.*, 1997a,b; MacKenzie and Westaway, 2001; Lindenbach *et al.*, 2007; Tan *et al.*, 2009; Welsch *et al.*, 2009) as well as co-localization of anti-prM or -E antibodies with ER marker proteins and lipids (MacKenzie and Westaway, 2001; Lorenz *et al.*, 2003; Lin and Wu, 2005; Miorin *et al.*, 2008; Limjindaporn *et al.*, 2009; Tan *et al.*, 2009; Welsch *et al.*, 2009). This concept has been validated, firstly by use of drugs that inhibit specific membrane trafficking and/or glycoprotein maturation (Courageot *et al.*, 2000; MacKenzie and Westaway, 2001; Lorenz *et al.*, 2003; Whitby *et al.*, 2005), and subsequently by direct visualization of virion budding at ER membranes directly opposed to VPs (Welsch *et al.*, 2009). It should be noted that previous to these studies, immuno-EM research involving the Sarafend strain of WNV revealed virion budding at the plasma membrane (Ng *et al.*, 1994). This process has not been observed for other members of family *Flaviviridae* and thus represents a possible alternative pathway of virion assembly and maturation to that described in Fig. 3.3. The site of SVP formation cannot be accurately differentiated from that of virions in these investigations; however it presumably also occurs within the ER. This is supported by visualization of SVP assembly at the ER in cells transfected with a prM–E gene cassette (Lorenz *et al.*, 2003). Studding the ER-derived lipid envelope are the two viral structural glycoproteins prM (~26 kDa) and E (~53 kDa), which associate as 60 heterotrimeric spikes on the surface of immature virus particles (Mukhopadhyay *et al.*, 2005; Lindenbach *et al.*, 2007; Li *et al.*, 2008). Correct heterodimer formation between prM and E on the luminal side of the ER membrane is crucial for the proper assembly of both SVPs and complete viral particles (Konishi and Mason, 1993). The primary roles attributed to prM are to assist the correct folding of E (Lorenz *et al.*, 2002). Early stage high-mannose glycosylation of the pr peptide has been implicated as an important facilitator of prMs chaperone duties, potentially via recruitment of the lectin-like chaperones calnexin and calreticulin (Goto *et al.*, 2005; Hanna *et al.*, 2005) which, in association with another ER resident chaperone human immunoglobulin heavy chain binding protein (Bip), have been shown to interact with flavivirus E (Limjindaporn *et al.*, 2009). Specific regions within the envelope proteins that mediate virion assembly/secretion are not known, though evidence is mounting for a role of the transmembrane domains (TMDs) in both proteins (Op De Beeck *et al.*, 2003), specifically TMD2 of the E protein (Orlinger *et al.*, 2006) and the semi-conserved His99 residue of the prM protein (Lin and Wu, 2005). The conserved Tyr78 of the prM ectodomain and DENV-conserved His39 of the M protein have also been implicated as critical for virus assembly/maturation (Pryor *et al.*, 2004; Tan *et al.*, 2009).

Virion maturation and release

Following budding of immature virus particles and their accumulation in ER cisternae, transport vesicles pinch off from the ER and transit to the Golgi apparatus for glycoprotein maturation (MacKenzie and Westaway, 2001; Welsch *et al.*, 2009). The mechanism driving this is as yet unknown; however involvement of the src family kinase c-Yes has been implicated as crucial (Hirsch *et al.*, 2005; Chu and Yang, 2007). The glycosylation status of the prM and E proteins has also been demonstrated to play a role in Golgi trafficking and virion secretion. N-linked glycosylation of prM and Asn153/154 of the E protein have been shown to greatly enhance the release of virus particles/SVPs into the media of cultured cells, depending on cell type (Goto *et al.*, 2005; Hanna *et al.*, 2005; Mondotte *et al.*, 2007; Kim *et al.*, 2008; Lee *et al.*, 2010). The role of glycosylation at Asn66/67 sites in E protein is more controversial with different groups variably finding it to be beneficial (Lee *et al.*, 2010), have minimal influence (Goto *et al.*, 2005), or be detrimental (Mondotte *et al.*, 2007) to secretion depending on cell type. Interestingly, the simultaneous presence of carbohydrate chains at both Asn residues in E appears to negatively affect infectivity of virus particles though promote particle secretion (Lee *et al.*, 2010).

Following Golgi trafficking immature virions localize in endosomal compartment and the *trans*-Golgi network (MacKenzie and Westaway, 2001). These compartments undergo acidification, initiating a conformational change in the prM–E heterodimers in which the trimeric spikes dissociate and the dimers lie flat on the virion surface, an arrangement analogous to that found in the mature virion (Mukhopadhyay *et al.*, 2005; Li *et al.*, 2008; Yu *et al.*, 2008). This conformational rearrangement exposes the furin cleavage site at the pr–M junction allowing the host enzyme to liberate the pr peptide, thus exposing the fusion loop on E (Stadler *et al.*, 1997; Mukhopadhyay *et al.*, 2005; Li *et al.*, 2008; Yu *et al.*, 2008). Mature virus particles are then assumed to be secreted at the plasma membrane immediately as no M containing particles have ever been found sequestered in cellular compartments of the secretory pathway.

Coupling between replication and packaging and the role of NS proteins in assembly

No packaging signal has so far been identified in the flavivirus genomic RNA. Thus another mechanism is likely to exist in its place to ensure the specificity of genome encapsidation. Various *trans*-packaging experiments that supply flavivirus structural proteins either expressed transiently from a plasmid (Yoshii *et al.*, 2005; Ansarah-Sobrinho *et al.*, 2008), as part of an alpha-virus replicon (Khromykh *et al.*, 1998b; Varnavski and Khromykh, 1999; Scholle *et al.*, 2004a; Patkar and Kuhn, 2008), or from a stable cell line (Gehrke *et al.*, 2003; Harvey *et al.*, 2004; Fayzulin *et al.*, 2006; Yun *et al.*, 2007; Lai *et al.*, 2008; Leung *et al.*, 2008) as a means to encapsidate replicon RNA have consistently failed to show any recovery of particles containing the RNA encoding structural gene cassette or other non-flavivirus RNA. Importantly, this demonstrates that RNA encapsidation by flavivirus structural proteins is specific for the flavivirus replicon/genome. To further analyse packaging mechanisms which ensure encapsidation of flavivirus RNA and its functional relationship to RNA replication, a CMV promoter-based plasmid DNA encoding replication-deficient (deletion of RNA polymerase active motif GDD) full-length KUNV infectious cDNA was employed (Khromykh *et al.*, 2001b). The CMV promoter allowed continuous transcription and accumulation of genomic RNA, as well as translation and processing of all viral proteins, but the GDD-deletion prevented viral RNA replication. This plasmid DNA was incapable of producing virus particles when transfected into normal cells. However, when RNA replication was rescued in KUNV replicon expressing cells, secreted infectious virus particles were readily detected. These experiments demonstrated that RNA packaging into virus particles is functionally coupled to RNA replication.

The coupling between replication and assembly suggests a potential role for the NS proteins in RNA packaging and/or virus assembly. Studies showing that point mutations in the NS2A protein of KUNV (I59N) (Leung *et al.*, 2008; Liu *et al.*, 2003) and YFV (K190S) (Kummerer and Rice, 2002) abrogated assembly/secretion of infectious virions without dramatically affecting RNA replication, established the role for NS2A protein in virus assembly/secretion. Importantly, virion assembly was restored upon complementation providing NS2A expressed *in trans* (Kummerer and Rice, 2002; Liu *et al.*, 2003). In addition, prolonged replication of mutant replicons/viruses resulted in frequent reversions to the wild type residue, or mutations to residues of similar to wild type residue properties, or to the appearance of compensatory mutations at other residues in the NS2A or NS3 (Kummerer and Rice, 2002; Leung *et al.*, 2008). Interestingly, one of the compensatory mutations identified in the YFV mutant was located in the NS3 helicase domain (D343V, A or G) (Kummerer and Rice, 2002), suggesting that packaging requires a concerted interaction between these two members of the RC. Indeed, subsequent alanine scanning mutations in the YFV E338-D354 loop of NS3 (in the absence of NS2A mutation) demonstrated that mutation of W349 blocked assembly of virions without affecting its protease or helicase activity (Patkar and Kuhn, 2008). A requirement for *cis*-encoding of NS3 to facilitate packaging of the KUNV genome has also been reported (Liu *et al.*, 2002; Pijlman *et al.*, 2006a), although *trans*-complementation of NS3 enzymatic functions can also not be excluded (Khromykh *et al.*, 2000). This requirement for *cis*-encoding does not appear to be universal among the *Flaviviridae* as YFV does not require NS3 produced in *cis* for packaging (Jones *et al.*, 2005). Although NS2A and NS3 each appear to have a functional role in virion assembly, the specific mechanisms they exert in this regard have not yet been elucidated.

Gaps in knowledge and future directions

Despite the elucidation of many processes in flavivirus replication leading to the generation of relatively comprehensive models of flavivirus replication and assembly (Fig. 3.2 and 3.3), many gaps in the scientific knowledge in this area still exist. Generally, these gaps concern a refinement in understanding of the roles of individual participants. For example, the exact viral composition of the RC remains unresolved with controversy surrounding the involvement of NS2B in RC. The specific mechanisms by which some viral components of RC are involved in formation of virus induced membranes (NS4A/NS4B) or in RNA synthesis (NS1, NS1′ and NS2A) are not fully elucidated. The molecular mechanisms allowing switches from translation to (–) strand RNA synthesis as well as from (–) strand RNA synthesis to (+) strand RNA synthesis have not yet been discovered. An *in vitro* translation/replication system similar to that developed for poliovirus (Molla *et al.*, 1991) would be highly beneficial for defining fine details of these processes; however such *in vitro* system has yet to be developed for flaviviruses. Viral proteins NS2A and NS3, shown to be an integral part of RC, are also involved in virion assembly via yet unknown mechanism. Although the RNA structures and in some cases specific sequences in the UTRs involved in binding to viral and host proteins and in long-range RNA interactions have been determined it is still not clear in most cases how exactly these interactions contribute to facilitating RNA replication, RNA packaging, and RNA/nucleocapsid transport to the sites of virus assembly. Many host proteins have been shown to be localized to the sites of RNA replication and assembly, to bind viral RNAs, to be involved in viral RNA degradation, and to play a role in other various aspects of flavivirus replication cycle; however the specific mechanisms of their involvement in these processes are only beginning to be uncovered. Filling in these and other knowledge gaps will allow to gain much better understanding of the details of flavivirus replication machinery and refine the models presented here.

Acknowledgements

The authors are grateful to Edwin Westaway (deceased) and Theodore Pierson for critical reading of the chapter and helpful suggestions.

References

Ackermann, M., and Padmanabhan, R. (2001). De novo synthesis of RNA by the dengue virus RNA-dependent RNA polymerase exhibits temperature dependence at the initiation but not elongation phase. J. Biol. Chem. *276*, 39926–39937.

Allison, S.L., Mandl, C.W., Kunz, C., and Heinz, F.X. (1994). Expression of cloned envelope protein genes from the flavivirus tick-borne encephalitis virus in mammalian cells and random mutagenesis by PCR. Virus Genes *8*, 187–198.

Allison, S.L., Stadler, K., Mandl, C.W., Kunz, C., and Heinz, F.X. (1995). Synthesis and secretion of recombinant tick-borne encephalitis virus protein E in soluble and particulate form. J. Virol. *69*, 5816–5820.

Allison, S.L., Tao, Y.J., O'Riordain, G., Mandl, C.W., Harrison, S.C., and Heinz, F.X. (2003). Two distinct size classes of immature and mature subviral particles from tick-borne encephalitis virus. J. Virol. *77*, 11357–11366.

Alvarez, D.E., De Lella Ezcurra, A.L., Fucito, S., and Gamarnik, A.V. (2005). Role of RNA structures present at the 3′ UTR of dengue virus on translation, RNA synthesis, and viral replication. Virology *339*, 200–212.

Alvarez, D.E., Filomatori, C.V., and Gamarnik, A.V. (2008). Functional analysis of dengue virus cyclization sequences located at the 5′ and 3′ UTRs. Virology *375*, 223–235.

Anraku, I., Mokhonov, V.V., Rattanasena, P., Mokhonova, E.I., Leung, J., Pijlman, G., Cara, A., Schroder, W.A., Khromykh, A.A., and Suhrbier, A. (2008). Kunjin replicon-based simian immunodeficiency virus gag vaccines. Vaccine *26*, 3268–3276.

Ansarah-Sobrinho, C., Nelson, S., Jost, C.A., Whitehead, S.S., and Pierson, T.C. (2008). Temperature-dependent production of pseudoinfectious dengue reporter virus particles by complementation. Virology *381*, 67–74.

Ashour, J., Laurent-Rolle, M., Shi, P.Y., and Garcia-Sastre, A. (2009). NS5 of dengue Virus mediates STAT2 binding and degradation. J. Virol. *83*, 5408–5418.

Bazan, J.F., and Fletterick, R.J. (1989). Detection of a trypsin-like serine protease domain in flaviviruses and pestiviruses. Virology *171*, 637–639.

Best, S.M., Morris, K.L., Shannon, J.G., Robertson, S.J., Mitzel, D.N., Park, G.S., Boer, E., Wolfinbarger, J.B., and Bloom, M.E. (2005). Inhibition of interferon-stimulated JAK-STAT signaling by a tick-borne flavivirus and identification of NS5 as an interferon antagonist. J. Virol. *79*, 12828–12839.

Bhuvanakantham, R., and Ng, M.L. (2005). Analysis of self-association of West Nile virus capsid protein and the crucial role played by Trp 69 in homodimerization. Biochem. Biophys. Res. Commun. *329*, 246–255.

Blackwell, J.L., and Brinton, M.A. (1995). BHK cell proteins that bind to the 3′ stem–loop structure of the West Nile virus genome RNA. J. Virol. *69*, 5650–5658.

Blackwell, J.L., and Brinton, M.A. (1997). Translation elongation factor-1 alpha interacts with the 3′ stem–loop region of West Nile virus genomic RNA. J. Virol. *71*, 6433–6444.

Blaney, J.E., Hanson, C.T., Firestone, C.Y., Hanley, K.A., Murphy, B.R., and Whitehead, S.S. (2004a). Genetically modified, live attenuated dengue virus type 3 vaccine candidates. Am. J. Trop. Med. Hyg. *71*, 811–821.

Blaney, J.E., Jr., Hanson, C.T., Hanley, K.A., Murphy, B.R., and Whitehead, S.S. (2004b). Vaccine candidates derived from a novel infectious cDNA clone of an American genotype dengue virus type 2. BMC Infect. Dis. *4*, 39.

Blaney, J.E., Sathe, N.S., Goddard, L., Hanson, C.T., Romero, T.A., Hanley, K.A., Murphy, B.R., and Whitehead, S.S. (2008). Dengue virus type 3 vaccine candidates generated by introduction of deletions in the 3′ untranslated region (3′-UTR) or by exchange of the DENV-3 3′-UTR with that of DENV-4. Vaccine *26*, 817–828.

Blitvich, B.J., Scanlon, D., Shiell, B.J., Mackenzie, J.S., and Hall, R.A. (1999). Identification and analysis of truncated and elongated species of the flavivirus NS1 protein. Virus Res. *60*, 67–79.

Brawner, I.A., Trousdale, M.D., and Trent, D.W. (1979). Cellular localization of Saint Louis encephalitis virus replication. Acta Virol. *23*, 284–294.

Bredenbeek, P.J., Kooi, E.A., Lindenbach, B., Huijkman, N., Rice, C.M., and Spaan, W.J.M. (2003). A stable full-length yellow fever virus cDNA clone and the role of conserved RNA elements in flavivirus replication. J. Gen. Virol. *84*, 1261–1268.

Brierley, I., Pennell, S., and Gilbert, R.J. (2007). Viral RNA pseudoknots: versatile motifs in gene expression and replication. Nat. Rev. Microbiol. *5*, 598–610.

Brinton, M.A., Fernandez, A.V., and Dispoto, J.H. (1986). The 3′-nucleotides of flavivirus genomic RNA form a conserved secondary structure. Virology *153*, 113–121.

Brion, P., and Westhof, E. (1997). Hierarchy and dynamics of RNA folding. Annu. Rev. Biophys. Biomol. Struct. *26*, 113–137.

Buckley, A., Gaidamovich, S., Turchinskaya, A., and Gould, E.A. (1992). Monoclonal antibodies identify the ns5 yellow-fever virus nonsturctural protein in the nuclei of infected cells. J. Gen. Virol. *73*, 1125–1130.

Bukreyev, A., Whitehead, S.S., Murphy, B.R., and Collins, P.L. (1997). Recombinant respiratory syncytial virus from which the entire SH gene has been deleted grows efficiently in cell culture and exhibits site-specific attenuation in the respiratory tract of the mouse. J. Virol. *71*, 8973–8982.

Castle, E., Nowak, T., Leidner, U., and Wengler, G. (1985). Sequence-analysis of the viral core protein and the membrane-associated protein-V1 and protein-NV2 of the flavivirus West Nile Viru and of the genome sequence for these proteins. Virology *145*, 227–236.

Castle, E., Leidner, U., Nowak, T., and Wengler, G. (1986). Primary structure of the West Nile flavivirus genome region coding for all nonstructural proteins. Virology *149*, 10–26.

Chambers, T.J., and Nickells, M. (2001). Neuroadapted yellow fever virus 17D: Genetic and biological characterization of a highly mouse-neurovirulent virus and its infectious molecular clone. J. Virol. *75*, 10912–10922.

Chambers, T.J., Hahn, C.S., Galler, R., and Rice, C.M. (1990). Flavivirus genome organization, expression, and replication. Annu. Rev. Microbiol. *44*, 649–688.

Chambers, T.J., Grakoui, A., and Rice, C.M. (1991). Processing of the Yellow-Fever Virus nonstructural polyprotein – a catalytically active NS3-proteinase domain and NS2B are required for cleavages at dibasic sites. J. Virol. *65*, 6042–6050.

Chang, D.C., Liu, W.J., Anraku, I., Clark, D.C., Pollitt, C.C., Suhrbier, A., Hall, R.A., and Khromykh, A.A. (2008). Single-round infectious particles enhance immunogenicity of a DNA vaccine against West Nile virus. Nat. Biotechnol. *26*, 571–577.

Chen, C.J., Kuo, M.D., Chien, L.J., Hsu, S.L., Wang, Y.M., and Lin, J.H. (1997). RNA–protein interactions: involvement of NS3, NS5, and 3′ noncoding regions of Japanese encephalitis virus genomic RNA. J. Virol. *71*, 3466–3473.

Chiu, W.W., Kinney, R.M., and Dreher, T.W. (2005). Control of translation by the 5′- and 3′-terminal regions of the dengue virus genome. J. Virol. *79*, 8303–8315.

Chu, J.J., and Ng, M.L. (2004). Infectious entry of West Nile virus occurs through a clathrin-mediated endocytic pathway. J. Virol. *78*, 10543–10555.

Chu, P.W., and Westaway, E.G. (1985). Replication strategy of Kunjin virus: evidence for recycling role of replicative form RNA as template in semiconservative and asymmetric replication. Virology *140*, 68–79.

Chu, J.J. H., and Yang, P.L. (2007). c-Src protein kinase inhibitors block assembly and maturation of dengue virus. Proc. Natl. Acad. Sci. U.S.A. *104*, 3520–3525.

Chu, J.J., Leong, P.W., and Ng, M.L. (2006). Analysis of the endocytic pathway mediating the infectious entry of mosquito-borne flavivirus West Nile into Aedes albopictus mosquito (C6/36) cells. Virology *349*, 463–475.

Chu, P.W., and Westaway, E.G. (1987). Characterization of Kunjin virus RNA-dependent RNA polymerase: reinitiation of synthesis *in vitro*. Virology *157*, 330–337.

Chu, P.W., and Westaway, E.G. (1992). Molecular and ultrastructural analysis of heavy membrane fractions associated with the replication of Kunjin virus RNA. Arch. Virol. *125*, 177–191.

Chung, K.M., Liszewski, M.K., Nybakken, G., Davis, A.E., Townsend, R.R., Fremont, D.H., Atkinson, J.P., and Diamond, M.S. (2006). West Nile virus nonstructural protein NS1 inhibits complement activation by binding the regulatory protein factor H. Proc. Natl. Acad. Sci. U.S.A. *103*, 19111–19116.

Chung, K.M., Thompson, B.S., Fremont, D.H., and Diamond, M.S. (2007). Antibody recognition of cell surface-associated NS1 triggers Fc-gamma receptor-mediated phagocytosis and clearance of West Nile virus-infected cells. J. Virol. *81*, 9551–9555.

Clark, D.C., Lobigs, M., Lee, E., Howard, M.J., Clark, K., Blitvich, B.J., and Hall, R.A. (2007). *In situ* reactions of monoclonal antibodies with a viable mutant of Murray Valley encephalitis virus reveal an absence of dimeric NS1 protein. J. Gen. Virol. *88*, 1175–1183.

Cleaves, G.R., and Dubin, D.T. (1979). Methylation status of intracellular dengue type 2 40 S RNA. Virology *96*, 159–165.

Cleaves, G.R., Ryan, T.E., and Schlesinger, R.W. (1981). Identification and characterization of type 2 dengue virus replicative intermediate and replicative form RNAs. Virology *111*, 73–83.

Clyde, K., and Harris, E. (2006). RNA secondary structure in the coding region of dengue virus type 2 directs translation start codon selection and is required for viral replication. J. Virol. *80*, 2170–2182.

Clyde, K., Barrera, J., and Harris, E. (2008). The capsid-coding region hairpin element (cHP) is a critical determinant of dengue virus and West Nile virus RNA synthesis. Virology *379*, 314–323.

Coia, G., Parker, M.D., Speight, G., Byrne, M.E., and Westaway, E.G. (1988). Nucleotide and complete amino-acid sequences of Kunjin virus – definitive gene order and characteristics of the virus-specified proteins. J. Gen. Virol. *69*, 1–21.

Corver, J., Ortiz, A., Allison, S.L., Schalich, J., Heinz, F.X., and Wilschut, J. (2000). Membrane fusion activity of tick-borne encephalitis virus and recombinant subviral particles in a liposomal model system. Virology *269*, 37–46.

Corver, J., Lenches, E., Smith, K., Robison, R.A., Sando, T., Strauss, E.G. and Strauss, J.H. (2003). Fine mapping of a cis-acting sequence element in yellow fever virus RNA that is required for RNA replication and cyclization. J. Virol. *77*, 2265–2270.

Courageot, M.P., Frenkiel, M.P., Santos, C.D.D., Deubel, V., and Despres, P. (2000). alpha-glucosidase inhibitors reduce dengue virus production by affecting the initial steps of virion morphogenesis in the endoplasmic reticulum. J. Virol. *74*, 564–572.

Crooks, A.J., Lee, J.M., Easterbrook, L.M., Timofeev, A.V., and Stephenson, J.R. (1994). The NS1 protein of tick-borne encephalitis-virus forms multimeric species upon secretion from the host-cell. J. Gen. Virol. *75*, 3453–3460.

Cui, T., Sugrue, R.J., Xu, Q., Lee, A.K., Chan, Y.C., and Fu, J. (1998). Recombinant dengue virus type 1 NS3 protein exhibits specific viral RNA binding and NTPase activity regulated by the NS5 protein. Virology *246*, 409–417.

Dalgarno, L., Trent, D.W., Strauss, J.H., and Rice, C.M. (1986). Partial nucleotide-sequence of the Murray Valley encephalitis virus genome – comparison of the encoded polypeptides with yellow fever virus structural and nonstructural proteins. J. Mol. Biol. *187*, 309–323.

De Nova-Ocampo, M., Villegas-Sepuveda, N., and del Angel, R.M. (2002). Translation elongation factor-1 alpha, La, and PTB interact with the 3′ untranslated region of dengue 4 virus RNA. Virology *295*, 337–347.

Dokland, T., Walsh, M., Mackenzie, J.M., Khromykh, A.A., Ee, K.H., and Wang, S. (2004). West Nile virus core protein: tetramer structure and ribbon formation. Structure *12*, 1157–1163.

Dong, H.P., Zhang, B., and Shi, P.Y. (2008). Flavivirus methyltransferase: A novel antiviral target. Antiviral Res. *80*, 1–10.

Edward, Z., and Takegami, T. (1993). Localization and functions of Japanese enephalitis-virus nonstructural proteins NS3 and NS5 for VIRAL-RNA synthesis in the infected-cells. Microbiol. Immunol. *37*, 239–243.

Elghonemy, S., Davis, W.G., and Brinton, M.A. (2005). The majority of the nucleotides in the top loop of the genomic 3′ terminal stem loop structure are cis-acting in a West Nile virus infectious clone. Virology *331*, 238–246.

Emara, M.M., Liu, H.A., Davis, W.G., and Brinton, M.A. (2008). Mutation of mapped TIA-1/TIAR binding sites in the 3′ terminal stem–loop of West Nile Virus minus-strand RNA in an infectious clone negatively affects genomic RNA amplification. J. Virol. *82*, 10657–10670.

Evans, J.D., and Seeger, C. (2007). Differential effects of mutations in NS4B on West Nile virus replication and inhibition of interferon signaling. J. Virol. *81*, 11809–11816.

Falgout, B., Pethel, M., Zhang, Y.M., and Lai, C.J. (1991). Both nonstructural proteins NS2B and NS3 are required for the proteolytic processing of dengue virus nonstructural proteins. J. Virol. *65*, 2467–2475.

Fayzulin, R., Scholle, F., Petrakova, O., Frolov, I., and Mason, P.W. (2006). Evaluation of replicative capacity and genetic stability of West Nile virus replicons using highly efficient packaging cell lines. Virology *351*, 196–209.

Ferlenghi, I., Clarke, M., Ruttan, T., Allison, S.L., Schalich, J., Heinz, F.X., Harrison, S.C., Rey, F.A., and Fuller, S.D. (2001). Molecular organization of a recombinant subviral particle from tick-borne encephalitis. Mol. Cell *7*, 593–602.

Filomatori, C.V., Lodeiro, M.F., Alvarez, D.E., Samsa, M.M., Pietrasanta, L., and Gamarnik, A.V. (2006). A 5′ RNA element promotes dengue virus RNA synthesis on a circular genome. Genes Dev. *20*, 2238–2249.

Firth, A.E., and Atkins, J.F. (2009). A conserved predicted pseudoknot in the NS2A-encoding sequence of West Nile and Japanese encephalitis flaviviruses suggests NS1 may derive from ribosomal frameshifting. Virol. J. *6*, 6.

Firth, A.E., Blitvich, B.J., Wills, N.M., Miller, C.L., and Atkins, J.F. (2009). Evidence for ribosomal frameshifting and a novel overlapping gene in the genomes of insect-specific flaviviruses. Virology *399*, 153–166.

Flamand, M., Megret, F., Mathieu, M., Lepault, J., Rey, F.A., and Deubel, V. (1999). Dengue virus type 1 nonstructural glycoprotein NS1 is secreted from mammalian cells as a soluble hexamer in a glycosylation-dependent fashion. J. Virol. *73*, 6104–6110.

Friebe, P., and Harris, E. (2010). The Interplay of RNA Elements in the dengue Virus 5′ and 3′ Ends Required for Viral RNA Replication. J Virol.

Funk, A., Truong, K., Nagasaki, T., Torres, S., Floden, N., Melian, E.B., Edmonds, J., Dong, H., Shi, P.Y., and Khromykh, A.A. (2010). RNA structures required for production of subgenomic flavivirus RNA. J. Virol. *84*, 11407–11417.

Gehrke, R., Ecker, M., Aberle, S.W., Allison, S.L., Heinz, F.X., and Mandl, C.W. (2003). Incorporation of tick-borne encephalitis virus replicons into virus-like particles by a packaging cell line. J. Virol. *77*, 8924–8933.

Gehrke, R., Heinz, F.X., Davis, N.L., and Mandl, C.W. (2005). Heterologous gene expression by infectious and replicon vectors derived from tick-borne encephalitis virus and direct comparison of this flavivirus system with an alphavirus replicon. J. Gen. Virol. *86*, 1045–1053.

Gillespie, L.K., Hoenen, A., Morgan, G., and Mackenzie, J.M. (2010). The Endoplasmic reticulum provides the membrane platform for biogenesis of the flavivirus replication complex. J. Virol. *84*, 10438–10447.

Gorbalenya, A.E., Koonin, E.V., Donchenko, A.P., and Blinov, V.M. (1989). Two related superfamilies of putative helicases involved in replication, recombination, repair and expression of DNA and RNA genomes. Nucleic Acids Res. *17*, 4713–4730.

Goto, A., Yoshii, K., Obara, M., Ueki, T., Mizutani, T., Kariwa, H., and Takashima, I. (2005). Role of the N-linked glycans of the prM and E envelope proteins in tick-borne encephalitis virus particle secretion. Vaccine *23*, 3043–3052.

Gritsun, T.S., Venugopal, K., Zanotto, P.M.D., Mikhailov, M.V., Sall, A.A., Holmes, E.C., Polkinghorne, I., Frolova, T.V., Pogodina, V.V., Lashkevich, V.A., *et al.* (1997). Complete sequence of two tick-borne flaviviruses isolated from Siberia and the UK: analysis and significance of the 5′ and 3′-UTRs. Virus Res. *49*, 27–39.

Grun, J.B., and Brinton, M.A. (1986). Characterization of West Nile virus RNA-dependent RNA-polymerase and cellular terminal adenylyl and uridylyl transferases in cell-free-extracts. J. Virol. *60*, 1113–1124.

Gualano, R.C., Pryor, M.J., Cauchi, M.R., Wright, P.J., and Davidsen, A.D. (1998). Identification of a major determinant of mouse neurovirulence of dengue virus type 2 using stably cloned genomic-length cDNA. J. Gen. Virol. *79*, 437–446.

Guirakhoo, F., Weltzin, R., Chambers, T.J., Zhang, Z.X., Soike, K., Ratterree, M., Arroyo, J., Georgakopoulos, K., Catalan, J., and Monath, T.P. (2000). Recombinant chimeric yellow fever-dengue type 2 virus is immunogenic and protective in nonhuman primates. J. Virol. *74*, 5477–5485.

Hahn, C.S., Hahn, Y.S., Rice, C.M., Lee, E., Dalgarno, L., Strauss, E.G. and Strauss, J.H. (1987). Conserved elements in the 3′ untranslated region of flavivirus RNAs and potential cyclization sequences. J. Mol. Biol. *198*, 33–41.

Hall, R.A., Khromykh, A.A., Mackenzie, J.M., Scherret, J.H., Khromykh, T.I., and Mackenzie, J.S. (1999). Loss of dimerisation of the nonstructural protein NS1 of Kunjin virus delays viral replication and reduces virulence in mice, but still allows secretion of NS1. Virology *264*, 66–75.

Hall, R.A., Nisbet, D.J., Pham, K.B., Pyke, A.T., Smith, G.A., and Khromykh, A.A. (2003). DNA vaccine coding for the full-length infectious Kunjin virus RNA protects mice against the New York strain of West Nile virus. Proc. Natl. Acad. Sci. U.S.A. *100*, 10460–10464.

Hanna, S.L., Pierson, T.C., Sanchez, M.D., Ahmed, A.A., Murtadha, M.M., and Doms, R.W. (2005). N-linked glycosylation of west nile virus envelope proteins influences particle assembly and infectivity. J. Virol. *79*, 13262–13274.

Harvey, T.J., Anraku, I., Linedale, R., Harrich, D., Mackenzie, J., Suhrbier, A., and Khromykh, A.A. (2003). Kunjin virus replicon vectors for human immunodeficiency virus vaccine development. J. Virol. *77*, 7796–7803.

Harvey, T.J., Liu, W.J., Wang, X.J., Linedale, R., Jacobs, M., Davidson, A., Le, T.T., Anraku, I., Suhrbier, A., Shi, P.Y., *et al.* (2004). Tetracycline-inducible packaging cell line for production of flavivirus replicon particles. J. Virol. *78*, 531–538.

Hayasaka, D., Gritsun, T.S., Yoshii, K., Ueki, T., Goto, A., Mizutani, T., Kariwa, H., Iwasaki, T., Gould, E.A., and Takashima, I. (2004a). Amino acid changes responsible for attenuation of virus neurovirulence in an infectious cDNA clone of the Oshima strain of tick-borne encephalitis virus. J. Gen. Virol. *85*, 1007–1018.

Hayasaka, D., Yoshii, K., Ueki, T., Iwasaki, T., and Takashima, I. (2004b). Sub-genomic replicons of tick-borne encephalitis virus – Brief report. Arch. Virol. *149*, 1245–1256.

Herd, K.A., Harvey, T., Khromykh, A.A., and Tindle, R.W. (2004). Recombinant Kunjin virus replicon vaccines induce protective T-cell immunity against human papillomavirus 16 E7-expressing tumour. Virology *319*, 237–248.

Hirsch, A.J., Medigeshi, G.R., Meyers, H.L., DeFilippis, V., Fruh, K., Briese, T., Lipkin, W.I., and Nelson, J.A. (2005). The src family kinase c-Yes is required for maturation of West Nile virus particles. J. Virol. *79*, 11943–11951.

Hoang-Le, D., Smeenk, L., Anraku, I., Pijlman, G.P., Wang, X.J., de Vrij, J., Liu, W.J., Le, T.T., Schroder, W.A., Khromykh, A.A., *et al.* (2009). A Kunjin replicon vector encoding granulocyte macrophage colony-stimulating factor for intra-tumoral gene therapy. Gene Ther. *16*, 190–199.

Hoenen, A., Liu, W., Kochs, G., Khromykh, A.A., and Mackenzie, J.M. (2007). West Nile virus-induced cytoplasmic membrane structures provide partial protection against the interferon-induced antiviral MxA protein. J. Gen. Virol. *88*, 3013–3017.

Hoenninger, V.M., Rouha, H., Orlinger, K.K., Miorin, L., Marcello, A., Kofler, R.M., and Mandl, C.W. (2008). Analysis of the effects of alterations in the tick-borne encephalitis virus 3′-noncoding region on translation and RNA replication using reporter replicons. Virology *377*, 419–430.

Hori, H., and Lai, C.J. (1990). Cleavage of dengue virus NS1-NS2A requires an octapeptide sequence at the C-terminus of NS1. J. Virol. *64*, 4573–4577.

Hurrelbrink, R.J., Nestorowicz, A., and McMinn, P.C. (1999). Characterization of infectious Murray Valley encephalitis virus derived from a stably cloned genome-length cDNA. J. Gen. Virol. *80*, 3115–3125.

Ishikawa, T., Widman, D.G., Bourne, N., Konishi, E., and Mason, P.W. (2008). Construction and evaluation of a chimeric pseudoinfectious virus vaccine to prevent Japanese encephalitis. Vaccine *26*, 2772–2781.

Issur, M., Geiss, B.J., Bougie, I., Picard-Jean, F., Despins, S., Mayette, J., Hobdey, S.E., and Bisaillon, M. (2009). The flavivirus NS5 protein is a true RNA guanylyltransferase that catalyzes a two-step reaction to form the RNA cap structure. RNA-Publ. RNA Soc. *15*, 2340–2350.

Ivanyi-Nagy, R., Lavergne, J.P., Gabus, C., Ficheux, D., and Darlix, J.L. (2008). RNA chaperoning and intrinsic disorder in the core proteins of Flaviviridae. Nucleic Acids Res. *36*, 712–725.

Johansson, M., Brooks, A.J., Jans, D.A., and Vasudevan, S.G. (2001). A small region of the dengue virus-encoded RNA-dependent RNA polymerase, NS5, confers interaction with both the nuclear transport receptor importin-ss and the viral helicase, NS3. J. Gen. Virol. *82*, 735–745.

Jones, C.T., Patkar, C.G., and Kuhn, R.J. (2005). Construction and applications of yellow fever virus replicons. Virology *331*, 247–259.

Kamer, G., and Argos, P. (1984). Primary structural comparison of RNA-dependent polymerases from plant, animal and bacterial viruses. Nucleic Acids Res. *12*, 7269–7282.

Kapoor, M., Zhang, L., Ramachandra, M., Kusukawa, J., Ebner, K.E., and Padmanabhan, R. (1995a). Association between NS3 and NS5 proteins of dengue virus type 2 in the putative RNA replicase is linked to differential phosphorylation of NS5. J. Biol. Chem. *270*, 19100–19106.

Kapoor, M., Zhang, L.W., Mohan, P.M., and Padmanabhan, R. (1995b). Synthesis and characterization of an infectious dengue virus type-2 RNA genome (New-Guinea C-strain). Gene *162*, 175–180.

Kent, S.J., De Rose, R., Mokhonov, V.V., Mokhonova, E.I., Fernandez, C.S., Alcantara, S., Rollman, E., Mason, R.D., Loh, L., Peut, V., *et al.* (2008). Evaluation of recombinant Kunjin replicon SIV vaccines for protective efficacy in macaques. Virology *374*, 528–534.

Khromykh, A.A. (2000). Replicon-based vectors of positive strand RNA viruses. Curr. Opin. Mol. Ther. *2*, 555–569.

Khromykh, A.A., and Westaway, E.G. (1994). Completion of Kunjin virus-RNA sequence and recovery of an infectious RNA transcribed from stably cloned full-length cDNA. J. Virol. *68*, 4580–4588.

Khromykh, A.A., and Westaway, E.G. (1996). RNA binding properties of core protein of the flavivirus Kunjin. Arch. Virol. *141*, 685–699.

Khromykh, A.A., and Westaway, E.G. (1997). Subgenomic replicons of the flavivirus Kunjin: construction and applications. J. Virol. *71*, 1497–1505.

Khromykh, A.A., Kenney, M.T., and Westaway, E.G. (1998a). trans-Complementation of flavivirus RNA polymerase gene NS5 by using Kunjin virus replicon-expressing BHK cells. J. Virol. *72*, 7270–7279.

Khromykh, A.A., Varnavski, A.N., and Westaway, E.G. (1998b). Encapsidation of the flavivirus kunjin replicon RNA by using a complementation system providing kunjin virus structural proteins in trans. J. Virol. *72*, 5967–5977.

Khromykh, A.A., Sedlak, P.L., Guyatt, K.J., Hall, R.A., and Westaway, E.G. (1999a). Efficient trans-complementation of the flavivirus kunjin NS5 protein but not of the NS1 protein requires its coexpression with other components of the viral replicase. J. Virol. *73*, 10272–10280.

Khromykh, A.A., Sedlak, P.L., and Westaway, E.G. (1999b). trans-Complementation analysis of the flavivirus Kunjin ns5 gene reveals an essential role for translation of its N-terminal half in RNA replication. J. Virol. *73*, 9247–9255.

Khromykh, A.A., Sedlak, P.L., and Westaway, E.G. (2000). cis- and trans-acting elements in flavivirus RNA replication. J. Virol. *74*, 3253–3263.

Khromykh, A.A., Meka, H., Guyatt, K.J., and Westaway, E.G. (2001a). Essential role of cyclization sequences in flavivirus RNA replication. J. Virol. *75*, 6719–6728.

Khromykh, A.A., Varnavski, A.N., Sedlak, P.L., and Westaway, E.G. (2001b). Coupling between replication and packaging of flavivirus RNA: evidence derived from the use of DNA-based full-length cDNA clones of Kunjin virus. J. Virol. *75*, 4633–4640.

Khromykh, A.A., Kondratieva, N., Sgro, J.Y., Palmenberg, A., and Westaway, E.G. (2003). Significance in replication of the terminal nucleotides of the flavivirus genome. J. Virol. *77*, 10623–10629.

Kiermayr, S., Kofler, R.M., Mandl, C.W., Messner, P., and Heinz, F.X. (2004). Isolation of capsid protein dimers from the tick-borne encephalitis flavivirus and *in vitro* assembly of capsid-like particles. J. Virol. *78*, 8078–8084.

Kim, J.M., Yun, S.I., Song, B.H., Hahn, Y.S., Lee, C.H., Oh, H.W., and Lee, Y.M. (2008). A single N-linked glycosylation site in the Japanese encephalitis virus prM protein is critical for cell type-specific prM protein biogenesis, virus particle release, and pathogenicity in mice. J. Virol. *82*, 7846–7862.

Kinney, R.M., and Huang, C.Y.H. (2001). Development of new vaccines against dengue fever and Japanese encephalitis. Intervirology *44*, 176–197.

Kinney, R.M., Butrapet, S., Chang, G.J.J., Tsuchiya, K.R., Roehrig, J.T., Bhamarapravati, N., and Gubler, D.J. (1997). Construction of infectious cDNA clones for dengue 2 virus: Strain 16681 and its attenuated vaccine derivative, strain PDK-53. Virology *230*, 300–308.

Kofler, R.M., Heinz, F.X., and Mandl, C.W. (2002). Capsid protein C of tick-borne encephalitis virus tolerates large internal deletions and is a favorable target for attenuation of virulence. J. Virol. *76*, 3534–3543.

Kofler, R.M., Leitner, A., O'Riordain, G., Heinz, F.X., and Mandl, C.W. (2003). Spontaneous mutations restore the viability of tick-borne encephalitis virus mutants with large deletions in protein C. J. Virol. *77*, 443–451.

Kofler, R.M., Aberle, J.H., Aberle, S.W., Allison, S.L., Heinz, F.X., and Mandl, C.W. (2004a). Mimicking live flavivirus immunization with a noninfectious RNA vaccine. Proc. Natl. Acad. Sci. U.S.A. *101*, 1951–1956.

Kofler, R.M., Heinz, F.X., and Mandl, C.W. (2004b). A novel principle of attenuation for the development of new generation live flavivirus vaccines. Arch. Virol. Suppl, 191–200.

Kofler, R.M., Hoenninger, V.M., Thurner, C., and Mandl, C.W. (2006). Functional analysis of the tick-borne encephalitis virus cyclization elements indicates major differences between mosquito-borne and tick-borne flaviviruses. J. Virol. *80*, 4099–4113.

Konishi, E., and Mason, P.W. (1993). Proper maturation of the Japanese encephalitis virus envelope glycoprotein requires cosynthesis with the premembrane protein. J. Virol. *67*, 1672–1675.

Konishi, E., Fujii, A., and Mason, P.W. (2001). Generation and characterization of a mammalian cell line continuously expressing Japanese encephalitis virus subviral particles. J. Virol. *75*, 2204–2212.

Konishi, E., Pincus, S., Paoletti, E., Shope, R.E., Burrage, T., and Mason, P.W. (1992). Mice immunized with a subviral particle containing the Japanese encephalitis virus prM/M and E proteins are protected from lethal JEV infection. Virology *188*, 714–720.

Koonin, E.V. (1991). The phylogeny of RNA-dependent RNA polymerases of positive-strand RNA viruses. J. Gen. Virol. *72*, 2197–2206.

Koonin, E.V. (1993). Computer-assisted identification of a putative methyl transferase domain in NS5 protein of flaviviruses and lambda-2 protein of Reovirus. J. Gen. Virol. *74*, 733–740.

Kopek, B.G., Perkins, G., Miller, D.J., Ellisman, M.H., and Ahlquist, P. (2007). Three-dimensional analysis of a viral RNA replication complex reveals a virus-induced mini-organelle. PLoS Biol. *5*, 2022–2034.

Kroeger, M.A., and McMinn, P.C. (2002). Murray Valley encephalitis virus recombinant subviral particles protect mice from lethal challenge with virulent wild-type virus. Arch. Virol. *147*, 1155–1172.

Kummerer, B.M., and Rice, C.M. (2002). Mutations in the yellow fever virus nonstructural protein NS2A selectively block production of infectious particles. J. Virol. *76*, 4773–4784.

Lai, C.J., Zhao, B., Hori, H., and Bray, M. (1991). Infectious RNA transcribed from stably cloned full-length cDNA of dengue type-4 virus. Proc. Natl. Acad. Sci. U.S.A. *88*, 5139–5143.

Lai, C.Y., Hu, H.P., King, C.C., and Wang, W.K. (2008). Incorporation of dengue virus replicon into virus-like particles by a cell line stably expressing precursor membrane and envelope proteins of dengue virus type 2. J. Biomed. Sci. *15*, 15–27.

Laurent-Rolle, M., Boer, E.F., Lubick, K.J., Wolfinbarger, J.B., Carmody, A.B., Rockx, B., Liu, W., Ashour, J., Shupert, W.L., Holbrook, M.R., *et al.* (2010). The NS5 protein of the virulent West Nile virus NY99 strain is a potent antagonist of type I interferon-mediated JAK-STAT signaling. J. Virol. *84*, 3503–3515.

Leary, K., and Blair, C.D. (1980). Sequential events in the morphogenesis of Japanese Encephalitis virus. J. Ultrastruct. Res. *72*, 123–129.

Lee, E., Leang, S.K., Davidson, A., and Lobigs, M. (2010). Both E Protein Glycans Adversely Affect dengue Virus Infectivity but Are Beneficial for Virion Release. J. Virol. *84*, 5171–5180.

Leung, J.Y., Pijlman, G.P., Kondratieva, N., Hyde, J., Mackenzie, J.M., and Khromykh, A.A. (2008). Role of nonstructural protein NS2A in flavivirus assembly. J. Virol. *82*, 4731–4741.

Li, H.T., Clum, S., You, S.H., Ebner, K.E., and Padmanabhan, R. (1999). The serine protease and RNA-stimulated nucleoside triphosphatase and RNA helicase functional domains of dengue virus type 2 NS3 converge within a region of 20 amino acids. J. Virol. *73*, 3108–3116.

Li, L., Lok, S.M., Yu, I.M., Zhang, Y., Kuhn, R.J., Chen, J., and Rossmann, M.G. (2008). The flavivirus precursor membrane-envelope protein complex: Structure and maturation. Science *319*, 1830–1834.

Li, W., Li, Y., Kedersha, N., Anderson, P., Emara, M., Swiderek, K.M., Moreno, G.T., and Brinton, M.A. (2002). Cell proteins TIA-1 and TIAR interact with the 3′ stem–loop of the West Nile virus complementary minus-strand RNA and facilitate virus replication. J. Virol. *76*, 11989–12000.

Li, X.F., Jiang, T., Yu, X.D., Deng, Y.Q., Zhao, H., Zhu, Q.Y., Qin, E.D., and Qin, C.F. (2010). RNA elements within the 5′ untranslated region of the West Nile virus genome are critical for RNA synthesis and virus replication. J. Gen. Virol. *91*, 1218–1223.

Limjindaporn, T., Wongwiwat, W., Noisakran, S., Srisawat, C., Netsawang, J., Puttikhunt, C., Kasinrerk, W.,

Avirutnan, P., Thiemmeca, S., Sriburi, R., *et al.* (2009). Interaction of dengue virus envelope protein with endoplasmic reticulum-resident chaperones facilitates dengue virus production. Biochem. Biophys. Res. Commun. *379*, 196–200.

Lin, R.J., Chang, B.L., Yu, H.P., Liao, C.L., and Lin, Y.L. (2006). Blocking of interferon-induced Jak-Stat signaling by Japanese encephalitis virus NS5 through a protein tyrosine phosphatase-mediated mechanism. J. Virol. *80*, 5908–5918.

Lin, Y.J., and Wu, S.C. (2005). Histidine at residue 99 and the transmembrane region of the precursor membrane prM protein are important for the prM–E heterodimeric complex formation of Japanese encephalitis virus. J. Virol. *79*, 8535–8544.

Lin, Y.L., Chen, L.K., Liao, C.L., Yeh, C.T., Ma, S.H., Chen, J.L., Huang, Y.L., Chen, S.S., and Chiang, H.Y. (1998). DNA immunization with Japanese encephalitis virus nonstructural protein NS1 elicits protective immunity in mice. J. Virol. *72*, 191–200.

Lindenbach, B.D., and Rice, C.M. (1997). trans-Complementation of yellow fever virus NS1 reveals a role in early RNA replication. J. Virol. *71*, 9608–9617.

Lindenbach, B.D., and Rice, C.M. (1999). Genetic interaction of flavivirus nonstructural proteins NS1 and NS4A as a determinant of replicase function. J. Virol. *73*, 4611–4621.

Lindenbach, B.D., Thiel, H.-J., and Rice, C.M. (2007). *Flaviviridae*: the viruses and their replication. In Fields Virology, 5th edn, Knipe, D.M., and Howley, P.M., eds (Philidelphia, Lippincott-Raven Publishers), pp. 1101–1152.

Liu, W.J., Sedlak, P.L., Kondratieva, N., and Khromykh, A.A. (2002). Complementation analysis of the flavivirus Kunjin NS3 and NS5 proteins defines the minimal regions essential for formation of a replication complex and shows a requirement of NS3 in cis for virus assembly. J. Virol. *76*, 10766–10775.

Liu, W.J., Chen, H.B., and Khromykh, A.A. (2003). Molecular and functional analyses of Kunjin virus infectious cDNA clones demonstrate the essential roles for NS2A in virus assembly and for a nonconservative residue in NS3 in RNA replication. J. Virol. *77*, 7804–7813.

Liu, W.J., Chen, H.B., Wang, X.J., Huang, H., and Khromykh, A.A. (2004). Analysis of adaptive mutations in Kunjin virus replicon RNA reveals a novel role for the flavivirus nonstructural protein NS2A in inhibition of beta interferon promoter-driven transcription. J. Virol. *78*, 12225–12235.

Liu, W.J., Wang, X.J., Mokhonov, V.V., Shi, P.Y., Randall, R., and Khromykh, A.A. (2005). Inhibition of interferon signaling by the New York 99 strain and Kunjin subtype of West Nile virus involves blockage of STAT1 and STAT2 activation by nonstructural proteins. J. Virol. *79*, 1934–1942.

Liu, Y., Wimmer, E., and Paul, A.V. (2009). Cis-acting RNA elements in human and animal plus-strand RNA viruses. Biochim. Biophys. Acta – Gene Regulatory Mechanisms *1789*, 495–517.

Lo, M.K., Tilgner, M., Bernard, K.A., and Shi, P.Y. (2003). Functional analysis of mosquito-borne flavivirus conserved sequence elements within 3′ untranslated region of West Nile virus by use of a reporting replicon that differentiates between viral translation and RNA replication. J. Virol. *77*, 10004–10014.

Lodeiro, M.F., Filomatori, C.V., and Gamarnik, A.V. (2009). Structural and functional studies of the promoter element for dengue virus RNA replication. J. Virol. *83*, 993–1008.

Lopez, C., Gil, L., Lazo, L., Menendez, I., Marcos, E., Sanchez, J., Valdes, I., Falcon, V., de la Rosa, M.C., Marquez, G., *et al.* (2009). *in vitro* assembly of nucleocapsid-like particles from purified recombinant capsid protein of dengue-2 virus. Arch. Virol. *154*, 695–698.

Lorenz, I.C., Allison, S.L., Heinz, F.X., and Helenius, A. (2002). Folding and dimerization of tick-borne encephalitis virus envelope proteins prM and E in the endoplasmic reticulum. J. Virol. *76*, 5480–5491.

Lorenz, I.C., Kartenbeck, J., Mezzacasa, A., Allison, S.L., Heinz, F.X., and Helenius, A. (2003). Intracellular assembly and secretion of recombinant subviral particles from tick-borne encephalitis virus. J. Virol. *77*, 4370–4382.

Ma, L.X., Jones, C.T., Groesch, T.D., Kuhn, R.J., and Post, C.B. (2004). Solution structure of dengue virus capsid protein reveals another fold. Proc. Natl. Acad. Sci. U.S.A. *101*, 3414–3419.

McAda, P.C., Mason, P.W., Schmaljohn, C.S., Dalrymple, J.M., Mason, T.L., and Fournier, M.J. (1987). Partial nucleotide sequence of the Japanese encephalitis virus genome. Virology *158*, 348–360.

McElroy, K.L., Tsetsarkin, K.A., Vanlandingham, D.L., and Higgs, S. (2005). Characterization of an infectious clone of the wild-type yellow fever virus Asibi strain that is able to infect and disseminate in mosquitoes. J. Gen. Virol. *86*, 1747–1751.

MacKenzie, J.M., and Westaway, E.G. (2001). Assembly and maturation of the flavivirus Kunjin virus appear to occur in the rough endoplasmic reticulum and along the secretory pathway, respectively. J. Virol. *75*, 10787–10799.

Mackenzie, J.M., Jones, M.K., and Young, P.R. (1996a). Immunolocalization of the dengue virus nonstructural glycoprotein NS1 suggests a role in viral RNA replication. Virology *220*, 232–240.

Mackenzie, J.M., Jones, M.K., and Young, Y.R. (1996b). Immunolocalization of the dengue virus nonstructural glycoprotein NS1 suggests a role in viral RNA replication. Virology *220*, 232–240.

Mackenzie, J.M., Khromykh, A.A., Jones, M.K., and Westaway, E.G. (1998). Subcellular localization and some biochemical properties of the flavivirus Kunjin nonstructural proteins NS2A and NS4A. Virology *245*, 203–215.

Mackenzie, J.M., Jones, M.K., and Westaway, E.G. (1999). Markers for trans-Golgi membranes and the intermediate compartment localize to induced membranes with distinct replication functions in flavivirus-infected cells. J. Virol. *73*, 9555–9567.

Mackenzie, J.M., Khromykh, A.A., and Westaway, E.G. (2001). Stable expression of noncytopathic Kunjin replicons simulates both ultrastructural and biochemical characteristics observed during replication of Kunjin virus. Virology *279*, 161–172.

Mackenzie, J.M., Kenney, M.T., and Westaway, E.G. (2007). West Nile virus strain Kunjin NS5 polymerase is a phosphoprotein localized at the cytoplasmic site of viral RNA synthesis. J. Gen. Virol. *88*, 1163–1168.

Malet, H., Egloff, M.P., Selisko, B., Butcher, R.E., Wright, P.J., Roberts, M., Gruez, A., Sulzenbacher, G., Vonrhein, C., Bricogne, G., *et al.* (2007). Crystal structure of the RNA polymerase domain of the West Nile virus non-structural protein 5. J. Biol. Chem. *282*, 10678–10689.

Mandl, C.W. (2004). Flavivirus immunization with capsid-deletion mutants: basics, benefits, and barriers. Viral Immunol *17*, 461–472.

Mandl, C.W., Holzmann, H., Kunz, C., and Heinz, F.X. (1993). Complete genomic sequence of Powassan virus: evaluation of genetic elements in tick-borne versus mosquito-borne flaviviruses. Virology *194*, 173–184.

Mandl, C.W., Ecker, M., Holzmann, H., Kunz, C., and Heinz, F.X. (1997). Infectious cDNA clones of tick-borne encephalitis virus European subtype prototypic strain Neudoerfl and high virulence strain Hypr. J. Gen. Virol. *78* (Pt 5), 1049–1057.

Mandl, C.W., Aberle, J.H., Aberle, S.W., Holzmann, H., Allison, S.L., and Heinz, F.X. (1998a). *In vitro*-synthesized infectious RNA as an attenuated live vaccine in a flavivirus model. Nat. Med. *4*, 1438–1440.

Mandl, C.W., Holzmann, H., Meixner, T., Rauscher, S., Stadler, P.F., Allison, S.L., and Heinz, F.X. (1998b). Spontaneous and engineered deletions in the 3′ noncoding region of tick-borne encephalitis virus: construction of highly attenuated mutants of a flavivirus. J. Virol. *72*, 2132–2140.

Markoff, L. (2003). 5′- and 3′-noncoding regions in flavivirus RNA. Adv. Virus Res. *59*, 177–228.

Mason, P.W. (1989). Maturation of Japanese encephalitis virus glycoproteins produced by infected mammalian and mosquito cells. Virology *169*, 354–364.

Mason, P.W., Shustov, A.V., and Frolov, I. (2006). Production and characterization of vaccines based on flaviviruses defective in replication. Virology *351*, 432–443.

Mazzon, M., Jones, M., Davidson, A., Chain, B., and Jacobs, M. (2009). Dengue Virus NS5 inhibits interferon-alpha signaling by blocking signal transducer and activator of transcription 2 phosphorylation. J. Infect. Dis. *200*, 1261–1270.

Melian, E.B., Hinzman, E., Nagasaki, T., Firth, A.E., Wills, N.M., Nouwens, A.S., Blitvich, B.J., Leung, J., Funk, A., Atkins, J.F., *et al.* (2010). NS1′ of flaviviruses in the Japanese encephalitis virus serogroup is a product of ribosomal frameshifting and plays a role in viral neuroinvasiveness. J. Virol. *84*, 1641–1647.

Men, R., Bray, M., Clark, D., Chanock, R.M., and Lai, C.J. (1996). Dengue type 4 virus mutants containing deletions in the 3′ noncoding region of the RNA genome: analysis of growth restriction in cell culture and altered viremia pattern and immunogenicity in rhesus monkeys. J. Virol. *70*, 3930–3937.

Miller, S., Sparacio, S., and Bartenschlager, R. (2006). Subcellular localization and membrane topology of the dengue virus type 2 non-structural protein 4B. J. Biol. Chem. *281*, 8854–8863.

Miller, S., Kastner, S., Krijnse-Locker, J., Buhler, S., and Bartenschlager, R. (2007). The non-structural protein 4A of dengue virus is an integral membrane protein inducing membrane alterations in a 2K-regulated manner. J. Biol. Chem. *282*, 8873–8882.

Miorin, L., Maiuri, P., Hoenninger, V.M., Mandl, C.W., and Marcello, A. (2008). Spatial and temporal organization of tick-borne encephalitis flavivirus replicated RNA in living cells. Virology *379*, 64–77.

Miyanari, Y., Atsuzawa, K., Usuda, N., Watashi, K., Hishiki, T., Zayas, M., Bartenschlager, R., Wakita, T., Hijikata, M., and Shimotohno, K. (2007). The lipid droplet is an important organelle for hepatitis C virus production. Nat. Cell Biol. *9*, 1089-U1074.

Mohan, P.M., and Padmanabhan, R. (1991). Detection of stable secondary structure at the 3′ terminus of dengue virus type-2 RNA. Gene *108*, 185–191.

Molla, A., Paul, A.V., and Wimmer, E. (1991). Cell-free, *de novo* synthesis of poliovirus. Science *254*, 1647–1651.

Mondotte, J.A., Lozach, P.Y., Amara, A., and Gamarnik, A.V. (2007). Essential role of dengue virus envelope protein *N* glycosylation at asparagine-67 during viral propagation. J. Virol. *81*, 7136–7148.

Mosimann, A.L.P., de Borba, L., Bordignon, J., Mason, P.W., and dos Santos, C.N.D. (2010). Construction and characterization of a stable subgenomic replicon system of a Brazilian dengue virus type 3 strain (BR DEN3 290–02). J. Virol. Methods *163*, 147–152.

Mukhopadhyay, S., Kuhn, R.J., and Rossmann, M.G. (2005). A structural perspective of the flavivirus life cycle. Nat. Rev. Microbiol. *3*, 13–22.

Munoz-Jordan, J.L., Sanchez-Burgos, G.G., Laurent-Rolle, M., and Garcia-Sastre, A. (2003). Inhibition of interferon signaling by dengue virus. Proc. Natl. Acad. Sci. U.S.A. *100*, 14333–14338.

Munoz-Jordan, J.L., Laurent-Rolle, M., Ashour, J., Martinez-Sobrido, L., Ashok, M., Lipkin, W.I., and Garcia-Sastre, A. (2005). Inhibition of alpha/beta interferon signaling by the NS4B protein of flaviviruses. J. Virol. *79*, 8004–8013.

Murphy, F.A., Harrison, A.K., Gary, G.W., Jr., Whitfield, S.G., and Forrester, F.T. (1968). St. Louis encephalitis virus infection in mice. Electron microscopic studies of central nervous system. Lab. Invest. *19*, 652–662.

Murray, C.L., Jones, C.T., and Rice, C.M. (2008). Opinion – architects of assembly: roles of Flaviviridae non-structural proteins in virion morphogenesis. Nat. Rev. Microbiol. *6*, 699–708.

Muylaert, I.R., Chambers, T.J., Galler, R., and Rice, C.M. (1996). Mutagenesis of the N-linked glycosylation sites of the yellow fever virus NS1 protein: effects on virus replication and mouse neurovirulence. Virology *222*, 159–168.

Muylaert, I.R., Galler, R., and Rice, C.M. (1997). Genetic analysis of the yellow fever virus NS1 protein: identification of a temperature-sensitive mutation which blocks RNA accumulation. J. Virol. *71*, 291–298.

Netherton, C., Moffat, K., Brooks, E., and Wileman, T. (2007). A guide to viral inclusions, membrane rearrangements, factories, and viroplasm produced during virus replication. Adv. Virus Res. *70*, 101–182.

Ng, C.Y., Gu, F., Phong, W.Y., Chen, Y.L., Lim, S.P., Davidson, A., and Vasudevan, S.G. (2007). Construction and characterization of a stable subgenomic dengue virus type 2 replicon system for antiviral compound and siRNA testing. Antiviral Res. *76*, 222–231.

Ng, M.L., and Hong, S.S. (1989). Flavivirus infection – essential ultrastructural-changes and association of Kunjin virus NS3 protein with microtubules. Arch. Virol. *106*, 103–120.

Ng, M.L., Pedersen, J.S., Toh, B.H., and Westaway, E.G. (1983). Immunofluorescent sites in Vero cells infected with the flavivirus Kunjin. Arch. Virol. *78*, 177–190.

Ng, M.L., Howe, J., Sreenivasan, V., and Mulders, J.J. L. (1994). Flavivirus West Nile (Sarafend) egress at the plasma membrane. Arch. Virol. *137*, 303–313.

Novoa, R.R., Calderita, G., Arranz, R., Fontana, J., Granzow, H., and Risco, C. (2005). Virus factories: associations of cell organelles for viral replication and morphogenesis. Biol. Cell *97*, 147–172.

Nowak, T., Farber, P.M., and Wengler, G. (1989). Analyses of the terminal sequences of West Nile virus structural proteins and of the *in vitro* translation of these proteins allow the proposal of a complete scheme of the proteolytic cleavages involved in their synthesis. Virology *169*, 365–376.

Olsthoorn, R.C., and Bol, J.F. (2001). Sequence comparison and secondary structure analysis of the 3′ noncoding region of flavivirus genomes reveals multiple pseudoknots. RNA *7*, 1370–1377.

Op De Beeck, A., Molenkamp, R., Caron, M., Ben Younes, A., Bredenbeek, P., and Dubuisson, J. (2003). Role of the transmembrane domains of prM and E proteins in the formation of yellow fever virus envelope. J. Virol. *77*, 813–820.

Orlinger, K.K., Hoenninger, V.M., Kofler, R.M., and Mandl, C.W. (2006). Construction and mutagenesis of an artificial bicistronic tick-borne encephalitis virus genome reveals an essential function of the second transmembrane region of protein e in flavivirus assembly. J. Virol. *80*, 12197–12208.

Överby, A.K., Popov, V.L., Niedrig, M., and Weber, F. (2010). Tick-borne encephalitis virus delays interferon induction and hides its double-stranded RNA in intracellular membrane vesicles. J. Virol. *84*, 8470–8483.

Patkar, C.G., and Kuhn, R.J. (2008). Yellow fever virus NS3 plays an essential role in virus assembly independent of its known enzymatic functions. J. Virol. *82*, 3342–3352.

Pierson, T.C., Sanchez, M.D., Puffer, B.A., Ahmed, A.A., Geiss, B.J., Valentine, L.E., Altamura, L.A., Diamond, M.S., and Doms, R.W. (2006). A rapid and quantitative assay for measuring antibody-mediated neutralization of West Nile virus infection. Virology *346*, 53–65.

Pijlman, G.P., Kondratieva, N., and Khromykh, A.A. (2006a). Translation of the flavivirus kunjin NS3 gene in cis but not its RNA sequence or secondary structure is essential for efficient RNA packaging. J. Virol. *80*, 11255–11264.

Pijlman, G.P., Suhrbier, A., and Khromykh, A.A. (2006b). Kunjin virus replicons: an RNA-based, non-cytopathic viral vector system for protein production, vaccine and gene therapy applications. Exp. Opin. Biol. Ther. *6*, 135–145.

Pijlman, G.P., Funk, A., Kondratieva, N., Leung, J., Torres, S., van der Aa, L., Liu, W.J., Palmenberg, A.C., Shi, P.Y., Hall, R.A., *et al.* (2008). A highly structured, nuclease-resistant, noncoding RNA produced by flaviviruses is required for pathogenicity. Cell Host Microbe *4*, 579–591.

Pletnev, A.G. (2001). Infectious cDNA clone of attenuated langat tick-borne flavivirus (strain E5) and a 3′ deletion mutant constructed from it exhibit decreased neuroinvasiveness in immunodeficient mice. Virology *282*, 288–300.

Poch, O., Sauvaget, I., Delarue, M., and Tordo, N. (1989). Identification of 4 conserved motifs among the RNA-dependent polymerase encoding elements. EMBO J. *8*, 3867–3874.

Polacek, C., Friebe, P., and Harris, E. (2009). Poly(A)-binding protein binds to the non-polyadenylated 3′ untranslated region of dengue virus and modulates translation efficiency. J. Gen. Virol. *90*, 687–692.

Polo, S., Ketner, G., Levis, R., and Falgout, B. (1997). Infectious RNA transcripts from full-length dengue virus type 2 cDNA clones made in yeast. J. Virol. *71*, 5366–5374.

Preugschat, F., Yao, C.W., and Strauss, J.H. (1990). Invitro processing of dengue virus type-2 nonstructural proteins NS2A, NS2B, and NS3. J. Virol. *64*, 4364–4374.

Proutski, V., Gould, E.A., and Holmes, E.C. (1997). Secondary structure of the 3′ untranslated region of flaviviruses: similarities and differences. Nucleic Acids Res. *25*, 1194–1202.

Proutski, V., Gritsun, T.S., Gould, E.A., and Holmes, E.C. (1999). Biological consequences of deletions within the 3′-untranslated region of flaviviruses may be due to rearrangements of RNA secondary structure. Virus Res. *64*, 107–123.

Pryor, M.J., and Wright, P.J. (1994). Glycosylation mutants of dengue virus NS1 protein. J. Gen. Virol. *75*, 1183–1187.

Pryor, M.J., Azzola, L., Wright, P.J., and Davidson, A.D. (2004). Histidine 39 in the dengue virus type 2 M protein has an important role in virus assembly. J. Gen. Virol. *85*, 3627–3636.

Puig-Basagoiti, F., Qing, M., Dong, H.P., Zhang, B., Zou, G., Yuan, Z.M., and Shi, P.Y. (2009). Identification and characterization of inhibitors of West Nile virus. Antiviral Res. *83*, 71–79.

Puri, B., Polo, S., Hayes, C.G., and Falgout, B. (2000). Construction of a full length infectious clone for dengue-1 virus Western Pacific, 74 strain. Virus Genes *20*, 57–63.

Qing, M., Yang, F., Zhang, B., Zou, G., Robida, J.M., Yuan, Z.M., Tang, H.L., and Shi, P.Y. (2009). Cyclosporine inhibits flavivirus replication through blocking the interaction between host cyclophilins and viral NS5 protein. Antimicrob. Agents Chemother. *53*, 3226–3235.

Rauscher, S., Flamm, C., Mandl, C.W., Heinz, F.X., and Stadler, P.F. (1997). Secondary structure of the 3′-noncoding region of flavivirus genomes: comparative analysis of base pairing probabilities. RNA *3*, 779–791.

Ray, D., Shah, A., Tilgner, M., Guo, Y., Zhao, Y., Dong, H., Deas, T.S., Zhou, Y., Li, H., and Shi, P.Y. (2006). West Nile virus 5′-cap structure is formed by sequential guanine N-7 and ribose 2′-O methylations by nonstructural protein 5. J. Virol. *80*, 8362–8370.

Rice, C.M., Lenches, E.M., Eddy, S.R., Shin, S.J., Sheets, R.L., and Strauss, J.H. (1985). Nucleotide sequence of yellow fever virus: implications for flavivirus gene expression and evolution. Science *229*, 726–733.

Roosendaal, J., Westaway, E.G., Khromykh, A., and Mackenzie, J.M. (2006). Regulated cleavages at the West Nile virus NS4A–2K-NS4B junctions play a major role in rearranging cytoplasmic membranes and Golgi trafficking of the NS4A protein. J. Virol. *80*, 4623–4632.

Rossi, S.L., Zhao, Q., O'Donnell, V.K., and Mason, P.W. (2005). Adaptation of West Nile virus replicons to cells in culture and use of replicon-bearing cells to probe antiviral action. Virology *331*, 457–470.

Samsa, M.M., Mondotte, J.A., Iglesias, N.G., Assuncao-Miranda, I., Barbosa-Lima, G., Da Poian, A.T., Bozza, P.T., and Gamarnik, A.V. (2009). Dengue virus capsid protein usurps lipid droplets for viral particle formation. PLoS Pathog. *5*, 14.

van der Schaar, H.M., Rust, M.J., Chen, C., van der Ende-Metselaar, H., Wilschut, J., Zhuang, X.W., and Smit, J.M. (2008). Dissecting the cell entry pathway of dengue virus by single-particle tracking in living cells. PLoS Pathog. *4*, 9.

van der Schaar, H.M., Rust, M.J., Waarts, B.L., van der Ende-Metselaarl, H., Kuhn, R.J., Wilschut, J., Zhuang, X.W., and Smit, J.M. (2007). Characterization of the early events in dengue virus cell entry by biochemical assays and single-virus tracking. J. Virol. *81*, 12019–12028.

Schalich, J., Allison, S.L., Stiasny, K., Mandl, C.W., Kunz, C., and Heinz, F.X. (1996). Recombinant subviral particles from tick-borne encephalitis virus are fusogenic and provide a model system for studying flavivirus envelope glycoprotein functions. J. Virol. *70*, 4549–4557.

Scholle, F., Girard, Y.A., Zhao, Q., Higgs, S., and Mason, P.W. (2004a). trans-Packaged West Nile virus-like particles: infectious properties *in vitro* and in infected mosquito vectors. J. Virol. 78, 11605–11614.

Scholle, F., Girard, Y.A., Zhao, Q.Z., Higgs, S., and Mason, P.W. (2004b). trans-packaged West Nile virus-like particles: Infectious properties *in vitro* and in infected mosquito vectors. J. Virol. 78, 11605–11614.

Seregin, A., Nistler, R., Borisevich, V., Yamshchikov, G., Chaporgina, E., Kwok, C.W., and Yamshchikov, V. (2006). Immunogenicity of West Nile virus infectious DNA and its noninfectious derivatives. Virology *356*, 115–125.

Shi, P.Y., Brinton, M.A., Veal, J.M., Zhong, Y.Y., and Wilson, W.D. (1996). Evidence for the existence of a pseudoknot structure at the 3′ terminus of the flavivirus genomic RNA. Biochemistry *35*, 4222–4230.

Shi, P.Y., Tilgner, M., Lo, M.K., Kent, K.A., and Bernard, K.A. (2002). Infectious cDNA clone of the epidemic West Nile virus from New York City. J. Virol. *76*, 5847–5856.

Shustov, A.V., Mason, P.W., and Frolov, I. (2007). Production of pseudoinfectious yellow fever virus with a two-component genome. J. Virol. *81*, 11737–11748.

Silva, P.A., Molenkamp, R., Dalebout, T.J., Charlier, N., Neyts, J.H., Spaan, W.J., and Bredenbeek, P.J. (2007). Conservation of the pentanucleotide motif at the top of the yellow fever virus 17D 3′ stem–loop structure is not required for replication. J. Gen. Virol. *88*, 1738–1747.

Silva, P.A., Pereira, C.F., Dalebout, T.J., Spaan, W.J., and Bredenbeek, P.J. (2010). An RNA pseudoknot is required for the production of yellow fever virus sfRNA by the host nuclease XRN1. J Virol.

Song, B.H., Yun, S.I., Choi, Y.J., Kim, J.M., Lee, C.H., and Lee, Y.M. (2008). A complex RNA motif defined by three discontinuous 5-nucleotide-long strands is essential for flavivirus RNA replication. RNA-Publ. RNA Soc. *14*, 1791–1813.

Speight, G., and Westaway, E.G. (1989a). Carboxy-terminal ananlysis of 9 proteins specified by the flavivirus Kunjin – Evidence that only the intracellular core protein is truncated. J. Gen. Virol. *70*, 2209–2214.

Speight, G., and Westaway, E.G. (1989b). Positive identification of NS4A, the last of the hypothetical nonstructural proteins of flaviviruses. Virology *170*, 299–301.

Speight, G., Coia, G., Parker, M.D., and Westaway, E.G. (1988). Gene-mapping and positive identification of the non-structural proteins NS2A, NS2B, NS3, NS4B and NS5 of the flavivirus Kunjin and their cleavage sites. J. Gen. Virol. *69*, 23–34.

Sriburi, R., Keelapang, P., Duangchinda, T., Pruksakorn, S., Maneekarn, N., Malasit, P., and Sittisombut, N. (2001). Construction of infectious dengue 2 virus cDNA clones using high copy number plasmid. J. Virol. Methods *92*, 71–82.

Stadler, K., Allison, S.L., Schalich, J., and Heinz, F.X. (1997). Proteolytic activation of tick-borne encephalitis virus by furin. J. Virol. *71*, 8475–8481.

Stollar, V., Schlesin.Rw, and Stevens, T.M. (1967). Studies on nature of dengue Viruses.3. RNA synthesis in cells infected with type 2 dengue virus. Virology *33*, 650-&.

Sumiyoshi, H., Hoke, C.H., and Trent, D.W. (1992). Infectious Japanese encephalitis-virus RNA can be synthesized from invitro-ligated cDNA templates. J. Virol. *66*, 5425–5431.

Suzuki, R., de Borba, L., dos Santos, C.N.D., and Mason, P.W. (2007). Construction of an infectious cDNA clone for a Brazilian prototype strain of dengue virus type 1: Characterization of a temperature-sensitive mutation in NS1. Virology *362*, 374–383.

Suzuki, R., Winkelmann, E.R., and Mason, P.W. (2009). Construction and characterization of a single-cycle chimeric flavivirus vaccine candidate that protects mice against lethal challenge with dengue virus type 2. J. Virol. *83*, 1870–1880.

Tan, T.T.T., Bhuvanakantham, R., Li, J., Howe, J., and Ng, M.L. (2009). Tyrosine 78 of premembrane protein is essential for assembly of West Nile virus. J. Gen. Virol. *90*, 1081–1092.

Uchil, P.D., Kumar, A.V.A., and Satchidanandam, V. (2006). Nuclear localization of flavivirus RNA synthesis in infected cells. J. Virol. *80*, 5451–5464.

Varnavski, A.N., and Khromykh, A.A. (1999). Noncytopathic flavivirus replicon RNA-based system for expression and delivery of heterologous genes. Virology *255*, 366–375.

Villordo, S.M., and Gamarnik, A.V. (2009). Genome cyclization as strategy for flavivirus RNA replication. Virus Res. *139*, 230–239.

Wallner, G., Mandl, C.W., Kunz, C., and Heinz, F.X. (1995). The flavivirus 3′-noncoding region: extensive size heterogeneity independent of evolutionary relationships among strains of tick-borne encephalitis virus. Virology *213*, 169–178.

Wang, P.G., Kudelko, M., Lo, J., Siu, L.Y.L., Kwok, K.T.H., Sachse, M., Nicholls, J.M., Bruzzone, R., Altmeyer, R.M., and Nal, B. (2009). Efficient assembly and secretion of recombinant subviral particles of the four dengue serotypes using native prM and E proteins. Plos One *4*, 13.

Welsch, S., Miller, S., Romero-Brey, I., Merz, A., Bleck, C.K.E., Walther, P., Fuller, S.D., Antony, C., Krijnse-Locker, J., and Bartenschlager, R. (2009). Composition and three-dimensional architecture of the dengue virus replication and assembly sites. Cell Host Microbe *5*, 365–375.

Wengler, G. (1991). The carboxy-terminal part of the NS 3 protein of the West Nile flavivirus can be isolated as a soluble protein after proteolytic cleavage and represents an RNA-stimulated NTPase. Virology *184*, 707–715.

Wengler, G. (1993). The NS 3 nonstructural protein of flaviviruses contains an RNA triphosphatase activity. Virology *197*, 265–273.

Wengler, G., and Castle, E. (1986). Analysis of structural-properties which possibly are characteristic for the 3′-terminal sequence of the genome RNA of flaviviruses. J. Gen. Virol. *67*, 1183–1188.

Wengler, G., and Gross, H.J. (1978). Studies on virus-specific nucleic-acids synthesized in vertebrate and mosquito cells infected with flaviviruses. Virology *89*, 423–437.

Wengler, G., Castle, E., Leidner, U., and Nowak, T. (1985). Sequence-analysis of the membrane-protein V3 of the flavivirus West Nile Virus and of its gene. Virology *147*, 264–274.

Wengler, G., Nowak, T., and Castle, E. (1990). Description of a procedure which allows isolation of viral nonstrucural proteins from BHK vertebrate cells infected with the West Nile flavivirus in a state which allows their direct chemical characterization. Virology *177*, 795–801.

Werme, K., Wigerius, M., and Johansson, M. (2008). Tick-borne encephalitis virus NS5 associates with membrane protein scribble and impairs interferon-stimulated JAK-STAT signalling. Cell. Microbiol. *10*, 696–712.

Westaway, E.G., Khromykh, A.A., Kenney, M.T., Mackenzie, J.M., and Jones, M.K. (1997a). Proteins C and NS4B of the flavivirus Kunjin translocate independently into the nucleus. Virology *234*, 31–41.

Westaway, E.G., Mackenzie, J.M., Kenney, M.T., Jones, M.K., and Khromykh, A.A. (1997b). Ultrastructure of Kunjin virus-infected cells: colocalization of NS1 and NS3 with double-stranded RNA, and of NS2B with NS3, in virus-induced membrane structures. J. Virol. *71*, 6650–6661.

Westaway, E.G., Khromykh, A.A., and Mackenzie, J.M. (1999). Nascent flavivirus RNA colocalized *in situ* with double-stranded RNA in stable replication complexes. Virology *258*, 108–117.

Westaway, E.G., Mackenzie, J.M., and Khromykh, A.A. (2002). Replication and gene function in Kunjin virus. Curr. Top. Microbiol. Immunol. *267*, 323–351.

Westaway, E.G., Mackenzie, J.M., and Khromykh, A.A. (2003). Kunjin RNA replication and applications of Kunjin replicons. Adv. Virus Res. *59*, 99–140.

Whitby, K., Pierson, T.C., Geiss, B., Lane, K., Engle, M., Zhou, Y., Doms, R.W., and Diamond, M.S. (2005). Castanospermine, a potent inhibitor of dengue virus infection *in vitro* and *in vivo*. J. Virol. *79*, 8698–8706.

Widman, D.G., Frolov, I., and Mason, P.W. (2008a). Third-generation flavivirus vaccines based on single-cycle, encapsidation-defective viruses. Adv. Virus Res. *72*, 77–126.

Widman, D.G., Ishikawa, T., Fayzulin, R., Bourne, N., and Mason, P.W. (2008b). Construction and characterization of a second-generation pseudoinfectious West Nile virus vaccine propagated using a new cultivation system. Vaccine *26*, 2762–2771.

Widman, D.G., Ishikawa, T., Giavedoni, L.D., Hodara, V.L., Garza Mde, L., Montalbo, J.A., Travassos Da Rosa, A.P., Tesh, R.B., Patterson, J.L., Carrion, R., Jr., *et al.* (2010). Evaluation of RepliVAX WN, a single-cycle flavivirus vaccine, in a non-human primate model of West Nile virus infection. Am. J. Trop. Med. Hyg. *82*, 1160–1167.

Widman, D.G., Ishikawa, T., Winkelmann, E.R., Infante, E., Bourne, N., and Mason, P.W. (2009). RepliVAX WN, a single-cycle flavivirus vaccine to prevent West Nile disease, elicits durable protective immunity in hamsters. Vaccine *27*, 5550–5553.

Wilson, J.R., de Sessions, P.F., Leon, M.A., and Scholle, F. (2008). West Nile virus nonstructural protein 1 inhibits TLR3 signal transduction. J. Virol. *82*, 8262–8271.

Winkler, G., Randolph, V.B., Cleaves, G.R., Ryan, T.E., and Stollar, V. (1988). Evidence that the mature form of the flavivirus nonstructural protein NS1 is a dimer. Virology *162*, 187–196.

Wu, J.H., Bera, A.K., Kuhn, R.J., and Smith, J.L. (2005). Structure of the flavivirus helicase: Implications for catalytic activity, protein interactions, and proteolytic processing. J. Virol. *79*, 10268–10277.

Xu, T., Sampath, A., Chao, A., Wen, D.Y., Nanao, M., Chene, P., Vasudevan, S.G., and Lescar, J. (2005). Structure of the dengue virus helicase/nucleoside triphosphatase catalytic domain at a resolution of 2.4 angstrom. J. Virol. *79*, 10278–10288.

Yamshchikov, V.F., Wengler, G., Perelygin, A.A., Brinton, M.A., and Compans, R.W. (2001). An infectious clone of the West Nile flavivirus. Virology *281*, 294–304.

Yon, C., Teramoto, T., Mueller, N., Phelan, J., Ganesh, V.K., Murthy, K.H., and Padmanabhan, R. (2005). Modulation of the nucleoside triphosphatase/RNA helicase and 5′-RNA triphosphatase activities of dengue virus type 2 nonstructural protein 3 (NS3) by interaction with NS5, the RNA-dependent RNA polymerase. J. Biol. Chem. *280*, 27412–27419.

Yoshii, K., Hayasaka, D., Goto, A., Kawakami, K., Kariwa, H., and Takashima, I. (2005). Packaging the replicon RNA of the Far-Eastern subtype of tick-borne encephalitis virus into single-round infectious particles: development of a heterologous gene delivery system. Vaccine *23*, 3946–3956.

Yoshii, K., Ikawa, A., Chiba, Y., Omori, Y., Maeda, J., Murata, R., Kariwa, H., and Takashima, I. (2009). Establishment of a neutralization test involving reporter gene-expressing virus-like particles of tick-borne encephalitis virus. J. Virol. Methods *161*, 173–176.

You, S., and Padmanabhan, R. (1999). A novel *in vitro* replication system for dengue virus – Initiation of RNA synthesis at the 3′-end of exogenous viral RNA templates requires 5′- and 3′-terminal complementary sequence motifs of the viral RNA. J. Biol. Chem. *274*, 33714–33722.

Yu, I.M., Zhang, W., Holdaway, H.A., Li, L., Kostyuchenko, V.A., Chipman, P.R., Kuhn, R.J., Rossmann, M.G., and Chen, J. (2008). Structure of the immature dengue virus at low pH primes proteolytic maturation. Science *319*, 1834–1837.

Yun, S.I., Choi, Y.J., Yu, X.F., Song, J.Y., Shin, Y.H., Ju, Y.R., Kim, S.Y., and Lee, Y.M. (2007). Engineering the Japanese encephalitis virus RNA genome for the expression of foreign genes of various sizes: Implications for packaging capacity and RNA replication efficiency. J. Neurovirol. *13*, 522–535.

Yun, S.I., Choi, Y.J., Song, B.H., and Lee, Y.M. (2009). 3′ cis-acting elements that contribute to the competence and efficiency of Japanese encephalitis virus genome replication: functional importance of sequence duplications, deletions, and substitutions. J. Virol. *83*, 7909–7930.

Zeng, L.L., Falgout, B., and Markoff, L. (1998). Identification of specific nucleotide sequences within the conserved 3 ‘-SL in the dengue type 2 virus genome required for replication. J. Virol. *72*, 7510–7522.

Zhang, B., Dong, H., Stein, D.A., Iversen, P.L., and Shi, P.Y. (2008). West Nile virus genome cyclization and RNA replication require two pairs of long-distance RNA interactions. Virology *373*, 1–13.

Zhang, Y., Kostyuchenko, V.A., and Rossmann, M.G. (2007). Structural analysis of viral nucleocapsids by subtraction of partial projections. J. Struct. Biol. *157*, 356–364.

Zhao, Z.J., Date, T., Li, Y.H., Kato, T., Miyamoto, M., Yasui, K., and Wakita, T. (2005). Characterization of the E-138 (Glu/Lys) mutation in Japanese encephalitis virus by using a stable, full-length, infectious cDNA clone. J. Gen. Virol. *86*, 2209–2220.

Zhou, Y., Ray, D., Zhao, Y., Dong, H., Ren, S., Li, Z., Guo, Y., Bernard, K.A., Shi, P.Y., and Li, H. (2007). Structure and function of flavivirus NS5 methyltransferase. J. Virol. *81*, 3891–3903.

The Many Faces of the Flavivirus Non-structural Glycoprotein NS1

4

David A. Muller and Paul R. Young

Abstract

The flavivirus non-structural protein, NS1 is an enigmatic protein whose structure and mechanistic function has remained elusive since it was first identified in 1970 as a viral antigen circulating in the sera of dengue infected patients. All flavivirus NS1 genes share a high degree of homology, encoding a 352 amino acid polypeptide that has a molecular weight of between 46 and 55 kDa depending on its glycosylation status. NS1 exists in multiple oligomeric forms and is found in different cellular locations; cell membrane associated in vesicular compartments or on the cell surface and as a secreted extracellular hexamer. Intracellular NS1 has been shown to co-localize with dsRNA and other components of the viral replication complex and plays an essential co-factor role in virus replication. However, the precise function of this protein in viral replication has yet to be elucidated. Secreted and cell surface-associated NS1 are highly immunogenic and both the protein and the antibodies it elicits have been implicated in either protection or pathogenesis in infected hosts. It is also an important biomarker for early diagnosis. In this review we provide an overview of these somewhat disparate areas of research.

Introduction

Flaviviruses are small, enveloped viruses with a positive-sense RNA genome. The flavivirus genus comprises many important human pathogens including dengue (DENV), yellow fever (YFV), Japanese encephalitis (JEV), West Nile (WNV), tick-borne encephalitis (TBEV), St Louis encephalitis (SLEV) and Murray Valley encephalitis (MVEV) viruses. Disease associated with these viruses varies greatly from asymptomatic infection and self-limiting febrile illness through to encephalitis or meningitis, haemorrhage and shock which can be fatal (Malavige *et al.*, 2004; Chappell *et al.*, 2008; Guzman *et al.*, 2010; Ross, 2010; Rossi *et al.*, 2010). The flavivirus genome encodes for three structural (C, prM and E) and seven non-structural proteins (NS1, NS2A, NS2B, NS3, NS4A, NS4B and NS5) (Lindenbach and Rice, 2003). The first non-structural protein NS1 is an enigmatic protein whose structure and mechanistic function has remained elusive since it was first identified in 1970 as a soluble complement fixing (SCF) antigen (Brandt *et al.*, 1970b; Smith and Wright, 1985). The SCF antigen, originally referred to as gp48 based on its molecular weight, was later renamed NS1 after the sequencing of the YFV genome in 1985 placed the gene encoding this protein as the first of the non-structural proteins (Rice *et al.*, 1985). All flavivirus NS1 genes share a high degree of homology and are 1056 nucleotides in length, encoding a 352 amino acid polypeptide (Mackow *et al.*, 1987; Deubel *et al.*, 1988; Mandl *et al.*, 1989; Wright *et al.*, 1989). NS1 has a molecular weight of between 46 and 55 kDa depending on its glycosylation status, exists in multiple oligomeric forms and is found at different cellular locations; either cell membrane associated (mNS1) in vesicular compartments within the cell or on the cell surface and as a secreted extracellular (non-virion) form (sNS1) (Smith and Wright, 1985; Westaway and Goodman, 1987; Winkler *et al.*, 1988; Mason, 1989). Intracellular NS1 plays an essential co-factor role in virus replication and has been shown to co-localize with dsRNA and other components of viral replication complexes (Mackenzie *et al.*, 1996; Westaway *et al.*, 1997). However, the precise function of this protein in viral replication has yet to be elucidated. Secreted and cell surface-associated NS1 are highly immunogenic and both the protein and the antibodies it elicits have been implicated in disease pathogenesis (Schlesinger *et al.*, 1987; Henchal *et al.*, 1988; Falgout *et al.*, 1990; Avirutnan *et al.*, 2006; Sun *et al.*, 2007). This review aims to bring together an extensive, and sometimes confusing body of literature on this unusual flavivirus protein. It has been implicated in a multitude of roles ranging from eliciting a protective immune response in infected hosts to playing a role in pathogenesis. Specifically, we will review the current state of knowledge of the structure and trafficking of this protein within and from the infected cell, its proposed role in viral replication, potential as a vaccine candidate, value in diagnostic applications and its role in pathogenesis *in vivo* through the interaction of the protein itself or the antibodies it elicits, with an ever increasing number of host cell targets.

Expression, post-translational processing and trafficking

Following flavivirus entry and uncoating, the viral genome provides the template for the first round of translation from a single open reading frame (Fig. 4.1). NS1 is translocated into the lumen of the ER via a signal sequence corresponding to the final 24 amino acids of E (Falgout *et al.*, 1989) and is released from E at its amino terminus via cleavage by the ER resident host signal peptidase (Nowak *et al.*, 1989). NS1 is cleaved at its C-terminus from the downstream NS2A by an as yet, unidentified ER resident host cell protease (Falgout and Markoff, 1995). The last eight amino acids of NS1 have been reported to be necessary for cleavage to take place by an ER resident protease recognizing the octapeptide motif L/M-V-X-S-X-V-X-A at the end of the NS1 protein (Chambers *et al.*, 1990; Hori and Lai, 1990; Pethel *et al.*, 1992; Falgout and Markoff, 1995). In a recombinant vaccinia virus expression system it was found that 70% of NS2A was required to mediate effective cleavage (Hori and Lai, 1990; Leblois and Young, 1995). However, only 26 amino acids at the N-terminus of NS2A were found to be required for NS1/2A cleavage using recombinant baculovirus expression in insect (*Sf9*) cells (Hori and Lai, 1990; Leblois and Young, 1995). Further studies are required to fully explore this apparent anomaly but conformational constraints imposed by variably truncated NS2A may be responsible. We had earlier thought that an ideal candidate for cleavage activity at the NS1–NS2A junction might be the ER resident glycosyl-phosphatidylinositol (GPI) transamidase following our finding of a GPI anchored form of NS1 (Jacobs *et al.*, 2000). However, while virus replication was indeed reduced in mutant cell lines defective for GPI addition, the NS1–NS2A junction was still efficiently cleaved (H. E. White and P. R. Young, unpublished observations). The protease responsible and the exact role of downstream NS2A sequences in the efficiency of cleavage remain to be determined.

The hydrophilic monomer that is released from the viral polyprotein contains 12 cysteines that form six discrete disulfide bonds that are thought to be important for both the structure and function of NS1 (Mason *et al.*, 1987; Leblois and Young, 1993; Blitvich *et al.*, 2001; Wallis *et al.*, 2004). The role of these disulfide bonds in stabilization and correct folding of the monomer is reflected in mutagenesis studies that showed that the last three cysteines were essential for NS1 maturation, secretion and the formation of oligomeric species (Pryor and Wright, 1993). Using mass spectrometry the first three and all six disulphide bonds were determined for MVE (Blitvich *et al.*, 2001) and DENV-2 (Wallis *et al.*, 2004) NS1, respectively. The disulfide bonds were determined to link in the following arrangements: C1/C2, C3/C4, C5/C6, C7/C12, C8/C10 and C9/C11 (Fig. 4.2A).

Following cleavage in the ER, NS1 is glycosylated by the addition of high mannose carbohydrates (Winkler *et al.*, 1988; Pryor and Wright, 1994) (Fig. 4.1, step 2). This hydrophilic monomer rapidly dimerizes (20–40 min) (Winkler *et al.*, 1989; Winkler *et al.*, 1988) acquiring a partial hydrophobic nature as demonstrated by the separation of dimeric NS1 into both membrane and aqueous phases in Triton X 114 phase separation experiments (Winkler *et al.*, 1989).

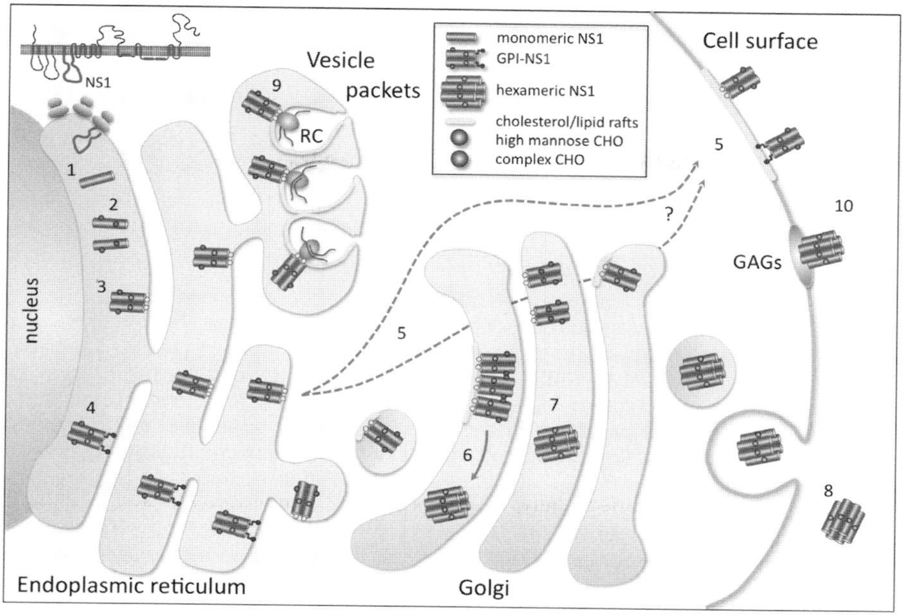

Figure 4.1 Schematic summary of NS1 trafficking in mammalian cells. NS1 is initially expressed in association with the endoplasmic reticulum (ER) as part of a long polyprotein that is encoded by a single open reading frame of the viral RNA genome. A signal sequence at the C-terminus of the virion glycoprotein E targets NS1 to the lumen of the ER and co-translational cleavage within the ER lumen at both its N- and C-terminus generates a hydrophilic monomeric subunit (1). This monomer is modified by the addition of high mannose carbohydrate moieties at multiple sites (2) and rapidly forms a dimeric species leading to the acquisition of a hydrophobic character (mNS1), resulting in membrane association (3). A subset of dengue virus NS1 acquires a glycosyl-phosphatidylinositol (GPI) anchor in the ER as a consequence of the recognition of a GPI addition signal at the N-terminus of NS2A (4). Both mNS1 and GPI-anchored NS1 are trafficked to the cell surface via an unknown pathway where they have been shown to associate with lipid rafts (yellow membrane highlights) (5). A proportion of NS1 traffics from the ER through the Golgi where dimeric units associate to form soluble hexamers (6), although it is also possible that these hexamers are formed earlier in the ER. Passage of this soluble hexameric species (sNS1) through the Golgi results in one of the high mannose carbohydrate moieties on each monomer being trimmed and processed to a complex carbohydrate form (7) and is then secreted from the cell (8). An alternative pathway for mNS1 from the ER sees a subset of the high mannose form becoming associated with vesicle packets (VP) where it co-localises with other non-structural viral proteins that comprise the viral replication complex (RC) (9). Some of the cell-surface-associated NS1 may be previously secreted NS1 that has bound directly to cell surface glycosaminoglycans (GAGs) (10) The oligomeric nature of the various cell-bound forms of NS1 remains largely unknown. A colour version of this figure is located in the plate section at the back of the book.

This newly acquired hydrophobicity is thought to be the major factor in NS1 becoming associated with the ER membrane (Fig. 4.1, step 3). However, the nature and location of this hydrophobic component has yet to be fully identified. Oligomerization of NS1 is a common feature of all flaviviruses with the stable dimeric form of NS1 first identified in SDS-PAGE analyses of infected mammalian and mosquito cell cultures (Winkler *et al.*, 1988; Chambers *et al.*, 1989; Mason, 1989). This dimer is resistant to treatment with both non-ionic and ionic detergents, however they can be dissociated by heat or acid (< pH 2.2–3) treatment (Falconar and Young, 1990; Winkler *et al.*, 1988). Recombinant expression studies have shown that multimeric species spontaneously form in the absence of other viral proteins, indicating that NS1 contains all the information needed to drive oligomerization (Parrish *et al.*, 1991; Pryor and Wright, 1993; Leblois and Young, 1995). Mutations of the octapeptide cleavage motif that leave the NS1/NS2A junction intact do

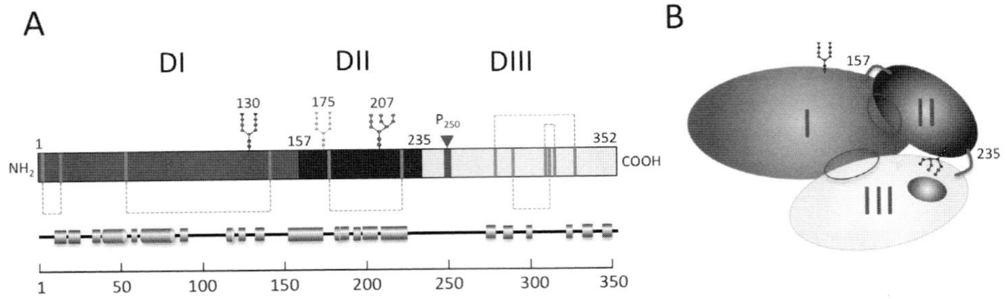

Figure 4.2 Antigenic and secondary structure models of NS1. (A) Linear representation of NS1 highlighting the three structural fragments, domain I (red), domain II (blue) and domain III (yellow) identified for WNV NS1. Disulfide linkages for all 12 conserved cysteines are shown in grey and the two conserved N-linked glycosylation sites are shown in black (complex CHO at Asn130 and high mannose CHO at Asn207). A third carbohydrate addition site at position 175 and found in the NS1 species of the JE serocomplex of flaviviruses is shown in grey. Destabilization of dimers by a naturally occurring mutation of residue 250 (green arrowhead) in WNV and MVE suggests that the dimerization domain is located in this region. A consensus secondary structure prediction based on an alignment of more than 40 flavivirus sequences and showing putative β sheets (green cylinders) and α helices (red cylinders) is shown below. (B) Proposed tertiary arrangement of the three structural domains of NS1 based on overlapping epitope reactivity by monoclonal antibodies (schematically represented by the green ovals). A colour version of this figure is located in the plate section at the back of the book.

not affect dimer formation (Parrish *et al.*, 1991; Pryor and Wright, 1993) and indeed the unique NS1′ of some encephalitic flaviviruses that carries a carboxy-terminal extension derived via a ribosomal frameshift is also dimeric (Mason, 1989; Melian *et al.*, 2009). In Kunjin virus and MVE, a single amino acid substitution at residue 250 from proline to leucine has been shown to result in a loss of detectable dimers, suggesting a role for this C-terminal region of the protein in the dimerization process (Fig. 4.2A). Despite the apparent loss of dimer formation, the resulting monomeric NS1 was still secreted. While these findings suggest that dimerization may not be essential for viral infectivity, this mutation did correlate with retarded virus growth and reduced virulence in mice (Hall *et al.*, 1999; Clark *et al.*, 2007). However, some caution needs to be made in drawing conclusions about the oligomeric nature of this mutant form of NS1 from these results. The presence of dimers was assessed by either SDS-PAGE analysis without sample heating or by reactivity in fixed infected cell monolayers with a MAb previously characterized as binding only to dimeric NS1 on immunoblots. While native NS1 normally retains its dimeric status when separated on SDS-PAGE in the absence of heating, the possibility that the proline to leucine mutation merely results in lower affinity interactions between individual monomers that are then disrupted by exposure to SDS treatment cannot be excluded. Furthermore, the reactivity of a MAb to a fixed cell substrate may not entirely reflect the native form of the protein. Gel filtration and/or cross-linking studies of untreated secreted NS1 would need to be performed on this mutant to adequately answer the question of whether or not the oligomeric form of NS1 is essential for viral replication.

Following dimerization, NS1 is trafficked to three separate destinations; sites of viral replication within the cell, the infected cell surface and secreted into the extracellular space (see Fig. 4.1). The majority of cell-associated NS1 co-localizes with dsRNA and other non-structural proteins involved in genome replication in structures referred to as vesicle packets (Fig. 4.1, step 9) (Mackenzie *et al.*, 1996; Lindenbach and Rice, 1997, 1999; Khromykh *et al.*, 1999), while a small proportion of cell-associated NS1 is also found at the infected cell surface (Fig. 4.1, step 5) (Winkler *et al.*, 1989; Schlesinger *et al.*, 1990). In mammalian cells, another component of expressed NS1 is trafficked through the Golgi via the secretory pathway, where exposed carbohydrate moieties are trimmed and processed to more complex sugars (Fig. 4.1, step 7) and then secreted from the cell as a soluble hexamer (Fig. 4.1, step 8) (Crooks *et al.*, 1990, 1994; Flamand *et al.*, 1999). Notably, NS1 is not secreted from infected insect cells (Mason, 1989) despite a recent report suggesting that it is (Ludert *et al.*, 2008). This report is based on NS1 detection in media harvests using a sensitive NS1 capture ELISA assay (Ludert *et al.*, 2008). These experiments were not quantitative and were performed relatively late in infection suggesting that detection of NS1 in the media was most probably the result of liberation from infected cells undergoing lysis. Pulse-chase experiments early during infection, where minimal cell lysis has occurred failed to demonstrate any significant secretion from insect cells (Mason, 1989). It is also worth noting that while intracellular, membrane-associated NS1 has been represented in the schematic presented in Fig. 4.1 as a dimeric species, and indeed this is the form often attributed to it in the literature, there is still no direct experimental evidence for this assumption. It has previously been shown that the hexameric form of NS1 is held together by weak hydrophobic interactions that are readily disrupted by detergent treatment (Flamand *et al.*, 1999) and so the detergent based lysis that is required to solubilize infected and/or transfected cells for analysis would not be compatible with the recovery of these higher oligomer forms. Therefore, while it is clear that NS1 is at least dimeric within the infected cell, the exact higher order oligomeric nature of intra-cellular NS1 awaits experimental confirmation. Despite this reservation, one early study employing cross-linking of intact yellow fever virus infected cells indicated that surface-associated NS1 probably exists in a dimeric, and not hexameric form (Schlesinger *et al.*, 1990).

As noted above, membrane-association probably follows the acquisition of a hydrophobic character following dimerization early after synthesis (Mason, 1989; Winkler *et al.*, 1989; Noisakran *et al.*, 2008a). Through unknown mechanisms, perhaps mediated by the local concentration of this form of NS1 on cholesterol rich lipid rafts (Noisakran *et al.*, 2008a), it is possible that a proportion of dimers are able to dissociate from the Golgi membrane and associate instead with two other dimeric units via this same hydrophobic domain. This association would sequester their inherent hydrophobicity to form the soluble hexameric species (Fig. 4.1, step 6) (Flamand *et al.*, 1999). However, it is also possible that both membrane-association and soluble hexamer formation occur immediately following dimerization and acquisition of hydrophobicity, with the fate of individual dimers being solely dependent on NS1 concentration. As the infection progresses, higher local concentrations of NS1 may lead to a greater likelihood of dimeric units partnering with others rather than the ER membrane. Pulse-chase labelling experiments have shown that the formation and secretion of sNS1 is significantly delayed following the initial synthesis of mNS1, a finding that would support either of these scenarios (Mason, 1989). Further studies are required to clarify the details of this early stage in NS1 maturation.

Glycosylation of a viral non-structural protein is somewhat unusual, given that this post-translational modification is usually restricted to virion surface proteins. Furthermore, a range of different glycosylation patterns are seen for NS1 that are dependant on the infecting flavivirus, the host cells they infect as well as their different cellular locations. NS1 from all serotypes of DENV, JEV and YFV contain two conserved glycosylation sites, at positions Asn 130 and Asn 207 (Fig. 4.2A) (Smith and Wright, 1985; Mason *et al.*, 1987; Zhao *et al.*, 1987; Flamand *et al.*, 1992; Pryor and Wright, 1994). With the addition of carbohydrate moieties, monomeric NS1 migrates on SDS-PAGE with molecular weights of between 49–55 kDa, depending on the level of processing and complex sugar addition. In mammalian cells, NS1 exists in two major forms. The membrane-associated form of NS1 (mNS1) migrates with a MW of approximately 49 kDa as a sharp band on SDS-PAGE as it contains only high mannose carbohydrate additions (Post *et al.*, 1991). The second, secreted form of NS1 (sNS1) migrates on SDS-PAGE some 3–6 kDa larger than mNS1 as a smear from 52–55 kDa, owing to the additional trimming and processing of the high mannose carbohydrate at Asn 130 with a heterogeneous mix of complex sugars. For some members of the JEV subgroup, including WNV, SLEV and MVEV an additional linkage site at Asn 175 is also processed to a complex form (Dalgarno *et al.*, 1986; Trent *et al.*, 1987; Coia *et al.*, 1988; Mandl *et al.*,

1989; Blitvich *et al.*, 1999, 2001). The addition of complex carbohydrates indicates the passage of secreted NS1 through the Golgi compartment where trimming of high mannose and the addition of more complex sugars occurs (Winkler *et al.*, 1988; Mason, 1989). Insect cells do not possess the required glycosylation machinery to process NS1 to the complex carbohydrate form. In these cells, and as noted above, NS1 is not secreted but instead accumulates in infected cells, suggesting an association between complex carbohydrate addition and secretion (Mason, 1989; Flamand *et al.*, 1999). This is supported by mutagenesis studies that have found that removal of either or both glycosylation sites in DENV, WNV or YFV results in decreased NS1 secretion, as well as reduced neurovirulence in mice, small plaque phenotype, decreased virus yields, reduced cytopathology and depressed RNA accumulation (Pryor and Wright, 1994; Muylaert *et al.*, 1996; Crabtree *et al.*, 2005; Tajima *et al.*, 2008). NS1 devoid of carbohydrate additions nevertheless appears to be trafficked efficiently to the cell surface (Youn *et al.*, 2010). Taken together, these results indicate that glycosylation is important for NS1 maturation, at least in terms of its secretion, virus RNA replication and virulence of disease.

The hexameric nature of the secreted form of NS1 (sNS1) was first identified in the media harvests of TBEV infected mammalian cells and later confirmed in DENV and WNV-infected cells (Crooks *et al.*, 1990; Crooks *et al.*, 1994; Flamand *et al.*, 1999; Chung *et al.*, 2006a). Secretion kinetics appear to be different for different flaviviruses with TBEV NS1 secreted within 45 minutes of expression, whereas NS1 from JEV and YFV has been shown to take up to 2 hours before it is detected in media harvests (Lee *et al.*, 1989; Mason, 1989). A short motif at the N-terminus of NS1 (residues 10 and 11) has recently been identified in a comparative study of WNV and DENV that may explain some of this variation in secretion versus cellular retention (Youn *et al.*, 2010). WNV NS1 was shown to accumulate at higher relative levels on the infected cell surface than DENV NS1 and revealed a distinctive reticular staining pattern in immunofluorescence analyses, with DENV NS1 showing a more diffuse surface distribution. In contrast, DENV NS1 was more efficiently secreted into the infected cell media. The authors suggest that this motif may mediate the differential binding of the respective NS1 species to an ER resident host cell protein and so influence the subsequent pathway of maturation to predominantly cell membrane association or secretion (Youn *et al.*, 2010). However, no direct evidence for such a host protein interaction was presented in this report. It could equally be argued that differences in specific residues at this location may directly influence the efficiency of hexamer formation and so increased cell membrane association versus secretion may simply be a loss-of-function mutation rather than reflecting binding to a host cell membrane protein. Further studies are required to clarify the mechanistic role of this motif in flavivirus replication, but it certainly adds to the growing list of phenotypic differences now identified between the various flavivirus NS1 species.

Gel filtration studies have shown that hexameric sNS1 has a molecular weight of 310 kDa and a Stokes radius of 64.4 Å. This form of NS1 is held together by weak hydrophobic interactions and will dissociate in the presence of non-ionic detergents to the more stable dimeric subunits that can only be dissociated to monomers following heat or acid treatment (Crooks *et al.*, 1994; Flamand *et al.*, 1999). Cross-linking experiments, using dimethylsuberimidate (DMS) and SDS-PAGE analysis have shown that hexameric NS1 denatures preferentially to tetramers, dimers and monomers, suggesting that the hexamer is made up of a trimer of dimers (Flamand *et al.*, 1999). As noted above, secretion of NS1 has been attributed in part to the differential glycosylation processing that occurs in mammalian cells. After NS1 dimerizes and moves through the secretory pathway, the high mannose carbohydrate at Asn-130 is trimmed and processed to a complex carbohydrate. The second carbohydrate addition site at Asn-207 is sterically protected from processing in the oligomeric form and so retains its high mannose carbohydrate moiety (Flamand *et al.*, 1999). Negative stain electron microscopy has shown that hexameric sNS1 is a roughly spherical macromolecule with 32-point symmetry and 2-fold symmetry along its axis (Flamand *et al.*, 1999). However, very little further structural information is available, with on-going efforts by a number of groups, including ours, to crystallize this species being unsuccessful to date. Nevertheless, some interpretation of the overall structural topography of sNS1 can be inferred from antigenic epitope competition mapping with monoclonal antibodies (Henchal *et al.*, 1987; Hall *et al.*, 1990; Young, 1990; Chung *et al.*, 2006b), as well as the localization of their binding site using synthetic linear peptides (Falconar *et al.*, 1994; Huang *et al.*, 1999; Wang *et al.*, 2009; Chen *et al.*, 2010) and recombinantly expressed fragments (Putnak and Schlesinger, 1990; Chung *et al.*, 2006b). The most comprehensive of these to date was an analysis of WNV NS1 using a panel of 22 MAbs that identified what appear to be three separate structural domains (schematically represented in Fig. 4.2), a finding that is entirely consistent with the earlier reports. Furthermore, antibody competition mapping and binding to recombinantly expressed subfragments showed that some epitopes overlap more than one fragment suggesting that domains at either end of the NS1 sequence may be in physical proximity (Fig. 4.2B). The final resolution of the quaternary structure of sNS1 is keenly awaited.

Dengue virus NS1 has been shown to bind to a wide variety of cells via a charge interaction with glycosaminoglycans (GAG), heparin sulphate and chondroitin sulphate E, although the amino acid sequence of flavivirus NS1 does not contain any obvious GAG-binding motifs (Avirutnan *et al.*, 2007). NS1 binds strongly to epithelial and fibroblast cells in culture with considerable variability in binding to endothelial cells (human dermal, lung microvascular and aortic endothelial cells) (Avirutnan *et al.*, 2007). Dengue virus sNS1 has also been shown to display a tropism for hepatocytes, both *in vitro* and *in vivo* (Alcon-LePoder *et al.*, 2005). Following internalization by endocytosis, it accumulates within late endosomes and can be detected for at least 48 h without degradation (Alcon-LePoder *et al.*, 2005). The pH of late endosomes is reported to be around pH 5.5 and at this pH, NS1 has been shown to be stable in its dimeric form (Winkler *et al.*, 1988; Falconar and Young, 1990; Alcon-LePoder *et al.*, 2005). It was also found that treatment of hepatocytes with NS1 leads to enhanced virus production (Alcon-LePoder *et al.*, 2005). Whether cellular binding and/or endosomal accumulation are responsible were not addressed and hence a mechanism for this enhanced infection remains unclear. The range of cells identified as substrates for NS1 binding and the number of protein partners characterized as binding different flavivirus NS1 species suggests that NS1 may be an inherently 'sticky' protein that forms interactions via non-specific as well as specific charged and hydrophobic interactions (Chua *et al.*, 2005; Chung *et al.*, 2006a; Avirutnan *et al.*, 2007; Kurosu *et al.*, 2007; Noisakran *et al.*, 2008b; Wilson *et al.*, 2008). The acquisition of hydrophobicity by the dimeric form and the relative fragility of hexameric NS1 suggests that interpretation of some of this binding data needs to be treated with some caution with many of the interactions identified requiring further confirmation. There is also growing evidence that many of the host cell derived partners identified for NS1 may be specific to individual flaviviruses suggesting a greater diversity in how different flaviviruses utilize their respective NS1 species than previously thought (Chung *et al.*, 2006a; Krishna *et al.*, 2009).

NS1 encoded by DENV-1 to 4 and JEV have recently been shown to associate with detergent resistant lipid rafts in infected cells (Lee *et al.*, 2008; Noisakran *et al.*, 2008a). Similar observations were made for NS1 expressed in cells transfected with recombinant NS1 constructs (Lee *et al.*, 2008; Noisakran *et al.*, 2008a). Mammalian cell membranes comprise a lipid bilayer made up mainly of three different types of lipids; phosphoglycerides, sphingolipids, and sterols (Simons and Ikonen, 1997, 2000; Ikonen, 2001). Lipid rafts are detergent-resistant microdomains (DRM) that are enriched in cholesterol and sphingolipids, accumulating in a liquid-ordered arrangement (Simons and Ikonen, 1997, 2000; Ikonen, 2001). DRM's have the ability to both include as well as exclude proteins and are known to cluster macromolecules involved in signal transduction, cholesterol homeostasis, lipid sorting and protein trafficking (Simons and Ikonen, 1997, 2000; Ikonen, 2001). Recently, cholesterol has been implicated in flavivirus entry, RNA uncoating and replication with NS1, NS3 and NS5 all being found to be associated with lipid rafts (Lee *et al.*, 2008; Heaton *et al.*, 2010).

Sequence analysis clearly reveals that NS1 is essentially a hydrophilic protein that lacks a traditional membrane spanning anchor domain. So the molecular basis of NS1 membrane-association, lipid raft-associated or otherwise, has remained an intriguing and unanswered question since this association was first identified two decades ago (Winkler *et al.*, 1989). The first report of a possible mechanism for membrane-association came with the identification of a glycosyl-phosphatidylinositol (GPI)-linked form of dengue NS1 (Jacobs *et al.*, 2000) that was subsequently confirmed by others (Jacobs *et al.*, 2000; Noisakran *et al.*, 2007, 2008a). This was the first viral encoded protein identified to be expressed with a GPI-anchor but has so far only been identified for DENV (Jacobs *et al.*, 2000). In this study, the amino terminus of NS2A was shown to contain a hydrophobic region which could act as a signal sequence for GPI anchor addition (Jacobs *et al.*, 2000). This post-translational modification takes place in the ER following cleavage of a carboxy-terminal hydrophobic signal sequence and the covalent addition of a preformed GPI precursor. GPI anchored NS1 is then targeted to the cell surface where it is lipid raft-associated (Jacobs *et al.*, 2000; Noisakran *et al.*, 2007, 2008a). Intriguingly, the addition of anti-NS1 antibodies that bind to GPI anchored NS1 at the cell surface results in signal transduction, as evidenced by tyrosine phosphorylation of cellular proteins (Jacobs *et al.*, 2000). While the consequences of the induction of this particular signal transduction pathway on virus replication remains unknown, the interaction of a cell surface-bound, GPI-anchored viral antigen with specific antibodies that are anamnestically elicited *in vivo* during secondary infection suggests a novel mechanism of cellular activation that may contribute to the pathogenesis of human dengue disease (Jacobs *et al.*, 2000).

However this and other reports have shown that NS1 expression on its own, in the absence of downstream sequences, retains its hydrophobic properties and is still found in association with the cell surface (Fan and Mason, 1990; Jacobs *et al.*, 2000; Youn *et al.*, 2010), indicating that GPI anchoring is not the only mechanism of membrane association. Furthermore, the observation that phospholipase C digestion of intact dengue virus-infected cells removes only a small component of NS1 displayed at the cell surface suggests that GPI anchoring most probably contributes only a small fraction of cell surface-associated NS1 (Jacobs *et al.*, 2000). The region of NS1 that acquires a hydrophobic character after dimerization and that is most probably responsible for the majority of membrane association remains unknown. As noted above, sequence variation in a short motif (specifically at residues 10 and 11) in the N-terminal region of WNV and DENV NS1 has recently been identified as influencing the differential targeting of NS1 to either the cell surface or for secretion (Youn *et al.*, 2010). While the contribution to membrane-association of this motif has yet to be fully explored, the authors clearly demonstrate that variation within this motif strongly influences the level of cell surface expression.

Flaviviruses in the JE serocomplex express an additional form of NS1 with a carboxy-terminal extension, designated NS1′ (Mason, 1989; Blitvich *et al.*, 1995, 1999; Chen *et al.*, 1996; Melian *et al.*, 2009). NS1′ is a 52–53 kDa species that is expressed in both mosquito and mammalian cells as a cell-associated oligomer. As with their NS1 counter-parts, NS1′ from mosquito cells is retained in the cell while NS1′ from infected mammalian cells is found in both cell-associated and secreted forms (Mason, 1989; Blitvich *et al.*, 1999). NS1′ has the same glycosylation pattern as NS1, with NS1′ retained in mammalian cells comprising high mannose carbohydrate additions, while secreted NS1′ contains additional complex carbohydrate additions (Mason, 1989). For many years NS1′ was thought to be the product of alternative cleavage of the viral polyprotein, consisting of full-length NS1 fused to the N-terminus of NS2A. Recently however, bioinformatics analysis suggested that expression of NS1′ may actually be the result of the presence of a conserved pseudoknot in the 5′ end of the NS2A nucleotide sequence that is followed by a conserved slippery heptanucleotide motif (Firth and Atkins, 2009). This possibility was quickly confirmed experimentally (Melian *et al.*, 2009). This pseudoknot and slippery heptanucleotide sequence is conserved in the JE serocomplex flaviviruses and is a classical −1 ribosomal frameshift motif. The ribosome frameshift occurs between codons 8 and 9 of NS2A and results in the addition of 52 extra amino acids. While the function of this novel NS1 species remains unknown, there is experimental evidence suggesting involvement in neurovirulence with ablation of NS1′ resulting in partial attenuation of viral neuroinvasiveness (Melian *et al.*, 2009). The frameshift, leading to the alternative NS1′ is terminated by a stop codon and therefore no further expression of downstream genes occurs. Whether the observed association with neuroinvasiveness is due to NS1′ itself or a change in the ratio of structural to non-structural proteins remains to be elucidated (Melian *et al.*, 2009).

NS1 as a co-factor in virus replication

While the exact functional involvement of NS1 in the viral replication cycle remains elusive, many studies have identified that NS1 plays an essential co-factor role in viral RNA replication (Lindenbach and Rice, 1997, 1999; Westaway *et al.*, 1997; Khromykh *et al.*, 2000). NS1 was initially thought to be involved in virus assembly and maturation as its secretion profile largely mirrored that of E and prM (Rice *et al.*, 1986; Lee *et al.*, 1989; Mason, 1989). However, it was also noted that in pulse-chase experiments a substantial component of expressed NS1 after extended chase periods remained cell-associated. Since these initial findings, this cell-associated NS1 was shown to localize to sites of viral RNA replication, in close association with vesicle packets and cytoplasmic vacuoles in Vero and C6/36 cells respectively (Mackenzie *et al.*, 1996). This co-localization with dsRNA and not to sites of virus assembly as initially suspected, suggested a role in RNA replication as a component of the viral replication complex (Mackenzie *et al.*, 1996; Westaway *et al.*, 1997). However, NS1 is expressed on the luminal side of the ER derived vesicular membrane while the viral replication machinery is found on the cytoplasmic side (see Fig. 4.1, step 9). Given this physical separation, NS1 is unlikely to contribute to RNA replication directly. Rather, it has been suggested that along with transmembrane replicase components it may fulfil a structural role, helping to anchor the replication complex to the membrane. *Trans*-complementation and mutagenesis experiments have further shown that whatever role NS1 does play it does so early in RNA replication (Muylaert *et al.*, 1996; Lindenbach and Rice, 1997; Muylaert *et al.*, 1997; Butrapet *et al.*, 2000; Liu *et al.*, 2006; Suzuki *et al.*, 2007). The complementation studies found that homologous NS1 supplied *in trans* could complement a defective YFV or WNV (Kunjun) genome resulting in recovered viral RNA synthesis and replication (Lindenbach and Rice, 1997; Khromykh *et al.*, 1999). This *trans*-complementation of NS1 was found to be species specific, as DENV-2 NS1 was not able to complement a defective YFV genome (Lindenbach and Rice, 1999). However, a genetic screen for suppressor mutants that were able to overcome this species specific interaction identified a single point mutation, Asn-42-Tyr in the NS4A gene that then enabled rescue of the defective YF genome (Lindenbach and Rice, 1999).

This rescue mutation was the first evidence of a genetic interaction between NS1 and NS4A. The mutation in NS4A is located on the cytoplasmic side of induced viral membranes while NS1 is found on the luminal side. Therefore it has been proposed that either this mutation in NS4A has an effect on the conformation of the luminally displayed regions of NS4A or that NS1 and the luminal peptide of NS4A may induce a conformational change in the region around amino acid 42 of NS4A resulting in recovery of RNA replication (Lindenbach and Rice, 1999). Further supportive biochemical evidence was identified using recombinant NS4A fused to glutathione-S-transferase. Column-bound NS4A was found to interact with all the proteins that are proposed to make up the flavivirus replication complex including NS1 (Mackenzie *et al.*, 1998; Welsch *et al.*, 2009). A reasonable hypothesis is that NS1 fulfils its membrane stabilizing structural role for the replication complex via a physical interaction with regions of NS4A displayed within the lumen of the ER.

NS1 engagement with host innate and adaptive immunity

Secreted NS1 has been found to circulate in the sera of flavivirus-infected individuals during the acute phase of disease (Young *et al.*, 2000; Alcon *et al.*, 2002; Macdonald *et al.*, 2005; Chung and Diamond, 2008). Indeed, during dengue virus infection, sNS1 may accumulate to very high levels, with up to 50 μg/ml being detected in some patient sera (Young *et al.*, 2000; Libraty *et al.*, 2002; Alcon-LePoder *et al.*, 2006). The *in vivo* function of this secreted viral protein has been the subject of many investigations, with emerging evidence that NS1 engages with the host in a multitude of different ways. These include the paradoxical ability to elicit both a protective (Schlesinger *et al.*, 1985, 1986, 1987) and a potentially pathogenic autoimmune response (Henchal *et al.*, 1987; Falconar, 1997; Chang *et al.*, 2002) as well as contributing directly to the disease process through its interaction with different cell types (Falconar, 1997; Chang *et al.*, 2002; Alcon-LePoder *et al.*, 2005; Avirutnan *et al.*, 2006, 2007) or via binding to a range of specific host proteins (Chung *et al.*, 2006a; Kanlaya *et al.*, 2010). Circulating NS1 has also been identified as an important diagnostic marker of infection (Young *et al.*, 2000; Libraty *et al.*, 2002; Alcon *et al.*, 2002). Collectively, these studies have revealed that NS1 displays a remarkable diversity of engagement with various components of the innate and adaptive arms of the host immune response. It is also becoming clear that many of these interactions may be specific to individual flaviviruses, suggesting that they may each have evolved separate strategies to utilize this secreted and cell surface bound protein. A schematic summary of some of these host interactions is presented in Fig. 4.3.

NS1 as a vaccine immunogen

NS1 is a major viral immunogen, which is not a surprising observation for a protein that circulates in relatively high concentrations in the sera of individuals during the acute phase of infection. In primary dengue infection, relatively low anti-NS1 IgM and IgG responses are elicited from 2 and 9 days respectively, post onset of symptoms (Huang *et al.*, 1999; Shu *et al.*, 2003, 2004), while in secondary dengue infection, an anamnestic antibody response results in a rapid rise in anti-NS1 antibodies early during the acute phase of disease (Falkler *et al.*, 1973; Kuno *et al.*, 1990; Churdboonchart *et al.*, 1991). As NS1 is not a component of the virion, these are not neutralizing antibodies. However, in the mid-1980s using some of the first flavivirus specific monoclonal antibodies (MAbs) to be generated, the somewhat surprising discovery was made that passive administration of selected MAbs against this non-structural protein were able to provide protection to mice against a lethal viral challenge (Schlesinger *et al.*, 1985; Gould *et al.*, 1986). A direct correlation was noted between those MAbs that fixed complement and those that afforded protection, suggesting that the probable mechanism of protection was via complement mediated lysis of infected cells following antibody recognition of cell surface bound NS1 (Schlesinger *et al.*, 1985) (Fig. 4.3, step 7). Since these initial experiments with YFV, passive administration of NS1 specific MAbs against a range of flaviviruses has also been shown to provide varying levels of protection against homologous virus challenge (Henchal *et al.*, 1988; Schlesinger *et al.*, 1985, 1986). Although there is a wealth of evidence confirming a role for complement mediated lysis of infected cells (Schlesinger *et al.*, 1985, 1986, 1987; Falgout *et al.*, 1990; Krishna *et al.*, 2009), protection does not always correlate with the ability of a MAb to fix complement, indicating that other mechanisms are also involved (Henchal *et al.*, 1988; Young, 1990; Schlesinger *et al.*, 1993; Jacobs *et al.*, 1994; Chung *et al.*, 2006b, 2007; Diamond *et al.*, 2008). Recent studies using complement and specific Fc-γ receptor knock-out mice along with WNV specific anti-NS1 MAbs of varying isotype, have shown that protection in mice can be mediated by phagocytosis and clearance of infected cells through Fc-γ receptor I and/or IV recognition of cell surface NS1 bound antibodies of the IgG2a subclass (Chung *et al.*, 2006b, 2007; Diamond *et al.*, 2008) (Fig. 4.3, step 8). These studies provide an explanation for the earlier observation of complement-independent protection afforded mice challenged with YFV, specifically by IgG2a isotype anti-NS1 MAbs (Schlesinger *et al.*, 1993) along with a more complete understanding of the role of anti-NS1 antibodies in providing protection against flaviviruses in general.

One of the major concerns for any vaccine strategy against the dengue viruses is the potential

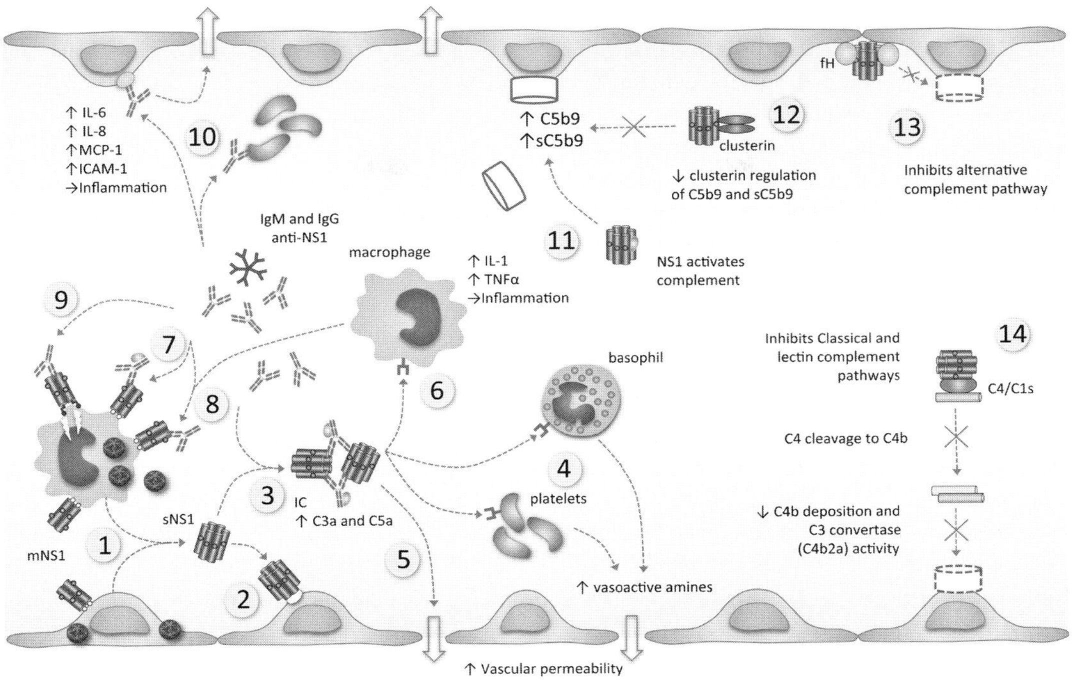

Figure 4.3 Schematic of flavivirus NS1 engagement with host cell components. The production of sNS1 and mNS1 from flavivirus infected cells and their interaction with selected host cell components is depicted. NS1 is expressed on the surface of, and secreted from, flavivirus infected cells (1). Circulating sNS1 can subsequently bind to the surface of both infected and uninfected cells via charged interactions with GAGs, heparin sulphate and chondroitin sulphate E (2). The consequences of that binding are yet to be fully determined. In dengue virus secondary infections, an anamnestic IgM and IgG antibody response to NS1 during the acute phase of disease can lead to the formation of immune complexes (3) that are capable of triggering a range of inflammatory processes including the activation of complement (green) to generate the anaphylatoxins C3a and C5a. ICs also act on basophils and platelets via Fc receptor engagement to release vasoactive amines (4) that can cause endothelial cell retraction and increase vascular permeability. This in turn may lead to IC deposition (5) inducing platelet aggregation and further complement activation. Binding of ICs to macrophages leads to their activation and the release of cytokines, further increasing the inflammatory response (6). NS1 specific antibody binding to cell surface-exposed NS1 targets infected cells for complement mediated lysis (7) and/or complement-independent phagocytosis (8). NS1-specific antibody binding of cell surface-exposed GPI anchored NS1 may also mediate activation of infected cells (9). A role for auto-immune, cross-reactive anti-NS1 antibodies in pathogenesis has also been proposed. These auto-antibodies have been shown to bind to host determinants on the surface of both platelets and endothelial cells resulting in the release of inflammatory cytokines and nitric oxide leading to inflammation and/or apoptosis. In the absence of antibody, circulating sNS1 has been shown to modulate/inhibit complement pathways through its interaction with various complement components (11–14). sNS1 has been shown to activate complement directly resulting in increased formation of the membrane attack complex C5b9 and sC5b9 (11) while its interaction with the complement inhibitory factor, clusterin is thought to result in the increased formation of sC5b9 in the serum of infected patients (12). sNS1 has also been shown to display an immune evasion function through its binding to the alternative complement pathway regulatory protein, fH (13) and its interaction with the classical complement pathway components C4 and C1s (14) with the resulting decrease in the deposition of C4b and C3 convertase leading to a decrease in the terminal membrane attack complex. The diverse array of ways in which NS1 and NS1-specific antibodies engage with the host as depicted in this schematic is by no means exhaustive. A colour version of this figure is located in the plate section at the back of the book.

priming of antibody-dependent enhancement (ADE), an important and accepted risk factor for the development of the more severe disease outcomes of dengue infection, dengue haemorrhagic fever (DHF) and dengue shock syndrome (DSS). The potential benefit to vaccine design of a protective immunogen that circumvents the induction of virion reactive antibodies and therefore avoids the potential risks of ADE was recognized early (Gibson *et al.*, 1988). Consequently, there was an immediate explosion in reports describing various immunization strategies using NS1 either alone or in combination with other viral proteins, each of which provided evidence of either partial or complete protection of mice or monkeys, from lethal virus challenge (Schlesinger *et al.*, 1985, 1986, 1987; Cane and Gould, 1988; Despres *et al.*, 1991; Falconar and Young, 1991; Lieberman *et al.*, 2007; Lin *et al.*, 2008b). Although overtaken to some degree by live attenuated chimeric virus vaccine candidates, there is renewed interest in NS1 as a component of second generation subunit vaccines for dengue as well as other flaviviruses (Krishna *et al.*, 2009; Miller, 2010). Many strategies have been employed for the delivery of NS1 as a vaccine candidate including immunization with recombinant or native protein subunit (Schlesinger *et al.*, 1985, 1987; Gould *et al.*, 1986; Gibson *et al.*, 1988; Zhang *et al.*, 1988; Qu *et al.*, 1993; Eckels *et al.*, 1994; Calvert *et al.*, 2006), live recombinant vaccinia virus (Falgout *et al.*, 1990; Hall *et al.*, 1996), defective recombinant adenovirus (Jacobs *et al.*, 1992, 1994; Timofeev *et al.*, 1998), naked DNA (Lin *et al.*, 1998; Wu *et al.*, 2003; Timofeev *et al.*, 2004; Costa *et al.*, 2007) and peptide (Volpina *et al.*, 2005) based approaches. In addition to complement-dependent and independent antibody mediated protection, NS1 has also been identified as a target of cell-mediated immunity, indicating that both arms of the immune response are likely to play a role in protection (Murali-Krishna *et al.*, 1995; Green and Rothman, 2006; Gao *et al.*, 2008).

Progress towards the development of NS1 as a viable vaccine candidate will not be without significant hurdles. One of the major stumbling blocks will be the fact that at least for dengue NS1, in addition to eliciting protective antibody responses as outlined above, it also appears to display epitopes that elicit auto-antibodies (discussed below) that cross-react with platelets and components of the extracellular matrix (Falconar, 1997; Chang *et al.*, 2002) as well as bind to, and induce damage to endothelial cells (Lin *et al.*, 2002). Indeed in mouse model studies, some passively administered MAbs have been shown to act either alone or synergistically to enhance disease (Henchal *et al.*, 1988). Whether these limitations apply to all of the flaviviruses remains to be determined, although preliminary evidence suggests that they may be flavivirus specific (Krishna *et al.*, 2009). Nevertheless, any vaccine based on NS1 will probably require a delivery and formulation strategy that ensures the induction of antibodies of the correct subclass (Chung *et al.*, 2007; Diamond *et al.*, 2009) as well as a degree of epitope specificity that avoids the induction of auto-antibodies (Cheng *et al.*, 2009).

NS1, complement and host cell protein binding partners

It has been 40 years since a series of landmark papers first described the dengue virus soluble complement fixing (SCF) antigen in infected mice and cell culture (Brandt *et al.*, 1970a,b; McCloud *et al.*, 1970; Russell *et al.*, 1970; Smith *et al.*, 1970). This SCF antigen was quickly identified as a secreted non-structural viral protein with a molecular weight of 39–43 kDa (Cardiff *et al.*, 1971), later to be confirmed as sNS1, and was immediately seen as a potential player in the pathogenesis of severe disease given the reported association between high levels of complement consumption and dengue shock syndrome (DSS) (Russell *et al.*, 1969). These studies lead to an expansion of interest in the role of complement pathway engagement by dengue viruses in infected patients and in particular, the role of immune complexes in potentiating the severity of disease (Russell, 1971; Bokisch *et al.*, 1973a; Sobel *et al.*, 1975; Theofilopoulos *et al.*, 1976; Ruangjirachuporn *et al.*, 1979). The measurement of a wide range of complement factors in comparative studies of both DHF/DSS and dengue fever patients highlighted the involvement of augmented complement activation in the pathophysiology of disease with decreased levels of C3 found in patients with more severe disease (Bokisch *et al.*, 1973b). Despite the reported observations of a clear and direct correlation between disease severity, complement consumption and a rise in the complement split products, C3a and C5a with known effects on vascular integrity, much of the on-going research effort from the mid-1980s was re-focused on the role of activated lymphocytes and the overproduction of vasoactive cytokines (Kurane *et al.*, 1994). As a consequence, the underlying *in vivo* mechanism of complement activation remained a matter of conjecture. However, the recent development of NS1 capture assays (Young *et al.*, 2000; Alcon *et al.*, 2002) and the discovery of remarkably high levels of circulating NS1 in patient sera that correlate with disease severity (Libraty *et al.*, 2002; Avirutnan *et al.*, 2006) has returned research efforts to the soluble complement fixing antigen first identified in 1970. Although other viral proteins may contribute, secreted NS1 is probably the major viral antigen responsible for immune complex formation and an important trigger of complement activation (Avirutnan *et al.*, 2006; Nascimento *et al.*, 2009) (Fig. 4.3, step 3).

In addition to contributing significantly to immune complex formation, NS1 has now been shown to bind a number of different complement pathway components as well as other host cell regulatory proteins. These include the complement regulation protein factor H (fH), complement inhibitory factor clusterin, complement proteins C4 and proC1s/C1s, hnRNP C1/C2, STAT3β and has been shown to trigger the generation of C5b-9 and SC5b-9 complexes (Chua *et al.*, 2005; Chung *et al.*, 2006a; Avirutnan *et al.*, 2006, 2010; Schlesinger, 2006; Kurosu *et al.*, 2007; Noisakran *et al.*, 2008b; Wilson *et al.*, 2008; Krishna *et al.*, 2009; Baronti *et al.*, 2010). It has also been suggested that WNV NS1 is able to antagonize TLR3 signalling (Wilson *et al.*, 2008), although this finding has recently been questioned in a study that failed to confirm inhibition of TLR3 signalling by NS1 from three different flaviviruses; DENV-2, YFV and WNV (Baronti *et al.*, 2010). In dengue infection, sNS1 has been shown to activate complement directly (Fig. 4.3, step 11), while binding of NS1 specific antibodies to mNS1 on infected cell surfaces directs complement attack (Avirutnan *et al.*, 2006). The consequent generation of membrane attack complexes (C5b-9) can trigger cellular activation and the production of inflammatory cytokines and along with soluble membrane attack complexes (sC5b-9) (Fig. 4.3, step 11) are likely to contribute to the pathogenesis of severe dengue (Avirutnan *et al.*, 2006). Indeed sC5b-9 levels in patient sera were found to follow similar kinetics to those of sNS1 and like NS1, were found to correlate with disease severity (Avirutnan *et al.*, 2006). Further evidence for NS1 involvement in the generation of soluble membrane attack complexes was suggested when NS1 was found to bind directly to the complement inhibitory factor clusterin, which inhibits the formation of the membrane attack complex (Kurosu *et al.*, 2007). It was proposed that NS1 binding results in a reduction in circulating free clusterin and hence to an increase in sC5b-9 formation in the serum of infected patients (Kurosu *et al.*, 2007) (Fig. 4.3, step 12). However, in the case of dengue, this hypothesis needs to be considered in the context of a secondary infection where peak NS1 levels are observed early during the acute phase of disease with free NS1 rapidly disappearing from circulation as a result of the anamnestic rise in cross-reactive anti-NS1antibodies. Consequently, high levels of free NS1 are not co-incident with the onset of severe disease (Libraty *et al.*, 2002). Furthermore, given that these cross-reactive anti-NS1 antibodies were originally elicited to the NS1 of a different serotype they are likely to recognize only a few epitopes in common between the primary and secondary infecting viruses. They may also be of low affinity to the NS1 of the secondary infecting virus. Both of these antibody characteristics are known to result in the formation of relatively small immune complexes (ICs) that are not efficiently cleared. They are more likely to be deposited in tissues and at the endothelial cell wall, potentially contributing to endothelial cell dysfunction and vascular permeability. At a time of rising anti-NS1 antibody levels and the presence of NS1 in ICs, IC deposition and complement activation is more likely to contribute to severe disease outcome, at least for dengue, than is free sNS1 binding to complement regulating proteins.

Recently it has been shown that NS1 may also display an immune evasion function through the inhibition of the classical/lectin complement pathways (Avirutnan *et al.*, 2010). In co-immunoprecipitation studies, sNS1 was found to co-precipitate with the complement proteins C4 and C4b (Avirutnan *et al.*, 2010). The authors found that sNS1 binds to proC1s/C1s and C4 which results in increased cleavage of C4 to C4b. They hypothesize that this limits the amount of C4 available, thus protecting virus from neutralization (Avirutnan *et al.*, 2010) (Fig. 4.3, step 14). Another proposed mechanism of viral mediated protection against complement directed killing of infected cells was suggested by the observation that WNV NS1 binds to factor H (fH), a circulating regulator of the alternative complement pathway (Chung *et al.*, 2006a). Binding of circulating fH by sNS1 may lead to accelerated breakdown of C3bBb convertase with consequent reduced C3b deposition and a resulting reduction in the formation of the terminal C5b-9 membrane attack complex (Chung *et al.*, 2006b; Schlesinger, 2006) (Fig. 4.3, step 13). However, it should be noted that this is not a generic property of flavivirus NS1 species as it has recently been reported that JEV NS1 does not bind to fH (Krishna *et al.*, 2009). While the antagonism of complement pathways by NS1 appears to be a common strategy employed by flaviviruses in engagement with their host, the underlying differences in specific interacting host partners suggests that its contributing role in the pathogenicity of infection probably varies between different flaviviruses.

NS1 induction of autoantibodies and a potential role in pathogenesis

Despite the fact that most antibodies directed against NS1 have been found to provide some level of passive protection to mice from a lethal flavivirus challenge (Schlesinger, 1985, 1986; Henchal *et al.*, 1988), a small number have been shown to increase morbidity (Henchal *et al.*, 1988). Since these early observations, anti-NS1 antibodies have been shown to cross-react with a wide range of host cell components including extracellular matrix, blood clotting and integrin/adhesion proteins, as well as to ATP synthase β-chain, protein disulfide isomerise (PDI), vimentin and heat shock protein on platelets and/or endothelial cells (Falconar, 1997, 2007; Lin *et al.*, 2001, 2002, 2003, 2006; Sun *et al.*, 2007; Cheng *et al.*, 2009). The induction of auto-antibodies relatively early in acute secondary dengue infections as part of the anamnestic antibody response, that can bind platelets and uninfected endothelial cells has suggested a possible role for these antibodies in the endothelial cell dysfunction that underlies the haemorrhage and vascular leak in DHF/DSS patients (Sun *et al.*, 2007).

Falconar (1997) was the first to show that anti-NS1 antibodies raised in mice were able to bind to common epitopes on human blood clotting, integrin/adhesion proteins as well as directly to human endothelial cells (Falconar, 1997). This binding to endothelial cells was subsequently found to induce apoptosis in a caspase-dependent manner with the up-regulation of p53 and Bax inducing nitric oxide leading to cell death (Lin *et al.*, 2002). It has also been demonstrated that binding to endothelial cells by cross-reactive NS1 MAbs, can lead to protein tyrosine phosphorylation and activation of NF-κB, resulting in an inflammatory response producing IL-6, IL-8, MCP-1 and increased expression of ICAM-1, followed by increased adhesion of PBMCs to endothelial cells (Lin *et al.*, 2005, 2006). Lin *et al.* (2003) have also shown that dengue patient antibodies react with endothelial cells and that there is an increase in anti-endothelial cell activity in patients suffering from acute DHF/DSS when compared with patients with acute dengue fever (Lin *et al.*, 2003).

In murine models, both passive administration of anti-NS1 antibodies and active immunization with purified DENV NS1 were shown to damage liver endothelial cells resulting in increased serum levels of aspartate aminotransferase (AST) and alanine aminotransferase (ALT) (Lin *et al.*, 2008a). Antibodies that were elicited by active immunization with DENV NS1 were found by histology to bind to liver vessel endothelium and passive administration of anti-NS1 antibodies was shown to lead to endothelial cell damage and monocyte infiltration (Lin *et al.*, 2008a). This immune-mediated liver injury in mouse models provides supporting evidence that anti-NS1 antibody responses may also play a role in the liver damage characteristically seen in human dengue virus disease. Using a proteomic approach, a sequence motif located between amino acid residues 311–330 of NS1 has been identified that is shared with a number of host components including the ATP synthase β chain, PDI, vimentin and heat shock protein 60 (Cheng *et al.*, 2009). Furthermore, sera from patients with DHF were shown to bind to these proteins (Cheng *et al.*, 2009). So there is now ample evidence in the literature that anti-NS1 antibodies not only have the potential to provide protection against flavivirus infection, but through their interaction with host cell components, may also exacerbate flavivirus disease. These findings have clear implications for vaccine design, with strategies that include NS1 as a component immunogen needing to take into account the potential for inducing auto-antibodies. Some progress towards this goal has been made with the recent report that a C-terminally truncated recombinant NS1 with the cross-reactive motif identified above removed, induced antibodies in mice that had lower platelet binding characteristics than antibodies induced by its full-length NS1 counterpart (Chen *et al.*, 2009).

Although a role in the pathogenesis of flavivirus infections for auto-antibodies elicited by epitopes on NS1 that are cross-reactive with host cell components is compelling, the dynamics of antibody kinetics over the course of the infection also needs to be taken into account. Most reports have provided solid *in vitro* and/or *in vivo* evidence for the binding of cross-reactive anti-NS1 antibodies to platelets and endothelial cells, as well as subsequent cellular damage and inflammatory activation (Falconar, 1997, 2007; Lin *et al.*, 2001, 2002, 2003, 2006, 2008a; Sun *et al.*, 2007; Cheng *et al.*, 2009). However, few of these studies have commented on the apparent paradox of disease recovery during convalescence in the presence of the on-going circulation of these otherwise damaging auto-antibodies. For example, auto-antibodies elicited by NS1 in dengue patients have been proposed as playing a role in the endothelial dysfunction and consequent vascular leak that is characteristic of DHF/DSS patients experiencing secondary infections (Lin *et al.*, 2008a). However, patient recovery from symptoms of vascular leak can often be quite dramatic (Mairuhu *et al.*, 2004) and does not coincide with a sudden drop in circulating antibody. Furthermore, in primary infections of young infants experiencing DHF/DSS, high levels of circulating anti-NS1 antibody are not likely to be found during the acute phase of the disease, despite the acquisition of maternal antibodies at birth. Thus, while platelet and endothelial cell reactive antibodies may indeed play some role in the pathogenesis of flavivirus disease, their activity needs to be viewed more in the context of a wide range of other modulating host and viral risk factors.

NS1 as a diagnostic biomarker

Correct serological diagnosis of a flavivirus infection can be challenging. Particularly in regions of the world where more than one flavivirus co-circulates, as the traditional serological approaches suffer from the relatively high antigenic cross-reactivity of the major virion envelope protein, E against which the majority of the antibody response is targeted. Furthermore, serological approaches, mostly based on IgM capture ELISA, not only suffer specificity problems from antigenic cross-reactivity but as they measure the patient immune response to infection, detection is not available early in the course of disease. Until recently, alternative laboratory diagnosis relied on virus isolation or RT-PCR, both of which present particular problems. Virus isolation requires lengthy culture, is expensive and has relatively poor sensitivity. RT-PCR, while accurate and ideal for early diagnosis requires specialized equipment not always available in tropical settings where many flaviviruses predominate. As NS1 is secreted from infected cells it was reasoned that this viral protein would be a suitable early surrogate biomarker for viraemia and/or infected cell mass in patients (Young *et al.*, 2000). An NS1 antigen capture ELISA for DENV was developed which revealed that NS1 is secreted from the onset of symptoms in some infected individuals at high levels of up to 50 μg/ml (Young *et al.*, 2000). Early assessment of this assay by a number of groups (Young *et al.*, 2000; Alcon *et al.*, 2002; Libraty *et al.*, 2002) lead to the commercial development of NS1 capture ELISAs by several companies. These have now been subjected to a large number of field evaluations that have proven the value of the assay in the early diagnosis of dengue infection (Lemes *et al.*, 2005; Xu *et al.*, 2006; Kumarasamy *et al.*, 2007a,b; Simmons *et al.*, 2007; Bessoff *et al.*, 2008; Blacksell *et al.*, 2008; Dussart *et al.*, 2008; Lapphra *et al.*, 2008; Ludert *et al.*, 2008; Chaiyaratana *et al.*, 2009; Hsieh and Chen, 2009; McBride, 2009; Phuong *et al.*, 2009; Shu *et al.*, 2009; Zainah *et al.*, 2009). Second generation rapid assays are now also available for point-of-care diagnosis (Chaiyaratana *et al.*, 2009; Hang *et al.*, 2009; Zainah *et al.*, 2009). The commercial development and application of NS1 detection as a diagnostic tool has revolutionized dengue diagnosis as it now provides a simple and relatively low-cost assay for early diagnosis that has high sensitivity and specificity (Castro-Jorge *et al.*, 2010; Lima Mda *et al.*, 2010; Wang and Sekaran, 2010b). More recent studies have shown that NS1 detection may also be applicable to the diagnosis of other flavivirus infections (Chung and Diamond, 2008).

What makes NS1 such an ideal diagnostic marker is the fact that it is found at high levels in the blood of infected individuals very early in infection, typically from, or before the onset of symptoms which is typically 2 to 3 days post-infection (Young *et al.*, 2000; Libraty *et al.*, 2002; Alcon *et al.*, 2002; McBride, 2009; Bessoff *et al.*, 2010; Thomas *et al.*, 2010). It is detected before an antibody response is mounted and as early as viral RNA, with the latter leading to NS1 being referred to as a surrogate marker for viraemia. However, closer examination of the kinetics of NS1 and viraemia levels in individual patients often reveals differential profiles suggesting a more complex association (Libraty *et al.*, 2002; McBride, 2009; Zainah *et al.*, 2009; Tricou *et al.*, 2010). Although NS1 should more accurately be referred to as a surrogate marker of infected cell mass, these differences may in part be due to differing mechanisms of clearance operating on virions and free NS1 protein.

In primary dengue infection, NS1 can be detected in patient serum or plasma samples taken as much as 9–12 days post disease onset (Alcon *et al.*, 2002; Libraty *et al.*, 2002; Xu *et al.*, 2006). However, a complicating factor for NS1 detection in secondary dengue infections is the rapid anamnestic rise of serotype cross-reactive anti-NS1 antibodies. As a consequence of the formation of immune complexes and their probable clearance from circulation, NS1 is rarely detected in secondary infected patients beyond 5–7 days post onset of symptoms (Vazquez *et al.*, 2010). Therefore, depending on the timing of clinical presentation, this may result in a dengue infected patient being tested as NS1 negative. This has led to some confusion in the field in the application of this assay for diagnosis of secondary infected individuals, by far the majority of patients seen clinically in endemic countries. A negative result under these circumstances is not a failure of the assay, as defined by poor sensitivities reported in a number of recent publications, but merely an accurate assessment of the circulation of low levels of free NS1 (Blacksell *et al.*, 2008; Datta and Wattal, 2010). While some efforts have been made to improve the sensitivity of the assay under circumstances where immune complex formation is likely to have occurred by incorporating methods for immune complex disruption (Lapphra *et al.*, 2008) it is imperative for accurate diagnosis of secondary infected dengue patients that the NS1 assay is not used alone but complemented with the detection of dengue specific IgM antibodies (Blacksell *et al.*, 2008; Lima Mda *et al.*, 2010; Osorio *et al.*, 2010; Wang and Sekaran, 2010a). Diagnostic assays incorporating both these markers are now making their way into practical use.

An observation made relatively early in the determination of NS1 levels in dengue-infected patients was that high levels of NS1 early in infection appeared to correlate with the later onset of severe disease (Libraty *et al.*, 2002). This finding has subsequently been confirmed in other studies (Avirutnan *et al.*, 2006) and offers the exciting prospect of including NS1 as an early prognostic biomarker of severe disease. In addition to the detection of NS1 itself, the antibodies that it elicits have also been used in developing useful diagnostic assays. The relative type specificity of the antibody responses elicited has led to the development of ELISA based assays that are capable of determining infecting serotype, primary or secondary infection and able to differentiate JEV from dengue virus, a particular problem in many countries of South East Asia (Shu *et al.*, 2000, 2001, 2002, 2003, 2004).

Conclusions

For a viral protein that has been the focus of research for more than two decades, it is remarkable that there is still so much that we do not know about its function in viral replication and disease. The post-translational modifications of NS1 that give rise to multiple species of both cell-associated and secreted forms have been identified. However, we do not know the molecular mechanism of membrane association nor the manner in which the hexamer is generated. Unequivocal evidence of a role for NS1 in viral RNA replication and an association with replication complexes has been documented. However, we do not know how NS1 is physically associated with the replication complex nor its specific role in RNA replication. Specific interactions with a wide range of host cell components have been reported for both sNS1 and mNS1 that may play a significant role in the pathogenesis of flavivirus disease. However, the molecular basis of these interactions and the level to which they are shared by different flaviviruses remains unclear. Dengue virus NS1 has been shown to display cross-reactive epitopes that are shared with a number of host cell components and the auto-antibodies that are induced by these determinants in secondary infections are thought to contribute to the platelet and endothelial cell damage that leads to the vascular permeability characteristic of severe DHF/DSS. However, little is known about the dynamics of the appearance and interactions between these different species during the course of disease. The presence of both NS1 and the antibodies it induces during the acute phase of disease is unusual and further studies are required to determine the modulating effect each has on the other, and of course the contribution of their interaction, immune complexes, to the overall disease process.

We have sought to provide in this review, a cohesive overview of the biochemical features of NS1, its application in the development of early diagnostics and its role in pathogenesis. Our hope is that it will provide insights to further research directions that may answer some of the outstanding questions posed above. Of paramount importance in our view, is the resolution of the structure of NS1. The three dimensional structure of monomeric and/or the biologically relevant dimeric and hexameric forms should provide extremely useful insight not only into the primary replicative function of this species but also into the reasons it is found in partnership with such a wide range of host cell components.

Despite the many gaps in our knowledge of the structure and function of NS1, a picture has emerged of a key viral protein that is involved in many stages of the virus life cycle, from its essential role in viral RNA replication to its somewhat contradictory contribution to the induction of both protection and pathogenesis in the infected host. Its detection in infected patients has also proven to be a very useful tool in early diagnosis. Perhaps one of the most intriguing aspects of this viral gene product can be found in a comparison with the genome coding strategy of the two related members of the *Flaviviridae* family, the hepaciviruses and pestiviruses. Virus members of these two genera do not encode an equivalent NS1 species. For a protein that is so essential for replication in the flaviviruses, how is this function compensated in these other *Flaviviridae* members? Is this a gene that has been lost during the separate evolution of these genera or have the *flaviviruses* picked up this gene from their host during their evolution? It is certainly worth noting that unlike the hepaciviruses and pestiviruses, most of the flaviviruses cycle between two quite different hosts, insects and mammals. Does NS1 play a role in bridging the separate replicative requirements of the two different cellular environments? Studies that directly compare the replication of these separate members of the *Flaviviridae* may be quite revealing.

On-going research into the flavivirus NS1 is bound to reveal more surprises and additions to the many faces of this somewhat enigmatic protein and we eagerly await further developments.

References

Alcon, S., Talarmin, A., Debruyne, M., Falconar, A., Deubel, V., and Flamand, M. (2002). Enzyme-linked immunosorbent assay specific to dengue virus type 1 nonstructural protein NS1 reveals circulation of the antigen in the blood during the acute phase of disease in patients experiencing primary or secondary infections. J. Clin. Microbiol. *40*, 376–381.

Alcon-LePoder, S., Drouet, M.T., Roux, P., Frenkiel, M.P., Arborio, M., Durand-Schneider, A.M., Maurice, M., Le Blanc, I., Gruenberg, J., and Flamand, M. (2005). The secreted form of dengue virus nonstructural protein NS1 is endocytosed by hepatocytes and accumulates in late endosomes: implications for viral infectivity. J. Virol. *79*, 11403–11411.

Alcon-LePoder, S., Sivard, P., Drouet, M.T., Talarmin, A., Rice, C., and Flamand, M. (2006). Secretion of flaviviral non-structural protein NS1: from diagnosis to pathogenesis. Novartis Found. Symp. *277*, 233–247.

Avirutnan, P., Punyadee, N., Noisakran, S., Komoltri, C., Thiemmeca, S., Auethavornanan, K., Jairungsri, A., Kanlaya, R., Tangthawornchaikul, N., Puttikhunt, C., *et al.* (2006). Vascular leakage in severe dengue virus infections: a potential role for the nonstructural viral protein NS1 and complement. J. Infect. Dis. *193*, 1078–1088.

Avirutnan, P., Zhang, L., Punyadee, N., Manuyakorn, A., Puttikhunt, C., Kasinrerk, W., Malasit, P., Atkinson, J.P., and Diamond, M.S. (2007). Secreted NS1 of dengue virus attaches to the surface of cells via interactions with heparan sulfate and chondroitin sulfate E. PLoS Pathog. *3*, 1798–1812.

Avirutnan, P., Fuchs, A., Hauhart, R.E., Somnuke, P., Youn, S., Diamond, M.S., and Atkinson, J.P. (2010). Antagonism of the complement component C4 by flavivirus nonstructural protein NS1. J. Exp. Med. *207*, 793–806.

Baronti, C., Sire, J., de Lamballerie, X., and Querat, G. (2010). Nonstructural NS1 proteins of several mosquito-borne flavivirus do not inhibit TLR3 signaling. Virology *404*, 319–330.

Bessoff, K., Delorey, M., Sun, W., and Hunsperger, E. (2008). Comparison of two commercially available dengue virus (DENV) NS1 capture enzyme-linked immunosorbent assays using a single clinical sample for diagnosis of acute DENV infection. Clin. Vaccine Immunol. *15*, 1513–1518.

Bessoff, K., Phoutrides, E., Delorey, M., Acosta, L.N., and Hunsperger, E. (2010). Utility of a commercial nonstructural protein 1 antigen capture kit as a dengue virus diagnostic tool. Clin. Vaccine Immunol. *17*, 949–953.

Blacksell, S.D., Mammen, M.P., Jr., Thongpaseuth, S., Gibbons, R.V., Jarman, R.G., Jenjaroen, K., Nisalak, A., Phetsouvanh, R., Newton, P.N., and Day, N.P. (2008). Evaluation of the Panbio dengue virus nonstructural 1 antigen detection and immunoglobulin M antibody enzyme-linked immunosorbent assays for the diagnosis of acute dengue infections in Laos. Diagn. Microbiol. Infect. Dis. *60*, 43–49.

Blitvich, B.J., Mackenzie, J.S., Coelen, R.J., Howard, M.J., and Hall, R.A. (1995). A novel complex formed between the flavivirus E and NS1 proteins: analysis of its structure and function. Arch. Virol. *140*, 145–156.

Blitvich, B.J., Scanlon, D., Shiell, B.J., Mackenzie, J.S., and Hall, R.A. (1999). Identification and analysis of truncated and elongated species of the flavivirus NS1 protein. Virus Res. *60*, 67–79.

Blitvich, B.J., Scanlon, D., Shiell, B.J., Mackenzie, J.S., Pham, K., and Hall, R.A. (2001). Determination of the intramolecular disulfide bond arrangement and biochemical identification of the glycosylation sites of the nonstructural protein NS1 of Murray Valley encephalitis virus. J. Gen. Virol. *82*, 2251–2256.

Bokisch, V.A., Muller-Eberhard, H.J., and Dixon, F.J. (1973a). The role of complement in hemorrhagic shock syndrome (dengue). Trans. Assoc. Am. Phys. *86*, 102–110.

Bokisch, V.A., Top, F.H., Jr., Russell, P.K., Dixon, F.J., and Muller-Eberhard, H.J. (1973b). The potential pathogenic role of complement in dengue hemorrhagic shock syndrome. N. Engl. J. Med. *289*, 996–1000.

Brandt, W.E., Cardiff, R.D., and Russell, P.K. (1970a). Dengue virions and antigens in brain and serum of infected mice. J. Virol. *6*, 500–506.

Brandt, W.E., Chiewslip, D., Harris, D.L., and Russell, P.K. (1970b). Partial purification and characterization of a dengue virus soluble complement-fixing antigen. J. Immunol. *105*, 1565–1568.

Butrapet, S., Huang, C.Y., Pierro, D.J., Bhamarapravati, N., Gubler, D.J., and Kinney, R.M. (2000). Attenuation markers of a candidate dengue type 2 vaccine virus, strain 16681 (PDK-53), are defined by mutations in the 5′ noncoding region and nonstructural proteins 1 and 3. J. Virol. *74*, 3011–3019.

Calvert, A.E., Huang, C.Y., Kinney, R.M., and Roehrig, J.T. (2006). Non-structural proteins of dengue 2 virus offer limited protection to interferon-deficient mice after dengue 2 virus challenge. J. Gen. Virol. *87*, 339–346.

Cane, P.A., and Gould, E.A. (1988). Reduction of yellow fever virus mouse neurovirulence by immunization with a bacterially synthesized non-structural protein (NS1) fragment. J. Gen. Virol. *69*, 1241–1246.

Cardiff, R.D., Brandt, W.E., McCloud, T.G., Shapiro, D., and Russell, P.K. (1971). Immunological and biophysical separation of dengue-2 antigens. J. Virol. *7*, 15–23.

Castro-Jorge, L.A., Machado, P.R., Favero, C.A., Borges, M.C., Passos, L.M., de Oliveira, R.M., and Fonseca, B.A. (2010). Clinical evaluation of the NS1 antigen-capture ELISA for early diagnosis of dengue virus infection in Brazil. J. Med. Virol. *82*, 1400–1405.

Chaiyaratana, W., Chuansumrit, A., Pongthanapisith, V., Tangnararatchakit, K., Lertwongrath, S., and Yoksan, S. (2009). Evaluation of dengue nonstructural protein 1 antigen strip for the rapid diagnosis of patients with dengue infection. Diagn. Microbiol. Infect. Dis. *64*, 83–84.

Chambers, T.J., McCourt, D.W., and Rice, C.M. (1989). Yellow fever virus proteins NS2A, NS2B, and NS4B:

identification and partial N-terminal amino acid sequence analysis. Virology *169*, 100–109.

Chambers, T.J., McCourt, D.W., and Rice, C.M. (1990). Production of yellow fever virus proteins in infected cells: identification of discrete polyprotein species and analysis of cleavage kinetics using region-specific polyclonal antisera. Virology *177*, 159–174.

Chang, H.H., Shyu, H.F., Wang, Y.M., Sun, D.S., Shyu, R.H., Tang, S.S., and Huang, Y.S. (2002). Facilitation of cell adhesion by immobilized dengue viral nonstructural protein 1 (NS1): arginine-glycine-aspartic acid structural mimicry within the dengue viral NS1 antigen. J. Infect. Dis. *186*, 743–751.

Chappell, K.J., Stoermer, M.J., Fairlie, D.P., and Young, P.R. (2008). West Nile Virus NS2B/NS3 protease as an antiviral target. Curr. Med. Chem. *15*, 2771–2784.

Chen, L.K., Liao, C.L., Lin, C.G., Lai, S.C., Liu, C.I., Ma, S.H., Huang, Y.Y., and Lin, Y.L. (1996). Persistence of Japanese encephalitis virus is associated with abnormal expression of the nonstructural protein NS1 in host cells. Virology *217*, 220–229.

Chen, M.C., Lin, C.F., Lei, H.Y., Lin, S.C., Liu, H.S., Yeh, T.M., Anderson, R., and Lin, Y.S. (2009). Deletion of the C-terminal region of dengue virus nonstructural protein 1 (NS1) abolishes anti-NS1-mediated platelet dysfunction and bleeding tendency. J. Immunol. *183*, 1797–1803.

Chen, Y., Pan, Y., Guo, Y., Qiu, L., Ding, X., and Che, X. (2010). Comprehensive mapping of immunodominant and conserved serotype- and group-specific B-cell epitopes of nonstructural protein 1 from dengue virus type 1. Virology *398*, 290–298.

Cheng, H.J., Lin, C.F., Lei, H.Y., Liu, H.S., Yeh, T.M., Luo, Y.H., and Lin, Y.S. (2009). Proteomic analysis of endothelial cell autoantigens recognized by anti-dengue virus nonstructural protein 1 antibodies. Exp. Biol. Med. *234*, 63–73.

Chua, J.J., Bhuvanakantham, R., Chow, V.T., and Ng, M.L. (2005). Recombinant non-structural 1 (NS1) protein of dengue-2 virus interacts with human STAT3beta protein. Virus Res. *112*, 85–94.

Chung, K.M., Liszewski, M.K., Nybakken, G., Davis, A.E., Townsend, R.R., Fremont, D.H., Atkinson, J.P., and Diamond, M.S. (2006a). West Nile virus nonstructural protein NS1 inhibits complement activation by binding the regulatory protein factor H. Proc. Natl. Acad. Sci. U.S.A. *103*, 19111–19116.

Chung, K.M., Nybakken, G.E., Thompson, B.S., Engle, M.J., Marri, A., Fremont, D.H., and Diamond, M.S. (2006b). Antibodies against West Nile Virus nonstructural protein NS1 prevent lethal infection through Fc gamma receptor-dependent and -independent mechanisms. J. Virol. *80*, 1340–1351.

Chung, K.M., Thompson, B.S., Fremont, D.H., and Diamond, M.S. (2007). Antibody recognition of cell surface-associated NS1 triggers Fc-gamma receptor-mediated phagocytosis and clearance of West Nile Virus-infected cells. J. Virol. *81*, 9551–9555.

Chung, K.M., and Diamond, M.S. (2008). Defining the levels of secreted non-structural protein NS1 after West Nile virus infection in cell culture and mice. J. Med. Virol. *80*, 547–556.

Churdboonchart, V., Bhamarapravati, N., Peampramprecha, S., and Sirinavin, S. (1991). Antibodies against dengue viral proteins in primary and secondary dengue hemorrhagic fever. Am. J. Trop. Med. Hyg. *44*, 481–493.

Clark, D.C., Lobigs, M., Lee, E., Howard, M.J., Clark, K., Blitvich, B.J., and Hall, R.A. (2007). *in situ* reactions of monoclonal antibodies with a viable mutant of Murray Valley encephalitis virus reveal an absence of dimeric NS1 protein. J. Gen. Virol. *88*, 1175–1183.

Coia, G., Parker, M.D., Speight, G., Byrne, M.E., and Westaway, E.G. (1988). Nucleotide and complete amino acid sequences of Kunjin virus: definitive gene order and characteristics of the virus-specified proteins. J. Gen. Virol. *69 (Pt 1)*, 1–21.

Costa, S.M., Azevedo, A.S., Paes, M.V., Sarges, F.S., Freire, M.S., and Alves, A.M. (2007). DNA vaccines against dengue virus based on the ns1 gene: the influence of different signal sequences on the protein expression and its correlation to the immune response elicited in mice. Virology *358*, 413–423.

Crabtree, M.B., Kinney, R.M., and Miller, B.R. (2005). Deglycosylation of the NS1 protein of dengue 2 virus, strain 16681: construction and characterization of mutant viruses. Arch. Virol. *150*, 771–786.

Crooks, A.J., Lee, J.M., Dowsett, A.B., and Stephenson, J.R. (1990). Purification and analysis of infectious virions and native non-structural antigens from cells infected with tick-borne encephalitis virus. J. Chromatogr. *502*, 59–68.

Crooks, A.J., Lee, J.M., Easterbrook, L.M., Timofeev, A.V., and Stephenson, J.R. (1994). The NS1 protein of tick-borne encephalitis virus forms multimeric species upon secretion from the host cell. J. Gen. Virol. *75 (Pt 12)*, 3453–3460.

Dalgarno, L., Trent, D.W., Strauss, J.H., and Rice, C.M. (1986). Partial nucleotide sequence of the Murray Valley encephalitis virus genome. Comparison of the encoded polypeptides with yellow fever virus structural and non-structural proteins. J. Mol. Biol. *187*, 309–323.

Datta, S., and Wattal, C. (2010). Dengue NS1 antigen detection: a useful tool in early diagnosis of dengue virus infection. Indian J. Med. Microbiol. *28*, 107–110.

Despres, P., Dietrich, J., Girard, M., and Bouloy, M. (1991). Recombinant baculoviruses expressing yellow fever virus E and NS1 proteins elicit protective immunity in mice. J. Gen. Virol. *72*, 2811–2816.

Deubel, V., Kinney, R.M., and Trent, D.W. (1988). Nucleotide sequence and deduced amino acid sequence of the nonstructural proteins of dengue type 2 virus, Jamaica genotype: comparative analysis of the full-length genome. Virology *165*, 234–244.

Diamond, M.S., Pierson, T.C., and Fremont, D.H. (2008). The structural immunology of antibody protection against West Nile virus. Immunol. Rev. *225*, 212–225.

Diamond, M.S., Mehlhop, E., Oliphant, T., and Samuel, M.A. (2009). The host immunologic response to West Nile encephalitis virus. Front. Biosci. *14*, 3024–3034.

Dussart, P., Petit, L., Labeau, B., Bremand, L., Leduc, A., Moua, D., Matheus, S., and Baril, L. (2008). Evaluation of two new commercial tests for the diagnosis of acute dengue virus infection using NS1 antigen detection in human serum. PLoS Negl. Trop. Dis. *2*, e280.

Eckels, K.H., Dubois, D.R., Summers, P.L., Schlesinger, J.J., Shelly, M., Cohen, S., Zhang, Y.M., Lai, C.J., Kurane, I., Rothman, A., *et al.* (1994). Immunization of monkeys with baculovirus-dengue type-4 recombinants containing envelope and nonstructural proteins: evidence of priming and partial protection. Am. J. Trop. Med. Hyg. *50*, 472–478.

Falconar, A.K. (1997). The dengue virus nonstructural-1 protein (NS1) generates antibodies to common epitopes on human blood clotting, integrin/adhesin proteins and binds to human endothelial cells: potential implications in haemorrhagic fever pathogenesis. Arch. Virol. *142*, 897–916.

Falconar, A.K. (2007). Antibody responses are generated to immunodominant ELK/KLE-type motifs on the nonstructural-1 glycoprotein during live dengue virus infections in mice and humans: implications for diagnosis, pathogenesis, and vaccine design. Clin. Vaccine Immunol. *14*, 493–504.

Falconar, A.K. (2008). Monoclonal antibodies that bind to common epitopes on the dengue virus type 2 nonstructural-1 and envelope glycoproteins display weak neutralizing activity and differentiated responses to virulent strains: implications for pathogenesis and vaccines. Clin. Vaccine Immunol. *15*, 549–561.

Falconar, A.K., and Young, P.R. (1990). Immunoaffinity purification of native dimer forms of the flavivirus non-structural glycoprotein, NS1. J. Virol. Methods *30*, 323–332.

Falconar, A.K., and Young, P.R. (1991). Production of dimer-specific and dengue virus group cross-reactive mouse monoclonal antibodies to the dengue 2 virus non-structural glycoprotein NS1. J. Gen. Virol. *72*, 961–965.

Falconar, A.K., Young, P.R., and Miles, M.A. (1994). Precise location of sequential dengue virus subcomplex and complex B cell epitopes on the nonstructural-1 glycoprotein. Arch. Virol. *137*, 315–326.

Falgout, B., and Markoff, L. (1995). Evidence that flavivirus NS1–NS2A cleavage is mediated by a membrane-bound host protease in the endoplasmic reticulum. J. Virol. *69*, 7232–7243.

Falgout, B., Chanock, R., and Lai, C.J. (1989). Proper processing of dengue virus nonstructural glycoprotein NS1 requires the N-terminal hydrophobic signal sequence and the downstream nonstructural protein NS2a. J. Virol. *63*, 1852–1860.

Falgout, B., Bray, M., Schlesinger, J.J., and Lai, C.J. (1990). Immunization of mice with recombinant vaccinia virus expressing authentic dengue virus nonstructural protein NS1 protects against lethal dengue virus encephalitis. J. Virol. *64*, 4356–4363.

Falkler, W.A., Jr., Diwan, A.R., and Halstead, S.B. (1973). Human antibody to dengue soluble complement-fixing (SCF) antigens. J. Immunol. *111*, 1804–1809.

Fan, W.F., and Mason, P.W. (1990). Membrane association and secretion of the Japanese encephalitis virus NS1 protein from cells expressing NS1 cDNA. Virology *177*, 470–476.

Firth, A.E., and Atkins, J.F. (2009). A conserved predicted pseudoknot in the NS2A-encoding sequence of West Nile and Japanese encephalitis flaviviruses suggests NS1′ may derive from ribosomal frameshifting. Virol. J. *6*, 14.

Flamand, M., Deubel, V., and Girard, M. (1992). Expression and secretion of Japanese encephalitis virus nonstructural protein NS1 by insect cells using a recombinant baculovirus. Virology *191*, 826–836.

Flamand, M., Megret, F., Mathieu, M., Lepault, J., Rey, F.A., and Deubel, V. (1999). Dengue virus type 1 nonstructural glycoprotein NS1 is secreted from mammalian cells as a soluble hexamer in a glycosylation-dependent fashion. J. Virol. *73*, 6104–6110.

Gao, G., Wang, Q., Dai, Z., Calcedo, R., Sun, X., Li, G., and Wilson, J.M. (2008). Adenovirus-based vaccines generate cytotoxic T lymphocytes to epitopes of NS1 from dengue virus that are present in all major serotypes. Hum. Gene Ther. *19*, 927–936.

Gibson, C.A., Schlesinger, J.J., and Barrett, A.D. (1988). Prospects for a virus non-structural protein as a subunit vaccine. Vaccine *6*, 7–9.

Gould, E.A., Buckley, A., Barrett, A.D., and Cammack, N. (1986). Neutralizing (54K) and non-neutralizing (54K and 48K) monoclonal antibodies against structural and non-structural yellow fever virus proteins confer immunity in mice. J. Gen. Virol. *67 (Pt 3)*, 591–595.

Green, S., and Rothman, A. (2006). Immunopathological mechanisms in dengue and dengue hemorrhagic fever. Curr. Opin. Infect. Dis. *19*, 429–436.

Guzman, M.G., Halstead, S.B., Artsob, H., Buchy, P., Farrar, J., Gubler, D.J., Hunsperger, E., Kroeger, A., Margolis, H.S., Martinez, E., *et al.* (2010). Dengue: a continuing global threat. Nat. Rev. Microbiol. *8*, S7-S16.

Hall, R.A., Kay, B.H., Burgess, G.W., Clancy, P., and Fanning, I.D. (1990). Epitope analysis of the envelope and non-structural glycoproteins of Murray Valley encephalitis virus. J. Gen. Virol. *71 (Pt 12)*, 2923–2930.

Hall, R.A., Brand, T.N., Lobigs, M., Sangster, M.Y., Howard, M.J., and Mackenzie, J.S. (1996). Protective immune responses to the E and NS1 proteins of Murray Valley encephalitis virus in hybrids of flavivirus-resistant mice. J. Gen. Virol. *77*, 1287–1294.

Hall, R.A., Khromykh, A.A., Mackenzie, J.M., Scherret, J.H., Khromykh, T.I., and Mackenzie, J.S. (1999). Loss of dimerisation of the nonstructural protein NS1 of Kunjin virus delays viral replication and reduces virulence in mice, but still allows secretion of NS1. Virology *264*, 66–75.

Hang, V.T., Nguyet, N.M., Trung, D.T., Tricou, V., Yoksan, S., Dung, N.M., Van Ngoc, T., Hien, T.T., Farrar, J., Wills, B., *et al.* (2009). Diagnostic Accuracy of NS1 ELISA and Lateral Flow Rapid Tests for dengue Sensitivity, Specificity and Relationship to Viraemia and Antibody Responses. PLoS Negl. Trop. Dis. 3, 20.

Heaton, N.S., Perera, R., Berger, K.L., Khadka, S., Lacount, D.J., Kuhn, R.J., and Randall, G. (2010). Dengue virus nonstructural protein 3 redistributes fatty acid synthase to sites of viral replication and increases cellular fatty acid synthesis. Proc. Natl. Acad. Sci. U.S.A. *107*, 17345–17350.

Henchal, E.A., Henchal, L.S., and Thaisomboonsuk, B.K. (1987). Topological mapping of unique epitopes on the dengue-2 virus NS1 protein using monoclonal antibodies. J. Gen. Virol. *68*, 845–851.

Henchal, E.A., Henchal, L.S., and Schlesinger, J.J. (1988). Synergistic interactions of anti-NS1 monoclonal antibodies protect passively immunized mice from lethal challenge with dengue 2 virus. J. Gen. Virol. *69*, 2101–2107.

Hori, H., and Lai, C.J. (1990). Cleavage of dengue virus NS1-NS2A requires an octapeptide sequence at the C terminus of NS1. J. Virol. *64*, 4573–4577.

Hsieh, C.J., and Chen, M.J. (2009). The commercial dengue NS1 antigen-capture ELISA may be superior to IgM detection, virus isolation and RT-PCR for rapid laboratory diagnosis of acute dengue infection based on a single serum sample. J. Clin. Virol. 2009; *44*, 102.

Huang, J.H., Wey, J.J., Sun, Y.C., Chin, C., Chien, L.J., and Wu, Y.C. (1999). Antibody responses to an immunodominant nonstructural 1 synthetic peptide in patients with dengue fever and dengue hemorrhagic fever. J. Med. Virol. *57*, 1–8.

Ikonen, E. (2001). Roles of lipid rafts in membrane transport. Curr. Opin. Cell Biol. *13*, 470–477.

Jacobs, M.G., Robinson, P.J., Bletchly, C., Mackenzie, J.M., and Young, P.R. (2000). Dengue virus nonstructural protein 1 is expressed in a glycosyl-phosphatidylinositol-linked form that is capable of signal transduction. Faseb J. *14*, 1603–1610.

Jacobs, S.C., Stephenson, J.R., and Wilkinson, G.W. (1992). High-level expression of the tick-borne encephalitis virus NS1 protein by using an adenovirus-based vector: protection elicited in a murine model. J. Virol. *66*, 2086–2095.

Jacobs, S.C., Stephenson, J.R., and Wilkinson, G.W. (1994). Protection elicited by a replication-defective adenovirus vector expressing the tick-borne encephalitis virus non-structural glycoprotein NS1. J. Gen. Virol. *75*, 2399–2402.

Kanlaya, R., Pattanakitsakul, S.N., Sinchaikul, S., Chen, S.T., and Thongboonkerd, V. (2010). Vimentin interacts with heterogeneous nuclear ribonucleoproteins and dengue nonstructural protein 1 and is important for viral replication and release. Mol. Biosyst. *6*, 795–806.

Khromykh, A.A., Sedlak, P.L., Guyatt, K.J., Hall, R.A., and Westaway, E.G. (1999). Efficient trans-complementation of the flavivirus kunjin NS5 protein but not of the NS1 protein requires its coexpression with other components of the viral replicase. J. Virol. *73*, 10272–10280.

Khromykh, A.A., Sedlak, P.L., and Westaway, E.G. (2000). cis- and trans-acting elements in flavivirus RNA replication. J. Virol. *74*, 3253–3263.

Krishna, V.D., Rangappa, M., and Satchidanandam, V. (2009). Virus-specific cytolytic antibodies to nonstructural protein 1 of Japanese encephalitis virus effect reduction of virus output from infected cells. J. Virol. *83*, 4766–4777.

Kumarasamy, V., Chua, S.K., Hassan, Z., Wahab, A.H., Chem, Y.K., Mohamad, M., and Chua, K.B. (2007a). Evaluating the sensitivity of a commercial dengue NS1 antigen-capture ELISA for early diagnosis of acute dengue virus infection. Singapore Med. J. *48*, 669–673.

Kumarasamy, V., Wahab, A.H., Chua, S.K., Hassan, Z., Chem, Y.K., Mohamad, M., and Chua, K.B. (2007b). Evaluation of a commercial dengue NS1 antigen-capture ELISA for laboratory diagnosis of acute dengue virus infection. J. Virol. Methods *140*, 75–79.

Kuno, G., Vorndam, A.V., Gubler, D.J., and Gomez, I. (1990). Study of anti-dengue NS1 antibody by western blot. J. Med. Virol. *32*, 102–108.

Kurane, I., Rothman, A.L., Livingston, P.G., Green, S., Gagnon, S.J., Janus, J., Innis, B.L., Nimmannitya, S., Nisalak, A., and Ennis, F.A. (1994). Immunopathologic mechanisms of dengue hemorrhagic fever and dengue shock syndrome. Arch. Virol. Suppl. *9*, 59–64.

Kurosu, T., Chaichana, P., Yamate, M., Anantapreecha, S., and Ikuta, K. (2007). Secreted complement regulatory protein clusterin interacts with dengue virus nonstructural protein 1. Biochem. Biophys. Res. Commun. *362*, 1051–1056.

Lapphra, K., Sangcharaswichai, A., Chokephaibulkit, K., Tiengrim, S., Piriyakarnsakul, W., Chakorn, T., Yoksan, S., Wattanamongkolsil, L., and Thamlikitkul, V. (2008). Evaluation of an NS1 antigen detection for diagnosis of acute dengue infection in patients with acute febrile illness. Diagn. Microbiol. Infect. Dis. *60*, 387–391.

Leblois, H., and Young, P.R. (1993). Sequence of the dengue virus type 2 (strain PR-159) NS1 gene and comparison with its vaccine derivative. Nucleic Acids Res. *21*.

Leblois, H., and Young, P.R. (1995). Maturation of the dengue-2 virus NS1 protein in insect cells: effects of downstream NS2A sequences on baculovirus-expressed gene constructs. J. Gen. Virol. *76 (Pt 4)*, 979–984.

Lee, C.J., Lin, H.R., Liao, C.L., and Lin, Y.L. (2008). Cholesterol effectively blocks entry of flavivirus. J. Virol. *82*, 6470–6480.

Lee, J.M., Crooks, A.J., and Stephenson, J.R. (1989). The synthesis and maturation of a non-structural extracellular antigen from tick-borne encephalitis virus and its relationship to the intracellular NS1 protein. J. Gen. Virol. *70*, 335–343.

Lemes, E.M., Miagostovicsh, M.P., Alves, A.M., Costa, S.M., Fillipis, A.M., Armoa, G.R., and Araujo, M.A. (2005). Circulating human antibodies against dengue NS1 protein: potential of recombinant D2V-NS1 proteins in diagnostic tests. J. Clin. Virol. *32*, 305–312.

Libraty, D.H., Young, P.R., Pickering, D., Endy, T.P., Kalayanarooj, S., Green, S., Vaughn, D.W., Nisalak, A., Ennis, F.A., and Rothman, A.L. (2002). High circulating levels of the dengue virus nonstructural protein NS1 early in dengue illness correlate with the development of dengue hemorrhagic fever. J. Infect. Dis. *186*, 1165–1168.

Lieberman, M.M., Clements, D.E., Ogata, S., Wang, G., Corpuz, G., Wong, T., Martyak, T., Gilson, L., Coller, B.A., Leung, J., *et al.* (2007). Preparation and immunogenic properties of a recombinant West Nile subunit vaccine. Vaccine *25*, 414–423.

Lima Mda, R., Nogueira, R.M., Schatzmayr, H.G., and dos Santos, F.B. (2010). Comparison of three commercially available dengue NS1 antigen capture assays for acute diagnosis of dengue in Brazil. PLoS Negl. Trop. Dis. *4*, e738.

Lin, C.F., Lei, H.Y., Liu, C.C., Liu, H.S., Yeh, T.M., Wang, S.T., Yang, T.I., Sheu, F.C., Kuo, C.F., and Lin, Y.S. (2001). Generation of IgM anti-platelet autoantibody in dengue patients. J. Med. Virol. *63*, 143–149.

Lin, C.F., Lei, H.Y., Shiau, A.L., Liu, H.S., Yeh, T.M., Chen, S.H., Liu, C.C., Chiu, S.C., and Lin, Y.S. (2002). Endothelial cell apoptosis induced by antibodies against dengue virus nonstructural protein 1 via production of nitric oxide. J. Immunol. *169*, 657–664.

Lin, C.F., Lei, H.Y., Shiau, A.L., Liu, C.C., Liu, H.S., Yeh, T.M., Chen, S.H., and Lin, Y.S. (2003). Antibodies from dengue patient sera cross-react with endothelial cells and induce damage. J. Med. Virol. *69*, 82–90.

Lin, C.F., Chiu, S.C., Hsiao, Y.L., Wan, S.W., Lei, H.Y., Shiau, A.L., Liu, H.S., Yeh, T.M., Chen, S.H., Liu, C.C., *et al.* (2005). Expression of cytokine, chemokine, and adhesion molecules during endothelial cell activation induced by antibodies against dengue virus nonstructural protein 1. J. Immunol. *174*, 395–403.

Lin, C.F., Wan, S.W., Cheng, H.J., Lei, H.Y., and Lin, Y.S. (2006). Autoimmune pathogenesis in dengue virus infection. Viral Immunol. *19*, 127–132.

Lin, C.F., Wan, S.W., Chen, M.C., Lin, S.C., Cheng, C.C., Chiu, S.C., Hsiao, Y.L., Lei, H.Y., Liu, H.S., Yeh, T.M., *et al.* (2008). Liver injury caused by antibodies against dengue virus nonstructural protein 1 in a murine model. Lab. Invest. *88*, 1079–1089.

Lin, C.W., Liu, K.T., Huang, H.D., and Chen, W.J. (2008). Protective immunity of E. coli-synthesized NS1 protein of Japanese encephalitis virus. Biotechnol. Lett. *30*, 205–214.

Lin, Y.L., Chen, L.K., Liao, C.L., Yeh, C.T., Ma, S.H., Chen, J.L., Huang, Y.L., Chen, S.S., and Chiang, H.Y. (1998). DNA immunization with Japanese encephalitis virus nonstructural protein NS1 elicits protective immunity in mice. J. Virol. *72*, 191–200.

Lindenbach, B.D., and Rice, C.M. (1997). trans-Complementation of yellow fever virus NS1 reveals a role in early RNA replication. J. Virol. *71*, 9608–9617.

Lindenbach, B.D., and Rice, C.M. (1999). Genetic interaction of flavivirus nonstructural proteins NS1 and NS4A as a determinant of replicase function. J. Virol. *73*, 4611–4621.

Lindenbach, B.D., and Rice, C.M. (2003). Molecular biology of flaviviruses. Adv. Virus Res. *59*, 23–61.

Liu, X., Cao, S., Zhou, R., Xu, G., Xiao, S., Yang, Y., Sun, M., Li, Y., and Chen, H. (2006). Inhibition of Japanese encephalitis virus NS1 protein expression in cell by small interfering RNAs. Virus Genes *33*, 69–75.

Ludert, J.E., Mosso, C., Ceballos-Olvera, I., and del Angel, R.M. (2008). Use of a commercial enzyme immunoassay to monitor dengue virus replication in cultured cells. Virol. J. *5*, 51.

McBride, W.J. (2009). Evaluation of dengue NS1 test kits for the diagnosis of dengue fever. Diagn. Microbiol. Infect. Dis. *64*, 31–36.

McCloud, T.G., Brandt, W.E., and Russell, P.K. (1970). Molecular size and charge relationships of the soluble complement-fixing antigens of dengue viruses. Virology *41*, 569–572.

Macdonald, J., Tonry, J., Hall, R.A., Williams, B., Palacios, G., Ashok, M.S., Jabado, O., Clark, D., Tesh, R.B., Briese, T., *et al.* (2005). NS1 protein secretion during the acute phase of West Nile virus infection. J. Virol. *79*, 13924–13933.

Mackenzie, J.M., Jones, M.K., and Young, P.R. (1996). Immunolocalization of the dengue virus nonstructural glycoprotein NS1 suggests a role in viral RNA replication. Virology *220*, 232–240.

Mackenzie, J.M., Khromykh, A.A., Jones, M.K., and Westaway, E.G. (1998). Subcellular localization and some biochemical properties of the flavivirus Kunjin nonstructural proteins NS2A and NS4A. Virology *245*, 203–215.

Mackow, E., Makino, Y., Zhao, B.T., Zhang, Y.M., Markoff, L., Buckler-White, A., Guiler, M., Chanock, R., and Lai, C.J. (1987). The nucleotide sequence of dengue type 4 virus: analysis of genes coding for nonstructural proteins. Virology *159*, 217–228.

Mairuhu, A.T., Wagenaar, J., Brandjes, D.P., and van Gorp, E.C. (2004). Dengue: an arthropod-borne disease of global importance. Eur J. Clin. Microbiol. Infect Dis *23*, 425–433.

Malavige, G.N., Fernando, S., Fernando, D.J., and Seneviratne, S.L. (2004). Dengue viral infections. Postgrad. Med. J. *80*, 588–601.

Mandl, C.W., Heinz, F.X., Stockl, E., and Kunz, C. (1989). Genome sequence of tick-borne encephalitis virus (Western subtype) and comparative analysis of nonstructural proteins with other flaviviruses. Virology *173*, 291–301.

Mason, P.W. (1989). Maturation of Japanese encephalitis virus glycoproteins produced by infected mammalian and mosquito cells. Virology *169*, 354–364.

Mason, P.W., McAda, P.C., Mason, T.L., and Fournier, M.J. (1987). Sequence of the dengue-1 virus genome in the region encoding the three structural proteins and the major nonstructural protein NS1. Virology *161*, 262–267.

Melian, E.B., Hinzman, E., Nagasaki, T., Firth, A.E., Wills, N.M., Nouwens, A.S., Blitvich, B.J., Leung, J., Funk, A., Atkins, J.F., *et al.* (2009). NS1′ of flaviviruses in the Japanese encephalitis serogroup is a product of ribosomal frameshifting and plays a role in viral neuro-invasiveness. J. Virol. *84*, 1641–1647.

Miller, N. (2010). Recent progress in dengue vaccine research and development. Curr. Opin. Mol. Ther. *12*, 31–38.

Murali-Krishna, K., Ramireddy, B., Ravi, V., and Manjunath, R. (1995). Recognition of nonstructural protein peptides by cytotoxic T lymphocytes raised against Japanese encephalitis virus. Microbiol. Immunol. *39*, 1021–1024.

Muylaert, I.R., Chambers, T.J., Galler, R., and Rice, C.M. (1996). Mutagenesis of the N-linked glycosylation sites of the yellow fever virus NS1 protein: effects on virus replication and mouse neurovirulence. Virology *222*, 159–168.

Muylaert, I.R., Galler, R., and Rice, C.M. (1997). Genetic analysis of the yellow fever virus NS1 protein: identification of a temperature-sensitive mutation which blocks RNA accumulation. J. Virol. *71*, 291–298.

Nascimento, E.J., Silva, A.M., Cordeiro, M.T., Brito, C.A., Gil, L.H., Braga-Neto, U., and Marques, E.T. (2009). Alternative complement pathway deregulation is correlated with dengue severity. PLoS One *4*, e6782.

Noisakran, S., Dechtawewat, T., Rinkaewkan, P., Puttikhunt, C., Kanjanahaluethai, A., Kasinrerk, W., Sittisombut, N., and Malasit, P. (2007). Characterization of dengue virus NS1 stably expressed in 293T cell lines. J. Virol. Methods *142*, 67–80.

Noisakran, S., Dechtawewat, T., Avirutnan, P., Kinoshita, T., Siripanyaphinyo, U., Puttikhunt, C., Kasinrerk, W., Malasit, P., and Sittisombut, N. (2008a). Association of dengue virus NS1 protein with lipid rafts. J. Gen. Virol. *89*, 2492–2500.

Noisakran, S., Sengsai, S., Thongboonkerd, V., Kanlaya, R., Sinchaikul, S., Chen, S.T., Puttikhunt, C., Kasinrerk, W., Limjindaporn, T., Wongwiwat, W., *et al.* (2008b). Identification of human hnRNP C1/C2 as a dengue virus NS1-interacting protein. Biochem. Biophys. Res. Commun. *372*, 67–72.

Nowak, T., Farber, P.M., and Wengler, G. (1989). Analyses of the terminal sequences of West Nile virus structural proteins and of the *in vitro* translation of these proteins allow the proposal of a complete scheme of the proteolytic cleavages involved in their synthesis. Virology *169*, 365–376.

Osorio, L., Ramirez, M., Bonelo, A., Villar, L.A., and Parra, B. (2010). Comparison of the diagnostic accuracy of commercial NS1-based diagnostic tests for early dengue infection. Virol. J. *7*, 361.

Parrish, C.R., Woo, W.S., and Wright, P.J. (1991). Expression of the NS1 gene of dengue virus type 2 using vaccinia virus. Dimerisation of the NS1 glycoprotein. Arch. Virol. *117*, 279–286.

Pethel, M., Falgout, B., and Lai, C.J. (1992). Mutational analysis of the octapeptide sequence motif at the NS1–NS2A cleavage junction of dengue type 4 virus. J. Virol. *66*, 7225–7231.

Phuong, H.L., Thai, K.T., Nga, T.T., Giao, P.T., Hung le, Q., Binh, T.Q., Nam, N.V., Groen, J., and de Vries, P.J. (2009). Detection of dengue nonstructural 1 (NS1) protein in Vietnamese patients with fever. Diagn. Microbiol. Infect. Dis. *63*, 372–378.

Post, P.R., Carvalho, R., and Galler, R. (1991). Glycosylation and secretion of yellow fever virus nonstructural protein NS1. Virus Res. *18*, 291–302.

Pryor, M.J., and Wright, P.J. (1993). The effects of site-directed mutagenesis on the dimerization and secretion of the NS1 protein specified by dengue virus. Virology *194*, 769–780.

Pryor, M.J., and Wright, P.J. (1994). Glycosylation mutants of dengue virus NS1 protein. J. Gen. Virol. *75 (Pt 5)*, 1183–1187.

Putnak, J.R., and Schlesinger, J.J. (1990). Protection of mice against yellow fever virus encephalitis by immunization with a vaccinia virus recombinant encoding the yellow fever virus non-structural proteins, NS1, NS2a and NS2b. J. Gen. Virol. *71*, 1697–1702.

Qu, X., Chen, W., Maguire, T., and Austin, F. (1993). Immunoreactivity and protective effects in mice of a recombinant dengue 2 Tonga virus NS1 protein produced in a baculovirus expression system. J. Gen. Virol. *74*, 89–97.

Rice, C.M., Lenches, E.M., Eddy, S.R., Shin, S.J., Sheets, R.L., and Strauss, J.H. (1985). Nucleotide sequence of yellow fever virus: implications for flavivirus gene expression and evolution. Science *229*, 726–733.

Rice, C.M., Aebersold, R., Teplow, D.B., Pata, J., Bell, J.R., Vorndam, A.V., Trent, D.W., Brandriss, M.W., Schlesinger, J.J., and Strauss, J.H. (1986). Partial N-terminal amino acid sequences of three nonstructural proteins of two flaviviruses. Virology *151*, 1–9.

Ross, T.M. (2010). Dengue virus. Clin. Lab. Med. *30*, 149–160.

Rossi, S.L., Ross, T.M., and Evans, J.D. (2010). West Nile virus. Clin. Lab. Med. *30*, 47–65.

Ruangjirachuporn, W., Boonpucknavig, S., and Nimmanitya, S. (1979). Circulating immune complexes in serum from patients with dengue haemorrhagic fever. Clin. Exp. Immunol. *36*, 46–53.

Russell, P.K. (1971). Mechanisms in the dengue shock syndrome. In Progress in Immunology, Amos, D.B, ed. (New York, Academic Press Inc), pp. 831–838.

Russell, P.K., Intavivat, A., and Kanchanapilant, S. (1969). Anti-dengue immunoglobulins and serum beta 1 c-a globulin levels in dengue shock syndrome. J. Immunol. *102*, 412–420.

Russell, P.K., Chiewsilp, D., and Brandt, W.E. (1970). Immunoprecipitation analysis of soluble complement-fixing antigens of dengue viruses. J. Immunol. *105*, 838–845.

Schlesinger, J.J. (2006). Flavivirus nonstructural protein NS1: complementary surprises. Proc. Natl. Acad. Sci. U.S.A. *103*, 18879–18880.

Schlesinger, J.J., Brandriss, M.W., and Walsh, E.E. (1985). Protection against 17D yellow fever encephalitis in mice by passive transfer of monoclonal antibodies to the nonstructural glycoprotein gp48 and by active immunization with gp48. J. Immunol. *135*, 2805–2809.

Schlesinger, J.J., Brandriss, M.W., Cropp, C.B., and Monath, T.P. (1986). Protection against yellow fever in monkeys by immunization with yellow fever virus nonstructural protein NS1. J. Virol. *60*, 1153–1155.

Schlesinger, J.J., Brandriss, M.W., and Walsh, E.E. (1987). Protection of mice against dengue 2 virus encephalitis by immunization with the dengue 2 virus non-structural glycoprotein NS1. J. Gen. Virol. *68 (Pt 3)*, 853–857.

Schlesinger, J.J., Brandriss, M.W., Putnak, J.R., and Walsh, E.E. (1990). Cell surface expression of yellow fever virus non-structural glycoprotein NS1: consequences of interaction with antibody. J. Gen. Virol. *71*, 593–599.

Schlesinger, J.J., Foltzer, M., and Chapman, S. (1993). The Fc portion of antibody to yellow fever virus NS1 is a determinant of protection against YF encephalitis in mice. Virology *192*, 132–141.

Shu, P.Y., Chen, L.K., Chang, S.F., Yueh, Y.Y., Chow, L., Chien, L.J., Chin, C., Lin, T.H., and Huang, J.H. (2000). Dengue NS1-specific antibody responses: isotype distribution and serotyping in patients with dengue fever and dengue hemorrhagic fever. J. Med. Virol. *62*, 224–232.

Shu, P.Y., Chen, L.K., Chang, S.F., Yueh, Y.Y., Chow, L., Chien, L.J., Chin, C., Lin, T.H., and Huang, J.H. (2001). Antibody to the nonstructural protein NS1 of Japanese encephalitis virus: potential application of mAb-based indirect ELISA to differentiate infection from vaccination. Vaccine *19*, 1753–1763.

Shu, P.Y., Chen, L.K., Chang, S.F., Yueh, Y.Y., Chow, L., Chien, L.J., Chin, C., Yang, H.H., Lin, T.H., and Huang, J.H. (2002). Potential application of nonstructural protein NS1 serotype-specific immunoglobulin G enzyme-linked immunosorbent assay in the seroepidemiologic study of dengue virus infection: correlation of results with those of the plaque reduction neutralization test. J. Clin. Microbiol. *40*, 1840–1844.

Shu, P.Y., Chen, L.K., Chang, S.F., Yueh, Y.Y., Chow, L., Chien, L.J., Chin, C., Lin, T.H., and Huang, J.H. (2003). Comparison of capture immunoglobulin M (IgM) and IgG enzyme-linked immunosorbent assay (ELISA) and nonstructural protein NS1 serotype-specific IgG ELISA for differentiation of primary and secondary dengue virus infections. Clin. Diagn. Lab. Immunol. *10*, 622–630.

Shu, P.Y., Chen, L.K., Chang, S.F., Su, C.L., Chien, L.J., Chin, C., Lin, T.H., and Huang, J.H. (2004). Dengue virus serotyping based on envelope and membrane and nonstructural protein NS1 serotype-specific capture immunoglobulin M enzyme-linked immunosorbent assays. J. Clin. Microbiol. *42*, 2489–2494.

Shu, P.Y., Yang, C.F., Kao, J.F., Su, C.L., Chang, S.F., Lin, C.C., Yang, W.C., Shih, H., Yang, S.Y., Wu, P.F., *et al.* (2009). Application of the dengue virus NS1 antigen rapid test for on-site detection of imported dengue cases at airports. Clin. Vaccine Immunol. *16*, 589–591.

Simmons, C.P., Chau, T.N., Thuy, T.T., Tuan, N.M., Hoang, D.M., Thien, N.T., Lien le, B., Quy, N.T., Hieu, N.T., Hien, T.T., *et al.* (2007). Maternal antibody and viral factors in the pathogenesis of dengue virus in infants. J. Infect. Dis. *196*, 416–424.

Simons, K., and Ikonen, E. (1997). Functional rafts in cell membranes. Nature *387*, 569–572.

Simons, K., and Ikonen, E. (2000). How cells handle cholesterol. Science *290*, 1721–1726.

Smith, G.W., and Wright, P.J. (1985). Synthesis of proteins and glycoproteins in dengue type 2 virus-infected vero and Aedes albopictus cells. J. Gen. Virol. *66 (Pt 3)*, 559–571.

Smith, T.J., Brandt, W.E., Swanson, J.L., McCown, J.M., and Buescher, E.L. (1970). Physical and biological properties of dengue-2 virus and associated antigens. J. Virol. *5*, 524–532.

Sobel, A.T., Bokisch, V.A., and Muller-Eberhard, H.J. (1975). C1q deviation test for the detection of immune complexes, aggregates of IgG, and bacterial products in human serum. J. Exp. Med. *142*, 139–150.

Sun, D.S., King, C.C., Huang, H.S., Shih, Y.L., Lee, C.C., Tsai, W.J., Yu, C.C., and Chang, H.H. (2007). Antiplatelet autoantibodies elicited by dengue virus non-structural protein 1 cause thrombocytopenia and mortality in mice. J. Thromb. Haemost. *5*, 2291–2299.

Suzuki, R., de Borba, L., Duarte dos Santos, C.N., and Mason, P.W. (2007). Construction of an infectious cDNA clone for a Brazilian prototype strain of dengue virus type 1: characterization of a temperature-sensitive mutation in NS1. Virology *362*, 374–383.

Tajima, S., Takasaki, T., and Kurane, I. (2008). Characterization of Asn130-to-Ala mutant of dengue type 1 virus NS1 protein. Virus Genes *36*, 323–329.

Theofilopoulos, A.N., Brandt, W.E., Russell, P.K., and Dixon, F.T. (1976). Replication of dengue-2 virus in cultured human lymphoblastoid cells and subpopulations of human peripheral leukocytes. J. Immunol. *117*, 953–961.

Thomas, L., Najioullah, F., Verlaeten, O., Martial, J., Brichler, S., Kaidomar, S., Moravie, V., Cabie, A., and Cesaire, R. (2010). Relationship between nonstructural protein 1 detection and plasma virus load in dengue patients. Am. J. Trop. Med. Hyg. *83*, 696–699.

Timofeev, A.V., Ozherelkov, S.V., Pronin, A.V., Deeva, A.V., Karganova, G.G., Elbert, L.B., and Stephenson, J.R. (1998). Immunological basis for protection in a murine model of tick-borne encephalitis by a recombinant adenovirus carrying the gene encoding the NS1 non-structural protein. J. Gen. Virol. *79*, 689–695.

Timofeev, A.V., Butenko, V.M., and Stephenson, J.R. (2004). Genetic vaccination of mice with plasmids encoding the NS1 non-structural protein from tick-borne encephalitis virus and dengue 2 virus. Virus Genes *28*, 85–97.

Trent, D.W., Kinney, R.M., Johnson, B.J., Vorndam, A.V., Grant, J.A., Deubel, V., Rice, C.M., and Hahn, C. (1987). Partial nucleotide sequence of St. Louis encephalitis virus RNA: structural proteins, NS1, ns2a, and ns2b. Virology *156*, 293–304.

Tricou, V., Minh, N.N., Van, T.P., Lee, S.J., Farrar, J., Wills, B., Tran, H.T., and Simmons, C.P. (2010). A randomized controlled trial of chloroquine for the treatment of dengue in Vietnamese adults. PLoS Negl. Trop. Dis. *4*, e785.

Vazquez, S., Ruiz, D., Barrero, R., Ramirez, R., Calzada, N., del Rosario Pena, B., Reyes, S., and Guzman, M.G. (2010). Kinetics of dengue virus NS1 protein in dengue 4-confirmed adult patients. Diagn. Microbiol. Infect. Dis. *68*, 46–49.

Volpina, O.M., Volkova, T.D., Koroev, D.O., Ivanov, V.T., Ozherelkov, S.V., Khoretonenko, M.V., Vorovitch,

M.F., Stephenson, J.R., and Timofeev, A.V. (2005). A synthetic peptide based on the NS1 non-structural protein of tick-borne encephalitis virus induces a protective immune response against fatal encephalitis in an experimental animal model. Virus Res. *112*, 95–99.

Wallis, T.P., Huang, C.Y., Nimkar, S.B., Young, P.R., and Gorman, J.J. (2004). Determination of the disulfide bond arrangement of dengue virus NS1 protein. J. Biol. Chem. *279*, 20729–20741.

Wang, B., Hua, R.H., Tian, Z.J., Chen, N.S., Zhao, F.R., Liu, T.Q., Wang, Y.F., and Tong, G.Z. (2009). Identification of a virus-specific and conserved B-cell epitope on NS1 protein of Japanese encephalitis virus. Virus Res. *141*, 90–95.

Wang, S.M., and Sekaran, S.D. (2010a). Early diagnosis of dengue infection using a commercial Dengue Duo rapid test kit for the detection of NS1, IGM, and IGG. Am. J. Trop. Med. Hyg. *83*, 690–695.

Wang, S.M., and Sekaran, S.D. (2010b). Evaluation of a commercial SD dengue virus NS1 antigen capture enzyme-linked immunosorbent assay kit for early diagnosis of dengue virus infection. J. Clin. Microbiol. *48*, 2793–2797.

Welsch, S., Miller, S., Romero-Brey, I., Merz, A., Bleck, C.K., Walther, P., Fuller, S.D., Antony, C., Krijnse-Locker, J., and Bartenschlager, R. (2009). Composition and three-dimensional architecture of the dengue virus replication and assembly sites. Cell Host Microbe *5*, 365–375.

Westaway, E.G. and Goodman, M.R. (1987). Variation in distribution of the three flavivirus-specified glycoproteins detected by immunofluorescence in infected Vero cells. Arch. Virol. *94*, 215–228.

Westaway, E.G., Mackenzie, J.M., Kenney, M.T., Jones, M.K., and Khromykh, A.A. (1997). Ultrastructure of Kunjin virus-infected cells: colocalization of NS1 and NS3 with double-stranded RNA, and of NS2B with NS3, in virus-induced membrane structures. J. Virol. *71*, 6650–6661.

Wilson, J.R., de Sessions, P.F., Leon, M.A., and Scholle, F. (2008). West Nile virus nonstructural protein 1 inhibits TLR3 signal transduction. J. Virol. *82*, 8262–8271.

Winkler, G., Maxwell, S.E., Ruemmler, C., and Stollar, V. (1989). Newly synthesized dengue-2 virus nonstructural protein NS1 is a soluble protein but becomes partially hydrophobic and membrane-associated after dimerization. Virology *171*, 302–305.

Winkler, G., Randolph, V.B., Cleaves, G.R., Ryan, T.E., and Stollar, V. (1988). Evidence that the mature form of the flavivirus nonstructural protein NS1 is a dimer. Virology *162*, 187–196.

Wright, P.J., Cauchi, M.R., and Ng, M.L. (1989). Definition of the carboxy termini of the three glycoproteins specified by dengue virus type 2. Virology *171*, 61–67.

Wu, S.F., Liao, C.L., Lin, Y.L., Yeh, C.T., Chen, L.K., Huang, Y.F., Chou, H.Y., Huang, J.L., Shaio, M.F., and Sytwu, H.K. (2003). Evaluation of protective efficacy and immune mechanisms of using a non-structural protein NS1 in DNA vaccine against dengue 2 virus in mice. Vaccine *21*, 3919–3929.

Wu-Hsieh, B.A., Yen, Y.T., and Chen, H.C. (2009). Dengue hemorrhage in a mouse model. Ann. N.Y. Acad. Sci. *1171* (Suppl. 1), E42–47.

Xu, H., Di, B., Pan, Y.X., Qiu, L.W., Wang, Y.D., Hao, W., He, L.J., Yuen, K.Y., and Che, X.Y. (2006). Serotype 1-specific monoclonal antibody-based antigen capture immunoassay for detection of circulating nonstructural protein NS1: Implications for early diagnosis and serotyping of dengue virus infections. J. Clin. Microbiol. *44*, 2872–2878.

Yen, Y.T., Chen, H.C., Lin, Y.D., Shieh, C.C., and Wu-Hsieh, B.A. (2008). Enhancement by tumor necrosis factor alpha of dengue virus-induced endothelial cell production of reactive nitrogen and oxygen species is key to hemorrhage development. J. Virol. *82*, 12312–12324.

Youn, S., Cho, H., Fremont, D.H., and Diamond, M.S. (2010). A short N-terminal peptide motif on flavivirus nonstructural protein NS1 modulates cellular targeting and immune recognition. J. Virol. *84*, 9516–9532.

Young, P.R. (1990). Antigenic analysis of dengue virus using monoclonal antibodies. Southeast Asian J. Trop. Med. Pub. Health *21*, 646–651.

Young, P.R., Hilditch, P.A., Bletchly, C., and Halloran, W. (2000). An antigen capture enzyme-linked immunosorbent assay reveals high levels of the dengue virus protein NS1 in the sera of infected patients. J. Clin. Microbiol. *38*, 1053–1057.

Zainah, S., Wahab, A.H., Mariam, M., Fauziah, M.K., Khairul, A.H., Roslina, I., Sairulakhma, A., Kadimon, S.S., Jais, M.S., and Chua, K.B. (2009). Performance of a commercial rapid dengue NS1 antigen immunochromatography test with reference to dengue NS1 antigen-capture ELISA. J. Virol. Methods *155*, 157–160.

Zhang, Y.M., Hayes, E.P., McCarty, T.C., Dubois, D.R., Summers, P.L., Eckels, K.H., Chanock, R.M., and Lai, C.J. (1988). Immunization of mice with dengue structural proteins and nonstructural protein NS1 expressed by baculovirus recombinant induces resistance to dengue virus encephalitis. J. Virol. *62*, 3027–3031.

Zhao, B.T., Prince, G., Horswood, R., Eckels, K., Summers, P., Chanock, R., and Lai, C.J. (1987). Expression of dengue virus structural proteins and nonstructural protein NS1 by a recombinant vaccinia virus. J. Virol. *61*, 4019–4022.

The Flavivirus NS3 Protein: Structure and Functions

5

Dahai Luo, Siew Pheng Lim and Julien Lescar

Abstract

The non-structural protein 3 (NS3) of flaviviruses is the second most conserved amongst the viral proteins. It bears a molecular mass of 69 kDa and is endowed with multiple functions including proteolytic processing, nucleic acid duplexes unwinding, nucleoside triphosphatase (NTPase) and RNA nucleoside 5′ triphosphatase (RTPase). Besides these enzymatic activities which are essential for replication of viral genomic RNA, this protein also participates in other aspects of the viral life cycle. Numerous crystal structures of apo- and ligand-bound 3D structures for the protease and helicase domains of NS3, as well as for the full-length NS3 polypeptide (NS3FL) have been determined, providing insight at the atomic level into its various enzymatic activities. In this chapter, we summarize recent progress published regarding the function and structure of this fascinating viral non-structural protein including its recently uncovered dynamic properties.

Introduction

The flavivirus NS3 protein encodes a serine protease domain at its N-terminal end that is required during the virus life cycle for proteolytic processing of the viral polyprotein. To be catalytically active, it must associate with a hydrophilic domain of the membrane -associated NS2B protein that acts as a cofactor. The C-terminal domain of NS3 functions as a RNA helicase, a NTPase and also a RTPase and works in close association with another non-structural protein which is also expressed during the intracellular stage of the virus replication cycle, the NS5 methyltranferase-polymerase. In addition to its well-established enzymatic activities listed above, other biological roles related to induction of cell death and disease severity have also been attributed to this protein. Hence NS3 assumes important duties in the intracellular part of the viral life cycle and represents an important target for the development of specific antiviral inhibitors to treat diseases caused by flaviviruses. From knowledge gained from crystal structures of the protease, helicase domains and full length protein, we first discuss the structure and function of each domain and further propose that flexibility imparted by the linker region that connects both domains, is necessary for a proper regulation of their activities. Consequences for the assembly of the supramolecular machinery that performs viral polyprotein processing and genomic viral RNA replication are also discussed. We also describe biological roles played by NS3 in the pathogenesis induced by flavivirus infection. Some of these topics have also been covered by recent reviews on dengue (Lescar *et al.*, 2008; Sampath and Padmanabhan, 2009; Noble *et al.*, 2010) and West Nile virus (Chappell *et al.*, 2008) proteases.

Biological roles played by the NS3 protein

The protease domain

Cleavage specificity

The *N*-terminal domain of NS3 spanning amino acids 1–168 of the polypeptide encodes a trypsin-like serine protease with the canonical catalytic triad His-51, Asp-75, and Ser-135 (Valle and Falgout, 1998). For activity, it requires association with its cofactor, the membrane-bound NS2B protein (Chambers *et al.*, 1991; Falgout *et al.*, 1991; Wengler *et al.*, 1991; Zhang *et al.*, 1992; Yusof *et al.*, 2000). The NS2B-NS3 protease complex is pivotal for viral polyprotein processing because it is responsible for all cleavages between proteins at the cytoplasmic side of the ER membrane, including at the NS2A/NS2B, NS2B/NS3, NS3/NS4A, NS4A/NS4B and NS4B/NS5 junctions (Fig. 5.1) (Chambers *et al.*, 1990; Preugschat *et al.*, 1990; Lindenbach *et al.*, 2007). As a result of this proteolytic activity, mature non-structural proteins are released in the infected cell and, together with host-cell components assemble to form a viral replication complex associated to the membrane. Disruption of the NS2B-NS3 protease enzymatic activity was shown to be lethal to virus replication (Chambers *et al.*, 1990; Wengler *et al.*, 1991; Droll *et al.*, 2000). Additionally, the NS2B-NS3 protease cleaves within the C, NS2A, NS3H and NS4A polypeptides (Fig. 5.1). Cleavage at the C-terminal region of the C protein is necessary for efficient generation of the amino-terminus of prM by signal peptidase and also for secretion of viral particles (Chambers *et al.*, 1990; Lobigs, 1993; Amberg *et al.*, 1994; Amberg and Rice, 1999). Moreover, association of C-prM with the N-terminal part of NS2B was found to be important for prM–E secretion (Yamshchikov *et al.*, 1997). The internal cleavage near the carboxy-terminus of the NS4A protein is a prerequisite for the efficient cleavage at the downstream NS4A/NS4B junction by signalase (Lin *et al.*, 1993). Similarly, cleavage within NS2A results in the production of a 20-kDa protein that is important for virus production (Nestorowicz *et al.*, 1994). Autoprocessing occurs within the helicase region both intracellularly (Teo and Wright, 1997)and when expressed as a fusion protein together with

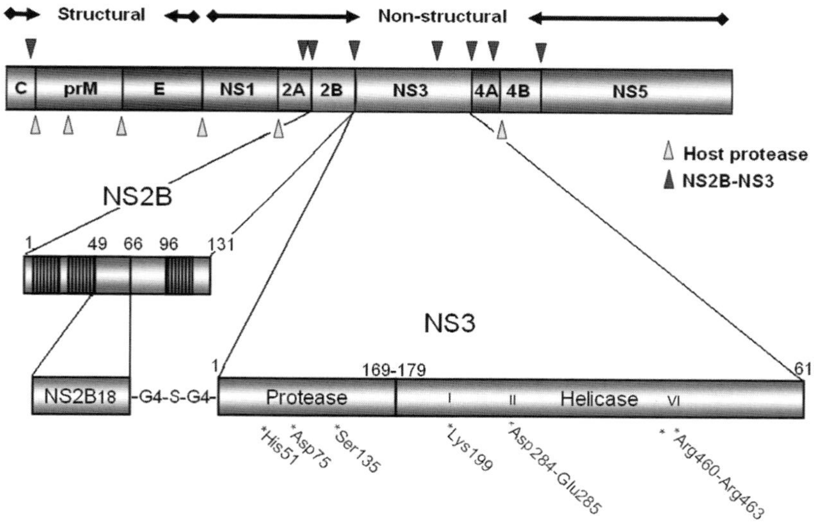

Figure 5.1 Cleavage sites proteolysed by the NS2B-NS3 protease between proteins at the cytoplasmic side of the ER membrane, including at the junctions of NS2A/NS2B, NS2B/NS3, NS3/NS4A, NS4A/NS4B and NS4B/NS5. Several internal cleavage sites are also indicated, as well as cleavage sites by host cell proteases. Also shown, the predicted distribution of hydrophilic (white) and hydrophobic (dark stripes) domains in the NS2B co-factor, as well as a strategy to design soluble constructs comprising parts of the NS2B hydrophilic domain and the NS3 polypeptide.

NS2B (Bera *et al.*, 2007; Lim *et al.*, unpublished results). The biological implications for this activity remain uncertain. Cellular processing of the viral polyprotein is believed to occur in virus-induced paracrystalline or convoluted membrane structures as both NS2B and NS3 have been co-localized to these sites (Westaway *et al.*, 1997). The NS2B protein is an integral membrane protein comprises three hydrophobic regions that span the membrane and a short – highly conserved – hydrophilic stretch of about 40 amino acids (aa) involved in the catalytic activation of the NS3 protease (Clum *et al.*, 1997). *in vitro* enzymatic activity of recombinant NS3 protease can be achieved through co-expression with the 40-aa hydrophilic sequence of NS2B as a fusion protein. NS3 protease can be linked to this 40 aa sequence either through C-terminal NS2B residues spanning the NS2B/NS3 cleavage site (Clum *et al.*, 1997; Yusof *et al.*, 2000; Mueller *et al.*, 2007) or via a flexible glycine-rich linker (Leung *et al.*, 2001; Nall *et al.*, 2004). Reducing the number of C-terminal NS2B residues cloned in-frame to the NS3 protease converts the recombinant NS2B–NS3 complex from an insoluble form to a soluble heterodimer that is non-covalently associated after undergoing auto-cleavage (Mueller *et al.*, 2007; Lim *et al.*, unpublished data). Single-chain constructs of NS2B-NS3FL have also been made and little influence of the helicase domain on protease activity was observed (Chappell *et al.*, 2007; Lim *et al.*, unpublished data). Except for the NS2B/NS3 junction, which contains a glutamine residue at the P2 position, the NS2B-NS3 protease preferentially cleaves the peptide bond downstream of a pair of highly conserved basic amino acids (Arg or Lys) at the P1 and P2 positions, followed by an amino-acid with a short side-chain (Gly, Ala, or Ser) at the P1′ site (Chambers *et al.*, 1990; Preugschat *et al.*, 1990). NS2B-NS3 proteases from all four dengue serotypes (DENV1–4) share 65–70% amino acid sequence identity and very similar peptide substrate recognition sites at both prime and non-prime sites (Li *et al.*, 2005). Hence, it should be possible to develop a single inhibitory agent targeting all four dengue NS3 proteases. Furthermore, NS3 from DENV shares greater than 50% amino-acid sequence homology with WNV and YFV proteases, suggesting that the design of a pan-flavivirus protease inhibitor may also be possible (Valle and Falgout, 1998; Knox *et al.*, 2006; Löhr *et al.*, 2007).

Enzymatic characterization

Several groups have characterized the enzymatic properties of NS2B-NS3 proteases with synthetic peptide substrates bearing endogenous dengue virus cleavage sites (Leung *et al.*, 2001; Khumthong *et al.*, 2002, 2003; Nall *et al.*, 2004; Li *et al.*, 2005). The best substrate from the DENV natural NS2A/2B cleavage site, Ac-RTSKKR-AMC measured a k_{cat}/K_m of $3366.4 \pm 1575.3\,M^{-1}s^{-1}$ whilst the optimal tetrapeptide substrate Bz-nKRR-ACMC, identified through peptide library profiling experiments, exhibited a specificity constant, k_{cat}/K_m, ranging from 51,780 to $112{,}10\,M^{-1}s^{-1}$ for four dengue NS2B-NS3 enzymes. Similarly, the best WNV natural cleavage site (NS3/4A), Ac-FASGKR-pNA, measured a k_{cat}/K_m of $4222 \pm 313\,M^{-1}s^{-1}$, whilst Bz-nKRR-AMC, gave k_{cat}/K_m, value of $33{,}389\,M^{-1}s^{-1}$ (D. Beer, unpublished results). These data suggest that natural cleavage sites are not necessarily occupied with optimal residues. In particular, synthetic peptides resembling the native cleavage sequence revealed much slower k_{cat}/K_m values for the substrate with NS2B/NS3 sequence (Leung *et al.*, 2001; Khumthong *et al.*, 2002, 2003; Li *et al.*, 2005). These data support the observation from the *in vitro* processing study that cleavage between NS2A/NS2B precedes cleavage between NS2B/NS3 (Preugschat *et al.*, 1990). The differential rates of cleavage at the four major cleavage sites may guide an ordered processing of dengue viral polypeptide needed to orchestrate viral replication and assembly (see below).

The helicase domain

Biological roles

The C-terminal domain (residues 180–618) of the flavivirus NS3 protein is a RNA-helicase bearing the DEAH sequence motif belonging to the SF2 superfamily, according to the classification of helicases based on the presence of certain conserved sequence signature motifs (Gorbalenya and Koonin, 1993; Tanner and Linder, 2001; Xu *et al.*, 2006). The helicase/NTPase activities and structures of NS3 protein from several members of the flavivirus have been characterized for WNV, YFV, JEV and MVEV (for a review see Lescar *et al.*, 2008). The flavivirus NS3 helicase domain harbours (1) RNA helicase activity required for unwinding duplex RNA and/or RNA secondary structures during genome replication by the NS5 RdRp, (2) nucleoside 5′-triphosphatase (NTPase) activity which is required to provide energy for the helicase unwinding activity, (3) 5′-terminal RNA triphosphatase activity (RTPase) which plays a role in the synthesis of the 5′-cap structure, and (4) a recently proposed role in virus assembly independent of its known enzymatic functions (Wengler *et al.*, 1991; Chambers *et al.*, 1993; Li *et al.*, 1999; Borowski *et al.*, 2000; Benarroch *et al.*, 2004; Lescar *et al.*, 2008; Patkar and Kuhn, 2008). Dengue viruses and bovine viral diarrhoea virus with impaired helicase activity are not able to replicate, demonstrating the importance of NS3 in the *Flaviviridae* life cycle and validating this enzyme as a target for the development of compounds with antiviral activity (Grassmann *et al.*, 1999; Matusan *et al.*, 2001). Below we briefly review what is known about these various activities.

The NTPase activity

The NTPase activity entails NTP binding and hydrolysis at its γ-phosphoric anhydride bond coupled with release of NDP and inorganic phosphate. Being common to many helicases and motor proteins, the conserved amino-acid sequence motifs directly involved in NTP hydrolysis are motif I (also known as 'Walker A' or 'Phosphate binding-loop or P-loop', involved in binding to the phosphate groups of the nucleotide substrate), motif II (also known as 'Walker B' that coordinates the divalent metal ion – Mg^{2+} or Mn^{2+} – needed for the catalytic activity and activation of the attacking water), and motif VI that participates in binding the nucleotide phosphates (Benarroch *et al.*, 2004; Enemark and Joshua-Tor, 2008; Li *et al.*, 1999; Wu *et al.*, 2005; Xu *et al.*, 2006). The NTPase activity is very basal and can be stimulated by addition of ssRNA like polyU (Li *et al.*, 1999; Wang *et al.*, 2009; Xu *et al.*, 2006). This RNA-stimulated NTPase activity is thought to be due to an allosteric effect, whereby conformational changes within the protein upon RNA binding result in a more kinetically favourable active site for interaction with the NTP substrate (Preugschat *et al.*, 1996). NS5 can also stimulate NTPase activity in a dose-dependent manner up to a 1:1 molar ratio (Yon *et al.*, 2005).

Helicase activity

Helicase activity entails mechanical separation of duplex oligonucleotides into individual strands. This activity is thought to utilize in part the energy released from NTP hydrolysis, as well as energy derived from binding the nucleic acid substrate itself. Interestingly, the HCV helicase is more active on DNA templates (Pang *et al.*, 2002), whereas the DENV helicase exhibits higher activity on RNA duplexes (Xu *et al.*, 2005). The flavivirus NS3 helicase has a 3′ to 5′ directionality and can unwind dsRNA with a 3′ overhang (Li *et al.*, 1999; Benarroch *et al.*, 2004; Xu *et al.*, 2005). Unlike the NS3 protein from HCV however, the flavivirus NS3 protein unwinds dsDNA very inefficiently (Tai *et al.*, 1996; Pang *et al.*, 2002; Frick *et al.*, 2004; Xu *et al.*, 2005; Chernov *et al.*, 2008; Wang *et al.*, 2009). Furthermore, in contrast to the NS3 protein from HCV and the NPH-II protein from Vaccinia virus, two other SF2 RNA helicases (Gross and Shuman, 1996; Jankowsky *et al.*, 2000; Pang *et al.*, 2002), the flavivirus NS3 helicase does not appear to be a processive RNA helicase. As the duplex region extends from 22 to 36 base-pairs, its unwinding ability decreases dramatically (Wang *et al.*, 2009), and ATP hydrolysis becomes absolutely required for activity (Li *et al.*, 1999; Benarroch *et al.*, 2004; Wang *et al.*, 2009). On the other hand, helicase assays using dsRNA substrates with a short duplex region of less than 20 bp show that the NTPase and helicase activities can be functionally uncoupled (Matusan *et al.*, 2001; Borowski *et al.*, 2002; Sampath *et al.*, 2006). Flavivirus NS3 helicase activity differs also from that of DEAD-box RNA helicases. The DEAD-box helicases usually demonstrate poor RNA unwinding activities and require assistance from cofactors (Cordin *et al.*, 2006; Bleichert and Baserga, 2007; Jankowsky and Fairman, 2007; Pyle, 2008). It is not known for certain how the energy from ATP hydrolysis is utilized for unwinding dsRNA and why the two enzymatic activities, NTPase/helicase, can be uncoupled in some RNA helicases but not others (Pyle, 2008).

RNA triphosphatase activity

RNA triphosphatase hydrolyses the γ-phosphoric anhydride bond of triphosphorylated RNA. This step is the first of the three sequential enzymatic reactions required for the addition of a 5′ cap to viral RNA, a process needed for proper translation of the viral genome (Lindenbach *et al.*, 2007; Dong *et al.*, 2008). Mutagenesis studies and competition experiments suggest that RTPase and NTPase activities share a common active site in the flavivirus NS3 protein (Bartelma and Padmanabhan, 2002; Benarroch *et al.*, 2004; Wang *et al.*, 2009). Not surprisingly, an exposed 5′-terminus makes the 5′-triphosphate group more accessible to the active site, and thus a better substrate for the RTPase activity (Wang *et al.*, 2009). The RTPase activity was shown to depend on divalent metal ions (Benarroch *et al.*, 2004; Wang *et al.*, 2009), although divalent metal ion independence was also reported (Bartelma and Padmanabhan, 2002). Furthermore, recent studies have shown that NS5 can also stimulate RTPase activity in a dose-dependent manner up to a 1:1 molar ratio (Yon *et al.*, 2005).

Structural studies

The NS2B47-NS3 protease

Overall structure

Crystal structures for the NS2B-NS3 protease from DENV and WNV have revealed a chymotrypsin-like fold with two β-barrels, each formed by six β-strands, with residues 51–57 of NS2B contributing an additional β-strand to the N-terminal β-barrel (Fig. 5.2) (Erbel *et al.*, 2006). At the interface of the two β-barrels, a cleft serves as the site for substrate binding and houses the conserved catalytic triad His-57, Asp-75 and Ser-135 (Fig. 5.2). For both protein constructs, a total of 47 residues from the central hydrophilic region of NS2B (abbreviated as '$NS2B_{47}$') were fused to the NS3 protease domain via a flexible and glycine-rich peptidic linker to generate a single-chain active NS2B-NS3 protease (Aleshin *et al.*, 2007; Chandramouli *et al.*, 2010; Erbel *et al.*, 2006; Robin *et al.*, 2009). In the absence of either ligands or inhibitors, the seventeen amino-terminal amino acids of $NS2B_{47}$ from both DENV and WNV wrap around the NS3 protease domain, forming one β-strand which associates with the N-terminal β-barrel of NS3, concealing

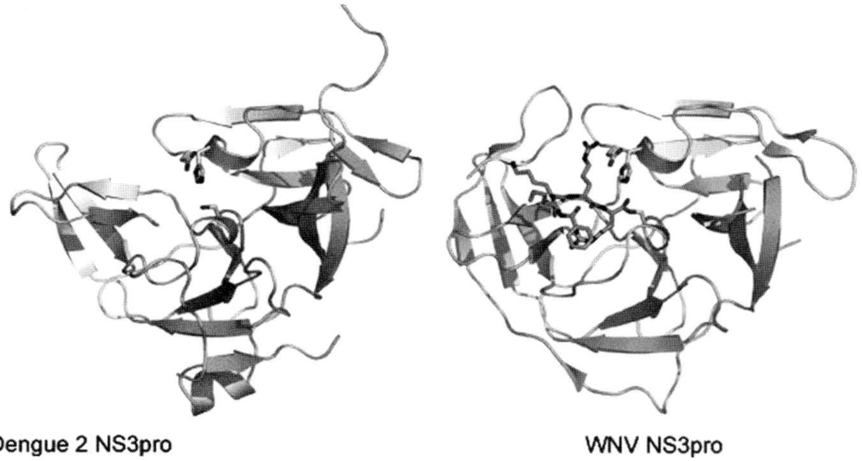

Figure 5.2 Left: The NS3 protease (white) from DENV2 has a chymotrypsin-like fold with two β-barrels, each formed by six β-strands, with residues 51–57 of NS2B (coloured in yellow) contributing an additional β-strand to the N-terminal β-barrel. Residues belonging to the catalytic triad are shown as sticks. Right: protease from WNV as a complex with the aldehyde-derived peptidic inhibitor Bzl-Nle-Lys-Arg-Arg-H in orange (Erbel *et al.*, 2006).

hydrophobic residues from the solvent, whilst several residues at the C-terminal end of NS2B$_{47}$ appear disordered (Fig. 5.2): in the case of the NS2B$_{47}$NS3 protease from DENV, electron density is discontinuous beyond aa76 suggesting that the C-terminal part of the NS2B co-factor can adopt multiple conformations in solution (Erbel *et al.*, 2006). Thus, the N-terminal region of NS2B$_{47}$ acts as a chaperone that stabilizes the fold of the protease. This is supported by observations that the protease domain is insoluble when expressed alone (Murthy *et al.*, 1999; Yusof *et al.*, 2000) and that addition of residues 49–66 at the N-terminal end of NS2B renders it soluble, yet catalytically inactive (Erbel *et al.*, 2006). Rather interestingly, the apo-structure of the NS2B$_{47}$-NS3 protease from DENV serotype 1 revealed a metal coordination site comprising His67 and His72 of NS2B and Glu74 of NS3, occupied by a Cadmium ion (Chandramouli *et al.*, 2010). This feature may be unique to DENV1 as neither His residues are conserved in other flaviviral NS2B sequences.

Structure of the protease bound to inhibitors

After years of efforts, structures for the NS2B$_{47}$-NS3 protease from WNV were determined as a complex either with the aldehyde-derived peptidic inhibitors Bzl-Nle-Lys-Arg-Arg-H (Erbel *et al.*, 2006), see Fig. 5.2 and Naph-Lys-Lys-Arg-H (Robin *et al.*, 2009) or with the pan-serine-protease inhibitor aprotinin/BPTI (Aleshin *et al.*, 2007). (Fig. 5.3). Complexed forms of dengue NS2B$_{47}$-NS3 remain elusive. Overall, the WNV protease complexes share a high level of structure similarity: Values for r.m.s. deviations of 0.51 and 0.60 Å, respectively, for 183 equivalent Cα atoms are returned when the protease from WNV bound to aldehyde-derived peptidic inhibitors is superimposed onto the protein bound to aprotinin. In these structures, the C-terminal part of NS2B adopts a 'closed' conformation that is markedly different from the unbound protease structures. In the bound enzyme, it encircles the C-terminal β-barrel of the NS3 protease by making contact with β-strands and ends up in a hairpin loop inserted into the active site. Notably, residues 83–86 (Erbel *et al.*, 2006; Aleshin *et al.*, 2007; Robin *et al.*, 2009) or 83–90 (Robin *et al.*, 2009) are directly involved in the formation of the S2 and S3 inhibitor binding pockets. Thus, it is clear from this work that NS2B not only stabilizes the NS3 protease by providing one additional β-strand to each of the two _-barrels, but also contributes to the substrate-binding site via the C-terminal region of its hydrophilic domain. This differs substantially from what is observed in the NS4A-NS3 protease from HCV, in which NS4A activates the NS3 protease by providing a more rigid framework and by orienting residues that form the substrate binding channel (Kim *et al.*, 1996; Love *et al.*, 1996). This accounts for the several orders of magnitude difference in enzyme activity measured *in vitro* between the NS3 polypeptide devoid of the co-factor and the NS2B47-NS3 protease (Yusof *et al.*, 2000; Erbel *et al.*, 2006). Nevertheless, in solution, both the 'open' (with the C-terminus of NS2B$_{47}$ being far from the substrate binding site) and 'closed' conformations of NS2B47-NS3 coexist, either in the presence or absence of inhibitors. This observation based on solution N.M.R. studies, suggests that the C-terminal region of NS2B47 is in constant dynamic equilibrium between different conformational states (Su *et al.*, 2009). This is also supported by the crystal structure of NS2B47-NS3 protease from DENV1, where part of the C-terminal end of NS2B47 spanning aa 63–71, forms a helix followed by a β-sheet that runs along the NS3 protease domain (Chandramouli *et al.*, 2010).

Aprotinin occupied more specificity pockets compared with the peptidic inhibitors, revealing the S2–S2′ residues involved in the interaction, whilst the latter structures allowed determination of the S4–S1 inhibitor binding sites. Binding of aprotinin and Naph-Lys-Lys-Arg-H resulted in a catalytically competent conformation of the oxyanion hole (comprising residues Gly133-Ser135) with the Arg P1 carbonyl atom (Arg-P1-O) interacting with the oxyanion hole. It points away from His51 and directly H-bonds with main chain N-atoms of Gly133 and Ser135. In contrast, in unbound and Bzl-Nle-Lys-Arg-Arg-H-bound structures, a flipped peptide bond between residues Thr132 and Gly133 results in the disruption of the oxyanion hole. The Arg-P1-O of

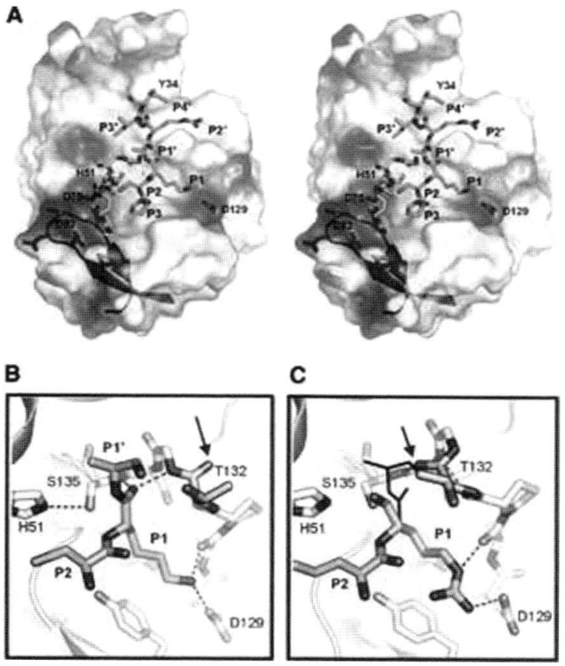

Figure 5.3 Structures for the NS2B$_{47}$-NS3 protease from DENV with the pan-serine-protease inhibitor aprotinin/BPTI. (A) Stereo view of WNV NS2B-NS3pro surface with selected aprotinin residues. The green and blue sticks are two parts of aprotinin (residues ^{13}PCKARII^{19} and ^{35}GGCR^{39}) that interact with the protease. The surface is coloured by electrostatic potential (negative, red; positive, blue). The magenta ribbons show the invading 2–3 hairpin of NS2B contributing to the active site. Selected residues of the protease are shown as sticks. (B, C) Comparison of the oxyanion hole conformations in the WNV NS2B–NS3pro complexed with (B) aprotinin and (C) a peptidic inhibitor (Erbel *et al.* 2006; PDB code 2FP7). Selected residues of the protease (yellow) and inhibitors (green) are shown. In (B), aprotinin induces the active conformation of the oxyanion hole, which is occupied by the P1 C=O. In *C*, the oxyanion hole is disrupted owing to a flip of the peptide bond at Gly-133–Thr-132, which creates a 3_{10} helical conformation (the C=O of Thr-132 is marked with an arrow in B and C). The P1 C=O and P1 side chain of aprotinin (shown as black stick in C) would clash with Thr-132 in this conformation. Selected hydrogen bonds are shown with dashed lines. From Aleshin *et al.* (2007).

Bzl-Nle-Lys-Arg-Arg-H was observed to directly H-bond with His51. The side chain of His51 is within hydrogen-bonding distance from the other two residues that form the catalytic triad, Ser125 and Asp75. Interestingly, two conformations of His51 were observed in NS2B$_{47}$-NS3 molecules bound with Naph-Lys-Lys-Arg-H (Robin *et al.*, 2009). In one case, His51 is orientated as in the Bzl-Nle-Lys-Arg-Arg-H-bound structure (active conformation), whilst in the other case, it is rotated out of the active site and interacts with NS2B Asp82 via a water molecule. It was suggested that an oxygen atom from water or the bound substrate would play an important role in maintaining the active conformation of His51 (Robin *et al.*, 2009). Similar rotations of His51 have been observed in crystal structures other serine proteases. For both bound peptidic inhibitors, a closed conformation was observed with either the benzoyl or naphthoyl groups forming intramolecular H-bond interactions with the P1-Arg residue (Erbel *et al.*, 2006; Robin *et al.*, 2009). It was proposed that removing this constraint via functional groups that allow the extension of the peptide inhibitors and occupation of the S4 pocket may be a way to increase inhibitor affinity (Robin *et al.*, 2009).

The NS3 NTPase/helicase domain

Overall structure

Crystal structures were first reported for the NS3 helicase domains from DENV2 (Fig. 5.4) and YFV (Wu *et al.*, 2005; Xu *et al.*, 2005). Subsequently, several structures for helicases from other flaviviruses or other DENV serotypes were published (Mastrangelo *et al.*, 2006, 2007; Mancini *et al.*, 2007; Lescar *et al.*, 2008; Luo *et al.*, 2008b; Speroni *et al.*, 2008; Yamashita *et al.*, 2008). The NS3 helicase comprises three subdomains. All known sequence motifs identified for the members of the SF2 superfamily of helicases are located within subdomains 1 and 2 (Fig. 5.4). Each of these two subdomains structurally resembles the DNA recombination protein RecA. The ATPase active site is housed between these two subdomains and is structurally well conserved compared with other SF2 helicases, including the one from HCV (Yao *et al.*, 1997; Xu *et al.*, 2005). It was proposed that ATP is primarily held through its triphosphate moiety via contacts mediated through the divalent metal ion and residues from motifs I, II and VI, a prediction that was confirmed by experimental crystallographic structures (Luo *et al.*, 2008b, 2010). Unlike DEAD-box RNA helicases, no 'Q motif' is found upstream of motif I, which is consistent with a lack of discrimination between various nucleotides as substrate (Locatelli *et al.*, 2002; Benarroch *et al.*, 2004). Like the NS3 protein from HCV, a tunnel is formed between subdomain 1, 2 on one side and subdomain 3 on the other side. This, in spite of the fact that subdomain 3 has the most divergent fold between the NS3 proteins from hepacivirus and

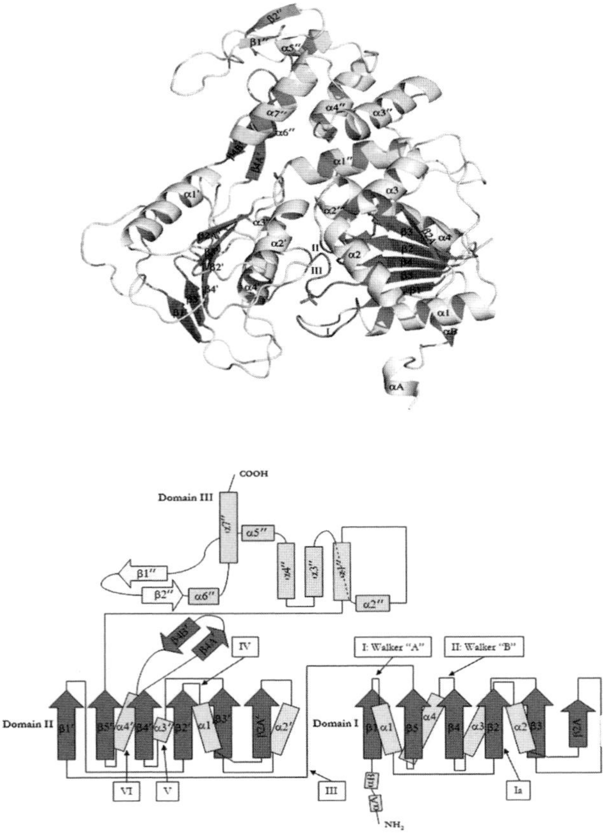

Figure 5.4 Crystal structure for the NS3 helicase domain from DENV2. Amino-acid sequence motifs important for the activity are highlighted with different colours.

flavivirus (Xu *et al.*, 2005). Residues belonging to the conserved motifs involved in ssRNA binding and duplex unwinding line the top of the two RecA-like domains and are facing towards this nucleic acid binding tunnel.

Structural comparison between flavivirus helicases

Superposition with other flavivirus helicases, including from YFV (Wu *et al.*, 2005), WNV (Mastrangelo *et al.*, 2007), MVEV (Mancini *et al.*, 2007) and DENV2 (Xu *et al.*, 2005), yields r.m.s. deviations of 1.77 Å (based on 431 equivalent Cα atoms), 1.53 Å (based on 415 Cα, using molecule A from the PDB), 1.05 Å (based on 429 Cα atoms) and 0.68 Å (based on 441 Cα atoms, using molecule B from the PDB), respectively. Among all three subdomains of the helicase, subdomain 1 is the most well conserved in structural terms. The main structural differences reside in subdomains 2 and 3, especially the two abutting helices, α1 of subdomain 2 and α7 of subdomain 3 that lie at the 5′ end of the ssRNA binding tunnel. In the helicase from YFV and MVEV, the distance between these two α helices, (measuring from the position of the Cα of Arg-371 and Gln-609 for YFV and from the Cα of Lys-367 and His-605 for MVEV) is 12.5 Å and 7.7 Å respectively. These values correspond to open and closed conformations of the ssRNA binding tunnel, demonstrating built-in flexibility of this functionally important region in the absence of ligand. Corresponding distances in WNV, DEN2 and DEN4 helicases are 11.8 Å, 10.0 Å and 8.9 Å, respectively. In addition, when comparing DENV4 NS3 helicase with that of YFV and WNV, another interesting structural variation is observed in subdomain 2, involving a short helix α3′ and a loop composed of residues Thr-409 to Phe-417 which belongs to motif V.

RNA recognition by NS3 helicase

Several structures of dengue NS3 helicase were resolved in the presence of a bound 12-mer ssRNA. This work provided for the first time a structural basis for sequence-independent RNA recognition and gave insight at atomic resolution into mechanochemical events that couple ATP hydrolysis and RNA unwinding by the flavivirus NS3 helicase. As anticipated, the ssRNA is accommodated in an extended conformation in the tunnel that separates subdomains 1 and 2 from subdomain 3 (Fig. 5.5). Subdomain 1 binds to the 3′ end of ssRNA while the 5′ end mainly interacts with subdomain 2. The RNA orientation conforms to that of a deoxyuridine octamer bound to the HCV helicase (Kim *et al.*, 1998). The path of the sugar-phosphate backbone, however, differs between these two structures: in the flavivirus helicase, the ssRNA adopts an 'A form'. Five bases of the RNA located at the 5′ end are stacked to each other. The RNA structure makes a sharp bend between bases 5 and 6. Only the sugar phosphate backbone is well ordered for nucleotides 6 and 7, and the corresponding bases appear mobile. At the 3′ end, the remaining five nucleotides are disordered. No aromatic side chain is in the right position for intercalating between the bases suggesting a different mechanism for unwinding than the one proposed for helicases from HCV or PcrA (Kim *et al.*, 1998; Soultanas *et al.*, 1999; Velankar *et al.*, 1999). Most of the protein RNA contacts are established with the sugar-phosphate backbone either directly or via water molecules. This lack of sequence specificity explains how helicases can accommodate diverse viral genomic sequences as it translocates from the 3′ towards the 5′ end.

Interestingly, the bound RNA moieties are closely superimposable between the structures from DENV helicase and eukaryotic DEAD box proteins eIF4AIII, Vasa or Mss116p in complex with polyU (Andersen *et al.*, 2006; Bono *et al.*, 2006; Sengoku *et al.*, 2006; Del Campo and Lambowitz, 2009), see Fig. 5.6. Superposition of these latter structures with the NS3 helicase reveals striking similarities in the molecular details of ssRNA recognition suggesting the existence of an ancestral ssRNA recognition module possibly borrowed by the *Flaviviridae* during evolution (Fig. 5.6). Through fusion with additional domains, this ancestral single strand nucleic acid recognition module has evolved to fulfil different molecular and cellular tasks. In the case of the DENV NS3 protein, both a protease domain and an original protein recognition module (subdomain 3 which binds to the NS5 polymerase) have been fused at its N and C-terminal ends, respectively.

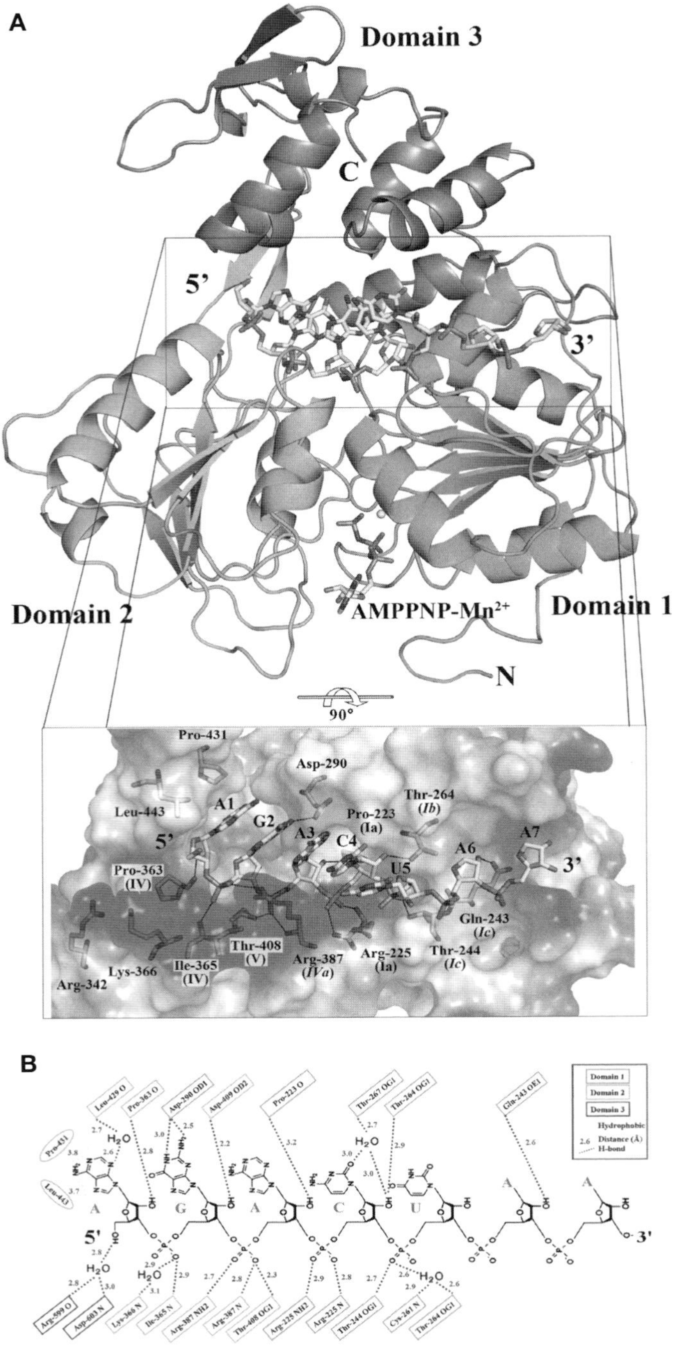

Figure 5.5 The ssRNA is accommodated in an extended conformation in the tunnel that separates subdomains 1 and 2 of NS3 helicase domain from its subdomain 3. (A) the NS3 helicase is shown as ribbons and the ssRNA, and non-hydrolysable ATP analogue AMPPNP as sticks. Manganese ion is depicted as a yellow sphere. Inset: Electrostatic surface of NS3 in the region contacting ssRNA. Positive charges: blue, negative charges: red. Amino acid residues in contact are displayed as coloured sticks and labelled based on the helicase domain sequence. Domain 3 was omitted for clarity. (B) Interaction network between NS3 helicase and ssRNA.

A

	I	**Ia**	**Ic**
DENV4NS3	193LHPGAGKTKRIL204	218TLILAPTRVVAAE230	239PIRYQTPAVKS249
eIF4AIII	SQSGTGKTATFS	ALILAPTRELAVQ	HACIGGTNVGE
Vasa	AQTGSGKTAAFL	VVIVSPTRELAIQ	GIVYGGTSFRH
HCVNS3	APTGSGKSTKVP	VLVLNPSVAATLG	PNIRTGVRTIT
Hel308	MPTAAGKTLLAE	SLYVVPLRALAGE	IGISTGDRDEH
	Ib	**II**	**III**
DENV4NS3	259LMCHATFTTRLLS271	279NLIVMDEAHFTDPC292	311AIFMTATPPGS321
eIF4AIII	VAGTPGRVFDMIR	KMLVLDEADEMLNK	VVLISATLPHE
Vasa	VIATPGRLLDFVD	RFVVLDEADRMLDM	TLMFSATFPEE
HCVNS3	TYSTYGKFLADGG	DIIICDECHSTDAT	VVLATATPPGS
Hel308	IVTTSEKADSLIR	SCLVVDEIHLLDSE	VIGLSATAPNV
	IV	**IVa/(QxxR)**	**V**
DENV4NS3	358TVWFVPSIKAGN369	384QLSRKTFDTEYPKT397	406VTTDISEMGAN416
eIF4AIII	AVIFCNTKRKVD	SMHGDMPQKERESI	ISTDVWARGLD
Vasa	TIVFVETKRGAD	SIHGDRLQSQREQA	IATSVASRGLD
HCVNS3	HLIFCHSKKKCD	AYYRGLDVSVIPTS	VATDALMTGFT
Hel308	VLVFESTRRGAE	FHHAGLLNGQRRVV	VATPTLAAGVN
	VI		
DENV4NS3	456QRRGRIGRNPAQ467		
eIF4AIII	HRIGRSGRYGRK		
Vasa	HRIGRTGRVGNN		
HCVNS3	QRRGRTGRGKPG		
Hel308	QMAGRAGRPGMD		

Motifs
ATP binding/Hydrolysis
RecA Domain Linker
2'OH (RNA Specific)
Backbone Pi

B

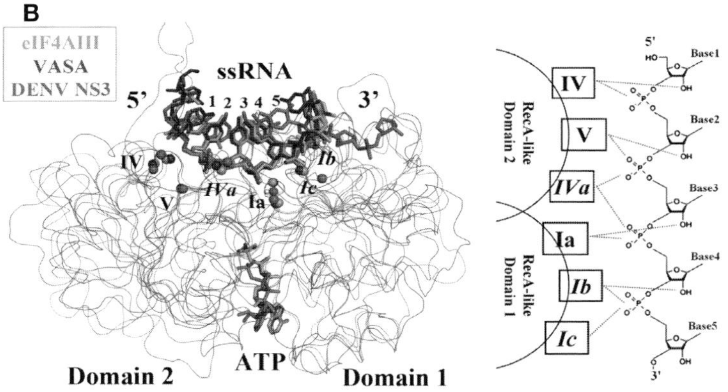

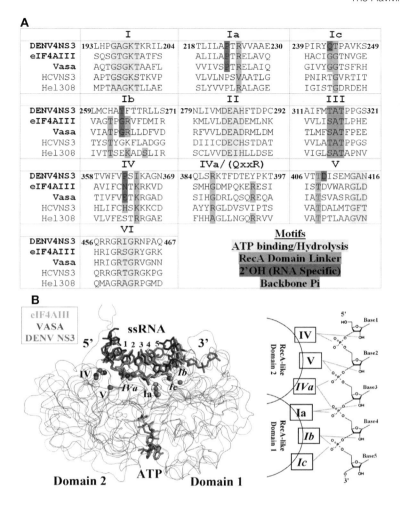

Figure 5.6 The bound RNA moieties are closely superim*posable between the structures of NS3 helicase from DENV and eukaryotic DEAD box proteins eIF4AIII, Vasa or Mss116p in complex with polyU. (A) Comparison of amino acids from DENV4, eIF4AIII, Vasa, HCV and Mss116p helicases of the region responsible for ATP binding and hydrolysis, the linker region between domains 1 and 2, and the ribose 2′-OH and phosphate backbone binding sequences. (B) Overlap of ssRNA bound-helicase structures from DENV4(red), eIF4AIII (green) and Vasa (blue).

Structural basis for RNA-stimulated NTPase activity

NS3 helicase undergoes several major conformational changes upon RNA binding (Fig. 5.7). A number of these may accompany RNA translocation: subdomain 3 rotates some 11° away from the ATP binding domains leading to a widening of the RNA binding tunnel and allowing the insertion of several water molecules in the groove. Helix α2′ which acts as the hinge for this movement, is deformed into a coil. There is closure between subdomains 1 and 2 by *ca* 12°, leading to a narrower ATP binding cleft (Fig. 5.7B). This closure leads to a translation of about 5 Å of the RNA binding motifs within subdomain 2 relative to subdomain 1. Thus, we proposed that one translocation step of the protein corresponds to the distance between two adjacent bases, in the 3′ to 5′ direction. Of note, such pronounced domain rearrangements have not been observed in the HCV NS3 helicase upon oligonucleotide binding (Kim *et al.*, 1998). For DENV4 helicase,

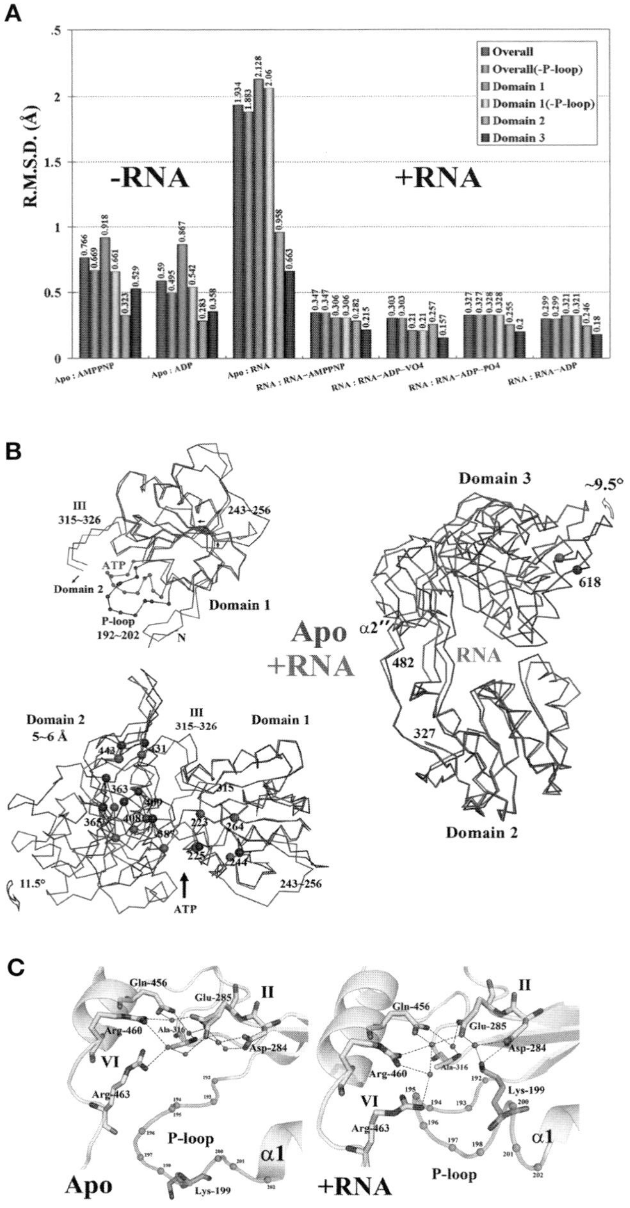

Figure 5.7 NS3 helicase undergoes major conformational changes upon RNA binding. (A) RMSDs between the various NS3 helicase structures or their individual domains after superimposition. (–P-loop) indicates values with residues 195–202 omitted from the calculation. A clear partition can be seen between the RNA-bound ('+RNA') and the RNA-free NS3 helicase structures ('–RNA'). (B) Comparison between the apo-NS3 helicase structure (blue) and RNA-bound structure (red). Top left: superposition of subdomain 1 highlighting P-loop movements. Left lower panel: closure of the ATP-binding cleft on RNA binding (both subdomains 1 were superposed to generate this figure). The position of the α-carbon atoms of residues that come in contact with RNA is shown as blue (apo-NS3 helicase) or red (+RNA) spheres and labelled. Right panel: re-orientation of subdomain 3 on RNA binding (subdomain 2 was superimposed). (C) Close-up view of the ATP-binding site highlighting conformational changes that occur in the P-loop (green) on RNA binding, in the absence of a bound nucleotide.

these movements in the quaternary structure are accompanied by an inward movement of the P-loop leading to a reduction of the volume of the ATP binding cleft and the formation of a salt bridge between Lys-199 and Asp-284 (Fig. 5.7C). Thus, allosteric conformational changes triggered by ssRNA binding, convert the P-loop into its substrate-bound conformation, poised to react with ATP (Luo *et al.*, 2008b).

The ATP hydrolysis cycle

Crystal structures of DENV4 helicase domain in complex with a non-hydrolysable ATP analogue, ADP/vanadate/Mn^{2+}, ADP/P_i/Mn^{2+}, and ADP/Mn^{2+}, provided snapshots along the hydrolytic cycle (Luo *et al.*, 2008b). The slowly hydrolysable ATP analogue 5′-adenylyl-β, γ-imidodiphosphate (AMPPNP) was used to trap an enzyme substrate ternary complex. In this complex, one water molecule hydrogen-bonded by residues from motif II and VI is positioned at right distance for an in-line nucleophilic attack of the γ-phosphate, probably following activation through proton transfer to Glu-285 and/or polarization by Gln-456. The geometry of the catalytic water relative to the γ-phosphate is more favourable with bound RNA, providing additional ground for the observed RNA-stimulated ATPase activity. The reaction intermediate is a trigonal bipyramidal pentavalent structure. A model for it was obtained using the ADP/Vanadate/Mn^{2+} ternary complex. Soaking NS3 helicase crystals with ATP led to the observation of an ADP-P_i/Mn^{2+} complex, indicating that the crystalline protein is active, opening the possibility of time-resolved studies. A small tunnel lined by Pro-195, Ala-316, Thr-317, Pro-326, Ala-455, and Gln-456 could serve as an exit route for the phosphate product of the hydrolytic reaction. The ATP hydrolysis cycle by the flavivirus helicase and the mechanism proposed for it are summarized in Fig. 5.8.

A model for RNA unwinding by NS3 helicase

Surprisingly, no changes in the structure of NS3 helicase domain were observed in the various altered nucleotide states, either for binary or for ternary (RNA-bound) complexes and RNA binding appears sufficient to convert the helicase to its closed conformation. This was surprising in the light of the unwinding mechanism proposed for the HCV NS3 protein, a DExH helicase, where the protein cycles between open and closed states, depending on the nucleotide-binding state and where the ATP hydrolysis cycle is coupled to loose versus tight RNA binding and translocation (Pyle, 2008). One explanation for this observation could relate to constraints imposed on the NS3 helicase protein by forces from the crystal lattice that would restrict conformational changes during the ATP hydrolytic cycle. It is quite possible that several conformational states of flavivirus NS3 helicase exist – some transient – that could not have been crystallized in our limited sampling of crystal structures. Two recent structural studies on DEAD box helicases (eIF4AIII and Mss116p) reveal similar observations, e.g. several nucleotide transition state analogues are able to bind to the protein without inducing conformational changes (Del Campo and Lambowitz, 2009; Nielsen *et al.*, 2009). Conversely, distinct conformations corresponding to different nucleotide binding states for the NS3 helicase from HCV have recently been reported, suggesting a ratchet translocation mechanism (Gu and Rice, 2010). It is therefore also quite possible that substantial mechanical differences exist between different helicases, mirroring subtle variations in their activities.

By extending the bound ssRNA strand into the 5′ direction and pairing it with a complementary strand to form a dsRNA substrate for NS3 helicase (Fig. 5.9), one β_hairpin element of DENV NS3 that joins strands β4A′ and β 4B′ is likely to play a crucial role for separating the two strands by disrupting base stacking and stabilizing the unwound duplex (Buttner *et al.*, 2007). In this model, one RNA strand would migrate through the tracking ssRNA tunnel, while the other strand would be forced towards the back of the protein by the β-hairpin. Interestingly, an equivalent structural device (christened 'separation pin') was proposed to play an active role in DNA unwinding by UvrD, a bacterial SF1 helicase (Lee and Yang, 2006). Thus this important functional feature is not limited to SF2 helicases but appears to be shared between helicases of these two superfamilies. How residues from the β-hairpin precisely interact with the fork and how this affect

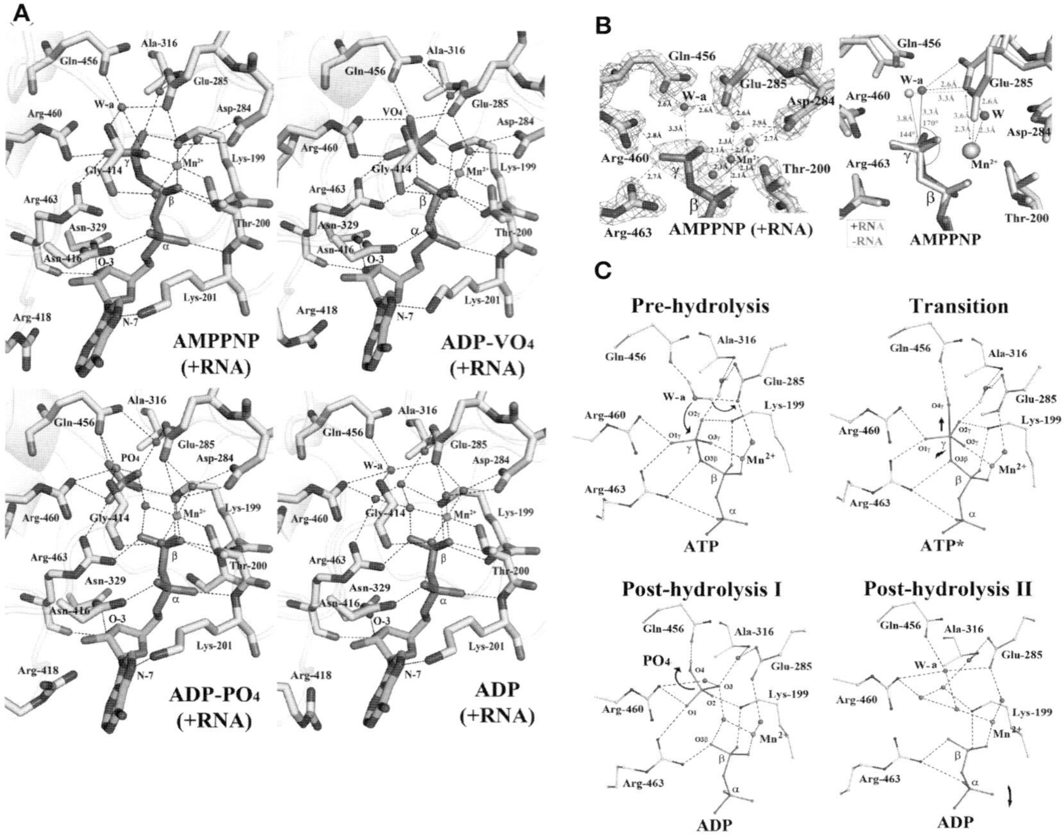

Figure 5.8 Dengue NS3 ATP hydrolysis. (A) Close-up view of NS3 ATP binding site with either AMPPNP, ADP-VO_4 (upper panel, left to right), ADP-PO_4 or ADP (lower panel; left to right) bound with in presence of RNA. Manganese ion (Mn) and water molecules (W) are shown as green and red spheres. H-bonds are depicted as dashed lines. (B) Comparison of amino acid interactions of AMPPNP bound in the NS3ATP binding site in the presence and absence of RNA. Manganese ion (Mn) and water molecules (W) are shown as grey and red spheres. H-bonds are depicted as dashed lines. (C) The mechanism proposed for the ATP hydrolysis cycle by the flavivirus helicase.

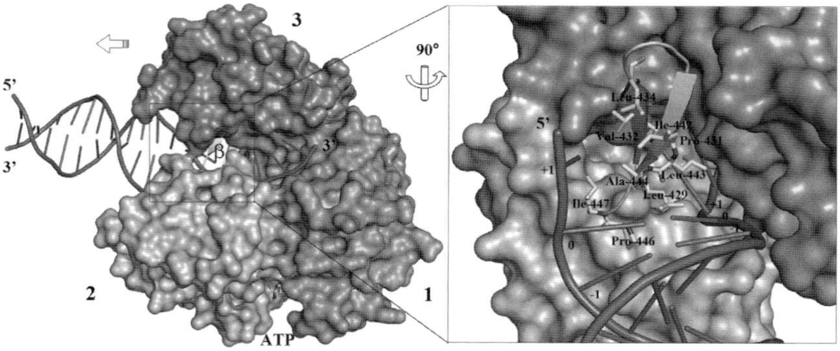

Figure 5.9 One β-hairpin element of DENV NS3 helicase that joins strands β4A′ and β 4B′ is likely to play a crucial role for separating the two strands by disrupting base stacking and stabilizing the unwound duplex.

RNA unwinding and processivity remains to be determined through mutagenesis and functional studies. Single-molecule studies would favour a second point of contact of the *Flaviviridae* helicase ahead of the fork, presumably in the form of a contact between dsRNA (Cheng *et al.*, 2007) and subdomain 3. The driving force for strand separation itself remains elusive (Betterton and Julicher, 2005). In an active mechanism, the β-hairpin would disrupt base pairs. For the HCV helicase, it was suggested that quaternary movements in the NS3 helicase subdomains accumulate increasing tension on the protein-nucleic acid complex, and also exert a torque on the duplex, (Myong *et al.*, 2007; Pyle, 2008). In a passive mechanism, the β-hairpin would merely act as an 'insulator' by taking advantage of the spontaneous thermal opening of the duplex at the fork and by stacking with the melted bases. In our model, base +1 would interact with the conserved Leu-443, that projects from the β-hairpin disrupting base stacking and thus splaying apart dsRNA at the fork region. We also noticed departure from A-form RNA between bases 5 and 6 at the 3′ end of the 12-mer bound to NS3 from DENV4. A distorted RNA conformation was indeed observed in the Vasa–polyU complex and was proposed to be important for the mechanism of RNA unwinding (Sengoku *et al.*, 2006). Another possibility is that this protein-induced strain is not directly related to RNA duplex unwinding. Rather, it would prevent NS3 from sliding backwards along the ssRNA substrate. The thermodynamics of energy transfer from ATP hydrolysis leading to base unpairing of the dsRNA macromolecular substrate remains unclear at the moment. It is commonly thought that all helicases use chemical energy released from ATP hydrolysis and convert it into mechanical energy, which in turn drives the enzyme directional movement along the nucleic acid. Models for helicase translocation along single stranded nucleic acid, coupled with nucleic acid duplex unwinding have been proposed based on studies of various helicases using different approaches (Theis *et al.*, 1999; Velankar *et al.*, 1999; Levin *et al.*, 2005; Yang and Jankowsky, 2006; Pyle, 2008). Apparently, no single model can explain all phenomena, a fact possibly linked to the use of different measurement techniques and methodologies. Furthermore, various helicases might have evolved to utilize different molecular mechanisms to perform different molecular tasks depending on the environments. For DENV NS3 helicase, the closest helicase that has been extensively studied is HCV NS3. Yet there is still no single model that can explain how HCV NS3 works. Both positive and passive unwinding models were proposed and each is supported by some experiments. Single-molecule kinetic studies are very powerful for the study of strand unwinding speed and step size (Dumont *et al.*, 2006; Myong *et al.*, 2007). The fact that DENV NS3 helicase is not processive and unwinds only RNA, not a DNA, highlights differences with the HCV NS3 helicase which is more processive and is a DNA/RNA dual-function helicase. Thus, further work is needed including a NS3–dsRNA complex structure and single molecule studies.

The full-length NS3 protease-helicase

Activity

NS3 carries in a single polypeptide chain two apparently disconnected enzymatic activities: a proteolytic activity needed for post-translational processing of the viral polyprotein and a helicase/ATPase/5′RNA triphosphatase activities required for viral RNA replication and 5′ RNA capping. The protease activity of the $NS2B_{47-}NS3pro$ domain and the full-length $NS2B_{47-}NS3$ show similar *in vitro* kinetics for both the WNV (Mueller *et al.*, 2007) and DENV protein constructs (Lim *et al.*, unpublished data). In contrast, although the helicase domain of NS3 is able to function independently of the protease domain, the full length NS3 (NS3FL) exhibits superior helicase/ATPase activity compared with the helicase domain alone (Yon *et al.*, 2005; Luo *et al.*, 2008b, 2010; Wang *et al.*, 2009). The active catalytic triad is dispensable for stimulatory effect suggesting that the overall fold of the N-terminal protease domain contributes to this enhancement (Yon *et al.*, 2005). Interestingly, the full-length NS3 enzymes also undergoes auto-cleavage in *cis* within its helicase domain the Arg-458–Gly-459 junction (Arias *et al.*, 1993; Teo and Wright, 1997; Bera *et al.*, 2007). How this activity is *in vivo* regulated as well as its physiological significance is unclear as cleavage at this site would result in inactivation of the helicase/NTPase. Since Arg-458 is buried inside the protein and inaccessible to proteases when the helicase adopts a native fold, cleavage at this position would require large conformational changes, re-orientations of the protein domain, and even partial unfolding of the helicase domain. These might be the rate limiting steps that are consistent with the fact that this internal cleavage is less efficient than the cleavage at the NS2B–NS3 junction. Thus, the N-terminal protease domain of dengue virus NS3 has a strong influence on its helicase and ATPase activity. The fact that NS3 protease activity requires integral membrane protein NS2B as co-factor also suggests that the NS3 protease-helicase functions close to the membrane surface (Luo *et al.*, 2008a, 2010; Assenberg *et al.*, 2009).

Structure

Structural and functional studies on NS3FL have been progressing rapidly. In the absence of the NS2B cofactor, the full-length NS3 protein is not very soluble and has no protease activity (Xu *et al.*, 2005; Sampath *et al.*, 2006). As detailed above, the central hydrophilic region of NS2B is essential for NS3 protease activity: the presence of this peptidic co-factor of 47 amino acids ($NS2B_{47}$) leads to a very soluble and active protease, and crystallographic structures of this $NS2B_{47}$–NS3Pro were determined (Erbel *et al.*, 2006). The presence of the first 18 amino acids from the NS2B co-factor region ($NS2B_{18}$) renders NS3 protease soluble but enzymatically inactive (Erbel *et al.*, 2006). Autolysis occurs within the $NS2B_{47}$–NS3Pro and $NS2B_{47}$–NS3FL proteins (Erbel *et al.*, 2006; Bera *et al.*, 2007). Based on these findings, a $NS2B_{18}$–NS3 protein using sequences from DENV4 was made (Fig. 5.1). By including the $NS2B_{18}$ segment in the construct, a soluble and proteolytically resistant NS3FL protein could be produced to large quantity needed for structural studies. Indeed, $NS2B_{18}$–NS3 showed no self-cleavage as its protease domain is inactive. This is in contrast to the MVEV NS2B–NS3FL construct produced by Assenberg and coworkers (2009) where the protein contained an active protease domain. Two distinct conformations of the $NS2B_{18}$–NS3 from DENV4 have been found using similar crystallization conditions (Luo *et al.*, 2008a, 2010). Together with the structure from Assenberg *et al.*, (2009), they shed new light on the dynamics and function of this multifunctional protein both *in vitro* and *in vivo* during flavivirus replication. In both DENV4 $NS2B_{18}$-NS3 conformations, the N-terminal protease domain resides next to the entrance of the ATPase active site between helicase subdomains 1 and 2. The relative orientation between the protease and helicase domain differs: there is 161° rotation of the protease domain relative to the helicase domain (Fig. 5.10). The interdomain contacts between the subdomain 2 of the helicase and the protease domain are absent in conformation II, leaving the ATPase pocket open. Both conformations give the protein an elongated shape, with approximate overall dimensions of $100\,Å \times 60\,Å \times 40\,Å$ that are compatible with small angle scattering studies in solution suggesting their coexistence in solution. A very similar elongated shape for a full-length NS3 protein from WNV Kunjin virus was obtained using small-angle X-ray scattering (Mastrangelo *et al.*, 2007). Interestingly, the crystallographic structure of the $NS2B_{47}$–NS3FL protein from MVEV adopts a similar but slightly extended conformation (Assenberg *et al.*, 2009). Moreover, a comparison of the $NS2B_{18}$–NS3 molecule with NS3–NS4A from HCV (Yao *et al.*, 1999) highlights a major difference for the relative orientations between the helicase and protease domains in the two proteins. The C-terminal end of NS3 from HCV participates actively in complex formation with the protease domain, being inserted into the active site of the protease domain as *cis* cleavage occurs at the NS3–NS4A junction. The $NS2B_{18}$-NS3 molecule from dengue, on the other hand, has its C terminus located far away from the protease active site.

Interdomain orientations

While no complex with nucleotide could be obtained with crystals where the protein adopted conformation I, ADP-Mn^{2+} was successfully soaked into the second crystal form of DENV4 $NS2B_{18}$–NS3 (conformation II), yielding a binary complex structure almost identical to that found

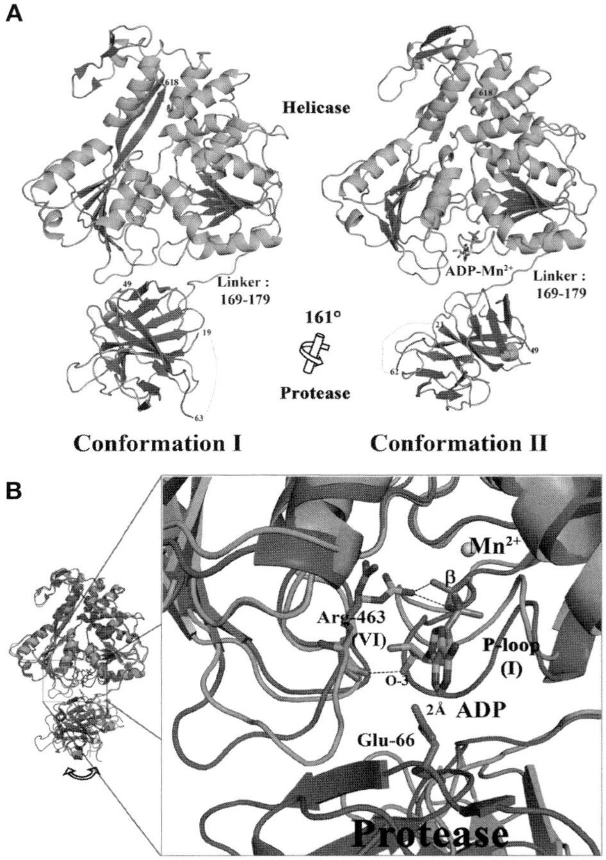

Figure 5.10 Alternative conformations for the flavivirus $NS2B_{18}$-NS3FL protein as revealed by X-ray crystallography. Both structures are compatible with studies in solution using X-ray scattering; see Luo *et al.*, 2010. The $NS2B_{18}$ co-factor that forms a β-strand is coloured in red. The region linking the protease and helicase (residues 169–179) is coloured in green. *N*-terminal residues are also labelled. The rotation axis that relates both protease domains is indicated with the corresponding angle. (B) ADP-Mn^{2+} was docked into $NS2B_{18}$-NS3 in conformation I (in magenta, PDB code 2VBC) and superimposed into $NS2B_{18}$-NS3 in conformation II (coloured in green, PDB code 2WHX) for comparison.

in NS3 helicase-ADP binary complex (Luo *et al.*, 2010). The closest distance between bound ADP and the protease domain is over 6Å, ruling out direct participation in binding by the protease domain. Interestingly, when the ADP-Mn^{2+} is docked into $NS2B_{18}$–NS3 conformation I, the adenine base is very close to the protease domain. The distance between the adenine ring to Glu-66 is 2 Å. In the helicase domain, P-loop (Motif I) and a segment comprising residues Arg-460 to Gln-471 (Motif VI) display different conformations in the two structures. In conformation I, the protein would have to undergo some local structural rearrangements in order to accommodate any nucleotide. The structural flexibility of $NS2B_{18}$-NS3 might thus be coupled to (or regulated by) its nucleotide binding states. Furthermore, since ATP hydrolysis is required to provide energy for effective RNA duplex unwinding, such structural switches could play a role in the regulation of $NS2B_{18}$–NS3 helicase activity. Mutagenesis studies suggests that the linker has evolved to its present length for an optimal coupling of both functions in the convoluted membranes or perinuclear vesicle

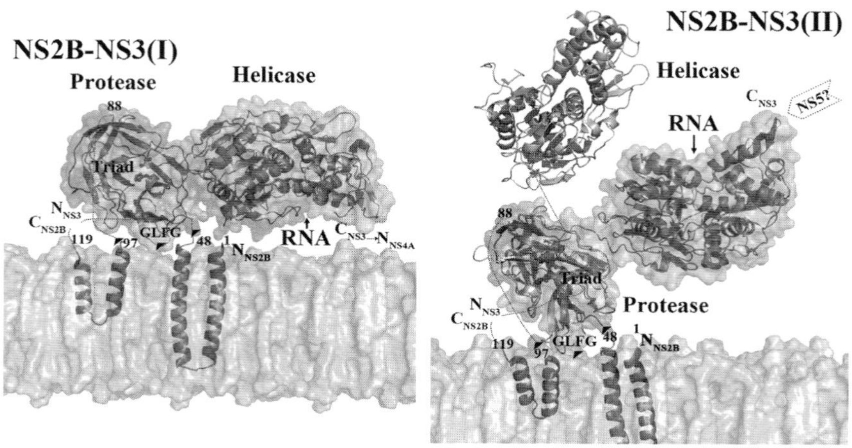

Figure 5.11 Model for the membrane-bound NS2B-NS3 protein complex. Left: Proposed association with the membrane of NS2B-NS3 when it adopts conformation I. Right: Membrane association of NS2B-NS3 (conformation II). See Luo *et al.*, 2010.

packets (Luo *et al.*, 2010). This is corroborated by our study in which a mutation introduced in the interdomain linker resulted in attenuation of viral replication (Luo *et al.*, 2010).

Based on secondary structure (Cserzo *et al.*, 1997; Hirokawa *et al.*, 1998) and deletion analyses of NS2B (Falgout *et al.*, 1993), residues N-terminal from 49 and C-terminal to 96 of NS2B are presumably inserted into the membrane forming two anchoring points. In addition, a hydrophobic loop (Gly-29–Gly-32) that projects from the NS3 protease domain is likely to face the lipid bilayer as well (Luo *et al.*, 2010), supporting a tripod-like structure for the NS2B–NS3 onto the membrane (Fig. 5.11). The latter loop which is conserved among flaviviruses, was proposed to play a role analogous to the amphipathic N-terminal helix α_0 of the HCV NS3-NS4A protease (Yan *et al.*, 1998; Aleshin *et al.*, 2007; Brass *et al.*, 2008). Thus, movements of the protease domain are probably quite restricted by a rather tight association with the lipid membrane (Fig. 5.11). The resulting orientation of the protease fully exposes its active site to the solvent and is compatible with unhindered proteolytic activity in the context of the lipid membrane of the infected cell. By contrast, the helicase domain of the NS2B-NS3 protein is connected to the protease domain through its highly flexible linker spanning aa 169–179 (Luo *et al.*, 2010). Thus, relative re-orientations between the protease and helicase domain, should allow a range of orientations for the helicase domain, where it would be either close to or away from the membrane (Fig. 5.11).

Intracellular localisation and interactions with other non-structural proteins

Several studies have shown that polyprotein processing and viral RNA replication take place at different intracellular locations in convoluted membranes (CM/PC) and vesicle packets (VP) respectively (Mackenzie *et al.*, 1998; Westaway *et al.*, 1999; Uchil and Satchidanandam, 2003; Denison, 2008). Thus, conformations adopted by NS2B-NS3 in CM/PC might be more favourable for viral polyprotein processing for which neither ATP hydrolysis, nor viral RNA binding and interaction with NS5 polymerase are needed. On the other hand, NS3 conformations within the viral replication complex in VP may be influenced by interactions with other viral non-structural proteins and be more compatible with the various functions performed by the helicase domain. Indeed, it has been reported that NS4B binds to NS3 helicase domain and enhances its unwinding activity (Umareddy *et al.*, 2006). In addition, NS4A was also proposed as a co-factor for NS3 helicase, sustaining its unwinding activity under conditions of ATP deficiency (Shiryaev *et al.*,

2009). The NS5 protein has also been shown to augment the NTPase and RTPase activities of NS3 (Yon *et al.*, 2005). Within infected cells, NS3 interacts with NS5 (Kapoor *et al.*, 1995; Uchil *et al.*, 2006). and this is probably mediated through the nuclear localization domain (β-NLS) of NS5 (Johansson *et al.*, 2001; Brooks *et al.*, 2002). NS3 and NS5 have been shown to colocalize in the nucleus of virally infected cells, although the precise significance of this observation is still unclear (Uchil *et al.*, 2006).

Non-enzymatic roles played by NS3

Whilst internal cleavage of NS2A by the NS2BNS3 protease is necessary for virus production (Nestorowicz *et al.*, 1994), a compensatory change at a second site in the helicase domain of NS3 (Asp 343 to an uncharged residue) restores infectious virus production in the absence of cleavage at the mutant NS2Aα site (Kummerer and Rice, 2002; Patkar and Kuhn, 2008). These findings suggest that physical association between NS3 and NS2A are essential for proper morphogenesis of infectious flavivirus particle and argue for a non-enzymatic role played by the NS3 protein. Moreover, mutation of the conserved residue Trp349 in the helicase domain abolishes production of viral particles without affecting viral protein and RNA production nor release of prM–E subviral particles, indicating that NS3 plays a role in virus assembly or secretion which is independent of its known enzymatic functions (Patkar and Kuhn, 2008). NS3 is also believed to play a role in inducing membranous structures observed in flavivirus-infected cells. Exogenous expression of DENV2 NS3 protein led to formation of membrane structures affiliated with the rough ER. NS3 was found to interact with nuclear receptor binding protein (NRBP), which influences trafficking between the endoplasmic reticulum (ER) and Golgi (Chua *et al.*, 2004) as well with microtubules and the tumour susceptibility gene 101 (TSG101) protein, which regulates microtubule networks (Chiou *et al.*, 2003). Thus, NS3 may function together with other non-structural proteins like NS2A and NS4, to form the vesicle packages that encapsidate the viral replication complexes.

Induction of apoptosis

Transfection of WNV full length NS2B-NS3, and either protease or helicase domains into cells triggered apoptotic pathways with activation of caspases-8 and -3. This consequently led to nuclear membrane ruptures and cleavage of the DNA-repair enzyme, PARP (Ramanathan *et al.*, 2006). Likewise, JEV NS2B-NS3pro was found to significantly inhibit AP-1 signalling pathway, and thus directly affect cell apoptosis (Lin *et al.*, 2006).

Viral pathogenicity associated with NS3

Recent observations have found an increasing frequency for neurological manifestations in dengue infections. Studies with two neurovirulent mice-adapted strains mapped mutations in E and NS3 helicase which enhanced their ability to replicate in the CNS of infected mice, causing extensive damage with leptomeningitis and encephalitis (Bordignon *et al.*, 2007). Interestingly, the same mutation in the helicase domain (Leu480Ser) was also previously implicated in the induction of apoptosis in dengue-infected cells (Duarte dos Santos *et al.*, 2000). Likewise, in JEV and TBE, a correlation between neurovirulence and NS2B-NS3pro mutations was proposed (McMinn, 1997; Chiou and Chen, 2001; Ruzek *et al.*, 2008). *in vitro* NS2B-NS3pro can directly cleave human protein proteins (Lin *et al.*, 2006), including the human myelin basic protein (Shiryaev *et al.*, 2006) causing disruptions in cellular functions.

Acknowledgements

We thank all our co-workers and collaborators whose named are listed in the original publications with a special mention to Drs Pei-Yong Shi, Ting Xu and Subhash G. Vasudevan. Work in the laboratory of J.L. has been generously supported over the years by the Singapore Biomedical Research Council and the National Research Foundation. Support from an ATIP grant of the CNRS is also gratefully acknowledged.

References

Aleshin, A.E., Shiryaev, S.A., Strongin, A.Y., and Liddington, R.C. (2007). Structural evidence for regulation and specificity of flaviviral proteases and evolution of the Flaviviridae fold. Protein Sci. *16*, 795–806.

Amberg, S.M., and Rice, C.M. (1999). Mutagenesis of the NS2B–NS3-mediated cleavage site in the flavivirus capsid protein demonstrates a requirement for coordinated processing. J. Virol. *73*, 8083–8094.

Amberg, S.M., Nestorowicz, A., McCourt, D.W., and Rice, C.M. (1994). NS2B-3 proteinase-mediated processing in the yellow fever virus structural region: *in vitro* and *in vivo* studies. J. Virol. *68*, 3794–3802.

Andersen, C.B., Ballut, L., Johansen, J.S., Chamieh, H., Nielsen, K.H., Oliveira, C.L., Pedersen, J.S., Seraphin, B., Le Hir, H., and Andersen, G.R. (2006). Structure of the exon junction core complex with a trapped DEAD-box ATPase bound to RNA. Science *313*, 1968–1972.

Arias, C.F., Preugschat, F., and Strauss, J.H. (1993). Dengue 2 virus NS2B and NS3 form a stable complex that can cleave NS3 within the helicase domain. Virology *193*, 888–899.

Assenberg, R., Mastrangelo, E., Walter, T.S., Verma, A., Milani, M., Owens, R.J., Stuart, D.I., Grimes, J.M., and Mancini, E.J. (2009). Crystal structure of a novel conformational state of the flavivirus NS3 protein: implications for polyprotein processing and viral replication. J. Virol. *83*, 12895–12906.

Bartelma, G., and Padmanabhan, R. (2002). Expression, purification, and characterization of the RNA 5′-triphosphatase activity of dengue virus type 2 nonstructural protein 3. Virology *299*, 122–132.

Benarroch, D., Selisko, B., Locatelli, G.A., Maga, G., Romette, J.L., and Canard, B. (2004). The RNA helicase, nucleotide 5′-triphosphatase, and RNA 5′-triphosphatase activities of dengue virus protein NS3 are Mg^{2+}-dependent and require a functional Walker B motif in the helicase catalytic core. Virology *328*, 208–218.

Bera, A.K., Kuhn, R.J., and Smith, J.L. (2007). Functional characterization of cis and trans activity of the flavivirus NS2B-NS3 protease. J. Biol. Chem. *282*, 12883–12892.

Betterton, M.D., and Julicher, F. (2005). Opening of nucleic-acid double strands by helicases: active versus passive opening. Phys. Rev. E. Stat. Nonlin. Soft Matter Phys. *71*, 011904.

Bleichert, F., and Baserga, S.J. (2007). The long unwinding road of RNA helicases. Mol. Cell *27*, 339–352.

Bono, F., Ebert, J., Lorentzen, E., and Conti, E. (2006). The crystal structure of the exon junction complex reveals how it maintains a stable grip on mRNA. Cell *126*, 713–725.

Bordignon, J., Strottmann, D.M., Mosimann, A.L., Probst, C.M., Stella, V., Noronha, L., Zanata, S.M., and Dos Santos, C.N. (2007). Dengue neurovirulence in mice: identification of molecular signatures in the E and NS3 helicase domains. J. Med. Virol. *79*, 1506–1517.

Borowski, P., Mueller, O., Niebuhr, A., Kalitzky, M., Hwang, L.H., Schmitz, H., Siwecka, M.A., and Kulikowsk, T. (2000). ATP-binding domain of NTPase/helicase as a target for hepatitis C antiviral therapy. Acta Biochim. Pol. *47*, 173–180.

Borowski, P., Niebuhr, A., Schmitz, H., Hosmane, R.S., Bretner, M., Siwecka, M.A., and Kulikowski, T. (2002). NTPase/helicase of Flaviviridae: inhibitors and inhibition of the enzyme. Acta Biochim. Pol. *49*, 597–614.

Brass, V., Berke, J.M., Montserret, R., Blum, H.E., Penin, F., and Moradpour, D. (2008). Structural determinants for membrane association and dynamic organization of the hepatitis C virus NS3–4A complex. Proc. Natl. Acad. Sci. U.S.A. *105*, 14545–14550.

Brooks, A.J., Johansson, M., John, A.V., Xu, Y., Jans, D.A., and Vasudevan, S.G. (2002). The interdomain region of dengue NS5 protein that binds to the viral helicase NS3 contains independently functional importin beta 1 and importin alpha/beta-recognized nuclear localization signals. J. Biol. Chem. *277*, 36399–36407.

Buttner, K., Nehring, S., and Hopfner, K.P. (2007). Structural basis for DNA duplex separation by a superfamily 2 helicase. Nat. Struct. Mol. Biol. *14*, 647–652.

Chambers, T.J., Weir, R.C., Grakoui, A., McCourt, D.W., Bazan, J.F., Fletterick, R.J., and Rice, C.M. (1990). Evidence that the N-terminal domain of nonstructural protein NS3 from yellow fever virus is a serine protease responsible for site-specific cleavages in the viral polyprotein. Proc. Natl. Acad. Sci. U.S.A. *87*, 8898–8902.

Chambers, T.J., Grakoui, A., and Rice, C.M. (1991). Processing of the yellow fever virus nonstructural polyprotein: a catalytically active NS3 proteinase domain and NS2B are required for cleavages at dibasic sites. J. Virol. *65*, 6042–6050.

Chambers, T.J., Nestorowicz, A., Amberg, S.M., and Rice, C.M. (1993). Mutagenesis of the yellow fever virus NS2B protein: effects on proteolytic processing, NS2B–NS3 complex formation, and viral replication. J. Virol. *67*, 6797–6807.

Chandramouli, S., Joseph, J.S., Daudenarde, S., Gatchalian, J., Cornillez-Ty, C., and Kuhn, P. (2010). Serotype-specific structural differences in the protease–co-factor complexes of the dengue virus family. J. Virol. *84*, 3059–3067.

Chappell, K.J., Stoermer, M.J., Fairlie, D.P., and Young, P.R. (2007). Generation and characterization of proteolytically active and highly stable truncated and full-length recombinant West Nile virus NS3. Protein Expr. Purif. *53*, 87–96.

Chappell, K.J., Stoermer, M.J., Fairlie, D.P., and Young, P.R. (2008). West Nile Virus NS2B/NS3 protease as an antiviral target. Curr. Med. Chem. *15*, 2771–2784.

Cheng, W., Dumont, S., Tinoco, I., Jr., and Bustamante, C. (2007). NS3 helicase actively separates RNA strands and senses sequence barriers ahead of the opening fork. Proc. Natl. Acad. Sci. U.S.A. *104*, 13954–13959.

Chernov, A.V., Shiryaev, S.A., Aleshin, A.E., Ratnikov, B.I., Smith, J.W., Liddington, R.C., and Strongin, A.Y. (2008). The two-component NS2B-NS3 proteinase represses DNA unwinding activity of the West Nile virus NS3 helicase. J. Biol. Chem. *283*, 17270–17278.

Chiou, S.S., and Chen, W.J. (2001). Mutations in the NS3 gene and 3′-NCR of Japanese encephalitis virus isolated from an unconventional ecosystem and implications for natural attenuation of the virus. Virology *289*, 129–136.

Chiou, C.T., Hu, C.C., Chen, P.H., Liao, C.L., Lin, Y.L., and Wang, J.J. (2003). Association of Japanese encephalitis virus NS3 protein with microtubules and tumour susceptibility gene 101 (TSG101) protein. J. Gen. Virol. *84*, 2795–2805.

Chua, J.J., Ng, M.M., and Chow, V.T. (2004). The non-structural 3 (NS3) protein of dengue virus type 2 interacts with human nuclear receptor binding protein and is associated with alterations in membrane structure. Virus Res. *102*, 151–163.

Clum, S., Ebner, K.E., and Padmanabhan, R. (1997). Cotranslational membrane insertion of the serine proteinase precursor NS2B-NS3(Pro) of dengue virus type 2 is required for efficient *in vitro* processing and is mediated through the hydrophobic regions of NS2B. J. Biol. Chem. *272*, 30715–30723.

Cordin, O., Banroques, J., Tanner, N.K., and Linder, P. (2006). The DEAD-box protein family of RNA helicases. Gene *367*, 17–37.

Cserzo, M., Wallin, E., Simon, I., von Heijne, G., and Elofsson, A. (1997). Prediction of transmembrane alpha-helices in prokaryotic membrane proteins: the dense alignment surface method. Protein Eng. *10*, 673–676.

Del Campo, M., and Lambowitz, A.M. (2009). Structure of the Yeast DEAD box protein Mss116p reveals two wedges that crimp RNA. Mol. Cell *35*, 598–609.

Denison, M.R. (2008). Seeking membranes: positive-strand RNA virus replication complexes. PLoS Biol. *6*, e270.

Dong, H., Zhang, B., and Shi, P.Y. (2008). Flavivirus methyltransferase: a novel antiviral target. Antiviral. Res. *80*, 1–10.

Droll, D.A., Krishna Murthy, H.M., and Chambers, T.J. (2000). Yellow fever virus NS2B-NS3 protease: charged-to-alanine mutagenesis and deletion analysis define regions important for protease complex formation and function. Virology *275*, 335–347.

Duarte dos Santos, C.N., Frenkiel, M.P., Courageot, M.P., Rocha, C.F., Vazeille-Falcoz, M.C., Wien, M.W., Rey, F.A., Deubel, V., and Despres, P. (2000). Determinants in the envelope E protein and viral RNA helicase NS3 that influence the induction of apoptosis in response to infection with dengue type 1 virus. Virology *274*, 292–308.

Dumont, S., Cheng, W., Serebrov, V., Beran, R.K., Tinoco, I., Jr., Pyle, A.M., and Bustamante, C. (2006). RNA translocation and unwinding mechanism of HCV NS3 helicase and its coordination by ATP. Nature *439*, 105–108.

Enemark, E.J., and Joshua-Tor, L. (2008). On helicases and other motor proteins. Curr. Opin. Struct. Biol. *18*, 243–257.

Erbel, P., Schiering, N., D'Arcy, A., Renatus, M., Kroemer, M., Lim, S.P., Yin, Z., Keller, T.H., Vasudevan, S.G., and Hommel, U. (2006). Structural basis for the activation of flaviviral NS3 proteases from dengue and West Nile virus. Nat. Struct. Mol. Biol. *13*, 372–373.

Falgout, B., Pethel, M., Zhang, Y.M., and Lai, C.J. (1991). Both nonstructural proteins NS2B and NS3 are required for the proteolytic processing of dengue virus nonstructural proteins. J. Virol. *65*, 2467–2475.

Falgout, B., Miller, R.H., and Lai, C.J. (1993). Deletion analysis of dengue virus type 4 nonstructural protein NS2B: identification of a domain required for NS2B-NS3 protease activity. J. Virol. *67*, 2034–2042.

Frick, D.N., Rypma, R.S., Lam, A.M., and Frenz, C.M. (2004). Electrostatic analysis of the hepatitis C virus NS3 helicase reveals both active and allosteric site locations. Nucleic Acids Res. *32*, 5519–5528.

Gorbalenya, A.E., and Koonin, E.V. (1993). Helicases: amino acid sequence comparisons and structure–function relationships. Curr. Opin. Struct. Biol. *3*, 419–429.

Grassmann, C.W., Isken, O., and Behrens, S.E. (1999). Assignment of the multifunctional NS3 protein of bovine viral diarrhea virus during RNA replication: an *in vivo* and *in vitro* study. J. Virol. *73*, 9196–9205.

Gross, C.H., and Shuman, S. (1996). Vaccinia virus RNA helicase: nucleic acid specificity in duplex unwinding. J. Virol. *70*, 2615–2619.

Gu, M., and Rice, C.M. (2010). Three conformational snapshots of the hepatitis C virus NS3 helicase reveal a ratchet translocation mechanism. Proc. Natl. Acad. Sci. U.S.A. *107*, 521–528.

Hirokawa, T., Boon-Chieng, S., and Mitaku, S. (1998). SOSUI: classification and secondary structure prediction system for membrane proteins. Bioinformatics *14*, 378–379.

Jankowsky, E., and Fairman, M.E. (2007). RNA helicases – one fold for many functions. Curr. Opin. Struct. Biol. *17*, 316–324.

Jankowsky, E., Gross, C.H., Shuman, S., and Pyle, A.M. (2000). The DExH protein NPH-II is a processive and directional motor for unwinding RNA. Nature *403*, 447–451.

Johansson, M., Brooks, A.J., Jans, D.A., and Vasudevan, S.G. (2001). A small region of the dengue virus-encoded RNA-dependent RNA polymerase, NS5, confers interaction with both the nuclear transport receptor importin-beta and the viral helicase, NS3. J. Gen. Virol. *82*, 735–745.

Kapoor, M., Zhang, L., Ramachandra, M., Kusukawa, J., Ebner, K.E., and Padmanabhan, R. (1995). Association between NS3 and NS5 proteins of dengue virus type 2 in the putative RNA replicase is linked to differential phosphorylation of NS5. J. Biol. Chem. *270*, 19100–19106.

Khumthong, R., Angsuthanasombat, C., Panyim, S., and Katzenmeier, G. (2002). *in vitro* determination of dengue virus type 2 NS2B-NS3 protease activity with fluorescent peptide substrates. J. Biochem. Mol. Biol. *35*, 206–212.

Khumthong, R., Niyomrattanakit, P., Chanprapaph, S., Angsuthanasombat, C., Panyim, S., and Katzenmeier, G. (2003). Steady-state cleavage kinetics for dengue virus type 2 ns2b-ns3(pro) serine protease with synthetic peptides. Protein Pept. Lett. *10*, 19–26.

Kim, J.L., Morgenstern, K.A., Lin, C., Fox, T., Dwyer, M.D., Landro, J.A., Chambers, S.P., Markland, W., Lepre, C.A., O'Malley, E.T., *et al.* (1996). Crystal structure of the hepatitis C virus NS3 protease domain complexed with a synthetic NS4A co-factor peptide. Cell *87*, 343–355.

Kim, J.L., Morgenstern, K.A., Griffith, J.P., Dwyer, M.D., Thomson, J.A., Murcko, M.A., Lin, C., and Caron, P.R. (1998). Hepatitis C virus NS3 RNA helicase domain with a bound oligonucleotide: the crystal structure provides insights into the mode of unwinding. Structure *6*, 89–100.

Knox, J.E., Ma, N.L., Yin, Z., Patel, S.J., Wang, W.L., Chan, W.L., Ranga Rao, K.R., Wang, G., Ngew, X., Patel, V., Beer, D., Lim S.P., Vasudevan, S.G., and Keller, T.H. (2006). Peptide inhibitors of West Nile NS3 protease: SAR study of tetrapeptide aldehyde inhibitors. J. Med. Chem *49*, 6585–90.

Kummerer, B.M., and Rice, C.M. (2002). Mutations in the yellow fever virus nonstructural protein NS2A selectively block production of infectious particles. J. Virol. *76*, 4773–4784.

Lee, J.Y., and Yang, W. (2006). UvrD helicase unwinds DNA one base pair at a time by a two-part power stroke. Cell *127*, 1349–1360.

Lescar, J., Luo, D., Xu, T., Sampath, A., Lim, S.P., Canard, B., and Vasudevan, S.G. (2008). Towards the design of antiviral inhibitors against flaviviruses: the case for the multifunctional NS3 protein from dengue virus as a target. Antiviral Res. *80*, 94–101.

Leung, D., Schroder, K., White, H., Fang, N.X., Stoermer, M.J., Abbenante, G., Martin, J.L., Young, P.R., and Fairlie, D.P. (2001). Activity of recombinant dengue 2 virus NS3 protease in the presence of a truncated NS2B co-factor, small peptide substrates, and inhibitors. J. Biol. Chem. *276*, 45762–45771.

Levin, M.K., Gurjar, M., and Patel, S.S. (2005). A Brownian motor mechanism of translocation and strand separation by hepatitis C virus helicase. Nat. Struct. Mol. Biol. *12*, 429–435.

Li, H., Clum, S., You, S., Ebner, K.E., and Padmanabhan, R. (1999). The serine protease and RNA-stimulated nucleoside triphosphatase and RNA helicase functional domains of dengue virus type 2 NS3 converge within a region of 20 amino acids. J. Virol. *73*, 3108–3116.

Li, J., Lim, S.P., Beer, D., Patel, V., Wen, D., Tumanut, C., Tully, D.C., Williams, J.A., Jiricek, J., Priestle, J.P., *et al.* (2005). Functional profiling of recombinant NS3 proteases from all four serotypes of dengue virus using tetrapeptide and octapeptide substrate libraries. J. Biol. Chem. *280*, 28766–28774.

Lin, C., Amberg, S.M., Chambers, T.J., and Rice, C.M. (1993). Cleavage at a novel site in the NS4A region by the yellow fever virus NS2B-3 proteinase is a prerequisite for processing at the downstream 4A/4B signalase site. J. Virol. *67*, 2327–2335.

Lin, C.W., Lin, K.H., Lyu, P.C., and Chen, W.J. (2006). Japanese encephalitis virus NS2B-NS3 protease binding to phage-displayed human brain proteins with the domain of trypsin inhibitor and basic region leucine zipper. Virus Res. *116*, 106–113.

Lindenbach, B.D., Thiel, H.J., and Rice, C.M. (2007). Flaviviridae: the viruses and their replication. (Lippincott-Raven Publishers, Philadelphia).

Lobigs, M. (1993). Flavivirus premembrane protein cleavage and spike heterodimer secretion require the function of the viral proteinase NS3. Proc. Natl. Acad. Sci. U.S.A. *90*, 6218–6222.

Locatelli, G.A., Spadari, S., and Maga, G. (2002). Hepatitis C virus NS3 ATPase/helicase: an ATP switch regulates the cooperativity among the different substrate binding sites. Biochemistry *41*, 10332–10342.

Löhr, K., Knox, J.E., Phong, W.Y., Ma, N.L., Yin, Z., Sampath, A., Patel, S.J., Wang, W., Chan, W., Ranga Rao, K.R., Wang, G., Vasudevan, S.G., Keller, T.H., and Lim, S.P. (2007). Yellow fever virus NS3 protease: peptide-inhibition studies. J. Gen. Virol. *88*, 2223–7.

Love, R.A., Parge, H.E., Wickersham, J.A., Hostomsky, Z., Habuka, N., Moomaw, E.W., Adachi, T., and Hostomska, Z. (1996). The crystal structure of hepatitis C virus NS3 proteinase reveals a trypsin-like fold and a structural zinc binding site. Cell *87*, 331–342.

Luo, D., Xu, T., Hunke, C., Gruber, G., Vasudevan, S.G., and Lescar, J. (2008a). Crystal structure of the NS3 protease-helicase from dengue virus. J. Virol. *82*, 173–183.

Luo, D., Xu, T., Watson, R.P., Scherer-Becker, D., Sampath, A., Jahnke, W., Yeong, S.S., Wang, C.H., Lim, S.P., Strongin, A., *et al.* (2008b). Insights into RNA unwinding and ATP hydrolysis by the flavivirus NS3 protein. EMBO J. *27*, 3209–3219.

Luo, D., Wei, N., Doan, D.N., Paradkar, P.N., Chong, Y., Davidson, A.D., Kotaka, M., Lescar, J., and Vasudevan, S.G. (2010). Flexibility between the protease and helicase domains of the dengue virus NS3 protein conferred by the linker region and its functional implications. J. Biol. Chem. *285*, 18817–18827.

Mackenzie, J.M., Khromykh, A.A., Jones, M.K., and Westaway, E.G. (1998). Subcellular localization and some biochemical properties of the flavivirus Kunjin nonstructural proteins NS2A and NS4A. Virology *245*, 203–215.

McMinn, P.C. (1997). The molecular basis of virulence of the encephalitogenic flaviviruses. J. Gen. Virol. *78 (Pt 11)*, 2711–2722.

Mancini, E.J., Assenberg, R., Verma, A., Walter, T.S., Tuma, R., Grimes, J.M., Owens, R.J., and Stuart, D.I. (2007). Structure of the Murray Valley encephalitis virus RNA helicase at 1.9 Angstrom resolution. Protein Sci. *16*, 2294–2300.

Mastrangelo, E., Bollati, M., Milani, M., Brisbarre, N., de Lamballerie, X., Coutard, B., Canard, B., Khromykh, A., and Bolognesi, M. (2006). Preliminary crystallographic characterization of an RNA helicase from Kunjin virus. Acta Crystallogr. Sect. F. Struct. Biol. Cryst. Commun. *62*, 876–879.

Mastrangelo, E., Milani, M., Bollati, M., Selisko, B., Peyrane, F., Pandini, V., Sorrentino, G., Canard, B., Konarev, P.V., Svergun, D.I., *et al.* (2007). Crystal structure and activity of Kunjin virus NS3 helicase;

protease and helicase domain assembly in the full length NS3 protein. J. Mol. Biol. *372*, 444–455.

Matusan, A.E., Pryor, M.J., Davidson, A.D., and Wright, P.J. (2001). Mutagenesis of the dengue virus type 2 NS3 protein within and outside helicase motifs: effects on enzyme activity and virus replication. J. Virol. *75*, 9633–9643.

Mueller, N.H., Yon, C., Ganesh, V.K., and Padmanabhan, R. (2007). Characterization of the West Nile virus protease substrate specificity and inhibitors. Int. J. Biochem. Cell Biol. *39*, 606–614.

Murthy, H.M., Clum, S., and Padmanabhan, R. (1999). Dengue virus NS3 serine protease. Crystal structure and insights into interaction of the active site with substrates by molecular modeling and structural analysis of mutational effects. J. Biol. Chem. *274*, 5573–5580.

Myong, S., Bruno, M.M., Pyle, A.M., and Ha, T. (2007). Spring-loaded mechanism of DNA unwinding by hepatitis C virus NS3 helicase. Science *317*, 513–516.

Nall, T.A., Chappell, K.J., Stoermer, M.J., Fang, N.X., Tyndall, J.D., Young, P.R., and Fairlie, D.P. (2004). Enzymatic characterization and homology model of a catalytically active recombinant West Nile virus NS3 protease. J. Biol. Chem. *279*, 48535–48542.

Nestorowicz, A., Chambers, T.J., and Rice, C.M. (1994). Mutagenesis of the yellow fever virus NS2A/2B cleavage site: effects on proteolytic processing, viral replication, and evidence for alternative processing of the NS2A protein. Virology *199*, 114–123.

Nielsen, K.H., Chamieh, H., Andersen, C.B., Fredslund, F., Hamborg, K., Le Hir, H., and Andersen, G.R. (2009). Mechanism of ATP turnover inhibition in the EJC. RNA *15*, 67–75.

Noble, C.G., Chen, Y.L., Dong, H., Gu, F., Lim, S.P., Schul, W., Wang, Q.Y., and Shi, P.Y. (2010). Strategies for development of dengue virus inhibitors. Antiviral Res. *85*, 450–62.

Pang, P.S., Jankowsky, E., Planet, P.J., and Pyle, A.M. (2002). The hepatitis C viral NS3 protein is a processive DNA helicase with co-factor enhanced RNA unwinding. EMBO J. *21*, 1168–1176.

Patkar, C.G., and Kuhn, R.J. (2008). Yellow Fever virus NS3 plays an essential role in virus assembly independent of its known enzymatic functions. J. Virol. *82*, 3342–3352.

Preugschat, F., Yao, C.W., and Strauss, J.H. (1990). *In vitro* processing of dengue virus type 2 nonstructural proteins NS2A, NS2B, and NS3. J. Virol. *64*, 4364–4374.

Preugschat, F., Averett, D.R., Clarke, B.E., and Porter, D.J. (1996). A steady-state and pre-steady-state kinetic analysis of the NTPase activity associated with the hepatitis C virus NS3 helicase domain. J. Biol. Chem. *271*, 24449–24457.

Pyle, A.M. (2008). Translocation and Unwinding Mechanisms of RNA and DNA Helicases. Annu. Rev. Biophys. *37*, 317–336.

Ramanathan, M.P., Chambers, J.A., Pankhong, P., Chattergoon, M., Attatippaholkun, W., Dang, K., Shah, N., and Weiner, D.B. (2006). Host cell killing by the West Nile Virus NS2B-NS3 proteolytic complex: NS3 alone is sufficient to recruit caspase-8-based apoptotic pathway. Virology *345*, 56–72.

Robin, G., Chappell, K., Stoermer, M.J., Hu, S.H., Young, P.R., Fairlie, D.P., and Martin, J.L. (2009). Structure of West Nile virus NS3 protease: ligand stabilization of the catalytic conformation. J. Mol. Biol. *385*, 1568–1577.

Ruzek, D., Bell-Sakyi, L., Kopecky, J., and Grubhoffer, L. (2008). Growth of tick-borne encephalitis virus (European subtype) in cell lines from vector and non-vector ticks. Virus Res. *137*, 142–146.

Sampath, A., Xu, T., Chao, A., Luo, D., Lescar, J., and Vasudevan, S.G. (2006). Structure-based mutational analysis of the NS3 helicase from dengue virus. J. Virol. *80*, 6686–6690.

Sampath A., and Padmanabhan, R. (2009). Molecular targets for flavivirus drug discovery. Antiviral Res. *81*, 6–15.

Sengoku, T., Nureki, O., Nakamura, A., Kobayashi, S., and Yokoyama, S. (2006). Structural basis for RNA unwinding by the DEAD-box protein Drosophila Vasa. Cell *125*, 287–300.

Shiryaev, S.A., Ratnikov, B.I., Chekanov, A.V., Sikora, S., Rozanov, D.V., Godzik, A., Wang, J., Smith, J.W., Huang, Z., Lindberg, I., *et al.* (2006). Cleavage targets and the D-arginine-based inhibitors of the West Nile virus NS3 processing proteinase. Biochem. J. *393*, 503–511.

Shiryaev, S.A., Chernov, A.V., Aleshin, A.E., Shiryaeva, T.N., and Strongin, A.Y. (2009). NS4A regulates the ATPase activity of the NS3 helicase: a novel co-factor role of the non-structural protein NS4A from West Nile virus. J. Gen. Virol. *90*, 2081–2085.

Soultanas, P., Dillingham, M.S., Velankar, S.S., and Wigley, D.B. (1999). DNA binding mediates conformational changes and metal ion coordination in the active site of PcrA helicase. J. Mol. Biol. *290*, 137–148.

Speroni, S., De Colibus, L., Mastrangelo, E., Gould, E., Coutard, B., Forrester, N.L., Blanc, S., Canard, B., and Mattevi, A. (2008). Structure and biochemical analysis of Kokobera virus helicase. Proteins *70*, 1120–1123.

Su, X.C., Ozawa, K., Qi, R., Vasudevan, S.G., Lim, S.P., and Otting, G. (2009). NMR analysis of the dynamic exchange of the NS2B co-factor between open and closed conformations of the West Nile virus NS2B-NS3 protease. PLoS Negl. Trop. Dis. *3*, e561.

Tai, C.L., Chi, W.K., Chen, D.S., and Hwang, L.H. (1996). The helicase activity associated with hepatitis C virus nonstructural protein 3 (NS3). J. Virol. *70*, 8477–8484.

Tanner, N.K., and Linder, P. (2001). DExD/H box RNA helicases: from generic motors to specific dissociation functions. Mol. Cell *8*, 251–262.

Teo, K.F., and Wright, P.J. (1997). Internal proteolysis of the NS3 protein specified by dengue virus 2. J. Gen. Virol. *78 (Pt 2)*, 337–341.

Theis, K., Chen, P.J., Skorvaga, M., Van Houten, B., and Kisker, C. (1999). Crystal structure of UvrB, a DNA helicase adapted for nucleotide excision repair. EMBO J. *18*, 6899–6907.

Uchil, P.D., and Satchidanandam, V. (2003). Architecture of the flaviviral replication complex. Protease, nuclease, and detergents reveal encasement within double-layered membrane compartments. J. Biol. Chem. *278*, 24388–24398.

Uchil, P.D., Kumar, A.V., and Satchidanandam, V. (2006). Nuclear localization of flavivirus RNA synthesis in infected cells. J. Virol. *80*, 5451–5464.

Umareddy, I., Chao, A., Sampath, A., Gu, F., and Vasudevan, S.G. (2006). Dengue virus NS4B interacts with NS3 and dissociates it from single-stranded RNA. J. Gen. Virol. *87*, 2605–2614.

Valle, R.P., and Falgout, B. (1998). Mutagenesis of the NS3 protease of dengue virus type 2. J. Virol. *72*, 624–632.

Velankar, S.S., Soultanas, P., Dillingham, M.S., Subramanya, H.S., and Wigley, D.B. (1999). Crystal structures of complexes of PcrA DNA helicase with a DNA substrate indicate an inchworm mechanism. Cell *97*, 75–84.

Wang, C.C., Huang, Z.S., Chiang, P.L., Chen, C.T., and Wu, H.N. (2009). Analysis of the nucleoside triphosphatase, RNA triphosphatase, and unwinding activities of the helicase domain of dengue virus NS3 protein. FEBS Lett *583*, 691–696.

Wengler, G., Czaya, G., Farber, P.M., and Hegemann, J.H. (1991). *In vitro* synthesis of West Nile virus proteins indicates that the amino-terminal segment of the NS3 protein contains the active centre of the protease which cleaves the viral polyprotein after multiple basic amino acids. J. Gen. Virol. *72 (Pt 4)*, 851–858.

Westaway, E.G., Mackenzie, J.M., Kenney, M.T., Jones, M.K., and Khromykh, A.A. (1997). Ultrastructure of Kunjin virus-infected cells: colocalization of NS1 and NS3 with double-stranded RNA, and of NS2B with NS3, in virus-induced membrane structures. J. Virol. *71*, 6650–6661.

Westaway, E.G., Khromykh, A.A., and Mackenzie, J.M. (1999). Nascent flavivirus RNA colocalized *in situ* with double-stranded RNA in stable replication complexes. Virology *258*, 108–117.

Westaway, E.G., Mackenzie, J.M., and Khromykh, A.A. (2003). Kunjin RNA replication and applications of Kunjin replicons. Adv. Virus Res. *59*, 99–140.

Wu, J., Bera, A.K., Kuhn, R.J., and Smith, J.L. (2005). Structure of the flavivirus helicase: implications for catalytic activity, protein interactions, and proteolytic processing. J. Virol. *79*, 10268–10277.

Xiao, M., Bai, Y., Xu, H., Geng, X., Chen, J., Wang, Y., Chen, J., and Li, B. (2008). Effect of NS3 and NS5B proteins on classical swine fever virus internal ribosome entry site-mediated translation and its host cellular translation. J. Gen. Virol. *89*, 994–999.

Xu, T., Sampath, A., Chao, A., Wen, D., Nanao, M., Chene, P., Vasudevan, S.G., and Lescar, J. (2005). Structure of the dengue virus helicase/nucleoside triphosphatase catalytic domain at a resolution of 2.4 A. J. Virol. *79*, 10278–10288.

Xu, T., Sampath, A., Chao, A., Wen, D., Nanao, M., Luo, D., Chene, P., Vasudevan, S.G., and Lescar, J. (2006). Towards the design of flavivirus helicase/NTPase inhibitors: crystallographic and mutagenesis studies of the dengue virus NS3 helicase catalytic domain. Novartis Found. Symp. *277*, 87–97; discussion *97–101*, 251–103.

Yamashita, T., Unno, H., Mori, Y., Tani, H., Moriishi, K., Takamizawa, A., Agoh, M., Tsukihara, T., and Matsuura, Y. (2008). Crystal structure of the catalytic domain of Japanese encephalitis virus NS3 helicase/nucleoside triphosphatase at a resolution of 1.8 A. Virology *373*, 426–436.

Yamshchikov, V.F., Trent, D.W., and Compans, R.W. (1997). Up-regulation of signalase processing and induction of prM–E secretion by the flavivirus NS2B-NS3 protease: roles of protease components. J. Virol. *71*, 4364–4371.

Yan, Y., Li, Y., Munshi, S., Sardana, V., Cole, J.L., Sardana, M., Steinkuehler, C., Tomei, L., De Francesco, R., Kuo, L.C., and Chen, Z. (1998). Complex of NS3 protease and NS4A peptide of BK strain hepatitis C virus: a 2.2 A resolution structure in a hexagonal crystal form. Protein Sci. *7*, 837–847.

Yang, Q., and Jankowsky, E. (2006). The DEAD-box protein Ded1 unwinds RNA duplexes by a mode distinct from translocating helicases. Nat. Struct. Mol. Biol. *13*, 981–986.

Yao, N., Hesson, T., Cable, M., Hong, Z., Kwong, A.D., Le, H.V., and Weber, P.C. (1997). Structure of the hepatitis C virus RNA helicase domain. Nat. Struct. Biol. *4*, 463–467.

Yao, N., Reichert, P., Taremi, S.S., Prosise, W.W., and Weber, P.C. (1999). Molecular views of viral polyprotein processing revealed by the crystal structure of the hepatitis C virus bifunctional protease-helicase. Structure *7*, 1353–1363.

Yon, C., Teramoto, T., Mueller, N., Phelan, J., Ganesh, V.K., Murthy, K.H., and Padmanabhan, R. (2005). Modulation of the nucleoside triphosphatase/RNA helicase and 5′-RNA triphosphatase activities of dengue virus type 2 nonstructural protein 3 (NS3) by interaction with NS5, the RNA-dependent RNA polymerase. J. Biol. Chem. *280*, 27412–27419.

Yusof, R., Clum, S., Wetzel, M., Murthy, H.M., and Padmanabhan, R. (2000). Purified NS2B/NS3 serine protease of dengue virus type 2 exhibits co-factor NS2B dependence for cleavage of substrates with dibasic amino acids *in vitro*. J. Biol. Chem. *275*, 9963–9969.

Zhang, L., Mohan, P.M., and Padmanabhan, R. (1992). Processing and localization of dengue virus type 2 polyprotein precursor NS3-NS4A-NS4B-NS5. J. Virol. *66*, 7549–7554.

Structure and Function of the Flavivirus NS5 Protein

6

Julien Lescar, Siew Pheng Lim and Pei-Yong Shi

Abstract

The non-structural protein 5 (NS5) of flaviviruses is the most conserved amongst the viral proteins. It is about 900 kDa and bears several enzymatic activities that play vital roles in virus replication. Its N-terminal domain encodes dual N7 and 2′-*O* methyltransferase activities (MTase), and possibly guanylyltransferase (GTase) involved in RNA cap formation. The C-terminal region comprises a RNA-dependent RNA polymerase (RdRp) required for viral RNA synthesis. Numerous crystal structures of the flavivirus MTase and RdRp domains have been solved. MTase in complex with S-adenosyl homocysteine (SAHC) or GTP analogues showed that the domain adopts a classical 2′-O MTase fold, however, the mechanism by which the protein can perform N7-methylation is still unknown. Besides its critical enzymatic activities, NS5 has also been implicated in viral pathogenesis through phosphorylation by host cell kinases, nucleus trafficking, and interference with interferon signalling and cytokine production. Here we summarize recent progress on this highly intriguing protein.

Introduction

The mRNA of flaviviruses contains a type 1 cap structure, followed by conserved AG dinucleotide at the 5′ end (m^7GpppAmG; Cleaves and Dublin, 1979; Wengler and Wengler, 1981) and is essential for mRNA stability and effective translation during infection (Furuchi and Shatkin, 2000; Shuman, 2001). Inside infected cells, a four-step cotranscriptional modification adds the 5′ cap onto flaviviral genomic RNAs (Dong *et al.*, 2008). First, the 5′ triphosphate of the plus-sense RNA is hydrolysed to a 5′ diphosphate by RNA triphosphatase (encoded by NS3) (Wengler and Wengler, 1981; Bartelma and Padmanabhan, 2002). Next, the GMP moiety from GTP is transferred to the 5′ diphosphate by an RNA guanylyltransferase (possibly resides in the N-terminal portion of NS5). The subsequent steps involve methylation at N-7 position of the guanine cap followed by methylation at the 2′-*O* position of the first nucleotide (adenosine) (Ray *et al.* 2006; Dong *et al.* 2008; Chung *et al.* 2010). Both N-7 and 2′-*O*-methylation are performed by a single MTase domain encoded within the first 300 amino acids (aa) of the NS5 protein (Koonin, 1993; Egloff *et al.*, 2002; Ray *et al.*, 2006; Dong *et al.*, 2008a) and utilize S-adenosyl-L-methionine (SAM) as the methyl donor. The methylation reaction generates S-adenosyl-L-homocysteine (SAHC) as a by-product.

The viral RNA comprises a long open reading frame flanked by 5′ and 3′ untranslated regions (UTR) which contain several conserved and complementary structures including several stem-loops (reviewed in Markoff 2003; Villordo and Gamarnik, 2009). Cyclization of these sequences is essential for viral RNA synthesis by the viral polymerase, which is present in the C-terminal 600 amino acids of the NS5 protein (RdRp; Ackermann and Padmanabhan, 2001). RdRp activity was first demonstrated *in vitro* with DENV2-infected cell lysates (Bartholomeusz *et al.*, 1993; You and Padmanabhan, 1999) and recombinant DENV1 RdRp protein (Tan *et al.*, 1996). Other flavivirus NS5 were subsequently expressed and *in vitro* activity was also shown

(Ackermann and Padmanabhan, 2001; Guyatt *et al.*, 2001; Nomaguchi *et al.*, 2003, 2004). NS5 alone exhibits specificity for the viral RNA and is able to initiate *de novo* minus-strand RNA synthesis without other viral or host cofactors (Ackermann and Padmanabhan, 2001; Nomaguchi *et al.*, 2004; Filomatori *et al.*, 2006; Selisko *et al.*, 2006). NS5 has been shown to interact with the viral RNA, other viral non-structural proteins, particularly NS3, as well as host proteins. These interactions are important for viral RNA replication, and formation of the viral replication complex. NS5 interaction with host proteins further controls of its subcellular localization, its phosphorylation states and its role in viral pathogenesis.

Thus, given its crucial role in the virus replication cycle, NS5 is a highly attractive therapeutic target for anti-dengue interventions. In the last few years, significant progresses have been made towards the understanding of flavivirus polymerases, including the three-dimensional structure, its enzymatic mechanism, the role of divalent metal ions in catalysis, and the interaction between the NS5 polymerase with other non-structural (NS) viral proteins as well as with the viral RNA genome. A number of reviews have recently been published on flavivirus NS5 (Davidson *et al.*, 2009), MTase (Dong *et al.*, 2008b) and RdRp (Lescar and Canard, 2009).

Biological roles played by NS5 protein

The methyltransferase domain (MTase)

Experimental evidence of dual N7 and 2′-*O* MTase activities

The flavivirus MTase domain when expressed as a recombinant protein contains the N7 methylation (Ray *et al.*, 2006) and the 2′-*O* methylation activities (Egloff *et al.*, 2002). The two methylation activities are independently of the polymerase activity. The presence of the RdRp domain does not enhance the MTase activity *in vitro* (Ray *et al.*, 2006). It is inactive on uncapped RNA and performs primarily 2′-*O* methylation when tested with GTP capped short RNA sequences *in vitro*, with a minimal requirement of six nucleotides (Bartelma and Padmanabhan, 2002; Benarroch *et al.*, 2004; Egloff *et al.*, 2007; Lim *et al.*, 2008; Selisko *et al.*, 2009). Methylation can occur on GTP capped RNA oligonucleotides bearing the conserved AG dinucleotides (Lim *et al.*, 2008) or with only the penultimate conserved A (Egloff *et al.*, 2007; Selisko *et al.*, 2009). In the presence of longer RNA templates corresponding to the authentic viral sequence, the same domain can catalyse both N7-and 2′-*O* methylation (Ray *et al.*, 2006; Kroschewski *et al.*, 2008; Milani *et al.*, 2009; Selisko *et al.*, 2009; Chung *et al.* 2010). This dual methylation function has been demonstrated for several flaviviruses including DENV1–4, WNV, YFV, Powassan virus (Dong *et al.*, 2007; Zhou *et al.*, 2007). N7 activity can only take place on the authentic virus 5′ UTR sequence and not unrelated capped RNA (Ray *et al.*, 2006). Interestingly, WNV N7 activity can occur on RNA templates comprising 74 nt of the viral 5′ UTR bearing the SLA sequence whilst the same activity in DENV requires at least 110 nt containing both SLA and SLB (Chung *et al.*, 2010). WNV N7 activity requires wild-type nucleotides at the second (G) and third (U) positions whilst 2′-*O* methylation requires wild-type nucleotides at the first (A) and second (G) positions (Dong *et al.*, 2007).

Characteristics of N7 and 2′-*O* MTase activities

Biochemical characterizations showed that *in vitro* N7 and 2′-*O* methylation activities require different optimal buffer conditions (Ray *et al.*, 2006; Dong *et al.*, 2007; Zhou *et al.*, 2007; Chung *et al.*, 2010). Specifically, N7 methylation is optimal around pH 7, and tolerated 0–250 mM NaCl, whilst 2′-*O* methylation is optimal at pH 9–10 with 5–10 mM $MgCl_2$ where N7 methylation is only at 30–50% compared with its activity at pH 7 (Zhou *et al.*, 2007). Addition of divalent cations were found to be detrimental to N7 activity (Zhou *et al.*, 2007; Chung *et al.*, 2010) whilst addition of NaCl inhibited 2′-*O* activity. Nevertheless, sequential N7 to 2′-*O* methylation, in the order GpppA-RNA → m7GpppA-RNA → m7GpppAm-RNA, can be obtained at neutral pH and with low NaCl (20 mM) and the reactions occurred via a random bi bi and processive mechanism that does not involve enzyme-RNA dissociation (Chung *et al.*, 2010). It is envisaged that following N7 methylation, repositioning of the N7-methylated cap structure into the GTP-binding pocket of the MTase domain enables 2′-*O* methylation to take place (Dong *et al.*, 2008a). Thus, it is highly probable that targeting flavivirus N7-activity, will also prevent 2′-*O* activity from proceeding. *In vitro* experiments demonstrated that N7 precedes 2′-*O* with GTP capped RNA being turned over faster than m7GTP capped RNA (Chung *et al.*, 2010). Flavivirus MTases bear a conserved KDKE motif that is the catalytic tetrad for all S-adenosyl methionine (SAM) dependent 2′-*O* MTases, including in vaccinia virus VP39 (Schnierle *et al.*, 1994). Mutagenesis studies show that the WNV enzyme requires the entire $K_{61}D_{146}K_{182}E_{218}$ motif for 2′-*O* activity, with K_{182} most probably participating in the deprotonation of ribose 2′-OH. On the other hand, only D_{146} is essential for N7 activity (Ray *et al.*, 2006). The exact mechanism of N7 methylation is still unknown. Structural studies with *Encephalitozoon cuniculi* (Ecm1) N7 MTase suggest that it is achieved through close proximity of methyl donor, SAM and the acceptor, N7 of guanine, without any nucleophilic attack (Fabregas *et al.*, 2004).

The NS5 polymerase domain

Experimental evidence of RdRp activity

Using Vero cell extracts after infection with DENV2 or WNV, an *in vitro* assay for flavivirus RdRp activity was established where RNA synthesis was triggered by an exogenous viral replicative form (RF) leading to conversion to the replicative intermediate (RI) (Bartholomeusz and Wright, 1993). Subsequently, an active recombinant form of NS5 from DENV1 was expressed in *E. coli* as a GST fusion protein. This construct could bind the RNA template and showed RdRp activity as detected by the incorporation of a radiolabel into a newly synthesized RNA strand (Tan *et al.*, 1996). Following these pioneering studies, the NS5 RdRp of flaviviruses has now been expressed in various systems in an active form (Ackermann and Padmanabhan, 2001; Guyatt *et al.*, 2001; Nomaguchi *et al.*, 2003, 2004; Selisko *et al.*, 2006) and various polymerase assays have been developed using homo- or hetero-polymeric RNA templates obtained via *in vitro* transcription. Thus, these functional studies established that recombinant versions of the flavivirus polymerase are catalytically competent even in the absence of other membrane, viral or host cell components.

Flavivirus RdRp initiates RNA synthesis *de novo*

Incubation with cytoplasmic extracts from DENV2-infected mosquito cells with a subgenomic DENV2 RNA template containing both 5′ and 3′ terminal regions of the genome allowed the formation of two RNA products (You and Padmanabhan, 1999). One enzymatic product had the same size as the RNA template, whilst the second product migrated faster. This second product had a size twice that of the template RNA; upon digestion with RNase A, this product could be converted to an RNA having the same mobility as the template RNA. Together, these results suggested that the latter product was a double-stranded hairpin RNA containing a short single-strand loop region. The authors therefore suggested that this hairpin species was synthesized by a 'copy back' mechanism. In this scheme, elongation of the 3′ end of the template takes place owing to intramolecular priming where the 3′ end of RNA anneals to a complementary sequence stretch of the RNA template. Likewise in other studies using a purified recombinant NS5 protein, similar RNA products were observed (Steffens *et al.*, 1999; Ackermann and Padmanabhan, 2001; Nomaguchi *et al.*, 2004; Selisko *et al.*, 2006). These reaction products were consistent with a *de novo* (also named 'primer-independent') mechanism of initiation of RNA synthesis. More evidence for a *de novo* model for initiation of RNA synthesis was provided by the following experiments: using subgenomic RNA templates from DENV2 and WNV with blocked 3′-OH, only template-size products were formed (Ackermann and Padmanabhan, 2001; Nomaguchi *et al.*, 2004). Accordingly, in the absence of a primer, both NS5 proteins from DENV2 or WNV were able to generate a poly(rG) product, using homopolymeric poly(rC) templates (Selisko *et al.*, 2006). Thus, the flavivirus RdRp is a primer-independent polymerase able to initiate RNA synthesis *de novo*. Based on these observations and given the lack of evidence for the formation of copy-back products *in vivo* (Chu and Westaway, 1985), it was proposed that *de novo* initiation of RNA synthesis corresponds to the RNA replication strategy adopted by flavivirus RdRps *in vivo*. This mechanism would also be in agreement with similar replication strategies used by other evolutionary-related viruses within the *Flaviviridae* family including the major human pathogen HCV (Oh *et al.*, 1999; Luo *et al.*, 2000; Zhong *et al.*, 2000) or the pestivirus bovine viral diarrhoea virus (Kao *et al.*, 1999).

Flavivirus RdRp interacts with a 5′ region of genomic template RNA

The 5′ and 3′ ends of the flavivirus RNA genome contain complementary ribonucleotide sequences that lead to cyclization of the genome (Rice *et al.*, 1985; You and Padmanabhan, 1999; You *et al.*, 2001). Using DENV2-infected cell lysates, these so-called cyclization sequences (CS) as well as a 3′ terminal stem–loop structure were shown to be important for RNA synthesis *in vitro*. These RNA structures are also essential for viral RNA replication in cultured cells, as shown by site-directed mutagenesis of infectious RNA from DENV (Men *et al.*, 1996; Zeng *et al.*, 1998) and from WNV (Khromykh *et al.*, 2001; Lo *et al.*, 2003) and YFV replicon RNA (Corver *et al.*, 2003). Accordingly, a purified recombinant DENV2 NS5 protein is able to synthesize the negative RNA strand using positive strand subgenomic RNA as a template containing intact CS stretches, but not using an RNA containing only the 3′ terminal region (Ackermann and Padmanabhan, 2001). Subsequent work by the Gamarnik group demonstrated that additional complementary sequence stretches at the 5′ and 3′ end of DENV genomic RNA were also important for DENV2 replication (Alvarez *et al.*, 2005). In the same series of elegant studies that revealed the importance of the structure of viral RNA itself for its replication in the infected cell, a physical interaction between the 5′ and 3′ ends of the DENV2 genome leading to its cyclization was actually visualized by atomic force microscopy, together with the interaction between the RdRp from DENV2 and the circular RNA genome. These studies showed that *de novo* RNA synthesis of the negative RNA strand relies on the presence of a promoter element located at the 5′ non-translated end of the genome. This sequence forms a large stem–loop structure christened 'SLA' (Filomatori *et al.*, 2006). Thus, how is the RNA polymerization initiated at the 3′ end, given that the RdRp binds to a promoter element located at the 5′ end of the viral genome? Following binding at the 5′ end of the genome, Gamarnik and collaborators proposed that long-range RNA–RNA interactions allow recruitment of the RdRp to the initiation site at the 3′ end of the flavivirus genome and that synthesis of the negative strand proceeds from there. Thus, as an aside, one possibility brought-up by this series of work is that disruption of either specific viral RNA structures important for RNA replication or the interaction between RdRp domain and SLA could lead to the development of new anti-flavivirus agents. Less is known for the synthesis of the positive RNA strand for which neither the minus strand 5′ end nor the 3′ CS motif are required (Nomaguchi *et al.*, 2004).

Guanylyltransferase

The guanylyltransferase activity in flavivirus remains poorly characterized. In general, the process is well-conserved across different species and involves formation of an intermediate with covalent linkage of GMP to the enzyme, usually via a phosphoamide to a Lys residue of the enzyme, followed by transfer of the GMP to the 5′-ppRNA. Binding of GTP to the N-terminal pocket of flavivirus MTases has been shown for several different members of the family (see 'Crystal structures of the methyltransferase domain', below). However, *in vitro* covalent attachment of GMP to flavivirus MTases has only been reported by two groups for Wesselsbron, DENV, YFV and WNV using radiolabelled GTP (Issur *et al.*, 2009; Bollati *et al.*, 2009). WNV GTase activity was reportedly stimulated by presence of full-length NS3. Furthermore, full-length WNV NS5-GMP intermediate transferred the GMP to pp-AG-RNA comprising the first 81 nt of the WNV 5′ UTR sequence but not to a similar RNA comprising non-viral sequences. Synthesis of the complete cap 1 structure on nascent triphosphorylated WNV viral RNA took place in the presence of full-length NS5 and NS3

(Issur *et al.*, 2009). Nevertheless, some questions remain over these findings. First, the observed WN GTase activity was low and required an input of large excess of radiolabelled GTP, suggesting that either conditions were not optimal or some co-factor is missing. Secondly, mutation of K29 to alanine retained residual GTase activity, casting doubt on its role in covalent attachment of GMP. That K29 is not conserved across the flavivirus MTases, further opens up more questions.

Structural studies

Crystal structures of the methyltransferase domain

N-terminal GTP binding pocket of MTase

Numerous crystal structures of flavivirus MTases have been solved, including from DENV (Egloff *et al.*, 2002; Geiss *et al.*, 2009), WNV (Zhou *et al.*, 2007), Murray valley virus (Assenberg *et al.*, 2007), Meabean virus (Mastroangelo *et al.*, 2006, 2007), TBV (Mastrangelo *et al.*, 2007), WESSV (Bollati *et al.*, 2009a), YFV (Geiss *et al.*, 2009) and two flavivirus without known vectors, YOKV (Bollati *et al.*, 2009b) and MODV (Jansson *et al.* 2009). Most structures were obtained with the active site occupied by the product inhibitor, S-adenosyl homocysteine (SAH) and are highly similar with RMSD values ranging from 0.6–0.7 Å. They display a conserved globular structure that comprises eight α-helices and 12 β-strands which collectively are organized into three motifs: an N-terminal domain (aa 1–58), a core domain that is found in all SAM-dependent MTases (aa 59–224; Fauman *et al.*, 1999) and a C-terminal domain (aa 225–265; Fig. 6.1A).

The N-terminal extension comprises a helix–turn–helix motif followed by a β-strand and an α-helix. It has a well conserved pocket that is formed with residues Lys13, Leu16, Asn17, Leu19, Phe24, Ser150, Ser152 (Pro152 in DENV), Arg197, Arg213, Ser215. It has been found to bind GTP (Fig. 6.1B) and its analogues (GDPMP, acyclovir, EICAR), ribavirin triphosphate and its analogues (RTP), and RNA cap analogues (GpppA, GpppG, m7GpppA, m7GpppG; Egloff *et al.*, 2002; Benarroch *et al.*, 2004; Assenberg *et al.*, 2007; Egloff *et al.*, 2007; Bollati *et al.*, 2009a; Geiss *et al.*, 2009). Binding of these ligands specifically inhibit 2′-O activity but not N7 activity (Benarroach *et al.*, 2004; Dong *et al.*, 2007; Lim *et al.*, 2008). Specifically, the guanosine moiety of GTP forms stacking interaction with the aromatic ring of Phe24 (in WNV; Phe25 in DENV). Accordingly, mutation of Phe24 to alanine is correlated with loss of GTP binding (Egloff *et al.*, 2002) and severe consequences on N7 and 2′-O methylation (Dong *et al.*, 2008a). Additionally, there is hydrogen bond formation of the guanosine with Leu16 and Leu19, whilst its 2′-OH of the ribose hydrogen bonds with Lys13 and Asn17 (in WNV; Lys14 and Asn18 in DENV). In most structures, only the alpha phosphate group is observed and lies close to Lys29 and Ser150. In YFV MTase, the β-phosphate has been observed to form hydrogen bonding with Ser150 (Geiss *et al.*, 2009). For all structures, the γ-phosphate of GTP or cap analogue is discontinuous and not observed. In structures with bound RNA cap analogues, the first nucleotide after the guanosine cap is either disordered, or flipped to form a stacking interaction with the cap (Fig. 6.1C) (Egloff *et al.*, 2007; Geiss *et al.*, 2009). This latter conformation has been proposed to mimic the guanylyl-transferase reaction, in which the GTP is covalently attached to 5′ penultimate A in the viral RNA.

Proposed sites for RNA binding in MTase

Interestingly a second triphosphate binding site has been observed for YFV MTase near residues Arg27 and Arg87, located above the SAH binding site whilst the same site in DENV MTase structures was occupied by a sulfate ion. This has led Geiss *et al.* (2009) to propose the region as an RNA channel in the MTase. Furthermore, based on this finding, a model has been hypothesized for the sequential N7 and 2′-*O* activities where binding of the capped RNA in this second site (also termed low binding site, LBS) positions the GTP cap in the SAM binding or active site and enables it to undergo N7-methylation. Subsequently, the methylated GTP cap docks into the N-terminal pocket (also termed high binding site, HBS), resulting in the re-positioning of the adenosine

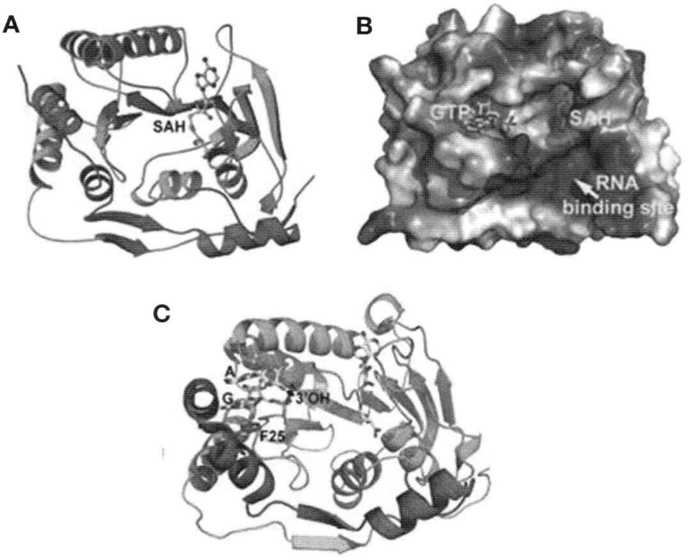

Figure 6.1 Crystal structure of flavivirus MTase. (A) DENV-2 MTase structure in complex of SAH (Egloff *et al.*, 2002). (B) Surface presentation of DENV-2 MTase. The binding sites for SAH, GTP, and RNA substrates are labelled. Electrostatic potential of the protein surface is shown with positive charges in blue, negative charges in red and neutral in white. (C) DENV-2 MTase structure in complex with SAH and cap analogue (Egloff *et al.*, 2007). The stacking interaction between F25 and the guanine and adenine of GpppA is depicted.

in the catalytic pocket and the resultant 2′-*O* methylation (Bollati *et al.*, 2009). Interestingly, Assenberg *et al.* observed that MVEV MTase forms dimers, with each molecule bearing two RNA cap analogues, binding separately in the HBS and LBS sites (2007). The binding of the four RNA cap analogues resemble a continuous strand of RNA, and the two LBS align to form a nearly continuous positively charged groove. Thus, a second model is proposed whereby dimerization of the MTase proteins creates a positively charged tunnel in which RNA binds. The GTP cap is first N7-methylated in the active site of one monomer, followed by translocation into the HBS of the second monomer via its LBS. This positions the adenosine into the active site of the second monomer and allows 2′-*O* methylation to proceed.

Comparative analyses of the sequences of the flavivirus MTases showed the presence of a conserved and almost continuous patch extending away from the active site to the back of the protein. It is made of one region comprising Tyr-89, Pro-113, Gly-120–Lys-127, Asp-131, Gly-263–Arg-265, and a second region with residues Ala-60, Trp6–4, Leu-207-Met-220, Arg-244, on the back of the protein (Bollati *et al.*, 200b). This region has been speculated to be involved in stabilization of the RNA chain during the methylation events. Anther putative RNA binding site comprising residues Arg-37, Arg-57, Arg-84 and Arg-213 has also been proposed as mutation of these residues have detrimental effects on N7 and or 2′-*O* methylation (Dong *et al.* 2008). Nevertheless how RNA interacts the enzyme and the mechanism by which the enzyme performs N7 methylation is still unclear and perhaps can be answered by a co-crystal structure of MTase and a long RNA template.

Core region of MTase structure

The core region comprises seven β-strands surrounded by four α-helices and two 310 helices, and is highly homologous to catalytic domains of other SAM-dependent MTases. Most flavivirus MTase structures contain the product inhibitor, SAH bound in its active site, and is likely to have derived from E. coli during the expression process. Additionally, Wesselsbron MTase has also been co-crystallized with SAM and its analogue, sinefungin (SF) (Bollati *et al.* 2009a). All three ligands bound in the active site in a very similar manner, and is stabilized by hydrogen bonds and van der Waals interactions with residues lining the pocket. The adenine ring is located within a hydrophobic pocket formed by side chains of Thr-104, Lys-105, Val-132, and Ile-147 and also hydrogen bonds with side chain of Asp-131 and main chain N of Val-132. The 2′-OH of the ribose moiety is bridged to Gly-106, H110 and/or Glu-111 through a water molecule and as well to Thr-104. Based on the Wesselsbron structures, loop 106–111 and E149 is present either as an open or closed gate in front of the SAM adenosine moiety, through movement of His110 towards or away from the solvent. This difference in the loop conformation may be functional in the association/dissociation of SAM and its product SAH during the co-factor turnover in the methylation process (Bollati *et al.*, 2009a). The homocysteine tail of the molecule forms hydrogen bonds with Ser-56, Gly-86, Asp-146, and via a water molecule to Asp-79, Arg-84, Trp-87 and Trp-219. In the SF bound structure, the additional –NH2 group hydrogen bonds to the main chain of Asp-146 and is also linked to two ordered water molecules (Bollati *et al.*, 2009b). Mutagenesis of selective residues in the active site have profound effect on N7 (H110, E149), 2′-O (D131, I147) or both methylation activities (S56, W87) (Dong *et al.*, 2008a).

C-terminal end of the MTase structure

The C-terminal subdomain is found between the N-terminal and core sequence and consists of an α-helix and two β-strands. Most crystals of flavivirus MTases are visible only up till about residues 265, with the stretch from aa 266–293 being disordered. Examination of the region from aa 266 to aa 285, which is visible in two structures of WESSV MTase domains, also shows that aa 267–269 are disordered. Expression of full-length NS5 *in vitro* leads to cleavage at Arg-264 or Cys-265 in TBEV Vasilchenko and Oshima strains (Lanciotti *et al.*, 2002; Bollati *et al.*, 2009b) or 272 in DENV3 (Yap *et al.*, 2007). Thus, the sequence from aa 266 to aa 272 could act as a linker between the MTase and RdRp domains which allows them to function in a concerted manner doing the process of viral RNA initiation/elongation and capping (Malet *et al.*, 2008).

Crystal structures of the NS5 polymerase domain from flaviviruses

Overall structure

Crystallographic structures for the RdRp domain of WNV and DENV (serotype 3) have been reported (Malet *et al.*, 2007; Yap *et al.*, 2007). Not surprisingly, these structures adopt a classical polymerase fold resembling a right-hand composed of fingers, palm, and thumb subdomains (Fig. 6.2). Unlike DNA-dependent DNA polymerases, which have a distinct U-shaped architecture, a characteristic feature of RdRps that is also found in the flavivirus RdRp architecture, consists in the formation of several interactions between the fingers and thumb subdomains, giving the flavivirus polymerase a more compact and heart-shaped appearance compared with their DNA polymerases counterparts. Unfortunately, several loops that connect the fingers and thumb subdomains are poorly visible in the electron density maps and a complete structure is still lacking for these regions. In addition, in these crystal forms, both proteins adopt a closed -presumably non-active-conformation (Fig. 6.2). However, several interesting insights can be gleaned from these crystal structures. In order to obtain protein constructs suitable for these crystallographic studies, two truncated fragments of the NS5 RdRp domain from WNV were designed, lacking either 272 or 316 residues from the N-terminus within the fingers subdomain. Only the larger fragment is enzymatically active. This demonstrates that residues 273–316 (WNV numbering) that were omitted from the shorter construct are essential for the structural integrity and the activity of the protein. Prior to structure determination, the segment spanning aa 322–407 was proposed to contain two nuclear localization signals (NLS) that could bind either to the importin from the host cell or to the helicase domain of the NS3 viral protein (Brooks *et al.*, 2002). Interestingly, this

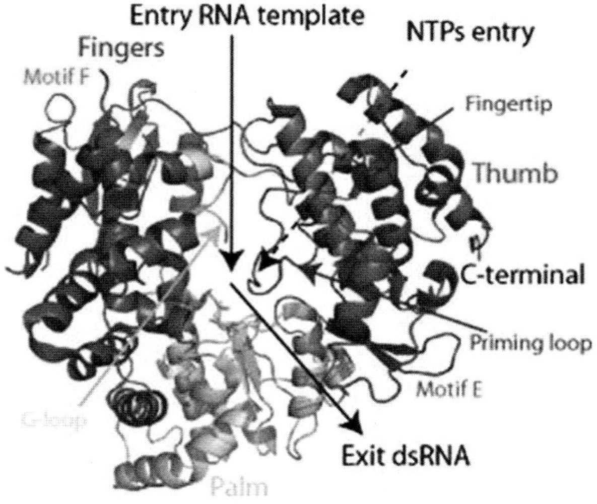

Figure 6.2 Ribbon representation of overall structure of DENV-3 RdRp showing three subdomains (fingers, thumb and palm), RNA template site, NTPs entry site, and dsRNA exit site (Yap *et al.*, 2007).

region forms an integral part of the folded RdRp structure by contributing six α-helices to the finger and thumb subdomains, respectively. However, the precise biological roles played by these two NLS regions within a functional importin/exportin system remain controversial (Mackenzie *et al.*, 2007). Moreover, flaviviruses like DENV and WNV seem to differ in that respect, a fact possibly related to differences in pathogenicity between these two viruses.

Little is known about the structure of the full-length NS5 protein. Limited proteolysis studies suggest that residues 260– 270 of the NS5 protein are accessible to the solvent and could form the linker region connecting the MTase and RdRp domains of NS5. Despite numerous efforts by several groups around the world, the full-length NS5 protein encompassing both its N-terminal MTase (residues 1–260) and C-terminal RdRp (271–905) domains has not been crystallized yet. The sensitivity of NS5 to proteolysis suggests flexibility between the two enzymatic domains that might explain the lack of success in crystallizing the complete protein so far. However, using reverse genetics and computer docking techniques, a model for the full-length NS5 protein from the 3D structures of its separated MTase and RdRp domains was put forward. This procedure positioned residues 512 of the RdRp domain in the vicinity of residues 46, 47, and 49 of the MTase domain. However, it is quite possible that no fixed orientation exist between the MTase and RdRp domain as was shown to be the case for the NS3 protease-helicase (see Chapter 5.).

Metal ion

An accurate identification of the location of metal ions in the crystal structures of RdRps is crucial given their various biological roles: Indeed, metal ions can favour primer-independent versus primer-dependent RNA polymerization, they can also modulate specificity and selectivity for nucleotide incorporation. Moreover, the binding of Mn^{2+} versus Mg^{2+} introduces differential structural flexibility into the RdRp structure. This increase in flexibility seems important for catalytic activity, both at the initiation and elongation stages. An enzymatic and biophysical study of the role of divalent cations was published for the flavivirus RdRp (Bougie *et al.*, 2009).

An unexpected observation in the structure of the RdRp from DENV was the presence of two zinc ions in the thumb and fingers subdomains respectively (Fig. 6.3). One zinc ion is coordinated by His-712, His-714, Cys-728 and Cys-847 of motif E. This ion is located at the interface between the palm and the thumb subdomains and could regulate conformational changes between

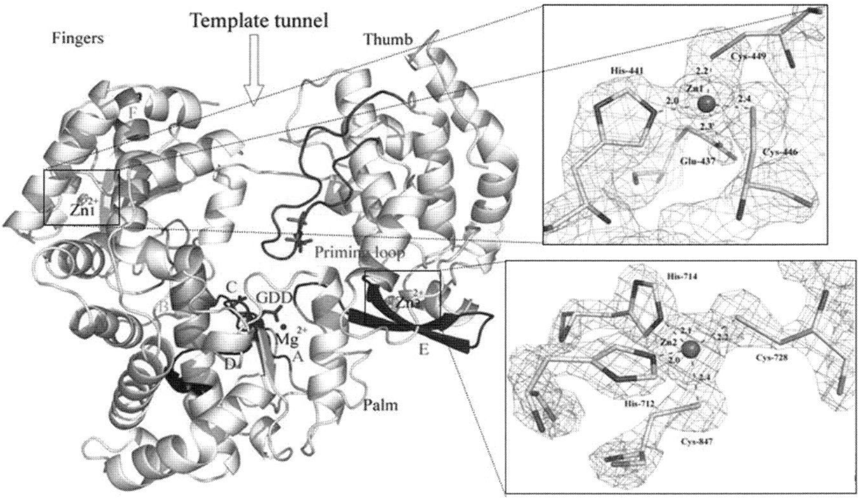

Figure 6.3 DENV-3 RdRp structure showing two Zn^{2+}-binding sites, GDD active site and priming loop (Yap *et al.*, 2007).

closed and open forms of the enzyme in its transition between initiation and elongation (Yap *et al.*, 2007). The residues that chelate this zinc are conserved in the four serotypes of DENV and also in the RdRp from YFV but not in the RdRp from WNV where His-714 is substituted by a Thr residue (Malet *et al.*, 2007). In the latter case, a disulfide bridge is observed instead. The second zinc ion, located in the fingers subdomain is coordinated by Cys-446, Cys-449 and His-441. This metal binding site is also found in the structure of the WNV RdRp.

The active site

The palm subdomain, which is the structurally most conserved, contains two catalytic aspartic-acid residues that coordinate two divalent metal ions essential for catalysis. The GDD catalytic motif C is located between strands β4 and β5 within this subdomain. In DENV RdRp, these two strands are shorter than in the corresponding enzyme from HCV. Moreover, only two β-strands are observed for the flavivirus RdRp, versus three β-strands for RdRps from HCV or BVDV. Following deprotonation by one of the three conserved carboxylate groups that acts as a general base at the catalytic site, the 3′-OH of the primer terminus emanating from a ribose moiety performs a nucleophilic attack on the nucleotide α-phosphate. Kinetic and mutagenesis studies performed by Cameron and coworkers (Castro *et al.*, 2009) have identified a conserved basic residue from motif D (Lys or Arg) that was proposed to act as an acid to assist PPi release after phosphodiester bond formation. The flavivirus RdRp catalytic site is located at the intersection of two tunnels. The first tunnel runs vertically between the fingers and thumb subdomains. In the conformation that was crystallized, this tunnel would be able to accommodate a ssRNA template strand of about six to seven ribonucleotides, but would not be wide enough to accommodate a RNA duplex. A second tunnel, approximately perpendicular to the first, traverses the entire protein, giving access to incoming rNTPs to the active site, by the back aperture as seen in Fig. 6.2. Following rNTP incorporation in the growing chain, exit of the nascent dsRNA product would proceed via the front part of the tunnel, looking at the protein in the orientation presented in Fig. 6.2. A key structural feature obstructs the flavivirus polymerase active site presumably hindering the exit of the dsRNA product. Interestingly, in the case of the flavivirus RdRp, this element, called 'priming loop' does not fold into a similar structure as the RdRp from HCV: It forms a β-hairpin in the HCV

RdRp while it forms an extended and potentially flexible structure in the flavivirus RdRp structures from both DENV and WNV. Recognition of ssRNA template must occur through the template channel that is large enough to accommodate five to six ribonucleotides. The 5′ portion of the template is likely to contact residues Trp-474 and Trp-477 from the fingers subdomain and its 3′-end would come in close contact with residues from motif C. Since the conformation of the priming loop and Trp-795 partially occlude the path of the template, it is probable that the apo-structure that has been captured represents a closed pre-initiation conformation and that the structure has to open-up substantially in order to accommodate a template and even more so in the elongation complex, where a RNA duplex has to fit in. Thus, it is probable that quaternary rearrangement occurs through concerted movements of the fingers subdomain away from the thumb, when compared with the corresponding subdomain in the RdRps from WNV or HCV. Efforts are under way to capture such structural states and one key requirement for success is probably the design a proper RNA template amenable to crystallization.

Mechanism for initiation of RNA synthesis

Much of the current conceptual framework to understand initiation of RNA-directed RNA polymerization derives from the pioneering work by Butcher *et al.* (2001) on the phi6 bacteriophage RdRp, a dsRNA virus. Comparable work was reported for the HCV RdRp by Bressanelli *et al.* (2002). In this model, three nucleotide-binding sites were identified within the polymerase and named (C) catalytic, (P) priming, and (I) interrogation. The substrate entry pore is denoted interrogation (I) site. The nucleotide at the 3′-end of the template is named as T1, while the nucleotides of the 5′-end daughter strand are named as D1 and D2. The template base T1 base pairs with D1, which is stacked against an aromatic residue: Tyr-630 in phi6, that could correspond to Trp-795 in DENV and Trp-800 in WNV, at the priming (P) site. The D2 base, which is stacked to D1 and base-pairs with T2, is positioned at the catalytic (C) site. The triphosphate moiety is stabilized by two Arg residues (Arg-734 and Arg-742 in WNV) that might assist the shuttling of the incoming rNTP between I and C sites and also ensure the fidelity and stability of the initiation complex (Fig. 6.4). Three spatially equivalent binding sites were found for the HCV RdRp in complex with various ribonucleotides (Bressanelli *et al.*, 2002). In addition a fourth site – solely for rGTP – was found at the surface of the HCV RdRp lying approximately 3 nm away from the catalytic site. However, no allosteric binding site for the DENV polymerase has been reported so far. A native DENV3 RdRp crystal soaked with the chain terminator 3′dGTP with an IC_{50} of 0.22 mM, revealed clear electron density for a triphosphate moiety coordinated by Ser-710, Arg-729, and Arg-737 near the priming loop. Neither the ribose nor the base was visible suggesting that the presence of a complementary base on a template strand is needed to stabilize them. The evolutionary conserved Trp-795 in DENV3 (Trp-800 in the RdRp from WNV) seems to be a crucial residue that provides a platform against which the priming base can stack to initiate RNA polymerization, playing a role equivalent from Tyr-630 in phi6, Tyr-448 in HCV, or Tyr-581 in BVDV RdRps. On the basis of these results, a model for flavivirus initiation complexes was proposed and is depicted in Fig. 6.4. As mentioned above, following initiation, a large conformational change is likely to occur to allow the elongation process, although not much is currently known about this structural state.

Drug discovery targeting the RdRp domain

Using the flavivirus RdRp structures Malet *et al.* (2007) performed a surface shape analysis in order to predict cavities that could potentially accommodate allosteric inhibitor binding sites. Two and five cavities were found in the RdRps from DENV3 and WNV RdRp, respectively. None was common with the HCV RdRp ruling out in principle the possibility of having a pan-*Flaviviridae* allosteric inhibitor. It was proposed that binding of a compound in some of these cavities could impair the conformational change of the priming loop necessary for the transition to elongation. Of note, cavity A and B that are

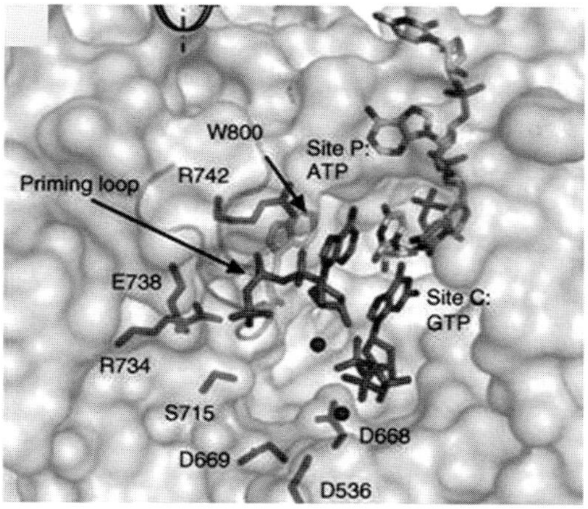

Figure 6.4 A proposed model for the initiation of flavivirus RNA synthesis (see text for details).

conserved between DENV3 and WNV RdRp might be used as a starting point for structure-based drug design. Inhibition, mutational and structural studies are under way to determine if these sites are druggable. The availability of a 3D crystallographic structure and of active soluble enzymes has also allowed the identification of several interesting nucleoside (Yin *et al.*, 2009a) and non-nucleoside inhibitors targeting this enzymatic activity (Yin *et al.*, 2009b). An adenosine analogue (named NITD008) containing a carbon substitution for N7 of the purine moiety and an acetylene group at the 2′ position of the ribose was synthesized (Yin *et al.*, 2009a). NITD008 is a chain terminator that stops RNA elongation. Treatment of cells with 9 μM of NITD008 reduced viral titres by more than 10^4-fold in various serotypes, including clinical DENV isolates. The *in vivo* efficacy of NITD008 tested using a mouse model of DENV infection, demonstrated that NITD008 could prevent death even with a delayed start of antiviral treatment (Yin *et al.*, 2009a) or a single-dose treatment (Chen *et al.*, 2010). Thus, these results provide a nice proof of concept that it is possible to use antiviral therapy to treat an acute viral disease like dengue. Unfortunately, this compound showed significant toxicity in rats and dogs. Further development of this compound was interrupted at this stage. (Yin *et al.*, 2009a).

Regulation of NS5

NS5 MTase and RdRp intramolecular interactions

Using reverse genetics mutagenesis, two studies found evidence for interaction between the MTase and RdRp domains. Mutation of three surface-exposed residues Lys-46/Arg-47/Glu-49 in WNV MTase which impaired virus growth, gave rise to a compensatory mutation in Leu-512 to Val in its RdRp domain and the restoration of wild-type virus growth kinetics (Malet *et al.*, 2007). *In silico* docking of the two domains indicated that that the MTase loop containing the three residues Lys-46/Arg-47/Glu-49 fitted into a groove formed by the fingers and thumb subdomains of the RdRp. Similarly, mutation of residue Asp-146 to Ser in WNV virus generated second site mutations in Lys-61 to Gln/Thr in the MTase domain and Trp-751 in the RdRp domain (Zhang *et al.*, 2008). Both changes led to enhanced activity of the N7-MTase and RdRp activities respectively. Although Trp-751 is surface-exposed, it is unlikely to participate in direct interaction with Asp-146 or Lys-61.

Interaction with viral RNA, NS3, and host proteins in viral RNA synthesis

Flavivirus RNA comprises 5′ and 3′ UTR sequences that contain complementary sequences that form either secondary hairpin loops (SLA,

SLB) and self-cyclize (3′SL with 5′UAR and 5′CS with 3′CS) (Brinton and Dispoto, 1988; Cahour *et al.*, 1995; Alveraz *et al.*, 2005, 2008; Filomatori *et al.*, 2006). Although both full-length WNV NS5 and DENV RdRp domains have been shown to bind mainly to SLA *in vitro* (Alveraz *et al.*, 2005; Filomatori *et al.*, 2006; Dong *et al.*, 2007), WNV RdRp domain alone was not observed to bind the viral 5′ UTR (Dong *et al.*, 2008c). Interaction between JEV NS5 and the terminal 83 nt of its 3′ UTR was detected in immunoprecipitation experiments which also pulled down NS3 (Chen *et al.*, 1997). Through its binding to viral 5′ and 3′ UTRs, La protein interacts with DENV NS5 and negatively regulates its RdRp activity (Yocupicio-Monroy *et al.*, 2003; Garcia-Montalvo *et al.*, 2004). In addition, glyceraldehyde-3-phosphate dehydrogenase (GAPDH) was found to bind JEV 3′ termini of plus- and minus-strand RNAs and colocalize with NS5 in the cytoplasm (Yang *et al.*, 2009). Its exact role in JEV replication is not determined.

Interaction of flavivirus NS3 and NS5 has been reported by several groups. They were shown by immunoprecipitation experiments using antibodies to either NS3 or NS5 with DENV-1, DENV-2, and JEV infected cell lysates and with recombinant expressed NS5 (Kapoor *et al.*,1995; Chen *et al.*, 1997; Cui *et al.*, 1998). NS5 interacts with NS3 and stimulates its NTPase, and RTPase activities (Cui *et al.*,1998; Yon *et al.*, 2005). Using yeast two-hybrid systems, aa 320–336 of NS3 were found to interact with the β-NLS located at aa 320–368 in NS5 (Johansson *et al.*, 2001). Interaction between MTase domain and the protease were also implicated by reverse genetics. Mutation of the MTase residues Glu-192, Lys-193 and Glu-195 resulted in a second site mutation in residue Gly70Ala in the protease sequence and restoration of virus growth (Davidson, 2009). These regions in NS5 and NS3 could not be trans-complemented in a Kunjin replicon system suggesting that they must interact in cis during viral polyprotein translation (Khromykh *et al.*, 1999). Two nuclear localization sequences (NLS) are found within aa 369–405 in NS5 RdRp domain (Forwood *et al.*, 1999). β-NLS located between aa 320–368 interacts with importin β besides NS3 (Johansson *et al.*, 2001), whilst αβ-NLS at aa 369–405 interacts with αβ-importin (Brooks *et al.*, 2002). Competitive binding between NS3 and importin β may thus determine its subcellular localization (See also Section 4.3).

Phosphorylation and nuclear trafficking

YFV, TBEV, DENV and Kunjin NS5 were shown to be phosphorylated by serine/threonine kinases in infected cells (Kapoor *et al.*, 1995; Morozova *et al.*, 1997; Reed *et al.*, 1998 MacKenzie *et al.*,2007). Phosphorylation occurred mainly in at least four distinct Ser residues (including Ser-56 and Thr-449) with low levels of Thr phosphorylation detected for YFV NS5. The exact identities of the kinases involved in regulating flavivirus NS5 phosphorylation have not been fully established. However, *in vitro* DEN,V NS5 is phosphorylated by casein kinase II (CK-II; Forwood *et al.*, 1997) or by protein kinase G (PKG) which in turn induces auto-phosphorylation of PKG (Bhattacharya *et al.*, 2009a). On the other hand, the alpha isoform of casein kinase 1 (CK-I) was reported to phosphorylate Ser-56 of YFV MTase (Bhattacharya *et al.*, 2008, 2009b). *In vitro* phosphorylation of YFV NS5 is inhibited by staurosporine, suggesting that a number of kinases may be involved (Reed *et al.*, 1998). Nevertheless, both inhibitors and activators of casein kinase CK-I as well as PKG modulate YFV and DENV replication in cell culture systems (Bhattacharya *et al.*, 2009a,b). Many workers have detected NS5 in the nucleus during infection of DENV (Kapoor *et al.*, 1995; Miller *et al.* 2006; MacKenzie *et al.*, 2007; Malet *et al.*, 2007; Pryor *et al.*, 2007) and YFV (Buckley *et al.*, 1992), but not for WNV or Kunjin virus (MacKenzie *et al.*, 2007; Malet *et al.*, 2007). Uchil *et al.* 2006 also showed that the host nucleus can function as a site for JEV replication complex, with presence of both NS3 and NS5 and newly synthesized viral RNA. NS5 phosphorylation states differed when the protein was located in the cytoplasm or nucleus, with nuclear NS5 mainly found in the hyperphosphorylated form (Kapoor *et al.*, 1995). It is possible that phosphorylation at the two NLS region in NS5 may determine if it binds NS3, importin β or αβ-importin (Johansson *et al.*, 2001; Brooks *et al.*, 2002) and thus determine its retention in the cytosol or nucleus. For more details on the mechanism of NS5 nuclear import (see Davidson, 2009). Mutation of two clusters of basic residues comprising aa Lys371/372 and Lys-387–389 in αβ-NLS (aa 369–405) also impaired nuclear accumulation of NS5 and led to reduction in virus growth and production (Pryor *et al.*, 2007). Conversely, inhibition of NS5 nuclear export with leptomycin B, resulted in its accumulation in the nucleus and enhanced virus production (Pryor *et al.*, 2006). This effect was mapped to a nuclear-export sequence (NES) in the β-NLS of NS5 and mutation of this sequence increased nuclear levels of NS5 and resistance to the effect of leptomycin B (Pryor *et al.*, 2006; Alvisi *et al.*, 2008). The exportin CRM1 binds also to this sequence and its inhibition coincided with altered kinetics of virus production (Rawlinson *et al.*, 2009).

Role of NS5 in viral pathogenesis

Several members of the flavivirus have been reported to inhibit cellular IFN α/β and γ responses, following viral infection. Inhibition is mediated by NS5 and other non-structural proteins through blockage of the JAK-STAT signalling. Immuno-precipitation studies revealed that LGTV NS5 interacted with IFN α/β receptor subunit FNAR2 and probably IFN γ receptor subunit IFNGR2, leading to inhibition of reporter genes driven by IFN α/β and γ-responsive promoter element. It did not directly bind STAT1 but inhibited STAT1 and JAK1 phosphorylation and nuclear translocation in cells treated with IFN α/β (Best *et al.*, 2005). Similarly JEV NS5 did not bind Tyk2 or JAK1 but also blocked phosphorylation of Tyk2 in JEV-infected cells, and the activation of downstream interferon-responsive genes (Lin *et al.*, 2004, 2006). Deletion analysis indicates that the C-terminal amino acid sequence of NS5 from region 584–762 and the N-terminal 166 amino acids contained domains that were important for STAT1 nuclear translocation, and the effect of NS5 on IFN-signalling was independent of its MTase and RdRp enzymatic activities (Lin *et al.*, 2006). For LGTV NS5 the region required for IFN-signalling was mapped to aa 374–380 (in particular Arg376, Asp380) and 624–647 (in particular Glu626, Glu628 and Trp647) having the most effect (Park *et al.*, 2007). Interestingly, these two regions are close to each other and surface exposed on the RdRp WNV structure. Amino acid 374–380 overlaps with the α/β-NLS identified in DENV-2 NS5 which was found to bind α/β-importin and involved in NS5 nuclear trafficking. NS5 from DENV-2 was shown to bind and inhibit STAT2 phosphorylation, resulting in inhibition of IFN-alpha, but not IFN-gamma, signalling. This effect was seen with the RdRp domain alone (Ashour *et al.*, 2009; Mazzon *et al.*, 2009). Degradation of STAT2 required ubiquitination and proteasome activity, and is linked to the proteolytic processing of the viral polyprotein during NS5 maturation (Ashour *et al.*, 2009). In contrast, α/β-IFN and γ-IFN signalling was inhibited by Kunjin virus non-structural proteins NS2A, NS2B, NS3, NS4A, and NS4B, but not by NS5 (Liu *et al.*, 2005). More recently, it was shown that NS5 from the virulent NY99 strain of WNV prevented STAT1 phosphorylation and suppressed IFN-dependent gene expression, suggesting that NS5 also function as an efficient IFN antagonist. In contrast, NS5 from KUN, a naturally attenuated subtype of WNV, was a poor suppressor of STAT1 phosphorylation. Furthermore, mutation of a single residue in KUN NS5 to the analogous residue in WNV-NY99 NS5 (S653F) rendered KUN NS5 an efficient inhibitor of pY-STAT1 (Laurent-Rolle *et al.*, 2010).

IL-8 belongs to the family of CXC chemokines and plays a vital role in the recruitment of monocytes and neutrophils in acute inflammatory responses. DENV-infected cells and serum from DENV-patients have been shown to contain increased levels of IL-8 (Raghupathy *et al.*, 1998; Juffrie *et al.*, 2000; Bosch *et al.*, 2002; Moreno-Alramirano *et al.*, 2004; Medlin *et al.*, 2005). Expression of NS5 in cells is sufficient to produce the same effect through the up-regulation of the IL-8 promoter via also the transcription factors c/EBP and NF-κB (Medlin *et al.*, 2005). However, expression of NS5 with mutation in the α/β-NLS sequence or treatment with inhibitors of CRM1, delayed nuclear accumulation of NS5 but increased IL-8 production suggesting NS5 also exerts its effect in the cytoplasm (Pryor *et al.*, 2007; Rawlinson *et al.*, 2009).

Acknowledgements

We thank all our co-workers and collaborators whose named are listed in the original publications in particular, Hongping Dong, Thai Leong Yap, Ka Yan Chung, Ting Xu and Subhash G. Vasudevan. Work in the laboratory of J.L. has been generously supported over the years by the Singapore Biomedical Research Council and the National Research Foundation. Support from an ATIP grant of the CNRS is also gratefully acknowledged.

References

Ackermann, M., and Padmanabhan, R. (2001). *De novo* synthesis of RNA by the dengue virus RNA-dependent RNA polymerase exhibits temperature dependence at the initiation but not elongation phase. J. Biol. Chem. *276*, 39926–39937.

Alvarez, D.E., Lodeiro, M.F., Luduena, S.J., Pietrasanta, L.I., and Gamarnik, A.V. (2005). Long-range RNA–RNA interactions circularize the dengue virus genome. J. Virol. *79*, 6631–6643.

Assenberg R., Mastrangelo, E., Walter, T.S., Verma, A., Milani, M., Owens, R.J., Stuart, D.I., and Grimes, J.M. and Mancini, E.J. (2009). Crystal structure of a novel conformational state of the flavivirus NS3 protein: implications for polyprotein processing and viral replication. J. Virol. *83*, 12895–906.

Barbosa, E., and Moss, B. (1978). mRNA(nucleoside-2′-)-methyltransferase from vaccinia virus. Characteristics and substrate specificity. J. Biol. Chem. *253*, 7698–7702.

Bartelma, G., and Padmanabhan, R. (2002). Expression, purification, and characterization of the RNA 5′-triphosphatase activity of dengue virus type 2 nonstructural protein 3. Virology *299*, 122–132.

Bartholomeusz, A., Tomlinson, E., Wright, P.J., Birch, C., Locarnini, S., Weigold, H., Marcuccio, S., and Holan, G. (1994). Use of a flavivirus RNA-dependent RNA polymerase assay to investigate the antiviral activity of selected compounds. Antiviral. Res. *24*, 341–350.

Bartholomeusz, A.I., and Wright, P.J. (1993). Synthesis of dengue virus RNA *in vitro*: initiation and the involvement of proteins NS3 and NS5. Arch. Virol. *128*, 111–121.

Benarroch, D., Egloff, M.-P., Mulard, L., Guerreiro, C., Romette, J. -L., and Canard, B. (2004). A structural basis for the inhibition of the NS5 dengue virus mRNA 2′-*O*-methyltransferase domain by ribavirin 5′-triphosphate. J. Biol. Chem. *279*, 35638–35643.

Bhattacharya, D., Hoover, S., Falk, S.P., Weisblum, B., Vestling, M., and Striker R. (2008). Phosphorylation of yellow fever virus NS5 alters methyltransferase activity. Virol. *380*, 276–84.

Bhattacharya, D., Mayuri, Best, S.M., Perera, R., Kuhn, R.J., Striker, R. (2009a). Protein kinase G phosphorylates mosquito-borne flavivirus NS5. J. Virol. *83*, 9195–205.

Bhattacharya, D., Ansari, I.H., and Striker, R. (2009b). The flaviviral methyltransferase is a substrate of casein kinase 1. Virus Res. *141*, 101–4.

Bollati, M., Milani, M., Mastrangelo, E., Ricagno, S., Tedeschi, G., Nonnis, S., Decroly, E., Selisko, B., de Lamballerie, X., Coutard, B., *et al.* (2009a). Recognition of RNA cap in the Wesselsbron virus NS5 methyltransferase domain: implications for RNA-capping mechanisms in flavivirus. J Mol Biol. *385*, 40–52.

Bollati, M., Milani, M., Mastrangelo, E., de Lamballerie, X., Canard, B., and Bolognesi, M. (2009b). Crystal structure of a methyltransferase from a no-known-vector flavivirus. Biochem. Biophys. Res. Commun. *382*, 200–4.

Bollati, M., Alvarez, K., Assenberg, R., Baronti, C., Canard, B., Cook, S., Coutard, B., Decroly, E., de Lamballerie, X., Gould, E.A., *et al.* (2010). Structure and functionality in flavivirus NS proteins: perspectives for drug design. Antiviral Res. *87*, 125–48.

Bougie, I., Bisaillon, M. (2009). Metal ion-binding studies highlight important differences between flaviviral RNA polymerases. Biochim. Biophys. Acta. *1794*, 50–60.

Bressanelli, S., Tomei, L., Rey, F.A., and De Francesco, R. (2002). Structural analysis of the hepatitis C virus RNA polymerase in complex with ribonucleotides. J. Virol. *76*, 3482–3492.

Brinton, M.A., and Dispoto, J.H. (1988). Sequence and secondary structure analysis of the 5′-terminal region of flavivirus genome RNA. Virology *162*, 290–9.

Brooks, A.J., Johansson, M., John, A.V., Xu, Y., Jans, D.A., and Vasudevan, S.G. (2002). The interdomain region of dengue NS5 protein that binds to the viral helicase NS3 contains independently functional importin beta 1 and importin alpha/beta-recognized nuclear localization signals. J. Biol. Chem. *277*, 36399–36407.

Butcher, S.J., Grimes, J.M., Makeyev, E.V., Bamford, D.H., and Stuart, D.I. (2001). A mechanism for initiating RNA-dependent RNA polymerization. Nature *410*, 235–240.

Cahour, A., Pletnev, A., Vazielle-Falcoz, M., Rosen, L., and Lai, C.J. (1995). Growth-restricted dengue virus mutants containing deletions in the 5′ noncoding region of the RNA genome. Virology *207*, 68–76.

Castro, C., Smidansky, E.D., Arnold, J.J., Maksimchuk, K.R., Moustafa, I., Uchida, A., Gotte, M., Konigsberg, W., and Cameron, C.E. (2009). Nucleic acid polymerases use a general acid for nucleotidyl transfer. Nat. Struct. Mol. Biol. *16*, 212–218.

Chambers, T.J., Hahn, C.S., Galler, R., and Rice, C.M. (1990). Flavivirus genome organization, expression, and replication. Ann. Rev. Microbiol. *44*, 649–688.

Chen, C.J., Kuo, M., D., Chien, L.J., Hsu, S.L., Wang Y.M., and Lin, J.H. (1997). RNA–protein interactions: involvement of NS3, NS5, and 3′ noncoding regions of Japanese encephalitis virus genomic RNA. J. Virol. *71*, 3466–73.

Chen, Y.L., Yin, Z., Lakshminarayana, S.B., Qing, M., Schul, W., Duraiswamy, J., Kondreddi, R.R., Goh, A., Xu, H.Y., Yip, A., *et al.* (2010). Inhibition of dengue virus by an ester prodrug of an adenosine analog. Antimicrob. Agents Chemother. *54*, 3255–61.

Chu, P.W., and Westaway, E.G. (1985). Replication strategy of Kunjin virus: evidence for recycling role of replicative form RNA as template in semiconservative and asymmetric replication. Virology *140*, 68–79.

Cleaves, G.R., Dubin, D.T. (1979). Methylation status of intracellular dengue type 2 40 S RNA. Virology *96*, 159–165.

Cui, T., Sugrue, R.J., Xu, Q., Lee, A.K., Chan, Y.C., and Fu, J. (1998). Recombinant dengue virus type 1 NS3 protein exhibits specific viral RNA binding and NTPase activity regulated by the NS5 protein. Virology *246*, 409–17.

Davidson A.D. (2009). New insights into flavivirus non structural protein 5. Adv. Virus Res. *74*, 41–101.

Dong, H., Ray, D., Ren, S., Zhang, B., Puig-Basagoiti, F., Takagi, Y., Ho, C.K., Li, H., and Shi, P.-Y. (2007). Distinct RNA Elements Confer Specificity to flavivirus RNA Cap Methylation Events. J. Virol. *81*, 4412–4421.

Dong, H., Ren, S., Zhang, B., Zhou, Y., Puig-Basagoiti, F., Li, H., and Shi, P.-Y. (2008a). West Nile Virus methyltransferase catalyzes two methylations of the viral RNA Cap through a substrate-repositioning mechanism. J. Virol. *82*, 4295–4307.

Dong, H., Zhang, B., and Shi, P.-Y. (2008b). Flavivirus methyltransferase: a novel antiviral target. Antiviral Res. *80*, 1–10.

Dong, H., Zhang, B., and Shi, P.-Y. (2008c). Terminal structures of West Nile virus genomic RNA and their interactions with viral NS5 protein. Virology *381*, 123–135.

Egloff, M.-P., Benarroch, D., Selisko, B., Romette, J. -L., and Canard, B. (2002). An RNA cap (nucleoside-2′-O-)-methyltransferase in the flavivirus RNA polymerase NS5: crystal structure and functional characterization. EMBO J. *21, 2757–2768.*

Egloff, M.-P., Decroly, E., Malet, H., Selisko, B., Benarroch, D., Ferron, F., and Canard, B. (2007). Structural and functional analysis of methylation and 5′-RNA sequence requirements of short capped RNAs by the methyltransferase domain of dengue virus NS5. J. Mol. Biol. *372*, 723–736.

Filomatori C.V., Lodeiro M.F., Alvarez D.E., Samsa M.M., Pietrasanta L., and Gamarnik A.V. (2006). A 5′ RNA element promotes dengue virus RNA synthesis on a circular genome. Genes Dev. *20*, 2238–49.

Forwood, J.K., Brooks, A., Briggs, L.J., Xiao, C.Y., Jans, D.A., and Vasudevan, S.G. (1997). The 37-amino-acid interdomain of dengue virus NS5 protein contains a functional NLS and inhibitory CK2 site. Biochem. Biophys. Res. Commun. *257*, 731–7.

Furuichi, Y., and Shatkin, A.J. (2000). Viral and cellular mRNA capping: past and prospects. Adv. Virus Res. *55*, 135–184.

García-Montalvo, B.M., Medina, F., del Angel, and R.M. (2004). La protein binds to NS5 and NS3 and to the 5′ and 3′ ends of dengue 4 virus RNA. Virus Res. *102*, 141–50.

Geiss, B.J., Thompson, A.A., Andrews, A.J., Sons, R.L., Gari, H.H., Keenan, S.M., and Peersen, O.B. (2009). Analysis of flavivirus NS5 methyltransferase cap binding. J. Mol. Biol. *385*, 1643–1654.

Gubler, D.J. (1998). Dengue and dengue hemorrhagic fever. Clin. Microbiol. Rev. *11*, 480–496.

Guyatt, K.J., Westaway, E.G. and Khromykh, A.A. (2001). Expression and purification of enzymatically active recombinant RNA-dependent RNA polymerase (NS5) of the flavivirus Kunjin. J. Virol. Methods *92*, 37–44.

Issur, M., Geiss, B.J., Bougie, I., Picard-Jean, F., Despins, S., Mayette, J., Hobdey, S.E., and Bisaillon, M. (2009). The flavivirus NS5 protein is a true RNA guanylyltransferase that catalyzes a two-step reaction to form the RNA cap structure. RNA *15*, 2340–2350

Jansson, A.M., Jakobsson, E., Johansson, P., Lantez, V., Coutard, B., de Lamballerie, X., Unge, T., and Jones. T.A. (2009). Structure of the methyltransferase domain from the Modoc virus, a flavivirus with no known vector. Acta Crystallogr. D. Biol. Crystallogr. *65*, 796–803.

Johansson, M., Brooks, A.J., Jans, D.A., and Vasudevan, S.G. (2001). A small region of the dengue virus-encoded RNA-dependent RNA polymerase, NS5, confers interaction with both the nuclear transport receptor importin-beta and the viral helicase, NS3. J. Gen. Virol. *82*, 735–45.

Kao, C.C., Del Vecchio, A.M., and Zhong, W. (1999). *De novo* initiation of RNA synthesis by a recombinant flaviviridae RNA-dependent RNA polymerase. Virology *253*, 1–7.

Kapoor, M., Zhang, L., Ramachandra, M., Kusukawa, J., Ebner, K.E., Padmanabhan, R. (1995). Association between NS3 and NS5 proteins of dengue virus type 2 in the putative RNA replicase is linked to differential phosphorylation of NS5. J. Biol. Chem. *270*, 19100–6.

Khromykh, A.A., Meka, H., Guyatt, K.J., and Westaway, E.G. (2001). Essential role of cyclization sequences in flavivirus RNA replication. J. Virol. *75*, 6719–6728.

Koonin, E.V. (1993). Computer-assisted identification of a putative methyltransferase domain in NS5 protein of flaviviruses and lambda 2 protein of reovirus. J. Gen. Virol. *74*, 733–40.

Kroschewski, H., Lim, S..P., Butcher, R.E., Yap, T.L., Lescar, J., Wright, P.J., Vasudevan S.G. and Davidson A.D. (2008). Mutagenesis of the dengue virus type 2 NS5 methyltransferase domain. J. Biol. Chem *283*, 19410–19421

Lim, S.P., Wen, D., Yap, T.L., Yan, C.K., Lescar, J., and Vasudevan, S.G. (2008). A scintillation proximity assay for dengue virus NS5 2′-O-methyltransferase–kinetic and inhibition analyses. Antiviral Res. *80*, 360–369.

Latour, D.R., Jekle, A., Javanbakht, H., Henningsen, R., Gee, P., Lee, I., Tran, P., Ren, S., Kutach, A.K., Harris, S.F., *et al.* (2010). Biochemical characterization of the inhibition of the dengue virus RNA polymerase by beta-d-2′-ethynyl-7-deaza-adenosine triphosphate. Antiviral Res. *87*, 213–222.

Laurent-Rolle, M., Boer, E.F., Lubick, K.J., Wolfinbarger, J.B., Carmody, A.B., Rockx, B., Liu, W., Ashour, J., Shupert, W.L., Holbrook, M.R., *et al.* (2010). The NS5 protein of the virulent West Nile virus NY99 strain is a potent antagonist of type I interferon-mediated JAK-STAT signaling. J. Virol. *84*, 3503–15.

Lescar, J. and Canard, B. (2009). RNA-dependent RNA polymerases from flaviviruses and picornaviridae. Curr. Opin. Struct. Biol. *19*, 759–767.

Lim, S.P., Wen, D., Yap, T.L., Chung, K.Y., Lescar, J. and Vasudevan. S.G. (2008). A scintillation proximity assay for dengue virus NS5 2′*O*-methyltransferase: Kinetic and inhibition analysis. Antiviral Res. *80*, 360–369

Luo, G., Hamatake, R.K., Mathis, D.M., Racela, J., Rigat, K.L., Lemm, J., and Colonno, R.J. (2000). *De novo* initiation of RNA synthesis by the RNA-dependent RNA poly merase (NS5B) of hepatitis C virus. J. Virol. *74*, 851–863.

Mackenzie, J.M., Kenney, M.T., and Westaway, E.G. (2007). West Nile virus strain Kunjin NS5 polymerase is a phosphoprotein localized at the cytoplasmic site of viral RNA synthesis. J. Gen. Virol. *88*, 1163–1168.

Malet, H., Massé, N., Selisko, B., Romette, J.-L., Alvarez, K., Guillemot, J.-C., Tolou, H., Yap, T.L., Vasudevan, S.G., Lescar, J. and Canard, B. (2008). The flavivirus polymerase as a target for drug discovery. Antiviral Res. *80*, 23–35.

Mastrangelo, E., Bollati, M., Milani, M., de Lamballerie, X., Brisbarre, N., Dalle, K., Lantez, V., Egloff, M.P., Coutard, B., Canard, B., *et al.* (2006). Preliminary characterization of (nucleoside-2′-O-)-methyltransferase crystals from Meaban and Yokose flaviviruses. Acta Crystallogr. Sect. F. Struct. Biol. Cryst. Commun. *62*, 768–70.

Mastrangelo, E., Bollati, M., Milani, M., Selisko, B., Peyrane, F., Canard, B., Grard, G., de Lamballerie, X., and Bolognesi, M. (2007). Structural bases for substrate recognition and activity in Meaban virus nucleoside-2′-O-methyltransferase. Protein Sci. *16*, 1133–45.

Men, R., Bray, M., Clark, D., Chanock, R.M., and Lai, C.J. (1999). Dengue type 4 virus mutants containing deletions in the 3′ noncoding region of the RNA genome: analysis of growth restriction in cell culture and altered viremia pattern and immunogenicity in rhesus monkeys J. Virol. *70*, 3930–3937.

Milani, M., Mastrangelo, E., Bollati, M., Selisko, B., Decroly, E., Bouvet, M., Canard, B., and Bolognesi, M. (2009). Flaviviral methyltransferase/RNA interaction: structural basis for enzyme inhibition. Antiviral Res. *83*, 28–34.

Noble, C.G., Chen, Y.L., Dong, H., Gu, F., Lim, S.P., Schul, W., Wang, Q.Y., and Shi, P.Y. (2010). Strategies for the development of dengue virus inhibitors. Antiviral Res. *85*, 450–462.

Nomaguchi, M., Ackermann, M., Yon, C., You, S., and Padmanabhan, R. (2003). *De novo* synthesis of negative-strand RNA by dengue virus RNA-dependent RNA polymerase *in vitro*: nucleotide, primer, and template parameters. J. Virol. *77*, 8831–8842.

Nomaguchi, M., Teramoto, T., Yu, L., Markoff, L., and Padmanabhan, R. (2004). Requirements for West Nile virus (−)- and (+)-strand subgenomic RNA synthesis *in vitro* by the viral RNA-dependent RNA polymerase expressed in *Escherichia coli*. J. Biol. Chem. *279*, 12141–12151.

Oh, J.W., Ito, T., and Lai, M.M. (1999). A recombinant hepatitis C virus RNA-dependent RNA polymerase capable of copying the full-length viral RNA. J. Virol. *73*, 7694–7702.

Ray, D., Shah, A., Tilgner, M., Guo, Y., Zhao, Y., Dong, H., Deas, T.S., Zhou, Y., Li, H., and Shi, P.-Y. (2006). West Nile Virus 5′-cap structure is formed by sequential guanine N-7 and ribose 2′-*O* methylations by nonstructural protein 5. J. Virol. *80*, 8362–8370.

Rawlinson, S.M., Pryor, M.J., Wright, P.J., and Jans. D.A. (2009). CRM1-mediated nuclear export of dengue virus RNA polymerase NS5 modulates interleukin-8 induction and virus production. J. Biol. Chem. *284*, 15589–97.

Rice, C.M., Lenches, E.M., Eddy, S.R., Shin, S.J., Sheets, R.L., and Strauss, J.H. (1985). Nucleotide sequence of yellow fever virus: implications for flavivirus gene expression and evolution. Science *229*, 726–33.

Schul, W., Liu, W., Xu, H.Y., Flamand, M., and Vasudevan. S.G. (2007). A dengue fever viremia model in mice shows reduction in viral replication and suppression of the inflammatory response after treatment with antiviral drugs. J. Infect. Dis. *195*, 665–674

Selisko, B., Dutartre, H., Guillemot, J.C., Debarnot, C., Benarroch, D., Khromykh, A., Despres, P., Egloff, M.P., and Canard, B. (2006). Comparative mechanistic studies of *de novo* RNA synthesis by flavivirus RNA-dependent RNA polymerases. Virology *351*, 145–158.

Selisko, B., Peyrane, F.F., Canard, B., Alvarez, K., and Decroly, E. (2010). Biochemical characterization of the (nucleoside-2′O)-methyltransferase activity of dengue virus protein NS5 using purified capped RNA oligonucleotides 7MeGpppACn and GpppACn. J. Gen. Virol. *91*, 112–21.

Shuman, S. (2001). Structure, mechanism, and evolution of the mRNA capping apparatus. Prog. Nucleic Acid Res. Mol. Biol. *66*, 1–40.

Steffens, S., Thiel, H.J., Behrens, S.E. (1999). The RNA-dependent RNA polymerases of different members of the family Flaviviridae exhibit similar properties *in vitro*. J. Gen. Virol. *80 (Pt 10)*, 2583–2590.

Stevens, A.J., Gahan, M.E., Mahalingam, S., and Keller, P.A. (2009). The medicinal chemistry of dengue fever. J. Med. Chem. *52*, 7911–7926.

Tan, B.H., Fu, J., Sugrue, R.J., Yap, E.H., Chan, Y.C., and Tan, Y.H. (1996). Recombinant dengue type 1 virus NS5 protein expressed in Escherichia coli exhibits RNA-dependent RNA polymerase activity. Virology *216*, 317–325.

Wengler, G. and Wengler, G. (1981). Terminal sequences of the genome and replicative-from RNA of the flavivirus West Nile virus: absence of poly(A) and possible role in RNA replication. Virology *113*, 544–55.

Yang, S.H., Liu, M.L., Tien, C.F., Chou, S.J., and Chang, R.Y. (2009). Glyceraldehyde-3-phosphate dehydrogenase (GAPDH) interaction with 3′ ends of Japanese encephalitis virus RNA and colocalization with the viral NS5 protein. J. Biomed. Sci. *15*, 16–40.

Yap, T.L., Xu, T., Chen, Y.L., Malet, H., Egloff, M.-P., Canard, B., Vasudevan, S.G. and Lescar. J. (2007a). The crystal structure of the dengue virus RNA-dependent RNA polymerase at 1.85 Å resolution. J. Virol. *81*, 4753–65

Yap, T.L., Chen, Y.L., Xu, T., Wen, D., Vasudevan, S.G. and Lescar. J. (2007b). A Multistep strategy to obtain crystals of the dengue virus RNA-dependent RNA polymerase that diffract to high resolution. Acta Cryst. F. *63*, 78–83.

Yin, Z., Chen, Y.L., Schul, W., Wang, Q.Y., Gu, F., Duraiswamy, J., Kondreddi, R.R., Niyomrattanakit, P., Lakshminarayana, S.B., Goh, A., *et al.* (2009a). An adenosine nucleoside inhibitor of dengue virus. Proc. Natl. Acad. Sci. U.S.A. *106*, 20435–9.

Yin, Z., Chen, Y.L., Kondreddi, R.R., Chan, W.L., Wang, G., Ng, R.H., Lim, J.Y., Lee, W.Y., Jeyaraj, D.A., Niyomrattanakit, P., *et al.* (2009b). N-sulfonylanthranilic acid derivatives as allosteric inhibitors of dengue viral RNA-dependent RNA polymerase. J. Med. Chem. 52, 7934–7.

You, S., and Padmanabhan, R. (1999). A novel *in vitro* replication system for dengue virus initiation of RNA synthesis at the 3′ -end of exogenous viral RNA templates requires 5′- and 3′-terminal complementary sequence motifs of the viral RNA. J. Biol. Chem. *274*, 33714–33722.

You, S., Falgout, B., Markoff, L., and Padmanabhan, R. (2001). *In vitro* RNA synthesis from exogenous dengue viral RNA templates requires long range interactions between 5′- and 3′-terminal regions that influence RNA structure. J. Biol. Chem. *276*, 15581–15591.

Yocupicio-Monroy, M., Padmanabhan, R., Medina, F., and del Angel, R.M. (2007). Mosquito La protein binds to the 3′ untranslated region of the positive and negative polarity dengue virus RNAs and relocates to the cytoplasm of infected cells. Virology *357*, 29–40.

Zhong, W., Ferrari, E., Lesburg, C.A., Maag, D., Ghosh, S.K., Cameron, C.E., Lau, J.Y., and Hong, Z. (2000). Template/primer requirements and single nucleotide incorporation by hepatitis C virus nonstructural protein 5B polymerase. J. Virol. *74*, 9134–9143.

Zhou, Y., Ray, D., Zhao, Y., Dong, H., Ren, S., Li, Z., Guo, Y., Bernard, K.A., Shi, P.-Y., and Li, H. (2007). Structure and function of flavivirus NS5 methyltransferase. J. Virol. *81*, 3891–3903.

Innate Immunity and Flavivirus Infection

7

Maudry Laurent-Rolle, Juliet Morrison and Adolfo García-Sastre

Abstract

Flaviviruses, along with the distantly related Hepacivirus and Pestiviruses, belong to the *Flaviviridae* family. Currently, more than 70 flaviviruses have been reported, including dengue virus serotypes 1 to 4 (DENV1–4), yellow fever virus (YFV), West Nile virus (WNV), Japanese encephalitis virus (JEV) and tick-borne encephalitis virus (TBEV). Flaviviruses are significant human and animal pathogens, creating a global public health challenge with more than 100 million people infected yearly. Typical manifestations of flaviviral disease in humans include acute febrile illness, jaundice, haemorrhagic disease, encephalitis, and even death. Currently, there are no specific antiviral treatments for infection with any of the flaviviruses. An understanding of the interplay between the virus and the host immune system would aid in the development of flaviviral therapeutics. The innate immune system is the host's first line of defence against invading pathogens. Critical components of the innate immune system include natural killer (NK) cells, the complement system, and the ability to recognize pathogens like viruses and induce antiviral cytokines. These components of the innate immune system play complementary roles in limiting viral replication and dissemination, as well as initiation of the adaptive immune response. While all flaviviruses examined thus far suppress host innate immune responses to viral infection, the mechanisms by which this occurs differ among viruses. In this chapter, we will examine the roles that the different arms of the innate immune system play in protecting the host against flavivirus infection. We will also discuss the mechanisms that flaviviruses use to subvert the innate immune system and establish infection.

Natural killer (NK) cells

Natural killer (NK) cells are one of the principal components of the innate immune response and serve to limit viral replication and dissemination even before the adaptive immune system can be activated (Trinchieri, 1989). They are a class of lymphocytes that contain cytoplasmic granules filled with chemicals and enzymes that are toxic to target cells. NK cells can kill target cells through a process known as granule exocytosis. NK cells recognize and kill abnormal cells such as cancer cells (Trinchieri, 1989) and virus-infected cells (Welsh, 1986), and produce a range of cytokines including IFNγ (Perussia, 1996). IFNγ activates macrophages to become more efficient at killing virus-infected cells. The different receptors expressed by NK cells can recognize various ligands and these receptors have both inhibitory and activating functions. Consequently, the fate of a target cell is determined by the balance between these two opposing functions of the receptors expressed on NK cells (Ljunggren and Karre, 1990). The activating receptors bind cell surface molecules that are commonly expressed on virus-infected cells and on immune cells containing intracellular microbes. The inhibitory receptors allow NK cells to recognize molecules expressed on the surface of normal host cells. Inhibitory signals prevent NK cells from becoming activated and killing normal cells.

The best characterized inhibitory signals are the major histocompatibility complex class I

(MHC-I) molecules, which are proteins expressed on all nucleated cells. These MHC-I molecules play an essential role in cellular immunity by presenting intracellular antigens produced from virus-infected cells to specific $CD8^+$ cytotoxic T lymphocytes (CTLs) of the immune system. $CD8^+$ CTLs are T lymphocytes that recognize and kill virus-infected cells by recognizing microbial peptides displayed by MHC-I on the infected cells (Shresta *et al.*, 1998). Therefore $CD8^+$ CTLs are restricted to recognize only cell-associated protein antigens that display the same MHC-I molecules, thus resulting in an antigen-specific immune response. However, engagement of an NK cell's inhibitory receptors with MHC-I molecules prevents the NK cell from eliminating the cell expressing the MHC-I molecules (Moretta *et al.*, 1996). Many viruses, including herpesviruses, poxviruses and adenoviruses, downregulate MHC-I molecules on their surface so as to escape elimination by virus-specific $CD8^+$ CTLs (Lodoen and Lanier, 2005). Consequently, this lack of cell surface MHC-I results in the activation of NK cells since the inhibitory receptors of the NK cells are not engaged. This results in the destruction of infected cells by NK cells (Moretta *et al.*, 1996; Lanier, 1998, 2008a,b).

Flaviviruses evade natural killer cell cytotoxicity

Virus-infected patients with NK-cell deficiencies exemplify the key role that NK cells play in the innate immune response. These patients tend to have severe and recurrent viral infections especially with members of the herpesvirus family (Biron *et al.*, 1989; Scalzo, 2002; Yokoyama and Scalzo, 2002; Lodoen and Lanier, 2006). Although NK cells are obviously important for controlling herpesvirus infections, the contribution of these cells to controlling flavivirus infection is less clear. Immunologic studies done in an immunocompetent inbred A/J mouse model showed a transient increase in NK cells in the spleen of dengue virus-infected mice at 3 days post infection (Shresta *et al.*, 2004a). Transient activation of NK cells was also observed in mice infected with WNV (Vargin and Semenov, 1986). These findings are supported by studies done in dengue-virus-infected patients. These patients showed a significant increase in the number and the activation of circulating NK cells during the early phase of dengue fever. These studies suggest that NK cells play an early role in controlling dengue virus infection and this has been associated with mild clinical dengue fever (Azeredo *et al.*, 2006). A study done with human liver tissue from patients who died of yellow fever showed increased numbers of NK cells in the area of the liver affected by YFV and points to the role that NK cells play in clearing virus-infected cells (Quaresma *et al.*, 2006, 2007). In support of these studies, early experiments by Kurane *et al.*, (1984) demonstrated that human NK cells are effective at lysing dengue virus-infected cells thus contributing to the elimination of infected cells (Kurane *et al.*, 1984).

More recent work demonstrated that activation of NK cells by flaviviruses might be mediated by a common mechanism through the interaction of the NK-activating receptor, NKp44, with the envelope (E) glycoprotein of flaviviruses (Hershkovitz *et al.*, 2009). NKp44 is one of the major activating receptors of NK cells and is involved in triggering the cytolytic activity of NK cells against virus-infected cells (Vitale *et al.*, 1998; Moretta *et al.*, 2001; Arnon *et al.*, 2006). Yet, studies done by an independent group suggest that NK cells do not play a significant role in controlling WNV infection in mice. Depletion of NK cells in mice using an antibody against the NK1.1 antigen showed no difference in morbidity and mortality compared to controls upon infection with WNV (Shrestha *et al.*, 2006). One plausible explanation to account for this difference in NK cell contribution to flavivirus protection in mice and humans is the lack of the murine homologue for the NKp44 receptor (Hershkovitz *et al.*, 2009). The results of this study add to the debate about the role that NK cells play in flavivirus infection.

While some viruses successfully evade immune detection by downregulating MHC-I molecules (Lodoen and Lanier, 2005), flaviviruses upregulate MHC-I molecules in vertebrate cells perhaps as a mechanism to escape early NK cell recognition of non-MHC-I-expressing cells (Lobigs *et al.*, 1996, 2003). Cells infected with WNV have increased expression of MHC-I. The increased expression of MHC-I in WNV-infected cells is likely due to an enhancement of peptide transport into the lumen of the endoplasmic reticulum (ER) by transporters associated with antigen presentation (TAP), and an increase in NF-κB-dependent gene transcription of MHC-I (Kesson and King, 2001; Momburg *et al.*, 2001; Kesson *et al.*, 2002;). Upregulation of MHC-I molecules appears to be independent of endogenous secreted cytokines like type I interferon, and might be dependent on some component of the virus (King and Kesson, 1988; Cheng *et al.*, 2004). It has been shown that the non-structural proteins of dengue virus are sufficient to enhance the upregulation of MHC-I. MHC-I is upregulated in dengue-virus-infected cells and in cells stably expressing dengue virus subgenomic replicons (Hershkovitz *et al.*, 2008). Whether this increased MHC-I expression in response to flavivirus infection is due to the presence and integrity of flaviviral RNA or to a specific flaviviral protein is unclear. Interestingly, JEV, a related flavivirus, induces MHC-I in an NF-κB-independent but type I interferon-dependent manner (Abraham *et al.*, 2010) reminding us of the inherent differences in the mechanisms that different flaviviruses use to ensure their success as pathogens.

It does seem paradoxical that flaviviruses would induce the upregulation of cellular factors (MHC-I) that would increase their susceptibility to elimination by MHC-class-I-restricted CTLs in order to inhibit NK cell cytolytic activity (Douglas *et al.*, 1994; Mullbacher and Lobigs, 1995). The answer may come down to timing. It has been proposed that the selective pressure of innate immunity has had a greater impact on the evolution of arthropod-borne flaviviruses than has adaptive immunity (Lobigs *et al.*, 2003a). Unlike the distantly-related hepatitis C virus, flaviviruses do not establish persistent infections in the vertebrate host. Additionally, infection of the arthropod vector is essential for transmission of the virus, which must occur during the short viremic period in the vertebrate host. This would result in the selection of viruses that can rapidly replicate to high titres in host cells so as to achieve an increased duration of viraemia (Monath, 1994; Lobigs *et al.*, 2003a). This would give flaviviruses an advantage in their transmission cycle between the mammalian host and the arthropod vector, thus enhancing their survival. Therefore, an effective adaptive immune response that clears flaviviruses at a later stage during infection would not exert significant evolutionary pressure on flaviviruses as their life cycle continues within the arthropod vector (Fernandez-Garcia *et al.*, 2009). It has been suggested that the upregulation of MHC-I molecules in flavivirus-infected cells may be a by-product of an elegant strategy of virus assembly involving the C-prM precursor rather than a mechanism of immune evasion (Lobigs *et al.*, 2004). However, studies have shown enhanced upregulation of MHC-I in cells stably expressing dengue virus replicons which do not express C-prM (Hershkovitz *et al.*, 2008). In summary, the ability of flaviviruses to inhibit NK-cell-mediated innate immunity by inducing an increase in MHC-I expression provides further evidence of the importance of the innate immune system in controlling flavivirus infection.

The complement system

The complement system is a group of more than thirty different soluble and membrane- associated proteins that work in concert to recognize altered self ligands, pathogens and immune complexes (Blue *et al.*, 2004; Avirutnan *et al.*, 2008, 2010). Activation of these proteins occurs sequentially upon encountering pathogens or antibodies, and triggers several antiviral responses including opsonization of the pathogen for phagocytosis, release of anaphylatoxins, which attract leucocytes and contribute to the host inflammatory response, and formation of the membrane attack complex (MAC), which lyses infected cells. The proteins of the complement system are produced predominantly by the liver, and are present in high concentrations in the blood and other tissues (Walport, 2001a,b; Blue *et al.*, 2004; Avirutnan *et al.*, 2010). The most abundant complement protein is C3, which is spontaneously hydrolysed in plasma into smaller unstable proteins (Lachmann and Hughes-Jones, 1984). One of the products of C3 hydrolysis is C3b, which plays a key role in the complement system. The complement cascade may be activated by one of three distinct pathways, the alternative, the classical, or the lectin pathway. These pathways are initiated with different recognition molecules but they all converge at the point of C3 cleavage and have the same effector functions (Walport, 2001a,b; Blue *et al.*, 2004).

The proteins of the alternative pathway are called factors, followed by a letter, such as factor B or factor D. The alternative pathway is spontaneous and is triggered when the product of C3 hydrolysis, C3b, is deposited on the surface of a microbe (Pangburn *et al.*, 1980). C3b binds either an amino or hydroxyl group found in proteins and carbohydrates on the surface of microbes, and activates downstream complement proteins. The C3b–microbe complex becomes a substrate for the binding of the Bb fragment. The Bb fragment is generated by cleavage of factor B by factor D. The Bb fragment remains attached to the C3b–microbe complex forming C3bBb, which functions as the alternative pathway C3 convertase. The C3 convertase can break down more C3 generating many more C3b and C3bBb molecules, which attach to the microbe. Some of the C3b molecules bind the C3bBb complex to form the C3bBb3b complex, which functions as the C5 convertase of the alternative pathway. This C5 convertase will cleave the complement protein C5 to form C5a, which is an anaphylatoxin, and C5b, which initiates the formation of C5b-C9, the membrane attack complex that lyses pathogens and infected cells (Walport, 2001a,b).

The proteins of the classical pathway are designated C1 through C9. The classical pathway is initiated when the C1 complement protein complex (which consists of C1q, two molecules of C1r and two molecules of C1s) binds to the Fc regions of an antibody bound to an antigen on the surface of microbes (Manderson *et al.*, 2001). C1 becomes enzymatically active (C1r and C1s) when bound to the antibody–antigen complex allowing for binding and cleavage of two other complement proteins, C4 and C2. The breakdown products C4b and C2a form the C4b2a complex, which functions as the classical pathway C3 convertase. The C3 convertase cleaves C3 and generates more C3b, some of which bind the C4b2a complex to form the C4b2a3b complex. The C4b2a3b complex functions as the C5 convertase of the classical and lectin pathways (Walport, 2001a,b).

The lectin pathway is initiated when the complex of mannose binding lectin (MBL) (Weis *et al.*, 1992) and the serine proteases, mannose-binding lectin-associated proteases 1 and 2 (MASP1 and MASP2), recognizes and binds carbohydrate structures on the surface of microbes or unhealthy apoptotic human cells (Matsushita and Fujita, 1992; Thiel *et al.*, 1997; Vorup-Jensen *et al.*, 1998). MBL is structurally similar to C1q and, upon binding its ligand, activates MASP1 and MASP2. MASP2 binds and cleaves C4 and C2 (Matsushita and Fujita, 1992; Thiel *et al.*, 1997; Matsushita *et al.*, 2000). The subsequent activation steps are shared with the classical pathway.

Regulation of the complement pathway is essential to prevent aberrant activation because this could lead to tissue damage. The host expresses regulatory proteins both in solution and on the surface of cells (Hourcade *et al.*, 1989). Some regulatory mechanisms include:

1. The catabolism of C3b and C4b to inactive products that is mediated by the plasma serine protease, factor I. This process involves the complement regulatory enzymes, factor I and factor H, and the co-factors, complement receptor type 1, CR1 (also called CD35) and membrane co-factor protein (also called CD46) (Kazatchkine *et al.*, 1979; Liszewski *et al.*, 2000; Fukui *et al.*, 2002).
2. The dissociation of the C3 and C5 convertases by the membrane protein decay-accelerating factor (DAF). DAF, along with complement receptor 1, C4 binding protein, C4BP and factor H, disrupts the binding of factor B to C3B and the binding of C4b2a to C3b, terminating both the alternative and classical pathways of complement (Weiler *et al.*, 1976; Scharfstein *et al.*, 1978; Krych-Goldberg *et al.*, 1999; Krych-Goldberg and Atkinson, 2001).
3. Inhibition of the formation of the MAC by the membrane regulator CD59 (also known as protectin) (Meri *et al.*, 1990; Davies and Lachmann, 1993).

Viruses have evolved several strategies to evade recognition by and activation of the complement cascade. These strategies include incorporation of complement regulators on the surface of the virion (Marschang *et al.*, 1995; Saifuddin *et al.*, 1995), secretion of viral products that functionally or structurally mimic complement regulatory proteins (Kotwal *et al.*, 1990), and upregulating complement regulatory proteins to prevent complement-dependent cell lysis (Spiller *et al.*, 1996; Nomura *et al.*, 2002; Lambris *et al.*, 2008).

Flaviviruses evade complement

Complement can inhibit flavivirus infection by stimulating the adaptive immune response, but it may also play a role in the pathogenesis of flaviviruses (Mehlhop *et al.*, 2005; Avirutnan *et al.*, 2008). WNV infection of mice deficient in components specific for each complement pathway showed increased viral burden and reduced antiviral antibody titres when compared to wild type mice (Mehlhop and Diamond, 2006). The non-structural protein (NS1) of dengue appears to be involved in activating complement and was associated with increased levels of C5b-9 in patient serum (Avirutnan *et al.*, 2006). This resulted in cell lysis and directly enhanced virus infection. Although the mechanism is unclear, DENV NS1 has been implicated in the pathogenesis of dengue haemorrhagic fever/dengue shock syndrome (DHF/DSS). The high levels of NS1 detected in the sera of dengue-virus-infected patients correlate with severe disease (Young *et al.*, 2000; Libraty *et al.*, 2002). It has been proposed that NS1 facilitates immune complex formation and causes endothelial cell damage (Avirutnan *et al.*, 1998; Young *et al.*, 2000; Lin *et al.*, 2002).

However, more recent studies show that the NS1 of flaviviruses plays a role in limiting complement activation to attenuate the host immune response. Protein interaction studies show that the soluble secreted form of WNV NS1 binds to and recruits the complement regulatory protein, factor H, resulting in increased proteolytic cleavage of C3b. Furthermore, cell surface NS1 binds factor H to decrease the deposition of C3 fragment and C5b-9 membrane attack complexes on cell surfaces. Thus WNV NS1 antagonizes the immune system by decreasing complement activation and complement-dependent cell lysis allowing for continued virus replication in the cell (Chung *et al.*, 2006). Other studies show that DENV NS1 binds to clusterin, a regulator of complement, and inhibits formation of the MAC (Kurosu *et al.*, 2007). This immune evasive function of NS1 appears to be conserved among several flaviviruses. The NS1 proteins of YFV, DENV, and WNV specifically inhibit both the classical and lectin pathway of complement activation by binding to C4 and C1s. This interaction reduced the C3 convertase (C4b2a) activity and the deposition of C4b by enhancing cleavage of C4 through the recruitment of the complement-specific protease, C1s, thus protecting virus-infected cells from complement-mediated lysis. The exact mechanism by which the interaction of the flavivirus NS1 protein with C1s results in an increased cleavage of C4 is unclear. One possibility is that binding of NS1 to C1s may cause a structural change that increases the proteolytic activity of C1. Another possibility may be that the interaction of NS1 may bring the C4 and C1s together and this interaction may enhance cleavage of the C4 to C4b by the C1s thereby decreasing the amount of C4 available for complement activation (Avirutnan *et al.*, 2010).

Cytokines of innate immunity

Type I interferon (IFNα/β)

The alpha/beta interferon response (IFNα/β) is one of the host's main innate immune mechanisms against viral infection and is present in all cell types. The IFNα/β response occurs during the early stages of viral infection and acts to limit viral replication and dissemination. In mammalian cells, innate immune responses to flavivirus infection are initiated when pathogen recognition receptors (PRRs) including the membrane-bound RNA sensors, Toll-like receptors 3 and 7 (TLR3 and TLR7), and the cytoplasmic RNA sensors, retinoic acid-inducible gene I (RIG-I) and melanoma differentiation-associated gene 5 (MDA5), bind conserved pathogen-associated molecular patterns (PAMPs) like double-stranded RNA formed during viral replication and 5′-triphosphate single-stranded RNA (Kumar *et al.*, 2011). Signalling through these sensors leads to the activation of downstream latent transcription factors including NF-κB, ATF2/c-Jun, and IRF3/IRF7, culminating in the subsequent induction of IFNβ and proinflammatory cytokines (Lenardo *et al.*, 1989; Du and Maniatis, 1992; Carey, 1998). (Fig. 7.1 and 7.2).

The innate antiviral response is amplified when secreted IFNα/β binds to a common multi-chain IFNα/β receptor (IFNAR1/IFNAR2) (Cleary *et al.*, 1994; Novick *et al.*, 1994) in an autocrine or paracrine fashion resulting in the activation of Jak1 and Tyk2. Activated Jak1 and Tyk2 then phosphorylate the intracellular domains of the receptor subunits creating a docking site for STAT2 (Colamonici *et al.*, 1994; Colamonici *et al.*, 1995; Domanski *et al.*, 1997; Schindler *et al.*, 2007; Zhao *et al.*, 2008). The Janus kinases phosphorylate STAT2 on tyrosine 689 (Y-689) allowing STAT2 to act as an adaptor, recruiting STAT1 to the receptor so it can be phosphorylated on tyrosine 701 (Y-701). Tyrosine-phosphorylated STAT1 and STAT2 heterodimerize and bind to a third component, IRF9/p48. Phosphorylated STAT1 and STAT2 along with the associated p48/IRF9 form a transcriptional complex called interferon-stimulated gene factor 3 (ISGF3). ISGF3 then translocates to the nucleus and binds to interferon-stimulated response element (ISRE) sequences found in the promoter/enhancer regions of interferon- stimulated genes (ISGs). This results

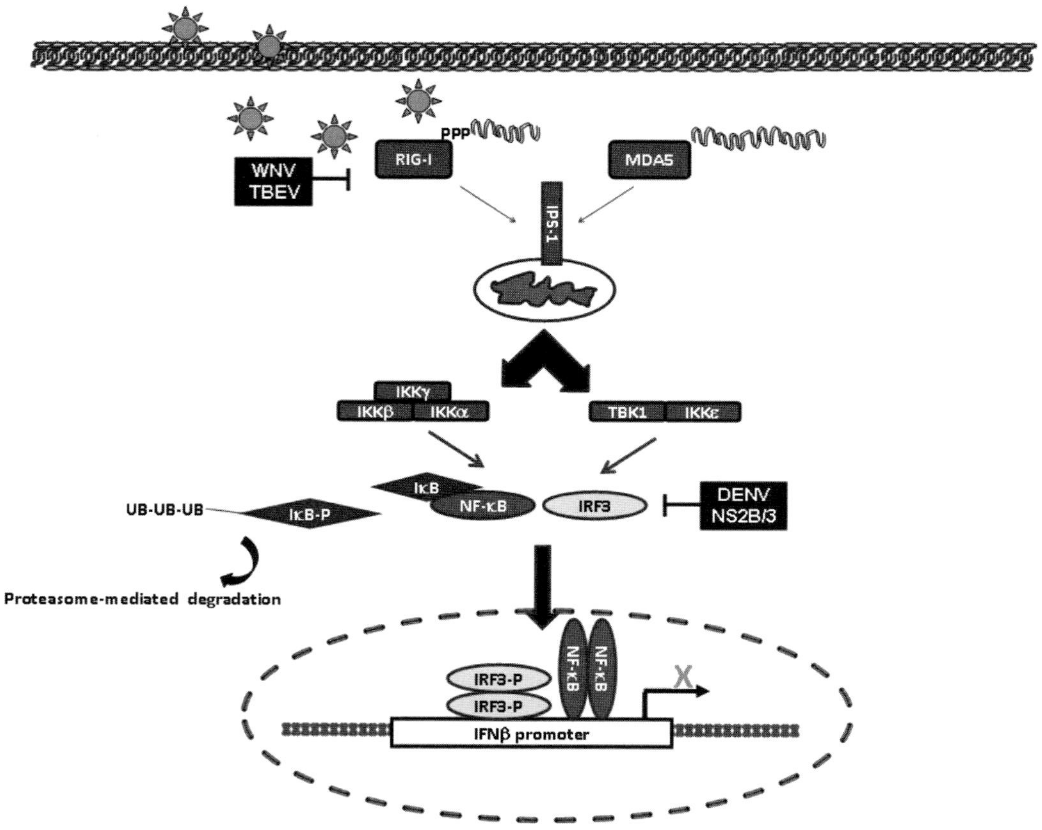

Figure 7.1 Flaviviruses disrupt RIG-I/MDA5 signaling pathways. The host innate immune response to flaviviruses is activated when RIG-I/MDA5 binds to conserved pathogen-associated molecular patterns (PAMPs) like double-stranded RNA intermediates formed during viral replication and 5'-triphosphate single-stranded RNA. Signaling through these sensors leads to the recruitment and activation of the adaptor protein IFNb promoter stimulator-1 (IPS-1) which is also known as MAVS, VISA or Cardif via CARD-CARD association with RIG-I/MDA5. This results in the activation and nuclear translocation of IRF3 and NF-kB and subsequent induction of IFNb and proinflammatory cytokines. Putative mechanisms of evasion of PRR signaling by flaviviruses include delay in activation of the RIG-I signaling pathway by WNV and TBEV, and inhibition of IRF3 activation by DENV NS2B/3.

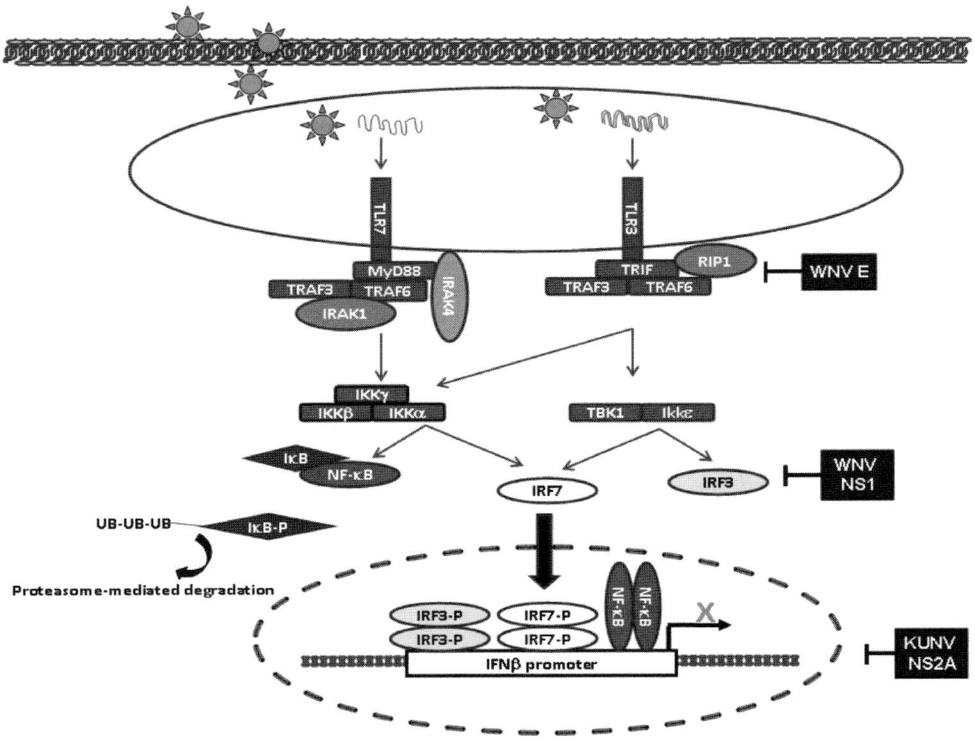

Figure 7.2 Flaviviruses disrupt TLR signaling. Flavivirus PAMPs which are incorporated into endosomal compartments of DCs and macrophages can be recognized by TLR3 and TLR7. RNA binding to these sensors leads to the recruitment and binding of the adaptor molecules TRIF and MyD88 respectively. TLR3-mediated TRIF signaling activates IRF3, IRF7, and NF-kB and results in the induction of the IFNb promoter. TLR7-mediated MyD88 signaling specifically activates IRF7 in addition to NF-kB. Putative mechanisms of evasion of TLR signaling by flaviviruses include inhibition of RIP1 signaling by WNV E protein, inhibition of IRF3 and NF-kB activation and nuclear translocation by WNV NS1, and inhibition of IFNb gene induction by KUN NS2A.

in transcription and translation of ISGs that can both directly or indirectly create an antiviral state within the cell (Platanias, 2005; Schindler *et al.*, 2007). The IFN system is summarized in Fig. 7.3. While hundreds of genes are induced by IFN, only a few encode proteins that have well characterized roles in the context of viral infection. These include dsRNA-dependent protein kinase (PKR), 2¢–5′ oligoadenylate synthase (2–5 OAS), RNase L, and Myxovirus resistant (Mx) GTPases. PKR is activated by double-stranded RNA and phosphorylates the transcription factor eIF-2α to prevent translation of viral proteins. Myxovirus resistant proteins are small GTPases which are believed to be involved in redistribution of viral capsid proteins thus inhibiting viral replication. 2¢5¢-OAS produces oligoadenylates which activate RNAse L leading to the degradation of viral RNAs and the induction of apoptosis (Samuel, 2001; Sadler and Williams, 2008).

Type I IFN also has potent immunomodulating activities, and serves to regulate the responses of cytotoxic T cells (CTLs) and natural killer (NK) cells. These cells eliminate virus-infected cells further curbing viral dissemination. Furthermore, type I IFN enhances the adaptive immune response by stimulating dendritic cells, and by activating B cells and T cells (Biron, 2001; Stetson and Medzhitov, 2006).

Type II interferon (IFNγ)

Interferon gamma (IFNγ) is produced by NK cells and T lymphocytes. It has potent antiviral and immunomodulating activities. The binding of IFNγ to the interferon gamma receptor, IFNGR, which consists of IFNGR1 and IFNGR2, leads to the activation of the Janus kinases, Jak1 and Jak2, through phosphorylation. Activated Jak1 phosphorylates the IFNGR1 chain, which serves as the binding site for STAT1. STAT1 subsequently becomes phosphorylated by Jak2. Phosphorylated STAT1 forms a homodimer called the gamma-activated factor (GAF) complex. GAF translocates to the nucleus and initiates the transcription of a subset of ISGs by binding to the gamma-activated sequence (GAS) within their promoters (Boehm *et al.*, 1997; Darnell, 1997; Stark *et al.*, 1998).

Type III interferon (IFNλ/IL-28)

Type III IFNs, IFNl1, IFNl2, and IFNl3, (also known as IL29, IL28A and IL28B, respectively) are cytokines with type-I-IFN-like antiviral activities. Like type I IFN, cells express type III IFN after virus infection or Toll-like receptor (TLR) stimulation (Onoguchi *et al.*, 2007; Osterlund *et al.*, 2007; Donnelly and Kotenko, 2010). These cytokines signal through a receptor complex consisting of IL10Rβ and IL28Rα chains (Kotenko *et al.*, 2003; Sheppard *et al.*, 2003). Like type I IFN, these cytokines signal through the Jak1/Tyk2 tyrosine kinases and ISGF3 complex, consequently leading to a type-I-IFN-like gene expression profile (Ank *et al.*, 2006; Zhou *et al.*, 2007). However, differences do exist between type I and type III IFN. A study that compared the antiviral effects of IFNα and IFNλ against WNV, demonstrated that IFNα is a more potent inhibitor of WNV (Ma *et al.*, 2009). Studies point to a role for type III IFN in resolution of HCV infection. IFNl1 inhibits HCV replication, induces ISGs and enhances the antiviral efficacy of IFNα (Marcello *et al.*, 2006). Furthermore, a genome-wide association study of chronically-infected HCV patients identified a single-nucleotide polymorphism (SNP) on chromosome 19q13 that maps upstream of the IL28B gene, which encodes IFNl3. This SNP was strongly associated with an enhanced positive response to HCV treatment in patients (Thomas *et al.*, 2009).

Negative regulation of the IFN signal transduction cascade

Several negative regulators of the JAK-STAT signalling pathway have been identified. Suppressors of cytokine signalling (SOCS), protein inhibitor of activated STATs (PIAS), and protein tyrosine phosphatases (PTPs) are involved in regulating the magnitude of cytokine-induced signal transduction (Wormald and Hilton, 2004). The cytoplasmic SOCS protein family is involved in regulating the JAK/STAT signalling cascade by directly preventing phosphorylation of the receptor or its associated kinases. All eight members of the SOCS protein family have a central SH2 domain and a C-terminal SOCS box (Alexander and Hilton, 2004; Akhtar and Benveniste, 2011). SOCS-1 and SOCS-3 have been shown to inhibit type I IFN signalling (Song and Shuai, 1998).

Protein tyrosine phosphatases are defined by the presence of a conserved signature motif (I/V)HCXAGXXR(S/T) at their active sites. PTPs regulate the signal transduction cascade by dephosphorylating phosphorylated signalling components (Wu *et al.*, 2002; Xu and Qu, 2008). CD45 and PTP1B, both PTPs, have been reported to inhibit IFN signalling by tyrosine dephosphorylation of the Janus kinases (Myers *et al.*, 2001).

Pattern recognition receptors that sense flaviviruses

Cytoplasmic RNA helicases: RIG-I and MDA5

Retinoic-acid-inducible gene I (RIG-I) and melanoma differentiation-associated protein 5 (MDA5) play a significant role in sensing viral infection and activating the host immune response. RIG-I and MDA5 both belong to the RIG-I-like receptor family (RLR) of DExD/H RNA helicases and contain an ATP-dependent helicase domain. Both RIG-I and MDA5 have two N-terminal CARD domains that are required for physically interacting with the adaptor protein, IFNβ-promoter stimulator-1 (IPS-1) also known as MAVS, VISA or CARDIF, and for downstream signalling (Yoneyama and Fujita, 2010; Kumar *et al.*, 2011). The regulatory/repressor domain (RD) at the C-terminus of RIG-I negatively regulates RIG-I activity in order to prevent aberrant signalling (Saito *et al.*, 2007). In contrast, MDA5 is negatively regulated by dihydroxacetone kinase (DAK) protein and not by its repressor domain (Diao *et al.*, 2007). Studies have also suggested that the RD is important in recognition of 5′-triphosphate RNA (Takahasi *et al.*, 2008, 2009). RIG-I and MDA5 are ubiquitously expressed in low abundance, and are IFN-inducible. Both sensors are activated by dsRNA, with RIG-I preferentially binding dsRNA with an exposed 5′-triphosphate group and MDA5 recognizing long dsRNA molecules. The binding of RNA to RIG-I/MDA5 activates these sensors leading to the recruitment and activation of IPS-1 via CARD–CARD association with RIG-I/MDA5 (Kawai *et al.*, 2005; Kumar *et al.*, 2011). This results in the activation of IRF3 and NF-κB and the subsequent induction of the IFNβ promoter (Yoneyama and Fujita, 2010) (Fig. 7.1). RIG-I and MDA5 have both been implicated in recognition of flavivirus RNA and thus contribute to the induction of host responses to flaviviruses (Chang *et al.*, 2006; Fredericksen and Gale, 2006; Kato *et al.*, 2006; Loo *et al.*, 2008).

Microarray analysis of WNV-infected mouse embryonic fibroblasts (MEFs) deficient in RIG-I revealed that the host response to WNV occurs in two distinct waves. The initial response, which includes the expression of IRF3-dependent genes and IFN-stimulated genes, was delayed at early times during infection of the $RIG^{-/-}$ MEFs compared to wild type MEFs. This suggests that RIG-I plays a significant role in the early detection of WNV infection in cells. However, no significant difference was observed between the $RIG^{-/-}$ MEFs and the WT MEFs in the second wave of IFN-dependent antiviral gene expression that occurred at later times during WNV infection. $RIG^{-/-}$ cells were still able to respond to WNV infection suggesting the contribution of another pattern recognition receptor (Fredericksen *et al.*, 2006). Further studies showed that MDA5 plays a considerable role in sensing WNV during the second phase of the innate immune response. WNV-infected MEFs deficient in IPS-1 demonstrated a severe phenotype including decreased IRF3 activation, delayed induction of host interferon production and ISG responses as well as an increase in viral replication. Thus in response to WNV infection, RIG-I and MDA5 operate in a concerted fashion to establish an antiviral state. More recent *in vivo* studies further demonstrate the essential role that IPS-1 plays in bridging the innate and adaptive immune response to effectively control WNV infection (Fredericksen *et al.*, 2008; Suthar *et al.*, 2010).

Cell-type-specific differences in RIG-I and MDA5 recognition of viruses exist. WNV-infected $MDA5^{-/-}$ myeloid dendritic cells produced IFN comparably to wild-type cells, suggesting that MDA5 is not essential for detecting WNV in these cell types. The molecular processes underlying the recognition of specific viruses in specific cells are unknown. They may be dependent on the ability of the virus to replicate to high enough levels in certain cells to be detected by the sensors or they may be dependent on the expression levels of the sensors. Furthermore, the role of other unknown PRRs as well as transcriptional activators like IRF7 in response to flaviviruses have to be considered since WNV-infected $IRF3^{-/-}$ animals produce levels of IFNα and IFNβ similar to wild type animals (Bourne *et al.*, 2007; Daffis *et al.*, 2007).

TLR3

TLR3 is one of the 13 mammalian TLRs identified to date. TLR3, along with TLR7 and TLR8, was identified as a sensor of RNA viruses (Alexopoulou *et al.*, 2001; Diebold *et al.*, 2004). TLR3 is expressed on the surface of fibroblasts and on the endosomes of myeloid cells. The extracellular domain of TLR3 consists of a leucine-rich repeat motif which is involved in pathogen recognition. The intracellular cytoplasmic Toll/interleukin 1 receptor (TIR) domain is involved in downstream signalling through interactions with adaptor molecules (Rock *et al.*, 1998; Bell *et al.*, 2003). TLR3 was the first PRR shown to induce IFN production in response to dsRNA. Initial studies done in $TLR3^{-/-}$ macrophages and $TLR3^{-/-}$ mice showed reduced IFN induction in response to the synthetic dsRNA polyriboinosinic:polyrib ocytidylic acid, poly (I:C) (Alexopoulou *et al.*, 2001). Following activation by dsRNA, TLR3 signals through its adaptor protein, TIR-domain-containing adaptor inducing IFN (TRIF). The recruitment and activation of the non-canonical IKK kinases lead to the phosphorylation and nuclear translocation of IRF3, activator protein 1 (AP1) and p50/p65 (NF-κB). These transcription factors act in a concerted fashion to initiate the transcriptional activation of the IFNβ promoter (Fitzgerald *et al.*, 2003). This leads to the production of type I IFN as well as proinflammatory cytokines such as IL-6, IL-8 and RANTES (Alexopoulou *et al.*, 2001) (Fig. 7.2).

The physiological importance of TLR3 in responding to WNV infection is not clear in light of contradictory studies. TRIF-deficient MEFs had similar WNV titres compared to WT MEFs suggesting that TLR3 may be dispensable for recognition of flavivirus replication in cells (Fredericksen and Gale, 2006). In keeping with the role that TLR3 signalling plays in inducing an inflammatory response, WNV-infected mice deficient in TLR3 showed impaired cytokine production and enhanced virus titres in the periphery but decreased viral load and inflammatory response in the brain compared with wild-type mice. This presumably occurred because of the reduced circulating levels of cytokines like TNFα that normally facilitate WNV entry into the brain (Wang *et al.*, 2004). Another group showed that $TLR3^{-/-}$ mice have increased WNV burden in the brain as well as enhanced mortality suggesting that TLR3 protects against WNV (Daffis *et al.*, 2008a). Currently it is not clear why the two groups observed these differences. One plausible explanation for these contradictory reports is the WNV stock used in the different experiments. It is possible that the different preparations of the virus contained additional factors or defective particles that could modulate the response observed in either experiment (Daffis *et al.*, 2008a). It appears that with functional RIG-I/MDA5 sensors, IFN production does not require TLR3 signalling. However it is clear that this receptor plays a role in controlling some viral infections since patients with a TLR3 deficiency are predisposed to more severe infections from HSV-1 (Zhang *et al.*, 2007) and to influenza A virus-induced encephalopathy (Hidaka *et al.*, 2006).

TLR7

TLR7 is a PRR predominantly found in the endosomal compartment of plasmacytoid dendritic cells (pDCs), and is involved in the induction of IFNα. It detects ssRNA rich in guanosine and uridine (Diebold *et al.*, 2004; Diebold, 2008). TLR7 transmits signals via the adaptor protein Myd88 and recruits downstream signalling molecules including IRF7 to form a signalling complex leading to the induction of IFN and proinflammatory cytokines (Honda *et al.*, 2004; Kawai *et al.*, 2004) (Fig. 7.2). TLR7 and Myd88 are essential for IFN production in pDCs following infection with RNA viruses. Dengue virus and YFV have been shown to activate DCs by triggering TLR7 (Querec *et al.*, 2006; Wang *et al.*, 2006). IRF7 plays a key role in inducing antiviral genes upon flavivirus infection. $IRF7^{-/-}$ mice demonstrate uncontrolled WNV replication, and decreased systemic IFN production. They rapidly succumb to WNV infection when compared to control mice (Daffis *et al.*, 2008b).

Interferon modulates flavivirus infection

The rapid host response to flaviviruses mediated by type I IFN occurs quickly and results in the induction of hundreds of ISGs that serve to limit flavivirus infection (Der *et al.*, 1998). Recent work has shown that several ISGs mediate antiviral effector functions against flaviviruses. The 2¢,5′-oligoadenylate synthetase (OAS) proteins are a group of interferon-inducible enzymes which become active upon binding dsRNA. Enzymatically active OAS produces 2′- to 5′-linked oligoadenylates. 2′- to-5′linked oligoadenylates bind and activate RNAse L. Activated RNase L degrades viral RNA in the early stages of the virus life cycle, thus preventing viral RNA accumulation in infected cells (Zhou *et al.*, 1993, 1997; Samuel, 2002; Kajaste-Rudnitski *et al.*, 2006; Scherbik *et al.*, 2006). Two independent groups identified the Oas1b isoform of the *Oas* gene as the gene that confers resistance to flaviviruses in mice (Mashimo *et al.*, 2002; Perelygin *et al.*,

2002). All mouse strains susceptible to flavivirus infection contain a nonsense mutation in exon 4 of the Oas1b gene that results in the expression of a truncated OAS isoform which lacks enzymatic activity. Furthermore, knock-in of the full length Oas1b allele into a flavivirus-susceptible mouse strain confers protection against yellow fever virus (Scherbik *et al.*, 2007). Therefore, Oas1b is a key host restriction factor for flaviviruses in mice. Interestingly, patients with a single nucleotide polymorphism (SNP) in exon 2 of OasL were more susceptible to severe WNV disease compared to control subjects. This polymorphism contains a splice enhancer sequence that increases splicing of RNA transcripts resulting in a truncated mutant form of Oas similar to that found in mice (Yakub *et al.*, 2005).

PKR is another antiviral effector molecule that has antiviral functions against flaviviruses. PKR is activated by double-stranded RNA that is generated during viral replication and then phosphorylates the transcription factor eIF-2α to prevent translation of host and viral proteins (Chong *et al.*, 1992; Meurs *et al.*, 1992). Initial studies with PKR indicated that it may play a contributory role in IFN production (Hu and Conway, 1993; Kumar *et al.*, 1997; Marcus and Sekellick, 1988; Zinn *et al.*, 1988). Subsequent studies have shown that PKR plays a role in the production of IFN in bone-marrow derived dendritic cells (BM-DCs) in response to poly (I:C) (Diebold *et al.*, 2003). More recent studies show that $PKR^{-/-}$ MEFs infected with WNV-derived virus-like particles (VLPs) produced less IFN than WT MEFs. Furthermore, treatment of a number of different human cell lines and MEFs with a PKR pharmacological inhibitor blocked IFN production in WNV-VLP-infected cells but not in Sendai-virus-infected cells indicating a specificity of PKR for WNV (Gilfoy and Mason, 2007). Both *in vitro* and *in vivo* studies show that PKR plays a significant role in controlling WNV infection. WNV-infected, PKR-deficient mice showed increased lethality with high viral burdens in the periphery (Gilfoy and Mason, 2007). The precise mechanism by which PKR exerts its antiviral function against WNV is not clear and may be due to inhibition of viral protein translation or to its role in IFN induction.

A number of other ISGs have been shown to play a role in the antiviral response against flaviviruses including: the interferon-inducible transmembrane proteins, IFITM1, 2 and 3 (Brass *et al.*, 2009); tumour necrosis factor-related apoptosis inducing ligand (TRAIL) (Warke *et al.*, 2008) and interferon-inducible gene 20 (ISG20) (Jiang *et al.*, 2010). It is likely that future studies will define the specific antiviral effector mechanisms used by other ISGs to limit flavivirus infection.

Flaviviruses disrupt the IFN response

Virus infection induces the secretion of IFNα/β (Katze *et al.*, 2002; Sen, 2001). However, upon infection of cells, flaviviruses counteract the antiviral action of IFN to increase intracellular viral replication and consequently disease severity. Treating cells with IFNα/β before infection protects human cells against flavivirus infection (Diamond *et al.*, 2000). Interestingly, treatment with IFN after infection does not inhibit infection illustrating that viral replication renders infected cells refractory to the antiviral action of IFN (Diamond *et al.*, 2000). Although it is now evident that flaviviruses evade the host IFN response, IFN still restricts viral replication and spread *in vivo*. Pre-treating mice and hamsters with IFN attenuates infection by flaviviruses (Brooks and Phillpotts, 1999; Leyssen *et al.*, 2001, 2003; Morrey *et al.*, 2004; Julander *et al.*, 2007; Diamond, 2009). Furthermore, type I IFN receptor-deficient mice infected with flaviviruses show enhanced flaviviral replication in normally resistant cell populations and enhanced lethality, suggesting that IFN restricts viral dissemination *in vivo* (Lobigs *et al.*, 2003b; Shresta *et al.*, 2004b; Samuel and Diamond, 2005; Meier *et al.*, 2009; Diamond, 2009). It is now evident that all flaviviruses examined have evolved means to successfully evade the type I IFN response at different levels of the pathway including (a) suppression of IFNβ gene induction (see Fig. 7.1 and Fig. 7.2), (b) inhibition of the IFN-signalling cascade and (c) inhibition of antiviral activities of IFN-stimulated genes (see Fig. 7.3).

Flaviviruses inhibit IFNβ production

Pathogenic strains of WNV delay the induction of type I IFN without actively inhibiting the host defence signalling pathway. Rather than inhibiting the activation of IRF3, WNV NY replication delays activation of IRF3 through an unknown mechanism. This delayed activation of IRF3 gives the virus the opportunity to replicate and achieve

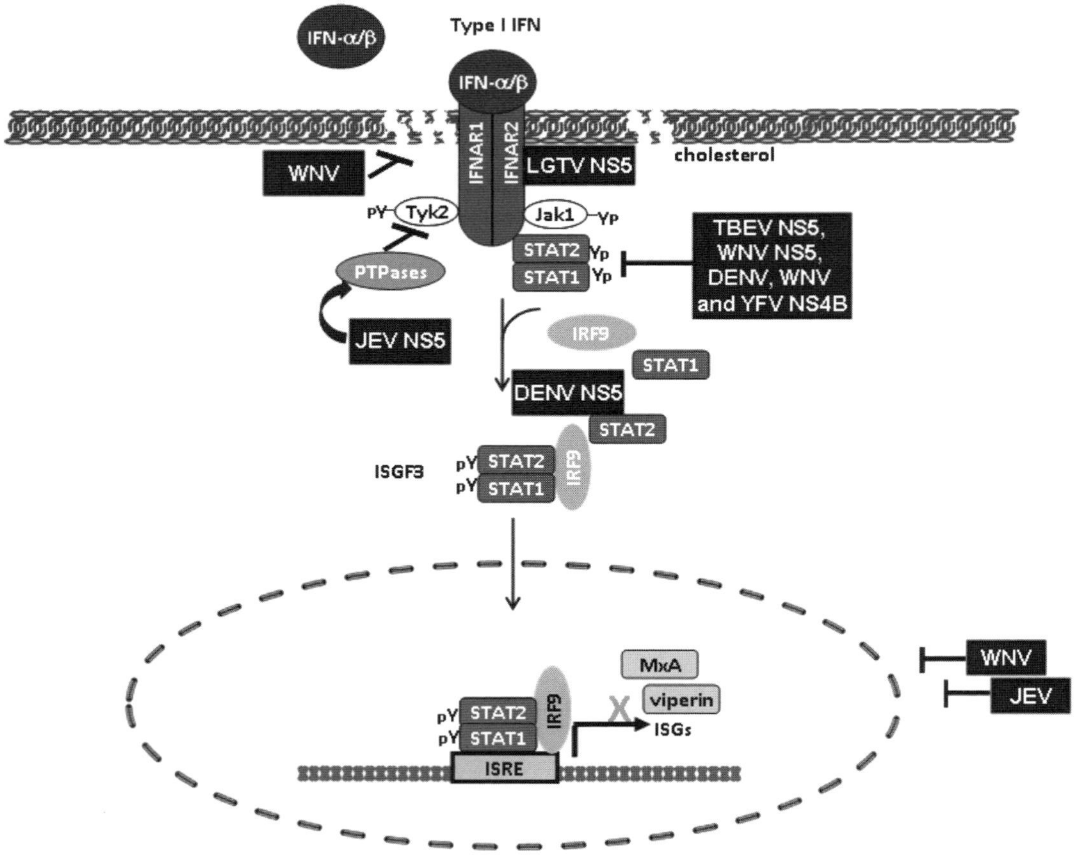

Figure 7.3 Flaviviruses antagonize type I IFN signaling. Flavivirus infection induces the secretion of IFN, which binds the IFNAR comprising of IFNAR1 and IFNAR2 subunits in an autocrine and paracrine fashion. Binding of IFN to the IFNAR results in activation of the JAKs, Jak1 and Tyk2, and the cytoplasmic tail of the IFNAR via tyrosine phosphorylation. STAT2 and STAT1 then get phosphorylated by the JAKs and form heterodimers, which then associate with IRF9 forming the ISGF3 complex. The ISGF3 complex translocates to the nucleus and binds the ISRE sequences found in the promoter region of ISGs to induce expression of hundreds of ISGs to create an antiviral state. Putative mechanisms of evasion of the type I IFN signaling cascade and gene induction by flaviviruses include WNV-induced redistribution of cholesterol from the plasma membrane to sites of viral replication which results in down-regulation of the IFNAR; association with the IFNAR and blocking phosphorylation of JAKs and STATs by LGTV NS5; modulating protein tyrosine phosphatase(s), which negatively regulate JAK/STAT signaling, by JEV NS5; proteasome-mediated degradation of STAT2 by DENV NS5; inhibition of STAT1 phosphorylation and nuclear import by the NS5 proteins of WNV and TBEV, and the NS4B protein of WNV, YFV and DENV; proteasome-mediated degradation of viperin by JEV; and WNV evasion of detection by MxA. Abbreviations: IFN, interferon; IFNAR, interferon α receptor; JAKs, Janus Kinases; Jak1, Janus kinase 1; Tyk2, protein tyrosine kinase 2; ISRE, interferon-stimulated response element.

maximum virus titres, thus establishing infection in the host before the induction of IFN (Fredericksen and Gale, 2004, 2006). A similar delay in the activation of IRF3 and the onset of IFN induction was observed in TBEV-infected cells (Overby *et al.*, 2010).

The contribution of the flavivirus NS1 protein to suppressing IFN induction is unclear owing to conflicting reports by two independent groups. Studies by Wilson *et al.* (2008) demonstrate that the NS1 protein of WNV may contribute to the inhibitory effect of WNV on TLR3 signalling. Expression of individual WNV proteins in mammalian cells revealed that WNV NS1 inhibits TLR3-induced transcriptional activation of the IFNβ promoter as well as IL6 production. Further analysis of the impact of WNV NS1 on the TLR3-signalling pathway demonstrated that WNV NS1 inhibited TLR3-induced nuclear translocation of IRF3 and NF-κB in response to poly (I:C) treatment. Currently, the mechanism of NS1 inhibition of the TLR3 pathway is unknown (Wilson *et al.*, 2008). However, Baronti *et al.* (2010) reported that transiently or stably expressed NS1 proteins of several flaviviruses do not play a role in suppressing TLR3 signalling. Protein interaction assays, subcellular localization studies and functional studies demonstrate that NS1 does not suppress TLR3 signalling (Baronti *et al.*, 2010). These discrepancies may be explained by the different experimental approaches used by the two groups. The two groups used different WNV strains. Furthermore, the ectopic expression of TLR3 in one study versus endogenous expression of TLR3 in the other study may also have contributed to differences in experimental results.

More than one viral protein can contribute to a virus' ability to inhibit the host protective immune response. The glycosylated WNV envelope protein (WNV E) inhibits dsRNA-induced production of antiviral and proinflammatory cytokines in murine macrophages. This effect was not directly dependent on TLR3 or its adaptor molecule TRIF but occurred downstream at the level of the serine-threonine kinase receptor-interacting protein 1 (RIP1) which is common to both the TLR3 and RIG-I pathways. WNV E protein is believed to blunt responses to viral RNA by inhibiting ubiquitination of RIP1. Interestingly, only mosquito cell-derived and not mammalian cell-derived WNV E protein has this effect on dsRNA-induced cytokine production. This suggests that WNV evolved a mechanism to evade innate immune responses during the initial stages of infection in the mammalian host (Arjona *et al.*, 2007).

The NS2A protein of Kunjin virus, a WNV strain, also plays a role in dampening the host IFN response by inhibiting IFNβ gene transcription (Liu *et al.*, 2004). Cells stably expressing a Kunjin virus subgenomic replicon showed a dramatic decrease in activation of the transcription of a reporter gene from the IFNβ promoter. Expression of the NS2A protein of Kunjin virus was able to inhibit transcription from the IFNβ promoter in response to Semliki Forest virus stimulation. A single amino acid substitution in NS2A at amino acid position 30 dramatically reduced its ability to inhibit transcription from the IFNβ promoter (Liu *et al.*, 2004). A mutant Kunjin virus containing this mutation elicited rapid and sustained type I IFN synthesis in both *in vitro* and *in vivo* infections. Consequently, this mutant virus was highly attenuated in weanling mice due to the robust induction of type I IFN which curbed viral dissemination and spread (Liu *et al.*, 2006). The cellular target of the NS2A protein is unknown as is the step in the IFN induction pathway that this protein exerts its effect.

More recent studies show that monocyte-derived dendritic cells (moDCs) are activated upon infection with dengue virus and release high levels of chemokines and proinflammatory cytokines with the exception of IFNα/β. It was observed that dengue-virus-infected moDCs lack phosphorylated IRF3 up to 24 hours post infection compared to control cells infected with Newcastle disease virus (NDV). Functionally, this translated to the inability of dengue-virus-infected moDCs to prime T cells thus dampening the adaptive immune response (Rodriguez-Madoz *et al.*, 2010). Furthermore, DENV-infected DCs had a reduced ability to produce type I IFN in response to several IFN inducers. Expression of the individual DENV proteins revealed that a catalytically active NS2B/3 protease is required for inhibition of type I IFN production in DENV-infected cells (Rodriguez-Madoz *et al.*, 2010).

Work done by Ishikawa *et al.* (2009) demonstrated a contribution by the YFV NS4B protein to suppressing type I IFN induction. Inhibition of IFN induction occurred possibly through a direct interaction of YFV NS4B with STING protein (also called MITA and MPYS) (Jin *et al.*, 2008; Zhong *et al.*, 2008). STING was discovered in a screen for constructs capable of inducing the IFNβ promoter. Overexpression of STING in IPS-1- or RIG-I-deficient MEFs resulted in the induction of an IFNβ-driven luciferase reporter construct placing STING downstream of IPS-1. Protein interaction assays demonstrate that STING interacts with both IPS-1 and RIG-I and facilitates transmission of virus-induced signals through these molecules (Ishikawa and Barber, 2008; Zhong *et al.*, 2008). It is currently unknown whether this function of NS4B is conserved among flaviviruses.

Flaviviruses antagonize interferon signalling

The importance of the IFN-signalling pathway has been demonstrated in mice deficient in components of the ISGF3 complex. These mice succumb to viral infection (Park *et al.*, 2000; Shresta *et al.*, 2005). The very existence of successful pathogenic viruses that encode proteins that block IFN-mediated signalling emphasizes the importance of type I IFN in the host defence against flaviviruses.

Tick-borne encephalitis virus serocomplex

Langat virus

Both the type I and type II IFN-signalling pathways are compromised in cells infected with Langat virus. Analysis of the effects of LGTV on IFN signalling showed that phosphorylation of Tyk2, Jak1, STAT1 and STAT2 was inhibited in LGTV-infected cells. However, the total levels of the Janus kinases and the STATs were unaffected. Analysis of the phosphorylation and nuclear translocation of STAT1 in response to type I IFN in cells expressing the individual proteins of LGTV identified the NS5 protein as an IFN signalling antagonist. Using IFNα/β– and IFNγ-responsive luciferase reporter gene constructs to further characterize the effects of the NS5 protein on type I and type II IFN signalling demonstrated that NS5 expression inhibited luciferase gene activity in response to either IFNα/β or IFNγ. Protein interaction assays in mammalian epithelial cells were used to examine the cellular interactors of the NS5 protein, and revealed that the NS5 protein interacted with the type I IFN receptor subunit, IFNAR2, and the type II IFN receptor subunit, IFNGR1. The interaction of the NS5 protein with the type I and type II IFN receptors was further confirmed in LGTV-infected human moDCs, which are target cells for virus replication (Best *et al.*, 2005). The minimal region of NS5 that was required for inhibiting IFN signalling was between amino acids 355 and 735, which lie within the RdRP domain of the NS5 protein. Further experiments identified two regions of amino acids, 374 to 380 and 624 to 647, within the RdRP that were required for inhibiting IFN signalling. Despite significant separation on the linear NS5 sequence, when these residues where modelled on the WNV RdRP crystal structure, they were shown to localize adjacent to each other and it is likely that these amino acids interact with the IFN receptors (Park *et al.*, 2007).

Tick-borne encephalitis virus

Tick-borne encephalitis virus NS5 has also been shown to inhibit IFN signalling. TBEV NS5 was shown to interact with the human scribble protein in a yeast two-hybrid assay. Scribble protein is localized at epithelial cell junctions and is involved in maintaining cell polarity (Dow *et al.*, 2003). Pulldown assays and immunofluorescence assays demonstrated an enrichment of NS5 at the cell–cell junctions where scribble localizes (Werme *et al.*, 2008). The association of TBEV NS5 with scribble and its recruitment to the plasma membrane seem to be essential for IFN antagonism. Localization of NS5 at the plasma membrane may allow it to interact with the IFN receptors.

Japanese encephalitis virus serogroup

Japanese encephalitis virus

In cells infected with JEV, Tyk2, STAT1 and STAT2 do not get phosphorylated in response to IFN. However, neither the IFNα/β receptor levels nor the expression levels of the proteins of the signalling cascade are affected (Lin *et al.*, 2004). Expression of the individual JEV proteins in mammalian cells revealed that only the NS5 protein inhibited phosphorylation and nuclear translocation of STAT1 in response to type I IFN. The negative effect of JEV and JEV NS5 on the IFN signalling pathway was shown by their ability to inhibit the induction of several IFN-inducible gene products in response to type I IFN and their ability to facilitate the replication of an IFN-sensitive virus in the presence of IFN. Protein interaction assays using a mammalian two-hybrid system revealed no interaction between NS5 and Tyk2 or Jak1. A potential mechanism for IFN signalling inhibition by NS5 involves modulating protein tyrosine phosphatases which are negative regulators of the IFN signalling pathway. Pre-treatment of JEV NS5 expressing cells with sodium orthovanadate, a broad spectrum protein tyrosine phosphatase inhibitor, resulted in an inability of NS5 to suppress IFN signalling. In order to determine the domain of NS5 required for this function, the ability of several truncated NS5 constructs to inhibit the phosphorylation and nuclear translocation of STAT1 in response to type I IFN was examined and it was revealed that amino acids 1–762 were sufficient for suppressing IFN signalling (Lin *et al.*, 2006).

West Nile virus

Analysis of the effects of WNV or WNV subgenomic replicons on the JAK-STAT pathway using reporter plasmid assays, immunofluorescence and biochemical assays demonstrated that phosphorylation of Tyk2 and Jak1 was suppressed. Consequently, activation and phosphorylation of STAT1 and STAT2 were also inhibited (Guo *et al.*, 2005). Analysis of the nuclear translocation of STAT2 in response to type I IFN in cells expressing the individual proteins of KUN showed that NS2A, NS2B3, NS4A and NS4B of Kunjin virus contribute to inhibition of IFN signalling. The proteins also inhibited gene activity from IFNα/β-responsive luciferase reporter gene constructs in cells treated with IFNα/β (Liu *et al.*, 2005). Work done by other groups supports the data that NS4B contributes to antagonism of the JAK-STAT pathway (Munoz-Jordan *et al.*, 2003, 2005). Further mutagenesis studies of the NS4B protein identified two amino acids (E22/K24) that were associated with IFN resistance in cells expressing subgenomic replicons. Surprisingly, in WNV-infected cells these two amino acids did not regulate the IFN response which suggests a role for the virus structural proteins in modulating the IFN response (Evans and Seeger, 2007). More recent studies using an IFNα/β responsive CAT-GFP reporter gene construct assay and an NDV-GFP complementation assay have identified the NS5 protein of virulent NY99 strain of WNV as an IFN antagonist. Immunofluorescence assays and flow cytometry were used to compare the WNV-NY99 NS5, NS4B and Kunjin NS5 proteins ability to inhibit the phosphorylation of STAT1 in response to type I IFN. Results showed that WNV-NY99 NS5 is a potent suppressor of the IFN signalling cascade in comparison to the NS4B and the KUN NS5. This study also showed that a single residue in KUN NS5 at position 653 was associated with a reduced ability to inhibit the IFN-signalling cascade (Laurent-Rolle *et al.*, 2010). This work suggests that the IFN antagonist function of NS5 may influence virus virulence in humans, as KUN does not induce severe disease. The ability of flaviviruses to encode multiple proteins to disarm the IFN signalling pathway may contribute to their pathogenicity.

Another potential strategy utilized by WNV and possibly other flaviviruses to evade the host antiviral response is manipulation of host cholesterol-biosynthesis pathways. Cells infected with WNV exhibit a marked relocalization of cholesterol and the key cholesterol-synthesizing enzyme, hydroxy-methylglutaryl-CoA reductase (HMGCR), to sites of viral replication. This redistribution of cholesterol diminishes the cholesterol-rich lipid rafts in the plasma membrane and is associated with the inhibition of the JAK-STAT pathway. Restoring plasma membrane concentrations by treating WNV-infected cells with exogenous cholesterol can partially overcome the dampening effects of WNV on the JAK-STAT pathway. Manipulation of cholesterol and the cholesterol synthesizing machinery appear to be crucial for virus survival since RNAi-mediated knockdown of the HMGCR gene results in a strong reduction in viral replication (Mackenzie *et al.*, 2007).

Dengue virus

Dengue virus inhibits type I IFN signalling by reducing STAT2 expression (Jones *et al.*, 2005). Other studies have also shown that the virus inhibits the phosphorylation of Tyk2 kinase (Ho *et al.*, 2005). Initial studies attributed dengue virus' ability to inhibit IFN signalling to several dengue virus proteins: NS2A, NS4A and NS4B (Munoz-Jordan *et al.*, 2003, 2005). Subsequent studies identified the NS5 protein as a potent inhibitor of the IFN-signalling pathway (Ashour *et al.*, 2009; Mazzon *et al.*, 2009). The ability of the NS5 protein of dengue virus to bind to STAT2 contributes to its IFN antagonism function since mutant NS5 proteins which lack the ability to bind STAT2 were unable to inhibit an IFNα/β-responsive reporter gene construct in response to IFNα/β. The expression of DEN-2 NS5 in the context of the viral polyprotein, which can be proteolytically processed for maturation, was shown to be necessary to mediate the proteasome-dependent degradation of human STAT2. Deletion analysis studies showed that the first 10 amino acids of NS5 are required for the loss of STAT2 but not for binding to STAT2 since a truncated NS5 of up to the first 202 amino acids could still bind STAT2 and inhibit IFN signalling (Ashour *et al.*, 2009). This ability of dengue virus to inhibit IFN signalling occurs in a species-specific manner since dengue NS5 is unable to bind and degrade mouse STAT2. This may be one reason why dengue virus does not naturally establish infections in mice (Ashour *et al.*, 2010). These findings are supported by studies done by Perry *et al* (2011).

Inhibition of antiviral activities of interferon-stimulated genes

Multiple flavivirus proteins act in a concerted fashion to inhibit the JAK-STAT pathway and prevent the induction of ISGs with possible antiviral activities. To further enhance their chances of survival, flaviviruses also target individual downstream antiviral effector proteins. Viperin is an IFN-stimulated gene with an ability to alter lipid raft formation and has been shown to have antiviral activity against a number of viruses including the related hepatitis C virus (Helbig *et al.*, 2005; Jiang *et al.*, 2008; Hinson and Cresswell, 2009). Japanese encephalitis virus counteracts the antiviral effects of viperin by promoting its rapid proteasome-mediated degradation. The molecular mechanism involved in this virus-mediated loss of viperin is not clear as transfection of individual JEV proteins failed to induce the loss of viperin. This suggests that viral RNA replication or a combined effect of more than one viral protein is required for proteasome-mediated viperin degradation (Chan *et al.*, 2008).

WNV employs a different mechanism to evade the antiviral effects of another IFN-stimulated gene, MxA. The type I and type III IFN-induced MxA belongs to the dynamin superfamily of large GTPases (Arnheiter *et al.*, 1996; Haller and Kochs, 2002; Holzinger *et al.*, 2007). MxA functions as a potent antiviral protein against both negative- and positive-strand RNA viruses (Landis *et al.*, 1998; Chieux *et al.*, 2001; Haller and Kochs, 2002; Holzinger *et al.*, 2007;). During flavivirus infection, characteristic membrane structures are induced to facilitate virus replication (Westaway *et al.*, 1997; Mackenzie *et al.*, 1998, 1999). Studies suggest that upregulation of these membranous structures in WNV-infected cells provides some protection for the virus against MxA antiviral action (Hoenen *et al.*, 2007).

Perspectives

We sincerely apologize to all authors whose work could not be cited. Our understanding of the intricate balance between flavivirus pathogenesis and the host innate immune response is progressing. Much has been learned from work done in both animal and cell culture systems. The host innate immune response occurs very quickly and the different arms of the innate immune system function cooperatively to limit virus infection. Yet, pathogenic flaviviruses have developed elegant ways to counteract the antiviral effects of the innate immune response so as to establish infection and increase survival in the host cell. One thing that is clear from these studies is that the flaviviruses have devoted a significant part of their genomes to evading the different arms of the innate immune response. The details of the molecular mechanisms involved in this process are largely unknown. More effort is needed to understand how the ability of flaviviruses to disarm the innate immune response contributes to their virulence. Answers to these questions will provide us with valuable information about the pathogenesis of these viruses and will be useful in the design of effective vaccines and therapeutic strategies to control flaviviral diseases.

Acknowledgements

Work in flaviviruses in A.G.-S.'s group is supported by NIAID (U54 AI057158). M.L-R is the recipient of an NIH fellowship (FAI077333A).

References

Abraham, S., Nagaraj, A.S., Basak, S., and Manjunath, R. (2010). Japanese encephalitis virus utilizes the canonical pathway to activate NF-kappaB but it utilizes the type I interferon pathway to induce major histocompatibility complex class I expression in mouse embryonic fibroblasts. J. Virol. *84*, 5485–5493.

Akhtar, L.N., and Benveniste, E.N. (2011). Viral exploitation of host SOCS protein functions. J. Virol. *85*, 1912–1921.

Alexander, W.S., and Hilton, D.J. (2004). The role of suppressors of cytokine signaling (SOCS) proteins in regulation of the immune response. Annu. Rev. Immunol. *22*, 503–529.

Alexopoulou, L., Holt, A.C., Medzhitov, R., and Flavell, R.A. (2001). Recognition of double-stranded RNA and activation of NF-kappaB by Toll-like receptor 3. Nature *413*, 732–738.

Anderson, J.F., and Rahal, J.J. (2002). Efficacy of interferon alpha-2b and ribavirin against West Nile virus *in vitro*. Emerg. Infect. Dis. *8*, 107–108.

Ank, N., West, H., Bartholdy, C., Eriksson, K., Thomsen, A.R., and Paludan, S.R. (2006). Lambda interferon (IFN-lambda), a type III IFN, is induced by viruses and IFNs and displays potent antiviral activity against select virus infections *in vivo*. J. Virol. *80*, 4501–4509.

Arjona, A., Ledizet, M., Anthony, K., Bonafe, N., Modis, Y., Town, T., and Fikrig, E. (2007). West Nile virus envelope protein inhibits dsRNA-induced innate immune responses. J. Immunol. *179*, 8403–8409.

Arnheiter, H., Frese, M., Kambadur, R., Meier, E., and Haller, O. (1996). Mx transgenic mice--animal models of health. Curr. Top. Microbiol. Immunol. *206*, 119–147.

Arnon, T.I., Markel, G., and Mandelboim, O. (2006). Tumor and viral recognition by natural killer cells receptors. Semin. Cancer Biol. *16*, 348–358.

Ashour, J., Laurent-Rolle, M., Shi, P.Y., and Garcia-Sastre, A. (2009). NS5 of dengue virus mediates STAT2 binding and degradation. J. Virol. *83*, 5408–5418.

Ashour, J., Morrison, J., Laurent-Rolle, M., Belicha-Villanueva, A., Plumlee, C.R., Bernal-Rubio, D., Williams, K.L., Harris, E., Fernandez-Sesma, A., Schindler, C., *et al.* (2010). Mouse STAT2 restricts early dengue virus replication. Cell Host Microbe *8*, 410–421.

Avirutnan, P., Malasit, P., Seliger, B., Bhakdi, S., and Husmann, M. (1998). Dengue virus infection of human endothelial cells leads to chemokine production, complement activation, and apoptosis. J. Immunol. *161*, 6338–6346.

Avirutnan, P., Punyadee, N., Noisakran, S., Komoltri, C., Thiemmeca, S., Auethavornanan, K., Jairungsri, A., Kanlaya, R., Tangthawornchaikul, N., Puttikhunt, C., *et al.* (2006). Vascular leakage in severe dengue virus infections: a potential role for the nonstructural viral protein NS1 and complement. J. Infect. Dis. *193*, 1078–1088.

Avirutnan, P., Mehlhop, E., and Diamond, M.S. (2008). Complement and its role in protection and pathogenesis of flavivirus infections. Vaccine *26 (Suppl 8)*, 1100–1107.

Avirutnan, P., Fuchs, A., Hauhart, R.E., Somnuke, P., Youn, S., Diamond, M.S., and Atkinson, J.P. (2010). Antagonism of the complement component C4 by flavivirus nonstructural protein NS1. J. Exp. Med. *207*, 793–806.

Azeredo, E.L., De Oliveira-Pinto, L.M., Zagne, S.M., Cerqueira, D.I., Nogueira, R.M., and Kubelka, C.F. (2006). NK cells, displaying early activation, cytotoxicity and adhesion molecules, are associated with mild dengue disease. Clin. Exp. Immunol. *143*, 345–356.

Baronti, C., Sire, J., de Lamballerie, X., and Querat, G. (2010). Nonstructural NS1 proteins of several mosquito-borne Flavivirus do not inhibit TLR3 signaling. Virology *404*, 319–330.

Bell, J.K., Mullen, G.E., Leifer, C.A., Mazzoni, A., Davies, D.R., and Segal, D.M. (2003). Leucine-rich repeats and pathogen recognition in Toll-like receptors. Trends Immunol. *24*, 528–533.

Best, S.M., Morris, K.L., Shannon, J.G., Robertson, S.J., Mitzel, D.N., Park, G.S., Boer, E., Wolfinbarger, J.B., and Bloom, M.E. (2005). Inhibition of interferon-stimulated JAK-STAT signaling by a tick-borne flavivirus and identification of NS5 as an interferon antagonist. J. Virol. *79*, 12828–12839.

Biron, C.A. (2001). Interferons alpha and beta as immune regulators--a new look. Immunity *14*, 661–664.

Biron, C.A., Byron, K.S., and Sullivan, J.L. (1989). Severe herpesvirus infections in an adolescent without natural killer cells. N. Engl. J. Med. *320*, 1731–1735.

Blue, C.E., Spiller, O.B., and Blackbourn, D.J. (2004). The relevance of complement to virus biology. Virology *319*, 176–184.

Boehm, U., Klamp, T., Groot, M., and Howard, J.C. (1997). Cellular responses to interferon-gamma. Annu. Rev. Immunol. *15*, 749–795.

Bourne, N., Scholle, F., Silva, M.C., Rossi, S.L., Dewsbury, N., Judy, B., De Aguiar, J.B., Leon, M.A., Estes, D.M., Fayzulin, R., *et al.* (2007). Early production of type I interferon during West Nile virus infection: Role for lymphoid tissues in IRF3-independent interferon production. J. Virol. *81*, 9100–9108.

Brass, A.L., Huang, I.C., Benita, Y., John, S.P., Krishnan, M.N., Feeley, E.M., Ryan, B.J., Weyer, J.L., van der Weyden, L., Fikrig, E., *et al.* (2009). The IFITM proteins mediate cellular resistance to influenza A H1N1 virus, West Nile virus, and dengue virus. Cell *139*, 1243–1254.

Brooks, T.J.G., and Phillpotts, R.J. (1999). Interferon-alpha protects mice against lethal infection with St Louis encephalitis virus delivered by the aerosol and subcutaneous routes. Antivir. Res. *41*, 57–64.

Carey, M. (1998). The enhanceosome and transcriptional synergy. Cell *92*, 5–8.

Chan, Y.L., Chang, T.H., Liao, C.L., and Lin, Y.L. (2008). The Cellular Antiviral Protein Viperin Is Attenuated by Proteasome-Mediated Protein Degradation in Japanese Encephalitis Virus-Infected Cells. J. Virol. *82*, 10455–10464.

Chang, T.H., Liao, C.L., and Lin, Y.L. (2006). Flavivirus induces interferon-beta gene expression through a pathway involving RIG-I-dependent IRF-3 and PI3K-dependent NF-kappaB activation. Microbes Infect. *8*, 157–171.

Cheng, Y., King, N.J., and Kesson, A.M. (2004). Major histocompatibility complex class I (MHC-I) induction by West Nile virus: involvement of 2 signaling pathways in MHC-I up-regulation. J. Infect. Dis. *189*, 658–668.

Chieux, V., Chehadeh, W., Harvey, J., Haller, O., Wattre, P., and Hober, D. (2001). Inhibition of coxsackievirus B4 replication in stably transfected cells expressing human MxA protein. Virology *283*, 84–92.

Chong, K.L., Feng, L., Schappert, K., Meurs, E., Donahue, T.F., Friesen, J.D., Hovanessian, A.G., and Williams, B.R.G. (1992). Human P68 Kinase Exhibits Growth Suppression in Yeast and Homology to the Translational Regulator Gcn2. Embo J. *11*, 1553–1562.

Chung, K.M., Liszewski, M.K., Nybakken, G., Davis, A.E., Townsend, R.R., Fremont, D.H., Atkinson, J.P., and Diamond, M.S. (2006). West Nile virus nonstructural protein NS1 inhibits complement activation by binding the regulatory protein factor H. Proc. Natl. Acad. Sci. U. S. A. *103*, 19111–19116.

Cleary, C.M., Donnelly, R.J., Soh, J., Mariano, T.M., and Pestka, S. (1994). Knockout and reconstitution of a functional human type I interferon receptor complex. J. Biol. Chem. *269*, 18747–18749.

Colamonici, O., Yan, H., Domanski, P., Handa, R., Smalley, D., Mullersman, J., Witte, M., Krishnan, K., and Krolewski, J. (1994). Direct binding to and tyrosine phosphorylation of the alpha subunit of the type I interferon receptor by p135tyk2 tyrosine kinase. *14*, 8133–8142.

Colamonici, O.R., Platanias, L.C., Domanski, P., Handa, R., Gilmour, K.C., Diaz, M.O., Reich, N., and Pitha-Rowe, P. (1995). Transmembrane signaling by the alpha subunit of the type I interferon receptor is essential for activation of the JAK kinases and the transcriptional factor ISGF3. J. Biol. Chem. *270*, 8188–8193.

Daffis, S., Samuel, M.A., Keller, B.C., Gale, M., Jr., and Diamond, M.S. (2007). Cell-specific IRF-3 responses protect against West Nile virus infection by interferon-dependent and -independent mechanisms. Plos Pathog. *3*, e106.

Daffis, S., Samuel, M.A., Suthar, M.S., Gale, M., and Diamond, M.S. (2008a). Toll-Like Receptor 3 Has a Protective Role against West Nile Virus Infection. J. Virol. *82*, 10349–10358.

Daffis, S., Samuel, M.A., Suthar, M.S., Keller, B.C., Gale, M., Jr., and Diamond, M.S. (2008b). Interferon regulatory factor IRF-7 induces the antiviral alpha interferon response and protects against lethal West Nile virus infection. J. Virol. *82*, 8465–8475.

Darnell, J.E., Jr. (1997). STATs and gene regulation. Science *277*, 1630–1635.

Davies, A., and Lachmann, P.J. (1993). Membrane defence against complement lysis: the structure and biological properties of CD59. Immunol. Res. *12*, 258–275.

Der, S.D., Zhou, A., Williams, B.R., and Silverman, R.H. (1998). Identification of genes differentially regulated by interferon alpha, beta, or gamma using oligonucleotide arrays. Proc. Natl. Acad. Sci. U. S. A. *95*, 15623–15628.

Diamond, M.S. (2009). Mechanisms of evasion of the type I interferon antiviral response by flaviviruses. J. Interferon. Cytokine Res. *29*, 521–530.

Diamond, M.S., Roberts, T.G., Edgil, D., Lu, B., Ernst, J., and Harris, E. (2000). Modulation of Dengue virus infection in human cells by alpha, beta, and gamma interferons. J. Virol. *74*, 4957–4966.

Diao, F., Li, S., Tian, Y., Zhang, M., Xu, L.G., Zhang, Y., Wang, R.P., Chen, D., Zhai, Z., Zhong, B., *et al.* (2007). Negative regulation of MDA5- but not RIG-I-mediated innate antiviral signaling by the dihydroxyacetone kinase. Proc. Natl. Acad. Sci. U. S. A. *104*, 11706–11711.

Diebold, S.S. (2008). Recognition of viral single-stranded RNA by Toll-like receptors. Adv. Drug Deliv. Rev. *60*, 813–823.

Diebold, S.S., Montoya, M., Unger, H., Alexopoulou, L., Roy, P., Haswell, L.E., Al-Shamkhani, A., Flavell, R., Borrow, P., and Reis e Sousa, C. (2003). Viral infection switches non-plasmacytoid dendritic cells into high interferon producers. Nature *424*, 324–328.

Diebold, S.S., Kaisho, T., Hemmi, H., Akira, S., and Reis e Sousa, C. (2004). Innate antiviral responses by means of TLR7-mediated recognition of single-stranded RNA. Science *303*, 1529–1531.

Domanski, P., Fish, E., Nadeau, O.W., Witte, M., Platanias, L.C., Yan, H., Krolewski, J., Pitha, P., and Colamonici, O.R. (1997). A region of the beta subunit of the interferon alpha receptor different from Box 1 interacts with Jak1 and is sufficient to activate the Jak-Stat pathway and induce an antiviral state. J. Biol. Chem. *272*, 26388–26393.

Donnelly, R.P., and Kotenko, S.V. (2010). Interferon-lambda: a new addition to an old family. J. Interferon. Cytokine Res. *30*, 555–564.

Douglas, M.W., Kesson, A.M., and King, N.J. (1994). CTL recognition of west Nile virus-infected fibroblasts is cell cycle dependent and is associated with virus-induced increases in class I MHC antigen expression. Immunology *82*, 561–570.

Drickamer, K. (1993). Recognition of complex carbohydrates by Ca(2+)-dependent animal lectins. Biochem. Soc. Trans. *21*, 456–459.

Du, W., and Maniatis, T. (1992). An ATF/CREB binding site is required for virus induction of the human interferon beta gene [corrected]. Proc. Natl. Acad. Sci. U. S. A. *89*, 2150–2154.

Evans, J.D., and Seeger, C. (2007). Differential effects of mutations in NS4B on west nile virus replication and inhibition of interferon signaling. J. Virol. *81*, 11809–11816.

Fernandez-Garcia, M.D., Mazzon, M., Jacobs, M., and Amara, A. (2009). Pathogenesis of flavivirus infections: using and abusing the host cell. Cell Host Microbe *5*, 318–328.

Fitzgerald, K.A., McWhirter, S.M., Faia, K.L., Rowe, D.C., Latz, E., Golenbock, D.T., Coyle, A.J., Liao, S.M., and Maniatis, T. (2003). IKKepsilon and TBK1 are essential components of the IRF3 signaling pathway. Nat. Immunol. *4*, 491–496.

Fredericksen, B.L., and Gale, M., Jr. (2006). West Nile virus evades activation of interferon regulatory factor 3 through RIG-I-dependent and -independent pathways without antagonizing host defense signaling. J. Virol. *80*, 2913–2923.

Fredericksen, B.L., Smith, M., Katze, M.G., Shi, P.Y., and Gale, M., Jr. (2004). The host response to West Nile Virus infection limits viral spread through the activation of the interferon regulatory factor 3 pathway. J. Virol. *78*, 7737–7747.

Fredericksen, B.L., Keller, B.C., Fornek, J., Katze, M.G., and Gale, M. (2008). Establishment and maintenance of the innate antiviral response to west nile virus involves both RIG-I and MDA5 signaling through IPS-1. J. Virol. *82*, 609–616.

Fukui, A., Yuasa-Nakagawa, T., Murakami, Y., Funami, K., Kishi, N., Matsuda, T., Fujita, T., Seya, T., and Nagasawa, S. (2002). Mapping of the sites responsible for factor I-cofactor activity for cleavage of C3b and C4b on human C4b-binding protein (C4bp) by deletion mutagenesis. J. Biochem. *132*, 719–728.

Gaucher, D., Therrien, R., Kettaf, N., Angermann, B.R., Boucher, G., Filali-Mouhim, A., Moser, J.M., Mehta, R.S., Drake, D.R., 3rd, Castro, E., *et al.* (2008). Yellow fever vaccine induces integrated multilineage and polyfunctional immune responses. J. Exp. Med. *205*, 3119–3131.

Gilfoy, F.D., and Mason, P.W. (2007). West nile virus-induced interferon production is mediated by the double-stranded RNA-dependent protein kinase PKR. J. Virol. *81*, 11148–11158.

Guo, J.T., Hayashi, J., and Seeger, C. (2005). West Nile virus inhibits the signal transduction pathway of alpha interferon. J. Virol. *79*, 1343–1350.

Haller, O., and Kochs, G. (2002). Interferon-induced mx proteins: dynamin-like GTPases with antiviral activity. Traffic *3*, 710–717.

Helbig, K.J., Lau, D.T.Y., Semendric, L., Harley, H.A.J., and Beard, M.R. (2005). Analysis of ISG expression in chronic hepatitis C identifies viperin as a potential antiviral effector. Hepatology *42*, 702–710.

Hershkovitz, O., Zilka, A., Bar-Ilan, A., Abutbul, S., Davidson, A., Mazzon, M., Kummerer, B.M., Monsoengo, A., Jacobs, M., and Porgador, A. (2008). Dengue virus replicon expressing the nonstructural proteins suffices to enhance membrane expression of HLA class I and inhibit lysis by human NK cells. J. Virol. *82*, 7666–7676.

Hershkovitz, O., Rosental, B., Rosenberg, L.A., Navarro-Sanchez, M.E., Jivov, S., Zilka, A., Gershoni-Yahalom, O., Brient-Litzler, E., Bedouelle, H., Ho, J.W., *et al.* (2009). NKp44 receptor mediates interaction of the envelope glycoproteins from the West Nile and dengue viruses with NK cells. J. Immunol. *183*, 2610–2621.

Hidaka, F., Matsuo, S., Muta, T., Takeshige, K., Mizukami, T., and Nunoi, H. (2006). A missense mutation of the Toll-like receptor 3 gene in a patient with influenza-associated encephalopathy. Clin. Immunol. *119*, 188–194.

Hinson, E.R., and Cresswell, P. (2009). The antiviral protein, viperin, localizes to lipid droplets via its N-terminal amphipathic alpha-helix. Proc. Natl. Acad. Sci. U. S. A. *106*, 20452–20457.

Ho, L.J., Hung, L.F., Weng, C.Y., Wu, W.L., Chou, P., Lin, Y.L., Chang, D.M., Tai, T.Y., and Lai, J.H. (2005). Dengue virus type 2 antagonizes IFN-alpha but not IFN-gamma antiviral effect via down-regulating Tyk2-STAT signaling in the human dendritic cell. J. Immunol. *174*, 8163–8172.

Hoenen, A., Liu, W., Kochs, G., Khromykh, A.A., and Mackenzie, J.M. (2007). West Nile virus-induced cytoplasmic membrane structures provide partial protection against the interferon-induced antiviral MxA protein. J. Gen. Virol. *88*, 3013–3017.

Holzinger, D., Jorns, C., Stertz, S., Boisson-Dupuis, S., Thimme, R., Weidmann, M., Casanova, J.L., Haller, O., and Kochs, G. (2007). Induction of MxA gene expression by influenza A virus requires type I or type III interferon signaling. J. Virol. *81*, 7776–7785.

Honda, K., Yanai, H., Mizutani, T., Negishi, H., Shimada, N., Suzuki, N., Ohba, Y., Takaoka, A., Yeh, W.C., and Taniguchi, T. (2004). Role of a transductional-transcriptional processor complex involving MyD88 and IRF-7 in Toll-like receptor signaling. Proc. Natl. Acad. Sci. U. S. A. *101*, 15416–15421.

Hourcade, D., Holers, V.M., and Atkinson, J.P. (1989). The regulators of complement activation (RCA) gene cluster. Adv. Immunol. *45*, 381–416.

Hu, Y., and Conway, T.W. (1993). 2-Aminopurine inhibits the double-stranded RNA-dependent protein kinase both *in vitro* and *in vivo*. J. Interferon Res. *13*, 323–328.

Ishikawa, H., and Barber, G.N. (2008). STING is an endoplasmic reticulum adaptor that facilitates innate immune signalling. Nature *455*, 674–678.

Jiang, D., Guo, H.T., Xu, C.X., Chang, J.H., Gu, B.H., Wang, L.J., Block, T.M., and Guo, J.T. (2008). Identification of three interferon-inducible cellular enzymes that inhibit the replication of hepatitis C virus. J. Virol. *82*, 1665–1678.

Jiang, D., Weidner, J.M., Qing, M., Pan, X.B., Guo, H., Xu, C., Zhang, X., Birk, A., Chang, J., Shi, P.Y., *et al.* (2010). Identification of five interferon-induced cellular proteins that inhibit west nile virus and dengue virus infections. J. Virol. *84*, 8332–8341.

Jin, L., Waterman, P.M., Jonscher, K.R., Short, C.M., Reisdorph, N.A., and Cambier, J.C. (2008). MPYS, a novel membrane tetraspanner, is associated with major histocompatibility complex class II and mediates transduction of apoptotic signals. Mol. Cell Biol. *28*, 5014–5026.

Jones, M., Davidson, A., Hibbert, L., Gruenwald, P., Schlaak, J., Ball, S., Foster, G.R., and Jacobs, M. (2005). Dengue virus inhibits alpha interferon signaling by reducing STAT2 expression. J. Virol. *79*, 5414–5420.

Julander, J.G., Morrey, J.D., Blatt, L.M., Shafer, K., and Sidwell, R.W. (2007). Comparison of the inhibitory effects of interferon alfacon-1 and ribavirin on yellow fever virus infection in a hamster model. Antiviral Res. *73*, 140–146.

Kajaste-Rudnitski, A., Mashimo, T., Frenkiel, M.P., Guenet, J.L., Lucas, M., and Despres, P. (2006). The 2′,5′-oligoadenylate synthetase 1b is a potent inhibitor of West Nile virus replication inside infected cells. J. Biol. Chem. *281*, 4624–4637.

Kato, H., Takeuchi, O., Sato, S., Yoneyama, M., Yamamoto, M., Matsui, K., Uematsu, S., Jung, A., Kawai, T., Ishii, K.J., *et al.* (2006). Differential roles of MDA5 and RIG-I helicases in the recognition of RNA viruses. Nature *441*, 101–105.

Katze, M.G., He, Y., and Gale, M., Jr. (2002). Viruses and interferon: a fight for supremacy. Nat. Rev. Immunol. *2*, 675–687.

Kawai, T., Sato, S., Ishii, K.J., Coban, C., Hemmi, H., Yamamoto, M., Terai, K., Matsuda, M., Inoue, J., Uematsu, S., *et al.* (2004). Interferon-alpha induction through Toll-like receptors involves a direct interaction of IRF7 with MyD88 and TRAF6. Nat. Immunol. *5*, 1061–1068.

Kawai, T., Takahashi, K., Sato, S., Coban, C., Kumar, H., Kato, H., Ishii, K.J., Takeuchi, O., and Akira, S. (2005). IPS-1, an adaptor triggering RIG-I- and Mda5-mediated type I interferon induction. Nat. Immunol. *6*, 981–988.

Kazatchkine, M.D., Fearon, D.T., and Austen, K.F. (1979). Human alternative complement pathway: membrane-associated sialic acid regulates the competition between B and beta1 H for cell-bound C3b. J. Immunol. *122*, 75–81.

Kesson, A.M., and King, N.J. (2001). Transcriptional regulation of major histocompatibility complex class I by flavivirus West Nile is dependent on NF-kappaB activation. J. Infect. Dis. *184*, 947–954.

Kesson, A.M., Cheng, Y., and King, N.J. (2002). Regulation of immune recognition molecules by flavivirus, West Nile. Viral Immunol. *15*, 273–283.

King, N.J., and Kesson, A.M. (1988). Interferon-independent increases in class I major histocompatibility complex antigen expression follow flavivirus infection. J. Gen. Virol. *69 (Pt 10)*, 2535–2543.

Kotenko, S.V., Gallagher, G., Baurin, V.V., Lewis-Antes, A., Shen, M., Shah, N.K., Langer, J.A., Sheikh, F., Dickensheets, H., and Donnelly, R.P. (2003). IFN-lambdas mediate antiviral protection through a distinct class II cytokine receptor complex. Nat. Immunol. *4*, 69–77.

Kotwal, G.J., Isaacs, S.N., McKenzie, R., Frank, M.M., and Moss, B. (1990). Inhibition of the complement cascade by the major secretory protein of vaccinia virus. Science *250*, 827–830.

Krych-Goldberg, M., and Atkinson, J.P. (2001). Structure–function relationships of complement receptor type 1. Immunol. Rev. *180*, 112–122.

Krych-Goldberg, M., Hauhart, R.E., Subramanian, V.B., Yurcisin, B.M., 2nd, Crimmins, D.L., Hourcade, D.E., and Atkinson, J.P. (1999). Decay accelerating activity of complement receptor type 1 (CD35). Two active sites are required for dissociating C5 convertases. J. Biol. Chem. *274*, 31160–31168.

Kumar, A., Yang, Y.L., Flati, V., Der, S., Kadereit, S., Deb, A., Haque, J., Reis, L., Weissmann, C., and Williams, B.R. (1997). Deficient cytokine signaling in mouse embryo fibroblasts with a targeted deletion in the PKR gene: role of IRF-1 and NF-kappaB. Embo J. *16*, 406–416.

Kumar, H., Kawai, T., and Akira, S. (2011). Pathogen recognition by the innate immune system. Int. Rev. Immunol. *30*, 16–34.

Kurane, I., Hebblewaite, D., Brandt, W.E., and Ennis, F.A. (1984). Lysis of dengue virus-infected cells by natural cell-mediated cytotoxicity and antibody-dependent cell-mediated cytotoxicity. J. Virol. *52*, 223–230.

Kurosu, T., Chaichana, P., Yamate, M., Anantapreecha, S., and Ikuta, K. (2007). Secreted complement regulatory protein clusterin interacts with dengue virus nonstructural protein 1. Biochem. Biophys. Res. Commun. *362*, 1051–1056.

Lachmann, P.J., and Hughes-Jones, N.C. (1984). Initiation of complement activation. Springer Semin. Immunopathol. *7*, 143–162.

Lambris, J.D., Ricklin, D., and Geisbrecht, B.V. (2008). Complement evasion by human pathogens. Nat. Rev. Microbiol. *6*, 132–142.

Landis, H., Simon-Jodicke, A., Kloti, A., Di Paolo, C., Schnorr, J.J., Schneider-Schaulies, S., Hefti, H.P., and Pavlovic, J. (1998). Human MxA protein confers resistance to Semliki Forest virus and inhibits the amplification of a Semliki Forest virus-based replicon in the absence of viral structural proteins. J. Virol. *72*, 1516–1522.

Lanier, L.L. (1998). NK cell receptors. Annu. Rev. Immunol. *16*, 359–393.

Lanier, L.L. (2008a). Evolutionary struggles between NK cells and viruses. Nat. Rev. Immunol. *8*, 259–268.

Lanier, L.L. (2008b). Up on the tightrope: natural killer cell activation and inhibition. Nat. Immunol. *9*, 495–502.

Laurent-Rolle, M., Boer, E.F., Lubick, K.J., Wolfinbarger, J.B., Carmody, A.B., Rockx, B., Liu, W., Ashour, J., Shupert, W.L., Holbrook, M.R., *et al.* (2010). The NS5 protein of the virulent West Nile virus NY99 strain is a potent antagonist of type I interferon-mediated JAK-STAT signaling. J. Virol. *84*, 3503–3515.

Lenardo, M.J., Fan, C.M., Maniatis, T., and Baltimore, D. (1989). The involvement of NF-kappa B in beta-interferon gene regulation reveals its role as widely inducible mediator of signal transduction. Cell *57*, 287–294.

Leyssen, P., Van Lommel, A., Drosten, C., Schmitz, H., De Clercq, E., and Neyts, J. (2001). A novel model for the study of the therapy of flavivirus infections using the Modoc virus. Virology *279*, 27–37.

Leyssen, P., Drosten, C., Paning, M., Charlier, N., Paeshuyse, J., De Clercq, E., and Neyts, J. (2003). Interferons, interferon inducers, and interferon-ribavirin in treatment of flavivirus-induced encephalitis in mice. Antimicrob. Agents Chemother. *47*, 777–782.

Li, H., Gade, P., Xiao, W., and Kalvakolanu, D.V. (2007). The interferon signaling network and transcription factor C/EBP-beta. Cell Mol. Immunol. *4*, 407–418.

Libraty, D.H., Young, P.R., Pickering, D., Endy, T.P., Kalayanarooj, S., Green, S., Vaughn, D.W., Nisalak, A., Ennis, F.A., and Rothman, A.L. (2002). High circulating levels of the dengue virus nonstructural protein NS1 early in dengue illness correlate with the development of dengue hemorrhagic fever. J. Infect. Dis. *186*, 1165–1168.

Lin, C.F., Lei, H.Y., Shiau, A.L., Liu, H.S., Yeh, T.M., Chen, S.H., Liu, C.C., Chiu, S.C., and Lin, Y.S. (2002). Endothelial cell apoptosis induced by antibodies against dengue virus nonstructural protein 1 via production of nitric oxide. J. Immunol. *169*, 657–664.

Lin, R.J., Liao, C.L., Lin, E., and Lin, Y.L. (2004). Blocking of the alpha interferon-induced Jak-Stat signaling pathway by Japanese encephalitis virus infection. J. Virol. *78*, 9285–9294.

Lin, R.J., Chang, B.L., Yu, H.P., Liao, C.L., and Lin, Y.L. (2006). Blocking of interferon-induced Jak-Stat signaling by Japanese encephalitis virus NS5 through a protein tyrosine phosphatase-mediated mechanism. J. Virol. *80*, 5908–5918.

Liszewski, M.K., Leung, M., Cui, W., Subramanian, V.B., Parkinson, J., Barlow, P.N., Manchester, M., and Atkinson, J.P. (2000). Dissecting sites important for complement regulatory activity in membrane cofactor protein (MCP; CD46). J. Biol. Chem. *275*, 37692–37701.

Liu, W.J., Chen, H.B., Wang, X.J., Huang, H., and Khromykh, A.A. (2004). Analysis of adaptive mutations in Kunjin virus replicon RNA reveals a novel role for the flavivirus nonstructural protein NS2A in inhibition of beta interferon promoter-driven transcription. J. Virol. *78*, 12225–12235.

Liu, W.J., Wang, X.J., Mokhonov, V.V., Shi, P.Y., Randall, R., and Khromykh, A.A. (2005). Inhibition of interferon signaling by the New York 99 strain and Kunjin subtype of West Nile virus involves blockage of STAT1 and STAT2 activation by nonstructural proteins. J. Virol. *79*, 1934–1942.

Liu, W.J., Wang, X.J., Clark, D.C., Lobigs, M., Hall, R.A., and Khromykh, A.A. (2006). A single amino acid substitution in the west nile virus nonstructural protein NS2A disables its ability to inhibit alpha/beta interferon induction and attenuates virus virulence in mice. J. Virol. *80*, 2396–2404.

Ljunggren, H.G., and Karre, K. (1990). In search of the 'missing self': MHC molecules and NK cell recognition. Immunol. Today *11*, 237–244.

Lobigs, M., Blanden, R.V., and Mullbacher, A. (1996). Flavivirus-induced up-regulation of MHC class I antigens; implications for the induction of CD8+ T-cell-mediated autoimmunity. Immunol. Rev. *152*, 5–19.

Lobigs, M., Mullbacher, A., and Regner, M. (2003a). MHC class I up-regulation by flaviviruses: Immune interaction with unknown advantage to host or pathogen. Immunol Cell Biol *81*, 217–223.

Lobigs, M., Mullbacher, A., Wang, Y., Pavy, M., and Lee, E. (2003b). Role of type I and type II interferon responses in recovery from infection with an encephalitic flavivirus. J. Gen. Virol. *84*, 567–572.

Lobigs, M., Mullbacher, A., and Lee, E. (2004). Evidence that a mechanism for efficient flavivirus budding upregulates MHC class I. Immunol Cell Biol *82*, 184–188.

Lodoen, M.B., and Lanier, L.L. (2005). Viral modulation of NK cell immunity. Nat. Rev. Microbiol. *3*, 59–69.

Lodoen, M.B., and Lanier, L.L. (2006). Natural killer cells as an initial defense against pathogens. Curr. Opin. Immunol. *18*, 391–398.

Loo, Y.M., Fornek, J., Crochet, N., Bajwa, G., Perwitasari, O., Martinez-Sobrido, L., Akira, S., Gill, M.A., Garcia-Sastre, A., Katze, M.G., *et al.* (2008). Distinct RIG-I and MDA5 signaling by RNA viruses in innate immunity. J. Virol. *82*, 335–345.

Ma, D., Jiang, D., Qing, M., Weidner, J.M., Qu, X., Guo, H., Chang, J., Gu, B., Shi, P.Y., Block, T.M., *et al.* (2009). Antiviral effect of interferon lambda against West Nile virus. Antiviral Res. *83*, 53–60.

Mackenzie, J.M., Khromykh, A.A., Jones, M.K., and Westaway, E.G. (1998). Subcellular localization and some biochemical properties of the flavivirus Kunjin nonstructural proteins NS2A and NS4A. Virology *245*, 203–215.

Mackenzie, J.M., Jones, M.K., and Westaway, E.G. (1999). Markers for trans-Golgi membranes and the intermediate compartment localize to induced membranes with distinct replication functions in flavivirus-infected cells. J. Virol. *73*, 9555–9567.

Mackenzie, J.M., Khromykh, A.A., and Parton, R.G. (2007). Cholesterol manipulation by West Nile virus perturbs the cellular immune response. Cell Host Microbe *2*, 229–239.

Manderson, A.P., Pickering, M.C., Botto, M., Walport, M.J., and Parish, C.R. (2001). Continual low-level activation of the classical complement pathway. J. Exp. Med. *194*, 747–756.

Marcello, T., Grakoui, A., Barba-Spaeth, G., Machlin, E.S., Kotenko, S.V., MacDonald, M.R., and Rice, C.M. (2006). Interferons alpha and lambda inhibit hepatitis C virus replication with distinct signal transduction and gene regulation kinetics. Gastroenterology *131*, 1887–1898.

Marcus, P.I., and Sekellick, M.J. (1988). Interferon induction by viruses. XVI. 2-Aminopurine blocks selectively and reversibly an early stage in interferon induction. J. Gen. Virol. *69 (Pt 7)*, 1637–1645.

Marschang, P., Sodroski, J., Wurzner, R., and Dierich, M.P. (1995). Decay-accelerating factor (CD55) protects human immunodeficiency virus type 1 from inactivation by human complement. Eur. J. Immunol. *25*, 285–290.

Mashimo, T., Lucas, M., Simon-Chazottes, D., Frenkiel, M.P., Montagutelli, X., Ceccaldi, P.E., Deubel, V., Guenet, J.L., and Despres, P. (2002). A nonsense mutation in the gene encoding 2′–5′-oligoadenylate synthetase/L1 isoform is associated with West Nile virus susceptibility in laboratory mice. Proc. Natl. Acad. Sci. U. S. A. *99*, 11311–11316.

Matsushita, M., and Fujita, T. (1992). Activation of the classical complement pathway by mannose-binding protein in association with a novel C1s-like serine protease. J. Exp. Med. *176*, 1497–1502.

Matsushita, M., Thiel, S., Jensenius, J.C., Terai, I., and Fujita, T. (2000). Proteolytic activities of two types of mannose-binding lectin-associated serine protease. J. Immunol. *165*, 2637–2642.

Mazzon, M., Jones, M., Davidson, A., Chain, B., and Jacobs, M. (2009). Dengue virus NS5 inhibits interferon-alpha signaling by blocking signal transducer and activator of transcription 2 phosphorylation. J. Infect. Dis. *200*, 1261–1270.

Mehlhop, E., and Diamond, M.S. (2006). Protective immune responses against West Nile virus are primed by distinct complement activation pathways. J. Exp. Med. *203*, 1371–1381.

Mehlhop, E., Whitby, K., Oliphant, T., Marri, A., Engle, M., and Diamond, M.S. (2005). Complement activation is required for induction of a protective antibody response against West Nile virus infection. J. Virol. *79*, 7466–7477.

Meier, K.C., Gardner, C.L., Khoretonenko, M.V., Klimstra, W.B., and Ryman, K.D. (2009). A mouse model for studying viscerotropic disease caused by yellow fever virus infection. Plos Pathog. *5*, e1000614.

Meri, S., Morgan, B.P., Davies, A., Daniels, R.H., Olavesen, M.G., Waldmann, H., and Lachmann, P.J. (1990). Human protectin (CD59), an 18,000–20,000 MW complement lysis restricting factor, inhibits C5b-8 catalysed insertion of C9 into lipid bilayers. Immunology *71*, 1–9.

Meurs, E.F., Watanabe, Y., Kadereit, S., Barber, G.N., Katze, M.G., Chong, K., Williams, B.R., and Hovanessian, A.G. (1992). Constitutive expression of human double-stranded RNA-activated p68 kinase in murine cells mediates phosphorylation of eukaryotic initiation factor 2 and partial resistance to encephalomyocarditis virus growth. J. Virol. *66*, 5805–5814.

Momburg, F., Mullbacher, A., and Lobigs, M. (2001). Modulation of transporter associated with antigen processing (TAP)-mediated peptide import into the endoplasmic reticulum by flavivirus infection. J. Virol. *75*, 5663–5671.

Monath, T.P. (1994). Dengue: the risk to developed and developing countries. Proc. Natl. Acad. Sci. U. S. A. *91*, 2395–2400.

Moretta, A., Bottino, C., Pende, D., Tripodi, G., Tambussi, G., Viale, O., Orengo, A., Barbaresi, M., Merli, A., Ciccone, E., *et al.* (1990). Identification of four subsets of human CD3-CD16+ natural killer (NK) cells by the expression of clonally distributed functional surface molecules: correlation between subset assignment of NK clones and ability to mediate specific alloantigen recognition. J. Exp. Med. *172*, 1589–1598.

Moretta, A., Bottino, C., Vitale, M., Pende, D., Biassoni, R., Mingari, M.C., and Moretta, L. (1996). Receptors for HLA class-I molecules in human natural killer cells. Annu. Rev. Immunol. *14*, 619–648.

Moretta, A., Bottino, C., Vitale, M., Pende, D., Cantoni, C., Mingari, M.C., Biassoni, R., and Moretta, L. (2001). Activating receptors and coreceptors involved in human natural killer cell-mediated cytolysis. Annu. Rev. Immunol. *19*, 197–223.

Morrey, J.D., Day, C.W., Julander, J.G., Blatt, L.M., Smee, D.F., and Sidwell, R.W. (2004). Effect of interferon-alpha and interferon-inducers on West Nile virus in mouse and hamster animal models. Antivir. Chem. Chemother. *15*, 101–109.

Mullbacher, A., and Lobigs, M. (1995). Up-regulation of MHC class I by flavivirus-induced peptide translocation into the endoplasmic reticulum. Immunity 3, 207–214.

Munoz-Jordan, J.L., Sanchez-Burgos, G.G., Laurent-Rolle, M., and Garcia-Sastre, A. (2003). Inhibition of interferon signaling by dengue virus. Proc. Natl. Acad. Sci. U. S. A. *100*, 14333–14338.

Munoz-Jordan, J.L., Laurent-Rolle, M., Ashour, J., Martinez-Sobrido, L., Ashok, M., Lipkin, W.I., and Garcia-Sastre, A. (2005). Inhibition of alpha/beta interferon signaling by the NS4B protein of flaviviruses. J. Virol. *79*, 8004–8013.

Myers, M.P., Andersen, J.N., Cheng, A., Tremblay, M.L., Horvath, C.M., Parisien, J.P., Salmeen, A., Barford, D., and Tonks, N.K. (2001). TYK2 and JAK2 are substrates of protein-tyrosine phosphatase 1B. J. Biol. Chem. *276*, 47771–47774.

Nomura, M., Kurita-Taniguchi, M., Kondo, K., Inoue, N., Matsumoto, M., Yamanishi, K., Okabe, M., and Seya, T. (2002). Mechanism of host cell protection from complement in murine cytomegalovirus (CMV) infection: identification of a CMV-responsive element in the CD46 promoter region. Eur J. Immunol. *32*, 2954–2964.

Novick, D., Cohen, B., and Rubinstein, M. (1994). The human interferon alpha/beta receptor: characterization and molecular cloning. Cell 77, 391–400.

Onoguchi, K., Yoneyama, M., Takemura, A., Akira, S., Taniguchi, T., Namiki, H., and Fujita, T. (2007). Viral infections activate types I and III interferon genes through a common mechanism. J. Biol. Chem. *282*, 7576–7581.

Osterlund, P.I., Pietila, T.E., Veckman, V., Kotenko, S.V., and Julkunen, I. (2007). IFN regulatory factor family members differentially regulate the expression of type III IFN (IFN-lambda) genes. J. Immunol. *179*, 3434–3442.

Overby, A.K., Popov, V.L., Niedrig, M., and Weber, F. (2010). Tick-borne encephalitis virus delays interferon induction and hides its double-stranded RNA in intracellular membrane vesicles. J. Virol. *84*, 8470–8483.

Pangburn, M.K., and Muller-Eberhard, H.J. (1980). Relation of putative thioester bond in C3 to activation of the alternative pathway and the binding of C3b to biological targets of complement. J. Exp. Med. *152*, 1102–1114.

Pangburn, M.K., Morrison, D.C., Schreiber, R.D., and Muller-Eberhard, H.J. (1980). Activation of the alternative complement pathway: recognition of surface structures on activators by bound C3b. J. Immunol. *124*, 977–982.

Park, C., Li, S., Cha, E., and Schindler, C. (2000). Immune response in Stat2 knockout mice. Immunity *13*, 795–804.

Park, G.S., Morris, K.L., Hallett, R.G., Bloom, M.E., and Best, S.M. (2007). Identification of residues critical for the interferon antagonist function of Langat virus NS5 reveals a role for the RNA-dependent RNA polymerase domain. J. Virol. *81*, 6936–6946.

Perelygin, A.A., Scherbik, S.V., Zhulin, I.B., Stockman, B.M., Li, Y., and Brinton, M.A. (2002). Positional cloning of the murine flavivirus resistance gene. Proc. Natl. Acad. Sci. U. S. A. *99*, 9322–9327.

Perussia, B. (1996). The cytokine profile of resting and activated NK cells. Methods *9*, 370–378.

Platanias, L.C. (2005). Mechanisms of type-I- and type-II-interferon-mediated signalling. Nat. Rev. Immunol. *5*, 375–386.

Quaresma, J.A., Barros, V.L., Pagliari, C., Fernandes, E.R., Guedes, F., Takakura, C.F., Andrade, H.F., Jr., Vasconcelos, P.F., and Duarte, M.I. (2006). Revisiting the liver in human yellow fever: virus-induced apoptosis in hepatocytes associated with TGF-beta, TNF-alpha and NK cells activity. Virology *345*, 22–30.

Quaresma, J.A., Barros, V.L., Pagliari, C., Fernandes, E.R., Andrade, H.F., Jr., Vasconcelos, P.F., and Duarte, M.I. (2007). Hepatocyte lesions and cellular immune response in yellow fever infection. Trans. R. Soc. Trop. Med. Hyg. *101*, 161–168.

Querec, T., Bennouna, S., Alkan, S., Laouar, Y., Gorden, K., Flavell, R., Akira, S., Ahmed, R., and Pulendran, B. (2006). Yellow fever vaccine YF-17D activates multiple dendritic cell subsets via TLR2, 7, 8, and 9 to stimulate polyvalent immunity. J. Exp. Med. *203*, 413–424.

Rock, F.L., Hardiman, G., Timans, J.C., Kastelein, R.A., and Bazan, J.F. (1998). A family of human receptors structurally related to Drosophila Toll. Proc. Natl. Acad. Sci. U. S. A. *95*, 588–593.

Rodriguez-Madoz, J.R., Bernal-Rubio, D., Kaminski, D., Boyd, K., and Fernandez-Sesma, A. (2010). Dengue virus inhibits the production of type I interferon in primary human dendritic cells. J. Virol. *84*, 4845–4850.

Sadler, A.J., and Williams, B.R. (2008). Interferon-inducible antiviral effectors. Nat. Rev. Immunol. *8*, 559–568.

Saifuddin, M., Parker, C.J., Peeples, M.E., Gorny, M.K., Zolla-Pazner, S., Ghassemi, M., Rooney, I.A., Atkinson, J.P., and Spear, G.T. (1995). Role of virion-associated glycosylphosphatidylinositol-linked proteins CD55 and CD59 in complement resistance of cell line-derived and primary isolates of HIV-1. J. Exp. Med. *182*, 501–509.

Saito, T., Hirai, R., Loo, Y.M., Owen, D., Johnson, C.L., Sinha, S.C., Akira, S., Fujita, T., and Gale, M., Jr. (2007). Regulation of innate antiviral defenses through a shared repressor domain in RIG-I and LGP2. Proc. Natl. Acad. Sci. U. S. A. *104*, 582–587.

Samuel, C.E. (2001). Antiviral actions of interferons. Clin. Microbiol. Rev. *14*, 778–809..

Samuel, C.E. (2002). Host genetic variability and West Nile virus susceptibility. Proc. Natl. Acad. Sci. U. S. A. *99*, 11555–11557.

Samuel, M.A., and Diamond, M.S. (2005). Alpha/beta interferon protects against lethal West Nile virus infection by restricting cellular tropism and enhancing neuronal survival. J. Virol. *79*, 13350–13361.

Scalzo, A.A. (2002). Successful control of viruses by NK cells--a balance of opposing forces? Trends Microbiol. *10*, 470–474.

Scharfstein, J., Ferreira, A., Gigli, I., and Nussenzweig, V. (1978). Human C4-binding protein. I. Isolation and characterization. J. Exp. Med. *148*, 207–222.

Scherbik, S.V., Kluetzman, K., Perelygin, A.A., and Brinton, M.A. (2007). Knock-in of the Oas1b(r) allele into a flavivirus-induced disease susceptible mouse generates the resistant phenotype. Virology *368*, 232–237.

Scherbik, S.V., Paranjape, J.M., Stockman, B.M., Silverman, R.H., and Brinton, M.A. (2006). RNase L plays a role in the antiviral response to West Nile virus. J. Virol. *80*, 2987–2999.

Schindler, C., Levy, D.E., and Decker, T. (2007). JAK-STAT signaling: from interferons to cytokines. J. Biol. Chem. *282*, 20059–20063.

Sen, G.C. (2001). Viruses and interferons. Annu. Rev. Microbiol. *55*, 255–281.

Seya, T., and Atkinson, J.P. (1989). Functional properties of membrane cofactor protein of complement. Biochem. J. *264*, 581–588.

Sheppard, P., Kindsvogel, W., Xu, W., Henderson, K., Schlutsmeyer, S., Whitmore, T.E., Kuestner, R., Garrigues, U., Birks, C., Roraback, J., *et al.* (2003). IL-28, IL-29 and their class II cytokine receptor IL-28R. Nat. Immunol. *4*, 63–68.

Shresta, S., Pham, C.T., Thomas, D.A., Graubert, T.A., and Ley, T.J. (1998). How do cytotoxic lymphocytes kill their targets? Curr Opin Immunol *10*, 581–587.

Shresta, S., Kyle, J.L., Robert Beatty, P., and Harris, E. (2004a). Early activation of natural killer and B cells in response to primary dengue virus infection in A/J mice. Virology *319*, 262–273.

Shresta, S., Kyle, J.L., Snider, H.M., Basavapatna, M., Beatty, P.R., and Harris, E. (2004b). Interferon-dependent immunity is essential for resistance to primary dengue virus infection in mice, whereas T- and B-cell-dependent immunity are less critical. J. Virol. *78*, 2701–2710.

Shresta, S., Sharar, K.L., Prigozhin, D.M., Snider, H.M., Beatty, P.R., and Harris, E. (2005). Critical roles for both STAT1-dependent and STAT1-independent pathways in the control of primary dengue virus infection in mice. J. Immunol. *175*, 3946–3954.

Shrestha, B., Samuel, M.A., and Diamond, M.S. (2006). CD8(+) T cells require perforin to clear West Nile virus from infected neurons. J. Virol. *80*, 119–129.

Song, M.M., and Shuai, K. (1998). The suppressor of cytokine signaling (SOCS) 1 and SOCS3 but not SOCS2 proteins inhibit interferon-mediated antiviral and antiproliferative activities. J. Biol. Chem. *273*, 35056–35062.

Spiller, O.B., Morgan, B.P., Tufaro, F., and Devine, D.V. (1996). Altered expression of host-encoded complement regulators on human cytomegalovirus-infected cells. Eur. J. Immunol. *26*, 1532–1538.

Stark, G.R., Kerr, I.M., Williams, B.R., Silverman, R.H., and Schreiber, R.D. (1998). How cells respond to interferons. Annu. Rev. Biochem. *67*, 227–264.

Stetson, D.B., and Medzhitov, R. (2006). Type I interferons in host defense. Immunity *25*, 373–381.

Summerfield, J.A., Sumiya, M., Levin, M., and Turner, M.W. (1997). Association of mutations in mannose binding protein gene with childhood infection in consecutive hospital series. BMJ *314*, 1229–1232.

Suthar, M.S., Ma, D.Y., Thomas, S., Lund, J.M., Zhang, N., Daffis, S., Rudensky, A.Y., Bevan, M.J., Clark, E.A., Kaja, M.K., *et al.* (2010). IPS-1 Is Essential for the Control of West Nile Virus Infection and Immunity. Plos Pathog. *6*, 1–15.

Takahasi, K., Yoneyama, M., Nishihori, T., Hirai, R., Kumeta, H., Narita, R., Gale, M., Jr., Inagaki, F., and Fujita, T. (2008). Nonself RNA-sensing mechanism of RIG-I helicase and activation of antiviral immune responses. Mol. Cell *29*, 428–440.

Takahasi, K., Kumeta, H., Tsuduki, N., Narita, R., Shigemoto, T., Hirai, R., Yoneyama, M., Horiuchi, M., Ogura, K., Fujita, T., *et al.* (2009). Solution structures of cytosolic RNA sensor MDA5 and LGP2 C-terminal domains: identification of the RNA recognition loop in RIG-I-like receptors. J. Biol. Chem. *284*, 17465–17474.

Thiel, S., Vorup-Jensen, T., Stover, C.M., Schwaeble, W., Laursen, S.B., Poulsen, K., Willis, A.C., Eggleton, P., Hansen, S., Holmskov, U., *et al.* (1997). A second serine protease associated with mannan-binding lectin that activates complement. Nature *386*, 506–510.

Thomas, D.L., Thio, C.L., Martin, M.P., Qi, Y., Ge, D., O'Huigin, C., Kidd, J., Kidd, K., Khakoo, S.I., Alexander, G., *et al.* (2009). Genetic variation in IL28B and spontaneous clearance of hepatitis C virus. Nature *461*, 798–801.

Trinchieri, G. (1989). Biology of natural killer cells. Adv. Immunol. *47*, 187–376.

Vargin, V.V., and Semenov, B.F. (1986). Changes of natural killer cell activity in different mouse lines by acute and asymptomatic flavivirus infections. Acta Virol. *30*, 303–308.

Vitale, M., Bottino, C., Sivori, S., Sanseverino, L., Castriconi, R., Marcenaro, E., Augugliaro, R., Moretta, L., and Moretta, A. (1998). NKp44, a novel triggering surface molecule specifically expressed by activated natural killer cells, is involved in non-major histocompatibility complex-restricted tumor cell lysis. J. Exp. Med. *187*, 2065–2072.

Vorup-Jensen, T., Jensenius, J.C., and Thiel, S. (1998). MASP-2, the C3 convertase generating protease of the MBLectin complement activating pathway. Immunobiology *199*, 348–357.

Walport, M.J. (2001a). Complement. First of two parts. N. Engl. J. Med. *344*, 1058–1066.

Walport, M.J. (2001b). Complement. Second of two parts. N. Engl. J. Med. *344*, 1140–1144.

Wang, J.P., Liu, P., Latz, E., Golenbock, D.T., Finberg, R.W., and Libraty, D.H. (2006). Flavivirus activation of plasmacytoid dendritic cells delineates key elements of TLR7 signaling beyond endosomal recognition. J. Immunol. *177*, 7114–7121.

Wang, T., Town, T., Alexopoulou, L., Anderson, J.F., Fikrig, E., and Flavell, R.A. (2004). Toll-like receptor 3 mediates West Nile virus entry into the brain causing lethal encephalitis. Nat. Med. *10*, 1366–1373.

Warke, R.V., Martin, K.J., Giaya, K., Shaw, S.K., Rothman, A.L., and Bosch, I. (2008). TRAIL is a novel antiviral protein against dengue virus. J. Virol. *82*, 555–564.

Weiler, J.M., Daha, M.R., Austen, K.F., and Fearon, D.T. (1976). Control of the amplification convertase of complement by the plasma protein beta1H. Proc. Natl. Acad. Sci. U. S. A. *73*, 3268–3272.

Weis, W.I., Drickamer, K., and Hendrickson, W.A. (1992). Structure of a C-type mannose-binding protein complexed with an oligosaccharide. Nature *360*, 127–134.

Welsh, R.M. (1986). Regulation of virus infections by natural killer cells. A review. Nat. Immun. Cell Growth Regul. *5*, 169–199.

Werme, K., Wigerius, M., and Johansson, M. (2008). Tick-borne encephalitis virus NS5 associates with membrane protein scribble and impairs interferon-stimulated JAK-STAT signalling. Cell Microbiol. *10*, 696–712.

Westaway, E.G., Mackenzie, J.M., Kenney, M.T., Jones, M.K., and Khromykh, A.A. (1997). Ultrastructure of Kunjin virus-infected cells: colocalization of NS1 and NS3 with double-stranded RNA, and of NS2B with NS3, in virus-induced membrane structures. J. Virol. *71*, 6650–6661.

Wilson, J.R., de Sessions, P.F., Leon, M.A., and Scholle, F. (2008). West Nile virus nonstructural protein 1 inhibits TLR3 signal transduction. J. Virol. *82*, 8262–8271.

Wormald, S., and Hilton, D.J. (2004). Inhibitors of cytokine signal transduction. J. Biol. Chem. *279*, 821–824.

Wu, T.R., Hong, Y.K., Wang, X.D., Ling, M.Y., Dragoi, A.M., Chung, A.S., Campbell, A.G., Han, Z.Y., Feng, G.S., and Chin, Y.E. (2002). SHP-2 is a dual-specificity phosphatase involved in Stat1 dephosphorylation at both tyrosine and serine residues in nuclei. J. Biol. Chem. *277*, 47572–47580.

Xu, D., and Qu, C.K. (2008). Protein tyrosine phosphatases in the JAK/STAT pathway. Front Biosci. *13*, 4925–4932.

Yakub, I., Lillibridge, K.M., Moran, A., Gonzalez, O.Y., Belmont, J., Gibbs, R.A., and Tweardy, D.J. (2005). Single nucleotide polymorphisms in genes for 2′-5′-oligoadenylate synthetase and RNase L inpatients hospitalized with West Nile virus infection. J. Infect. Dis. *192*, 1741–1748.

Yokoyama, W.M., and Scalzo, A.A. (2002). Natural killer cell activation receptors in innate immunity to infection. Microbes Infect. *4*, 1513–1521.

Yoneyama, M., and Fujita, T. (2010). Recognition of viral nucleic acids in innate immunity. Rev. Med. Virol. *20*, 4–22.

Young, P.R., Hilditch, P.A., Bletchly, C., and Halloran, W. (2000). An antigen capture enzyme-linked immunosorbent assay reveals high levels of the dengue virus protein NS1 in the sera of infected patients. J. Clin. Microbiol. *38*, 1053–1057.

Zhang, S.Y., Jouanguy, E., Ugolini, S., Smahi, A., Elain, G., Romero, P., Segal, D., Sancho-Shimizu, V., Lorenzo, L., Puel, A., *et al.* (2007). TLR3 deficiency in patients with herpes simplex encephalitis. Science *317*, 1522–1527.

Zhao, W., Lee, C., Piganis, R., Plumlee, C., de Weerd, N., Hertzog, P.J., and Schindler, C. (2008). A conserved IFN-alpha receptor tyrosine motif directs the biological response to type I IFNs. J. Immunol. *180*, 5483–5489.

Zhong, B., Yang, Y., Li, S., Wang, Y.Y., Li, Y., Diao, F., Lei, C., He, X., Zhang, L., Tien, P., *et al.* (2008). The adaptor protein MITA links virus-sensing receptors to IRF3 transcription factor activation. Immunity *29*, 538–550.

Zhou, A., Hassel, B.A., and Silverman, R.H. (1993). Expression cloning of 2–5A-dependent RNAase: a uniquely regulated mediator of interferon action. Cell *72*, 753–765.

Zhou, A., Paranjape, J., Brown, T.L., Nie, H., Naik, S., Dong, B., Chang, A., Trapp, B., Fairchild, R., Colmenares, C., *et al.* (1997). Interferon action and apoptosis are defective in mice devoid of 2′,5′-oligoadenylate-dependent RNase L. Embo J. *16*, 6355–6363.

Zhou, Z., Hamming, O.J., Ank, N., Paludan, S.R., Nielsen, A.L., and Hartmann, R. (2007). Type III interferon (IFN) induces a type I IFN-like response in a restricted subset of cells through signaling pathways involving both the Jak-STAT pathway and the mitogen-activated protein kinases. J. Virol. *81*, 7749–7758.

Zinn, K., Keller, A., Whittemore, L.A., and Maniatis, T. (1988). 2-Aminopurine selectively inhibits the induction of beta-interferon, c-fos, and c-myc gene expression. Science *240*, 210–213.

Host Responses During Mild and Severe Dengue

8

Mark Schreiber, F. Joel Leong and Martin L. Hibberd

Abstract

Dengue fever is an acute viral infection that can produce a wide spectrum of disease outcomes in patients, ranging from mild or even asymptomatic fever to severe manifestations including haemorrhagic fever and shock. With the incidence of the severe forms increasing in most tropical countries as well as an overall increase in dengue incidence, dengue fever is becoming a significant burden on the health systems of affected countries.

In this review, we examine the clinical definitions and presentation of mild and severe dengue as well as recent research into the underlying molecular mechanisms of the differential host response. Finally, we will examine how host responses from the early phase of the disease might be useful as biomarkers for predicting the eventual disease outcome.

Introduction

Dengue fever (DF) is an acute viral fever caused by infection with the one of four immunologically distinct but genetically similar dengue serotypes (DENV-1 to DENV-4) (SABIN, 1952; Halstead, 2007). While generally self-limiting and not fatal, the condition represents a major economic and public health burden on endemic tropical and subtropical regions (Okanurak *et al.*, 1997; Meltzer *et al.*, 1998; Clark, Jr., *et al.*, 2005; Anderson *et al.*, 2007; Garg *et al.*, 2008; Huy *et al.*, 2009; Beaute and Vong, 2010). Like many flaviviruses, mosquitoes, predominantly the *Aedes aegypti* species, transmit DENV. Although some other Aedes species can transmit the virus, their public health impact is not thought to be as significant (Halstead, 2008). The four serotypes are distributed throughout the tropical and subtropical regions of the world and in many places, particularly South East Asia, the four serotypes are hyperendemic.

Although generally 'mild', clinicians began to identify the first cases of severe version of the disease in the 1950s and 1960s (Hammon *et al.*, 1960; Halstead, 1965; Cohen and Halstead, 1966). The severe cases presented with haemorrhagic fever, characterized by vascular leakage, sometimes leading to shock. These conditions have been termed dengue Haemorrhagic Fever (DHF) and dengue Shock Syndrome (DSS). The incidence of both the mild and severe forms has been increasing steadily since their identification (Gubler, 1998, 2002; Ooi and Gubler, 2009). Typically, severe manifestations do not present early in the fever. What begins as a classical dengue fever may progress to severe manifestations later. Often the onset of severe vascular leak occurs at or around the point of defervescence (Libraty *et al.*, 2002b; Srikiatkhachorn *et al.*, 2007; Guilarde *et al.*, 2008).

In addition to the classical DF and DHF/DSS conditions, there is strong evidence that considerable 'silent' transmission of dengue occurs (Chen *et al.*, 1996; Steel *et al.*, 2010). Indeed, it is conceivable that very mild DF could easily be symptomatically confused with influenza infection. Symptom free infections may also occur. The incidence of these inapparent infections seems to vary by virus strain and host population, but in some cases, inapparent infections can greatly outnumber symptomatic infections suggesting that symptomatic dengue is sometimes the 'tip of the iceberg' (Capeding *et al.*, 2010).

The causes of these different host responses are not fully understood, although some risk factors are known. Perhaps the most important risk factor is secondary dengue infection (Guzman *et al.*, 2002b; Halstead, 2009). Because there are four serotypes, multiple infections are possible in a patient's lifetime. Antibodies specific to the infecting serotype provide lifelong protection to that serotype but only partial protection to the other serotypes. Upon subsequent infection with a different serotype, these antibodies may recognize but not neutralize the virus. These partially opsonized virus particles are efficiently taken up by monocytes but, as they are not neutralized, they proceed to infect the cell, leading to higher levels of viral infection and subsequently increased severity (Halstead *et al.*, 1984; Morens *et al.*, 1987; Halstead, 1989; Morens and Halstead, 1990). This phenomenon is known as antibody-dependent enhancement (ADE) and it is the dominant theory explaining the severity risk of secondary dengue (Halstead, 1989).

Although ADE is recognized as a major factor in severity, secondary infection does not always lead to severe outcomes and some apparently primary infections are associated with severe symptoms, although perhaps not full DHF/DSS. Other severity risk factors include patient haplotypes (Sakuntabhai *et al.*, 2005; Vejbaesya *et al.*, 2009), cell receptor mutations (Loke *et al.*, 2002; Chen *et al.*, 2008), viral genotype (Watts *et al.*, 1999; Halstead *et al.*, 2001; Barcelos *et al.*, 2008), and complicating chronic conditions such as diabetes, asthma and an immunocompromised state (Bravo *et al.*, 1987; Guzman *et al.*, 1992; Gonzalez *et al.*, 2005; Lahiri *et al.*, 2008; Figueiredo *et al.*, 2010). Age is also a significant factor, with younger people more at risk of DHF and DSS (Guzman *et al.*, 2002a).

Definitions of mild and severe dengue fever

For the past few decades, the most widely used severity and diagnostic criteria are those published by the WHO. The WHO criteria for DF/DHF and DSS are very specific and are paraphrased in the following paragraphs. The WHO defines cases of probable dengue fever as – an acute febrile illness with two or more of the following manifestations: headache, retro-orbital pain, myalgia, arthralgia, rash, haemorrhagic manifestations or leucopenia – and with supportive serology or occurrence at the same location and time as other confirmed cases of dengue fever. There are four criteria for confirming dengue in the laboratory: the isolation of dengue virus from serum or autopsy samples, a fourfold or greater change in anti dengue IgG or IgM in paired serum samples, detection of dengue antigens in tissue, serum or cerebrospinal fluid, or detection of dengue genomic material in tissue, serum or cerebrospinal fluid. Clinical confirmation is not required to diagnose probable dengue, which is important in settings where serology and molecular diagnostics are not available. The WHO classification further defines two forms of more severe dengue; DHF and DSS.

A dengue case is classified clinically as DHF when *all* of the following symptoms, signs and laboratory criteria are present: fever or biphasic fever lasting more than 2–7 days, haemorrhagic manifestations (at least a positive tourniquet test), thrombocytopenia ($<100{,}000$ platelets per ml of serum), haemoconcentration (rising haematocrit $\geq 20\%$ above baseline or normal for age/gender/ethnicity) or other evidence of plasma leakage (ascites, pleural effusions, low serum protein/albumin/cholesterol). Haemorrhagic manifestations include a positive tourniquet test (also known as the Hess test or Rumpel–Leede capillary fragility test), petechiae (1–3 mm diameter, red spots caused by minor haemorrhage), purpura (red or purple skin discoloration 3 mm to 1 cm in diameter), ecchymosis (bruising >1 cm diameter), and bleeding from mucosa, gastrointestinal tract or other locations (Kalayanarooj and Nimmannitya, 2004).

A diagnosis of DSS requires all of the DHF criteria with additional manifestations of shock such as a rapid and weak pulse, narrowing of pulse pressure to ≤ 20 mmHg, or hypotension, poor capillary refill time >2 seconds and cold clammy extremities or restlessness.

The WHO also publishes four grades of severity which are useful in public health reporting or retrospective analysis of outcome. Grade I is dengue haemorrhagic fever without shock and only minor haemorrhagic manifestations (positive tourniquet test). Grade II includes the manifestations of Grade I plus spontaneous bleeding. Grade III requires circulatory failure and Grade IV is profound shock including undetectable blood pressure or pulse. Grades III and IV would be considered DSS. The scale indicates increasing severity however, it is important to realize that dengue illness is heterogeneous in severity, and that severe clinical features can occur in patients who do not meet the criteria for DHF (Srikiatkhachorn, Gibbons *et al.*, 2010). It is plasma leakage which separates DHF/DSS from DF. A case of DHF with signs of vascular leakage but no spontaneous haemorrhage would be WHO Grade I and could be less severe than DF accompanied by spontaneous bleeding ('DF with unusual haemorrhage', under the current WHO classification).

While comprehensive and precise in its definition, the WHO severity criteria have been criticized for being difficult to understand and implement in clinical practice (Balasubramanian *et al.*, 2006; Rigau-Perez, 2006; Gupta *et al.*, 2010). Many clinicians and researchers dislike it because it requires demonstration of plasma leakage to distinguish dengue fever from dengue haemorrhagic fever. Yet it is plasma leakage and thrombocytopenia which contribute to the specificity (92%) of the WHO case definition and help identify dengue cases that require intervention (Srikiatkhachorn *et al.*, 2010). Despite the name, haemorrhagic tendency does not reliably differentiate DF from DHF.

Given that vascular leakage can appear transiently, accurate classification requires either skilled clinical observations for signs of leakage (pleural effusion, ascites) or frequent sampling to detect a 20% rise in haematocrit above baseline or expected reference range. Effective implementation of these standards does sometimes require changes in standard hospital practice. For example, to perform frequent haematocrit tests a testing station would need to be set up in the infectious diseases ward. Daily results from a central haematology laboratory would be inadequate. While perceived by some as inconvenient, one could argue that by forcing the physician to examine and monitor their patient carefully, the patient will receive better clinical management.

Modification and misinterpretation of the severity criteria is common and many clinicians will not use the tourniquet test (Cao *et al.*, 2002), which calls into question any DHF diagnosis in the absence of other demonstrations of plasma leakage. The use of the tourniquet test involves inflating a blood pressure cuff around the patient's arm between diastolic and systolic pressure for 5 minutes and counting the number of petechiae in a square inch of skin. Some clinicians are reluctant to perform the test, regarding it as time consuming and uncomfortable for the patient, while others have successfully integrated this into their screening workflow. Opinions are mixed, with some believing it is clinically useful for screening but non-specific with a high false-positive rate (78.7% positive predictive value and 28.6% negative predictive values in one study; Norlijah *et al.*, 2006). This test can also be positive in those with sun-damaged skin or scurvy. Petechiae can be difficult to see in individuals with dark skin, leading to false-negative interpretations.

The definition is also very difficult to apply during retrospective analysis of outbreaks unless clinicians practice thorough case documentation. A recent review of 37 published papers on dengue classification found that 27 did not use the full WHO definition or modified the definition. In cases where the strict definition was used, many clinically severe cases and even deaths could not be classified as DHF or DSS owing to a lack of data or an absence of characteristic symptoms and signs (Bandyopadhyay *et al.*, 2006).

Rigau-Perez (2006) has also questioned the use of an outdated definition of thrombocytopenia; a level of less than 150,000 platelets per ml is more current. Further, the interpretation of a positive test is inconsistent with many clinicians using 10 petechiae per square inch while the WHO definition calls for 20. He also pointed out that the definition is based on observations in South-East Asian populations – mainly children – in the 1960s that may not always be applicable in other settings. Further, he questioned the value of the diagnosis when early intervention with fluid treatment will prevent detection of the DHF/DSS leakage criteria unless X-rays or sonograms are used.

The current classification focuses heavily on narrowing of pulse pressure and vascular leak, which are two of the most dangerous complications of the disease. However, it does not account for rare but potentially life threatening complications such as encephalitis (Lum *et al.*, 1996; Garcia-Rivera and Rigau-Perez, 2002; Agarwal *et al.*, 2009), cardiomyopathy (Obeyesekere and Hermon, 1972; Neo *et al.*, 2006), liver failure (Osorio *et al.*, 2008; Chongsrisawat *et al.*, 2009) and renal impairment (George *et al.*, 1988; Kularatne *et al.*, 2005). While these are serious conditions, it is possible that they are outcomes of prolonged and untreated vascular leakage and not directly due to the viral infection. Rather than invalidating the definition, they may represent complications of DHF and DSS.

In its favour, the current WHO classification provides a high positive predictive value in endemic areas without requiring laboratory testing such as dengue virus PCR or NS1 detection. Two studies have indicated that the criteria, when strictly applied, have a specificity of 82% and a sensitivity of 94–100% (Kalayanarooj *et al.*, 1997; Phuong *et al.*, 2004).

As part of a WHO/TDR dengue control working group (DENCO) meeting, an expert group discussed the existing definition and suggested some simplifications and modifications. The proposed revised dengue classification consists of two classes: dengue and severe dengue, further subcategorizing dengue into dengue without warning signs and dengue with warning signs. A confirmed dengue diagnosis requires laboratory evidence whereas a probable dengue diagnosis requires living within or recent travel to a dengue-endemic area, fever and two of the following criteria: nausea or vomiting, rash, aches and pains, positive tourniquet test, leukopenia or any severity warning sign. The warning signs consist of abdominal pain, persistent vomiting, clinical fluid accumulation, mucosal bleed, lethargy or restlessness, liver enlargement > 2 cm, increase in haematocrit concurrent with a rapid decrease in platelet count. Any warning sign is considered prognostic of a poor clinical outcome and would indicate the need for strict observation and medical intervention. Severe dengue includes any of severe plasma leakage, severe bleeding or severe organ involvement (Deen *et al.*, 2006).

The revised classification makes tourniquet testing and frequent haematocrit measurements optional, which should make it easier to classify severity. It also incorporates potentially serious complications that may not fulfil parts of the older classification. The addition of warning signs would be of additional benefit to inexperienced clinicians and could alert them of a possible progression to severe dengue.

At the time of writing, the new classification proposal is being validated in several research sites. The outcome of these evaluations will determine the relative performance of both classification schemes. In the absence of published results, we can only speculate on the relative strengths and weaknesses.

The revised system allows easier classification and may improve consistency between physicians. We also anticipate that the new scheme would be much easier to apply to retrospective studies, as it would require less clinical documentation to validate the diagnosis of severity. However, the reduced criteria for probable dengue may result in a high false-positive rate of diagnosis. Because of this, clinicians may require more laboratory confirmation of dengue virus. This may lead to better research but is an added cost to the practising physician and may not be available at all in healthcare centres. If rapid laboratory diagnostics were inexpensive and readily accessible, the new classification could be superior to the current system. The proposed classification would also be very useful if an effective antiviral was available, owing to a reduced need for clinical intervention. However, the existing classification – based on detecting and managing vascular leakage – has utility that is more clinically relevant. One concern with the new proposal is that removing the need for close monitoring may mean clinicians miss the onset of vascular leakage and manage fluid balance incorrectly, either over or under compensating as the disease progresses through febrile, critical and convalescence phases.

There is a need for a clear and effective classification scheme to improve diagnosis and treatment, to allow unambiguous interpretation of experimental studies and to provide a convenient and feasible end-point for drug and vaccine trials. The scheme would also need to be easy to implement to prevent inconsistencies in classification between sites and ideally effective for use by clinicians with little experience of dengue.

The immune response to dengue

An immunocompetent dengue patient posses a number of antiviral defences that are somewhat overlapping and redundant. The two major immune responses are the innate and adaptive cellular responses (Navarro-Sanchez *et al.*, 2005; Flano, 2008). Limiting and clearing the viral infection is the role of both of these pathways. Triggered by the presence of molecules associated with pathogens, the innate system involves interferon and chemokine signals to coordinate immune cells such as monocytes and natural killer cells (Azeredo *et al.*, 2001; Libraty *et al.*, 2001). The innate response also induces fever and other symptoms commonly associated with viral infection. In non-immune patients, the innate response is rapidly activated and, unlike the innate immune response, does not require any previous exposure to the pathogen.

The adaptive response takes longer to develop but results, in the case of dengue, in lifelong immunity to the infecting serotype (Hotta *et al.*, 1968; Halstead, 1974). A population of T lymphocytes circulating in the body that recognize a dengue-specific antigen initiates the response. Lymphocytes that can recognize specific dengue serotype antigens will be present at very low concentrations but eventually one or more will encounter their specific antigen and become activated. This process is mediated by antigen presenting cells, predominantly dendritic cells (DCs). DCs circulating in the dermis and associated tissues encounter DENV particles shortly after infection and carry these particles, bound to their DC-SIGN receptors, to the thymus where they are presented to T lymphocytes for antigen recognition. Ironically, DCs are one of the cells that can be infected by DENV and this infection may be mediated by the DC-SIGN receptor (Navarro-Sanchez *et al.*, 2003; Tassaneetrithep *et al.*, 2003; Lozach *et al.*, 2005). The infected DC effectively becomes a Trojan horse and carries the infection to the lymphatic system.

T cells that recognize the presented antigens will become activated and begin to proliferate. Over time the activated T cells will proliferate and produce a large amount of antibody that will bind the antigen, potentially neutralizing circulating infectious DENV particles. Antibody bound viruses are then cleared by macrophages and other immune cells. After the infection subsides, the patient will be immune to both homologous and heterologous serotypes. This cross-reactive immunity lasts from three to four months (SABIN, 1952). Cross-immunity is gradually lost as the active T-cell population diminishes and specializes. Guzman *et al.* (2002b, 2007) observed that the incidence of DHF increases as the time between primary and secondary infection increases. This may be due to gradual reduction in cross-protective antibodies resulting in increased risk of ADE. However, it is impossible to rule out the influence of genetic variations in the infecting virus.

The ADE severity hypothesis

Under the ADE hypothesis (Halstead *et al.*, 1970), partially neutralizing or non-neutralizing antibodies present from a primary, heterologous infection, increase the infection rate of monocytes leading rapidly to very high levels of viraemia. The increased infection is due to Fcγ receptor-mediated uptake and subsequent escape of the non-neutralized virus–antibody complex (Littaua *et al.*, 1990). This enhanced infection results in a very strong innate immune response leading to high levels of inflammatory cytokines and increased vascular leakage. Many studies have linked secondary dengue infection to the incidence of DHF and high viraemia prior to defervescence is a strong risk factor for DHF at defervescence (Vaughn *et al.*, 2000; Libraty *et al.*, 2002a; Endy *et al.*, 2004; Wichmann *et al.*, 2004; Tantracheewathorn and Tantracheewathorn, 2007).

Antibodies to the unprocessed dengue M protein (prM) have limited neutralizing ability and are particularly prone to enhancement (Dejnirattisai *et al.*, 2010). They also greatly enhance the infectivity of immature viruses containing unprocessed prM protein in the capsid that are otherwise poorly infective, suggesting a role for these antibodies in the ADE of DENV (Dejnirattisai *et al.*, 2010; Rodenhuis-Zybert *et al.*, 2010). Approximately 30% of dengue virions contain prM protein, providing a pool of prM antigen in primary dengue and a source of infectious virus in secondary dengue (Zybert *et al.*, 2008).

ADE provides an intriguing explanation for cases of DHF in infants. Children who have not had a previous dengue exposure can be at risk of DHF. It is thought that the presence of heterologous maternal antibodies can create the immunological equivalent of a secondary dengue infection (Marchette *et al.*, 1979; Kliks *et al.*, 1988). These antibodies decline over time and become undetectable in the blood of infants after 6–12 months resulting in small window – approximately 4–12 months of age – in which the maternal antibodies have declined below a neutralizing titre but are sufficient to cause ADE (Marchette *et al.*, 1979; Kliks *et al.*, 1988; Simmons *et al.*, 2007a; Chau *et al.*, 2009).

Transcriptional and pathway responses in mild and severe dengue

Pathogen-induced host transcriptional responses are complex across the span of disease progression. These host transcriptional responses can be the cause or consequence of the severity of a disease (Katze *et al.*, 2008; Jenner and Young, 2005). Unravelling strong associations between these transcriptional responses and the disease states at different time course by means of genomic approaches can provide a better understanding of the molecular mechanisms that give rise to the outcome of disease, including dengue (Hibberd *et al.*, 2006).

At the early stages of dengue infection, it is believed that a transient response of the innate immunity plays a role in reducing the initial viral load. Among the immune-response pathways presented in Long *et al.* 2009, the interferon (IFN) pathway, interleukin (IL)-10, antigen presentation, and IL-6 signalling pathways were significantly represented (Long *et al.*, 2009). Several of these IFN-stimulated genes (ISGs) (OAS1, OAS2, OAS3, IRF7, SOCS1, STAT1, and viperin) were observed to be abundant in the transcriptome of mild dengue patient (Fink *et al.*, 2007; Long *et al.*, 2009). A general association of dengue severity and up regulation of genes involved in innate immune response (such as IFNG and ILI2A) during the acute phase of infection in children was also highlighted by de Kruif *et al.* (2008). In a comparison of acute dengue and DSS patients with identical duration of illness a diverse set of related processes such as death receptor, IFN, apoptosis, IL-6, NF-κB, and IL-10 signalling were over expressed in the acute patients whereas the DSS profile appeared to be somewhat attenuated (Long *et al.*, 2009). This shows that early innate immune response is of paramount importance during mild dengue presentation but may not as effective for DSS dengue presentation.

As innate immunity is transient, it is reasonable that this response may only be captured as early as the day of disease onset. However, it is clinically challenging to recruit this group of patients. As such, it is common for studies to characterize the transcriptome of acute mild dengue by a highly active metabolic state consistent with a host response to infection, instead of in combination with the activation of innate immunity response. Transcripts related to oxidative metabolism and protein ubiquitination were usually highly differentiated (Long *et al.*, 2009). Genes whose protein products are expressed in the ER are also identified to be the most significantly enriched specifically including UBE2 and the PSMB genes (Fink *et al.*, 2007; Simmons *et al.*, 2007b). In addition, the study by Loke *et al.* (2010) comparing all acute cases with convalescence found responses that appeared unrelated to the immune response, but mainly to metabolism-related processes such as oxidative phosphorylation, protein targeting, nucleic acid metabolism, DNA metabolism and replication, and protein metabolism and modification.

Ubol *et al.* (2008) found that innate immunity response is active during the early acute phase of the disease in mild patients but not in DHF patients. Loke *et al.* (2010) have shown that innate immune response such as IFN response are usually underrepresented in *in vivo* studies compared with *in vitro* studies, suggesting the observation of innate immune response may be less likely 3–6 days after disease onset (Loke *et al.*, 2010).

Interestingly, the majority of the transcriptome of patients with acute DHF is very similar to that of patients with acute DF. Both have an abundance of transcripts from cell cycle and endoplasmic reticulum (ER) related genes. These genes such as cyclins, cyclin-dependent kinases, and topoisomerase 2A, suggest progression of the mitotic cell cycle which may be due to immune cell proliferation. Several genes participate in the processing and folding of nascent proteins or the transport of proteins across the ER are also expressed, such as peptidylprolyl isomerase B (cyclophilin B; PPIB); signal sequence receptors a, g, and d (SSR1, SSR3, and SSR4); and translocation-associated membrane protein 1 (TRAM1). This suggests a proliferative response accompanied by ER stress. Furthermore, genes associated with proapoptotic pathways and immune responses were also elevated in acute samples, including: proapoptotic caspase adaptor protein (PACAP), apoptosis-inducing factor, mitochondrion-associated, 2 (AMID), BCL2-antagonist killer 1 (BAK1), programmed cell death 5 (PDCD5), serine/threonine kinase 17a (STK17A), cell surface markers, immunoglobulin, innate response elements and neutrophil-derived antimicrobial peptides (lactotransferrin and defensin A1/A3) (Simmons *et al.*, 2007b; Loke *et al.*, 2010). This implies that the innate immune response is involved in the transition from DF to DHF during the acute phase of the disease. From the transcriptional analysis of peripheral blood mononuclear cells (PBMCs), it is interesting to note that similar results was observed except for the decrease of innate immune response with development of acute DHF (Ubol *et al.*, 2008; Nascimento *et al.*, 2009), which was similar in patients with DSS (Long *et al.*, 2009).

The transcriptome of patients with DSS was surprisingly 'benign', particularly with regard to transcripts derived from apoptotic and type I interferon pathways. Under-represented transcripts included those from the death receptor, apoptotic, and IL-10 signalling pathways. In addition, there was a relatively low abundance of transcripts from interferon-stimulated genes (ISGs) among paediatric patients with DSS at the time of cardiovascular decompensation, even in the presence of a measurable viraemia (Long *et al.*, 2009). Similarly, many of the gene transcripts associated with DSS in adults, which are induced by type I interferons, were less abundant in patients with DSS than in those without DSS, such as canonical IFN-stimulated genes (ISGs), including MX1, MX2, ISG15 (G1P2), IFIT-2, and OAS3 (Simmons *et al.*, 2007b). Enriched genes that were differentially expressed by DSS patients were found to be unrelated to the immune response, but were involved in protein biosynthesis and, protein metabolism and modification (Loke *et al.*, 2010). Interestingly, it was hypothesized by Devignot *et al.* (2010) that a second inflammatory response loop may occur in acute DSS patients that results in impairment of T and NK lymphocyte transcriptional responses globally and overexpression of genes implicated in compensatory anti-inflammatory and repair/remodelling responses and in innate immune responses. The 'severe' transcriptional profile of DSS may be transient and may have occurred earlier than the 'benign' profile of DSS described by Long *et al.* (2009). Owing to the difference in sample collection period and the 'no fold change' statistical analysis used by Devignot *et al.* (2010). This unique phenomenon may reflect the beginning of recovery phase of DSS. However, more studies are required to confirm this.

When interpreting whole blood transcription profiles one must proceed with caution as others have found that some differential expression in whole blood can be explained by changes in the relative populations of blood cells (Chaussabel *et al.*, 2008), which are known to change in severe and mild dengue infections (Tanner *et al.*, 2008). Future gene-expression studies should attempt to consider this.

Notably absent from many of these studies is the differential expression of genes encoding chemokines. This may be due to the timing of the studies or the types of samples taken; possibly the majority of the cytokines seen in blood are not produced by PBMCs and may be coming from other tissues such as the endothelium. Another explanation may be that differential expression is not easily detected, as cytokine expression is high in both DF and DHF patients. Whatever the cause of the absence, severe dengue is associated with high levels of chemokines.

Chemokines associated with severity

Secreted by cells as part of the innate and adaptive immune responses, cytokines are involved in the modulation of the inflammatory response and the activation and migration of immune cells. Chemokines that are elevated in DHF/DSS include many interleukins: IL-2, IL-4, IL-6, IL-7, IL-8, IL-10, IL-13, IL-18 (Hober *et al.*, 1993; Raghupathy *et al.*, 1998; Juffrie *et al.*, 2001; Mustafa *et al.*, 2001; Gagnon *et al.*, 2002; Huang *et al.*, 2003; Chen *et al.*, 2006, 2007; Bozza *et al.*, 2008; Houghton-Trivino *et al.*, 2010; Priyadarshini *et al.*, 2010). Produced by diverse cell types, these interleukins would result in a comprehensive proliferation, differentiation and chemotaxis of B, T and NK cells.

The T cell-attracting cytokines, MIG, IP-10 and I-TAC (CXCL9, CXCL10, and CXCL11) are also highly expressed in patients with severe dengue (Fink *et al.*, 2007; Dejnirattisai *et al.*, 2008; Becerra *et al.*, 2009). CXCL10 (IP10) is the most up regulated chemokine found in the blood of DENV infected patients (Fink *et al.*, 2007). It is also a strong predictor of progression to severe disease and correlates tightly with DENV genome copy number in blood (Tanner *et al.*, 2008). Strongly stimulated by IFNγ, all three cytokines bind to the G-protein coupled receptor CXCLR3 and act as T-Cell attractants and inhibitors or endothelial cell growth (Weng *et al.*, 1998; Tensen *et al.*, 1999; Lasagni *et al.*, 2003). CXCR3($^{-}/^{-}$) and CXCL10($^{-}/^{-}$) knockout mice are more susceptible to intracerebral infection by DENV and have higher mortality than wild-type mice (Hsieh *et al.*, 2006) demonstrating a direct role for both, in controlling DENV infection. While CXCL10 promotes immune cell infiltration into tissues CXCL10($^{-}/^{-}$) mice do not display this deficiency, presumably owing to the presence of several other chemokines, most obviously CXCL9 and CXCL11 (Ip and Liao, 2010). However, others have found mice with anti-CXCL10 showed reduced NK cell infiltration and reduced production of NK related signals (Chen *et al.*, 2006). Further, CXCL10 has been shown to compete with DENV for binding to the cell surface co-receptor, heparan sulphate and mutations in CXCL10 that prevent CXCL10-heperan sulphate binding also fail to block DENV binding (Chen *et al.*, 2006).

The activation of T lymphocytes in DHF patients is greater than that in DF and peaks just before defervescence and plasma leakage begins (Kurane *et al.*, 1991). In a unique follow up study Kurane *et al.* (1995) monitored the levels of these same proteins in two volunteers who contracted dengue fever following vaccination with a candidate live-viral vaccine. This allowed them to collect samples throughout the period of infection and convalescence, including a baseline sample before inoculation. During the viraemic period, they detected elevated levels of sIL-2R, IL-2, sCD4 and IFN-γ. Following viraemia sCD8 increased while IFN γ and sIL-2R declined. This is consistent with an early activation of T helper cells followed by activation of cytotoxic T cells and NK cells at or around the time of defervescence; the period when decompensation usually occurs.

Bozza *et al.* (2008) performed extensive statistical analysis of patient chemokine profiles and found increased levels of IL-1β, IFN-γ, IL-4, IL-6, IL-13, IL-7, and GM-CSF in patients with severe dengue. Thrombocytopenia was associated with IL-1β, IL-8, TNF-α, and MCP-1 levels, while MCP-1 and GM-CSF correlated with hypotension. MIP-1β was elevated in mild dengue and associated with CD56^{+} NK cells numbers. MIP-1β and IFN-γ are associated with patient outcome.

Recent *ex vivo* experiments on PBMCs from previously infected and naive patients suggest that secondary dengue infections may produce different molecular responses in the presence of non-neutralizing antibody. (Sierra *et al.*, 2010b), DENV infected *ex vivo* PBMCs from previously infected dengue patients produced more MIP-1a (CCL3) and MCP-1 (CCL2) than PBMCs from non-immune patients, although this was dependant on the serotype of the previous infection. Some exposure sequences also resulted in elevated levels of IFNγ, TNFα and IL-10. In the presence of heterologous antibodies MCP-1 and MIP-1a were elevated in dengue immune individuals with the sequence of infection again being important. No difference was found in non-immune PBMCs. In a related *ex vivo* PBMC study, Sierra *et al.* (2010a) found that the regulatory cytokines IL-10 and TGFβ where increased in a primary dengue challenge but not in a secondary, heterotypic dengue challenge. IL-10 decreases the expression of Th1 cytokines and enhances B-cell proliferation and antibody production (Moore *et al.*, 2001). TGF-β plays a role in reducing the short-lived $CD8^+$ T-cell population and reduces IFN-γ production, which may help prevent pathogenesis caused by an unrestrained T-cell response. The reduction of IL-10 and TGF-β suggest that in secondary dengue the regulation of the T-cell response is not adequate to prevent a cytokine storm (Sierra *et al.*, 2010a).

Contradicting this result is a study by Ubol *et al.* (2010), which found that the binding of a DENV non-neutralizing antibody complex increased IL-10 expression in THP-1 monocytes, while suppressing IFN-b production. Both of these immune suppressive effects could be blocked with CD32 antibody suggesting that of the binding of the DENV antibody complex to the Fcγ receptor IIA is responsible for triggering the response. They also found increased IL-10 in patients with a severe secondary dengue infection but not in those with a mild secondary infection. RIG-I and MDA5 where reduced in both DENV-ADE infected THP-1 cells and also in PBMCs from patients with secondary DHF but not in DENV infected THP-1 or PBMCs from mild secondary patients.

MCP-1 (CCL2) and MIP-1a (CCL3) are produced by T cells (Wu *et al.*, 2009), hinting that memory T cells may play an important role in producing elevated levels of pro-inflammatory cytokines during secondary dengue infection. Several other non-PBMC cell lines can also produce CCL2 and CCL3 (Wu *et al.*, 2009) and many DENV infects many of these, at least *in vitro*. In addition, many of the immune cells that contain the CCR2 receptor – the receptor of CCL2 – are permissive to dengue infection and Sierra *et al.* (2010b) have suggested that dengue may benefit from the migration of permissive cells to the site of infection. Together these studies suggest a pathogenic role of non-neutralizing DENV antibodies in addition to ADE.

Algorithms for the prognosis of severity

Tanner *et al.* (2008) published the results of a large study that showed that differential haematology and cytokine measurements could be used to train a decision tree classifier that could accurately differentiate between dengue fever and other non-specific fever with samples taken at 48 hours. Interestingly, they also showed that it was possible to predict the eventual onset of thrombocytopenia using a similar approach, albeit in a smaller subset of patients (Tanner *et al.*, 2008). Although thrombocytopenia is not in itself severe, the study did indicate that it might be possible to detect patients who may develop severe manifestations before they develop them. Since this observation, others have published similar studies using different populations and approaches (Faisal *et al.*, 2010a,b; Gomes *et al.*, 2010; Ibrahim *et al.*, 2010; Potts *et al.*, 2010). Taken together, they provide some confidence that the approach is possible.

Unfortunately, many of the studies published thus far have involved data that includes multiple clinical or molecular measurements from small populations. Although standard cross-validation techniques are used, the selection of biomarkers is highly vulnerable to the 'curse of dimensionality'. This problem has been encountered in other biomarker discovery efforts, particularly in the field of oncology (Ioannidis, 2005a,b; Michiels *et al.*, 2005). In essence, when the number of independent observations (patient samples) is not substantially larger than the number of potential biomarkers, there is a very real danger that some biomarkers will be significantly different by chance alone. The result is that the accuracy of the trained classifier is grossly overestimated and one would expect that it would not generalize to new patient populations, a problem known as over fitting. The problem is often at its worst when biomarkers are identified using microarrays as around 20,000 genes are simultaneously measured in very much smaller populations.

Another, less well-known problem is overoptimization. Most machine learning and model building approaches have multiple tunable parameters; often tens or hundreds of combinations are possible. In addition, researchers will typically explore several types of model such as decision trees, neural networks, support vector machines, etc., before settling on the one that gives the best performance for their data. Further optimization is possible in selecting which features will be used as biomarkers and how these features might be transformed or combined to increase accuracy. All of these approaches will result in overestimation of accuracy unless the approach can be validated in a completely new population. Cross-validation of the parameter and feature optimization, in addition to cross-validation of the model, will help give estimates that are more realistic but unseen data are always best.

Pleasingly, many of the published biomarkers have a reasonable biological link with severe dengue. This suggests that people are at least on the right track although their estimates of accuracy should be discounted until larger multipopulation studies are conducted.

Several biological factors can also confound small studies. For example, if there is a significant strain or serotype effect on severity, owing to pathogenic strains or increased ADE risk due to previous population exposure, this will be missed in a one off study. Different ethnic populations may experience different responses to infection, which means that some biomarkers may be less valuable in certain populations. For example, African Americans in Cuba are reported to be more resistant to DHF (Guzman *et al.*, 1984). Patient age is a known severity risk factor, and adults seem to have a different clinical presentation (Binh *et al.*, 2009; Ramos *et al.*, 2009; Lye *et al.*, 2010; Thomas *et al.*, 2010). Prognostic markers that work in children may not be appropriate for adults. The relevance of some of these differences can only be assessed by larger, multicentre, multiyear studies.

A reliable dengue severity prognostic could have very real value in patient triage. Considerable health resources are consumed by the hospitalization of dengue patients who will only develop mild symptoms (Clark *et al.*, 2005; Khun and Manderson, 2008; Huy *et al.*, 2009). Unfortunately, in the absence of reliable prognostic tests clinicians will naturally want to exercise caution and, where possible, admit patients for monitoring. To be effective the prognostic would need to incorporate, or be combined with, a reliable diagnostic. Both the diagnostic and prognostic tests would need to be fast so that patients can be triaged early in fever, before the onset of severity. In addition, the patient population would need to seek medical care early in the febrile phase. A rapid and cheap test administered in outpatient settings or community clinics would be the most valuable.

While any good test requires a high level of accuracy, there is usually some trade-off between sensitivity (detection of true positives) and specificity (avoidance of false negatives). For a dengue prognostic, intuition suggests that we would want sensitivity to be as high as possible. Missing a patient that will develop DHF or DSS is potentially fatal. However, high sensitivity can always be achieved by allowing for a large number of false positives. At the extreme, we can classify all patients as likely to be severe, thereby guaranteeing 100% sensitivity. Of course, specificity will be very low owing to false positives. The effect is enhanced because the incidence of severe dengue is much lower than mild. Whatever our reservations about missing severe patients there is a competing interest in terms of public health and health economics. If there are too many false positives due to an unwarranted emphasis on specificity, then too many patients will be hospitalized, and too many health resources, which could be used in other ways, will be needlessly consumed. High sensitivity at all costs is not always a good thing. On the other hand, if sensitivity is too low, fewer resources will be consumed but more patients will be incorrectly treated resulting increased morbidity and possible deaths. Ideally, sensitivity and specificity will both be very high but the relative value of each is ultimately a question of health economics and health policy. If clinicians and public health officials can agree on relative priorities, there are well-established techniques for the incorporation of 'cost based sensitivity' into prognostic models (Yang *et al.*, 2009).

The study of Potts *et al.* (2010) is arguably the best prognostic algorithm published to date. The study enrolled 1384 paediatric patients from two Thai hospitals from 1994 to 2007. Using a CART based decision tree, a diagnostic algorithm was able to predict progression to DSS with 97% sensitivity. A second algorithm could predict the progression of a patient to DSS or serious pleural effusion with 99% sensitivity. Interestingly, the prognostic algorithms did not rely on a molecularly confirmed dengue diagnosis showing that this might not be required in all cases. One would imagine it could be applied to all cases of suspected dengue. Unfortunately, the algorithms suffer from a high rate of false positives, which reduce their specificity. While the high sensitivity would certainly assist clinical decisions, the low specificity may not result in a significant reduction in hospitalization or utilization of health resources. Cost-based training was used, and a misclassification penalty of 10:1 for false negatives vs. false positives was assigned. This would explain the imbalance between specificity and sensitivity. Although the authors mention the use of five times cross-validation they do not report any statistics on this or the confidence intervals of the accuracy estimates making it difficult to estimate how well the approach would generalize to other populations.

Ibrahim and colleagues have produced an interesting body of work using engineering approaches to determine prognosis in at-risk patients using non-invasive measures (Faisal *et al.*, 2010a,c; Ibrahim *et al.*, 2005a,b). They considered 35 clinical and biophysical parameters in a population of 210 patients, synchronized by day of defervescence. Highly correlated parameters where filtered for redundancy and linear and multiple regressions were used to find significant descriptors. They found nine descriptors that could be used to train a neural network to identify 'at risk' dengue patients, were at risk was defined as two or more of the following criteria: platelet counts less ≤ 40,000 cells per ml, haematocrit concentration greater than or equal to 25%, a fivefold increase in aspartate aminotransferase or alanine aminotransferase. Using body mass index, resistance, reactance, body cell mass, extracellular water/intracellular water ratio, abdominal pain, bleeding tendency, petechiae rash, weakness of the lower limbs, their model achieved a prognostic accuracy of 75% with a good balance of sensitivity and specificity.

It appears that prognostic markers of severe dengue exist and may be of value in a clinical setting. However, the existing approaches have been inadequate to make confident estimates of the possible impact of a prognostic test. There is a clear need for one or more, large biomarker discovery campaigns with large and diverse patient populations classified using strict clinical definitions. These discovery campaigns would need to be clinically validated in multiple locations and in multiple outbreaks.

Conclusions

Molecular studies have greatly enhanced our understanding of the contributions of the innate and adaptive immune responses to controlling viral replication and to the induction of vascular leakage. Generally, innate immunity plays a key role during the acute phase of dengue. It produces a transient response and may be the 'cause' or 'consequence' of the severity of disease. Severe dengue is also associated with attenuation of selected aspects of the innate host response. However, there may be only small differences in these innate immunity expression profiles between the different severities of disease during the acute phase, which would be statistically and clinically challenging to differentiate. More prospective studies with daily sample collection have to be carried out to accurately differentiate these differences.

The production of cytokines indicates activation of an appropriate, protective immune response to dengue virus infection. However, overwhelming production of cytokines and activation of pathological T-Cell responses could be causing vascular leakage, shock and death. Without significantly enriched adaptive immune related genes responding during acute DSS, it is hypothesized that the high intensity of innate immune response, influenced by the previously discussed risk factors, may initiate the 'cytokine storm' to reduce the virus load. At the same time, without proper control of anti-inflammatory molecules as observed by Devignot *et al.* (2010), it may result in severe dengue.

Emerging evidence suggests that ADE may be playing a dual role in increasing cytokine levels. Certain cytokines (e.g. IP-10/CXCL10) are tightly correlated with viraemia which is increased by ADE (Fink *et al.*, 2007; Tanner *et al.*, 2008). It also appears that complexes of DENV with non-neutralizing antibodies may increase inflammatory cytokine levels via binding to the Fcγ receptor (Sierra *et al.*, 2010a,b; Ubol *et al.*, 2010).

Cytokine stimulated immune cell migration and infiltration may explain many of the observed features of vascular leakage. CCL2, CCL3 and CXCL9, CXCL10 and CXCL11 strongly promote the migration of monocytes, T Lymphocytes and NK cells across the endothelium (Carr *et al.*, 1994; Shields *et al.*, 1999; Luther and Cyster, 2001). They appear to do this by stimulating the separation of tight junctions, allowing the immune cells to infiltrate the underlying tissues. Supernatants from DENV infected macrophages have been shown to be able to induce endothelial permeability in-vitro implying that soluble factors mediate the process and that viral infection of the endothelium or macrophage destruction of endothelial cells is not required for leakage (Carr *et al.*, 2003). In addition, the high levels of TNF-α seen in DENV infection would be expected to increase endothelial permeability. Large increases in endothelial permeability could lead to leakage of vascular fluid in a manner that would probably not cause any apparent cellular damage (Trung and Wills, 2009). The effect would also be rapidly reversed when cytokine levels fall, which would be consistent with observations of rapid recompensation in DHF/DSS cases (Trung and Wills, 2009). The depletion of many types of white blood cells in the blood of infected patients may also be partially due to their migration out of the vasculature and may not be a result of their reduction by infection.

Although considerable progress has been made in understanding the causes and pathology of severe dengue several challenges remain. There is currently a paucity of studies that serially patients from the onset of fever till the period when severity occurs and resolves. This makes it difficult to link early clinical and molecular findings with patient outcome. Comparisons between studies are also challenging as the point of sampling is not always consistent and the nature of the virus, host population and the history of infection varies between outbreaks. Controlled *in vivo* experiments are not possible owing to the lack of suitable and well-validated animal models of severe dengue. Large, multicentre studies in diverse but well-characterized populations and using well-defined and relevant severity classification criteria would be invaluable in increasing our understanding of severe dengue.

References

Agarwal, J.P., Bhattacharyya, P.C., Das, S.K., Sharma, M., and Gupta, M. (2009). Dengue encephalitis. Southeast Asian J. Trop. Med. Public Health *40*, 54–55.

Anderson, K.B., Chunsuttiwat, S., Nisalak, A., Mammen, M.P., Libraty, D.H., Rothman, A.L., Green, S., Vaughn, D.W., Ennis, F.A., and Endy, T.P. (2007). Burden of symptomatic dengue infection in children at primary school in Thailand: a prospective study. Lancet *369*, 1452–1459.

Azeredo, E.L., Zagne, S.M., Santiago, M.A., Gouvea, A.S., Santana, A.A., Neves-Souza, P.C., Nogueira, R.M., Miagostovich, M.P., and Kubelka, C.F. (2001). Characterisation of lymphocyte response and cytokine patterns in patients with dengue fever. Immunobiology *204*, 494–507.

Balasubramanian, S., Janakiraman, L., Kumar, S.S., Muralinath, S., and Shivbalan, S. (2006). A reappraisal of the criteria to diagnose plasma leakage in dengue hemorrhagic fever. Indian Pediatr. *43*, 334–339.

Bandyopadhyay, S., Lum, L.C., and Kroeger, A. (2006). Classifying dengue: a review of the difficulties in using the WHO case classification for dengue haemorrhagic fever. Trop. Med. Int. Health *11*, 1238–1255.

Barcelos, F.L., Batista, C.A., Portela, F.G., Paiva, D.B., Germano, d. O., Bonjardim, C.A., Peregrino Ferreira, P.C., and Kroon, E.G. (2008). Dengue virus 3 genotype 1 associated with dengue fever and dengue hemorrhagic fever, Brazil. Emerg. Infect. Dis. *14*, 314–316.

Beaute, J., and Vong, S. (2010). Cost and disease burden of dengue in Cambodia. BMC. Public Health *10*, 521.

Becerra, A., Warke, R.V., Martin, K., Xhaja, K., de, B.N., Rothman, A.L., and Bosch, I. (2009). Gene expression profiling of dengue infected human primary cells identifies secreted mediators *in vivo*. J. Med. Virol. *81*, 1403–1411.

Binh, P.T., Matheus, S., Huong, V.T., Deparis, X., and Marechal, V. (2009). Early clinical and biological features of severe clinical manifestations of dengue in Vietnamese adults. J. Clin. Virol. *45*, 276–280.

Bozza, F.A., Cruz, O.G., Zagne, S.M., Azeredo, E.L., Nogueira, R.M., Assis, E.F., Bozza, P.T., and Kubelka, C.F. (2008). Multiplex cytokine profile from dengue patients: MIP-1beta and IFN-gamma as predictive factors for severity. BMC. Infect. Dis. *8*, 86.

Bravo, J.R., Guzman, M.G., Kouri, G.P. (1987). Why dengue haemorrhagic fever in Cuba? 1. Individual risk factors for dengue haemorrhagic fever/dengue shock syndrome (DHF/DSS). Trans. R. Soc. Trop. Med. Hyg. *81*, 816–820.

Cao, X.T., Ngo, T.N., Wills, B., Kneen, R., Nguyen, T.T., Ta, T.T., Tran, T.T., Doan, T.K., Solomon, T., Simpson, J.A., *et al.* (2002). Evaluation of the World Health

Organization standard tourniquet test and a modified tourniquet test in the diagnosis of dengue infection in Vietnam. Trop. Med. Int. Health 7, 125–132.

Capeding, R.Z., Brion, J.D., Caponpon, M.M., Gibbons, R.V., Jarman, R.G., Yoon, I.K., and Libraty, D.H. (2010). The incidence, characteristics, and presentation of dengue virus infections during infancy. Am. J. Trop. Med. Hyg. *82*, 330–336.

Carr, J.M., Hocking, H., Bunting, K., Wright, P.J., Davidson, A., Gamble, J., Burrell, C.J., and Li, P. (2003). Supernatants from dengue virus type-2 infected macrophages induce permeability changes in endothelial cell monolayers. J. Med. Virol. *69*, 521–528.

Carr, M.W., Roth, S.J., Luther, E., Rose, S.S., and Springer, T.A. (1994). Monocyte chemoattractant protein 1 acts as a T-lymphocyte chemoattractant. Proc. Natl. Acad. Sci. U.S.A. *91*, 3652–3656.

Chau, T.N., Hieu, N.T., Anders, K.L., Wolbers, M., Lien, l. B., Hieu, L.T., Hien, T.T., Hung, N.T., Farrar, J., Whitehead, S., *et al.* (2009). Dengue virus infections and maternal antibody decay in a prospective birth cohort study of Vietnamese infants. J. Infect. Dis. *200*, 1893–1900.

Chaussabel, D., Quinn, C., Shen, J., Patel, P., Glaser, C., Baldwin, N., Stichweh, D., Blankenship, D., Li, L., Munagala, I., *et al.* (2008). A modular analysis framework for blood genomics studies: application to systemic lupus erythematosus. Immunity *29*, 150–164.

Chen, J.P., Lu, H.L., Lai, S.L., Campanella, G.S., Sung, J.M., Lu, M.Y., Wu-Hsieh, B.A., Lin, Y.L., Lane, T.E., Luster, A.D., *et al.* (2006). Dengue virus induces expression of CXC chemokine ligand 10/IFN-gamma-inducible protein *10*, which competitively inhibits viral binding to cell surface heparan sulfate. J. Immunol. *177*, 3185–3192.

Chen, L.C., Lei, H.Y., Liu, C.C., Shiesh, S.C., Chen, S.H., Liu, H.S., Lin, Y.S., Wang, S.T., Shyu, H.W., and Yeh, T.M. (2006). Correlation of serum levels of macrophage migration inhibitory factor with disease severity and clinical outcome in dengue patients. Am. J. Trop. Med. Hyg. *74*, 142–147.

Chen, R.F., Yang, K.D., Wang, L., Liu, J.W., Chiu, C.C., and Cheng, J.T. (2007). Different clinical and laboratory manifestations between dengue haemorrhagic fever and dengue fever with bleeding tendency. Trans. R. Soc. Trop. Med. Hyg. *101*, 1106–1113.

Chen, S.T., Lin, Y.L., Huang, M.T., Wu, M.F., Cheng, S.C., Lei, H.Y., Lee, C.K., Chiou, T.W., Wong, C.H., and Hsieh, S.L. (2008). CLEC5A is critical for dengue-virus-induced lethal disease. Nature *453*, 672–676.

Chen, W.J., Chen, S.L., Chien, L.J., Chen, C.C., King, C.C., Harn, M.R., Hwang, K.P., and Fang, J.H. (1996). Silent transmission of the dengue virus in southern Taiwan. Am. J. Trop. Med. Hyg. *55*, 12–16.

Chongsrisawat, V., Hutagalung, Y., and Poovorawan, Y. (2009). Liver function test results and outcomes in children with acute liver failure due to dengue infection. Southeast Asian J. Trop. Med. Public Health *40*, 47–53.

Clark, D.V., Mammen, M.P., Jr., Nisalak, A., Puthimethee, V., and Endy, T.P. (2005). Economic impact of dengue fever/dengue hemorrhagic fever in Thailand at the family and population levels. Am. J. Trop. Med. Hyg. *72*, 786–791.

Cohen, S.N., and Halstead, S.B. (1966). Shock associated with dengue infection. I. Clinical and physiologic manifestations of dengue hemorrhagic fever in Thailand, (1964). J. Pediatr. *68*, 448–456.

Deen, J.L., Harris, E., Wills, B., Balmaseda, A., Hammond, S.N., Rocha, C., Dung, N.M., Hung, N.T., Hien, T.T., and Farrar, J.J. (2006). The WHO dengue classification and case definitions: time for a reassessment. Lancet *368*, 170–173.

Dejnirattisai, W., Duangchinda, T., Lin, C.L., Vasanawathana, S., Jones, M., Jacobs, M., Malasit, P., Xu, X.N., Screaton, G., and Mongkolsapaya, J. (2008). A complex interplay among virus, dendritic cells, T cells, and cytokines in dengue virus infections. J. Immunol. *181*, 5865–5874.

Dejnirattisai, W., Jumnainsong, A., Onsirisakul, N., Fitton, P., Vasanawathana, S., Limpitikul, W., Puttikhunt, C., Edwards, C., Duangchinda, T., Supasa, S., *et al.* (2010). Cross-reacting antibodies enhance dengue virus infection in humans. Science *328*, 745–748.

Devignot, S., Sapet, C., Duong, V., Bergon, A., Rihet, P., Ong, S., Lorn, P.T., Chroeung, N., Ngeav, S., Tolou, H.J., *et al.* (2010). Genome-wide expression profiling deciphers host responses altered during dengue shock syndrome and reveals the role of innate immunity in severe dengue. PLoS ONE *5*, e11671.

Endy, T.P., Nisalak, A., Chunsuttitwat, S., Vaughn, D.W., Green, S., Ennis, F.A., Rothman, A.L., and Libraty, D.H. (2004). Relationship of preexisting dengue virus (DV) neutralizing antibody levels to viremia and severity of disease in a prospective cohort study of DV infection in Thailand. J. Infect. Dis. *189*, 990–1000.

Faisal, T., Ibrahim, F., and Taib, M.N. (2010a). A noninvasive intelligent approach for predicting the risk in dengue patients. Expert Syst. App. *37*, 2175–2181.

Faisal, T., Taib, M.N., and Ibrahim, F. (2010b). Neural network diagnostic system for dengue patients risk classification. J. Med. Syst.

Faisal, T., Taib, M.N., and Ibrahim, F. (2010c). Reexamination of risk criteria in dengue patients using the self-organizing map. Med. Biol. Eng. Comput. *48*, 293–301.

Figueiredo, M.A., Rodrigues, L.C., Barreto, M.L., Lima, J.W., Costa, M.C., Morato, V., Blanton, R., Vasconcelos, P.F., Nunes, M.R., and Teixeira, M.G. (2010). Allergies and diabetes as risk factors for dengue hemorrhagic fever: results of a case control study. PLoS. Negl. Trop. Dis. *4*, e699.

Fink, J., Gu, F., Ling, L., Tolfvenstam, T., Olfat, F., Chin, K.C., Aw, P., George, J., Kuznetsov, V.A., Schreiber, M., Vasudevan, S.G., *et al.* (2007). Host gene expression profiling of dengue virus infection in cell lines and patients. PLoS. Negl. Trop. Dis. *1*, e86.

Flano, E. (2008). Viral immunity: it takes two to tango. Viral Immunol. *21*, 281–283.

Gagnon, S.J., Mori, M., Kurane, I., Green, S., Vaughn, D.W., Kalayanarooj, S., Suntayakorn, S., Ennis, F.A., and Rothman, A.L. (2002). Cytokine gene expression and protein production in peripheral blood mononuclear cells of children with acute dengue virus infections. J. Med. Virol. *67*, 41–46.

Garcia-Rivera, E.J., and Rigau-Perez, J.G. (2002). Encephalitis and dengue. Lancet *360*, 261.

Garg, P., Nagpal, J., Khairnar, P., and Seneviratne, S.L. (2008). Economic burden of dengue infections in India. Trans. R. Soc. Trop. Med. Hyg. *102*, 570–577.

George, R., Liam, C.K., Chua, C.T., Lam, S.K., Pang, T., Geethan, R., and Foo, L.S. (1988). Unusual clinical manifestations of dengue virus infection. Southeast Asian J. Trop. Med. Public Health *19*, 585–590.

Gomes, A.L., Wee, L.J., Khan, A.M., Gil, L.H., Marques, E.T., Jr., Calzavara-Silva, C.E., and Tan, T.W. (2010). Classification of dengue fever patients based on gene expression data using support vector machines. PLoS ONE *5*, e11267.

Gonzalez, D., Castro, O.E., Kouri, G., Perez, J., Martinez, E., Vazquez, S., Rosario, D., Cancio, R., and Guzman, M.G. (2005). Classical dengue hemorrhagic fever resulting from two dengue infections spaced 20 years or more apart: Havana, dengue 3 epidemic, 2001–2002. Int. J. Infect. Dis. *9*, 280–285.

Gubler, D.J. (1998). Dengue and dengue hemorrhagic fever. Clin. Microbiol. Rev. *11*, 480–96.

Gubler, D.J. (2002). Epidemic dengue/dengue hemorrhagic fever as a public health, social and economic problem in the 21st century. Trends Microbiol. *10*, 100–103.

Guilarde, A.O., Turchi, M.D., Siqueira, J.B., Jr., Feres, V.C., Rocha, B., Levi, J.E., Souza, V.A., Boas, L.S., Pannuti, C.S., and Martelli, C.M. (2008). Dengue and dengue hemorrhagic fever among adults: clinical outcomes related to viremia, serotypes, and antibody response. J. Infect. Dis. *197*, 817–824.

Gupta, P., Khare, V., Tripathi, S., Nag, V.L., Kumar, R., Khan, M.Y., and Dhole, T.K. (2010). Assessment of World Health Organization definition of dengue hemorrhagic fever in North India. J. Infect. Dev. Ctries. *4*, 150–155.

Guzman, M.G., Kouri, G.P., Bravo, J., Calunga, M., Soler, M., Vazquez, S., and Venereo, C. (1984). Dengue haemorrhagic fever in Cuba. I. Serological confirmation of clinical diagnosis. Trans. R. Soc. Trop. Med. Hyg. *78*, 235–238.

Guzman, M.G., Kouri, G., Soler, M., Bravo, J., Rodriguez, d. L., V., Vazquez, S., and Mune, M. (1992). Dengue 2 virus enhancement in asthmatic and non asthmatic individual. Mem. Inst. Oswaldo Cruz *87*, 559–564.

Guzman, M.G., Kouri, G., Bravo, J., Valdes, L., Vazquez, S., and Halstead, S.B. (2002a). Effect of age on outcome of secondary dengue 2 infections. Int. J. Infect. Dis. *6*, 118–124.

Guzman, M.G., Kouri, G., Valdes, L., Bravo, J., Vazquez, S., and Halstead, S.B. (2002b). Enhanced severity of secondary dengue-2 infections: death rates in 1981 and 1997 Cuban outbreaks. Rev. Panam. Salud Publica *11*, 223–227.

Guzman, M.G., Alvarez, M., Rodriguez-Roche, R., Bernardo, L., Montes, T., Vazquez, S., Morier, L., Alvarez, A., Gould, E.A., Kouri, G., *et al.* (2007). Neutralizing antibodies after infection with dengue 1 virus. Emerg. Infect. Dis. *13*, 282–286.

Halstead, S.B. (1965). Dengue and hemorrhagic fevers of Southeast Asia. Yale J. Biol. Med. *37*, 434–454.

Halstead, S.B. (1974). Etiologies of the experimental dengues of Siler and Simmons. Am. J. Trop. Med. Hyg. *23*, 974–982.

Halstead, S.B. (1989). Antibody, macrophages, dengue virus infection, shock, and hemorrhage: a pathogenetic cascade. Rev. Infect. Dis. *11* (*Suppl 4*), S830–S839.

Halstead, S.B. (2007). Dengue. Lancet *370*, 1644–1652.

Halstead, S.B. (2008). Dengue virus–mosquito interactions. Annu. Rev. Entomol. *53*, 273–291.

Halstead, S.B. (2009). Antibodies determine virulence in dengue. Ann. N.Y. Acad. Sci. *1171* (*Suppl 1*), E48–E56.

Halstead, S.B., Nimmannitya, S., and Cohen, S.N. (1970). Observations related to pathogenesis of dengue hemorrhagic fever. IV. Relation of disease severity to antibody response and virus recovered. Yale J. Biol. Med. *42*, 311–328.

Halstead, S.B., Venkateshan, C.N., Gentry, M.K., and Larsen, L.K. (1984). Heterogeneity of infection enhancement of dengue 2 strains by monoclonal antibodies. J. Immunol. *132*, 1529–1532.

Halstead, S.B., Streit, T.G., Lafontant, J.G., Putvatana, R., Russell, K., Sun, W., Kanesa-Thasan, N., Hayes, C.G., and Watts, D.M. (2001). Haiti: absence of dengue hemorrhagic fever despite hyperendemic dengue virus transmission. Am. J. Trop. Med. Hyg. *65*, 180–183.

Hammon, W.M., Rudnick, A., and Sather, G.E. (1960). Viruses associated with epidemic hemorrhagic fevers of the Philippines and Thailand. Science *131*, 1102–1103.

Hibberd, M.L., Ling, L., Tolfvenstam, T., Mitchell, W., Wong, C., Kuznetsov, V.A., George, J., Ong, S.H., Ruan, Y., Wei, C.L., *et al.* (2006). A genomics approach to understanding host response during dengue infection. Novartis. Found. Symp. *277*, 206–214.

Hober, D., Poli, L., Roblin, B., Gestas, P., Chungue, E., Granic, G., Imbert, P., Pecarere, J.L., Vergez-Pascal, R., and Wattre, P. (1993). Serum levels of tumor necrosis factor-alpha (TNF-alpha), interleukin-6 (IL-6), and interleukin-1 beta (IL-1 beta) in dengue-infected patients. Am. J. Trop. Med. Hyg. *48*, 324–331.

Hotta, S., Yamamoto, M., Tokuchi, M., and Sakakibara, S. (1968). Long persistence of anti-dengue antibodies in the serum of native residents of some Japanese main islands. Kobe J. Med. Sci. *14*, 149–153.

Houghton-Trivino, N., Salgado, D.M., Rodriguez, J.A., Bosch, I., and Castellanos, J.E. (2010). Levels of soluble ST2 in serum associated with severity of dengue due to tumour necrosis factor alpha stimulation. J. Gen. Virol. *91*, 697–706.

Hsieh, M.F., Lai, S.L., Chen, J.P., Sung, J.M., Lin, Y.L., Wu-Hsieh, B.A., Gerard, C., Luster, A., and Liao, F. (2006). Both CXCR3 and CXCL10/IFN-inducible protein 10 are required for resistance to primary infection by dengue virus. J. Immunol. *177*, 1855–1863.

Huang, Y.H., Lei, H.Y., Liu, H.S., Lin, Y.S., Chen, S.H., Liu, C.C., and Yeh, T.M. (2003). Tissue plasminogen activator induced by dengue virus infection of human endothelial cells. J. Med. Virol. *70*, 610–616.

Huy, R., Wichmann, O., Beatty, M., Ngan, C., Duong, S., Margolis, H.S., and Vong, S. (2009). Cost of dengue and other febrile illnesses to households in rural Cambodia: a prospective community-based case-control study. BMC Public Health *9*, 155.

Ibrahim, F., Taib, M.N., Abas, W.A.B.W., Guan, C.C., and Sulaiman, S. (2005a). A novel approach to classify risk in dengue hemorrhagic fever (DHF) using bioelectrical impedance analysis (BIA). IEEE Trans. Inst. Measure. *54*, 237–244.

Ibrahim, F., Taib, M.N., Abas, W.A.W., Guan, C.C., and Sulaiman, S. (2005b). A novel dengue fever (DF) and dengue haemorrhagic fever (DHF) analysis using artificial neural network (ANN). Comp. Methods Prog. Biomed. *79*, 273–281.

Ibrahim, F., Faisal, T., Mohamad Salim, M.I., and Taib, M.N. (2010). Non-invasive diagnosis of risk in dengue patients using bioelectrical impedance analysis and artificial neural network. Med. Biol. Eng. Comput. *48*, 1141–1148.

Ioannidis, J.P. (2005a). Microarrays and molecular research: noise discovery? Lancet *365*, 454–455.

Ioannidis, J.P. (2005b). Why most published research findings are false. PLoS Med. *2*, e124.

Ip, P.P., and Liao, F. (2010). Resistance to dengue virus infection in mice is potentiated by CXCL10 and is independent of CXCL10-mediated leukocyte recruitment. J. Immunol. *184*, 5705–5714.

Jenner, R.G., and Young, R.A. (2005). Insights into host responses against pathogens from transcriptional profiling. Nat. Rev. Microbiol. *3*, 281–294.

Juffrie, M., Meer, G.M., Hack, C.E., Haasnoot, K., Sutaryo, Veerman, A.J., and Thijs, L.G. (2001). Inflammatory mediators in dengue virus infection in children: interleukin-6 and its relation to C-reactive protein and secretory phospholipase A2. Am. J. Trop. Med. Hyg. *65*, 70–75.

Kalayanarooj, S., and Nimmannitya, S. (2004). Guidelines for dengue hemorrhagic fever case management. Bangkok Medical Publisher.

Kalayanarooj, S., Vaughn, D.W., Nimmannitya, S., Green, S., Suntayakorn, S., Kunentrasai, N., Viramitrachai, W., Ratanachu-eke, S., Kiatpolpoj, S., Innis, B.L., *et al.* (1997). Early clinical and laboratory indicators of acute dengue illness. J. Infect. Dis. *176*, 313–321.

Katze, M.G., Fornek, J.L., Palermo, R.E., Walters, K.A., and Korth, M.J. (2008). Innate immune modulation by RNA viruses: emerging insights from functional genomics. Nat. Rev. Immunol. *8*, 644–654.

Khun, S., and Manderson, L. (2008). Poverty, user fees and ability to pay for health care for children with suspected dengue in rural Cambodia. Int. J. Equity Health *7*, 10.

Kliks, S.C., Nimmanitya, S., Nisalak, A., and Burke, D.S. (1988). Evidence that maternal dengue antibodies are important in the development of dengue hemorrhagic fever in infants. Am. J. Trop. Med. Hyg. *38*, 411–419.

de Kruif, M.D., Setiati, T.E., Mairuhu, A.T., Koraka, P., Aberson, H.A., Spek, C.A., Osterhaus, A.D., Reitsma, P.H., Brandjes, D.P., Soemantri, A., *et al.* (2008). Differential gene expression changes in children with severe dengue virus infections. PLoS. Negl. Trop. Dis. *2*, e215.

Kularatne, S.A., Gawarammana, I.B., and Kumarasiri, P.R. (2005). Epidemiology, clinical features, laboratory investigations and early diagnosis of dengue fever in adults: a descriptive study in Sri Lanka. Southeast Asian J. Trop. Med. Public Health *36*, 686–692.

Kurane, I., Innis, B.L., Nimmannitya, S., Nisalak, A., Meager, A., Janus, J., and Ennis, F.A. (1991). Activation of T lymphocytes in dengue virus infections. High levels of soluble interleukin 2 receptor, soluble CD4, soluble CD8, interleukin 2, and interferon-gamma in sera of children with dengue. J. Clin. Invest *88*, 1473–1480.

Kurane, I., Innis, B.L., Hoke, C.H., Jr., Eckels, K.H., Meager, A., Janus, J., and Ennis, F.A. (1995). T cell activation *in vivo* by dengue virus infection. J. Clin. Lab. Immunol. *46*, 35–40.

Lahiri, M., Fisher, D., and Tambyah, P.A. (2008). Dengue mortality: reassessing the risks in transition countries. Trans. R. Soc. Trop. Med. Hyg. *102*, 1011–1016.

Lasagni, L., Francalanci, M., Annunziato, F., Lazzeri, E., Giannini, S., Cosmi, L., Sagrinati, C., Mazzinghi, B., Orlando, C., Maggi, E., *et al.* (2003). An alternatively spliced variant of CXCR3 mediates the inhibition of endothelial cell growth induced by IP-10, Mig, and I-TAC, and acts as functional receptor for platelet factor 4. J. Exp. Med. *197*, 1537–1549.

Libraty, D.H., Pichyangkul, S., Ajariyakhajorn, C., Endy, T.P., and Ennis, F.A. (2001). Human dendritic cells are activated by dengue virus infection: enhancement by gamma interferon and implications for disease pathogenesis. J. Virol. *75*, 3501–3508.

Libraty, D.H., Endy, T.P., Houng, H.S., Green, S., Kalayanarooj, S., Suntayakorn, S., Chansiriwongs, W., Vaughn, D.W., Nisalak, A., Ennis, F.A., *et al.* (2002a). Differing influences of virus burden and immune activation on disease severity in secondary dengue-3 virus infections. J. Infect. Dis. *185*, 1213–1221.

Libraty, D.H., Endy, T.P., Kalayanarooj, S., Chansiriwongs, W., Nisalak, A., Green, S., Ennis, F.A., and Rothman, A.L. (2002b). Assessment of body fluid compartment volumes by multifrequency bioelectrical impedance spectroscopy in children with dengue. Trans. R. Soc. Trop. Med. Hyg. *96*, 295–299.

Littaua, R., Kurane, I., and Ennis, F.A. (1990). Human IgG Fc receptor II mediates antibody-dependent enhancement of dengue virus infection. J. Immunol. *144*, 3183–3186.

Loke, H., Bethell, D., Phuong, C.X., Day, N., White, N., Farrar, J., and Hill, A. (2002). Susceptibility to dengue hemorrhagic fever in vietnam: evidence of an association with variation in the vitamin d receptor and Fc gamma receptor IIa genes. Am. J. Trop. Med. Hyg. *67*, 102–106.

Loke, P., Hammond, S.N., Leung, J.M., Kim, C.C., Batra, S., Rocha, C., Balmaseda, A., and Harris, E. (2010). Gene expression patterns of dengue virus-infected children from nicaragua reveal a distinct signature of increased metabolism. PLoS. Negl. Trop. Dis. *4*, e710.

Long, H.T., Hibberd, M.L., Hien, T.T., Dung, N.M., Ngoc, T.V., Farrar, J., Wills, B., and Simmons, C.P. (2009). Patterns of gene transcript abundance in the blood of children with severe or uncomplicated dengue highlight differences in disease evolution and host response to dengue virus infection. J. Infect. Dis. *199*, 537–546.

Lozach, P.Y., Burleigh, L., Staropoli, I., Navarro-Sanchez, E., Harriague, J., Virelizier, J.L., Rey, F.A., Despres, P., Arenzana-Seisdedos, F., and Amara, A. (2005). Dendritic cell-specific intercellular adhesion molecule 3-grabbing non-integrin (DC-SIGN)-mediated enhancement of dengue virus infection is independent of DC-SIGN internalization signals. J. Biol. Chem. *280*, 23698–23708.

Lum, L.C., Lam, S.K., Choy, Y.S., George, R., and Harun, F. (1996). Dengue encephalitis: a true entity? Am. J. Trop. Med. Hyg. *54*, 256–259.

Luther, S.A., and Cyster, J.G. (2001). Chemokines as regulators of T cell differentiation. Nat. Immunol. *2*, 102–107.

Lye, D.C., Lee, V.J., Sun, Y., and Leo, Y.S. (2010). The benign nature of acute dengue infection in hospitalized older adults in Singapore. Int.J. Infect. Dis. *14*, e410–e413.

Marchette, N.J., Halstead, S.B., O'Rourke, T., Scott, R.M., Bancroft, W.H., and Vanopruks, V. (1979). Effect of immune status on dengue 2 virus replication in cultured leukocytes from infants and children. Infect. Immun. *24*, 47–50.

Meltzer, M.I., Rigau-Perez, J.G., Clark, G.G., Reiter, P., and Gubler, D.J. (1998). Using disability-adjusted life years to assess the economic impact of dengue in Puerto Rico: 1984–1994. Am. J. Trop. Med. Hyg. *59*, 265–71.

Michiels, S., Koscielny, S., and Hill, C. (2005). Prediction of cancer outcome with microarrays: a multiple random validation strategy. Lancet *365*, 488–492.

Moore, K.W., de Waal, M.R., Coffman, R.L., O'Garra, A. (2001). Interleukin-10 and the interleukin-10 receptor. Annu. Rev. Immunol. *19*, 683–765.

Morens, D.M., Venkateshan, C.N., and Halstead, S.B. (1987). Dengue 4 virus monoclonal antibodies identify epitopes that mediate immune infection enhancement of dengue 2 viruses. J. Gen. Virol. *68 (Pt 1)*, 91–98.

Morens, D.M., and Halstead, S.B. (1990). Measurement of antibody-dependent infection enhancement of four dengue virus serotypes by monoclonal and polyclonal antibodies. J. Gen. Virol. *71 (Pt 12)*, 2909–2914.

Mustafa, A.S., Elbishbishi, E.A., Agarwal, R., and Chaturvedi, U.C. (2001). Elevated levels of interleukin-13 and IL-18 in patients with dengue hemorrhagic fever. FEMS Immunol. Med. Microbiol. *30*, 229–233.

Nascimento, E.J., Braga-Neto, U., Calzavara-Silva, C.E., Gomes, A.L., Abath, F.G., Brito, C.A., Cordeiro, M.T., Silva, A.M., Magalhaes, C., Andrade, R., *et al.* (2009). Gene expression profiling during early acute febrile stage of dengue infection can predict the disease outcome. PLoS ONE *4*, e7892.

Navarro-Sanchez, E., Altmeyer, R., Amara, A., Schwartz, O., Fieschi, F., Virelizier, J.L., Arenzana-Seisdedos, F., and Despres, P. (2003). Dendritic-cell-specific ICAM3-grabbing non-integrin is essential for the productive infection of human dendritic cells by mosquito-cell-derived dengue viruses. EMBO Rep. *4*, 723–728.

Navarro-Sanchez, E., Despres, P., and Cedillo-Barron, L. (2005). Innate immune responses to dengue virus. Arch. Med. Res. *36*, 425–435.

Neo, H.Y., Wong, R.C., Seto, K.Y., Yip, J.W., Yang, H., Ling, L.H. (2006). Noncompaction cardiomyopathy presenting with congestive heart failure during intercurrent dengue viral illness: importance of phenotypic recognition. Int. J. Cardiol. *107*, 123–125.

Norlijah, O., Khamisah, A.N., Kamarul, A., Paeds, M., and Mangalam, S. (2006). Repeated tourniquet testing as a diagnostic tool in dengue infection. Med. J. Malaysia *61*, 22–27.

Obeyesekere, I., and Hermon, Y. (1972). Myocarditis and cardiomyopathy after arbovirus infections (dengue and chikungunya fever). Br. Heart J. *34*, 821–827.

Okanurak, K., Sornmani, S., and Indaratna, K. (1997). The cost of dengue hemorrhagic fever in Thailand. Southeast Asian J. Trop. Med. Public Health *28*, 711–717.

Ooi, E.E., and Gubler, D.J. (2009). Dengue in Southeast Asia: epidemiological characteristics and strategic challenges in disease prevention. Cad. Saude Publica *25 (Suppl 1)*, S115–S124.

Osorio, J., Carvajal, C., Sussman, O., Buitrago, R., and Franco-Paredes, C. (2008). Acute liver failure due to dengue virus infection. Int.J. Infect. Dis. *12*, 444–445.

Phuong, C.X., Nhan, N.T., Kneen, R., Thuy, P.T., van Thien, C., Nga, N.T., Thuy, T.T., Solomon, T., Stepniewska, K., and Wills, B. (2004). Clinical diagnosis and assessment of severity of confirmed dengue infections in Vietnamese children: is the world health organization classification system helpful? Am. J. Trop. Med. Hyg. *70*, 172–179.

Potts, J.A., Gibbons, R.V., Rothman, A.L., Srikiatkhachorn, A., Thomas, S.J., Supradish, P.O., Lemon, S.C., Libraty, D.H., Green, S., and Kalayanarooj, S. (2010). Prediction of dengue disease severity among pediatric Thai patients using early clinical laboratory indicators. PLoS Negl. Trop. Dis. *4*, e769.

Priyadarshini, D., Gadia, R.R., Tripathy, A., Gurukumar, K.R., Bhagat, A., Patwardhan, S., Mokashi, N., Vaidya, D., Shah, P.S., and Cecilia, D. (2010). Clinical findings and pro-inflammatory cytokines in dengue patients in Western India: a facility-based study. PLoS ONE. *5*, e8709.

Raghupathy, R., Chaturvedi, U.C., Al Sayer, H., Elbishbishi, E.A., Agarwal, R., Nagar, R., Kapoor, S., Misra, A., Mathur, A., Nusrat, H., *et al.* (1998). Elevated levels of IL-8 in dengue hemorrhagic fever. J. Med. Virol. *56*, 280–285.

Ramos, M.M., Tomashek, K.M., Arguello, D.F., Luxemburger, C., Quinones, L., Lang, J., and Munoz-Jordan, J.L. (2009). Early clinical features of dengue infection in Puerto Rico. Trans. R. Soc. Trop. Med. Hyg. *103*, 878–884.

Rigau-Perez, J.G. (2006). Severe dengue: the need for new case definitions. Lancet Infect. Dis. *6*, 297–302.

Rodenhuis-Zybert, I.A., van der Schaar, H.M., da Silva Voorham, J.M., Ende-Metselaar, H., Lei, H.Y., Wilschut, J., and Smit, J.M. (2010). Immature dengue virus: a veiled pathogen? PLoS Pathog. *6*, e1000718.

SABIN, A.B. (1952). Research on dengue during World War II. Am. J. Trop. Med. Hyg. *1*, 30–50.

Sakuntabhai, A., Turbpaiboon, C., Casademont, I., Chuansumrit, A., Lowhnoo, T., Kajaste-Rudnitski, A., Kalayanarooj, S.M., Tangnararatchakit, K., Tangthawornchaikul, N., Vasanawathana, S., *et al.* (2005). A variant in the CD209 promoter is associated with severity of dengue disease. Nat.Genet. *37*, 507–513.

Shields, P.L., Morland, C.M., Salmon, M., Qin, S., Hubscher, S.G., and Adams, D.H. (1999). Chemokine and chemokine receptor interactions provide a mechanism for selective T cell recruitment to specific liver compartments within hepatitis C-infected liver. J. Immunol. *163*, 6236–6243.

Sierra, B., Perez, A.B., Vogt, K., Garcia, G., Schmolke, K., Aguirre, E., Alvarez, M., Kern, F., Kouri, G., Volk, H.D., *et al.* (2010a). Secondary heterologous dengue infection risk: Disequilibrium between immune regulation and inflammation? Cell Immunol. *262*, 134–140.

Sierra, B., Perez, A.B., Vogt, K., Garcia, G., Schmolke, K., Aguirre, E., Alvarez, M., Volk, H.D., and Guzman, M.G. (2010b). MCP-1 and MIP-1alpha expression in a model resembling early immune response to dengue. Cytokine *52*, 175–183.

Simmons, C.P., Chau, T.N., Thuy, T.T., Tuan, N.M., Hoang, D.M., Thien, N.T., Lien, l. B., Quy, N.T., Hieu, N.T., Hien, T.T., *et al.* (2007a). Maternal antibody and viral factors in the pathogenesis of dengue virus in infants. J. Infect. Dis. *196*, 416–424.

Simmons, C.P., Popper, S., Dolocek, C., Chau, T.N., Griffiths, M., Dung, N.T., Long, T.H., Hoang, D.M., Chau, N.V., Thao, l. T., *et al.* (2007b). Patterns of host genome-wide gene transcript abundance in the peripheral blood of patients with acute dengue hemorrhagic fever. J. Infect. Dis. *195*, 1097–1107.

Srikiatkhachorn, A., Krautrachue, A., Ratanaprakarn, W., Wongtapradit, L., Nithipanya, N., Kalayanarooj, S., Nisalak, A., Thomas, S.J., Gibbons, R.V., Mammen, M.P., Jr., *et al.* (2007). Natural history of plasma leakage in dengue hemorrhagic fever: a serial ultrasonographic study. Pediatr. Infect. Dis. J. *26*, 283–290.

Srikiatkhachorn, A., Gibbons, R.V., Green, S., Libraty, D.H., Thomas, S.J., Endy, T.P., Vaughn, D.W., Nisalak, A., Ennis, F.A., Rothman, A.L., *et al.* (2010). Dengue hemorrhagic fever: the sensitivity and specificity of the world health organization definition for identification of severe cases of dengue in Thailand, 1994–2005. Clin. Infect. Dis. *50*, 1135–1143.

Steel, A., Gubler, D.J., Bennett, S.N. (2010). Natural attenuation of dengue virus type-2 after a series of island outbreaks: a retrospective phylogenetic study of events in the South Pacific three decades ago. Virology *405*, 505–512.

Tanner, L., Schreiber, M., Low, J.G., Ong, A., Tolfvenstam, T., Lai, Y.L., Ng, L.C., Leo, Y.S., Thi, P.L., Vasudevan, S.G., *et al.* (2008). Decision tree algorithms predict the diagnosis and outcome of dengue Fever in the early phase of illness. PLoS Negl. Trop. Dis. *2*, e196.

Tantracheewathorn, T., and Tantracheewathorn, S. (2007). Risk factors of dengue shock syndrome in children. J. Med. Assoc. Thai. *90*, 272–277.

Tassaneetrithep, B., Burgess, T.H., Granelli-Piperno, A., Trumpfheller, C., Finke, J., Sun, W., Eller, M.A., Pattanapanyasat, K., Sarasombath, S., Birx, D.L., *et al.* (2003). DC-SIGN (CD209) mediates dengue virus infection of human dendritic cells. J. Exp. Med. *197*, 823–829.

Tensen, C.P., Flier, J., Van Der Raaij-Helmer E.M., Sampat-Sardjoepersad, S., Van Der Schors, R.C., Leurs, R., Scheper, R.J., Boorsma, D.M., and Willemze, R. (1999). Human IP-9: A keratinocyte-derived high affinity CXC-chemokine ligand for the IP-10/Mig receptor (CXCR3). J. Invest. Dermatol. *112*, 716–722.

Thomas, L., Brouste, Y., Najioullah, F., Hochedez, P., Hatchuel, Y., Moravie, V., Kaidomar, S., Besnier, F., Abel, S., Rosine, J., *et al.* (2010). Predictors of severe manifestations in a cohort of adult dengue patients. J. Clin. Virol. *48*, 96–99.

Trung, D.T., and Wills, B. (2009). Systemic vascular leakage associated with dengue infections – the clinical perspective. In: Dengue Virus, Rothman, A.L., ed. (Springer Berlin Heidelberg), pp. 57–66.

Ubol, S., Masrinoul, P., Chaijaruwanich, J., Kalayanarooj, S., Charoensirisuthikul, T., and Kasisith, J. (2008). Differences in global gene expression in peripheral blood mononuclear cells indicate a significant role of the innate responses in progression of dengue fever but not dengue hemorrhagic fever. J. Infect. Dis. *197*, 1459–1467.

Ubol, S., Phuklia, W., Kalayanarooj, S., and Modhiran, N. (2010). Mechanisms of immune evasion induced by a complex of dengue virus and preexisting enhancing antibodies. J. Infect. Dis. *201*, 923–935.

Vaughn, D.W., Green, S., Kalayanarooj, S., Innis, B.L., Nimmannitya, S., Suntayakorn, S., Endy, T.P., Raengsakulrach, B., Rothman, A.L., Ennis, F.A., *et al.* (2000). Dengue viremia titer, antibody response pattern, and virus serotype correlate with disease severity. J. Infect. Dis. *181*, 2–9.

Vejbaesya, S., Luangtrakool, P., Luangtrakool, K., Kalayanarooj, S., Vaughn, D.W., Endy, T.P., Mammen, M.P., Green, S., Libraty, D.H., Ennis, F.A., *et al.* (2009). TNF and LTA gene, allele, and extended HLA haplotype associations with severe dengue virus infection in ethnic Thais. J. Infect. Dis. *199*, 1442–1448.

Watts, D.M., Porter, K.R., Putvatana, P., Vasquez, B., Calampa, C., Hayes, C.G., and Halstead, S.B. (1999). Failure of secondary infection with American genotype dengue 2 to cause dengue haemorrhagic fever. Lancet *354*, 1431–1434.

Weng, Y., Siciliano, S.J., Waldburger, K.E., Sirotina-Meisher, A., Staruch, M.J., Daugherty, B.L., Gould, S.L., Springer, M.S., and DeMartino, J.A. (1998). Binding and functional properties of recombinant and endogenous CXCR3 chemokine receptors. J. Biol. Chem. *273*, 18288–18291.

Wichmann, O., Hongsiriwon, S., Bowonwatanuwong, C., Chotivanich, K., Sukthana, Y., and Pukrittayakamee, S. (2004). Risk factors and clinical features associated with severe dengue infection in adults and children during the 2001 epidemic in Chonburi, Thailand. Trop. Med. Int. Health *9*, 1022–1029.

Wu, C., Orozco, C., Boyer, J., Leglise, M., Goodale, J., Batalov, S., Hodge, C.L., Haase, J., Janes, J., Huss, J.W., III, *et al.* (2009). BioGPS: an extensible and customizable portal for querying and organizing gene annotation resources. Genome Biol. *10*, R130.

Yang, F., Wang, H.Z., Mi, H., Lin, C.D., and Cai, W.W. (2009). Using random forest for reliable classification and cost-sensitive learning for medical diagnosis. BMC Bioinformatics *10*, S22.

Zybert, I.A., Ende-Metselaar, H., Wilschut, J., and Smit, J.M. (2008). Functional importance of dengue virus maturation: infectious properties of immature virions. J. Gen. Virol. *89*, 3047–3051.

Flavivirus Fitness and Transmission

9

Gregory D. Ebel and Laura D. Kramer

Abstract

Flavivirus fitness is inextricably linked to the ability of a particular agent to be efficiently transmitted among relevant hosts in natural transmission cycles. Thus, fitness is an inherent component of the virus–host relationship. The mechanisms through which virus fitness is maximized are poorly understood, but have recently been examined in increasing detail. This chapter examines recent developments in the study of flavivirus fitness from both observational and experimental studies, highlighting important emergent and/or resurgent tick- and mosquito-borne members of the flavivirus genus.

Introduction

RNA viruses are the most abundant molecular pathogens infecting humans, animals, and plants (Murphy, 1994); and more than 50 RNA viruses have been classified as emerging pathogens in the past 30 years (Domingo and Holland, 1997). The ever increasing activity of *dengue virus* (DENV) serotypes 1–4, and the establishment and successful expansion of range of *West Nile virus* (WNV) in North America have highlighted the success of RNA viruses. Their adaptability, i.e. the capacity for success in variable environments, is achieved through the exploitation of high mutation rates (Domingo and Holland, 1994), high levels of viral replication, and large populations size (Drake and Holland, 1999), leading to viral plasticity. Yet despite this capacity, it is known that arboviruses possess genetic and antigenic stability in nature (Kuno *et al.*, 1998). This chapter will address the concept of viral fitness and evolution with a focus on the family *Flaviviridae,* genus *Flavivirus.* First, we will introduce flaviviruses, provide a definition of terms, and explore the significance of the interdependence of viral fitness and viral transmission. Then we will discuss how these concepts are important in the biology of four selected flaviviruses – one tick-borne virus [*tick-borne encephalitis virus* (TBEV)], one *Culex*-borne virus [*West Nile virus* (WNV)], and two *Aedes*-borne viruses [DENV *and Yellow fever virus* (YFV)]. The first two of these viruses depend on an enzootic cycle to be maintained in nature, although they are transmitted by different arthropod orders, *Ixodidae* (TBEV) and *Diptera* (WNV). Both DENV and YFV, also transmitted by *Diptera,* have sylvatic cycles, but also can be transmitted from human to human by *Aedes* mosquitoes. DENV does not depend on a sylvatic cycle for perpetuation while YFV does. The consequences of these differences in strategy on viral fitness and viral evolution will be explored.

The family *Flaviviridae* consists of three genera – *Flavivirus, Hepacivirus* and *Pestivirus.* Members of the family *Flaviviridae* are spherical, enveloped virions, and approximately 50 nm in diameter (Rice, 1996). The virion contains a host-derived lipid bilayer surrounding a nucleocapsid core consisting of the viral RNA complexed with multiple copies of the capsid protein (Mukhopadhyay *et al.*, 2003). The viral genome is linear, positive sense, single-stranded RNA of approximately 11 kilobases in length with a 5′ untranslated region (UTR), followed by a single, long open-reading frame (ORF), and a 3′ UTR. The 5′ end of the genome is capped but is not polyadenylated. The ORF encodes a polyprotein that is co- and post-translationally processed by viral and cellular proteases into three structural proteins [C, premembrane (prM) or M and E] and seven non-structural proteins (NS1, NS2A, NS2B, NS3, NS4A, NS4B and NS5). Structural proteins are mainly involved in viral particle formation, while non-structural proteins function in viral replication, virion assembly, and evasion of host innate immune response. Although the flaviviruses are closely related, they have minor differences in their genome organization (Fig. 9.1) (Weaver and Vasilakis, 2009). There are approximately 70 known viruses in the genus *Flavivirus,* the genus that is the focus of this chapter.

Medically important mosquito-borne viruses in this genus either cause encephalitis in humans and are maintained in an enzootic transmission cycle by *Culex* species mosquitoes and avian hosts; or they cause haemorrhagic fever in humans and are maintained by *Aedes* species mosquitoes and primates (Gaunt *et al.*, 2001). The tick-borne flaviviruses form two distinct groups separate from mosquito-borne flaviviruses. The phylogenetic tree structure of the two arthropod-transmitted groups differ, with continual and asymmetric branching in the tick-transmitted tree compared with explosive radiation in the last 200 years of mosquito-transmitted flaviviruses (Zanotto *et al.*, 1996). Differing evolutionary dynamics between these two viral groups has been hypothesized to be a consequence of different modes of dispersal, propagation, and changes in the size of host populations.

Viral fitness

Mutation rates have been reported to range from 10^{-3} to 10^{-5} per nucleotide per round of replication (Domingo and Holland, 1994; Drake and Holland, 1999) for many RNA viruses. This translates to 0.1–10 mutations in each newly generated genome (Domingo and Holland, 1997) in the case of a 10-kb template. As a consequence of these high mutation rates, RNA virus populations are composed of dynamic mutant swarms, i.e. a mixture of genetic and phenotypic variants that are constantly undergoing selection, competition, and complementation upon viral replication.

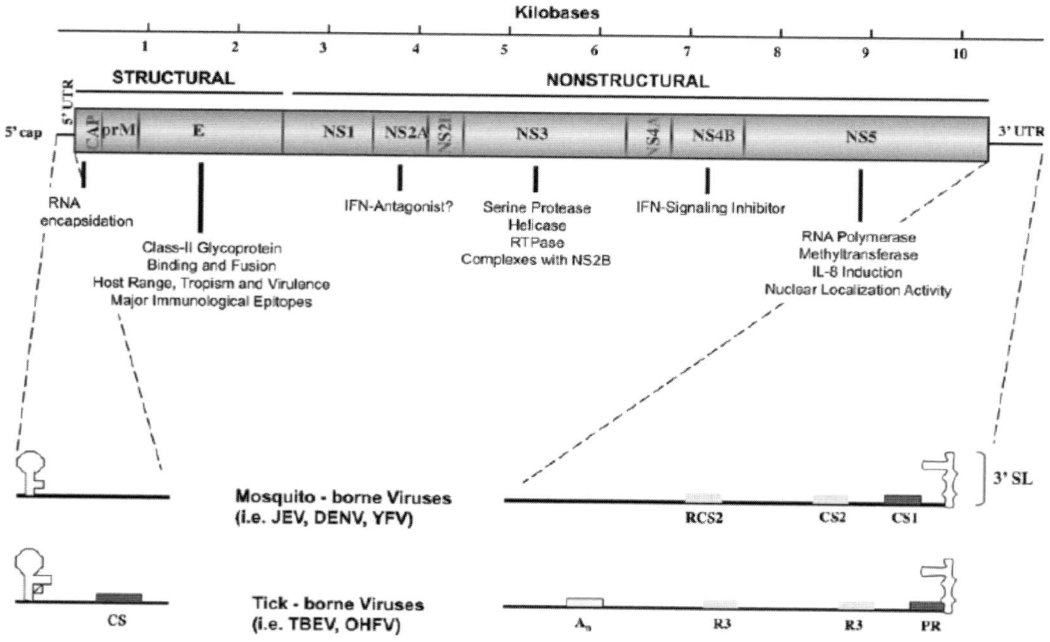

Figure 9.1 Genome organization of flaviviruses. Reproduced with permission from Weaver and Vasilakis (2009).

Interactions among the variants may lead to an alteration in viral phenotype. This mutant swarm is also referred to as the viral 'quasispecies', i.e. the diverse mutant spectrum surrounding a master genotype that presumably possesses the highest fitness value (Eigen and Biebricher, 1988; Eigen, 1993; Domingo *et al.*, 1998; Holmes and Moya, 2002), although essential to quasispecies theory is the concept that selection acts on viral populations as a whole, rather than on individual virions. The presence of genetic variants within a given host following infection with flaviviruses has been confirmed (Wang *et al.*, 2002; Lin *et al.*, 2004; Jerzak *et al.*, 2005; Chao *et al.*, 2005). Studies with WNV demonstrate that genetic diversity is host-dependent (Jerzak *et al.*, 2005, 2007; Ciota *et al.*, 2007c), as did studies with the plant viruses *Tobacco mosaic virus* and *Cucumber mosaic virus* (Schneider and Roossinck, 2001).

Each genotype within the swarm has a relative 'fitness', referring to the replicative competitiveness of a virus under defined conditions. The fittest viruses are those that most readily adapt to the environment in which they are replicating, and the genotype of the fittest virus may change as conditions change. Changes in viral fitness occur when large population passages of RNA viruses allow either competitive optimization of mutant distributions (gain of fitness) or when repeated bottlenecks lead to Muller's ratchet (loss of fitness). In addition, fitness assessment may occur on more than one level, i.e. virus populations compete, as do individual viruses within a population. Most evaluations of viral fitness have been conducted *in vitro* in tissue culture, with results varying with the virus, as well as with the types of cells that are used for viral propagation. A search of the literature indicates that among flaviviruses, experimental fitness studies have been conducted to a significant extent only with DENV and WNV. Furthermore, the relevance of the *in vitro* findings to the *in vivo* situation has not been well studied with some exceptions (Ciota *et al.*, 2008, 2009; Coffey *et al.*, 2008; Jerzak *et al.*, 2008; Fitzpatrick *et al.*, 2010).

The relationships between virus fitness and pathogenic potential are of longstanding interest but are relatively poorly understood. Importantly, diminished fitness may not equate with diminished pathogenesis or virulence, as recently demonstrated with *foot-and-mouth-disease virus* (Herrera *et al.*, 2007). Therefore, it appears that in some circumstances, viruses that replicate poorly may still cause serious disease. In addition, compensatory mutations may emerge in a viral genome that result in increased replication efficiency.

'Host radiation' refers to the ability of parasites to adapt to new environments/hosts and expand or change their niches. The presence of minority variants in the quasispecies of RNA viruses allows them to adapt rapidly to new landscapes. However, adaptation to one particular environment may involve fitness losses in an alternative environment, which is referred to as a 'fitness trade-off'. Bottlenecks may occur as RNA viruses replicate in different hosts or in diverse tissues of a single host, i.e. a significant proportion of the viral population might not contribute to the following generations. Such bottlenecks allow genetic drift to play a more significant role in virus population biology, which would generally decrease fitness since most mutations are probably deleterious (Muller, 1964; Domingo *et al.*, 1996).

Genome recombination also leads to increased variation for many viruses (Strauss, 1993), yet there is little evidence for this phenomenon in flaviviruses and it is clearly not a prerequisite for adaptability. Adaptability of virus populations and success in new and dynamic environments are clear benefits of diverse quasispecies structures. In studies with *human immunodeficiency virus* (HIV) and *hepatitis C virus,* avoidance of host immune responses has been attributed to the diverse quasispecies populations (Weiner *et al.*, 1995; Farci *et al.*, 2000, 2002; Frost *et al.*, 2005). Studies with *foot-and-mouth-disease virus* (FMDV), HIV (Briones *et al.*, 2006), and *vesicular stomatitis virus* (Novella *et al.*, 2007) have identified the presence of molecular memory of passage history in intrahost populations. In these studies, minority mutations well adapted to certain conditions are available for swift adaptation to new hosts or changing environments. Evidence for this phenomenon also was found in studies with WNV, in which specific variants from mosquito cell adapted virus were quickly selected following infection in avian cell culture (Ciota *et al.*, 2007c).

A study with multiple plant viruses has shown a correlation between host range and the size of the mutant spectrum (Schneider and Roossinck, 2000). These findings demonstrate the importance of identifying differences in levels of genetic diversity among RNA viruses in order to assess the potential for virus adaptability.

Intrahost genetic diversity also has phenotypic implications beyond population adaptability. Quasispecies have been shown to be important in viral fitness (Martinez *et al.*, 1991; Duarte *et al.*, 1994; Ruiz-Jarabo *et al.*, 2002). Findings from a study with FMDV identified genetically diverse populations following sequential cell culture passage (Arias *et al.*, 2001). In studies with VSV, fitness increases have been measured with no changes in consensus sequences (Novella and Ebendick-Corp, 2004), and genetic diversity was correlated with cell culture adaptation of WNV in a previous study (Ciota *et al.*, 2007c). This study also demonstrated that consensus genetic change was not solely responsible for the adaptive phenotype measured. Interestingly, identical consensus mutations were identified in a separate lineage of adapted WNV (Ciota *et al.*, 2007b). This indirectly implicated aspects of the mutant spectrum in adaptation and replication of WNV, and suggested interactions between genotypes could be necessary for conferring the adaptive phenotype. A recent study with DENV suggests that a defective genome can be maintained within the mutant spectrum by complementation of functional genomes (Aaskov *et al.*, 2006). A separate study with *Poliovirus* and WNV provides evidence for the role of complementation and cooperative interactions in pathogenesis (Vignuzzi *et al.*, 2006). The phenotypic importance of the mutant spectrum has also been noted in previous reports with other viruses (Farci *et al.*, 2002; Sauder *et al.*, 2006) and further clarification of the intrahost role of non-consensus variants is essential if the implications of genetic change in RNA viruses are to be accurately assessed.

Although all RNA viruses have significant potential for adaptation, variability in terms of both mutation and replication rates clearly exists among them. Studies with YFV indicate that the production of errors is much lower than the generally accepted error rate for RNA viruses (Pugachev *et al.*, 2004). Studies with the RNA bacteriophage $\phi 6$ show that mutation rates can even differ significantly among different strains of the same virus, and that mutation rate itself is a trait under selection (Burch and Chao, 2000). Decreased rate of replication is often coupled with less mutation. This is for the obvious reason that less copying means less error, but also because slower replication means more accuracy despite the lack of proofreading (Arnold *et al.*, 2005). It has been suggested that emerging pathogens tend to have relatively high mutation rates (LeClerc *et al.*, 1996). This is not a surprise given that it is often mutation that results in the capacity for emergence. North American strains of the flavivirus *St. Louis encephalitis virus* (SLEV) tend to replicate slower than those of the closely related WNV (Ciota *et al.*, 2007a). Given that SLEV has a relatively long evolutionary history in North America (Reisen, 2003), and WNV is a recently emerged pathogen which continues to establish in new areas, replication and mutation rate may be indirectly tied to evolutionary history with these two closely related pathogens.

Quasispecies theory has often been used to predict how genetically diverse populations may adapt and evolve (Eigen, 1993; Wilke *et al.*, 2001; Wilke, 2005). Although experimental evidence is generally lacking, many theoretical models and digital simulations predict that, under high mutation rates, adaptation of a mutant swarm will result in a mutational robustness (Wilke *et al.*, 2001; Montville *et al.*, 2005; Forster *et al.*, 2006), defined specifically as resistance to phenotypic change in the face of mutation (Montville *et al.*, 2005). This is important because in such a case, fitness is not linked to replicative ability alone. Specifically, a population composed of variants with equal replicative ability could be favoured over a population consisting of a single variant with a higher replicative rate surrounded by low fitness variants. A previous laboratory finding demonstrating that the majority of subclones from a highly diverse, high fitness variant of VSV obtained from cell culture passage possessed fitness values lower than that of the parental population (Duarte *et al.*, 1994) did not demonstrate this theoretical possibility. A digital simulation of evolving RNA sequences demonstrated that selection for robustness can be predicted as a product of mutation rate and population size (Forster *et al.*, 2006). More extensive laboratory experimentation is necessary to clarify factors important in determining how selection will act on newly emerging and preexisting pathogens which exist as genetically diverse quasispecies.

Transmission of arboviruses

As noted above, flaviviruses are a diverse group of viruses that are maintained in nature in a wide array of transmission cycles. With the exception of *Hepatitis C virus* (genus *Hepacivirus*; family *Flaviviridae*), the most important flaviviruses from a human health perspective involve arthropod vectors, mainly mosquitoes and ticks. Flaviviruses with no known vector and insect-only flaviviruses will not be discussed because they appear to pose minimal risk to public health and because relatively little is known about their mode(s) of perpetuation in nature.

Two critical concepts, vector competence and vectorial capacity, are central to our understanding of flavivirus transmission, evolution and epidemiology. Vector competence refers to the intrinsic ability of an arthropod species to acquire, support replication of, and transmit an infectious organism. In competent arthropod vectors, virus is ingested at the time of bloodfeeding and infects midgut epithelial cells. After initial replication in epithelial cells, virus is released into the haemocoel where it infects parenteral tissues, ultimately including the salivary gland. Replication in salivary gland cells leads to virus release into the salivary secretion where it may be inoculated during subsequent episodes of bloodfeeding. Competent vectors are clearly a biological requirement for maintenance of flaviviruses and their emergence as health threats.

Factors that influence vector competence have been studied extensively. Seminal studies demonstrated the existence of midgut infection and escape barriers (Kramer *et al.*, 1981; Houk *et al.*, 1986). More recent studies showed that vector competence is at least partly genetically determined by identifying quantitative trait loci for DEN vector competence in *Aedes aegypti* mosquitoes (Bosio *et al.*, 1998, 2000). Currently, several investigators have sought to profile transcriptional and small RNA responses to virus infection in important vector species (Sanchez-Vargas *et al.*, 2004; Myles *et al.*, 2008; Brackney *et al.*, 2009; Girard *et al.*, 2010). Nonetheless, the molecular determinants of mosquito vector competence for flaviviruses are relatively poorly understood.

Not surprisingly, vector competence is rarely a deterministic trait of a vector species, with all or almost all individuals of a given species equally competent. It is more accurately represented as a probability function associated with a particular arthropod species under specific conditions. Extreme variation has long been noted in free-ranging populations of mosquito species known to be important vectors of DENV and WNV (Boromisa *et al.*, 1987; Goddard *et al.*, 2002; Bennett *et al.*, 2002). Further, even in long standing mosquito colonies, some individuals become infected following feeding on an infectious blood meal, while others do not. Indeed, outbreaks of flavivirus disease have been attributed to 'incompetent' vectors (Miller *et al.*, 1989). Thus, although vector competence is relatively straightforward to measure in the laboratory and to conceptualize, it seems to poorly predict epidemiological or epizootiological transmission patterns. Other factors discussed at length below are therefore at least as critical as vector competence in determining the extent to which an arthropod species is an important flavivirus vector.

Many of these factors are incorporated conceptually under the term 'vectorial capacity.' The importance of vectorial capacity to the dynamics of flavivirus infection in mosquitoes has been reviewed extensively (Kramer and Ebel, 2003). The concept dates to the middle of the 20th century and has its origins in early mathematical models generated to formalize factors influencing malaria transmission and control (MacDonald, 1961). Briefly, vectorial capacity incorporates variables including vector abundance, degree of host focus, daily probability of vector survival and the extrinsic incubation period of the agent of interest. These variables are themselves influenced by factors such as rainfall, temperature, host availability, etc.. A vast literature documents the impact of all of these factors in the transmission dynamics of arthropod-borne infections. Vectorial capacity and vector competence represent two critical concepts in understanding the transmission of arthropod-borne flaviviruses.

The relationship between virus fitness and transmission is complex and the two are interdependent. Virus fitness influences transmission. For example, in North America, changes to the WNV genome (Davis *et al.*, 2005b) that increased fitness in *Culex pipiens* and *Cx. tarsalis* mosquitoes occurred rapidly. The fitness increase was apparent as a reduced extrinsic incubation period that effectively increased the vectorial capacity of local mosquitoes for WNV (Ebel *et al.*, 2004a). Ultimately more fit WNV variants displaced the existing WNV strain. Notably, the fitness benefit was entirely dependant on transmission by mosquitoes, highlighting their interrelatedness and the complex ways that the relationship can work. Transmission also influences fitness. It has long been assumed that transmission by arthropods imposes population bottlenecks on arbovirus populations. These bottlenecks, if sufficiently tight, can reduce virus fitness through the action of Muller's ratchet (Muller, 1964). Alternatively, bottlenecks might diminish or remove the modulating effects of the genetically complex population structure that RNA viruses tend to adopt (see introduction to viral fitness), potentially leading to the emergence of highly pathogenic virus variants (de la Torre and Holland, 1990). The complexity of the relationship(s) between transmission and virus fitness leads to inherent difficulty in making generalizations that are broadly applicable across the diverse mechanisms through which flaviviruses perpetuate. Therefore, in this chapter we will consider the relationships between fitness and transmission in specific cases, providing descriptions of how fitness and transmission are related.

WNV

Introduction to the agent

West Nile virus (WNV) is currently the most widely distributed arbovirus in the world, found on all continents except Antarctica (Kramer *et al.*, 2008). WNV is a member of the Japanese encephalitis (JE) serogroup of the flaviviruses, and shares coding and replication strategies with other flaviviruses. WNV perpetuates in nature in an enzootic cycle involving mosquitoes (mainly *Culex* spp.) and birds (Komar, 2000). Other arthropod species such as *Ochlerotatus* spp. mosquitoes and ticks, and various mammals may be infected, but these hosts contribute to perpetuation of the agent only minimally, if at all (Bernard *et al.*, 2001; Bunning *et al.*, 2002; Anderson *et al.*, 2003; Austgen *et al.*, 2004; Davis *et al.*, 2005a; Teehee *et al.*, 2005). The inadvertent introduction of WNV into the New York City area in 1999 (Lanciotti *et al.*, 1999) and the dramatic range expansion that followed have triggered intense interest into the evolutionary dynamics of this agent, including determinants of fitness and detailed studies of transmission mechanisms. Recent reviews have covered the epidemiology (Kramer *et al.*, 2008), molecular epidemiology and diversity (Ebel and Kramer, 2009), and ecology (Kramer, 2001; Komar, 2006; among others) of this virus. As has been pointed out by several authors, the introduction of WNV into North America at a well-defined time and place provided a rare opportunity to observe the adaptation of an exotic pathogen within a naive ecosystem prospectively. WNV thus has been an ideal system for studying the relationships between transmission and fitness because laboratory-based studies have been informed by an on-going natural experiment, providing a constant external control against which experimental results can be compared. The results from observational studies of WNV molecular epidemiology and evolution, and experimental studies informed by these observations, provide an excellent example of how tightly arbovirus transmission and fitness are associated in nature. Finally, because the WNV transmission cycle is relatively tractable in the laboratory – it can be modelled using colonized mosquitoes and chickens – the experimental literature on WNV fitness is rich in comparison with many flaviviruses.

Dominance and extinction of new WNV lineages

Convention holds that arboviruses are remarkably stable in nature, and tend to evolve more slowly than other RNA viruses because they undergo replication in cells of two radically different host types (arthropods and vertebrates) (Weaver *et al.*, 1992). This results in an evolutionary trade-off wherein optimal replication in either host is traded for adequate replication in both. It follows that arboviruses are subject to very strong evolutionary constraints and would be expected to change little in a relatively static environment. Molecular epidemiological studies of WNV nucleotide sequences were thus undertaken with considerable interest in the years following its introduction.

The first studies tended to confirm the prediction that WNV would remain stable in the Americas (Anderson *et al.*, 2001; Ebel *et al.*, 2001a; Beasley *et al.*, 2003). These studies found remarkable genetic conservation among sampled strains and formed a reference dataset against which subsequent studies could be compared. The observed genetic stasis was unsurprising because the relevant hosts involved in WNV transmission (*Culex* spp. mosquitoes and birds) were similar to those important in WNV transmission in Africa and Eurasia. By 2003, however, novel sequence variants and attenuated mutants had been detected (Beasley *et al.*, 2003; Davis *et al.*, 2003, 2004), and reviewed extensively elsewhere (Ebel and Kramer, 2009). Among these were attenuated strains isolated from mosquitoes and birds in Texas, and a genotype that clustered along a coastal region to the southwest of metropolitan Houston (Davis *et al.*, 2005b). These variants were generally sampled during only one transmission season, demonstrating both the propensity for new WNV lineages to be generated through mutation, and that many such variants, even those that undergo sustained transmission, ultimately become extinct. Molecular epidemiological studies of WNV in New York (Ebel *et al.*, 2004a) and a focus of transmission in suburban Chicago (Bertolotti *et al.*, 2007) demonstrated that pairwise genetic distance between strains increased with time (Fig. 9.2), suggesting a growing pool of WNV genotypes. From this pool of genotypes emerged one strain carrying a signature mutation (V to A at position 159 of the E protein) that increased in frequency from the time that it was first detected in 2001, when it represented approximately 10% of strains sampled, until 2004, when it represented 100% of strains sampled (Fig. 9.3) (Davis *et al.*, 2005b; Snapinn *et al.*, 2006). This WNV genotype (termed the WN02 or North American genotype) is now dominant throughout North America: strains lacking this mutation (i.e. the introduced or NY99 genotype) have not been detected since 2003. In contrast to conventional wisdom, then, WNV represents a dynamic virus population, with new lineages arising and becoming extinct frequently.

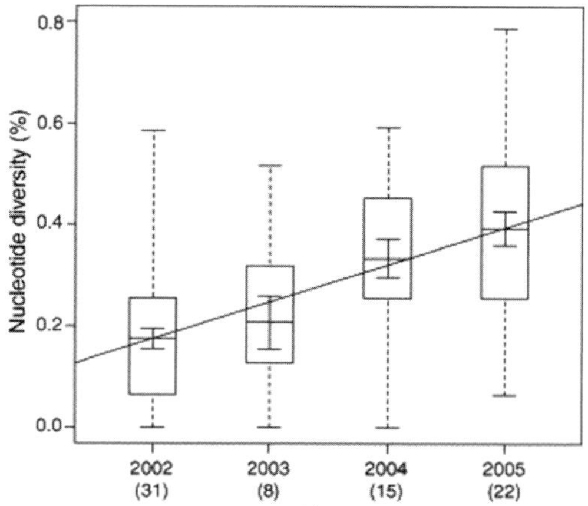

Figure 9.2 Increasing pairwise genetic diversity of WNV in North America, 2002–2005. Reproduced with permission from Bertolotti *et al.* (2007).

The establishment and dominance of the WN02 genotype provided an ideal opportunity to test which aspects of the host–virus interaction determined the fitness of the virus. Studies comparing the NY99 and WN02 genotypes found no evidence in the ability of several strains of each genotype to infect and produce viraemia in chickens or to infect mosquitoes. However, WN02 strains were transmitted after two to four fewer days extrinsic incubation than NY99 strains (Ebel *et al.*, 2004a). This difference in the time to transmission effectively reduces the extrinsic incubation period required for WN02 transmission by mosquitoes, increasing the basic reproductive rate (i.e. vectorial capacity, reviewed in Kramer and Ebel, 2003) of WN02 genotype viruses relative to NY99 viruses. Subsequent studies showed that WN02 replicates more efficiently outside of the mosquito midgut, providing a plausible mechanism for the increased transmissibility of WN02 by mosquitoes, and its observed dominance in nature (Moudy *et al.*, 2007). Collectively, this

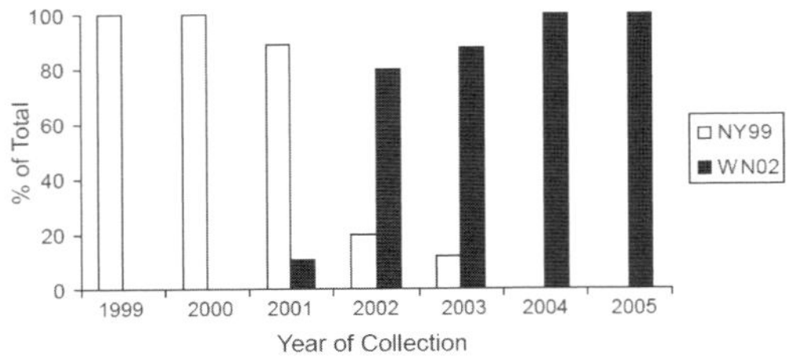

Figure 9.3 Displacement of the introduced (NY99) genotype WNV by derived (WN02) genotype, 1999–2005. WN02 is currently the dominant genotype in North America. Reproduced with permission from Snappin *et al.* (2007).

demonstrates clearly how transmission and fitness are tightly linked in the WNV system: faster replication in mosquitoes leads to increased transmissibility which leads to increased fitness. Interestingly, this also provides an example of how context-specific replication in one host might lead to the emergence of disease in another host. Notably, the largest epidemic of WNV encephalitis ever recorded was coincident with the emergence of the WN02 genotype in 2002 and 2003. Flavivirus fitness and transmission are therefore proximal determinants of public health risk in this system.

Effect of host alternation on fitness

One hypothesis for the considerable genetic stability of arboviruses is that the alternative cycles of viral replication in vertebrate hosts and arthropod vectors, which are required for arbovirus perpetuation, constrain evolution (Strauss *et al.*, 1990; Weaver *et al.*, 1992, 1994; Scott *et al.*, 1994). This implies that compromises must be made in replicative ability of virus populations in both the arthropod vectors and vertebrate hosts owing to differential selection in each. Specifically, mutations exclusively advantageous or neutral in one host may be purged by purifying selection in the other. Genetic change then only results from the infrequent mutations resulting in no cost in fitness in either host or in coadaptation. Reduced positive selection and increased purifying selection in vector-borne RNA viruses have been reported previously (Woelk and Holmes, 2002; Holmes, 2003; Jerzak *et al.*, 2005). Many studies measuring viral fitness changes resulting from cell culture passage have attempted to characterize the contribution of individual host cell types and the consequences of cycling. Viral fitness represents the propensity of an individual genome to make a contribution to the gene pool of the next generation, i.e. its ability to survive and reproduce in a particular specified environment and population (Mayr, 1994). Phenotypic and genotypic evidence for adaptive compromises resulting from cycling have been provided for the flavivirus dengue-2 (Chen *et al.*, 2003) and togaviruses *Sindbis virus* (Greene *et al.*, 2005) and *eastern equine encephalitis virus* (Cooper and Scott, 2001; Weaver *et al.*, 1999), yet a lack of evolutionary stasis with cycling was reported for studies with *vesicular stomatitis virus* (VSV) (Novella *et al.*, 1999; Zarate and Novella, 2004). Two flaviviruses, WNV and *St. Louis encephalitis virus* (SLEV), have been demonstrated to be capable of significant cell-specific adaptation resulting from sequential passaging in mosquito cell culture (Ciota *et al.*, 2007b). These adaptations did not generally result in decreased replicative fitness in vertebrate cell culture, yet, for WNV, one passage in avian cell culture significantly decreased both the mosquito cell adapted phenotype and genetic variation (Ciota *et al.*, 2007c). With a few exceptions; e.g. *duck hepatitis B virus* (Carrillo *et al.*, 1998), *foot-and-mouth disease virus* (Lenhoff *et al.*, 1998), Venezuelan equine encephalitis virus (Coffey *et al.*, 2008) and WNV (Ciota *et al.*, 2008, 2009; Fitzpatrick *et al.*, 2010), *in vivo* viral fitness studies are largely lacking. Studies in intact vertebrates and arthropods, while significantly more complex, recognize immune and antiviral responses and other selective pressures that are critical to our understanding of how within-host ecology impacts viral evolution.

Effect of mutational diversity on fitness

WNV, like other RNA viruses, exists within hosts as a genetically diverse swarm of genotypes that differ in their mutational distance from the consensus sequence of the group as a whole (Jerzak *et al.*, 2005). Intrahost genetic diversity of WNV is greater in mosquitoes than in avians, both in field-collected samples, and following laboratory passage (Jerzak *et al.*, 2005, 2007, 2008). Two main factors lead to the increase in genetic diversity in mosquitoes. First, analysis of synonymous and non-synonymous variation in WNV passed in only mosquitoes or only in birds demonstrated that purifying selection is strong in birds but relaxed in mosquitoes (Jerzak *et al.*, 2008). Second, the RNA interference-based response to WNV infection in *Culex* mosquitoes creates an intracellular environment where rare virus genotypes are favoured because they do not match existing guide siRNA sequences (Brackney *et al.*, 2009). This latter observation led to the prediction that more genetically diverse WNV populations would have a fitness advantage in mosquitoes. Fitness competition studies conducted *in vivo* in mosquitoes and chickens demonstrated that WNV populations that are more genetically diverse are indeed more fit in mosquitoes, but not in chickens (Fig. 9.4) (Fitzpatrick *et al.*, 2010). Therefore, intrahost genetic diversity provides a fitness benefit to WNV populations. Additional studies are required to characterize the importance of this diversity on transmission, where population bottlenecks may become more important.

Yellow fever virus and dengue virus

Introduction to the agents

Yellow fever virus (YFV) is the prototype member of the *Flaviviridae*, with the virus family taking its name from the jaundiced condition (flavus) of the patients infected with this virus. *Aedes* spp. mosquitoes are the predominant vectors of both DENV and YFV. Unlike *Culex*-borne flaviviruses, mosquitoes may be competent vectors following feeding on humans infected with these two *Aedes*-borne flaviviruses, and virus amplification within human reservoir populations is critical to outbreaks. Consequently the fitness of YFV and DENV may be affected by adaptation to human as well as non-human primates in addition to the mosquito vector. However, DENV and YFV differ in a number of ways. Whereas YFV is largely sylvatic with only occasional spillover to humans, DENV is maintained predominantly through a human-mosquito cycle.

YFV is transmitted to nonhuman primates by several species of mosquitoes in tropical and subtropical sylvatic areas in South America and Africa. *Aedes* spp. transmit YFV in Africa; forest canopy mosquitoes of the genera *Haemagogus* and *Sabethes* transmit YFV in South America (Barrett and Higgs, 2007). YFV has not been found in Asia. DENV is transmitted in a sylvatic cycle in Asia and Africa. *Ae. aegypti* is the predominant vector of both YFV and DENV in urban settings, although *Ae. albopictus* also has been implicated in DENV outbreaks.

Other differences between these two *Aedes*-transmitted flaviviruses are that a single serotype of YFV exists, while four serotypes of DENV (DENV-1, -2, -3, -4) may co-circulate. Since one serotype of DENV does not confer full protection against the others, and cross-reactive non-neutralizing antibodies may facilitate virus entry into host cells through the phenomenon known as antibody dependent enhancement (Burke and Kliks, 2006), different pressures exist on this virus than on YFV where a single infection is assumed to confer life-long immunity against re-infection. Furthermore, a vaccine has been available providing protection against YFV since the 1950s, while DENV vaccine is still in the developmental and trial stages.

The evolutionary and epidemiological histories of both DENV (Vasilakis and Weaver, 2008) and YFV (Barrett and Monath, 2003) have been well documented and won't be repeated here for lack of space. Recent YFV outbreaks have been relatively small, with fewer than 5000 human cases reported in Africa and South America from 2000 to 2005 (Barrett and Higgs, 2007). However, YFV virus has a long history of re-emerging as a public health threat, highlighting the need to better understand how virus fitness relates to transmission and emergence in this system. Further, DENV outbreaks continue to increase owing to an unplanned urban expansion, increased vector density owing to inconsistent control effort, and an increase in air travel facilitating the rapid movement of viraemic humans since the end of the Second World War. Evidence supporting the case for human dispersal as the single most important factor shaping the genetic structure of DENV populations comes from numerous studies which have demonstrated that the evolutionary history of DENV is a relatively recent one (Zanotto *et al.*, 1996; Twiddy *et al.*, 2003; Dunham and Holmes, 2007). Dengue

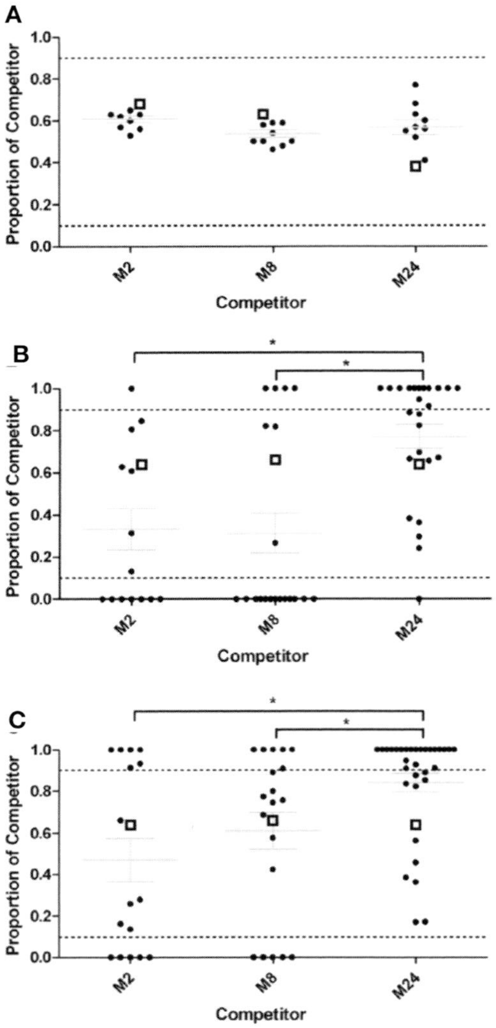

Figure 9.4 Genetic diversity of WNV leads to fitness increases within mosquitoes, but not chickens. WNV populations of low (M2), medium (M8) and high (M24) genetic diversity were competed against a genetically marked reference virus *in vivo* in day-old chickens (panel A) and in mosquitoes for either seven (panel B) or 14 days (panel C). Observations from individual chickens and mosquitoes are shown as dark circles and the proportions of reference and test viruses in the input mixture are shown as open squares. Grey bars indicate mean and standard error, dashed line indicates the approximate linear range of the assay, and asterisks indicate statistically significant ($p<0.05$) results. Reproduced with permission from Fitzpatrick *et al.* (2010).

viruses are responsible for the highest incidence of human disease among all mosquito-borne viruses, including 100 million infections annually (Gubler, 1998), resulting in approximately 500,000 cases of dengue haemorrhagic fever (DHF), a serious manifestation of dengue disease (Halstead *et al.*, 2007). The reported number of dengue cases has doubled in the current decade (Kroeger and Nathan, 2006). Severe disease is largely a consequence of hyperendemnicity of the four serotypes of DENV.

Transmission cycles

Three transmission cycles have been associated with YFV. A sylvatic or jungle cycle occurs in Africa and South America, and in addition, an intermediate or savannah cycle also is found in Africa. The urban cycle found in Africa has not been evident in South America since 1942. DENV also has a sylvatic cycle in Asia and Africa, but it is not clear whether significant numbers of human cases are associated with it. There are those who think that an insufficient number and density of susceptible human hosts are available at this time for sylvatic dengue to establish itself (Dunham and Holmes, 2007; Rico-Hesse and Mota, 2010); however, others argue that it may be an important and unrecognized cause of morbidity (Vasilakis *et al.*, 2010). Several mosquito species are involved in the sylvatic and savannah DENV cycles, with *Aedes aegypti* being the main vector in the urban cycle. *Aedes aegypti* have several important feeding and biological features that have facilitated its geographical expansion: the ability of larvae to develop rapidly in standing water around domestic dwellings, the ability of the eggs to survive drying and a preference for feeding on humans almost exclusively as well as daily (Harrington *et al.*, 2001; Barrett and Higgs, 2007). Efficient transmission between humans of YFV and DENV by *Ae. aegypti* allows for explosive outbreaks to occur in urban settings. *Ae. aegypti* are not highly competent vectors of DENV, selecting for those strains that replicate to the highest titres in humans. Nonetheless, the virulence of DENV is still lower than that of YFV, the latter having a mortality rate approaching 64% in some African locations (Sall *et al.*, 2010).

YFV is continually present in the forests of east and central Africa, but outbreaks do not occur in these two regions except in times of disruption such as war or following movement of large numbers of people. At these times, different selection pressures on the virus would take place. West African genotype I, which is more frequently associated with outbreaks, is more genetically heterogeneous, and consequently may be more adaptable to changing conditions. Nonetheless, YFV appears to have a substitution rate significantly lower than DENV (Sall *et al.*, 2010).

Genetic variation

YFV is maintained in the jungles of South America most probably by spread of the virus by infected non-human primates along river banks from one location to a contiguous one. Epizootics occur intermittently when susceptible populations of primates become large enough to sustain transmission (Vasconcelos *et al.*, 1997). The mobility of the reservoir host in the virus transmission cycle has an impact on viral evolution. This becomes evident in comparisons of evolutionary patterns of such flaviviruses as WNV and JEV, where avian hosts are amplification vertebrates, and other flaviviruses such as YFV where non-human primates act as amplification hosts in the enzootic sylvatic cycles. Since non-human primates live in distinct geographically limited habitats, the viruses infecting these hosts tend to be similarly genetically partitioned, thus genetic differentiation is characteristic of YFV (Deubel *et al.*, 1986). In contrast, birds acting as reservoir hosts lead to broadly distributed viral lineages (MacKenzie *et al.*, 1995). DENV, which is predominantly transmitted among humans by vector mosquitoes, demonstrates complex patterns of evolution depending on travel behaviour of the human host, as well as immune status of the population. Genetic shifts and virus strain replacements occur frequently under such circumstances. YFV, on the other hand, appears to have a lower rate of evolutionary change (Bryant *et al.*, 2007; Sall *et al.*, 2010) as it is more dependent on the less dynamic sylvatic cycle, especially in South America, where non-human primates often suffer mortality rather than survive with immunity following infection. It also has been hypothesized that YFV may evolve more slowly than DENV because it is dependent on vertical transmission of the virus by the mosquito vector. This, in addition to the consequence of the maintenance of the virus in a mosquito–primate system, may partially explain the different evolutionary rates observed in Peru between geographically separate locations (Bryant *et al.*, 2003). Similarly, there is substantial genetic variability among YFV variants in Brazil which may be connected to human migration throughout the country (Vasconcelos *et al.*, 2004). Studies of YFV nucleotide sequences have identified seven genotypes (Barrett and Higgs, 2007). Each of the genotypes corresponds to a specific geographic region: East and Central African genotype, East African genotype, Angola genotype, South America genotype I, South America genotype II, West Africa genotype I and West Africa genotype II. In addition, only slight sequence variation among YFV strains isolated across years was detected, demonstrating the stability of genotypes over time (Barrett and Higgs, 2007).

In contrast with YFV, DENV is expanding its range and increasingly causing major outbreaks. The severity of the outbreaks varies temporally both seasonally and annually depending on immune status of the population, environmental factors such as rainfall and temperature, vector variability, i.e. competence, population density, etc. (Strickman and Kittayapong, 2002, 2003; Honorio *et al.*, 2009; Bennett *et al.*, 2010). In general, a three to five year periodicity has been noted (Bennett *et al.*, 2010), although this number varies depending on the approach taken to define activity, i.e. serotype-dependent calculations, disease manifestation, sequence change. As virus transmission increases in response to a convergence of favourable virus–vector–vertebrate factors, nucleotide changes accumulate, and consequently genetic diversity and potentially viral evolution increase. Genetic diversity of DENV sequences of isolates from Puerto Rico were used to estimate effective population size (N_e) using a coalescent approach (Bennett *et al.* 2010). It was demonstrated that when DENV-4 was introduced into the country, it underwent a dramatic expansion in population size (Carrington *et al.*, 2005) as demonstrated by generation of significantly greater genetic diversity than had been seen with other dengue serotypes, doubling effective population size every 2 weeks. Phylogenetic analysis indicated diversity was not due to introduction of new viral strains, but rather diversity generated *in situ*.

Examination of the full genome of the four dengue serotypes indicated that selection is acting on amino acid positions of the genes and serotypes differentially, however sampling bias may have contributed to the differences observed, particularly for DENV-1. These results differ from findings for other flaviviruses discussed in this chapter where no evidence of positive selection has been found in the envelope gene (Twiddy *et al.*, 2002). Nonetheless, selection pressure is considerably weaker than for non-vector-borne viruses, perhaps as a consequence of the need to replicate in arthropod hosts.

Strains comprising the four serotypes of DENV have been extensively sequenced allowing for greater understanding of the viruses' population dynamics. One explanation offered for the increased rate of viral evolution of DENV than YFV is that DENV epidemics lead to higher transmission rates than endemic situations. However, a comparison of evolutionary rates between regions of Brazil with epidemic and endemic DENV-3 strains showed no differences in overall rates (Araujo *et al.*, 2009). In other studies, DEN-4 was found to undergo greater increase in the number of viral lineages than DENV-2, indicating faster viral population growth or geographic dispersal in the Americas. Thus one may not be able to generalize from one DENV serotype to another; however some studies analysed the E gene only, while others analysed both structural and non-structural genes, therefore direct comparisons between datasets may be problematic. Clearly, the E gene is under greater immunological selection pressure imposed by the human immune system (Innis *et al.*, 1989; Gritsun *et al.*, 1995). Nonetheless, differences have been observed within genotypes, between genotypes, and between serotypes. In summary, DENV evolution appears to be shaped by a combination of random genetic drift and fixation of genetic changes, periodic replacement of viral clades, the increasing number of infections globally, and the complex interactions among serotypes as a consequence of serotype-specific host immunity (Kyle and Harris, 2008).

TBEV

Introduction to the agents

The tick-borne encephalitis virus (TBEV) serological complex comprises an ecologically diverse set of organisms that share the requirement for a zoonotic transmission cycle because no ticks feed preferentially on human beings (Calisher and Gresikova, 1989). In nature these viruses are maintained by ticks that feed on a wide array of vertebrate hosts. Several members have been described, and include *tick-borne encephalitis virus* (TBEV), *Omsk haemorrhagic fever virus*, *Kyasanur Forest disease virus*, *Langat virus*, the seabird-associated *Saumarez Reef virus* and *Tyuleniy virus* and others. The most important member of the complex and the focus of this discussion, TBEV, is maintained in Europe and Asia by ticks of the *Ixodes ricinus* species complex, including *Ix. ricinus* and *Ix. persulcatus*. TBEV includes several subtypes, including Central European, Siberian and Far Eastern genotypes, and Louping Ill virus. These agents vary in pathogenic potential from the relatively mild Central European (~0.5% case fatality rate) to the relatively severe Far Eastern (~10% case fatality rate) subtypes. The North American member of the TBE complex is *Powassan virus* (POWV), which is maintained by the North American member of the *Ix. ricinus* complex, *Ix. scapularis*, among others. POWV causes severe encephalitis and has a case-fatality rate of approximately 12%. Recent and past reviews discuss the epidemiology, replication, pathogenesis and ecology of TBEV and POWV in detail (Labuda and Nuttall, 2004; Charrel *et al.*, 2004; Ebel, 2010). Tick-borne flaviviruses, then, constitute severe and/or emerging health threats in areas where they are enzootic.

Two general features of TBEV fitness and transmission should be noted because they markedly differ from mosquito-borne flaviviruses. The first (rather obviously) is their requirement for transmission by ticks. Ticks differ from mosquitoes in several important characteristics of their life history and feeding behaviour/strategy that probably influence the population biology and fitness of tick-borne viruses. Second, TBEV and related viruses tend to persist in nature in discrete, extremely stable foci that have been termed 'nidi'. The nidality of tick-borne infection has important epidemiological implications: whereas mosquito-borne agents may emerge in explosive short-lived epidemics, those transmitted by ticks tend to reside within a particular landscape with epidemics building much more slowly and intractably. The reliance on ticks for perpetuation in nature and the apparent nidality of tick-borne infections thus differentiate them from mosquito-borne infections.

Therefore, in the following sections, we discuss these shared characteristics of tick-borne flaviviruses and provide a discussion of how they relate to virus fitness. In general, however, experimental literature on TBEV fitness is non-existent. A PubMed search including the terms '*Tick-borne encephalitis virus* fitness' conducted on in June 2010 returned zero records. Observational studies that may inform understanding of TBEV fitness consist of molecular epidemiological studies (Grard *et al.*, 2007; Hartemink *et al.*, 2008b; Leonova *et al.*, 2009) and a single study of intrahost genetic diversity of POWV in tick hosts (Brackney *et al.*, 2010). Given the severe, intractable and emerging nature of tick-borne flaviviruses, rigorous studies aimed at understanding how reliance on ticks influences TBEV populations are important and may shed light on other tick-borne viruses that present serious health threats, e.g. *Crimean Congo haemorrhagic fever virus*.

Horizontal transmission

Whereas the mosquito-borne flaviviruses are maintained in transmission cycles mainly by horizontal (mosquito-vertebrate-mosquito) transmission, and nominally by vertical transmission, the mechanisms by which tick-borne flaviviruses are maintained are more complicated. Hard (Ixodid) ticks only feed once during each of their three active life stages (larvae, nymph and adult). This has two important implications for virus transmission and presumably fitness. First, virus must be maintained within infected ticks between life stages (i.e. trans-stadial transmission, discussed below). The duration of the tick life cycle, and in particular the several weeks to months that occur between episodes of tick feeding owing to moulting, etc., releases the virus from the requirement for extremely rapid replication during an extrinsic incubation period, which is critical for mosquito-borne flaviviruses (Ebel *et al.*, 2004), but irrelevant for tick-borne flaviviruses. Second, since infection with tick-borne flaviviruses in vertebrates generally does not result in a lengthy viraemia, the timing of feeding of the different life stages on a common host must ensure that later (infected nymphal) stages will cofeed along with earlier (uninfected larval) stages for 'horizontal' transmission to occur. In horizontal transmission, susceptible vertebrates (i.e. those that are frequently host to competent ticks) become infected by the bite of an infected tick and develop viraemia sufficient to establish infection in another tick. It has been demonstrated for alphaviruses that in order to be transmitted horizontally, the virus makes an evolutionary trade-off: replication must be adequate in both vertebrates and invertebrates, presumably at the cost of being highly efficient in either (Greene *et al.*, 2005; Weaver, 2006; Coffey *et al.*, 2008). However, this may not be the case for flaviviruses as demonstrated by *in vivo* assays rather than in cell culture systems (Ciota 2008). In addition, this hypothesis has not been examined to date with tick-borne flaviviruses, and fitness determinants of flaviviruses that may influence horizontal transmission (i.e. the ability to replicate well in taxonomically diverse cell types) are poorly understood. Evidence from several studies using TBEV and POWV in various animal models have generally found modest titres (Kokernot *et al.*, 1969; Timoney, 1971; Chunikhin and Kurenkov, 1979; Kozuch *et al.*, 1981; Zarnke and Yuill, 1981) compared with reported viral loads attained in ticks (Ebel and Kramer, 2004), suggesting that for TBE viruses, replication efficiency within ticks may be more closely correlated with overall virus fitness than is replication within vertebrates.

Trans-stadial transmission

Because ticks feed only once during each of their active life stages, tick-borne flaviviruses must be passed between life stages (i.e. transstadially). The requirement for transstadial transmission is relatively unique to tick-borne agents. Between stages, extensive tissue reorganization and remodelling occur prior to ecdysis [for a comprehensive treatment of the physiological changes associated with moulting in the acari, (Sonenshine, 1993)]. Tick-borne flaviviruses presumably posses specific adaptations that facilitate transstadial transmission since non tick-borne members of the genus are not efficiently transmitted in this manner (Anderson *et al.*, 2003), but the viral determinants that might facilitate transstadial transmission have not been described. For several decades, however, it has been appreciated that transstadial transmission may influence vector competence and virus replication. Chernesky (1969) documented steep declines in POW titres shortly after *D. andersoni* larvae moulted, followed by rapid increases in the week immediately following (Chernesky, 1969). Additional studies using POW and *Ix. scapularis* found similar results and suggested that significant virus replication (an ~2 $\log_{10}$ increase in titre) was associated with metamorphosis between larval and nymphal stages (Ebel and Kramer, 2004). The efficiency of trans-stadial transmission has been estimated by two studies, which independently determined that at least 22% of infected larvae transmit virus transstadially (Costero and Grayson, 1996; Ebel and Kramer, 2004). Trans-stadial transmission, therefore, presents a potential population bottleneck because the number and diversity of virus genomes may be drastically reduced within hosts during metamorphosis and because the overall rate of transstadial transmission may be relatively low. The potential influence of the bottlenecks associated with trans-stadial transmission on TBEV fitness include the strengthening of founder's effects which could either lead to stochastic loss of fitness through the action of Muller's ratchet (Duarte *et al.*, 1992) or the emergence of high fitness variants (de la Torre and Holland, 1990). Studies of POWV mutational diversity in field-collected adult *Ix. scapularis* demonstrated relatively low intrahost diversity levels, which would be consistent with the action of population bottlenecks. Supporting this, most ticks contained a unique consensus sequence (Fig. 9.5) (Brackney *et al.*, 2010), which would also be expected to occur if founder's effects during trans-stadial transmission between nymphal and adult stages were strong. Therefore, trans-stadial transmission is a unique characteristic of tick-borne flaviviruses that seems likely to influence their fitness by imposing population bottlenecks.

Vertical transmission

Evidence has been presented from several studies of flaviviruses (Miller *et al.*, 2000; Baqar *et al.*, 1993; Goddard *et al.*, 2003; Anderson *et al.*, 2007), including tick-borne agents (Costero and Grayson, 1996), to suggest that vertical transmission (i.e. transmission from an infected female to her offspring) occurs. The extent to which this mode of transmission contributes to enzootic perpetuation of tick-borne flaviviruses is not currently clear: Costero and Grayson found that ~15% of infected females transmitted virus to their offspring (Costero and Grayson, 1996). Modelling studies suggest that although prevalence of infection in ticks is relatively insensitive to variation in vertical transmission efficiency, vertical transmission becomes a crucial intergenerational link for virus persistence when larval and nymphal feeding do not overlap. Very high vertical transmission rates can sustain virus prevalence, but the rates required seem unrealistic (Nonaka *et al.*, 2010).

Transmission among cofeeding ticks

Several studies have demonstrated that TBEV can be transmitted between cofeeding ticks without the production of systemic vertebrate viraemia (Labuda *et al.*, 1993a–d). This mode of transmission is often referred to as 'non-viraemic transmission' (NVT). The mechanisms leading to this mode of transmission have been reviewed extensively elsewhere (Nuttall and Labuda, 2003). Briefly, ticks possess potent anti-inflammatory, anti-haemostatic and other immune-modulating compounds that are inoculated into the feeding site during the several days/weeks that ticks feed (Wang and Nuttall, 1995; Hajnicka *et al.*, 1998; Schoeler *et al.*, 2000; Hajnicka *et al.*, 2005). These compounds interact with cellular components of the localized immune response to tick feeding that occurs within the skin to enhance transmission between cofeeding ticks (Labuda *et al.*, 1996b). Similar studies of mosquito-borne flaviviruses have demonstrated that 'non-viraemic' transmission among cofeeding mosquitoes occurs (Higgs *et al.*, 2005; McGee *et al.*, 2007) but the contribution of this mode of transmission to the epidemiology of the mosquito-borne agents has not been evaluated. In contrast, several modelling studies have suggested that cofeeding transmission is required for the perpetuation, and thus the epidemiology, of tick-borne agents (Labuda *et al.*, 1993e; Ogden *et al.*, 2007; Hartemink *et al.*, 2008; Nonaka *et al.*, 2010). An important corollary of this model of TBEV transmission is that viraemia in vertebrates has been described as a 'red herring' of little relevance to the ability of tick-borne flaviviruses to perpetuate (Nuttall and Labuda, 2003). This would presumably have significant implications for the fitness of tick-borne flaviviruses because it would lessen the importance of the ability to produce systemic viraemia in vertebrates, although the ability to infect and replicate in skin at the site of tick attachment and/or in Langerhans cells would still be required (Labuda *et al.*, 1996a; Nuttall and Labuda, 2003). Thus, transmission among cofeeding ticks seems to be a major mechanism for the perpetuation of tick-borne flaviviruses in nature, with other transmission routes of somewhat lesser importance.

Nidality of infection

It has been noted for decades that tick-borne pathogens tend to persist in spatially discrete, highly stable enzootic foci. This has been noted for TBEV in Eurasia (Babenko *et al.*, 1958), and POWV in North America (Brackney *et al.*, 2008), and distinguishes tick-borne flaviviruses from their mosquito-borne counterparts such as *DENV* and *West Nile virus*, which have rapidly emerged and expanded their ranges, leading to explosive epidemics (Rico-Hesse *et al.*, 1998; Kramer *et al.*, 2008; Balmaseda *et al.*, 2010). The pattern for tick-borne viruses suggests that conditions favourable to their long-term perpetuation occur relatively infrequently in a given landscape and that favourable conditions are themselves a stable characteristic of a landscape. Molecular epidemiological studies have shown that tick-borne flaviviruses follow a cline across the northern hemisphere (Zanotto *et al.*, 1995) These studies suggest that the TBE serogroup may diversify through colonization of new permissive environments by a small population of 'founders' followed by selection of the most fit variants in that environment and subsequent long-term perpetuation. In support of this, POWV nucleotide sequence data from ticks within a single transmission focus in Wisconsin showed that most individual ticks had a unique virus sequence. In addition, intrahost variation within ticks was low, and purifying selection was strong, suggesting tight evolutionary constraint in addition to the action of founders effects. Further, two studies have suggested that adaptation to tick species may contribute to the natural selection of TBEV variants in nature. A genotype of POWV, deer tick virus (Telford, III *et al.*, 1997), may have specifically adapted to *Ix. scapularis* vectors in North America (Ebel *et al.*, 2001b). Similarly, adaptation of Siberian TBEV strains to *Ix. persulcatus* may be driving their emergence and contributing to their ability to produce severe disease in human beings (Khasnatinov *et al.*, 2009). Relevant fitness experiments addressing how these different mechanisms may influence virus transmission are currently lacking.

Conclusions

Flaviviruses persist in complex and highly divergent transmission cycles, involving mosquitoes, ticks, and a wide array of vertebrate hosts, including human beings. The need to perpetuate in the highly divergent transmission cycles, detailed above, has led to complex evolutionary dynamics that appear to be driven by the unique factors within each of these cycles. The determinants that allow virus persistence and maximize fitness, however, are poorly understood and have not been adequately studied. While molecular epidemiological studies have provided an important framework for generating hypotheses about what determines virus fitness in a given environment, extremely few experiments have been conducted to assess flavivirus fitness. This shortcoming is striking given the rapidity of the emergence of several flaviviruses, the threats they pose to public health, and the wealth of observational data available for hypothesis generation. Significant barriers to conducting the required research include lack of inexpensive and tractable model systems, the

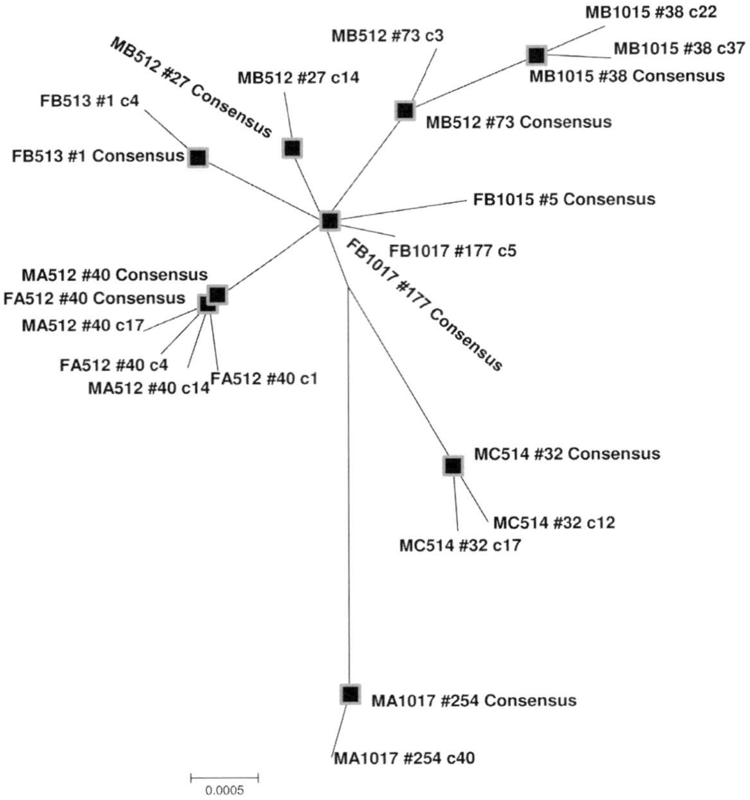

Figure 9.5 Genetic diversity differs by scale in a tick-borne flavivirus. Phylogenetic analysis of all consensus and haplotype sequences. Neighbour-joining analysis of consensus sequences (black squares) and all unique haplotypes detected in POWV populations infecting adult *Ixodes scapularis*. Presented is the 50% majority rule consensus tree of 1000 bootstrap replicates. Intrahost genetic diversity is extremely low in POW-infected *Ix. scapularis*, but each tick tends to have a unique consensus sequence. Reproduced with permission from Brackney *et al.* (2010).

variability inherent in working with outbred (or even wild-caught) hosts, and the technical and analytic complexity of relevant experiments. Recent technical advances in measuring fitness and in creating relevant reference viruses should facilitate further experimental studies into flavivirus transmission and fitness. This research should lead to a more detailed understanding of the host–virus relationship, which seems to be a prerequisite for effective suppression of virus transmission.

References

Aaskov, J., Buzacott, K., Thu, H.M., Lowry, K., and Holmes, E.C. (2006). Long-term transmission of defective RNA viruses in humans and *Aedes* mosquitoes. Science *311*, 236–238.

Anderson, J.F., Vossbrinck, C.R., Andreadis, T.G., Iton, A., Beckwith, W.H., and Mayo, D.R. (2001). A phylogenetic approach to following West Nile virus in Connecticut. Proc. Natl. Acad. Sci. U.S.A. *98*, 12885–12889.

Anderson, J.F., Main, A.J., Andreadis, T.G., Wikel, S.K., and Vossbrinck, C.R. (2003). Transstadial transfer of West Nile virus by three species of ixodid ticks (Acari: Ixodidae). J. Med. Entomol. *40*, 528–533.

Anderson, J.F., Main, A.J., Delroux, K., and Fikrig, E. (2007). Culex pipiens amplifies west nile virus in northeastern united states by vertical and horizontal transmission. Am. J. Trop. Med. Hyg. *77*, 185.

Araujo, J.M., Nogueira, R.M., Schatzmayr, H.G., Zanotto, P.M., and Bello, G. (2009). Phylogeography and evolutionary history of dengue virus type 3. Infect. Genet. Evol. *9*, 716–725.

Arias, A., Lazaro, E., Escarmis, C., and Domingo, E. (2001). Molecular intermediates of fitness gain of an RNA virus: characterization of a mutant spectrum by biological and molecular cloning. J. Gen. Virol. *82*, 1049–1060.

Arnold, J.J., Vignuzzi, M., Stone, J.K., Andino, R., and Cameron, C.E. (2005). Remote site control of an active site fidelity checkpoint in a viral RNA-dependent RNA polymerase. J. Biol. Chem. *280*, 25706–25716.

Austgen, L.E., Bowen, R.A., Bunning, M.L., Davis, B.S., Mitchell, C.J., and Chang, G.J. (2004). Experimental infection of cats and dogs with West Nile virus. Emerg. Infect. Dis. *10*, 82–86.

Babenko, L.V., Davydova, M.S., Zakorkina, T.N., Blokhin, Voronkov, N.A., Naumov, R.L., and Khizhinskii, P.G. (1958). [Characteristics of a nidus of tick-borne encephalitis in the construction zone of the Krasnoyarsk Hydroelectric Station and development of measures for protection of workers against ticks; preliminary report.]. Med Parazitol. (Mosk) *27*, 6–14.

Balmaseda, A., Standish, K., Mercado, J.C., Matute, J.C., Tellez, Y., Saborio, S., Hammond, S.N., Nunez, A., Aviles, W., Henn, M.R., *et al.* (2010). Trends in patterns of dengue transmission over 4 years in a pediatric cohort study in Nicaragua. J. Infect. Dis. *201*, 5–14.

Baqar, S., Hayes, C.G., Murphy, J.R., and Watts, D.M. (1993). Vertical transmission of West Nile virus by *Culex* and *Aedes* species mosquitoes. Am. J. Trop. Med. Hyg. *48*, 757–762.

Barrett, A.D., and Higgs, S. (2007). Yellow fever: a disease that has yet to be conquered. Annu. Rev. Entomol. *52*, 209–229.

Barrett, A.D.T., and Monath, T.P. (2003). Epidemilolgy and ecology of yellow fever virus. Adv. Virus Res. *61*, 291–315.

Beasley, D.W., Davis, C.T., Guzman, H., Vanlandingham, D.L., Travassos da Rosa, A.P., Parsons, R.E., Higgs, S., Tesh, R.B., and Barrett, A.D. (2003). Limited evolution of West Nile virus has occurred during its southwesterly spread in the United States. Virology *309*, 190–195.

Bennett, K.E., Olson, K.E., Munoz, M.L., Fernandez-Salas, I., Farfan-Ale, J.A., Higgs, S., Black, W.C., and Beaty, B.J. (2002). Variation in vector competence for dengue 2 virus among 24 collections of *Aedes aegypti* from Mexico and the United States. Am. J. Trop. Med. Hyg. *67*, 85–92.

Bennett, S.N., Drummond, A.J., Kapan, D.D., Suchard, M.A., Munoz-Jordan, J.L., Pubus, O.G., Holmes, E.C., and Gubler, D.J. (2010). Epidemic dynamics revealed in dengue evolution. Mol. Biol. Evol. *27*, 811–818.

Bernard, K.A., Maffei, J., Jones, S.A., Kauffman, E.B., Ebel, G.D., Dupuis, A.P., Ngo, K., Nicholas, D., Young, D., Shi, P.Y., *et al.* (2001). West Nile virus infection in birds and mosquitoes, New York State, (2000). Emerg. Infect. Dis. *7*, 679–685,

Bertolotti, L., Kitron, U., and Goldberg, T.L. (2007). Diversity and evolution of West Nile virus in Illinois and the United States, 2002–2005. Virology *360*, 143–149.

Boromisa, R.D., Rai, K.S., and Grimstad, P.R. (1987). Variation in the vector competence of geographic strains of *Aedes albopictus* for dengue 1 virus. J. Am. Mosq. Control Assoc. *3*, 378–386.

Bosio, C.F., Beaty, B.J., and Black, W.C. (1998). Quantitative genetics of vector competence for dengue-2 virus in *Aedes aegypti*. Am. J. Trop. Med. Hyg. *59*, 965–970.

Bosio, C.F., Fulton, R.E., Salasek, M.L., Beaty, B.J., and Black, W.C. (2000). Quantitative trait loci that control vector competence for dengue-2 virus in the mosquito *Aedes aegypti*. Genetics *156*, 687–698.

Brackney, D.E., Nofchissey, R.A., Fitzpatrick, K.A., Brown, I.K., and Ebel, G.D. (2008). Stable prevalence of Powassan virus in *Ixodes scapularis* in a northern Wisconsin focus. Am. J. Trop. Med. Hyg. *79*, 971–973.

Brackney, D.E., Beane, J.E., and Ebel, G.D. (2009). RNAi targeting of West Nile virus in mosquito midguts promotes virus diversification. PLoS Pathog. *5*, e1000502.

Brackney, D.E., Brown, I.K., Nofchissey, R.A., Fitzpatrick, K.A., and Ebel, G.D. (2010). Homogeneity of Powassan virus populations in naturally infected Ixodes scapularis. Virology *402*, 366–371.

Briones, C., de Vicente, A., Molina-Paris, C., and Domingo, E. (2006). Minority memory genomes can influence the evolution of HIV-1 quasispecies *in vivo*. Gene *384*, 129–138.

Bryant, J., Wang, H., Cabezas, C., Ramirez, G., Watts, D., Russell, K., and Barrett, A. (2003). Enzootic transmission of yellow fever virus in Peru. Emerg. Infect. Dis. *9*, 926–933.

Bryant, J.E., Holmes, E.C., and Barrett, A.D. (2007). Out of Africa: a molecular perspective on the introduction of yellow fever virus into the Americas. PLoS Pathog. *3*, e75.

Bunning, M.L., Bowen, R.A., Cropp, C.B., Sullivan, K.G., Davis, B.S., Komar, N., Godsey, M.S., Baker, D., Hettler, D.L., Holmes, D.A., *et al.* (2002). Experimental infection of horses with West Nile virus. Emerg. Infect. Dis. *8*, 380–386.

Burch, C.L., and Chao, L. (2000). Evolvability of an RNA virus is determined by its mutational neighbourhood. Nature *406*, 625–628.

Burke, D.S., and Kliks, S. (2006). Antibody-dependent enhancement in dengue virus infections. J. Infect. Dis. *193*, 601–603.

Calisher, C.H. and Gresikova, M. (1989). Tick-borne encephalitis. In The Arboviruses: Epidemiology and Ecology, Monath, T.P., ed. (CRC Press, Boca Raton), pp. 177–202.

Carrillo, C., Borca, M., Moore, D.M., Morgan, D.O., and Sobrino, F. (1998). *In vivo* analysis of the stability and fitness of variants recovered from foot-and-mouth disease virus quasispecies. J. Gen. Virol. *79 (Pt 7)*, 1699–1706.

Chao, D.Y., King, C.C., Wang, W.K., Chen, W.J., Wu, H.L., and Chang, G.J. (2005). Strategically examining the full-genome of dengue virus type 3 in clinical isolates reveals its mutation spectra. Virol. J. *2*, 72.

Charrel, R.N., Attoui, H., Butenko, A.M., Clegg, J.C., Deubel, V., Frolova, T.V., Gould, E.A., Gritsun, T.S., Heinz, F.X., Labuda, M., *et al.* (2004). Tick-borne virus diseases of human interest in Europe. Clin. Microbiol. Infect. *10*, 1040–1055.

Chen, W.J., Wu, H.R., and Chiou, S.S. (2003). E/NS1 modifications of dengue 2 virus after serial passages in mammalian and/or mosquito cells. Intervirology *46*, 289–295.

Chernesky, M.A. (1969). Powassan virus transmission by ixodid ticks infected after feeding on viremic rabbits injected intravenously. Can. J. Microbiol. *15*, 521–526.

Chunikhin, S.P. and Kurenkov, V.B. (1979). Viraemia in Clethrionomys glareolus – a new ecological marker of tick-borne encephalitis virus. Acta Virol. *23*, 257–260.

Ciota, A.T., Lovelace, A.O., Jones, S.A., Payne, A., and Kramer, L.D. (2007a). Adaptation of two flaviviruses results in differences in genetic heterogeneity and virus adaptability. J. Gen. Virol. *88*, 2398–2406.

Ciota, A.T., Lovelace, A.O., Ngo, K.A., Le, A.N., Maffei, J.G., Franke, M.A., Payne, A.F., Jones, S.A., Kauffman, E.B., and Kramer, L.D. (2007b). Cell-specific adaptation of two flaviviruses following serial passage in mosquito cell culture. Virology 357, 165–174.

Ciota, A.T., Ngo, K.A., Lovelace, A.O., Payne, A.F., Zhou, Y., Shi, P.-Y., and Kramer, L.D. (2007c). Role of the mutant spectrum in adaptation and replication of West Nile virus. J. Gen. Virol. *88*, 865–874.

Ciota, A.T., Lovelace, A.O., Jia, Y., Davis, L.J., Young, D.S., and Kramer, L.D. (2008). Characterization of mosquito-adapted West Nile virus. J. Gen. Virol. *89*, 1633–1642.

Ciota, A.T., Jia, Y., Payne, A.F., Jerzak, G., Davis, L.J., Young, D.S., Ehrbar, D., and Kramer, L.D. (2009). Experimental passage of St. Louis encephalitis virus *in vivo* in mosquitoes and chickens reveals evolutionarily significant virus characteristics. PLoS. One. *4*, e7876.

Coffey, L.L., Vasilakis, N., Brault, A.C., Powers, A.M., Tripet, F., and Weaver, S.C. (2008). Arbovirus evolution *in vivo* is constrained by host alternation. Proc. Natl. Acad. Sci. U.S.A. *105*, 6970–6975.

Cooper, L.A. and Scott, T.W. (2001). Differential evolution of eastern equine encephalitis virus populations in response to host cell type. Genetics *157*, 1403–1412.

Costero, A., and Grayson, M.A. (1996). Experimental transmission of Powassan virus (Flaviviridae) by *Ixodes scapularis* ticks (Acari:Ixodidae). Am. J. Trop. Med. Hyg. 55, 536–546.

Costero, A., and Grayson, M.A. (1996b). Experimental transmission of Powassan virus (Flaviviridae) by *Ixodes scapularis* ticks (Acari:Ixodidae). Am. J. Trop. Med. Hyg. 55, 536–546.

Davis, A., Bunning, M., Gordy, P., Panella, N., Blitvich, B., and Bowen, R. (2005a). Experimental and natural infection of North American bats with West Nile virus. Am. J. Trop. Med. Hyg. 73, 467–469.

Davis, C.T., Beasley, D.W., Guzman, H., Raj, R., D'Anton, M., Novak, R.J., Unnasch, T.R., Tesh, R.B., and Barrett, A.D. (2003). Genetic variation among temporally and geographically distinct West Nile virus isolates, United States, 2001, 2002. Emerg. Infect. Dis. *9*, 1423–1429.

Davis, C.T., Beasley, D.W., Guzman, H., Siirin, M., Parsons, R.E., Tesh, R.B., and Barrett, A.D. (2004). Emergence of attenuated West Nile virus variants in Texas, 2003. Virology *330*, 342–350.

Davis, C.T., Ebel, G.D., Lanciotti, R.S., Brault, A.C., Guzman, H., Siirin, M., Lambert, A., Parsons, R.E., Beasley, D.W., Novak, R.J., *et al.* (2005). Phylogenetic analysis of North American West Nile virus isolates, 2001–2004: evidence for the emergence of a dominant genotype. Virology *342*, 252–265.

Deubel, V., Digoutte, J.P., Monath, T.P., and Girard, M. (1986). Genetic heterogeneity of yellow fever virus strains from Africa and the Americas. J. Gen. Virol. *67 (Pt 1)*, 209–213.

Domingo, E. and Holland, J.J. (1994). Mutation rates and rapid evolution of RNA viruses. In Evolutionary Biology of Viruses, Morse, S.S., ed. (Raven Press, New York), pp. 161–184.

Domingo, E., and Holland, J.J. (1997). RNA virus mutations and fitness for survival. Annu. Rev. Microbiol. *51*, 151–178.

Domingo, E., Escarmis, C., Sevilla, N., Moya, A., Elena, S.F., Quer, J., Novella, I.S., and Holland, J.J. (1996). Basic concepts in RNA virus evolution. FASEB J. *10*, 859–864.

Domingo, E., Escarmis, C., Sevilla, N., and Baranowski, E. (1998). Population dynamics in the evolution of RNA viruses. Adv. Exp. Med. Biol. *440*, 721–727.

Drake, J.W., and Holland, J.J. (1999). Mutation rates among RNA viruses. Proc. Natl. Acad. Sci. U.S.A. *96*, 13910–13913.

Duarte, E., Clarke, D., Moya, A., Domingo, E., and Holland, J. (1992). Rapid fitness losses in mammalian RNA virus clones due to Mueller's ratchet. Proc. Natl. Acad. Sci. U.S.A. *89*, 6015–6019.

Duarte, E.A., Novella, I.S., Ledesma, S., Clarke, D.K., Moya, A., Elena, S.F., Domingo, E., and Holland, J.J. (1994). Subclonal components of consensus fitness in an RNA virus clone. J. Virol. *68*, 4295–4301.

Dunham, E.J., and Holmes, E.C. (2007). Inferring the timescale of dengue virus evolution under realistic models of DNA substitution. J. Mol. Evol. *64*, 656–661.

Ebel, G.D. (2010). Update on Powassan virus: emergence of a North American tick-borne flavivirus. Annu. Rev. Entomol. 55, 95–110.

Ebel, G.D., and Kramer, L.D. (2004). Short report: duration of tick attachment required for transmission of powassan virus by deer ticks. Am. J. Trop. Med. Hyg. *71*, 268–271.

Ebel, G.D., and Kramer, L.D. (2009). West Nile virus: molecular epidemiology and diversity. In West Nile encephalitis virus infection: Viral pathogenesis and the host immune response, Diamond, M., ed. (New York: Springer), pp. 25–43.

Ebel, G.D., Dupuis, A.P., Ngo, K., Nicholas, D., Kauffman, E.B., Jones, S.A., Young, D., Maffei, J., Shi, P.Y., Bernard, K.A., *et al.* (2001a). Partial genetic characterization of WNV strains isolated in New York State during the 2000 transmission season. Emerg. Infect. Dis. *7*, 650–653.

Ebel, G.D., Spielman, A., and Telford, S.R., III (2001b). Phylogeny of North American Powassan virus. J Gen. Virol. *82*, 1657–1665.

Ebel, G.D., Carricaburu, J., Young, D., Bernard, K.A., and Kramer, L.D. (2004). Genetic and phenotypic variation of West Nile virus in New York, 2000–2003. Am. J. Trop. Med. Hyg. *71*, 493–500.

Eigen, M. (1993). Viral quasispecies. Sci. Am. *269*, 42–49.

Eigen, M., and Biebricher, D.K. (1988). Sequence space and quasispecies distribution. In RNA Genetics, Domingo, E., Holland, J.J., and Ahlquist, P., eds. (CRC Press, Boca Raton, FL), pp. 211–245.

Farci, P., Shimoda, A., Coiana, A., Diaz, G., Peddis, G., Melpolder, J.C., Strazzera, A., Chien, D.Y., Munoz, S.J., Balestrieri, A., *et al.* (2000). The outcome of acute hepatitis C predicted by the evolution of the viral quasispecies. Science *288*, 339–344.

Farci, P., Strazzera, R., Alter, H.J., Farci, S., Degioannis, D., Coiana, A., Peddis, G., Usai, F., Serra, G., Chessa, L., *et al.* (2002). Early changes in hepatitis C viral quasispecies during interferon therapy predict the therapeutic outcome. Proc. Natl. Acad. Sci. U.S.A. *99*, 3081–3086.

Fitzpatrick, K.A., Deardorff, E., Pesko, K., Brackney, D., Zhang, B., Bedrick, E., Shi, P.Y., and Ebel, G.D. (2010). Population variation of West Nile virus confers a host-specific fitness benefit in mosquitoes. Virology *404*, 89–95.

Forster, R., Adami, C., and Wilke, C.O. (2006). Selection for mutational robustness in finite populations. J. Theor. Biol. *243*, 181–190.

Frost, S.D.W., Wrin, T., Smith, D.M., Pond, S.L.K., Liu, Y., Paxinos, E., Chappey, C., Galovich, J., Beauchaine, J., Petropoulos, C.J., *et al.* (2005). Neutralizing antibody responses drive the evolution of human immunodeficiency virus type 1 envelope during recent HIV infection. Proc. Natl. Acad. Sci. U.S.A. *102*, 18514–18519.

Gaunt, M.W., Sall, A.A., de Lamballerie, X., Falconar, A.K., Dzhivanian, T.I., and Gould, E.A. (2001). Phylogenetic relationships of flaviviruses correlate with their epidemiology, disease association and biogeography. J. Gen. Virol. *82*, 1867–1876.

Girard, Y.A., Mayhew, G.F., Fuchs, J.F., Li, H., Schneider, B.S., McGee, C.E., Rocheleau, T.A., Helmy, H., Christensen, B.M., Higgs, S., *et al.* (2010). Transcriptome changes in Culex quinquefasciatus (Diptera: Culicidae) salivary glands during West Nile virus infection. J. Med. Entomol. *47*, 421–435.

Goddard, L.B., Roth, A.E., Reisen, W.K., and Scott, T.W. (2002). Vector competence of California mosquitoes for West Nile virus. Emerg. Infect. Dis. *8*, 1385–1391.

Goddard, L.B., Roth, A.E., Reisen, W.K., and Scott, T.W. (2003). Vertical transmission of West Nile Virus by three California Culex (Diptera: Culicidae) species. J. Med. Entomol. *40*, 743–746.

Grard, G., Moureau, G., Charrel, R.N., Lemasson, J.J., Gonzalez, J.P., Gallian, P., Gritsun, T.S., Holmes, E.C., Gould, E.A., and de Lamballerie, X. (2007). Genetic characterization of tick-borne flaviviruses: new insights into evolution, pathogenetic determinants and taxonomy. Virology *361*, 80–92.

Greene, I.P., Wang, E., Deardorff, E.R., Milleron, R., Domingo, E., and Weaver, S.C. (2005). Effect of alternating passage on adaptation of sindbis virus to vertebrate and invertebrate cells. J. Virol. *79*, 14253–14260.

Gubler, D.J. (1998). Dengue and dengue hemorrhagic fever. Clin. Microbiol. Rev. *11*, 480–496.

Hajnicka, V., Fuchsberger, N., Slovak, M., Kocakova, P., Labuda, M., and Nuttall, P.A. (1998). Tick salivary gland extracts promote virus growth *in vitro*. Parasitology *116 (Pt 6)*, 533–538.

Hajnicka, V., Vancova, I., Kocakova, P., Slovak, M., Gasperik, J., Slavikova, M., Hails, R.S., Labuda, M., and Nuttall, P.A. (2005). Manipulation of host cytokine network by ticks: a potential gateway for pathogen transmission. Parasitology *130*, 333–342.

Halstead, S.B., Suaya, J.A., and Shepard, D.S. (2007). The burden of dengue infection. Lancet *369*, 1410–1411.

Harrington, L.C., Edman, J.D., and Scott, T.W. (2001). Why do female *Aedes aegypti* (Diptera: Culicidae) feed preferentially and frequently on human blood? J. Med. Entomol. *38*, 411–422.

Hartemink, N.A., Randolph, S.E., Davis, S.A., and Heesterbeek, J.A. (2008). The basic reproduction number for complex disease systems: defining R(0) for tick-borne infections. Am. Nat. *171*, 743–754.

Herrera, M., Garcia-Arriaza, J., Pariente, N., Escarmis, C., and Domingo, E. (2007). Molecular basis for a lack of correlation between viral fitness and cell killing capacity. PLoS Pathog. *3*, e53.

Higgs, S., Schneider, B.S., Vanlandingham, D.L., Klingler, K.A., and Gould, E.A. (2005). Nonviremic transmission of West Nile virus. Proc. Natl. Acad. Sci. U.S.A. *102*, 8871–8874.

Holmes, E.C. (2003). Patterns of intra- and interhost nonsynonymous variation reveal strong purifying selection in dengue virus. J. Virol. *77*, 11296–11298.

Holmes, E.C. and Moya, A. (2002). Is the quasispecies concept relevant to RNA viruses? J. Virol. *76*, 460–465.

Honório, N.A., Nogueira, R.M., Codeço, C.T., Carvalho, M.S., Cruz, O.G., Magalhães Mde, A., de Araújo, J.M., de Araújo, E.S., Gomes, M.Q., Pinheiro, L.S., *et al.* (2009). Spatial evaluation and modeling of dengue seroprevalence and vector density in Rio de Janeiro, Brazil. PLoS Negl. Trop. Dis. *3*, e545.

Houk, E.J., Kramer, L.D., Hardy, J.L., and Presser, S.B. (1986). An interspecific mosquito model for the mesenteronal infection barrier to western equine encephalomyelitis virus (*Culex tarsalis* and *Culex pipiens*). Am. J. Trop. Med. Hyg *35*, 632–641.

Jerzak, G., Bernard, K.A., Kramer, L.D., and Ebel, G.D. (2005). Genetic variation in West Nile virus from naturally infected mosquitoes and birds suggests quasispecies structure and strong purifying selection. J Gen. Virol. *86*, 2175–2183.

Jerzak, G.V., Brown, I., Shi, P.Y., Kramer, L.D., and Ebel, G.D. (2008). Genetic diversity and purifying selection in West Nile virus populations are maintained during host switching. Virology *374*, 256–260.

Jerzak, G.V.S., Bernard, K., Kramer, L.D., Shi, P.Y., and Ebel, G.D. (2007). The West Nile virus-mutant spectrum is host-dependant and a determinant of mortality in mice. Virology *360*, 469–476.

Khasnatinov, M.A., Ustanikova, K., Frolova, T.V., Pogodina, V.V., Bochkova, N.G., Levina, L.S., Slovak, M., Kazimirova, M., Labuda, M., Klempa, B., *et al.* (2009). Non-hemagglutinating flaviviruses: molecular mechanisms for the emergence of new strains via adaptation to European ticks. PLoS One *4*, e7295.

Kokernot, R.H., Radivojevic, B., and Anderson, R.J. (1969). Susceptibility of wild and domesticated mammals to four arboviruses. Am. J. Vet. Res. *30*, 2197–2203.

Komar, N. (2000). West Nile Viral Encephalitis. Rev. Sci. Tech. *19*, 166–176.

Kozuch, O., Chunikhin, S.P., Gresikova, M., Nosek, J., Kurenkov, V.B., and Lysy, J. (1981). Experimental characteristics of viraemia caused by two strains of tick-borne encephalitis virus in small rodents. Acta Virol. *25*, 219–224.

Kramer, L.D., and Ebel, G.D. (2003). Dynamics of flavivirus infection in mosquitoes. Adv. Virus Res. *60*, 187–232.

Kramer, L.D., Hardy, J.L., Presser, S.B., and Houk, E.J. (1981). Dissemination barriers for western equine encephalomyelitis virus in *Culex tarsalis* infected after ingestion of low viral doses. Am. J. Trop. Med. Hyg. *30*, 190–197.

Kramer, L.D., Styer, L.M., and Ebel, G.D. (2008). A global perspective on the epidemiology of West Nile virus. Annu. Rev. Entomol. *53*, 61–81.

Kroeger, A., and Nathan, M.B. (2006). Dengue: setting the global research agenda. Lancet *368*, 2193–2195.

Kuno, G., Chang, G.J., Tsuchiya, K.R., Karabatsos, N., and Cropp, C.B. (1998). Phylogeny of the genus *Flavivirus*. J. Virol. *72*, 73–83.

Kyle, J.L., and Harris, E. (2008). Global spread and persistence of dengue. Annu. Rev. Microbiol. *62*, 71–92.

Labuda, M., and Nuttall, P.A. (2004). Tick-borne viruses. Parasitology *129* (*Suppl.*), S221–S245.

Labuda, M., Austyn, J.M., Zuffova, E., Kozuch, O., Fuchsberger, N., Lysy, J., and Nuttall, P.A. (1996). Importance of localized skin infection in tick-borne encephalitis virus transmission. Virology *219*, 357–366.

Labuda, M., Danielova, V., Jones, L.D., and Nuttall, P.A. (1993a). Amplification of tick-borne encephalitis virus infection during co-feeding of ticks. Med. Vet. Entomol. *7*, 339–342.

Labuda, M., Jones, L.D., Williams, T., Danielova, V., and Nuttall, P.A. (1993b). Efficient transmission of tick-borne encephalitis virus between cofeeding ticks. J. Med. Entomol. *30*, 295–299.

Labuda, M., Jones, L.D., Williams, T., and Nuttall, P.A. (1993c). Enhancement of tick-borne encephalitis virus transmission by tick salivary gland extracts. Med. Vet. Entomol. *7*, 193–196.

Labuda, M., Nuttall, P.A., Kozuch, O., Eleckova, E., Williams, T., Zuffova, E., and Sabo, A. (1993d). Non-viraemic transmission of tick-borne encephalitis virus: a mechanism for arbovirus survival in nature. Experientia *49*, 802–805.

Lanciotti, R.S., Roehrig, J.T., Deubel, V., Smith, J., Parker, M., Steele, K., Crise, B., Volpe, K.E., Crabtree, M.B., Scherret, J.H., *et al.* (1999). Origin of the West Nile virus responsible for an outbreak of encephalitis in the northeastern United States. Science *286*, 2333–2337.

LeClerc, J.E., Li, B., Payne, W.L., and Cebula, T.A. (1996). High mutation frequencies among *Escherichia coli* and *Salmonella* pathogens. Science *274*, 1208–1211.

Lenhoff, R.J., Luscombe, C.A., and Summers, J. (1998). Competition *in vivo* between a cytopathic variant and a wild-type duck hepatitis B virus. Virology *251*, 85–95.

Leonova, G.N., Kondratov, I.G., Ternovoi, V.A., Romanova, E.V., Protopopova, E.V., Chausov, E.V., Pavlenko, E.V., Ryabchikova, E.I., Belikov, S.I., and Loktev, V.B. (2009). Characterization of Powassan viruses from Far Eastern Russia. Arch. Virol. *154*, 811–820.

Lin, S.R., Hsieh, S.C., Yueh, Y.Y., Lin, T.H., Chao, D.Y., Chen, W.J., King, C.C., and Wang, W.K. (2004). Study of sequence variation of dengue type 3 virus in naturally infected mosquitoes and human hosts: implications for transmission and evolution. J. Virol. *78*, 12717–12721.

MacDonald, G. (1961). Epidemiologic models in the studies of vector-borne diseases. Public Health Rep. *76*, 753–764.

McGee, C.E., Schneider, B.S., Girard, Y.A., Vanlandingham, D.L., and Higgs, S. (2007). Nonviremic transmission of West Nile virus: evaluation of the effects of space, time, and mosquito species. Am. J Trop. Med Hyg. *76*, 424–430.

MacKenzie, J.S., Poidinger, M., Lindsay, M.D., Hall, R.A., and Sammels, L.M. (1995). Molecular epidemiology and evolution of mosquito-borne flaviviruses and alphaviruses enzootic in Australia. Virus Genes *11*, 225–237.

Martinez, M.A., Carrillo, C., Gonzalez-Candelas, F., Moya, A., Domingo, E., and Sobrino, F. (1991). Fitness alteration of foot-and-mouth disease virus mutants: measurement of adaptability of viral quasispecies. J. Virol. *65*, 3954–3957.

Mayr, E. (1994). Driving forces in evolution: an analysis of natural selection. In The Evolutionary Biology of Viruses, Morse, S.S., ed. (Raven Press, Ltd, New York), pp. 29–48.

Miller, B.R., Monath, T.P., Tabachnick, W.J., and Ezike, V.I. (1989). Epidemic yellow fever caused by an incompetent mosquito vector. Trop. Med Parasitol. *40*, 396–399.

Miller, B.R., Nasci, R.S., Godsey, M.S., Savage, H.M., Lutwama, J.J., Lanciotti, R.S., and Peters, C.J. (2000). First field evidence for natural vertical transmission of West Nile virus in Culex univittatus complex mosquitoes from Rift Valley province, Kenya. Am. J. Trop. Med. Hyg. *62*, 240–246.

Montville, R., Froissart, R., Remold, S.K., Tenaillon, O., and Turner, P.E. (2005). Evolution of mutational robustness in an RNA virus. PLoS Biol. *3*, e381.

Moudy, R.M., Meola, M.A., Morin, L.L., Ebel, G.D., and Kramer, L.D. (2007). A newly emergent genotype of West Nile virus is transmitted earlier and more efficiently by Culex mosquitoes. Am. J. Trop. Med. Hyg. *77*, 365–370.

Mukhopadhyay, S., Kim, B.S., Chipman, P.R., Rossmann, M.G., and Kuhn, R.J. (2003). Structure of West Nile virus. Science *302*, 248.

Muller, H.J. (1964). The relation of recombination to mutational advance. Mutat. Res. *106*, 2–9.

Murphy, F.A. (1994). New, emerging, and reemerging infectious diseases. Adv. Virus Res. *43*, 1–52.

Myles, K.M., Wiley, M.R., Morazzani, E.M., and Adelman, Z.N. (2008). Alphavirus-derived small RNAs modulate pathogenesis in disease vector mosquitoes. Proc. Natl. Acad. Sci. *105*, 19938–19943.

Nonaka, E., Ebel, G.D., and Wearing H.J. (2010). Persistence of pathogens with short infectious periods in seasonal tick populations: The relative importance of three transmission routes. PLoS. One. *5*, e11745.

Novella, I.S. and Ebendick-Corp (2004). Molecular basis of fitness loss and fitness recovery in vesicular stomatitis virus. J. Mol. Biol. *342*, 1423–1430.

Novella, I.S., Hershey, C.L., Escarmis, C., Domingo, E., and Holland, J.J. (1999). Lack of evolutionary stasis during alternating replication of an arbovirus in insect and mammalian cells. J. Mol. Biol. *287*, 459–465.

Novella, I.S., Ebendick-Corp, Zarate, S., and Miller, E.L. (2007). Emergence of mammalian-adapted vesicular stomatitis virus from persistent infections of insect-vector cells. J. Virol. JVI.

Nuttall, P.A. and Labuda, M. (2003). Dynamics of infection in tick vectors and at the tick–host interface. Adv. Virus Res. *60*, 233–272.

Ogden, N.H., Bigras-Poulin, M., O'callaghan, C.J., Barker, I.K., Kurtenbach, K., Lindsay, L.R., and Charron, D.F. (2007). Vector seasonality, host infection dynamics and fitness of pathogens transmitted by the tick Ixodes scapularis. Parasitology *134*, 209–227.

Pugachev, K.V., Guirakhoo, F., Ocran, S.W., Mitchell, F., Parsons, M., Penal, C., Girakhoo, S., Pougatcheva, S.O., Arroyo, J., Trent, D.W., *et al.* (2004). High fidelity of Yellow Fever Virus RNA polymerase. J. Virol. *78*, 1032–1038.

Reisen, W.K. (2003). Epidemiology of St. Louis encephalitis virus. Adv. Virus Res. *61*, 139–183.

Rice, C.M. (1996). Flaviviridae: the viruses and their replication. In: Fields Virology, Fields, B.N., Knipe, D.M., and Howley, P.M., eds. (Lippincott-Raven, Philadelphia), 931–960.

Rico-Hesse, R., and Mota, J. (2010). Authors reply to Vasilakis, N., Cardosa, J., Diallo, M., Sall A.A., Holmes, E.C., Hanley, K.A., and Weaver, S.C. (2010). Sylvatic dengue viruses share the pathogenic potential of urban/endemic dengue viruses. J. Virol. *84*, 3726–3728. Letter to the Editor.

Rico-Hesse, R., Harrison, L.M., Nisalak, A., Vaughn, D.W., Kalayanarooj, S., Green, S., Rothman, A.L., and Ennis, F.A. (1998). Molecular evolution of dengue type 2 virus in Thailand. Am. J. Trop. Med. Hyg. *58*, 96–101.

Ruiz-Jarabo, C.M., Arias, A., Molina-Paris, C., Briones, C., Baranowski, E., Escarmis, C., and Domingo, E. (2002). Duration and fitness dependence of quasispecies memory. J. Mol. Biol. *315*, 285–296.

Sall, A.A., Faye, O., Diallo, M., Firth, C., Kitchen, A., and Holmes, E.C. (2010). Yellow fever virus exhibits slower evolutionary dynamics than dengue virus. J. Virol. *84*, 765–772.

Sanchez-Vargas, I., Travanty, E.A., Keene, K.M., Franz, A.W.E., Beaty, B.J., Blair, C.D., and Olson, K.E. (2004). RNA interference, arthropod-borne viruses, and mosquitoes. Virus Res. *102*, 65–74.

Sauder, C.J., Vandenburgh, K.M., Iskow, R.C., Malik, T., Carbone, K.M., and Rubin, S.A. (2006). Changes in mumps virus neurovirulence phenotype associated with quasispecies heterogeneity. Virology *350*, 48–57.

Schneider, W.L., and Roossinck, M.J. (2000). Evolutionarily related Sindbis-like plant viruses maintain different levels of population diversity in a common host. J. Virol. *74*, 3130–3134.

Schneider, W.L., and Roossinck, M.J. (2001). Genetic diversity in RNA virus quasispecies is controlled by host–virus interactions. J. Virol. *75*, 6566–6571.

Schoeler, G.B., Bergman, D.K., Manweiler, S.A., and Wikel, S.K. (2000). Influence of soluble proteins from the salivary glands of ixodid ticks on the in-vitro proliferative responses of lymphocytes from BALB/c and C3H/HeN mice. Ann. Trop. Med. Parasitol. *94*, 507–518.

Scott, T.W., Weaver, S.C., and Mallampalli, V.L. (1994). Evolution of mosquito-borne viruses. In The Evolutionary Biology of Viruses, Morse, S.S., ed. (Raven Press, Ltd, New York), pp. 293–324.

Snapinn, K.W., Holmes, E.C., Young, D.S., Bernard, K.A., Kramer, L.D., and Ebel, G.D. (2006). Declining growth rate of West Nile Virus in North America. J. Virol. *81*, 2531–2534.

Sonenshine, D. (1993). Biology of Ticks. (Oxford University Press, New York).

Strauss, J.H. (1993). Recombination in the evolution of RNA viruses. In Emerging Viruses, Morse, S.S., ed. (Oxford Univ. Press, Oxford), pp. 241–251.

Strauss, E.G., Strauss, J.H., and Levine, A.J. (1990). Virus evolution. In Virology, Fields B.N., and Knipe, D.M., eds. (Raven Press, New York), pp. 167–190.

Strickman, D., and Kittayapong, P. (2002). Dengue and its vectors in Thailand: introduction to the study and seasonal distribution of *Aedes* larvae. Am. J. Trop. Med. Hyg. *67*, 247–259.

Strickman, D., and Kittayapong, P. (2003). Dengue and its vectors in Thailand: calculated transmission risk from total pupal counts of *Aedes aegypti* and association of wing-length measurements with aspects of the larval habitat. Am. J. Trop. Med. Hyg. *68*, 209–217.

Teehee, M.L., Bunning, M.L., Stevens, S., and Bowen, R.A. (2005). Experimental infection of pigs with West Nile virus. Arch. Virol. *150*, 1249–1256.

Telford, S.R., III, Armstrong, P.M., Katavolos, P., Foppa, I., Garcia, A.S., Wilson, M.L., and Spielman, A. (1997).

A new tick-borne encephalitis-like virus infecting New England deer ticks, Ixodes dammini. Emerg. Infect. Dis. *3*, 165–170.

Timoney, P. (1971). Powassan virus infection in the grey squirrel. Acta Virol. *15*, 429.

de la Torre, J.C., and Holland, J.J. (1990a). RNA virus quasispecies populations can suppress vastly superior mutant progeny. J. Virol. *64*, 6278–6281.

Twiddy, S.S., Woelk, C.H., and Holmes, E.C. (2002). Phylogenetic evidence for adaptive evolution of dengue viruses in nature. J. Gen. Virol. *83*, 1679–1689.

Twiddy, S.S., Holmes, E.C., and Rambaut, A. (2003). Inferring the rate and time-scale of dengue virus evolution. Mol. Biol. Evol. *20*, 122–129.

Vasconcelos, P.F., Rodrigues, S.G., Degallier, N., Moraes, M.A., da Rosa, J.F., da Rosa, E.S., Mondet, B., Barros, V.L., and da Rosa, A.P. (1997). An epidemic of sylvatic yellow fever in the southeast region of Maranhao State, Brazil, 1993–1994: epidemiologic and entomologic findings. Am. J. Trop. Med. Hyg. *57*, 132–137.

Vasconcelos, P.F., Bryant, J.E., da Rosa, T.P., Tesh, R.B., Rodrigues, S.G., and Barrett, A.D. (2004). Genetic divergence and dispersal of yellow fever virus, Brazil. Emerg. Infect. Dis. *10*, 1578–1584.

Vasilakis, N., and Weaver, S.C. (2008). The history and evolution of human dengue emergence. Adv. Virus Res. *72*, 1–76.

Vasilakis, N., Cardosa, J., Diallo, M., Sall A.A., Holmes, E.C., Hanley, K.A., and Weaver, S.C. (2010). Sylvatic dengue viruses share the pathogenic potential of urban/endemic dengue viruses. J. Virol. *84*, 3726–3728. Letter to the Editor.

Vignuzzi, M., Stone, J.K., Arnold, J.J., Cameron, C.E., and Andino, R. (2006). Quasispecies diversity determines pathogenesis through cooperative interactions in a viral population. Nature *439*, 344–348.

Wang, H., and Nuttall, P.A. (1995). Immunoglobulin G binding proteins in male *Rhipicephalus appendiculatus* ticks. Parasite Immunol. *17*, 517–524.

Wang, W.K., Lin, S.R., Lee, C.M., King, C.C., and Chang, S.C. (2002). Dengue type 3 virus in plasma is a population of closely related genomes: quasispecies. J. Virol. *76*, 4662–4665.

Weaver, S.C. (2006). Evolutionary influences in arboviral disease. Curr. Top. Microbiol. Immunol. *299*, 285–314.

Weaver, S.C., and Vasilakis, N. (2009). Molecular evolution of dengue viruses: contributions of phylogenetics to understanding the history and epidemiology of the preeminent arboviral disease. Infect. Genet. Evol. *9*, 523–540.

Weaver, S.C., Rico-Hesse, R., and Scott, T.W. (1992). Genetic diversity and slow rates of evolution in New World alphaviruses. Curr. Top. Microbiol. Immunol. *176*, 99–117.

Weaver, S.C., Hagenbaugh, A., Bellew, L.A., Gousset, L., Mallampalli, V., Holland, J.J., and Scott, T.W. (1994). Evolution of alphaviruses in the eastern equine encephalomyelitis complex. J. Virol. *68*, 158–169.

Weaver, S.C., Brault, A.C., Kang, W., and Holland, J.J. (1999). Genetic and fitness changes accompanying adaptation of an arbovirus to vertebrate and invertebrate cells. J. Virol. *73*, 4316–4326.

Weiner, A., Erickson, A.L., Kansopon, J., Crawford, K., Muchmore, E., Hughes, A.L., Houghton, M., and Walker, C.M. (1995). Persistent hepatitis C virus Infection in a chimpanzee is associated with emergence of a cytotoxic T lymphocyte escape variant. Proc. Natl. Acad. Sci. U.S.A. *92*, 2755–2759.

Wilke, C.O. (2005). Quasispecies theory in the context of population genetics. BMC Evol. Biol. *5*, 44.

Wilke, C.O., Wang, J.L., Ofria, C., Lenski, R.E., and Adami, C. (2001). Evolution of digital organisms at high mutation rates leads to survival of the flattest. Nature *412*, 331–333.

Woelk, C.H., and Holmes, E.C. (2002). Reduced positive selection in vector-borne RNA viruses. Mol. Biol. Evol. *19*, 2333–2336.

Zanotto, P.M., Gao, G.F., Gritsun, T., Marin, M.S., Jiang, W.R., Venugopal, K., Reid, H.W., and Gould, E.A. (1995). An arbovirus cline across the northern hemisphere. Virology *210*, 152–159.

Zanotto, P.M., Gould, E.A., Gao, G.F., Harvey, P.H., and Holmes, E.C. (1996). Population dynamics of flaviviruses revealed by molecular phylogenies. Proc. Natl. Acad. Sci. U.S.A. *93*, 548–553.

Zarate, S., and Novella, I.S. (2004). Vesicular stomatitis virus evolution during alternation between persistent infection in insect cells and acute infection in mammalian cells is dominated by the persistence phase. J. Virol. *78*, 12236–12242.

Zarnke, R.L., and Yuill, T.M. (1981). Powassan virus infection in snowshoe hares (*LepuSouth Americanus*). J. Wildl. Dis. *17*, 303–310.

Flavivirus Vaccines

Scott B. Halstead

10

Abstract

Eight flaviviruses cause significant morbidity and mortality around the globe: yellow fever (YF), Japanese encephalitis (JE), tick-borne encephalitis (TBE), dengue 1, 2, 3 and 4 and West Nile (WN). Four, YF, JE, TBE and WN, are zoonoses, with the consequence that vaccines are the only means of protecting humans. The successful YF 17D vaccine, introduced in 1937, produced dramatic reductions in epidemic activity. Effective killed JE and TBE vaccines were introduced in the middle of the 20th century. Unacceptable adverse events have prompted change from a mouse-brain killed JE vaccine to safer and more effective second-generation JE vaccines. These may come into wide use to effectively prevent this severe disease in the huge populations of Asia – North, South and South-East. The dengue viruses produce many millions of infections annually due to transmission by a successful global mosquito vector. As mosquito control has failed, several dengue vaccines are in varying stages of development. A tetravalent chimeric vaccine that splices structural genes of the four dengue viruses onto a 17D YF backbone is in phase III clinical testing. For each of the eight flaviviruses, clinical disease, epidemiology, vaccine development history, vaccine usage, precautions and adverse events are briefly presented.

Introduction

The flaviviruses are a family of more than 70 single-stranded RNA viruses, the majority of which are transmitted to their hosts by arthropod vectors such as mosquitoes or ticks. Flavivirions are spherical and composed of a lipid membrane surrounding a capsid (C) protein bound to a viral RNA (positive-sense), which together form the nucleocapsid core. The surface proteins are the glycosylated envelope (E) and membrane (M) proteins. The M protein is a mature form of the pre-membrane (prM) protein and is important for infectivity and pathogenicity. The E protein is responsible for viral attachment to cellular receptors and specific membrane fusion. The E protein is the major target for virus-neutralizing and haemagglutination-inhibiting antibodies.

Flaviviruses are responsible for much of the global morbidity and mortality caused by arthropod-borne viruses. All are zoonoses. Overt epidemics are caused by four dengue (DENV), Ilheus, Japanese encephalitis (JEV), Kyasanur Forest disease (KFDV), Murray Valley encephalitis (MVEV), Omsk haemorrhagic fever, St. Louis encephalitis (SLEV), tick-borne encephalitis (TBEV) – European and Russian subtypes, West Nile virus (WNV) and yellow fever virus (YFV). Other than early administration of specific antibodies, there is as yet no antiviral treatment for flavivirus infections. Arthropod vector control has proven difficult in the case of dengue and yellow fever and is impossible for viruses that are predominantly zoonotic. Further, no flavivirus can be eradicated given the complexity of their circulation in non-human reservoirs. Hence, vaccination is the most effective approach to disease control. Importantly, for this chapter, flavivirus vaccines are among a small family of effective vaccines. Licensed vaccines protect against YF, JE, and

TBE, second-generation vaccines have just been introduced for JE and a very large number of candidate vaccines against DEN and WN are in late stage development. Unfortunately, even with licensed vaccines affordable and available, the global burden of flavivirus disease continues to rise for one or more of the following reasons: incomplete immunization or ineffective vector control programmes, increasing populations of humans and vector arthropods or increasing business and recreational travel. To explain the requirement for vaccines and the attributes of effective immunization programmes special emphasis is given in this section to the epidemiology of vaccine-eligible flaviviruses.

Licensed vaccines

Yellow fever

Yellow fever (YF) was the first flavivirus isolated in the laboratory and is the prototype of the genera. The virus is enzootic in two cycles involving a number of African subhuman primates and one cycle in the American tropics. West African YFV has escaped from its zoonotic reservoir on numerous occasions into an urban transmission cycle involving the domesticated African variant of *Aedes aegypti*. This YFV was imported to the American hemisphere in the 1600s, where large urban outbreaks took an enormous toll; at the same time the virus escaped into a new sylvatic cycle. YFV was identified by Walter Reed in Cuba in 1900 and was attenuated in 1937 by workers at the Rockefeller Foundation, New York. YF vaccine is manufactured by developed as well as developing world manufacturers. Its use has substantially reduced the burden of this disease, although the circulation of virus in sylvatic cycles in Africa and the Americas requires continuous large scale vaccination of humans. While extensive use has shown the live attenuated 17D YF vaccine to be safe and effective, important low-frequency adverse events have been identified that merit close attention for those developing or contemplating use of flavivirus vaccines.

Disease

Severity is quite variable; a high proportion of cases are mild or moderate and cannot be distinguished from other viral syndromes. After the bite of an infected mosquito there is an incubation period of 3–6 days or more. Classical disease begins with abrupt onset of fever accompanied by headache, photophobia, dizziness, back pain, malaise, anorexia, gastrointestinal disturbances and irritability (Kerr, 1951). Patients often are unable to sleep, and are overalert. They may observe a macular blanching rash together with conjunctival congestion and exhibit a facies that suggests alcohol intoxication. On physical examination during the acute illness stage there may be mid-epigastric tenderness and an enlarged liver. An early pathognomonic sign is a white coated tongue, described as 'small' with red margins and tip (Berry and Kitchen, 1931; Beeuwkes, 1936). By the second day of fever and thereafter the patient has bradycardia relative to degree of fever (Faget's sign). Laboratory findings include a marked leucopenia with relative neutropenia, elevation of serum SGOT and SGPT. On day 4–6 after onset of fever there may be a brief remission. Next, during the so-called period of intoxication, hepatic and renal pathology accompanies a rebound in fever (saddle back fever curve) while approximately 15% of patients with serologically confirmed infections develop moderate or severe disease, characterized by vomiting, jaundice, haemorrhagic diathesis and oliguria. Alkaline phosphatise values are normal, while bilirubin values vary from 3 to 5 g/l. In some patients renal failure predominates. Most yellow fever patients exhibit albuminuria starting during the early acute stage and worsening in proportion to overall disease severity. During the intoxication period, patients may experience haematemesis, melaena, haematuria, metrorrhagia, petechiae, ecchymoses, epistaxis, oozing of blood from gums or from needle puncture sites and anuria. Imminent fatal outcome is sometimes heralded by delirium and intense agitation (Kerr, 1951). Clinical severity and lethality are higher in adults than children.

Yellow fever virus in Africa causes a wider spectrum of illness than in the Americas. In African adults, yellow fever is often a relatively mild disease. Case fatality rates are relatively lower than the terrifyingly high rates common in American outbreaks (Monath, 1997). These differences were sufficient to motivate the Rockefeller Foundation in 1926 to study YF in virus laboratories sited in Lagos, Nigeria and Accra, Ghana (Strode, 1951). The African patient, Asibi, who had a mild, self-limited disease, yielded the prototype YFV strain and parent to the 17D vaccine. Almost immediately thereafter, three Rockefeller researchers infected with the Asibi strain acquired yellow fever and died (Strode, 1951). There are extensive anecdotal reports in the medical literature over several centuries that suggest that disease attack and severity rates are lower in individuals of African ancestry than in whites or Asians. In the Caribbean and southern United States yellow fever outbreaks involving Africans who had lived in the Americas for generations were frequently mild with very low attack rates (Carter, 1931; Kiple and Kiple, 1977; Kiple, 1984; Kiple and Ornelas, 1996). Other reports document a lower incidence of overt YF disease and lower severity and death rates in children compared with adults. Beeuwkes reported that, among 91 individuals under the age of 30, the death rate was 8%, while of 31 persons 30 years and older, 32% died (Beeuwkes, 1936).

Epidemiology

In Africa, yellow fever is maintained in two sylvatic cycles. One is in West Africa in which virus is transmitted among a wide range of subhuman primate species by *Aedes furcifur-taylori* or *Aedes luteocephalus* and other species of *Stegomyia*. The relative stability of the yellow fever genome in Africa suggests that virus is also maintained by transovarial transmission (Sall *et al.*, 2010). Virus in this cycle is transmitted by these vector species to humans and then from human to human in rural and urban areas by *A aegypti*. This species in Africa occupies a wider range of habitats and has a broader host-feeding preference than outside Africa. Outbreaks occur in mid-rainy season (August) to early dry season (October). The other zoonotic cycle is in East and Central Africa, where virus is transmitted among subhuman primates by *Aedes africanus* and from monkeys to humans by *Aedes simpsoni*. Virus from this cycle produced the large 1960 Ethiopian outbreak and the smaller 1990 and 1996 outbreaks in Kenya (Monath, 1989, 1994, 1997). Despite the long-term presence of *Aedes aegypti* in urban centres of East Africa, no Central African YF virus has ever escaped to be transmitted in an urban cycle (Strode, 1951). African subhuman primates support high enough viraemia to sustain transmission, but infections in these animals are clinically inapparent. The selection for host resistance to fatal yellow fever in African subhuman primates is consistent with evidence that the virus evolved in Africa as a zoonosis.

Early genetic studies concluded that there were at least two or three genotypes of yellow fever virus in circulation: Type I in Central/East Africa, type IIA in West Africa and Type IIB in the Americas (Chang *et al.*, 1995). More recently, seven genotypes based on greater than 7% differences in nucleotides have been described from full-length sequences, two in the Americas, three in Angola, East and Central Africa and two in West Africa (Mutebi *et al.*, 2001; von Lindern *et al.*, 2006). The oldest viruses with the greatest genetic distance from West African and American viruses are transmitted in Central/East African sylvatic cycles (Lepiniec *et al.*, 1994; Chang *et al.*, 1995; Mutebi *et al.*, 2001; von Lindern *et al.*, 2006). A relatively close genetic relationship exists between American and West African genotypes, consistent with the hypothesized West African origin of American yellow fever viruses, which then adapted to a new group of vertebrate hosts and mosquito vectors. Little work has been done, however, to define phenotypic differences between West and Central/East African viruses. In fact, there is no formal evidence that the pylogenetically older Central/East African viruses are capable of being transmitted by *A aegypti*. However, many phenotypic differences are described for American sylvatic and West African urban YF viruses (Fitzgeorge and Bradish, 1980; Deubel *et al.*, 1987; Henderson *et al.*, 1970).

Based upon abundant evidence, historical and genetic, yellow fever virus moved out of the West African sylvatic cycle to cause African and then American urban outbreaks (Sall *et al.*, 2010). Yellow fever was recognized as a classic pandemic disease in the New World. This emergence event occurred at a date unknown but likely the early 1600s (Strode, 1951). Domesticated West African

A aegypti undoubtedly infested many of the sailing vessels which plied the trade in human beings which supplied sugar plantation workers to the British, French, Spanish and Dutch colonies in the Caribbean and to Portuguese in South America. On occasion, the human cargo and crew must have offered the combustible mixture of yellow fever-infecteds, -immunes and -susceptibles to support an epidemic during the 2- to 3-month journey across the Atlantic Ocean. It is also plausible that virus was transported in living female *A. aegypti*, infected either in Africa or on-board epidemics. World trade soon introduced yellow fever to many of the towns and cities on the coasts of North and South America and to the banks of the Mississippi and Amazon rivers.

Another remarkable event occurred silently and unheralded in the Americas: yellow fever virus escaped from the urban cycle to a forest cycle, infecting a wide range of South American subhuman primates as well as marsupials and transmitted by several species of *Haemagogus* mosquitoes. Studies have shown that YF epizootics travel up and down the vast tropical forests in the river basins of South America east of the Andean Cordillera. As evidence that the New World zoonotic cycle began relatively recently, in nature and in the laboratory, American subhuman primates experience high fatality rates when infected with wild-type yellow fever virus (Strode, 1951; Waddell and Taylor, 1945; Waddell and Taylor, 1947; Waddell and Taylor, 1948; Waddell and Taylor, 1946).

At the beginning of the 20th century, the urban yellow fever mosquito vector was identified and effective mosquito control methods designed and implemented (Strode, 1951). Using these methods, as early as 1934, urban yellow fever was eradicated from the Western Hemisphere and by 1960, *A. aegypti* had been eradicated from 13 large countries occupying 85% of the Central and South American land mass (Monath, 1994). Despite this show of competence, by the 1970s, *A. aegypti* began to reinvade and by 1990 had virtually attained its previous range. As early as the late 1800s, humans were infected with yellow fever virus repeatedly from the American sylvatic cycle, principally adults with occupational exposure to forest or forest fringe areas of Bolivia, Brazil, Venezuela, Colombia, Ecuador and Peru. At present, epizootics and human cases occur during months of high rainfall and humidity, January–May (Monath *et al.*, 2008). Over the past 50 years, in these six countries, the Pan American Health Organization reports that there have been between 50 and 500 cases each year with as many as 200–250 deaths.

The reintroduction of yellow fever into urban areas of the Americas from the sylvatic cycle has been repeatedly predicted. A few putative occurrences have been reported, but urban outbreaks have not occurred. The reason for the failure of this much heralded event is not clear. One must ask whether jungle *Haemagogus*-adapted IIB viruses have lost their ability to be transmitted by *A. aegypti*. In the 1940s, Waddell and Taylor demonstrated that South American yellow fever strains were capable of being serially transmitted among subhuman primates by *A. aegypti* (Waddell and Taylor, 1945, 1946, 1947, 1948). However the proper experiment has not been done. This would be to compare the competence of American *A aegypti* strains to transmit type IIA and IIB viruses. Fortunately for the world, but unfortunately for the experimental design, American urban yellow fever virus strains do not exist and only the ability of sylvatic YF strains to be transmitted by South American *A. aegypti* has been studied (Aitken *et al.*, 1977; Miller *et al.*, 1989; Miller and Ballinger, 1988; Tabachnik *et al.*, 1985; Lourenco de Oliveira *et al.*, 2002, 2004; Mutebi *et al.*, 2004).

The invasion of Asia by yellow fever has been the emergence event most widely predicted and feared during the 20th century (Dudley, 1934; Strode, 1951). Considering this fear, it is surprising that contingency plans do not seem to be in place to combat the introduction of yellow fever. The most likely reason for the failure of this emergence event is the lack of transmission of YF in urban cycles on the East Coast of Africa. It appears that the Central/East African YF zoonotic cycle poses an effective barrier to the spread of West African YF strains.

Vaccines

At present, only strains of 17D yellow fever virus are licensed as vaccines.

Origin of 17 D strains

Development of YF vaccine was initiated by sequential passage of prototype Asibi virus in cultures of minced mouse embryo tissues. After 18 subcultures, the virus was passed to minced whole chick embryo. After 58 passages, the virus, now designated 17D, was propagated in chick embryo cultures from which brain and spinal cord had been removed. Final passage of virus for use as a vaccine was (and is) made in embryonated hen's eggs (Lloyd *et al.*, 1936; Theiler and Smith, 1937a,b; Monath *et al.*, 2008). All live attenuated YF vaccines are made from strains designated YF 17DD or 17D 204. The lineages of derivation of these strains have been documented (Monath *et al.*, 1983). There are 20 amino acid changes in the genome and four nucleotide differences in the 3 non-coding region between wild-type Asibi strain and 17D vaccine viruses (dos Santos *et al.*, 1995; Rice *et al.*, 1985; Monath *et al.*, 2008). One amino acid difference is in the M protein, eight in the E protein and 11 in NS proteins. Wild-type YF viruses share amino acids at envelope protein positions 52, 200, 305 and 380, suggesting that amino acid differences observed at these sites between wild and vaccine viruses may be responsible for attenuation. Interestingly, a single or limited passage of wild-type YF virus in HeLa cells resulted in loss of neurotropism and viscerotropism (Dunster *et al.*, 1990, 1999). Five changes in amino acids were in E – E27, E 155, E 228, E 331 and E 390 – while one change was noted in NS2A, three in NS2B and one in NS4B. As neurotropism and viscerotropism are lost separately in genetic constructs, the stable attenuation of YF 17D is probably due to several interacting mutations. It is also possible that mutations in NS proteins convey attenuation as, the NS amino acid changes and nucleotide changes in the 3′ NCR are present in dengue/YF chimeras that have been made into tetravalent dengue vaccines. These NS and NCR changes appear to attenuate vaccines when dengue prM and E genes have been inserted into the YF 17D backbone (Monath *et al.*, 2002a; Guirakhoo *et al.*, 2006a). In addition to the changes already noted, there are eight amino acid changes between 17DD (Brazil) and 17 D 204 (USA, UK, France, Senegal) vaccines found in the E protein, eight differ in NS proteins and there are four nucleotide differences in the 3′ NCR (Jennings *et al.*, 1993; Rice *et al.*, 1985; Galler *et al.*, 1998; Dupuy *et al.*, 1989; Pugachev *et al.*, 2002).

Protective mechanisms

Because of its safety and efficacy, YF 17D virus has been given to human volunteers whose innate and acquired immune responses have been studied in detail (Pulendran, 2009). One dose of vaccine is known to protect more than 90% of vaccinees and antibodies circulate essentially for a lifetime (Monath *et al.*, 2008). Neutralizing antibodies are thought to be the primary correlate of protective immunity (Monath *et al.*, 2008). In addition, administration of YF 17D results in a massive expansion of $CD8^+$ T cells, peaking at 15 days (Miller *et al.*, 2008). Vaccination with YF 17D induces polyvalent adaptive immune responses, including the production of cytotoxic T cells, a mixed T helper 1 (T_H1) and T_H2 cell profile and robust neutralizing antibodies that can persist for up to 40 years after vaccination (Barrett and Teuwen, 2009). YF 17D signals through RIG1 and MDA5, activating transcription factors that regulate the expression of type I IFNs. The solute carrier family two genes, SLC2A6 and EIF2AK4, that mediate an integrated stress response in innate immune cells, predicted high CD8+ T cells and antibody responses (Querec *et al.*, 2009). In addition, with respect to antibody responses, TNF receptor superfamily, receptor 17, a receptor for B cell-activating factor (BAFF), was a key gene in predictive signatures (Querec *et al.*, 2009). Other workers have shown that infection with live respiratory syncytial virus (RSV) stimulates a different set of toll receptors (TLR) than does formalin-inactivated RSV (Delgado *et al.*, 2009).

Production facilities

YF vaccine is made in 10 facilities worldwide. Five of these are principal suppliers of vaccine – Sanofi Pasteur (USA), Bio-Manguinhos (Brazil), Chiron/Novartis (UK), Sanofi Pasteur, Marcy l'Etoile (France) and the Pasteur Institute of Dakar, Senegal (Africa). In 1966, yellow fever 17D vaccine was discovered to be contaminated with avian leucosis virus. (ALV). All vaccines produced at that time contained the agent, owing to the high prevalence of ALV infection in flocks used for egg production. New seeds free of avian leucosis virus were developed in the 1970s, and all manufacturers now employ leucosis-free YF virus seeds and leucosis-free flocks as stipulated by WHO standards.

Vaccine usage

Indications for use of 17D vaccine

Ideally, all individuals living in YF-enzootic countries should receive 17D vaccine. In practice, in Brazil, Venezuela, Colombia, Ecuador, Peru and Bolivia, residents who live near areas where sylvatic YF cases have occurred are routinely vaccinated, whereas the residents of large cities on the Atlantic, Pacific or Gulf coasts are not. In Brazil, it is in this unvaccinated majority that YF occurs during outings to areas of sylvatic YFV transmission. In West Africa, where the threat of urban YF is universal, all residents should be immunized. Immigrants, travellers, military personnel and expatriate residents and their dependants planning to live or travel in YF-enzootic countries should receive YF vaccine. For tourists to South America, anyone travelling to any part of the Amazon Basin should be immunized. Vaccine should be given 10 days prior to start of travel to enzootic countries. For conventional travel destinations travellers from the United States should consult recommendations provided by the CDC and the Advisory Committee on Immunization Practices (Staples *et al.*, 2010). Alternatively, the WHO provides a detailed listing of areas of YF activity and a description of countries that require YF vaccination certification on an interactive map accessible at www.who.int/ith/en/.

Dose and route of administration

17D vaccine is given by the subcutaneous route in a volume of 0.5 ml, generally in the upper arm. The minimum dose requirement is 1000 mouse LD_{50} or the equivalent of 10,000 plaque-forming units (pfu). Vaccines vary with respect to stabilizer additives and salt content. Some contain sodium chloride and buffer salts. All vaccines are lyophilized. As YF vaccines do not contain antibiotics, when reconstituted, vaccines should be administered within 1 hour. When maintained on ice, vaccines in multidose vials given under epidemic conditions are used for a period of 8 hours. Vaccines are supplied in single- and multiple-dose vials, up to 20 doses.

Stabilizers

Vaccines should meet a stability standard (World Health Organization, 1988). The lyophilized vaccine (1) held at 37°C for 14 days must maintain minimum potency (> 1000 MLD_{50} per 0.5-ml dose) and (2) show a mean loss of titre of less than 1.0 $log_{10}LD_{50}$. Without stabilizers yellow fever vaccine loses 1.5–2.5 log_{10}/dose during 14 days at 37°C. Stabilized vaccines lose only 0.3–0.5 log_{10}/dose during this interval. Stabilizers used in vaccines produced in the USA and UK are sorbitol and gelatin. The vaccine made in France contains sugars, amino acids and divalent cations (Monath *et al.*, 2008).

Composition of vaccine

Virologically, YF 17D vaccines are heterogeneous and presumably composed of quasispecies. Despite this heterogeneity, fully virulent viruses have not been found in vaccine mixtures (Monath *et al.*, 2008). A genetically defined monoclonal chimeric YF-dengue vaccine has been given to thousands of volunteers. The non-structural genes of YF 17D when given with DENV structural genes have retained the non-virulent phenotype (Monath *et al.*, 2002a).

Viraemia

YF 17 D vaccines, when given to susceptible rhesus monkeys or human beings, produce low levels of viraemia when compared with wild-type virus. In a recent study of the 17D 204 vaccine, approximately 30–60% of susceptible adult subjects were viraemic between days 4 and 6 after vaccination. The mean duration of viraemia was approximately 2 days. Viraemia levels were low (mean titres < 20 pfu) (Monath *et al.*, 2002a, 2003). However, in a recent study, no virus was isolated or detected by RT-PCR on days 2, 4 and 7 after vaccination of eight susceptible adults given 17D vaccine (dos Santos *et al.*, 2005).

Seroconversion rate

Neutralizing antibody seroconversion rates to YF 17D vaccines made before and after the introduction of avian leucosis-free virus seeds and production flocks have been measured many times, and in virtually all instances the neutralizing antibody seroconversion rate has approached 100% (Monath *et al.*, 2008). Neutralizing antibody responses can generally be detected within 14 days after vaccination (Theiler and Smith, 1937b; Smith *et al.*, 1938; Wisseman and Sweet, 1962) It is likely that protective efficacy develops very early following administration of vaccine as in rhesus monkeys neutralizing antibodies were detectable in serum on day 6 or 7 at which time animals were refractory to challenge.(Theiler and Smith, 1937b) Interestingly, antibody responses following YF 17D immunizations do not evolve in the usual manner with a IgM to IgG shift. Prolonged IgM responses have been observed (Monath, 1971).

Duration of immunity

Live attenuated YF 17D viruses share the property with wild-type viruses and other viral vaccines of lifetime circulation of neutralizing antibodies following a infection or vaccination (Amanna *et al.*, 2006, 2007, 2008; Poland *et al.*, 1981; Halstead, 1974; Bass *et al.*, 1976, 1978) Although the International Health Regulations of the World Health Organization stipulate that the yellow fever immunization certificate for international travel is valid for 10 years, it is likely that the duration of protective immunity to YF 17D vaccine is lifelong (World Health Organization, 2005). Many studies of booster vaccinations have been made. In general, only individuals with low levels of neutralizing antibodies will respond with an increase in titre of neutralizing antibodies (Monath *et al.*, 2008). It is doubtful that booster immunizations alter the protection status against yellow fever disease. Although it is possible that immunity to heterologous flaviviruses might reduce 'takes' of YF 17D, in a large study, Peruvian children with or without dengue antibodies had similar high seroconversion responses to two commercial YF 17D 204 vaccines (US and UK). (Belmusto-Worn *et al.*, 2005). Immunity from yellow fever vaccination was able to enhance seroconversion rates and viraemia following the administration of candidate live attenuated dengue vaccines (Bancroft *et al.*, 1984; Guirakhoo *et al.*, 2006b).

Precautions

Vaccine contraindications

The International Health Regulations of the World Health Organization stipulate that yellow fever 17D vaccines should not be administered to individuals with egg protein allergies, during pregnancy or to children under the age of 9 months (World Health Organization, 2005; Staples *et al.*, 2010). Over its use over nearly three-quarters of a century, YF 17D has been administered inadvertently or in planned studies to many pregnant women (Suzano *et al.*, 2006). These studies have not provided any evidence of danger to the fetus or the mother nor evidence of congenital anomalies in infants (Suzano *et al.*, 2006; Monath *et al.*, 2008). Prior to the adoption of the seed lot system of maintaining phenotypic consistency of 17D vaccines, cases of encephalitis and hepatitis in adult vaccine recipients were reported (Monath *et al.*, 2008). However, since the seed lot system was adopted in 1941, side-effects in susceptible adults have been minimal (Camacho *et al.*, 2005). In one recent study, in which 1440 subjects were monitored, half received YF-VAX and half Arilvax. Safety was monitored using diary notes and clinic visits (Monath *et al.*, 2002a). No serious side-effects were noted using either vaccine. However, non-serious side effects, such as local reactions, headache, asthenia, myalgia, malaise, fever and chills, were noted, and in 7–8% of vaccinees these were serious enough to result in absence from work.

Adverse events

Encephalitis The incidence of severe reactions to YF 17D vaccines has been estimated at 1.6 per 100,000 doses based upon adverse reports from the US, where 300,000 doses are administered annually. These occurred predominantly in persons over the age of 60 years (Lindsey *et al.*, 2008; Khromava *et al.*, 2005). 17D vaccine strains retain a degree of neurovirulence as demonstrated by intracerebral inoculation of mice or monkeys (Monath *et al.*, 2005). This appears to have a counterpart in humans as manifested in young infants and the elderly (vaccine-associated neurological disease – YEL-AND). Nonetheless, encephalitis in adults is rare. Between 1945 and 1991 only 21 adult cases of YF post-vaccination encephalitis were reported worldwide (Monath *et al.*, 2008). Between 1991 and March 2006, after institution of a formal adverse events reporting system, there were 29 post-vaccination encephalitis cases reported, mostly in individuals over the age of 60 years (McMahon *et al.*, 2007; Marfin *et al.*, 2005). In 1952–53, five cases of encephalitis were reported among 1800 infants who were vaccinated under the age of 1 year (Stuart, 1956). Since then restrictions were imposed on giving 17D vaccine to infants under 9 months of age. As a result, the incidence of encephalitis has been very low.

Viscerotropic disease In 2001, multiorgan failure was described in seven individuals receiving YF 17D vaccine; six cases were fatal (Vaccine-associated viscerotropic disease – YEL-AVD) (Martin *et al.*, 2001; Vasconcelos *et al.*, 2001). These were observed in individuals immunized with both 17DD and 17D 204 vaccine strains. The signs and symptoms of the syndrome resembled wild-type yellow fever disease, including rapid onset of fever, malaise and myalgia within 2–5 days of vaccination followed by jaundice, oliguria, cardiovascular signs and haemorrhage. Large amounts of YF antigen were found in liver and heart. Virus recovered did not differ significantly from dominant genome present in vaccines (Jennings *et al.*, 1994; Engel *et al.*, 2006). A partial genome sequence from a individual who died after receiving vaccine in 1975 revealed that only 17 DD vaccine virus was recovered from organs (Engel *et al.*, 2006). For a time, it was thought that individuals who had thymus resections were uniquely at risk to YEL-AVD, but as more cases have been described no important risk factors have been established. Individuals with YEL-AVD appear to mount unusual cytokine responses along with brisk and fairly normal acquired immune responses (Silva *et al.*, 2010).

Japanese encephalitis

Japanese encephalitis (JE), a mosquito-borne flaviviral zoonotic infection, is the leading recognized cause of viral encephalitis in Asia. JE viruses are transmitted by *Culex* spp. mosquitoes throughout Asia, a rice-growing region with an indigenous population of more than 3 billion people that is also a major tourist destination. For many years, only inactivated JE vaccines prepared from infected mouse brains were licensed for use by residents and travellers. Use of this vaccine resulted in an unacceptable level of adverse safety events and its production has been discontinued by major manufacturers. Recently, the JE vaccine landscape has changed. A safe and efficacious single-dose, live-attenuated vaccine produced in China has become available in many Asian countries. A new, inactivated JE alum-adjuvanted vaccine is now licensed for use in Europe, Australia and the United States and a yellow fever (YF)-JE chimeric vaccine candidate is nearing licensure in developed as well as developing countries.

Disease

Typically symptoms start suddenly following a variable incubation period of 2 days to 2 weeks, sometimes heralded by a non-specific viral prodrome. The earliest symptoms include lethargy, fever, headache, abdominal pain, nausea and vomiting (Tsai and Solomon, 2004). These may be followed by a combination of nuchal rigidity, photophobia, altered consciousness, hyperexcitability, masked facies, muscle rigidity, cranial nerve palsies, tremulous eye movements, tremors and involuntary movement of the extremities, paresis, incoordination and pathological reflexes. Neurological manifestations may include meningeal (meningitis), parenchymal (encephalitis) or spinal cord (myelitis) involvement. (Solomon and Vaughn, 2002) Sensory deficits are rare. In children, 50–85% develop focal or general seizures compared with 10% of adult cases (Tsai and Solomon, 2004; Halstead and Jacobson, 2008). Seizures have been associated with poor clinical outcome (Solomon *et al.*, 2002).

Peripheral leucocytosis with left shift and hyponatraemia may be observed. Opening pressure of cerebral spinal fluid (CSF) may be elevated in up to 50% of cases; protein levels are often normal or mildly elevated (Solomon and Vaughn, 2002). CSF pleocytosis ranges from 10 to a few thousand cells per cubic millimetre (median of several hundred), and are predominantly of lymphocytic origin (Halstead and Tsai, 2004). Electroencephalography demonstrates diffuse delta wave activity and, rarely, spike and seizure patterns (Tsai and Solomon, 2004). Imaging studies demonstrate diffuse white matter oedema and abnormal signals in the thalamus – often bilateral and haemorrhagic – basal ganglia, cerebellum, midbrain, pons and spinal cord (Kumar *et al.*, 1992, 1997; Misra *et al.*, 1994).

In non-fatal cases, clinical improvement begins after approximately 1 week, paralleling defervescence. Recovery of neurological function may take weeks to years. Seizure disorders, motor and cranial nerve paresis, and movement disorders may persist in up to one-third of patients. Persistent behavioural and/or psychological abnormalities occur in 45–75% of survivors and are more severe in children (Kumar *et al.*, 1993). There is no specific therapy for JE; supportive care focuses on controlling seizures, ventilator support of respiratory failure, and monitoring and reducing cerebral oedema (Tsai and Solomon, 2004). Anecdotal use of interferon-alpha and ribavirin has been reported with negative results (Kumar *et al.*, 1993; Harinasuta *et al.*, 1985). Fatality rates vary between 5% and 40% and are often reflective of the standard of medical care available.

Epidemiology

The rice paddy-breeding *Culex tritaeniorhynchus summarosus*, a night-biting mosquito that feeds preferentially on large domestic animals and birds and infrequently on humans, is the principal vector of zoonotic and human JE in northern Asia. A more complex ecology prevails in southern Asia, from Taiwan to India, where *Cx. tritaeniorhynchus* and members of the closely related *Cx. vishnui* group are vectors. Before the introduction of JE vaccine, summer outbreaks of JE occurred regularly in Japan, Korea, China, Okinawa and Taiwan. Over the past decade, there has been a pattern of steadily enlarging recurrent seasonal outbreaks in Vietnam, Thailand, Nepal, and India, with small outbreaks in the Philippines, Indonesia, and the northern tip of Queensland, Australia (Halstead and Jacobson, 2008; Kari *et al.*, 2006). Seasonal rains are accompanied by increases in mosquito populations and increased transmission. Pigs serve as amplifying hosts. In contrast, humans are likely dead-end hosts because they experience short-duration and low-level viraemia (Halstead and Jacobson, 2008). There is no direct human to human transmission.

In economically advanced Asian countries, such as Japan, Korea and Taiwan, and in moderate- and low-income countries, such as Thailand, Sri Lanka and Nepal, the recent integration of JE vaccine into routine immunization programmes has led to the near-elimination of JE (Halstead and Jacobson, 2008) (Nepal programme: J.B. Tandan, personal communication, December 2009). Despite widespread immunization, in 2007, over 9000 cases were reported in the South-East Asia and Western Pacific regions (http://www.who.int/whosis/whostat/EN_WHS09_Full.pdf). From the standpoint of risk, it is important to understand that reported cases vastly underestimate the infectious burden. The ratio of infections to symptomatic JE cases has been estimated to vary between 1:25 and 1:300; the lower rates (1:200–1:300) were observed in northern Asian persons, indigenous to the zoonotic heartland of JE, while higher rates were measured in non-indigenous military personnel (Halstead and Grosz, 1962; Benenson *et al.*, 1975; Gajanana *et al.*, 1995). Determinants for occurrence of overt neurological disease following infection are not understood. JE resembles West Nile infections in this respect (Gyure, 2009; Solomon and Vaughn, 2002).

In endemic areas, the incidence of JE disease is greater in the young; attack rates in the 3- to 15-year age group are 5–10 times higher than in older persons (Solomon, 2004). The higher disease rates in younger persons reflect high immunity rates in adults. There is some evidence that young children are intrinsically at greater risk of developing encephalitis than are susceptible adults (Hammon *et al.*, 1958). Numerous epidemiological observations document a weak protective effect of prior dengue virus infection on subsequent overt JE disease (Grossman *et al.*, 1973; Edelman *et al.*, 1975; Hammon *et al.*, 1958; Tarr and Hammon, 1974; Libraty *et al.*, 2000).

Vaccines

General principles

Antibodies directed against E protein neutralize the virus and play an important role in protection (Srivastava *et al.*, 1987; Konishi *et al.*, 1999). A JE neutralizing antibody titre ≥1:10 is commonly accepted as evidence of protection (Sukhavachana *et al.*, 1969; Markoff, 2000; Hombach *et al.*, 2005; Hoke *et al.*, 1988).

First-generation JE vaccines

Formalin-inactivated JE vaccines have been available since the 1950s. Two vaccines were widely used: (1) an inactivated mouse brain-derived vaccine and (2) an inactivated vaccine cultivated on primary hamster kidney (PHK) cells.

An inactivated mouse brain vaccine using either Nakayama or Beijing-1 virus strains was developed in Japan. Local production of this vaccine also contributed to a decrease in the incidence of JE in Thailand, India, Korea, Taiwan, Vietnam and areas of Malaysia and Sri Lanka (Halstead and Jacobson, 2008). For several decades this JE vaccine has been available for use by tourists or military personnel in developed countries (BIKEN, distributed by Sanofi Pasteur as JE-Vax®). The Korean Green Cross Vaccine Co. also produced a mouse brain JE vaccine. This was licensed for use in the UK and available in Europe but distributed by MASTA on a named-patient basis [http://www.nathnac.org/travel/factsheets/japanese_enc.htm; http://www.masta-travel-health.com/]. Seroconversion rates, quantitative neutralizing antibody titres following vaccination, and efficacy rates varied according to the population studied (indigenous versus non-indigenous) and number of doses administered (one, two or three doses in the primary immunization series) (Halstead and Jacobson, 2008). A single efficacy trial showed equivalent protection afforded by either Beijing-1 or Nakayama strains (Hoke *et al.*, 1988). For travellers, a three-dose immunization series has been recommended (Immunization Practices Advisory Committee (ACIP), 1993). Mild to moderate vaccine reactogenicity was accepted over years of use but the occurrence of a single case of acute disseminated encephalomyelitis (ADEM) temporally related to vaccination in Japan prompted the Japanese government (May 2005) to suspend routine childhood JE vaccination. Data from the United Kingdom, Australia, Canada, and the US estimate the rate of hypersensitivity reactions to be between 0.7 and 104 per 10,000 vaccinees (Halstead and Jacobson, 2008). The cause of temporally associated neurological or hypersensitivity reactions is not clearly understood; the presence of murine neural proteins, gelatin and/or thimerosal in vaccine preparations have all been implicated but none proven as causative. BIKEN ceased production of JE-VAX® in 2005; supplies may be exhausted in 2009.

A second inactivated vaccine, manufactured from virus grown in hamster kidney cells, has been in wide use. From 1968 to 1990, this was China's principal JE vaccine (Gu and Ding, 1987). Approximately 70 million doses of the PHK cell culture inactivated JE vaccine (Beijing-3, P-3 strain) were administered in China yearly until 2005. Urticarial allergic reactions were reported in approximately 1 in 15,000 vaccine recipients (Hoke *et al.*, 1988). Randomized field trials demonstrated vaccine efficacy ranging from 76% to 95% (Oya, 1988; Halstead and Tsai, 2004).

Second-generation JE vaccines

The development and licensure of second-generation, non-mouse brain derived JE vaccines provides improved safety profile and lower dosage requirements and opportunities for greatly expanded use.

SA14-14-2 strain of JEV was isolated from *C. pipiens* in Xian in 1954. To achieve attenuation, it was passaged in mice and then over 100 times in primary hamster kidney cells (PHK), chick embryo tissues and suckling mice (Halstead and Jacobson, 2008; Yu, 2010). The parental and vaccine viruses differ by four amino acids in the envelope and one each in NS2, NS3, and NS4B (Ni *et al.*, 1997). Studies suggest that a heparin-sensitive mutation at Glu-306 leads to a loss of neuroinvasiveness (Lee *et al.*, 2004). This remarkably non-neurovirulent live-attenuated vaccine has gradually been introduced into China, where it has demonstrated an excellent safety profile and very high efficacy (88–96%) in large-scale trials ($n > 200,000$ children) and effectiveness (Hennessy *et al.*, 1996; Liu *et al.*, 1997; Tsai *et al.*,

1998; Kumar *et al.*, 2009). In China, SA 14-14-2, currently administered as a two- or three-dose vaccine, is rapidly replacing the inactivated PHK vaccine (Yu Yong Xin, personal communication, 17 December 2009). Since its licensure in China in 1988, more than 300 million doses have been produced and administered to over 120 million children. Numerous large scale evaluations of vaccine safety demonstrate low rates (0.2–6%) of short-lived local and systemic (i.e. fever) reactogenicity and even lower (2.3 per 10,000) predicted rates of neurotoxicity (Halstead and Jacobson, 2008; Halstead and Thomas, 2010). Case–control studies of a large vaccine trial in Nepal showed rapid onset of protection followed by a 5-year efficacy of 96% after a single dose of vaccine (Bista *et al.*, 2001; Ohrr *et al.*, 2005; Tandan *et al.*, 2007). A similar result was observed in India (Kumar *et al.*, 2009). In a small, single study, SA 14-14-2 vaccine was co-administered with live measles vaccine in children; normal immune responses were retained to each vaccine (Gatchalian *et al.*, 2008). Recently, the vaccine has been licensed for use and millions of doses administered in Nepal, India, Sri Lanka and South Korea (Halstead and Jacobson, 2008; Yaich, 2009; Elias *et al.*, 2009). The Chinese manufacturer, Chengdu Institute of Biological Products, is seeking prequalification by the World Health Organization (WHO) (http://whqlibdoc.who.int/hq/2006/WHO_IVB_05.19_eng.pdf).

The Program for Advanced Technology in Health (PATH) has negotiated concessional prices for the use of SA 14-14-2 in India, Sri Lanka and Nepal for public health prevention programmes (http://www.path.org/projects/japanese_encephalitis_project.php). The SA 14-14-2 vaccine has recently been incorporated into the national extended programme for immunizations (EPI) in Nepal and China (personal communications from Dr Jay Tandan and Dr Yu Yong Xin, respectively, 17 December 2009).

IC51 (in US and Europe available as IXIARO®, in Australia and New Zealand available as JESPECT®)

The IC51 vaccine is a purified, formalin-inactivated, whole-virus JE vaccine developed by Intercell AG (Vienna, Austria); the product was licensed for use in individuals ≥ 17 years of age the US, Australia and Europe in the spring of 2009. Kollaritsch and colleagues have published a useful review of the product's development life cycle (Kollaritsch *et al.*, 2009). Stockpiled supplies of mouse brain vaccine are available for children 16 years and less.

The vaccine construct was developed at the Walter Reed Army Institute of Research (WRAIR, Silver Spring, MD, USA) based on the JE SA14-14-2 virus strain passaged eight times in primary dog kidney (PDK) cells, that was then cultivated in Vero cells in serum-free medium, formalin-inactivated and formulated with 0.1% aluminium hydroxide (Eckels *et al.*, 1988). The absence of serum allows for a simplified purification process and potentially, superior safety profile (Montagnon and Vincent-Falquet, 1998; Srivastava *et al.*, 2001). As safety advantage this vaccine does not require additional stabilizers or additives (see 'First-generation JE vaccines', above).

Several early-phase clinical studies established the safety and immunogenicity of the IC51 candidate (Kollaritsch *et al.*, 2009; Lyons *et al.*, 2007). Multinational phase III immunogenicity trials demonstrated the IC51 vaccine was well tolerated and elicited non-inferior immune responses compared with control (JE-VAX®) (Tauber *et al.*, 2007). There was no evidence of an increased incidence of rare adverse events (including anaphylaxis or anaphylactoid reactions) compared to placebo (Tauber *et al.*, 2008) .

Additional studies demonstrated high (83%) seroconversion rates 1 year following vaccination, superiority of the standard dosing regimen (2 × 6 μg) compared with using low- or high-dose strategies and the observation that the presence of TBE antibodies had no safety impact but heightened the JE immune response following a single dose. There was no adverse impact on safety or immunogenicity of either vaccine when IC51 was given together with hepatitis A vaccine, HAVRIX® 1440 (Schuller *et al.*, 2008a, 2009; Kaltenbock *et al.*, 2009; Dubischar-Kastner *et al.*, 2010).

Non-inferiority studies The vaccine was licensed after a trial in approximately 5000 adults of JEV, which found that neutralizing antibodies were equivalent to those produced by the licensed mouse brain vaccine. The pivotal non-inferiority immunogenicity study compared two doses of JE-IC51given on days 0 and 28 with three doses of JE-MB given on days 0, 7 and 28 to adults aged ≥ 18 years in the United States, Austria, and Germany (Tauber *et al.*, 2007). In the 'per protocol' analysis, 352 (96%) of 365 JE-IC51 recipients developed a $PRNT_{50} \geq 10$ compared with 347 (94%) of 370 JE-MB recipients at 28 days after the last dose (CDC, 2010). Two doses of vaccine are necessary as 28 days after receiving 1 dose of the standard 6-μg regimen, only 95 (41%) of 230 JE-IC51 recipients had seroconverted with a $PRNT_{50} \geq 10$. By contrast, 97% (110/113) of the subjects who had received two doses had a $PRNT_{50} \geq 10$ (Tauber *et al.*, 2008).

Co-administering vaccines Administering IC51 to subjects with pre-existing antibodies against tick-borne encephalitis virus (TBEV) enhanced JE neutralizing antibody response after the first dose but had no effect following the two-dose primary series (Schuller *et al.*, 2008b).

Duration of immunity In two studies performed in Europe, 95% and 83% of subjects who received two doses of JE-IC51 maintained protective neutralizing antibodies ($PRNT_{50} \geq 10$) at 6 months after receiving the first dose, while 83% and 58% still had protective levels of antibodies at 12 months after the first dose (Schuller *et al.*, 2008a; Dubischar-Kastner *et al.*, 2010). Among 44 subjects who no longer had protective antibodies (17 subjects at 6 months after their first dose and 27 subjects at 12 months after their first dose), all developed $PRNT_{50} \geq 10$ after receiving a booster dose (Dubischar-Kastner *et al.*, 2010).

Immunogenicity in children Currently, IC51 is not licensed for use in children. A phase II trial investigating safety and immunogenicity of JE-IC51 was performed in healthy children aged 1 and 2 years in India, using a standard (6 μg) or half (3 μg) dose (Kaltenbock *et al.*, 2010). Children in both groups received two doses of JE-IC51 administered 28 days apart. A third group of children received three doses (days 0, 7 and 28) of an inactivated mouse brain-derived JE vaccine produced by Korean Green Cross. Four weeks after the last dose, seroconversion rates in the 6-μg ($n = 21$) and 3-μg ($n = 23$) IC51 recipient groups and the inactivated mouse brain-derived group ($n = 11$) were 95%, 96%, and 91%, respectively. Differences in seroconversion rates or geometric mean titres were not statistically different.

The ACIP has issued recommendations for the use of XIARO vaccine on 12 March 2010 (CDC, 2010). Intercell AG has entered into distribution partnerships to supply vaccine to markets in the US, EU, Japan, South Korea, Latin America, parts of Asia India, Bhutan, Nepal, Bangladesh, Australia, New Zeeland, Papua New Guinea and Pacific Islands (http://www.intercell.com/main/forvaccperts/products/japanese-encephalitis-vaccine/, accessed 26 August 2009).

ChimeriVax™-JE (IMOJEV)

Sanofi Pasteur has licensed a JE vaccine developed by Acambis, Inc., Cambridge, MA, USA and based on a chimeric virus generated at the St. Louis University Health Sciences Center, St. Louis, MO, USA. ChimeriVax™-JE virus (JE-CV) is produced using infectious clone technology based on insertion of prM and E genes from SA JE SA 14-14-2 virus into the non-structural genes of YF 17D viral strain as viral 'backbone' (Chambers *et al.*, 1999). The resulting chimeric RNA was electroporated into Vero cells. Progeny virus particles contain JE-specific antigenic determinants that elicit neutralizing antibodies as well as cytotoxic T lymphocytes (Konishi *et al.*, 1998). YF 17D was chosen as backbone to the chimera because of its proven record of safety and efficacy.

A single dose of JE-CV was shown to protect against a virulent wt JEV challenge in mice (Guirakhoo *et al.*, 1999). Passive protection experiments also demonstrated that mouse serum raised against JE-CV was protective against all four JEV genotypes (Beasley *et al.*, 2004). Moreover, a computer analysis of human leucocyte antigen (HLA) class II-restricted T-cell epitopes in JEV E protein revealed a high degree of conservation among putative T helper epitopes in JE-CV and circulating JEV representing all four genotypes (De Groot *et al.*, 2007). Inoculation of ChimeriVax™-JE in non-human primates resulted in no illness, transient, low viraemia, followed by high titres of anti-JE neutralizing antibodies. A WHO monkey neurovirulence test scored only minimal brain and spinal cord lesions. Vaccinated monkeys were protected against IC or intranasal (IN) virulent JE virus challenge (Monath *et al.*, 2000).

Because it is a living agent, ChimeriVax™-JE virus has been evaluated for its ability to replicate in and to be transmitted by vector mosquitoes. Individual *Cx. tritaeniorhynchus, Aedes albopictus,* and *Ae aegypti* mosquitoes ingested a virus-laden blood meal or were inoculated intrathoracically (IT). ChimeriVax™-JE virus did not replicate following oral feeding in any of the three mosquito species. In *Cx. Tritaeniorhynchus* replication was not detected after IT inoculation of ChimeriVax™-JE. No genetic changes were associated with replication of vaccine virus in mosquitoes (Bhatt *et al.*, 2000) None of three Australian mosquitoes (*Cx. annulirostris, Cx. gelidus,* and *Ae. vigilax*) became infected after being fed orally with 6.1 log10 pfu/ml of ChimeriVax™-JE (JE-CV) (Reid *et al.*, 2006).

Safety and immunogenicity of ChimeriVax™-JE (high and low dose) was established in a phase I trial comparing immune responses in YF immune and flavivirus naive volunteer. Low-level viraemias were observed in both groups and the frequency of all adverse events was similar between groups. Overall, 96% of subjects who received ChimeriVax™-JE developed neutralizing antibodies to one or more of the wild-type JE strains tested (Beijing, P3, Nakayama) (Monath *et al.*, 2002a) In a phase II study, ChimeriVax™-JE was well tolerated with viraemias of short duration and low titre; 94% of subjects administered graded doses (1.8–5.8 $\log_{10}$) of ChimeriVax™-JE developed neutralizing antibodies. A second dose, administered 30 days later, had no booster effect. Previous inoculation with YF did not interfere with ChimeriVax™-JE, but there was a suggestion (not statistically significant) that ChimeriVax™-JE interfered with YF-VAX® administered 30 days later. In a further study in a larger number of subjects, ChimeriVax™-JE did not interfere with YF vaccine (Farshad Guirakhoo, personal communication). There is evidence that ChimeriVax™-JE induces sterilizing immunity in mice (Monath *et al.*, 2003).

As of October 2010, clinical studies involving 1400 children and 2400 adults had been conducted in the US, Australia, Thailand and the Philippines (Guy *et al.*, 2010) A single dose of ChimeriVax™-JE in adults achieved an immune response equivalent to that of three doses of JE-VAX®. A single dose of ChimeriVax™-JE elicited a protective level of neutralizing antibodies in 94% and 99% of subjects 14 days and 1 month after vaccine administration, respectively. A phase III non-inferiority study in toddlers showed that 95% of JE virus-naive subjects were seroprotected within one month following a single dose of ChimeriVax™-JE (T. Monath, personal communication, September 2010). No vaccine safety concerns were uncovered. Sanofi Pasteur has filed (7 July 2009) for marketing authorization with the Therapeutic Goods Administration in Australia (TGA) and the Food and Drug Administration in Thailand (TFDA) (http://www.sanofipasteur.com/sanofi-pasteur2/front/index.jsp?siteCode=SP_CORP&codeRubrique=34).

Vaccine usage

SA 14-14-2

Route of administration Subcutaneous.

Dosage 0.5 ml in children less than 6 years, 1.0 ml in older children and adults.

Immunization schedule In China, doses are given to children at 1 and 2 years of age and at school entry; large-scale immunization studies suggest that a single dose can be administered effectively at any time after 12 months of age (Tandan *et al.*, 2007).

IC51 vaccine

Each 0.5-ml dose contains approximately 6 μg of purified, inactivated JEV proteins and 0.1% aluminium hydroxide as an adjuvant (Jelinek, 2009). The finished product does not include gelatin stabilizers, antibiotics, or thimerosal. The vaccine should be stored at 35–46°F (2–8°C); it should not be frozen. The vaccine should be protected from light. During storage, the vaccine might appear as a clear liquid with a white precipitate. After agitation, it forms a cloudy white suspension.

Route of administration Intramuscular (i.m.).

Dosage and schedule 0.5 ml i.m. on days 0 and 28 (CDC, 2010).

Vaccination recommendations

On 12 March 2010, the US Centers for Disease Control and Prevention Advisory Committee on Immunization Practices (ACIP) published provisional recommendations for use of JE vaccine (CDC, 2010). The ACIP suggests that physicians and others should counsel travellers to be aware of the low level but almost unavoidable risk of acquiring JE when travelling to JE-enzootic countries. Risk can be reduced by using personal protective measures to avoid vector exposure. But JE vaccination should be considered as an important option to reduce the risk. Risk is a function of exposure, which in turn is a function of disease epidemiology. Travel during the rainy season, residence in rural or agricultural areas, residence near pig farms or long-duration visits all increase risk for JE. Currently, vaccination is recommended for such individuals; particularly high-risk activities are those that occur outdoors, near agricultural areas, during evening hours, and where lodging is in the open without use of bed nets. Travel to a JE-endemic area without a defined destination calls for vaccination. JE vaccine is not recommended for short-term travellers whose visit will be restricted to urban areas or times outside of a well-defined JEV transmission season. Lastly, JE vaccine is recommended for laboratory personnel who work with live, wild-type JEV strains. Vaccinated, at-risk laboratory personnel should receive appropriate booster doses of JE vaccine or be evaluated regularly for JEV-specific neutralizing antibodies to assure adequate titres (http://www.cdc.gov/vaccines/recs/provisional/#acip). Other groups have offered recommendations for the expanded use of JE vaccines in travellers and expatriates but data are currently lacking to support a consensus (Burchard *et al.*, 2009; Hatz *et al.*, 2009; Teitelbaum, 2009; Tsai *et al.*, 2009).

There is no other effective public health option for the prevention of Japanese encephalitis and reason to believe that new generation vaccines are safe and will result in long-lasting protection. This being true, the continued occurrence of cases of JE is intolerable. JE vaccines should be administered universally to all who live in or visit enzootic areas.

Precautions

Current-generation JE vaccines seem quite safe. Issues in selecting JE vaccines for use by private or public sector include valid licensing and commercial availability, cost and speed of development of protective immunity.

SA 14-14-2

For many residents of JE enzootic countries, SA 14-14-2 vaccine is by far the most appealing choice for national health authorities, due, in part to concessional prices and because large data sets provide evidence that a single dose is safe and sufficient to produce long-lasting protection. (Liu *et al.*, 1997; Tandan *et al.*, 2007). Also, SA 14-14-2 vaccine appears to be protective immediately after administration (Bista *et al.*, 2001).

IC51

In clinical trials, mild local and systemic adverse events caused by IC51 were similar to those reported for JE mouse brain vaccines or placebo adjuvant alone. No serious hypersensitivity reactions or neurological adverse events were identified among ~5000 recipients enrolled in the clinical trials, but, the possibility of rare serious adverse events cannot be excluded. Additional post-licensure studies and monitoring of surveillance data are planned to evaluate the safety of JE-IC51 in a larger population. In a pooled analysis of 6-month safety data from seven studies, severe injection-site reactions were reported by 3% of the 3558 JE-IC51 subjects, which was comparable to the 3% among 657 placebo adjuvant recipients but lower than the 14% among 435 individuals receiving mouse brain vaccine (Dubischar-Kastner *et al.*, 2010). Systemic symptoms were reported with similar frequency among subjects who received IC51 (40%), mouse brain (36%), or placebo (40%). Serious adverse events were reported by 1% of the subjects in the JE-IC51 group. Serious allergic reactions were not observed in any study group, including JE-IC51, JE-MB, or placebo recipients. The most common local reactions after IC51 administration were pain and tenderness. Two cases of urticaria were noted during the study. Angioedema was not observed. No serious neurological events were identified. In the non-inferiority immunogenicity trial, the frequency of adverse events reported following JE-IC51 vaccination (428 subjects) was similar to that reported by persons receiving JE-MB (435 subjects) (Tauber *et al.*, 2007).

For non-indigenous travellers and military personnel, the only option in much of the world is the ICI vaccine. The optimal vaccinations series is three doses administered over 3 weeks. For travellers, a two dose series administered over 1 week may provide protection.

JE-CV

Reported adverse events with ChimeraVax JE resemble those reported for 17 D YF vaccine. The long term events that may follow use of this chimeric live virus cannot be known. Despite the fact that ChimeraVax JE is a live attenuated virus, the current recommendation is to administer two doses.

Regardless of vaccination status, individuals travelling to endemic areas are well advised to reduce their risk of vector exposure and infection by wearing mosquito repellent and long-sleeved shirts and trousers, by avoiding outdoor activities in the evening, and by sleeping under permethrin-impregnated mosquito nets or in screened or air-conditioned rooms (Luo *et al.*, 1994).

Tick-borne encephalitis

TBEV circulates in nature as three subtypes, Western, Far Eastern and Siberian. The disease is also referred to as Central European encephalitis, Far Eastern encephalitis, Russian spring–summer encephalitis and diphasic milk fever. TBEV are related to a large group of zoonotic tick-borne flaviviruses. Some affect humans, producing encephalitis (Langat, Powassan, louping ill) or haemorrhagic fever (Kyasanur Forest disease, Omsk haemorrhagic fever). The Far Eastern and Siberian strains of TBEV are closely related, whereas Western TBEV is actually more closely related genetically to louping Ill virus, than to Central European and Siberian strains of TBEV.

Disease

The average incubation period between tick bite and onset of symptoms is 11 days (4–28 days, range). Infection with the Far Eastern subtype of TBEV results in a disease with a monophasic illness course whereas infection with the Western subtype usually produces a biphasic course. Nearly 75% of patients experienced a biphasic course of illness with the first phase lasting between 1 and 7 days. The incubation period is generally 7–14 days. During a biphasic infection symptoms initially include fever, fatigue, headache, vomiting and pain in the neck, shoulders or lower back. This may be followed by an asymptomatic period lasting 2–10 days and may progress to neurological involvement with symptoms attributable to the involvement of meninges, brain parenchyma or spinal cord. Acute TBE is characterized by encephalitic symptoms in 45–56% of patients. Symptoms range from mild meningitis to severe meningoencephalomyelitis, which is characterized by muscular weakness which develops 5–10 days after fever subsides. Severely affected patients may demonstrate altered consciousness and a poliomyelitis-like syndrome that may result in long-term disability (Dumpis *et al.*, 1999; Gritsun *et al.*, 2003). Severity of illness increases with the age of the patient. Pareses and lasting sequelae appear to occur less frequently in young patients (Kunze *et al.*, 2004). After approximately age 40, TBE patients increasingly develop the encephalitis form of the disease (Kunze *et al.*, 2005). In patients older than 60 years, TBE takes a severe course, often leading to paralysis and death. In Siberia, approximately 80% of cases present with fever but without neurological sequelae.

A chronic form of TBE is observed in patients from Siberia or far-eastern Russia and is thought to be associated with infections by the Siberian subtype of TBEV (Gritsun *et al.*, 2003). This occurs in two forms: (1) neurological symptoms that occur for years following bite of infected tick, including Kozhevnikov's epilepsy, progressive neuritis of the shoulder plexus, lateral and dispersed sclerosis and a Parkinson's-like disease with progressive muscle atrophy. This is often accompanied by mental deterioration; and (2) hyperkinesis or epileptoid syndrome. The case fatality rate is approximately 1–2% in infections with the Western subtype, but as high as 20–40% with infections by the Far-Eastern subtype. Infection with the Siberian subtype produces a mortality rate of 2–3%. There have been reports of increased severity of TBE illness in a few individuals, usually children, who were bitten by ticks and given human immune globulin as a method of immunoprophylaxis. This has been attributed by some as evidence of antibody-mediated enhanced disease (ADE), although in the absence of evidence of a central role for macrophages as infection target cells, it is more likely to be an example of antibody-mediated pathogenicity of viral infections in the brain (Webb *et al.*, 1967; Halstead, 1982; Gould and Buckley, 1989).

Epidemiology

Western TBEV is transmitted primarily by *Ixodes ricinus*, whereas the vector for the Siberian and Far-Eastern subtypes is *I. Persulcatus. I ricinus* is a three-host tick, with each parasitic stage (larva, nymph, adult) feeding for a period of a few days on a different host. Ticks become active when the mean outdoor temperatures are at or above 11°C. As each stage in the life cycle takes approximately 1 year to moult to the next stage, the entire life cycle may be completed in 2–6 years, depending on geographical location. Once infected with TBEV, ticks remain infected throughout their life cycle. Furthermore, TBEV is passed from adult transovarially to larvae. Larvae and nymphs feed predominantly on ground-dwelling rodents, particularly yellow-necked field mice and voles. Rodents can be chronically infected. In addition, uninfected larvae or nymphs can become infected when feeding on a host at the same time that infected nymphs or larvae are feeding (co-transmission). Although larger animals host ticks, these are largely adults. It is not believed that infections of large animals contribute substantially to maintain infectious cycles. In this respect, maintenance of virus may be somewhat similar to that of Lyme disease *Borellia;* infection in rodents supports the *Borellia* life cycle, while deer support and maintain the tick population. All three known subtypes of TBEV are capable of co-circulating in the same area, as is currently the situation in Estonia. The disease is widespread across Europe and Asia with recent extension to areas not previously affected, such as Norway (Mansfield *et al.*, 2009). About 3000 hospitalized cases of TBE are recorded annually in Europe (Heinz *et al.*, 2007). Russia and Western Siberia contribute the largest number of cases, 10,000–15,000 in some years. In Western Europe, formerly Austria and now the Czech Republic has the highest incidence rate, with 400–1000 cases annually (Mansfield *et al.*, 2009).

Historically, TBE has been transmitted to humans via contaminated unpasteurized milk or milk products made from chronically infected cows, goats or sheep. Although this mechanism of transmission is still possible, better standards for production and pasteurization of milk and milk products now widely practised throughout the enzootic area of TBE have greatly reduced transmission by the oral route.

There is evidence of an increase in reported cases of TBE over the past two decades. This has been attributed by some to warming temperatures in Europe and Asia, but a wide range of political, ecological, economic and demographic factors appear to play role in aiding the spread of tick-borne diseases (Kunze, 2010). This includes both increases and decreases in forestation. Newly planted fields may bring farmers into contact with infected ticks. There is a growing participation by residents and tourists in outdoor pursuits such as hiking and fishing. Also, the economic downturn may have resulted in increasing foraging for food in forests by persons who are unvaccinated. In Central Europe, two seasonal peaks of TBE occur, one in June/July and the second in September/October, corresponding to two waves of feeding by larvae and nymphs.

Vaccines

Four vaccines are used to prevent TBE disease. All are formalin-inactivated virus grown in tissue culture cells: (1) Austrian (Baxter) FSME-Immun 0.5 ml and FSME-Immun 0.25 ml Junior; (2) Encepur (Novartis) Adult and Children; (3) Moscow TBE vaccine; and (4) Tomsk TBE vaccine.(Barrett *et al.*, 2008). In Europe over 80 million doses of TBE vaccines have been administered. Austrian vaccine is prepared growing the Neudorfl virus strain in specific pathogen-free chick embryo cells with aluminium hydroxide as an adjuvant. Human albumin is added as a stabilizer to reduce tumour necrosis factor alpha and interleukin 1β production in vaccinees. The Novartis vaccine is produced using the K23 strain of TBEV grown on chick embryo cells. Following inactivation by formaldehyde, vaccine is purified by continuous-flow density-gradient ultracentrifugation. Alum is used as an adjuvant. The vaccine was initially stabilized using processed bovine gelatin. Reactions thought to be allergic phenomena to gelatin resulted in a reformulation to use sucrose as stabilizer. Of the four vaccines that are produced, that in Austria has been in use for the longest period. No formal prospective efficacy trial was conducted prior to licensing. This vaccine has been used since 1976 with upwards of 88% of the Austrian population now fully vaccinated. A prospective, randomized, phase II double-blind dose-finding study (0.6, 1.2 and 2.4 μg antigen) was carried out in volunteers, 16–65 years, using the Baxter vaccine ($n = 504$) who were bled 21 days following administration of two doses at an interval of 30 days. Similar dose-finding studies (0.6, 1.2 and 2.4 μg antigen) were carried out in children aged 1–5 years ($n = 639$) and 6–15 years ($n = 639$). Seroconversion rates varied from 85% to 97% of cases. A dose of 1.2 μg was selected as optimal. Subsequently, this dose was given to over 3000 adults in a large safety trial. Using data on the incidence of vaccination among cases of serologically proven TBEV infections, a field effectiveness of 99% has been calculated among fully immunized individuals (Heinz *et al.*, 2007).

Vaccine usage

The standard immunization schedule consists of three immunizations administered intramuscularly on day 0, a second dose administered 1–3 months later and a third dose administered 5–12 months after the second dose. Rapid immunization schedules have been introduced, three immunizations to be administered on days 0, 7 and 21. The Enceptor rapid immunization schedule includes a fourth vaccination after 12–18 months. For those under 60 years of age, booster doses are recommended at 5-year intervals; for those older than 60 years, booster doses are recommended at 3-year intervals. A retrospective study found a high level of protection following two doses of the Baxter vaccine (Heinz *et al.*, 2007). Vaccines are not recommended for use as post-exposure prophylaxis or as a therapeutic vaccine. Vaccination is recommended for persons living in TBEV enzootic areas, woodcutters, farmers, military personnel, laboratory workers and tourists who plan to hike, bicycle or camp in enzootic areas (Barrett *et al.*, 2008).

Precautions

An immune globulin concentrate containing IgG antibodies against TBEV was in use as a post-exposure treatment. However, reports of severe clinical outcomes in patients given TBE immune globulins plus evidence of *in vitro* ADE led to critical scrutiny and the abandonment of post-exposure treatment (Waldvogel *et al.*, 1996; Ozherelkov *et al.*, 2008; Aebi and Schaad, 1994). Recent studies in a mouse model of TBE failed to demonstrate antibody enhanced infection (Kreil *et al.*, 1998). Some workers continue to argue that during critical time windows administration of high levels of neutralizing antibodies are protective (Pen'evskaia and Rudakov, 2010; Broker and Kollaritsch, 2008).

As with all vaccines given intramuscularly, local reactions such as reddening, swelling and pain occur frequently, while fever occurs occasionally after the first injection. Reactions are less common accompanying second or third injections (Barrett *et al.*, 2008).

Vaccines under development

Dengue fever

Primary infections with dengue virus types 2 and 4 usually are inapparent in children and this is often the case in adults also (Halstead, 1997; Vaughn, 2000; Vaughn *et al.*, 2000; Guzman *et al.*, 2000). By contrast, while primary infections with dengue virus types 1 and 3 in some children are relatively mild, others are quite severe. Most adults infected with these viruses develop biphasic fever, rash and other characteristic features of the dengue fever syndrome (Sabin, 1952; Halstead, 1997; Vaughn, 2000; Vaughn *et al.*, 2000; Guzman *et al.*, 2000). In infants and young children, the disease may be undifferentiated or characterized by a 1- to 5-day fever, pharyngeal inflammation, rhinitis and mild cough.

In classic dengue fever, after an incubation period of 2–7 days, patients experience a sudden onset of fever, which rapidly rises to 39.5–41.4 °C (103–106° F) and is usually accompanied by frontal or retro-orbital headache. Occasionally, back pain precedes the fever. A transient, macular, generalized rash that blanches under pressure may be seen during the first 24–48 hours of fever. The pulse rate may be slow in proportion to the degree of fever. Myalgia or bone pain occurs soon after onset and increases in severity. During the second to sixth day of fever, nausea and vomiting are apt to occur, and generalized lymphadenopathy, cutaneous hyperaesthesia or hyperalgesia, aberrations in taste and pronounced anorexia may develop. One or 2 days after defervescence, a generalized, morbilliform, maculopapular rash appears, with sparing of the palms and soles. It disappears in 1–5 days. In some cases, oedema of the palms and soles may be noted, and desquamation may occur. About the time of appearance of this second rash, the body temperature, which has fallen to normal, may become elevated slightly and establish the biphasic temperature curve. Epistaxis, petechiae and purpuric lesions, though uncommon, may occur at any stage of the disease. Swallowed blood from epistaxis may be passed per rectum or be vomited and could be interpreted as bleeding of gastrointestinal origin. Gastrointestinal bleeding, menorrhagia and bleeding from other organs have been observed in some dengue fever outbreaks. Peptic ulcers predispose to gastrointestinal haemorrhage during dengue infections; some patients may exsanguinate during an otherwise normal dengue fever (Tsai *et al.*, 1991). This syndrome occurs without vascular permeability and contributes to a misunderstanding of the case definition of dengue haemorrhagic fever/dengue shock syndrome (DHF/DSS). After the febrile stage, prolonged asthenia, mental depression, bradycardia and ventricular extrasystoles are noted commonly in adults (Lumley and Taylor, 1943).

Dengue haemorrhagic fever/dengue shock syndrome (DHF/DSS)

DHF/DSS is an acute vascular permeability syndrome accompanied by abnormal haemostasis. The incubation period of DHF/DSS is unknown, but presumed to be the same as that of dengue fever. In children, progression of the illness is characteristic (Cohen and Halstead, 1966; Nimmannitya *et al.*, 1969; Nimmannitya, 1987; World Health Organization, 1997). A relatively mild first phase with an abrupt onset of fever, malaise, vomiting, headache, anorexia and cough may be followed after 2–5 days by rapid deterioration and physical collapse. In Thailand, the median day of admission to the hospital after the onset of fever is day 4. In this second phase, the patient usually has cold, clammy extremities, a warm trunk, a flushed face, and diaphoresis. Patients are restless and irritable and complain of midepigastric pain. Frequently, scattered petechiae appear on the forehead and extremities, spontaneous ecchymoses may develop, and easy bruisability and bleeding at sites of venepuncture are common findings. Circumoral and peripheral cyanosis may occur. Respirations are rapid and often laboured. The pulse is weak, rapid and thready, and the heart sounds are faint. The pulse pressure is frequently narrow ($\leq$ 20 mmHg); systolic and diastolic pressure may be low or unobtainable. The liver may become palpable two or three fingerbreadths below the costal margin and is usually firm and non-tender. Chest radiographs show unilateral (right) or bilateral pleural effusions. Approximately 10% of patients have gross ecchymosis or gastrointestinal bleeding. After a 24- or 36-hour period of crisis, convalescence is fairly rapid in children who recover. The temperature may return to normal before or during the stage of shock.

Epidemiology

The four dengue viruses evolved from a common ancestor in subhuman primates and moved into the 'urban' *Aedes aegypti*–human–*Aedes aegypti* cycle somewhere around 4–500 years ago (Wang *et al.*, 2000). It is likely that *A. aegypti* moved from Africa throughout the world during the era of European exploration and colonization. This mosquito provided an ecological niche quickly occupied by several human viral pathogens: yellow fever, chikungunya, and the dengue viruses. During the 18th and 19th centuries, largely because of the necessity for storage of domestic water in frontier areas, *A. aegypti*-borne epidemics occurred in newly settled lands. Isolated shipboard or garrison outbreaks, often confined to non-indigenous settlers or visitors, were reported in Africa, the Indian subcontinent, and South-East Asia (Siler *et al.*, 1926). Simmons and colleagues were the first to note that wild-caught *Macaca philippinensis* resisted dengue infection whereas *Macaca fuscatus* (Japanese macaque) individuals were susceptible. (Simmons *et al.*, 1931). Work by Rudnick in Malaysia revealed a jungle cycle of dengue transmission involving canopy-feeding monkeys and *Aedes niveus*, a species that feeds on both monkeys and humans (Rudnick, 1978). In the early 1980s, an extensive epizootic of DENV 2 involved subhuman primates over wide areas of West Africa (Roche *et al.*, 1983). The full geographic range of the subhuman primate zoonotic reservoir is not known for Asia or Africa. Although the urban human dengue and sylvatic dengue cycles are relatively compartmentalized, strains of sylvatic dengue cause overt disease in humans (Cardosa *et al.*, 2009; Wang *et al.*, 2000; Vasilakis *et al.*, 2010). Urban dengue is vectored by anthropophilic mosquitoes, and the virus travels along routes of transportation. *A. aegypti* and susceptible humans are so abundant and so widespread that detection of low-level exchange of DENV between humans and monkeys will be extremely difficult. Should urban dengue transmission be eliminated by vaccination but without eradicating *Aedes aegypti*, the reintroduction of virus from jungle cycles could become a health threat similar to that of yellow fever today.

Dengue outbreaks were commonly reported in Asia, Europe and the Americas during the 19th century. Of these, there is serological evidence that dengue 1 and 2 caused 1927–28 outbreaks in Greece and dengue 4 and 1 viruses, respectively, were used in the 1923 and 1929 experimental studies of Siler and Simmons in the Philippines (Halstead, 1974; Halstead and Papaevangelou, 1980). Dengue 2, American genotype, probably circulated in the Caribbean prior to the 1940s (Anderson and Downs, 1956; Rosen, 1958; Guzman *et al.*, 1990). During the Second World War, dengue 1 and 2 virus infections occurred commonly in combatants of the Pacific War and spread to staging areas not normally infected: Japan, Hawaii, and Polynesia (Sabin, 1952). Dengue 3 and 4 were identified in 1956 from the Philippines by Hammon and colleagues from a new disease called dengue haemorrhagic fever (DHF) (Hammon *et al.*, 1960). A DHF-like disease was described clinically in Thailand beginning in 1950 and in the Philippines from 1953 (Halstead and Yamarat, 1965; Quintos *et al.*, 1954). Cases were confirmed aetiologically as dengue in 1958 and 1956, respectively (Hammon *et al.*, 1960). The circulation of multiple dengue viruses and the occurrence of DHF was first described in Singapore and Malaysia in 1962, Vietnam in 1963, India in 1963, Ceylon (Sri Lanka) in 1965, Indonesia in 1969, Burma (Myanmar), in 1970, China in 1985, and Kampuchea and Laos from about 1985, and major outbreaks have occurred in Sri Lanka and India since 1988, in French Polynesia since 1990, in Pakistan since 1998, and in Bangladesh since 1999. (Halstead, 1980a, 1997). DHF occurs at consistently high endemicity in Thailand, Burma, Vietnam and Indonesia.

In the Americas, shortly after the end of the Second World War, a major effort was made to eradicate *A. aegypti*. After initial success in the 1970s this effort failed. One by one, each of the four dengue viruses was imported from South-East Asia to the Americas (Gubler, 1997b). After the introduction of DENV 1 in 1977, multiple dengue viruses are now endemic on the larger Caribbean islands and in coastal Central America and the tropical areas of Guyana, Venezuela, Colombia, Ecuador, Peru and Brazil (Halstead, 2006). A sharp DHF epidemic due to DENV 2 in Cuba in 1981 led to island-wide *A. aegypti* control and apparent eradication of the virus (Guzman *et al.*, 1990). In 1986 and 1987, dengue virus type 1 spread through most of coastal Brazil and from there to Paraguay and to Peru and Ecuador (1979; Gubler, 1989). In 1990, more than 9000 dengue cases were reported from Venezuela; 2600 of them were classified as DHF, and 74 deaths were associated with the epidemic (Ramirez-Ronda and Garcia, 1994). Dengue virus types 1, 2 and 4 were isolated. Shortly thereafter, DHF/DSS caused by dengue type 2 was reported from Brazil and French Guiana (Nogueira *et al.*, 1991; Reynes *et al.*, 1994). In 1995, dengue virus type 3 was introduced into the region. In 1997, dengue virus type 2 with a South-East Asian genotype was introduced into Santiago de Cuba and caused a sharp DHF/DSS outbreak observed only in individuals 20 years and older (Guzman *et al.*, 2000).

During the past 20 years, major epidemics of all four dengue serotypes have occurred on many Pacific islands (Gubler, 1989; Halstead, 1997).

In Africa, dengue virus types 1 and 2 were recovered from humans with mild clinical illness in Nigeria in the absence of epidemic disease (Carey *et al.*, 1971). In 1983, dengue virus type 3 was isolated in Mozambique (Gubler, 1997a). DHF/DSS has not been reported, and large dengue fever outbreaks are rare. In this respect, Africa resembles the situation in Haiti, where multiple dengue serotypes are transmitted at high rates among a predominantly black population, but severe disease is not recognized (Halstead *et al.*, 2001). Today, it is estimated that dengue viruses circulate in over 100 countries with a combined population of nearly 3 billion, with perhaps 50 million persons infected, up to10 million with DF and 0.5 million hospitalized with DHF annually.

Dengue vaccines – general consideration

The immunopathological role of dengue antibodies together with the co-circulation of multiple DENV dictates that effective vaccines must be tetravalent. The challenge is to show that vaccines protect against all four dengue viruses but do not directly cause symptoms in the vaccinated subject or provoke other safety problems. This section will review those vaccines that are or have been in human clinical trials or have been tested in subhuman primates. In assessing preclinical data, because administration of attenuated live viruses or inactivated virus preparation often has resulted in rapid waning of neutralizing antibody titres, those studies in subhuman primates in which neutralizing antibody titres were measured at least 6 months after completion of immunization series or trials that included challenge with wild-type DENV will receive special emphasis. Brief mention will be made of other DENV vaccine candidates. A number of excellent reviews on dengue vaccine development have been published recently (Webster *et al.*, 2009; Durbin and Whitehead, 2010; Blaney *et al.*, 2010; Guzman *et al.*, 2010; Swaminathan *et al.*, 2010).

Classical live-attenuated vaccines

Mahidol University vaccine

The discovery in the early 1980s that DENV were selected for attenuation attributes during serial passage in primary dog kidney (PDK) cells led to the development of a live-attenuated tetravalent dengue vaccine by investigators at the University of Hawaii and Mahidol University in Bangkok, Thailand (Halstead and Marchette, 2003). PDK-passaged viral strains DEN-1 16007, DEN-2 16681 and DEN-4 1036 were tested in small groups of flavivirus-susceptible adult Thai volunteers. Passage levels with acceptable reactogenicity and immunogenicity were identified for DEN 1, 2 and 4 viruses at passage levels PDK13, PDK53 and PDK48, respectively (Bhamarapravati and Sutee, 2000; Bhamarapravati *et al.*, 1987; Bhamarapravati and Yoksan, 1997). In the hands of these investigators DENV-3 viruses consistently failed to replicate in PDK cells and attenuation was attempted by serial passage in primary African green monkey kidney (GMK) cells. Based upon studies in flavivirus susceptible volunteers, an acceptable vaccine candidate was found for DENV-3 16562 at the 48th GMK cell passage with three final passages in fetal rhesus lung (FRhL) cells (S. Yoksan, personal communication, 1995). Serial passage selected for amino acid changes, DENV-1 16007, PDK 13 differed from parental virus by 14 amino acids, while DENV-2 16681, PDK 53 vaccine differed by nine amino acids. (Huang *et al.*, 2000, 2003).

A single dose of vaccine candidate DEN-2 16681 PDK53 in 10 US Army soldiers had acceptable reactogenicity and raised neutralizing antibodies in all subjects (Vaughn *et al.*, 1996). The four monovalent candidates elicited neutralizing antibody seroconversion in 3/5, 5/5, 5/5 and 5/5 American volunteers, respectively, after a single dose of approximately 10^3–10^4 pfu (Kanesa-thasan *et al.*, 2001). PDK vaccine strains when cloned and passaged in Vero cells retained phenotypic attributes including immunogenic attenuation and immunogenicity for human beings. (Kinney *et al.*, 2010a,b).

Bivalent and trivalent formulations using DEN-1, -2 and -4 vaccine candidates elicited uniform seroconversions in Thai subjects (Bhamarapravati and Yoksan, 1989). However, when all four serotypes were combined into a tetravalent vaccine, the predominant response was the development of DENV-3 viraemia with a predominant neutralizing antibody response to DENV-3 (Kanesa-thasan *et al.*, 2001; Sun *et al.*, 2003; Sabchareon *et al.*, 2004; Kitchener *et al.*, 2006). However, in Thai children, some of whom had been previously infected by dengue viruses, seroconversion rates were higher and neutralizing antibodies persisted for many years (Sabchareon *et al.*, 2002, 2004; Chanthavanich *et al.*, 2006). Further evidence of innate DENV 3 reactogenicity was obtained when a mutagenized and Vero-adapted strain of DENV-3 16562 produced clinical dengue fever in 15 of 15 susceptible adult volunteers (Sanchez *et al.*, 2006). This result led to the abandonment of further development of the Mahidol live-attenuated vaccine.

Walter Reed Army Institute of Research vaccine

The Walter Reed Army Institute of Research (WRAIR) also developed a tetravalent live-attenuated dengue vaccine based upon serial passage of wild-type DENV, including DENV-3 in PDK cells. Vaccine candidates (specific virus strains and PDK cell passage levels) were selected based on the results of a series of Phase 1 safety and immunogenicity studies conducted at the WRAIR and the University of Maryland School of Medicine Center for Vaccine Development (Kanesa-Thasan *et al.*, 2003). Selected candidate vaccines were produced with terminal passage in FRhL cells – DEN-1 45AZ5 PDK20/FRhL3, DEN-2 16803 PDK50/FRhL3, DEN-3 CH53489 PDK20/FRhL3 and DEN-4 341750 PDK20/FRhL4. The DENV-2 and DENV-4 candidates had been selected for attenuation at the University of Hawaii (Halstead and Marchette, 2003; Marchette *et al.*, 1990) and tested for clinical responses at WRAIR (Kanesa-Thasan *et al.*, 2003; Hoke *et al.*, 1990). Among the selected passage levels, the seroconversion rates were 100%, 92%, 46% and 58% for a single dose of DEN-1–4 respectively. The WRAIR DEN-2, -3 and- 4 vaccine viruses were well tolerated by volunteers. The DEN-1 PDK20/FRhL3 monovalent candidate was associated with increased reactogenicity, with 40% developing fever and generalized rash. Vaccine related reactions consisted of modified symptoms of dengue fever to include headache, myalgia, and rash (Edelman *et al.*, 1994, 2003; Mackowiak *et al.*, 1994; Sun *et al.*, 2003).

Sixteen different dose formulations of the WRAIR tetravalent vaccine were tested in 64 adult volunteers (Edelman *et al.*, 2003; Sun *et al.*, 2009). The formulations were derived by using undiluted vaccine (5–6 logs of virus) or a 1:30 dilution for each virus serotype. Seroconversion rates after a single dose of tetravalent vaccine were 83%, 65%, 57% and 25% to DENV-1–4 respectively, similar to that seen with monovalent vaccines. Few additional seroconversions were seen following a booster dose 1 month after the first dose. Though the sample size was small, a trend towards increased reactogenicity was observed when a full dose of the DENV-1 component was combined with lower doses of DENV-2 and/or DENV-4. With these viral strains, at the doses evaluated, viral interference did not affect antibody response but may have modified reactogenicity. In an attempt to reduce reactogenicity due to the DEN-1 component and to increase the immunogenicity of the DENV-4 component, a 17th formulation using DENV-1 45AZ5 PDK 27 and DENV-4341750 PDK 6 was evaluated. Expanded testing of vaccine formulations along with booster doses at an interval of 6 months yielded higher seroconversion rates along with acceptable reactogenicity in adults (Sun *et al.*, 2009).

Clinical testing of a battery of 16 different formulations of the WRAIR vaccine showed the DENV-1 component to be slightly over attenuated and the DENV-4 component to be slightly under-attenuated (Sun *et al.*, 2003; Edelman *et al.*, 2003). As a result a modified formulation (Formulation 17) was developed by replacing the DENV-1 and DENV-4 components with the corresponding viruses from later and earlier PDK passages. A phase II trial of Formulation 17 in adults in the US was reported to induce tetravalent seroconversion in 63% of volunteers who received two doses (Sun *et al.*, 2009). Although hospital-based studies in Thailand have shown that most primary dengue virus infections in young children are clinically silent (Vaughn *et al.*, 2000; Green *et al.*, 1999). DHF/DSS is regularly seen during secondary infections in infants 2 years of age and older. For this reason an evaluation of tetravalent vaccines in infants was required. The new formulation WRAIR vaccine was found to be well tolerated and immunogenic in a phase I paediatric study in Thailand (Simasathien *et al.*, 2008).

Following the publication of these studies a pause in clinical trials was announced (S. Thomas, personal communication, 2008) during which each DENV vaccine strain was rederived genetically. At present, further development work at WRAIR on the LAV vaccine has apparently stalled. It is not clear whether or not this pause is related to the report that passage of dengue viruses in FRhL cells resulted in a Glu237 to Gly substitution that enhanced interaction with heparan sulfate and reduced infectivity of both vaccine and wild-type DENV 4 strains for susceptible rhesus monkeys (Anez *et al.*, 2009).

Precautions

There are two main safety issues with live dengue vaccines. First, a principal concern is the theoretical risk of enhanced disease following incomplete tetravalent immunization. A tetravalent vaccine approach was adopted for the purpose of inducing primary-type immune responses to all four dengue viruses simultaneously, a phenomenon readily demonstrated when wild-type DENV were inoculated in susceptible rhesus monkeys (Halstead and Palumbo, 1973). In monkeys, these neutralizing antibodies yielded solid immunity to challenge. As discussed above, primary vaccine failures have been observed frequently to one or more components following tetravalent vaccination. It is possible, but not yet demonstrated in an experimental model or in human subjects, that these individuals are at risk to antibody-enhanced disease during a subsequent infection with a wild-type dengue virus. Enhanced seroconversions, but not enhanced disease, were observed when an underattenuated DENV-2 vaccine was given to yellow fever immune humans (Bancroft *et al.*, 1984). Duration of immunity following vaccination could be a problem, as it has been shown that the severity of disease accompanying second DENV-2 infections increased as the interval between first and second infections lengthened (Guzman, 2002).

Second, enhanced vaccine reactogenicity might occur when attenuated DENV viruses are administered to individuals circulating flavivirus antibodies. It has been observed that wild-type DENV-4 and DENV-2 viruses may not cause overt disease in flavivirus-naive humans (Vaughn, 2000; Guzman *et al.*, 1990). This implies that when inoculated with apparently 'attenuated' laboratory passaged dengue strains inoculated in partially immune volunteers these DENV may become 'virulent', producing antibody-enhanced infection and disease.

Because more than 95% of secondary dengue virus infections in humans are inapparent, safety tests on live-attenuated vaccines must be performed on large numbers of partially dengue-immune volunteers to adequately assess these two risks (Halstead, 1980b). As yet, increased reactogenicity has not been observed in clinical trials conducted in Thailand and in the United States in which 2 doses of tetravalent vaccines were given 6 months apart (R. Edelman and J.F. Saluzzo, personal communications) A better understanding of the immune mechanisms of vaccine protection and enhanced disease is needed to better determine risks associated with phase III field trials of LAV.

Other safety concerns with live-attenuated dengue viral vaccines include the presence of cell culture-derived adventitious agents in the final vaccine product, community spread of vaccine virus present in viraemic immunized humans by local vector mosquitoes, vaccine virus neurovirulence, and the effects of vaccine administration in immunocompromised hosts.

Interference

A mixture of DENV-1, -2 and -4 serially passaged in PDK cells with a DENV-3 that had been serially passaged in African green monkey kidney cells (AGMK) resulted in overgrowth of DENV-3 in flavivirus susceptible humans (Sabchareon *et al.*, 2004; Sun *et al.*, 2003; Kitchener *et al.*, 2006). Only DENV-3 viral RNA was recovered from volunteers and neutralizing antibody responses were predominantly to DENV-3 (Kanesa-thasan *et al.*, 2001). A similar interference or overgrowth phenomenon was not observed when all four dengue viruses that had been attenuated by passaging in PDK cells and then in FRhL cells were given to flavivirus-susceptible rhesus monkeys or to adult human volunteers (Sun *et al.*, 2003, 2006, 2009). This phenomenon has not been studied in vaccinated humans or reproduced in an animal model; accordingly, there are no data available for hypothesis-making. A possibly similar interference phenomenon was observed in humans and in monkeys given mixtures of the four yellow fever DENV chimeras (see below) (Guy, 2009; Guy *et al.*, 2009).

Chimeric vaccines

The flavivirus genome is a single-stranded, positive-sense RNA molecule of nearly 11 kb containing a single open reading frame. The RNA is translated into a polyprotein that is processed into at least 10 gene products: the three structural proteins nucleocapsid or core (C), premembrane (prM), and envelope (E) and seven non-structural (NS) proteins, NS1, NS2A, NS2B, NS3, NS4A, NS4B, and NS5. The untranslated regions of the genome at the 5′ and 3′ ends are crucial for protein translation and minus strand transcription (Henchal and Putnak, 1990).

Yellow fever virus as molecular backbone

Acambis/Sanofi Pasteur vaccine (CYD)

Chimeric DENV vaccines have been derived by coupling genes encoding dengue virus prM and E structural proteins with capsid and non-structural genes from an attenuated yellow fever vaccine virus. The hypothesis is that the attenuation characteristics of 17D YF vaccine virus will impart similar behaviour to the chimera (see 'JE-CV', above). Insertion of dengue genes into the cDNA backbone of yellow fever17D was developed commercially by Acambis, Inc. and licensed to Sanofi Pasteur (Rice *et al.*, 1989; Chambers *et al.*, 1999; Guirakhoo *et al.*, 2000, 2001). Chimeric viruses are replicated by the YF RNA polymerase. Vero cells serve as the substrate for vaccine virus production.

Dengue/yellow fever virus chimeras have been developed for each of the four dengue virus serotypes using the *PrM-E*-genes from DENV-1 PUO-359, DENV-2 PUO-218, DENV-3 PaH881, and DENV-4 1228 viruses (Guirakhoo *et al.*, 2000; Guirakhoo *et al.*, 2001). Neurovirulence in mice inoculated with dengue/YF chimeras was reduced compared to the yellow fever 17D vaccine virus (Monath *et al.*, 2005). Viraemia in rhesus macaques is similar to that of yellow fever 17D virus and greatly reduced compared to wild-type dengue viruses. Seroconversion in rhesus monkeys was 100% using a tetravalent formulation comprising 5 logs of DENV1–4 chimeras and 3 logs of DENV1–4 chimeras (Guirakhoo *et al.*, 2001, 2004). High-titred DENV 1–4 neutralizing antibody titres were seen with 5 logs of each component and somewhat lower titres with 3 logs of each component. Rhesus monkeys vaccinated with 5 logs of each component were protected from viraemia when challenged with all four wild-type dengue viruses but, with mixed doses, limited breakthrough viraemia was observed with DENV-1 and DENV-4 wild-type virus challenge (Guirakhoo *et al.*, 2001, 2004; Guirakhoo and Pugachev, 2002). Encouraging phase I data for the yellow fever 17D – Japanese encephalitis virus chimera (ChimeriVax-JE) suggested that this approach would also be effective for dengue. (Monath *et al.*, 2002a) Data in humans demonstrated high rates of seroconversion with a DENV-2/YF chimera (CYD2) with higher levels of neutralizing antibodies produced in individuals previously vaccinated with 17D YF (Guirakhoo *et al.*, 2006a).

Clinical studies of tetravalent ChimeriVax Dengue vaccines have progressively involved larger and larger groups of susceptible, yellow fever-immune and dengue-immune human volunteers (Morrison *et al.*, 2010; Lang, 2009; Guy, 2009; Guy *et al.*, 2010). As of April 2009, over 880 children and adults had received mono or tetravalent CYD in phase I or II trials in Asia and the Americas (Guy *et al.*, 2010). Phase IIb/phase III evaluations of tetravalent dengue vaccine were initiated in Thai children in 2009 (J. Lang, personal communication). In susceptible humans, low levels of viraemia were observed following administration of tetravalent vaccines, with DENV-4 being recovered most frequently (Guy and Almond, 2008; Guy, 2009). Viraemia was noted less frequently following administration of a second or third dose of vaccine. Approximately 12% of susceptible adults develop neutralizing antibodies to all four DENV after receiving a single dose of tetravalent vaccine formulated to contain 10^5 pfu/ml each (Guy, 2009; Morrison *et al.*, 2010). It was observed that administration of a second dose of live vaccine did not successfully boost antibody response unless delivered 4–6 months after the first dose; further, a third dose, also administered 6 months later, was require to achieve 100% tetravalent neutralizing antibody response (Guy *et al.*, 2010). Limited data suggest that by extending the interval between first and second dose of tetravalent vaccine to 12–15 months, nearly 100% tetravalent seroconversions can be achieved (Morrison *et al.*, 2010; Guy *et al.*, 2010). In individuals circulating YF antibodies or children circulating dengue antibodies, administration of two doses of CYD was sufficient to evoke full tetravalent neutralizing antibody responses (Guy *et al.*, 2010).

Interference

If virus replication depends upon non-structural genes and/or non-coding regions of the genome those chimeras using a common genetic backbone for all component viruses should not result in growth interference. However, with the yellow fever 17D–dengue virus chimera (ChimeriVax-DEN), different formulations of the four component viruses resulted in significantly different neutralizing antibody responses (Guirakhoo *et al.*, 2001). This suggests that interference may occur with chimeric vaccines. While the mechanism of interference is unknown, it is unlikely to be the same mechanism that resulted in DENV-3 overgrowth in the Mahidol/Sanofi Pasteur tetravalent LAV (see above). Recombination events might occur following administration of tetravalent chimeric vaccines but, such events are unlikely to result in reversion to a more virulent phenotype. Of interest, has been the ability to circumvent interference in susceptible rhesus monkeys by simultaneous inoculation of bivalent vaccines at different sites.(Guy *et al.*, 2009).

Dengue 2 virus as molecular backbone

CDC vaccine

The US Centers for Disease Control and Prevention (CDC) is developing a tetravalent chimeric dengue vaccine by introducing DENV-1, -3 and -4 C, prM and E genes into cDNA derived from the successfully attenuated DENV-2 component of the Mahidol University/Aventis Pasteur LAV vaccine (DENV-2 16681 PDK-53). (Vaughn *et al.*, 1996; Kinney *et al.*, 1997; Kinney and Huang, 2001)Previously, this group had reported that the attenuation markers for DENV-2 vaccine reside outside the structural gene region providing a justification for using this virus backbone to receive *C-PrM-E* genes from the other dengue virus types. (Butrapet *et al.*, 2000) Chimeras for DENV-1 have been produced using structural genes from both the Mahidol DENV-1 PDK-13 vaccine virus and the near wild-type DENV-1 16007 virus. The structural genes of the DENV-1 16007 virus appear to be more immunogenic in mice than those of the PDK-13 vaccine virus. (Huang *et al.*, 2000) Chimeras for DENV-3 and DENV-4 have been constructed using the wild-type virus structural genes, and have characterized in monkeys (D. Stinchcomb, Inviragen, personal communication).

Dengue 4 virus as molecular backbone

NIH vaccine

A recombinant DENV-4 described in the following paragraph served as a chimera to produce vaccines by substituting genes for DENV-1 and DENV-2 *C-PrM-E* to code for the respective viral structural proteins (Bray and Lai, 1991; Lai *et al.*, 1998). These chimeric viruses produced robust antibody responses and protection from challenge with wild-type viruses when used as monovalent vaccines or bivalent vaccines in rhesus monkeys (Bray *et al.*, 1996) Work with the DENV-3/DENV-4 chimera is in progress (Chen *et al.*, 1995).

DENV 2/4 chimera

To make recombinant (r) DENV2/4Delta30(ME) the prM and E structural proteins of the DEN4 candidate vaccine rDENV4Delta30 were those of DENV2 NGC. rDEN2/4Delta30(ME) was evaluated at a dose of 1000 pfu in 20 healthy dengue-naive adult volunteers. Low-level vaccine viraemia, transient asymptomatic rashes and mild neutropenia were noted in some vaccinees. All developed and maintained significant neutralizing antibody titres over six months observation. This virus was selected as a future component of a tetravalent formulation (see below) (Durbin *et al.*, 2006b).

DENV 3/4 and DENV1/4 chimeras

A vaccine candidate, rDENV3/4Delta30(ME), was made which contain the membrane (M) precursor and envelope (E) genes of DENV2 and DENV3, respectively, and a 30 nucleotide deletion (Delta30) in the 3′ untranslated region of the DENV4 backbone. Based on the promising preclinical phenotypes of rDENV2/4 and rDENV3/4 four antigenic chimeric DENV1/4 viruses were made DENV-1 Puerto Rico/94 strain. rDENV1/4Delta30(ME) was significantly attenuated in a SCID-HuH-7 mouse xenograft model with a 25-fold or greater reduction in replication compared to wild type DENV-1, produced reduced viraemia in rhesus monkeys followed by production on neutralizing antibodies and was non-infectious for *Aedes aegypti* mosquitoes and is scheduled for clinical evaluation in humans (Blaney *et al.*, 2008).

Tetravalent NIH vaccine

A phase I clinical trial of a mixture of all four attenuated chimeric or mutant (see below) DENV is scheduled for completion by the fall of 2010. Results from this clinical trial have not yet been reported (2011).

DENV2/WN chimera: RepliVAX

Yet another approach uses a capsid-deleted WNV vector to produce DENV pseudo-infectious derivatives (Suzuki *et al.*, 2009). When a RepliVax vector containing DENV 2 prM and E genes was propagated in a cell line that constituitively expressed WNV c protein a construct was produced that underwent one cycle of replication and successfully raised neutralizing antibodies in AG 129 mice. The DENV-WNV chimeras are expected to retain the wild-type WNV attribute of high replicative efficiency enhancing the productive potential of such an approach (Suzuki *et al.*, 2009).

Molecularly attenuated viruses

Dengue 1, 3, 4 NIH vaccine

The cloning of a DENV-4 (WRAIR 814669, Dominica, 1981) by C.J. Lai and colleagues at the National Institutes of Health permitted the introduction of selected genes in a dengue genome and opened a new era of dengue vaccine research (Lai *et al.*, 1991). In rhesus monkeys, a DENV-4 mutant (DENV-4 2AΔ30), transcribed from recombinant cDNA, with a 30-nucleotide deletion in the 3″ untranslated region (10478–10507) produced lower viraemia and slightly decreased neutralizing antibody responses compared with wild-type DENV 4 (Men *et al.*, 1996). After being grown in Vero cells, this virus evaluated in 20 volunteers (Durbin *et al.*, 2001). Following inoculation of 10^5 pfu in 0.5 ml subcutaneously there was no injection site soreness or swelling and only minimal systemic symptoms. Ten volunteers (50%) experienced a mild rash (unnoticed by the volunteers themselves). Seven volunteers reported mild headaches lasting less than 2 hours. One volunteer reported a fever of 38.6°C without other symptoms. A mild increase in serum ALT levels was seen in five volunteers (25%) with a transient increase to five times the upper limit of normal (238 IU/l) for one volunteer. Seventy per cent of volunteers had low-level viraemia and all developed neutralizing antibody (mean reciprocal neutralizing antibody titres of 662 for those who were viraemic and 426 for non-viraemic volunteers 28 days following a single dose of vaccine). To lessen ALT responses, additional mutations were introduced into rDENV4Delta30 to further attenuate the virus in animal models and ultimately human vaccinees. Based on the elevated liver enzymes associated with the 10^5 pfu dose of rDENV4Delta30 and the known involvement of liver infection in dengue virus pathogenesis, a large panel of mutant viruses was screened for level of replication in the HuH-7 human hepatoma cell line, a surrogate for human liver cells and selected viruses were further analysed for level of viraemia in SCID-HuH-7 mice. It was hypothesized that rDENV4Delta30 derivatives with restricted replication *in vitro* and *in vivo* in HuH-7 human liver cells would be restricted in replication in the liver of vaccinees. Two mutations identified by this screen, NS3 4995 and NS5 200,201, were separately introduced into rDENV4Delta30 and found to further attenuate the vaccine candidate for SCID-HuH-7 mice and rhesus monkeys while retaining sufficient immunogenicity in rhesus monkeys to confer protection (Blaney *et al.*, 2010). Clinical studies of rDENV4Delta30–4995 revealed a high incidence of an erythematous rash at the injection site, a low level of ALT elevation, no viraemia and adequate neutralizing antibody response (Wright *et al.*, 2009). In humans, the rDENV4Delta30-200,201 vaccine candidate administered at 10^5 pfu exhibited greatly reduced viraemia, high infectivity and lacked liver toxicity while inducing serum neutralizing antibody at a level comparable to that observed in volunteers immunized with rDENV4Delta30 (McArthur *et al.*, 2008) .

DENV-1 vaccine

A similar rDENV1Delta30 construct was made and evaluated in animal models and found to be attenuated and immunogenic. rDENV1Delta30 was given at a dose of 10^3 pfu to 20 healthy adult human volunteers. Neutropenia and a transient asymptomatic rash was observed in 40% of vaccinees while 95% developed homotypic neutralizing antibodies that persisted throughout 6 months of observation. This candidate was selected for inclusion in a tetravalent dengue vaccine formulation (Durbin *et al.*, 2006a).

Dengue 1 FDA vaccine

Investigators at the Center for Biologics Evaluation and Research at the US Food and Drug Administration (FDA) created a chimeric virus combining a DENV-2 with the terminal 3′ stem and loop structure of West Nile virus. This virus grew normally in mammalian LLC-MK2 cells but was severely restricted for growth in C6/36 insect cells and was designated ‘mutant F’ or ‘mutF’ (Zeng *et al.*, 1998). A DEN-1 mutF virus was created starting with an infectious clone of the DEN-1 West Pacific; this virus shared the phenotype of restricted growth in insect cells and was evaluated in rhesus macaque monkeys (Markoff *et al.*, 2002). Viraemia in rhesus monkeys was greatly reduced compared to DEN-1 West Pacific parent suggesting that the chimera will be less reactogenic in humans (Edelman *et al.*, 1994) .Immune responses were similar to the wild-type virus, all animals seroconverted with mean reciprocal neutralizing antibody titres of 320 and 240 for wild-type and DEN-1mutF virus recipients, respectively. Monkeys challenged 12 or 17 months following a single dose of vaccine were protected from viraemia. Phase I testing is planned.

The two approaches described may generate tetravalent dengue vaccines either by separately introducing 3′ terminus changes into each of the four dengue viruses or by making structural gene chimeras so as to include all four sets of dengue antigens. Such a vaccine may have advantages over empiric attenuation in cell culture because there is a known molecular basis to attenuation and there will be a reduced risk of adventitious agents resulting in lower product quality assurance costs. It remains to be determined if genetically altered viruses have lost dengue-like virulence and whether interference phenomena can occur when these viruses are given as a tetravalent formulation.

Vectored vaccines

With the advent of genetic engineering, many attempts have been made to develop vaccines by inserting the gene for structural proteins in replicative carriers. The virus most commonly used early in this era was vaccinia. The first such vaccine was constructed by inserting 80% of the E gene of DENV 2, NGC. Two or three vaccine doses given to rhesus monkeys intramuscularly reduced or eliminated viraemia following inoculation of parental DENV 2 (Men *et al.*, 2000).

Adenovirus as vector

At least two laboratories have developed second generation adenovirus complex vectors to express and present dengue antigen. Two constructs were developed, one with prM and E genes of DENV-1 and 2 and the other with the same genes of DENV-3 and -4 (Holman *et al.*, 2007). These vaccines were mixed and given to a large group of rhesus monkeys. Twenty per cent of animals raised tetravalent neutralizing antibodies following one dose but 100% developed high-titred neutralizing antibodies when boosted with a second dose 3 months later. When challenged with live DENV 1–4 at 1 or 6 months, all animals were protected against DENV 1–3 with moderate protection to DENV-4 challenge (Raviprakash *et al.*, 2008). A second laboratory has inserted domain III of DENV-2 (EDIII) into an adenovirus vector. Using an adenovirus prime followed by a booster immunization of EDIII raised neutralizing antibodies and dengue antigen responsive CD8+ T cells in mice (Khanam *et al.*, 2006, 2007). This work was extended by creating a tetravalent antigen derived by in-frame fusion of the EDIII encoding sequences of the four DENV. These were inserted into a replication-defective human adenovirus type 5. When inoculated into mice intraperitoneally and boosted with three doses of tetravalent EDIII plasmid this construct raised neutralizing antibodies to all four DENV (Khanam *et al.*, 2009).

Envelope domain III vaccines

Recently, EDIII has emerged as an antigen that has attracted considerable interest by dengue vaccine developers (Guzman *et al.*, 2010). The selection of this antigen is based upon the extensive literature documenting domain III as the site of attachment on dengue virions of mouse monoclonal DENV neutralizing antibodies (Roehrig *et al.*, 1998, 1990). Inhibition of fusion with endosomal membranes has been proposed as mechanism for neutralization of antibodies of EDIII specificity (Pierson *et al.*, 2008). An advantage of flavivirus DIIIs is that they can be synthesized *in vitro* to achieve normal three-dimensional configuration (Kuhn *et al.*, 2002). Cuban scientists have investigated the use of *Neisseria meningitidis* p64k protein as a fusion partner for two different EDIII molecules (Guzman *et al.*, 2010). The domain III of DENV-2 Jamaica, coding for amino acids 286–426 of dengue E glycoprotein was fused to the C-terminus carrier protein P64k to obtain the fusion protein PD5. When this vaccine was administered to African green monkeys in Freund's complete adjuvant, it raised neutralizing antibodies and provided good protection to wild-type DENV-2 challenge (Hermida *et al.*, 2006). Similar results were obtained with DENV-1 (Bernardo *et al.*, 2008). Antibodies raised were able to neutralize several genotypes of DENV-1 and -2 (Bernardo *et al.*, 2009). When alum was substituted for Freund's adjuvant, monkeys again raised specific neutralizing antibodies, but with less evidence of protection to challenge (Valdes *et al.*, 2009). Protection was greatly improved by administering PD5 to monkeys either before or after they were infected with a live DENV-2 (Valdes *et al.*, 2010). The authors suggest that a EDIII vaccine can be used in a prime-boost setting to boost immune responses raised following vaccination with live-attenuated viruses (mixtures of which often fail to raise tetravalent antibodies following a single dose – see discussion of interference above).

Inactivated or subunit vaccines

Tissue culture-based WRAIR vaccine

Purified, inactivated DENV-2 vaccines (D2-PIV) were prepared in Vero and fetal rhesus lung (FRhL) cells under good manufacturing practices by the Department of Biologics Research, WRAIR (Putnak *et al.*, 1996a,b). For preparation of D2-PIV, D2 strain S16803, Vero cell passage three, was propagated in certified Vero cells which were maintained in serum-free Eagle's Minimum Essential medium. Virus from the culture supernatant fluid was concentrated by ultrafiltration and purified on 15–60% sucrose gradients. The high-titre purified virus (approximately 9 $\log_{10}$ pfu/ml) was inactivated with 0.05% formalin at 22°C for 10 days. Following inactivation, residual formalin was neutralized with sodium bisulfite and the D2-PIV was stored at 4°C prior to formulating with adjuvant. Two doses of D2-PIV with alum induced high PRNT antibody levels and 100% protection against viraemia in a primate model (Putnak *et al.*, 1996a; Eckels and Putnak, 2003).

More recently, the immunogenicity and efficacy of D2-PIV formulated with four different adjuvants (alum and adjuvants SBAS4, SBAS5 and SBAS8 provided by GlaxoSmithKline Biologicals) was evaluated in rhesus monkeys in groups of three animals each (Putnak *et al.*, 2005). The WRAIR DENV2 S16803 PDK-50 LAV vaccine and saline served as positive and negative controls respectively. Two doses of D2-PIV, D2 LAV, or saline were given 3 months apart. The animals were challenged with wild-type D2 S16803 parent virus 3 months after the second dose, and their subsequent viraemia and antibody responses were measured. All but one vaccinated animal seroconverted after the first dose of vaccine. Moreover, all vaccinated animals had anamnestic antibody rises following the second dose of vaccine, with mean reciprocal PRNT antibody titres 30 days after the second dose of vaccine of 4600 (D2-PIV/SBAS5), 3800 (D2-PIV/SBAS8), 680 (D2-PIV/SBAS4), 630 (D2 LAV vaccine), and 200 (D2-PIV/alum group). After virus challenge, viraemia was detected by cell culture in the saline group (3–5 days of viraemia for all animals, mean 4 days), the D2-PIV/alum group (two animals without viraemia, one animal with 2 days of low-titre viraemia), and the D2-PIV/SBAS4 (two animals without viraemia, one animal with 2 days of low-titre viraemia). No viraemia was detected in the animals that received D2-PIV/SBAS5, D2-PIV/SBAS8 or D2 LAV. A subset of monkeys was re-challenged with wild-type virus 1 year later when PRNT antibody titres had sharply decreased. The animals were protected against viraemia (Putnak *et al.*, 2005). A phase I evaluation of the DEN-2-PIV vaccine is planned.

As a follow-up to these studies, tetravalent purified inactivated virus (TPIV) or tetravalent plasmid DNA vaccines expressing the structural prME gene region (TDNA) were boosted 2 months later with the WRAIR tetravalent live attenuated virus vaccine. TPIV raised neutralizing antibodies against all four serotypes (GMT 1:28 to 1:43). Boosting with TLAV led to an increase in the GMT for each serotype (1:500 to 1:1200 for DENVs 1, 3, and 4, and greater than 1:6000 for DENV 2). Titres declined by month 8 (GMT 1:62 for DENV 3, 1:154 for DENV 1, 1:174 for DENV 4, and 1:767 for DENV 2). After challenge with each the four DENV serotypes, vaccinated animals exhibited no viraemia but experienced anamnestic antibody responses to the challenge viruses (Simmons *et al.*, 2010).

Recombinant subunit vaccines

T and B cell epitopes have been mapped on dengue virus structural and non-structural proteins (Brinton *et al.*, 1998; Rey *et al.*, 1995). The right combination of epitopes expressed in protein subunit vaccines could be the basis for an effective and safe vaccine at moderate cost (Trent *et al.*, 1997). Structural and non-structural dengue virus proteins have been produced in adequate amounts in many expression systems including *E. coli* (Fonseca *et al.*, 1991; Srivastava *et al.*, 1995; Simmons *et al.*, 1998b), baculovirus in *Spodoptera frugiperda* insect cells and *Drosophila* cells (Delenda *et al.*, 1994a,b; Velzing *et al.*, 1999; Bielefeldt Ohmann *et al.*, 1997), yeast (Sugrue *et al.*, 1997) and vaccinia virus (Zhao *et al.*, 1987; Deubel *et al.*, 1988; Men *et al.*, 2000). The last approach, 80% *E* gene expression in *Drosophila* cells, was explored by the Hawaii Biotechnology, Inc. One microgram of DEN-2 E protein with the SBAS5 adjuvant protected rhesus monkeys from viraemia following challenge with wild-type virus. A monovalent formulation is undergoing phase I evaluation in 2010 (H. Margolis, personal communication). Results of this clinical trial have not yet been published (2011).

DNA vaccines

DNA plasmid

DNA vaccines consist of a plasmid or plasmids containing dengue genes reproduced to high copy number in bacteria such as *E. coli* (Whalen, 1996). The plasmid contains a eukaryotic promoter and termination sequence to drive transcription in the vaccine recipient. The transcribed RNA is translated to produce proteins to be processed and presented to the immune system in the context of MHC molecules. Additional genes such as intracellular trafficking and immunostimulatory sequences can be added to the plasmid. The target organism's immune system recognizes the expressed antigen, and generates antibodies and/or cell-mediated immune responses. DNA-based vaccine constructs can be modified without the need for subsequent viability as required when working with infectious clones. DNA vaccines afford numerous advantages over conventional vaccines including ease of production, stability and transport at room temperature, and they provide a possibility to immunize against multiple pathogens with a single vaccine. Dengue DNA vaccines offer the possibility of achieving protective immunity with reduced concern about reactogenicity or the interference seen with multivalent live virus vaccines. The immunology of DNA vaccines has been recently reviewed (Gurunathan *et al.*, 2000).

Workers at the Naval Medical Research Institute evaluated two eukaryotic plasmid expression vectors (pkCMVint-Polyli and pVR1012, Vical, Inc., San Diego, CA, USA) expressing the PrM protein and 92% of the E protein for DEN-2 virus (New Guinea C strain). Both constructs induced neutralizing antibody in all mice (Kochel *et al.*, 1997) with a subsequent improvement seen with the addition of immunostimulatory CpG motifs (pUC 19, Gibco BRL, Gaithersburg, MD, USA) (Porter *et al.*, 1998). Konishi and colleagues successfully immunized mice with a similar DEN-2 vaccine construct using *C-PrM-E* genes in the pcDNA3 vector (Invitrogen, Corp., San Diego, CA, USA) (Konishi *et al.*, 2000).

In subsequent experiments using genes from the West Pacific 74 strain of DEN-1 virus and the pVR1012 plasmid, it was determined that the full length *E* gene with *PrM* served as a better immunogen (Raviprakash *et al.*, 2000a) and was shown to reduced the frequency and duration of viraemia in rhesus macaques following challenge with wild-type virus (Raviprakash *et al.*, 2000b). At present, the DNA approach has produced modest neutralizing antibody levels in non-human primates with only a portion of animals fully protected from viraemia for varying lengths of time (K. Porter and R. Putnak, personal communication). With the recognition that skin dendritic cells are highly permissive for dengue virus replication (Wu *et al.*, 2000), on-going efforts may focused on vaccine delivery systems that target these cells. The DENV-1 DNA vaccine was used in a prime-boost model. When cynomolgous monkeys were given two doses of DNA DENV1 plasmid intramuscularly at an interval of 4 weeks and boosted 3 months later with a Venezuelan equine encephalitis construct in which DENV 1 Western Pacific 74 prM and E genes were inserted, all animals developed neutralizing antibodies and were protected against viraemia following wild-type DENV-1 challenge (Chen *et al.*, 2007).

DNA shuffling

Tetravalent dengue vaccines have been created by shuffling the envelope genes from the four dengue viruses. Selected shuffled DNA was transfected into human cells, subjected to flow cytometry and reacted with type-specific dengue antibodies. Antibody markers permitted rapid screening of libraries and identification of novel expressed chimeric antigens for all four dengue types. By DNA shuffling of codon-optimized dengue 1–4 E genes, a panel of novel chimeric clones expressing C-terminal truncated E antigens from all four dengue viruses was created. When mixtures of these DNA molecules were inoculated in mice they raised cross-reactive and cross-neutralizing antibodies. Immunized mice survived challenge with lethal intracerebral doses of all four dengue viruses (Apt *et al.*, 2006).

In follow-up studies, three shuffled constructs (sA, sB and sC) were evaluated in the rhesus macaque model. Constructs sA and sC expressed pre-membrane and envelope genes, whereas construct sB expressed only the ectodomain of envelope protein. Five of six, and four of six animals vaccinated with sA and sC, respectively, developed antibodies that neutralized all four dengue serotypes *in vitro*. Four of six animals vaccinated with construct sB developed neutralizing antibodies against three serotypes (DEN-1, -2 and -3). When challenged with live dengue-1 or dengue-2 virus, partial protection against dengue-1 was observed (Raviprakash *et al.*, 2006). These approaches have been supplemented successfully in susceptible monkeys priming with DNA vaccines and boosting with the WRAIR live-attenuated vaccine,(Simmons *et al.*, 2010) or with various DENV subunit proteins (Simmons *et al.*, 2006).

While the DNA approach offers unique advantages it also carries unique risks. (Klinman *et al.*, 1997) These include the theoretical risk of nucleic acid integration into the host's chromosomal DNA to potentially inactivate tumour suppressor genes or activate oncogenes. This risk appears to be well below the spontaneous mutation frequency for mammalian cells (Nichols *et al.*, 1995; Martin *et al.*, 1999). However, if a mutation due to DNA integration is a part of a multiple hit phenomenon leading to carcinogenesis, it could take many years before this problem became evident. Another concern is that foreign DNA might induce anti DNA antibodies leading to autoimmune diseases such as systemic lupus erythematosus. However, studies in lupus-prone mice, normal mice, rabbits, and people to date have not validated this concern (Parker *et al.*, 1999; Mor *et al.*, 1997) and, in fact, DNA vaccine are being proposed as an approach to the management of autoimmune diseases (Prud'homme *et al.*, 2001; Karin, 2000).

Vaccine constructs not tested in subhuman primates

Measles virus as vector

In this technology, genes coding for 100 amino acids of the envelope domain III (EDIII) fused to 40 amino acids of the ectodomain of the membrane protein (ectoM) from DENV-1 were inserted into the genome of the Schwarz strain of measles vaccine (MV). Immunization of mice resulted in production of DENV-1 neutralizing antibodies (Brandler and Tangy, 2008). The adjuvant property of the pro-apoptotic sequence ectoM correlated with the ability of this construct to promote the maturation of dendritic cells and the secretion of proinflammatory and antiviral cytokines and chemokines involved in adaptive immunity (Brandler *et al.*, 2007). A gene that expressed the four EDIII in the sequence, 1–4 were fused at the C terminal end to a single ectoM sequence from DENV-1 was inserted into the MV and, when injected intraperitoneally in two doses at a 4-week interval raised neutralizing antibodies to all four DENV in measles-susceptible mice (Brandler *et al.*, 2010).

Replication-incompetent vaccines

Several approaches in the patent literature have been described for the use of recombinant EDIII or genes that express EDIII as dengue vaccines (Swaminathan *et al.*, 2010). Amino acid residues can be modified at the prM junction to enhance host furin-mediated cleavage during virus maturation. This replication-competent mutant, which has a lesser proportion of prM in the virion surface compared to wild-type virus manifests a defect in release of progeny virus from infected cells. An alternative approach inserts a furin cleavage site on the membrane glycoprotein, such a mutant self destructs during virion maturation in the trans-Golgi network.

A new technology disclosed by Arbovax Inc. exploits the observation that the mammalian cell membrane is thicker than its insect counterpart because it contains cholesterol. The technology shortens the transmembrane domain of E that anchors it to host cell membrane. This generates host-range mutants that replicate and release infectious progeny from insect cells but fails to assemble into mature virions in mammalian cells due to the mutant E protein being unable to anchor itself into the thicker mammalian membrane (Swaminathan *et al.*, 2010).

A technology named RepliVAX inserts dengue PrM and E genes into a capsid-deleted West Nile virus that is capable of only a single cycle of translation and gene expression. A DENV-2 vaccine with mutations in prM and E permitting growth in capsid supplemented cells, successfully produced neutralizing antibodies in AG 129 mice which somewhat resisted lethal wild-type virus challenge (Suzuki *et al.*, 2009). Similar vaccines have been produced for West Nile and Japanese encephalitis (Widman *et al.*, 2008).

Envelope domain III-based vaccines

For this reason a number of EDIII DENV constructs synthesized in bacteria or yeast cells and presented with a fusion protein partner or as a virus-like particle are proposed in the patent literature (Swaminathan *et al.*, 2010). *E. coli* maltose binding protein has been used as a fusion partner with DENV-2 EDIII to elicit neutralizing and protective antibodies in mice.(Simmons *et al.*, 1998a) Indian scientists have designed a gene that expresses a single domain III protein using a bioinformatics approach that has antigenic sequences of all four DENV. A recombinant antigen made by linking the EDIII of each DENV-1–4 was expressed in *Pichia pastoris* cells in translational fusion raised neutralizing antibodies to all four DENV in mice (Etemad *et al.*, 2008).

Synthetic peptides

Synthetic peptides containing B and T cell epitopes are immunogenic in mice and combinations of peptides could be effective as subunit vaccines (Roehrig *et al.*, 1992; Becker, 1994). Antibodies directed to synthetic peptides have been detected in sera from patients convalescent from dengue infections. (Huang *et al.*, 1999; Garcia *et al.*, 1997). Peptides are particularly problematic as vaccines to protect against dengue infection because they provide fewer epitopes than do other candidate vaccines and lack conformational epitopes.

Inactivated whole virus or subunit vaccines are the simplest approach to vaccine development. Such vaccines have two potential advantages compared with live-attenuated vaccines: they cannot revert to a more pathogenic phenotype and when combined are unlikely to produce interference. Cell mediated as well as humoral immune responses have been demonstrated with an inactivated flavivirus vaccine (Aihara *et al.*, 2000). On the other hand, killed or subunit vaccines raise antibodies to only a portion of the structural proteins and normal virion-based structural conformation. Three dimensional structural antigens crucial to protection may not be presented to B cells. Other disadvantages include the requirement for multiple doses to fully immunize plus the need to prepare sufficient quantity of antigens that could result in increased cost per dose. Safety concerns in dengue may differ from those in viral encephalitis. Two inactivated vaccines have been licensed and are in wide use, safe and effective in preventing Japanese encephalitis, (Hoke *et al.*, 1988) and tick-borne encephalitis (Craig *et al.*, 1999).Because of antibody-dependent enhancement it is unlikely that inactivated or subunit dengue vaccines will ever be used alone, although they may be useful in a prime-boost strategy with live or DNA vaccines.

West Nile

WNV is a zoonotic pathogen which first appeared in the western hemisphere in 1999 in New York City, before spreading rapidly across North America, causing disease in wild birds, horses, and humans. The virus has also been isolated in the West Indies and neutralizing antibody-positive birds and horses have been identified in Mexico, Jamaica, and the Dominican Republic; West Nile virus also appeared recently in South America, including Brazil and Argentina. The spectrum of disease extends from a mild febrile exanthem to fatal encephalitis; neuroinvasive disease occurs in approximately 30% of reported WNV cases. Around 30,000 cases and more than 1000 deaths have been attributed to WNV infection since its introduction (Guy *et al.*, 2010). There is no effective drug treatment against this infection. While the introduction of vaccines for horses represents a potent tool to control the veterinary disease, a human vaccine against WNV may represents an important approach to the prevention and control of this emerging disease although recent epidemiological data show a sharp decline in the number WNV cases in the past years. (Monath, 2002; Monath *et al.*, 2001). West Nile virus (WNV) was isolated from a Ugandan patient with a mild fever in 1937. For many years WNV was thought to cause a dengue fever-like disease in humans and encephalitis in horses.

Disease

Neuroinvasive WN

After an incubation period of around 2–14 days, usually 1 week, patients develop a high fever, headache, neck stiffness which may progress to stupor, disorientation, coma, tremors, convulsions, muscle weakness, vision loss, numbness and paralysis. Patients with neuroinvasive disease often have a flu-like prodrome followed by development of neurological signs and symptoms. It is estimated that approximately 35–40% of patients with WNV neuroinvasive disease have meningitis, 55–60% have encephalitis and 5–10% acute flaccid paralysis, although numbers vary in different case series. Patients with encephalitis typically have fever (85–100%), headache (47–90%) and an altered mental status (46–74%) (Davis *et al.*, 2008). Following infection, many patients develop movement disorders which include tremors of various types, myoclonus and Parkinsonism. Cerebellar signs and symptoms occur in a variable number. Weakness is also common and may be generalized and non-specific or a lower motor neuron pattern of flaccid paralysis associated with absent or reduced reflexes and preserved sensation. Cranial nerve palsies have been reported in 10% or more of cases and most commonly involve the ophthalmic (2nd) and facial (7th) nerves.

Visual problems in patients with WNND have become increasingly recognized. Patients often complain of blurry vision, trouble seeing, and photophobia. Clinically, reduced vision, active non-granulomatous uveitis, vitreitis, multifocal chorioretinitis and optic neuritis have been described. Immunosuppressed patients, such as patients with organ transplants or AIDS, often develop a prolonged serious encephalitis during WNV infections (Davis *et al.*, 2008). Severe outcomes of WNV infection are more frequently observed in persons of advanced age or those with diabetes. A genetic factor, CCR5, leads to more serious complications.

Epidemiology

Serosurveys showed WNV antibodies to be widely distributed in sub-Saharan Africa, in parts of the Middle East including Egypt and in India and Pakistan. (Diamond, 2009) A strain of West Nile, Kunjin, circulates as a zoonosis in Australia. Until the virus was imported into the United States in 1999, it had been only associated with isolated outbreaks of encephalitis in elderly patients in Israel and Romania. The virus is transmitted in nature by various species of *Culex* mosquitoes predominantly between various species of birds. Mammals, reptiles and amphibians of many species are infected, sometimes with fatal outcome. Crows, blue jays and hawks are particularly at risk to fatal WNV infection. The virus was imported to the New York City area in 1999 causing encephalitis in humans, dogs, cats and horses with extensive deaths in birds, particularly crows. The US virus was very closely related to a lineage 1 strain isolated in Israel in 1998. The virus has spread to cause clinical disease in 47 US states (all except Maine, Alaska and Hawaii), Canada, Caribbean islands and Central America. In the United States, from 1999 through the summer of 2010, there were 30,301 cases or mild and severe disease. Recent WNV outbreaks have been reported in Russia, Israel, Greece and the Czech Republic. In the US WNV is transmitted by *Culex pipiens* (Eastern US), *Culex tarsalis* (Midwest and West) and *Culex quinquefasciatus* (Southeast). Infections in humans are predominantly (80%) asymptomatic. Perhaps 20% of infected persons will have mild symptoms including fever, headache, body aches, nausea, vomiting and sometimes a skin rash on the trunk. About one in 150 infected humans, higher in individuals 60 years or older, will develop encephalitis.

The number of dying birds in a locale during early summer months is used to monitor wildlife infection and predict possible human or equine cases (Diamond, 2009). Ecology studies suggest that *Culex pipiens*, the dominant enzootic (bird-to-bird) and bridge (bird-to-human) vector of WNV in urbanized areas in the northeast and north-central United States, shifts its feeding preferences from birds to humans during the late summer and early fall, coincident with the dispersal of its avian hosts. Most (~85%) of human infections in the US occur in the late summer with a peak number of cases in August and September. In warmer parts of the country, virtually year-round transmission has been observed.

The outbreak in the US revealed that WNV can be spread through blood transfusions, organ transplants, intrauterine exposure and breast feeding. Since 2003, blood banks in the US routinely screen plasma for virus.

Vaccines

Chimeric vaccines

Two ChimeriVax West Nile (WN-CV) viruses have been constructed.

In the first, WN01, the complete sequences of the prM/E genes of the YFV 17D virus were replaced with those of the WNV strain NY99 introduced in the USA, without any modification (Monath *et al.*, 2001; Arroyo *et al.*, 2001). WN01was developed by Intervet as a single-dose vaccine for horses (PreveNileTM, Schering-Plough Animal Health/Merck) and has been commercially available since 2006 in the USA. As part of the development of aWNV vaccine for human use (WN02), three mutations were introduced into the WN01 E gene [E107 (L→F), E316 (A→V) and E440 (K→R)]. These mutations are known to attenuate the closely related JEV. In the human WN02 vaccine, these mutations were shown to independently enhance neuroattenuation of the chimera, so that reversion at one residue or even two residues would maintain the attenuated phenotype (Arroyo *et al.*, 2001). Even in a very unlikely event that all three mutations simultaneously reverted to wt codons, the chimera, would be similar to the attenuated WN01 which is not neuroinvasive and is even less neurovirulent than YFV 17D vaccine by direct intracerebral (i.c.) inoculation of mice and monkeys (Arroyo *et al.*, 2004). A passage 5 (P5) WN02 vaccine was safe and highly immunogenic in a human phase I study in healthy adults (Monath *et al.*, 2006). However, this vaccine produced heterogeneous plaque populations: large plaque (LP) and small plaque (SP) were identified in the vaccine lot at a ratio of 50:50. The mutation at M66 (L→P) that is responsible for the SP phenotypic change appears to be an adaptation to propagation in serum-free (SF) Vero cells. The two subpopulations (with and without the M66 L→P mutation) were biologically and phenotypically identical, both being attenuated with respect to mouse neurovirulence. However, since the SP virus generated a lower viraemia than the LP in a hamster model of WN infection (see below), SP was selected for production of vaccine lots to be used in phase II clinical trials in humans.

Replication-incompetent vaccines

RepliVAX West Nile vaccine is propagated in C-expressing cells (or as a unique two-component virus) using methods similar to those used to produce today's economical and potent LAVs. Owing to deletion of most of the gene for the C protein, RepliVAX cannot spread between normal cells, and is unable to cause disease in vaccinated animals. RepliVAX provides an efficacious WN vaccine in animal models that demonstrate that it is potent, economical to produce, and safe for immunization of the immunocompromised host (Widman *et al.*, 2008; Suzuki *et al.*, 2009).

Measles vaccine as vector

Genes for WNV IS-98-ST1 envelop protein were inserted into the genome of live-attenuated Schwarz measles vaccine (MV). Inoculation of this construct into MV-susceptible mice induced high levels of specific anti-WNV neutralizing antibodies and protected mice from a lethal challenge with WNV. In addition, antibodies raised in mice were given to BALB/c mice which were protected against lethal outcome following a challenge with a high dose of WNV (Lorin *et al.*, 2005; Despres *et al.*, 2005).

References

Aebi, C., and Schaad, U.B. (1994). [TBE-immunoglobulins – a critical assessment of efficacy]. Schweiz Med. Wochenschr. *124*, 1837–1840.

Aihara, H., Takasaki, T., Toyosaki-Maeda, T., Suzuki, R., Okuno, Y., and Kurane, I. (2000). T-cell activatoin and induction of antibodies and memory T cells by immunization with inactivated Japanese encephalitis vaccine. Viral Immunol. *13*, 179–186.

Aitken, T.H.G., Downs, W.G., and Shope, R.E. (1977). *Aedes aegypti* strain fitness for yellow fever virus transmission. Am. J. Trop. Med. Hyg. *26*, 985–989.

Amanna, I.J., Carlson, N.E., and Slifka, M.K. (2007). Duration of humoral immunity to common viral and vaccine antigens. N. Engl. J. Med. *357*, 1903–15.

Amanna, I.J., Messaoudi, I., and Slifka, M.K. (2008). Protective immunity following vaccination: how is it defined? Hum. Vaccine *4*, 316–319.

Amanna, I.J., Slifka, M.K., and Crotty, S. (2006). Immunity and immunological memory following smallpox vaccination. Immunol. Rev. *211*, 320–337.

Anderson, C.R., and W.G., Downs. (1956). Isolation of dengue virus from a human being in Trinidad. Science *124*, 224–225.

Anez, G., Men, R., Eckels, K.H., and Lai, C.J. (2009). Passage of dengue virus type 4 vaccine candidates in fetal rhesus lung cells selects heparin-sensitive variants that result in loss of infectivity and immunogenicity in rhesus macaques. J. Virol. *83*, 10384–10394.

Anonymous. (1979). Dengue in the Caribbean, 1977, vol. PAHO Scientific Publication No. 375, Washington, DC.

Anonymous. (1995). Isolation of dengue type 3 virus prompts concern and action. Bull. Pan Am Health Org. *29*, 184–185.

Apt, D., Raviprakash, K., Brinkman, A., Semyonov, A., Yang, S., Skinner, C., Diehl, L., Lyons, R., Porter, K., and Punnonen, J. (2006). Tetravalent neutralizing antibody response against four dengue serotypes by a single chimeric dengue envelope antigen. Vaccine *24*, 335–344.

Arroyo, J., Miller, C., Catalan, J., Myers, G.A., Ratterree, M.S., Trent, D.W., and Monath, T.P. (2004). ChimeriVax-West Nile virus live-attenuated vaccine: preclinical evaluation of safety, immunogenicity, and efficacy. J. Virol. *78*, 12497–12507.

Arroyo, J., Miller, C.A., Catalan, J., and Monath, T.P. (2001). Yellow fever vector live-virus vaccines: West Nile virus vaccine development. Trends Mol. Med. *7*, 350–354.

Bancroft, W.H., Scott, R.M., Eckels, K.H., Hoke, C.H., Jr., Simms, T.E., Jesrani, K.D., Summers, P.L., Dubois, D.R., Tsoulos, D., and Russell, P.K. (1984). Dengue virus type 2 vaccine: reactogenicity and immunogenicity in soldiers. J. Infect.Dis. *149*, 1005–1010.

Barrett, A.D., and Teuwen, D.E. (2009). Yellow fever vaccine – how does it work and why do rare cases of serious adverse events take place? Curr. Opin. Immunol .*21*, 308–313.

Barrett, P.N., Plotkin, S.A., and Ehrlich, H.J. (2008). Tick-borne Encephalitis Virus Vaccines, pp. 841–856. In Plotkin, S.A., Orenstein, W.A., and Offit, P.A. (eds.), Vaccines, 5th edn. Elsevier, Philadelphia.

Bass, J.S., Halstead, S.B., Fischer, G.W., and Podgore, J.K. (1976). Booster vaccination with futher live attenuated measles vaccine. J. Am. Med. Assoc. *235*, 31–34.

Bass, J.W., Halstead, S.B., Fischer, G.W., and Podgore, J.K. (1978). Oral polio vaccine: effect of booster vaccination one to 14 years after primary series. J. Am. Med. Assoc. *239*, 2252–2255.

Beasley, D.W., Li, L., Suderman, M.T., Guirakhoo, F., Trent, D.W., Monath, T.P., Shope, R.E., and Barrett, A.D. (2004). Protection against Japanese encephalitis virus strains representing four genotypes by passive transfer of sera raised against ChimeriVax-JE experimental vaccine. Vaccine *22*, 3722–3726.

Becker, Y. (1994). Dengue fever virus and Japanese encephalitis virus synthetic peptides, with motifs to fit HLA class I haplotypes prevalent in human populations in endemic regions, can be used for application to skin Langerhans cells to prime antiviral CD8(+) cytotoxic T cells (CTLs) – A novel approach to the protection of humans. Virus Genes *9*, 33–45.

Beeuwkes, H. (1936). Clinical manifestations of yellow fever in the West African native as observed during four extensive epidemics of the disease in the gold Coast and Nigeria. Trans. R. Soc. Trop. Med. Hyg. *30*, 61–86.

Belmusto-Worn, V.E., Sanchez, J.L., McCarthy, K., Nichols, R., Bautista, C.T., Magill, A.J., Pastor-Cauna, G., Echevarria, C., Laguna-Torres, V.A., Samame, B.K., *et al.* (2005). Randomized, double-blind, phase III, pivotal field trial of the comparative immunogenicity, safety, and tolerability of two yellow fever 17D vaccines (Arilvax and YF-VAX) in healthy infants and children in Peru. Am. J. Trop. Med. Hyg. *72*, 189–97.

Benenson, M.W., Top, F.H., Jr., Gresso, W., Ames, C.W., and Altstatt, L.B. (1975). The virulence to man of Japanese encephalitis virus in Thailand. Am. J. Trop. Med. Hyg. *24*, 974–980.

Bernardo, L., Fleitas, O., Pavon, A., Hermida, L., Guillen, G., and Guzman, M.G. (2009). Antibodies induced by dengue virus type 1 and 2 envelope domain III recombinant proteins in monkeys neutralize strains with different genotypes. Clin. Vaccine Immunol. *16*, 1829–1831.

Bernardo, L., Izquierdo, A., Alvarez, M., Rosario, D., Prado, I., Lopez, C., Martinez, R., Castro, J., Santana, E., Hermida, L., Guillen, G., and Guzman, M.G. (2008). Immunogenicity and protective efficacy of a recombinant fusion protein containing the domain III of the dengue 1 envelope protein in non-human primates. Antiviral Res. *80*, 194–199.

Berry, G.P., and Kitchen, S.F. (1931). Yellow fever accidentally contracted in the laboratory: a study of seven cases. Am. J. Trop. Med. Hyg. *11*, 365–434.

Bhamarapravati, N., and Sutee, Y. (2000). Live attenuated tetravalent dengue vaccine. Vaccine *18*, Suppl. *2*, 44–74.

Bhamarapravati, N., and Yoksan, S. (1989). Study of bivalent dengue vaccine in volunteers. Lancet *1*, 1077.

Bhamarapravati, N., and Yoksan, S. (1997). Live attenuated tetravalent dengue vaccine, pp. 367–377. In Gubler, D.J. and Kuno, G. (eds.), Dengue and Dengue Hemorrhagic Fever. CAB International, New York.

Bhamarapravati, N., Yoksan, S., Chayaniyayothin, Angsubphakorn, T., S. and Bunyaratvej, A. (1987). Immunization with a live attenuated dengue-2-virus candidate vaccine (16681-PDK 53): clinical, immunological and biological responses in adult volunteers. Bull.World Health Org. *65*, 189–195.

Bhatt, T.R., Crabtree, M.B., Guirakhoo, F., Monath, T.P., and Miller, B.R. (2000). Growth characteristics of the chimeric Japanese encephalitis virus vaccine candidate, ChimeriVax-JE (YF/JE SA14-14-2), in *Culex tritaeniorhynchus*, *Aedes albopictus*, and *Aedes aegypti* mosquitoes. Am. J. Trop. Med. Hyg. *62*, 480–484.

Bielefeldt Ohmann, H., Beasley, D.W., Fitzpatrick, D.R., and Aaskov, J.G. (1997). Analysis of a recombinant dengue-2 virus–dengue-3 virus hybrid envelope protein expressed in a secretory baculovirus system. J. Gen. Virol. *78*, 2723–2733.

Bista, M.B., Banerjee, M.K., Shin, S.H., Tandan, J.B., Kim, M.H., Sohn, Y.M., Ohhr, H.C., and Halstead, S.B. (2001). Efficacy of single dose SA 14-14-2 vaccine against Japanese encephalitis: a case-control study. Lancet *358*, 791–795.

Blaney, J.E., Jr., Durbin, A.P., Murphy, B.R., and Whitehead, S.S. (2010). Targeted mutagenesis as a rational approach to dengue virus vaccine development. Curr. Top. Microbiol. Immunol. *338*, 145–158.

Blaney, J.E., Jr., Sathe, N.S., Goddard, L., Hanson, C.T., Romero, T.A., Hanley, K.A., Murphy, B.R., and Whitehead, S.S. (2008). Dengue virus type 3 vaccine candidates generated by introduction of deletions in the 3' untranslated region (3'-UTR) or by exchange of the DENV-3 3'-UTR with that of DENV-4. Vaccine *26*, 817–828.

Brandler, S., Lucas-Hourani, M., Moris, A., Frenkiel, M.P., Combredet, C., Fevrier, M., Bedouelle, H., Schwartz, O., Despres, P., and Tangy, F. (2007). Pediatric measles vaccine expressing a dengue antigen induces durable serotype-specific neutralizing antibodies to dengue virus. PLoS Negl. Trop. Dis. *1*, e96.

Brandler, S., Ruffie, C., Najburg, V., Frenkiel, M.P., Bedouelle, H., Despres, P., and Tangy, F. (2010). Pediatric measles vaccine expressing a dengue tetravalent antigen elicits neutralizing antibodies against all four dengue viruses. Vaccine *28*, 6730–6739.

Brandler, S., and Tangy, F. (2008). Recombinant vector derived from live attenuated measles virus: potential for flavivirus vaccines. Comp Immunol. Microbiol. Infect. Dis. *31*, 271–291.

Bray, M., and Lai, C.J. (1991). Construction of intertypic chimeric dengue viruses by substitution of structural protein genes. Proc. Natl. Acad. Sci. U.S.A. *88*, 10342–10346.

Bray, M., Men, R., and Lai, C.J. (1996). Monkeys immunized with intertypic chimeric dengue viruses are protected against wild-type virus challenge. J. Virol. *70*, 4162–4166.

Brinton, M.A., Kurane, I., Mathew, A., Zeng, L., Shi, P.Y., Rothman, A., and Ennis, F.A. (1998). Immune mediated and inherited defences against flaviviruses. Clin Diagn Virol *10*, 129–139.

Broker, M., and Kollaritsch, H. (2008). After a tick bite in a tick-borne encephalitis virus endemic area: current positions about post-exposure treatment. Vaccine *26*, 863–868.

Burchard, G.D., Caumes, E., Connor, B.A., Freedman, D.O., Jelinek, T., Jong, E.C., F. von Sonnenburg, Steffen, R., Tsai, T.F., Wilder-Smith, A., and Zuckerman, J. (2009). Expert opinion on vaccination of travelers against Japanese encephalitis. J. Travel Med. *16*, 204–216.

Butrapet, S., Huang, C.Y., Pierro, D.J., Bhamarapravati, N., Gubler, D.J., and Kinney, R.M. (2000). Attenuation markers of a candidate dengue type 2 vaccine virus, strain 16681 (PDK-53), are defined by mutations in the 5′ noncoding region and nonstructural proteins 1 and 3. J. Virol. *74*, 3011–3019.

Camacho, L.A., de Aguiar, S.G., Freire Mda, S., Leal Mda, L., do Nascimento , J.P., Iguchi, T., Lozana, J.A., and R Farias, H. (2005). Reactogenicity of yellow fever vaccines in a randomized, placebo-controlled trial. Rev. Saude Publica *39*, 413–420.

Cardosa, J., Ooi, M.H., Tio, P.H., Perera, D., Holmes, E.C., Bibi, K., and Abdul Manap, Z. (2009). Dengue virus serotype 2 from a sylvatic lineage isolated from a patient with dengue hemorrhagic Fever. PLoS Negl. Trop. Dis. *3*, e423.

Carey, D.E., Causey, O.R., Reddy, S., and Cooke, A.R. (1971). Dengue viruses from febrile patients in Nigeria, 1964–68. Lancet *1*, 105–106.

Carter, H.R. (1931). Yellow Fever, an Epidemiological and Historical Study of its Place of Origin. Williams and Wilkins, Baltimore.

CDC (2010). Japanese encephalitis vaccines: Recommendations of the Advisory Committee on Immunization Practices (ACIP). MMWR 59, No. RR-1.

Chambers, T.J., Nestorowicz, A., Mason, P.W., and Rice, C.M. (1999). Yellow fever/Japanese encephalitis chimeric viruses: construction and biological properties. J. Virol. *73*, 3095–3101.

Chang, G.J., Cropp, B.C., Kinney, R.M., Trent, D.W., and Gubler, D.J. (1995). Nucletide sequence variation of the envelope protein gene identifies two distinct genotypes of the yellow fever virus. J. Virol. *69*, 5773–5780.

Chanthavanich, P., Luxemburger, C., Sirivichayakul, C., Lapphra, K., Pengsaa, K., Yoksan, S., Sabchareon, A., and Lang, J. (2006). Short report: immune response and occurrence of dengue infection in thai children three to eight years after vaccination with live attenuated tetravalent dengue vaccine. Am. J. Trop. Med. Hyg. 75, 26–28.

Chen, L., Ewing, D., Subramanian, H., Block, K., Rayner, J., Alterson, K.D., Sedegah, M., Hayes, C., Porter, K., and Raviprakash, K. (2007). A heterologous DNA prime-Venezuelan equine encephalitis virus replicon particle boost dengue vaccine regimen affords complete protection from virus challenge in cynomolgus macaques. J. Virol. *81*, 11634–11639.

Chen, W., Kawano, H., Men, R., Clark, D., and Lai, C.J. (1995). Construction of intertypic chimeric dengue viruses exhibiting type 3 antigenicity and neurovirulence for mice. J. Virol. *69*, 5186–5190.

Cohen, S.N., and Halstead, S.B. (1966). Shock associated with dengue infection. I. Clinical and physiologic manifestations of dengue hemorrhagic fever in Thailand, 1964. J. Pediatr. *68*, 448–456.

Craig, S.C., Pittman, P.R., Lewis, T.E., Rossi, C.A., Henchal, E.A., Kuschner, R.A., Martinez, C., Kohlhase, K.F., Cuthie, J.C., Welch, G.E., and Sanchez, J.L. (1999). An accelerated schedule for tick-borne encephalitis vaccine: the American military experience in Bosnia. Am. J. Trop. Med. Hyg. *61*, 874–878.

Davis, L.E., Beckham, J.D., and Tyler, K.L. (2008). North American encephalitic arboviruses. Neurol Clin 26, 727–57, ix.

De Groot, A.S., Martin, W., Moise, L., Guirakhoo, F., and Monath, T. (2007). Analysis of ChimeriVax Japanese Encephalitis Virus envelope for T-cell epitopes and comparison to circulating strain sequences. Vaccine *25*, 8077–8084.

Delenda, C., Frenkiel, M.P., and Deubel, V. (1994). Protective efficacy in mice of a secreted form of recombinant dengue-2 virus envelope protein produced in baculovirus infected insect cells. Arch. Virol. *139*, 197–207.

Delenda, C., Staropoli, I., Frenkiel, M.P., Cabanie, L., and Deubel, V. (1994). Analysis of C-terminally truncated dengue 2 and dengue 3 virus envelope glycoproteins: Processing in insect cells and immunogenic properties in mice. J. Gen.Virol. *75*, 1569–1578.

Delgado, M.F., Coviello, S., Monsalvo, A.C., Melendi, G.A., Hernandez, J.Z., Batalle, J.P., Diaz, L., Trento, A., Chang, H.Y., Mitzner, W., Ravetch, J., Melero, J.A., Irusta, P.M., and Polack, F.P. (2009). Lack of antibody affinity maturation due to poor Toll-like receptor stimulation leads to enhanced respiratory syncytial virus disease. Nat. Med. *15*, 34–41.

Despres, P., Combredet, C., Frenkiel, M.P., Lorin, C., Brahic, M., and Tangy, F. (2005). Live measles vaccine expressing the secreted form of the West Nile virus envelope glycoprotein protects against West Nile virus encephalitis. J. Infect. Dis. (*191*, 207–214.

Deubel, V., Kinney, R.M., Esposito, J.J., Cropp, C.B., Vorndam, A.V., Monath, T.P., and Trent, D.W. (1988). Dengue 2 virus envelope protein expressed by a recombinant vaccinia virus fails to protect monkeys against dengue. J. Gen.Virol. *69*, 1921–1929.

Deubel, V., Schlesinger, J.J., and Digoutte, J.P. (1987). Comparative immunochemical and biological analysis of African and South American yellow fever viruses. Arch. Virol. *94*, 331–338.

Diamond, M.S. (2009). Progress on the development of therapeutics against West Nile virus. Antiviral Res. *83*, 214–227.

dos Santos, A.P., Bertho, A.L., and Dias, D.C. (2005). Lymphocyte subset analyses in healthy adults vaccinated with yellow fever 17DD virus. Mem. Inst. Oswaldo Cruz *100*, 331–337.

dos Santos, C.N., Post, P.R., Carvalho, R., Ferreira, I.I., Rice, C.M., and Galler, R. (1995). Complete nucleotide sequence of yellow fever virus vaccine strains 17DD and 17D-213. Virus Res. *35*, 35–41.

Dubischar-Kastner, K., Eder, S., Buerger, V., Gartner-Woelfl, G., Kaltenboeck, A., Schuller, E., Tauber, E., and Klade, C. (2010). Long-term immunity and immune response to a booster dose following vaccination with the inactivated Japanese encephalitis vaccine IXIARO, IC51. Vaccine *28*, 5197–5202.

Dudley, S.F. (1934). Can yellow fever spread into Asia? J. Trop. Med. Hyg. *37*, 273–278.

Dumpis, U., Crook, D., and Oksi, J. (1999). Tick-borne encephalitis. Clin. Infect. Dis. *28*, 882–890.

Dunster, L.M., Gibson, C.A., Stephenson, J.R., Minor, P.D., and Barrett, A.D. (1990). Attenuation of virulence of flaviviruses following passage in HeLa cells. J. Gen. Virol. *71*, 601–607.

Dunster, L.M., Wang, H., Ryman, K.D., Miller, B.R., Watowich, S.J., Minor, P.D., and Barrett, A.D. (1999). Molecular and biological changes associated with HeLa cell attenuation of wild-type yellow fever virus. Virology *261*, 309–18.

Dupuy, A., Despres, P., Cahour, A., Girard, M., and Bouloy, M. (1989). Nucleotide sequence comparison of the genome of two 17D-204 yellow fever vaccines. Nucleic Acids Res. *17*, 3989.

Durbin, A P., Karron, R.A., Sun, W., Vaughn, D.W., Reynolds, M.J., Perreault, J.R., Thumar, B., Men, R., Lai, C.-J., Elkins, W.R., Chanock, R.M., Murphy, B.R., and Whitehead, S.S. (2001). Attenuation and immunogenicity in humans of a live dengue virus type-4 vaccine candidate with a 30 nucleotide deletion in its 3′-untranslated region. Am. J. Trop. Med. Hyg. *65*, 405–413.

Durbin, A.P., McArthur, J., Marron, J.A., Blaney, J.E., Jr., Thumar, B., Wanionek, K., Murphy, B.R., and Whitehead, S.S. (2006). The live attenuated dengue serotype 1 vaccine rDEN1Delta30 is safe and highly immunogenic in healthy adult volunteers. Hum. Vaccine *2*, 167–173.

Durbin, A.P., McArthur, J.H., Marron, J.A., Blaney, J.E., Thumar, B., Wanionek, K., Murphy, B.R., and Whitehead, S.S. (2006). rDEN2/4Delta30(ME), a live attenuated chimeric dengue serotype 2 vaccine is safe and highly immunogenic in healthy dengue-naive adults. Hum Vaccine *2*, 255–260.

Durbin, A.P., and Whitehead, S.S. (2010). Dengue vaccine candidates in development. Curr. Top. Microbiol. Immunol. *338*, 129–143.

Eckels, K.H., and Putnak, R. (2003). Formalin-inactivated whole virus and recombinant subunit flavivirus vaccines. Adv. Virus Re.s *61*, 395–418.

Eckels, K.H., Yu, Y.X., Dubois, D.R., Marchette, N.J., Trent, D.W., and Johnson, A. (1988). Japanese encephalitis virus live-attenuated vaccine, Chinese strain SA14–14–2; adaptation to primary canine kidney cell cultures and preparation of a vaccine for human use. Vaccine *6*, 513–518.

Edelman, R., Schneider, R.J., Chieowanich, P., Pornpibul, R., and Voodhikul, P. (1975). The effect of dengue virus infection on the clinical sequelae of Japanese encephalitis: A one year follow-up study in Thailand. Southeast Asian J. Trop. Med. Publ. Hlth *6*, 308–315.

Edelman, R., Tacket, C.O., Wasserman, S.S., Vaughn, D.W., Eckels, K.H., Dubois, D.R., Summers, P.L., and Hoke, C.H., Jr. (1994). A live attenuated dengue-1 vaccine candidate (45AZ5) passaged in primary dog kidney cell culture is attenuated and immunogenic for humans. J. Infect. Dis. *170*, 1448–1455.

Edelman, R., Wasserman, S.S., Bodison, S.A., Putnak, R.J., Eckels, K.H., Tang, D., Kanesa-Thasan, N., Vaughn, D.W., Innis, B.L., and Sun, W. (2003). Phase I trial of 16 formulations of a tetravalent live-attenuated dengue vaccine. Am. J. Trop. Med. Hyg. *69*, 48–60.

Elias, C., Okwo-Bele, J.M., and Fischer, M. (2009). A strategic plan for Japanese encephalitis control by 2015. Lancet Infect. Dis. *9*, 7.

Engel, A.R., Vasconcelos, P.F., McArthur, M.A., and Barrett, A.D. (2006). Characterization of a viscerotropic yellow fever vaccine variant from a patient in Brazil. Vaccine *24*, 2803–2809.

Etemad, B., Batra, G., Raut, R., Dahiya, S., Khanam, S., Swaminathan, S., and Khanna, N. (2008). An envelope domain III-based chimeric antigen produced in Pichia pastoris elicits neutralizing antibodies against all four dengue virus serotypes. Am. J. Trop. Med. Hyg. *79*, 353–363.

Fitzgeorge, R., and Bradish, C.J. (1980). The in vivo differentiation of strains of yellow fever virus in mice. J. Gen. Virol. *46*, 1–13.

Fonseca, B.A., Khoshnood, K., Shope, R.E., and Mason, P.W. (1991). Flavivirus type-specific antigens produced from fusions of a portion of the E protein gene with the Escherichia coli trpE gene. Am. J. Trop. Med. Hyg. *44*, 500–508.

Gajanana, A., Thenmozhi, V., Samuel, P.P., and Reuben, R. (1995). A community-based study of subclinical flavivirus infections in children in an area of Tamil Nadu, India, where Japanese encephalitis is endemic. Bull.World Health Org. *73*, 237–244.

Galler, R., Post, P.R., Santos, C.N., and Ferreira, II. (1998). Genetic variability among yellow fever virus 17D substrains. Vaccine *16*, 1024–1028.

Garcia, G., Vaughn, D.W., and Del, Angel, R.M. (1997). Recognition of synthetic oligopeptides from nonstructural proteins NS1 and NS3 of dengue-4 virus by sera from dengue virus-infected children. Am. J. Trop. Med. Hyg. *56*, 466–470.

Gatchalian, S., Yao, Y., Zhou, B., Zhang, L., Yoksan, S., Kelly, K., Neuzil, K.M., Yaich, M., and Jacobson, J. (2008). Comparison of the immunogenicity and safety of measles vaccine administered alone or with live, attenuated Japanese encephalitis SA 14-14-2 vaccine in Philippine infants. Vaccine *26*, 2234–2241.

Gould, E.A., and Buckley, A. (1989). Antibody-dependent enhancement of yellow fever and Japanese encephalitis virus neurovirulence. J. Gen.Virol. *70*, 1605–1608.

Green, S., Vaughn, D.W., Kalayanarooj, S., Nimmannitya, S., Suntayakorn, S., Nisalak, A., Lew, R., Innis, B.L., Kurane, I., Rothman, A.L., and Ennis, F.A. (1999). Early immune activation in acute dengue illness is related to development of plasma leakage and disease severity. J. Infect. Dis. *179*, 755–762.

Gritsun, T.S., Lashkevich, V.A., and Gould, E.A. (2003). Tick-borne encephalitis. Antiviral Res. *57*, 129–146.

Grossman, R.A., Edelman, R., Chieowanich, P., Voodhikul, P., and Siriwan, C. (1973). Study of Japanese encephalitis virus in Chiangmai valley, Thailand. II. Human clinical infections. Am. J. Epidemiol. *98*, 121–132.

Gu, P.W., and Ding, Z.F. (1987). Inactivated Japanese encephalitis (JE) vaccine made from hamster kidney cell culture (a review). JE Hers Bull. *2*, 15–26.

Gubler, D.J. (1989). Aedes aegypti and Aedes aegypti-borne disease control in the 1990s: top down or bottom up. Charles Franklin Craig Lecture. Am. J. Trop. Med. Hyg. *40*, 571–578.

Gubler, D.J. (1997). Dengue and dengue hemorrhagic fever: its history and resurgence as a global public health problem, pp. 1–22. In Gubler, D.J. and Kuno, G. (eds.), Dengue and Dengue Hemorrhagic Fever. CAB International, New York.

Gubler, D.J. (1997). The emergence of dengue/dengue hemorrhagic fever as a global public health problem, pp. 83–92. In Saluzzo, J.F. and Dodet, B. (ed.), Factors in the Emergence of Arbovirus Diseases. Elisevier, Paris.

Guirakhoo, F., Arroyo, J., Pugachev, K.V., Miller, C., Zhang, Z.X., Weltzin, R., Georgakopoulos, K., Catalan, J., Ocran, S., Soike, K., Ratterree, M., and Monath, T.P. (2001). Construction, safety, and immunogenicity in nonhuman primates of a chimeric yellow fever-dengue virus tetravalent vaccine. J. Virol. *75*, 7290–7304.

Guirakhoo, F., Kitchener, S., Morrison, D., Forat, R., McCarthy, K., Nichols, R., and Yoksan, S. (2006). Live attenuated chimeric yellow fever dengue type 2 (ChimeriVax-DEN2) vaccine: Phase I clinical trial for safety and immunogenicity: effect of yellow fever pre-immunity in induction of cross neutralizing antibody responses to all 4 dengue serotypes. Hum. Vaccine *2*, 60–67.

Guirakhoo, F., Kitchener, S., Morrison, D., Forrat, R., McCarthy, K., Nichols, R., Yoksan, S., Duan, X., Ermak, T.H., Kanesa-Thasan, N., Bedford, P., Lang, J., Quentin-Millet, M.J., and Monath, T.P. (2006). Live attenuated chimeric yellow fever dengue type 2 (ChimeriVax-DEN2) vaccine: Phase I clinical trial for safety and immunogenicity: effect of yellow fever pre-immunity in induction of cross neutralizing antibody responses to all 4 dengue serotypes. Hum. Vaccine *2*, 60–67.

Guirakhoo, F., Pugachev, K., Zhang, Z., Myers, G., Levenbook, I., Draper, K., Lang, J., Ocran, S., Mitchell, F., Parsons, M., Brown, N., Brandler, S., Fournier, C., Barrere, B., Rizvi, F., Travassos, A., Nichols, R., Trent, D., and Monath, T. (2004). Safety and efficacy of chimeric yellow Fever-dengue virus tetravalent vaccine formulations in nonhuman primates. J. Virol. *78*, 4761–4775.

Guirakhoo, F., and Pugachev, K.V. (2002). Viremia and immunogenicity in nonhuman primates of a tetravalent yellow fever-dengue chimeric vaccine: genetic reconstructions, dose adjustment, and antibody responses against wild-type dengue virus isolates. Virology *298*, 146–159.

Guirakhoo, F., Weltzin, R., Chambers, T.J., Zhang, Z.X., Soike, K., Ratterree, M., Arroyo, J., Georgakopoulos, K., Catalan, J., and Monath, T.P. (2000). Recombinant chimeric yellow fever-dengue type 2 virus is immunogenic and protective in nonhuman primates. J. Virol. *74*, 5477–5485.

Guirakhoo, F., Zhang, Z.X., Chambers, T.J., Delagrave, S., Arroyo, J., Barrett, A.D., and Monath, T.P. (1999). Immunogenicity, genetic stability, and protective efficacy of a recombinant, chimeric yellow fever-Japanese encephalitis virus (ChimeriVax-JE) as a live, attenuated vaccine candidate against Japanese encephalitis. Virology *257*, 363–372.

Gurunathan, S., Klinman, D.M., and Seder, R.A. (2000). DNA vaccines: immunology, application, and optimization. Annu Rev Immunol *18*, 927–74.

Guy, B. (2009). Immunogenicity of sanofi pasteur tetravalent dengue vaccine. J. Clin. Virol. *46*, Suppl. *2*, S16–19.

Guy, B., and Almond, J.W. (2008). Towards a dengue vaccine: progress to date and remaining challenges. Comp Immunol Microbiol Infect Dis *31*, 239–52.

Guy, B., Barban, V., Mantel, N., Aguirre, M., Gulia, S., Pontvianne, J., Jourdier, T.M., Ramirez, L., Gregoire, V., Charnay, C., Burdin, N., Dumas, R., and Lang, J. (2009). Evaluation of interferences between dengue vaccine serotypes in a monkey model. Am. J. Trop. Med. Hyg. *80*, 302–311.

Guy, B., Guirakhoo, F., Barban, V., Higgs, S., Monath, T.P., and Lang, J. (2010). Preclinical and clinical development of YFV 17D-based chimeric vaccines against dengue, West Nile and Japanese encephalitis viruses. Vaccine *28*, 632–49.

Guzman, M.G., Hermida, L., Bernardo, L., Ramirez, R., and Guillen, G. (2010). Domain III of the envelope protein as a dengue vaccine target. Expert Rev Vaccines *9*, 137–147.

Guzman, M.G., Kouri, G., Valdes, L., Bravo, J., Alvarez, M., Vazquez, S., Delgado, I., and Halstead, S.B. (2000). Epidemiologic studies on dengue in Santiago de Cuba, 1997. Am. J. Epidemiol. *152*, 793–799.

Guzman, M.G., Kouri, G.P., Bravo, J., Soler, M., Vazquez, S., and Morier, L. (1990). Dengue hemorrhagic fever in Cuba, 1981: a retrospective seroepidemiologic study. Am. J. Trop.Med.Hyg. *42*, 179–184.

Guzman, M.G., Kouri, G., Valdes, L., Bravo, J., Vazquez, S., and Halstead, S.B. (2002). Enhanced severity of secondary dengue 2 infections occurring at an interval of 20 compared with 4 years after dengue 1 infection. PAHO J. Epidemiol. *81*, 223–227.

Gyure, K.A. (2009). West Nile virus infections. J. Neuropathol. Exp. Neurol. *68*, 1053–1060.

Halstead, S.B. (1974). Etiologies of the experimental dengues of Siler and Simmons. Am. J. Trop. Med. Hyg. *23*, 974–982.

Halstead, S.B. (1980a). Dengue haemorrhagic fever – a public health problem and a field for research. Bull. World Health Org. *58*, 1–21.

Halstead, S.B. (1980b). Immunological parameters of Togavirus disease syndromes. In The Togaviruses, Biology, Structure, Replication, Schlesinger, R.W. ed. (New York, Academic Press), pp. 107–173.

Halstead, S.B. (1982). Immune enhancement of viral infection. Prog. Allergy *31*, 301–364.

Halstead, S.B. (1997). Epidemiology of dengue and dengue hemorrhagic fever, p. 23–44. In Gubler, D.J. and Kuno, G. (eds.), Dengue and Dengue Hemorrhagic Fever. CAB, Wallingford, UK.

Halstead, S.B. (2006). Dengue in the Americas and Southeast Asia: Do they differ? Rev. Panam Salud Publica *20*, 407–415.

Halstead, S.B., and Grosz, C.R. (1962). Subclinical Japanese encephalitis I. Infection of Americans with limited residence in Korea. Am. J. Hyg. *75*, 190–201.

Halstead, S.B., and Jacobson, J. (2008). Japanese encephalitis vaccines, p. 311–352. In Plotkin, S.A., Orenstein, W.A., and Offit, P.A. (eds.), Vaccines, 5th edn. Elsevier, Philadelphia.

Halstead, S.B., and Marchette, N. (2003). Biological properties of dengue viruses following serial passage in primary dog kidney cells. Am. Trop, J. Med. Hyg. *69* (*Suppl. 6*), 5–11.

Halstead, S.B., and Palumbo, N.E. (1973). Studies on the immunization of monkeys against dengue: II. Protection following inoculation of combinations of viruses. Am. J. Trop. Med. Hyg. *22*, 375–381.

Halstead, S.B., and Papaevangelou, G. (1980). Transmission of dengue 1 and 2 viruses in Greece in 1928. Am. J. Trop. Med. Hyg. *29*, 635–637.

Halstead, S.B., and Thomas, S.J. (2010). Japanese encephalitis: new options for active immunization. Clin. Infect. Dis. *50*, 1155–1164.

Halstead, S.B., and Tsai, T.F. (2004). Japanese encephalitis vaccines, pp. 919–958. In Plotkin, S.A. and Orenstein, W.A. (ed.), Vaccines, 4th edn. Elsevier, Philadelphia.

Halstead, S.B., and Yamarat, C. (1965). Recent epidemics of hemorrhagic fever in Thailand. Observations related to pathogenesis of a "new" dengue disease. J. Am. Publ. Health Assoc. *55*, 1386–1395.

Halstead, S.B., Streit, T.G., Lafontant, J., Putvatana, R., Russell, K., Sun, W., Kanesa-Thasan, N., Hayes, C.G., and Watts, D.M. (2001). Haiti: Absence of dengue hemorrhagic fever despite hyperendemic dengue virus transmission. Am. J. Trop. Med. Hyg. *65*, 180–183.

Hammon, W.M., Rudnick, A., and Sather, G.E. (1960). Viruses associated with epidemic hemorrhagic fevers of the Philippines and Thailand. Science *131*, 1102–1103.

Hammon, W.M., Tigertt, W.D., Sather, G.E., Berge, T.O., and Meiklejohn, G. (1958). Epidemiologic studies of concurrent 'virgin' epidemics of Japanese B encephalitis and of mumps on Guam, 1947–1948, with subsequent observations including dengue, through 1957. Am. J. Trop. Med. Hyg. *7*, 441–468.

Harinasuta, C., Nimmanitya, S., and Titsyakorn, U. (1985). The effect of interferon-alpha A on two cases of Japanese encephalitis in Thailand. Southeast Asian J. Trop. Med. Publ. Hlth *16*, 332–336.

Hatz, C., Werlein, J., Mutsch, M., Hufnagel, M., and Behrens, R.H. (2009). Japanese encephalitis: defining risk incidence for travelers to endemic countries and vaccine prescribing from the UK and Switzerland. J. Travel Med. *16*, 200–203.

Heinz, F.X., Holzmann, H., Essl, A., and Kundi, M. (2007). Field effectiveness of vaccination against tick-borne encephalitis. Vaccine *25*, 7559–7567.

Henchal, E.A., and Putnak, J.R. (1990). The dengue viruses. Clin. Microbiol. Rev. *3*, 376–396.

Henderson, B.E., Cheshire, P.P., Kirya, G.B., and Lule, M. (1970). Immunological studies with yellow fever and selected African group B arboviruses in rhesus and vervet monkeys. Am. J. Trop. Med. Hyg. (*19*, 110–119.

Hennessy, S., Zhengle, L., Tsai, T.F., Strom, B.L., Wan, C.M., Liu, H.L., Wu, T.X., Yu, H.J., Liu, Q.M., Karabatsos, N., Bilker, W.B., and Halstead, S.B. (1996). Effectiveness of live-attenuated Japanese encephalitits vaccine (SA14–14–2): a case control study. Lancet *347*, 1 583–1571.

Hermida, L., Bernardo, L., Martin, J., Alvarez, M., Prado, I., Lopez, C., Sierra, Bde, L., Martinez, R., Rodriguez, R., Zulueta, A., Perez, A.B., Lazo, L., Rosario, D., Guillen, G., and Guzman, M.G. (2006). A recombinant fusion protein containing the domain III of the dengue-2 envelope protein is immunogenic and protective in nonhuman primates. Vaccine *24*, 3165–3171.

Hoke, C.H., Jr., Malinoski, F.J., Eckels, K.H., Scott, R.M., Dubois, D.R., Summers, P.L., Simms, T., Burrous, J., Hasty, S.E., and Bancroft, W.H. (1990). Preparation of an attenuated dengue 4 (341750 Carib) virus vaccine. II. Safety and immunogenicity in humans. Am. J. Trop. Med. Hyg. *43*, 219–226.

Hoke, C.H., Jr., Nisalak, A., Sangawhipa, N., Jatanasen, S., Laorakpongse, T., Innis, B.L., Kotchasenee, S., Gingrich, J.B., Latendresse, J., Fukai, K., *et al.* (1988). Protection against Japanese encephalitis by inactivated vaccines. N. Engl. J. Med. *319*, 608–614.

Holman, D.H., Wang, D., Raviprakash, K., Raja, N.U., Luo, M., Zhang, J., Porter, K.R., and Dong, J.Y. (2007). Two complex, adenovirus-based vaccines that together induce immune responses to all four dengue virus serotypes. Clin. Vaccine Immunol. *14*, 182–189.

Hombach, J., Solomon, T., Kurane, I., Jacobson, J., and Wood, D. (2005). Report on a WHO consultation on immunological endpoints for evaluation of new Japanese encephalitis vaccines, WHO, Geneva, 2–3 September, 2004. Vaccine *23*, 5205–5211.

Huang, C.Y., Butrapet, S., Pierro, D.J., Chang, G.J., Hunt, A.R., Bhamarapravati, N., Gubler, D.J., and Kinney, R.M. (2000). Chimeric dengue type 2 (vaccine strain PDK-53)/dengue type 1 virus as a potential candidate dengue type 1 virus vaccine. J. Virol. *74*, 3020–3028.

Huang, C.Y., Butrapet, S., Tsuchiya, K.R., Bhamarapravati, N., Gubler, D.J., and Kinney, R.M. (2003). Dengue

2 PDK-53 virus as a chimeric carrier for tetravalent dengue vaccine development. J. Virol. 77, 11436–11447.

Huang, J.H., Wey, J.J., Sun, Y.C., Chin, C., Chien, L.J., and Wu, Y.C. (1999). Antibody responses to an immunodominant nonstructural 1 synthetic peptide in patients with dengue fever and dengue hemorrhagic fever. J. Med. Virol. 57, 1–8.

Immunization Practices Advisory Committee (ACIP). (1993). Inactivated Japanese encepahltis virus vaccine: recommendations of the ACIP. MMWR Morb. Mortal. Wkly. Rep. 42, 1–15.

Jelinek, T. (2009). Ixiaro: a new vaccine against Japanese encephalitis. Exp. Rev. Vaccines 8, 1501–1511.

Jennings, A.D., Gibson, C.A., Miller, B.R., Mathews, J.H., Mitchell, C.J., Roehrig, J.T., Wood, D.J., Taffs, F., Sil, B.K., Whitby, S.N., *et al.* (1994). Analysis of a yellow fever virus isolated from a fatal case of vaccine-associated human encephalitis. J. Infect. Dis. 169, 512–518.

Jennings, A.D., Whitby, J.E., Minor, P.D., and Barrett, A.D.T. (1993). Comparison of the nucleotide and deduced amino acid sequences of the structural protein genes of the yellow fever 17DD vaccine strain from Senegal with those of other yellow fever vaccine viruses. Vaccine 11, 679–681.

Kaltenbock, A., Dubischar-Kastner, K., Eder, G., Jilg, W., Klade, C., Kollaritsch, H., Paulke-Korinek, M., F. von Sonnenburg, Spruth, M., Tauber, E., Wiedermann, U., and Schuller, E. (2009). Safety and immunogenicity of concomitant vaccination with the cell-culture based Japanese Encephalitis vaccine IC51 and the hepatitis A vaccine HAVRIX1440 in healthy subjects: A single-blind, randomized, controlled Phase 3 study. Vaccine 27, 4483–4489.

Kaltenbock, A., Dubischar-Kastner, K., Schuller, E., Datla, M., Klade, C.S., and Kishore, T.S. (2010). Immunogenicity and safety of IXIARO (IC51) in a Phase II study in healthy Indian children between 1 and 3 years of age. Vaccine 28, 834–839.

Kanesa-Thasan, N., Edelman, R., Tacket, C.O., Wasserman, S.S., Vaughn, D.W., Coster, T.S., Kim-Ahn, G.J., Dubois, D.R., Putnak, J.R., King, A., Summers, P.L., Innis, B.L., Eckels, K.H., and Hoke, C.H., Jr. (2003). Phase 1 studies of Walter Reed Army Institute of Research candidate attenuated dengue vaccines: selection of safe and immunogenic monovalent vaccines. Am. J. Trop. Med. Hyg. 69, 17–23.

Kanesa-thasan, N., Sun, W., Kim-Ahn, G., Van, Albert, S., Putnak, J.R., King, A., Raengsakulsrach, B., Christ-Schmidt, H., Gilson, K., Zahradnik, J.M., Vaughn, D.W., Innis, B.L., Saluzzo, J.F., and Hoke, C.H., Jr. (2001). Safety and immunogenicity of attenuated dengue virus vaccines (Aventis Pasteur) in human volunteers. Vaccine 19, 3179–3188.

Kari, K., Liu, W., Gautama, K., Mammen, M.P., Jr., Clemens, J.D., Nisalak, A., Subrata, K., Kim, H.K., and Xu, Z.Y. (2006). A hospital-based surveillance for Japanese encephalitis in Bali, Indonesia. BMC Med. 4, 8.

Karin, N. (2000). Gene therapy for T cell mediated autoimmunity: teaching the immune system how to restrain its own harmful activities by targeted DNA vaccines. Isr. Med. Assoc. J. 2S, 63–8.

Kerr, J.A. (1951). The clinical aspects and diagnosis of yellow fever, pp. 389–425. In Strode, G.K. (ed.), Yellow Fever. McGraw-Hill Book Company, Inc., New York.

Khanam, S., Khanna, N., and Swaminathan, S. (2006). Induction of neutralizing antibodies and T cell responses by dengue virus type 2 envelope domain III encoded by plasmid and adenoviral vectors. Vaccine 24, 6513–6525.

Khanam, S., Pilankatta, R., Khanna, N., and Swaminathan, S. (2009). An adenovirus type 5 (AdV5) vector encoding an envelope domain III-based tetravalent antigen elicits immune responses against all four dengue viruses in the presence of prior AdV5 immunity. Vaccine 27, 6011–6021.

Khanam, S., Rajendra, P., Khanna, N., and Swaminathan, S. (2007). An adenovirus prime/plasmid boost strategy for induction of equipotent immune responses to two dengue virus serotypes. BMC Biotechnol. 7, 10.

Khromava, A.Y., Eidex, R.B., Weld, L.H., Kohl, K.S., Bradshaw, R.D., Chen, R.T., and Cetron, M.S. (2005). Yellow fever vaccine: an updated assessment of advanced age as a risk factor for serious adverse events. Vaccine 23, 3256–63.

Kimura, K., Dosaka, A., Hashimoto, Y., Yasunaga, T., Uchino, M., and Ando, M. (1997). Single-photon emission CT findings in acute Japanese encephalitis. Am.Neuroradiol, J. (AJNR) 18, 465–469.

Kinney, R., Kinney, C.Y.H., Barban, V., Lang, J., and Guy, B. January 5, 2010. Dengue serotype 1 attenuated strain. US Patent.

Kinney, R., Kinney, C.Y.H., Barban, V., Lang, J., and Guy, B. January 5 2010. Dengue serotype 2 attenuated strain. US Patent.

Kinney, R.M., Butrapet, S., Chang, G.J., Tsuchiya, K.R., Roehrig, J.T., Bhamarapravati, N., and Gubler, D.J. (1997). Construction of infectious cDNA clones for dengue 2 virus: strain 16681 and its attenuated vaccine derivative, strain PDK-53. Virology 230, 300–308.

Kinney, R.M., and Huang, C.Y. (2001). Development of new vaccines against dengue fever and Japanese encephalitis. Intervirology 44, 176–197.

Kiple, K.F. (1984). The Caribbean Slave: A Biological History. Cambridge University Press, Cambridge.

Kiple, K.F., and Kiple, V.H. (1977). Black yellow fever immunities, innate and acquired. Soc. Sci. Hist. 2, 419–436.

Kiple, K.F., and Ornelas, K.C. (1996). Race, war and tropical medicine in the eighteenth-century Caribbean. Clio Med. 35, 65–79.

Kitchener, S., Nissen, M., Nasveld, P., Forrat, R., Yoksan, S., Lang, J., and Saluzzo, J.F. (2006). Immunogenicity and safety of two live-attenuated tetravalent dengue vaccine formulations in healthy Australian adults. Vaccine 24, 1238–1241.

Klinman, D.M., Takeno, M., Ichino, M., Gu, M., Yamschikov, G., Mor, G., and Conover, J. (1997). DNA vaccines: safety and efficacy issues. Springer Semin. Immunopathol. (*19*, 245–256.

Kochel, T., Wu, S.J., Raviprakash, K., Hobart, P., Hoffman, S., Porter, K., and Hayes, C. (1997). Inoculation of plasmids expressing the dengue-2 envelope gene elicit neutralizing antibodies in mice. Vaccine *15*, 547–552.

Kollaritsch, H., Paulke-Korinek, M., and Dubischar-Kastner, K. (2009). IC51 Japanese encephalitis vaccine. Exp. Opin. Biol. Ther. *9*, 921–31.

Konishi, E., Yamaoka, M., Khin, Sane, W., Kurane, I., and Mason, P.W. (1998). Induction of protective immunity against Japanese encephalitis in mice by immunization with a plasmid encoding Japanese encephalitis virus premembrane and envelope genes. J. Virol. *72*, 4925–4930.

Konishi, E., Yamaoka, M., Khin Sane, W., Kurane, I., Takada, K., and Mason, P.W. (1999). The anamnestic neutralizing antibody response is critical for protection of mice from challenge following vaccination with a plasmid encoding the Japanese encephalitis virus premembrane and envelope genes. J. Virol. *73*, 5527–5534.

Konishi, E., Yamaoka, M., Kurane, I., and Mason, P.W. (2000). A DNA vaccine expressing dengue type 2 virus premembrane and envelope genes induces neutralizing antibody and memory B cells in mice. Vaccine *18*, 1133–1139.

Kreil, T.R., Burger, I., Attakpah, E., Olas, K., and Eibl, M.M. (1998). Passive immunization reduces immunity that results from simultaneous active immunization against tick-borne encephalitis virus in a mouse model. Vaccine *16*, 955–959.

Kuhn, R.J., Zhang, W., Rossmann, M.G., Corver, J., Lenches, E., Jones, C.T., Mukhopadhyay, S., Chipman, P.R., Strauss, E.G., Baker, T.S., and Strauss, J.H. (2002). Structure of dengue virus. Implications for flavivirus organization, maturation and fusion. Cell *108*, 717–725.

Kumar, R., Kohli, N., Mathur, A., and Wakhlu, I. (1992). Use of the computed tomographic scan in Japanese encephalitis. Ann. Trop. Med. Parasitol. *86*, 77–81.

Kumar, R., Mathur, A., and Singh, K.B. (1993). Clinical sequelae of Japanese encephalitis in children. Ind. J. Med. Res. *97*, 9–13.

Kumar, R., Tripathi, P., and Rizvi, A. (2009). Effectiveness of one dose of live-attenuated SA 14-14-2 vaccine against Japanense encephalitis. N. Engl. J. Med. *360*, 1465–1466.

Kunze, U. (2010). TBE – awareness and protection: the impact of epidemiology, changing lifestyle, and environmental factors. Wien Med/ Wochenschr. *160*, 252–255.

Kunze, U., Asokliene, L., Bektimirov, T., Busse, A., Chmelik, V., Heinz, F.X., Hingst, V., Kadar, F., Kaiser, R., Kimmig, P., *et al.* (2004). Tick-borne encephalitis in childhood – consensus 2004. Wien Med. Wochenschr. *154*, 242–245.

Kunze, U., Baumhackl, U., Bretschneider, R., Chmelik, V., Grubeck-Loebenstein, B., Haglund, M., Heinz, F., Kaiser, R., Kimmig, P., Kunz, C., *et al.* (2005). The Golden Agers and Tick-borne encephalitis. Conference report and position paper of the International Scientific Working Group on Tick-borne encephalitis. Wien Med. Wochenschr. *155*, 289–294.

Lai, C.J., Bray, M., Men, R., Cahour, A., Chen, W., Kawano, H., Tadano, M., Hiramatsu, K., Tokimatsu, I., Pletnev, A., Arakai, S., Shameem, G., and Rinaudo, M. (1998). Evaluation of molecular strategies to develop a live dengue vaccine [In Process Citation]. Clin. Diagn. Virol. *10*, 173–179.

Lai, C.J., Zhao, B.T., Hori, H., and Bray, M. (1991). Infectious RNA transcribed from stably cloned full-length cDNA of dengue type 4 virus. Proc. Natl. Acad. Sci. U.S.A. *88*, 5139–5143.

Lang, J. (2009). Recent progress on sanofi pasteur's dengue vaccine candidate. J. Clin. Virol. *46*, Suppl. *2*, S20–S4.

Lee, E., Hall, R.A., and Lobigs, M. (2004). Common E protein determinants for attenuation of glycosaminoglycan-binding variants of Japanese encephalitis and West Nile viruses. J. Virol. *78*, 8271–8280.

Lepiniec, L., Dalgarno, L., Huong, V.T., Monath, T.P., Digoutte, J.P., and Deubel, V. (1994). Geographic distribution and evolution of yellow fever viruses based on direct sequencing of genomic cDNA fragments. J. Gen. Virol. *75*, 417–423.

Libraty, D.H., Nisalak, A., Endy, T.P., Suntayakorn, S., Vaughn, D.W., and Innis, B.L. (2000). Presented at the Annual Meeting of the American Society of Tropical Medicine and Hygiene, Houston, TX, 29 October–2 November.

Lindsey, N.P., Schroeder, B.A., Miller, E.R., Braun, M.M., Hinckley, A.F., Marano, N., Slade, B.A., Barnett, E.D., Brunette, G.W., Horan, K., Staples, J.E., Kozarsky, P.E., and Hayes, E.B. (2008). Adverse event reports following yellow fever vaccination. Vaccine *26*, 6077–6082.

Liu, Z.L., Hennessy, S., Strom, B.L., Tsai, T.F., Wan, C.M., Tang, C.S., Xiang, C.F., Bilker, W.B., Pan, X.P., Yao, Y.J., Xu, Z.W., and Halstead, S.B. (1997). Short-term safety of live attenuated Japanese encephalitis vaccine (SA14-14-2): Results of a randomized trial with 26, 239 subjects. J. Infect. Dis. *176*, 1366–1369.

Lloyd, W., Theiler, M., and Ricci, N.I. (1936). Modification of the virulence of yellow fever virus by cultivation in tissues in vitro. Trans. R. Soc. Trop. Med. Hyg. *29*, 481–529.

Lorin, C., Combredet, C., Labrousse, V., Mollet, L., Despres, P., and Tangy, F. (2005). A paediatric vaccination vector based on live attenuated measles vaccine. Therapie *60*, 227–233.

Lourenco de Oliveira, R., Vazeille, M., Bispo, A.M. de Filippis, and Failloux, A.B. (2002). Oral suasceptibility to yellow fever virus of Aedes aegypti from Brazil. Mem. Inst Oswaldo Cruz. *97*, 437–439.

Lourenco de Oliveira, R., M. vazeille, A.M. de Filippis, and Failloux, A.B. (2004). Aedes aegypti in Brazil: genetically differentiated populations with high susceptibility to dengue and yellow fever viruses. Trans. R. Soc. Trop. Med. Hyg. *98*, 43–54.

Lumley, G.F., and Taylor, F.H. (1943). Dengue, vol. 3. Glebe, N.S.W., Australia.

Luo, D.P., Yao, R.G., Song, J.D., Huo, H.G., and Wang, Z. (1994). The effect of DDT spraying and bed nets impregnated with pyrethroid insecticide on the incidence of Japanese encephalitis virus infection. Trans. R. Soc. Trop. Med. Hyg. *88*, 629–631.

Lyons, A., Kanesa-thasan, N., Kuschner, R.A., Eckels, K.H., Putnak, R., Sun, W., Burge, R., Towle, A.C., Wilson, P., Tauber, E., and Vaughn, D.W. (2007). A Phase 2 study of a purified, inactivated virus vaccine to prevent Japanese encephalitis. Vaccine *25*, 3445–3453.

Mackowiak, P.A., Wasserman, S.S., Tacket, C.O., Vaughn, D.W., Eckels, K.H., Dubois, D.R., Hoke, C.H., Jr., and Edelman, R. (1994). Quantitative relationship between oral temperature and severity of illness following inoculation with candidate attenuated dengue virus vaccines. Clin. Infect. Dis. (19948–19950.

Mansfield, K.L., Johnson, N., Phipps, L.P., Stephenson, J.R., Fooks, A.R., and Solomon, T. (2009). Tick-borne encephalitis virus – a review of an emerging zoonosis. J. Gen. Virol. *90*, 1781–1794.

Marchette, N.J., Dubois, D.R., Larsen, L.K., Summers, P.L., Kraiselburd, E.G., Gubler, D.J., and Eckels, K.H. (1990). Preparation of an attenuated dengue 4 (341750 Carib) virus vaccine. I. Pre-clinical studies. Am. J. Trop. Med. Hyg. *43*, 212–218.

Marfin, A.A., Eides, R.S., Kozarsky, P.E., and Cetron, M.S. (2005). Yellow fever and Japanese encephalitis vaccines: indications and complications. Infect. Dis. Clin. North Am. (*19*, 151–168.

Markoff, L. (2000). Points to consider in the development of a surrogate for efficacy of novel Japanese encephalitis virus vaccines. Vaccine *18*, Suppl. *2*, 26–32.

Markoff, L., Pang, X., Houng Hs, H.S., Falgout, B., Olsen, R., Jones, E., and Polo, S. (2002). Derivation and characterization of a dengue type 1 host range-restricted mutant virus that is attenuated and highly immunogenic in monkeys. J. Virol. *76*, 3318–3328.

Martin, M., Tsai, T.F., Cropp, B., Chang, G.J., Holmes, D.A., Tseng, J., Shieh, W., Zaki, S.R., Al-Sanouri, I., Cutrona, A.F., Ray, G., Weld, L.H., and Cetron, M.S. (2001). Fever and multisystem organ failure associated with 17D-204 yellow fever vaccination: a report of four cases. Lancet *358*, 98–104.

Martin, T., Parker, S.E., Hedstrom, R., Le, T., Hoffman, S.L., Norman, J.E., Hobart, P., and Lew, D. (1999). Plasmid DNA malaria vaccine: the potential for genomic integration after intramuscular injection. Hum. Gene Ther. *10*, 759–768.

McArthur, J.H., Durbin, A.P., Marron, J.A., Wanionek, K.A., Thumar, B., Pierro, D.J., Schmidt, A.C., Blaney, J.E., Jr., Murphy, B.R., and Whitehead, S.S. (2008). Phase I clinical evaluation of rDEN4Delta30-200, 201: a live attenuated dengue 4 vaccine candidate designed for decreased hepatotoxicity. Am. J. Trop. Med. Hyg. *79*, 678–684.

McMahon, A.W., Eidex, R.B., Marfin, A.A., Russell, M., Sejvar, J.J., Markoff, L., Hayes, E.B., Chen, R.T., Ball, R., Braun, M.M., and Cetron, M. (2007). Neurologic disease associated with 17D-204 yellow fever vaccination: a report of 15 cases. Vaccine *25*, 1727–1734.

Men, R., Bray, M., Clark, D., Chanock, R.M., and Lai, C.J. (1996). Dengue type 4 virus mutants containing deletions in the 3' noncoding region of the RNA genome: analysis of growth restriction in cell culture and altered viremia pattern and immunogenicity in rhesus monkeys. J. Virol. *70*, 3930–3937.

Men, R., Wyatt, L., Tokimatsu, I., Arakaki, S., Shameem, G., Elkins, R., Chanock, R., Moss, B., and Lai, C.J. (2000). Immunization of rhesus monkeys with a recombinant of modified vaccinia virus Ankara expressing a truncated envelope glycoprotein of dengue type 2 virus induced resistance to dengue type 2 virus challenge. Vaccine *18*, 3113–3122.

Miller, B.R., and Ballinger, M.E. (1988). *Aedes albopictus* mosquitoes introduced into Brazil: vector competence for yellow fever and dengue viruses. Trans. R. Soc. Trop. Med. Hyg. *82*, 476–477.

Miller, B.R., Monath, T.P., Tabachnik, W.J., and Ezike, V.I. (1989). Epidemic yellow fever caused by an incompetent mosquito vector. Tro.p Med. Parasitol. *40*, 396–399.

Miller, J.D., R.G. van der Most, Akondy, R.S., Glidewell, J.T., Albott, S., Masopust, D., Murali-Krishna, K., Mahar, P.L., Edupuganti, S., Lalor, S., *et al.* (2008). Human effector and memory CD8+ T cell responses to smallpox and yellow fever vaccines. Immunity *28*, 710–722.

Misra, U.K., Kalita, J., Jain, S.K., and Mathur, A. (1994). Radiological and neurophysiological changes in Japanese encephalitis. J. Neurol. Neurosurg. Psychiatr. *57*, 1484–1487.

Monath, T.P. (1971). Neutralizing antibody responses in the major immunoglobulin classes to yellow fever 17D vaccination of humans. Am. J. Epidemiol. *93*, 122–129.

Monath, T.P. (1989). Recent epidemics of yellow fever in Africa and the risk of future urbanization and spread. In Uren, M.F., Blok, J., and Manderson, L.H. (ed.), Arbovirus Research in Australia. Queensland Institute of Medical Research, Brisbane.

Monath, T.P. (1994). Yellow fever and dengue – The interactions of virus, vector and host in the re-emergence of epidemic disease. Semin. Virol. *5*, 133–145.

Monath, T.P. (1997). Epidemiology of yellow fever: current status and speculations on future trends, pp. 143–156. In Saluzzo, J.F. and Dodet, B. (eds.), Factors in the Emergence of Arbovirus Diseases. Elsevier, Paris.

Monath, T.P. (2002). Editorial: jennerian vaccination against West Nile virus. Am. J. Trop. Med. Hyg. *66*, 113–114.

Monath, T.P., Kinney, R.M., Schlesinger, J.J., Brandriss, M.W., and Bres, P. (1983). Ontogeny of yellow fever 17D vaccine: RNA oligonucliotide fingerprint and monoclonal antibody analyses of vaccines produced world-wide. J. Gen. Virol. *64*, 627–637.

Monath, T.P., Levenbook, I., Soike, K., Zhang, Z.X., Ratterree, M., Draper, K., Barrett, A.D., Nichols, R., Weltzin, R., Arroyo, J., and Guirakhoo, F. (2000).

Chimeric yellow fever virus 17D-Japanese encephalitis virus vaccine: dose-response effectiveness and extended safety testing in rhesus monkeys. J. Virol. *74*, 1742–1751.

Monath, T.P., Arroyo, J., Miller, C., and Guirakhoo, F. (2001). West Nile virus vaccine. Curr. Drug Targets Infect. Disord. *1*, 37–50.

Monath, T.P., Nichols, R., Archambault, W.T., Moore, L., Marchesani, R., Tian, J., Shope, R.E., Thomas, N., Schrader, R., Furby, D., and Bedford, P. (2002a). Comparative safety and immunogenicity of two yellow fever 17D vaccines (ARILVAX and YF-VAX) in a phase III multicenter, double-blind clinical trial. Am. J. Trop. Med. Hyg. *66*, 533–541.

Monath, T.P., McCarthy, K., Bedford, P., Johnson, C.T., Nichols, R., Yoksan, S., Marchesani, R., Knauber, M., Wells, K.H., Arrroyo, J., and Guirakhoo, F. (2002b). Clinical proof of principle for ChimeriVax: recombinant live, attenuated vaccines against flavivirus infections. Vaccine *20*, 1004–1018.

Monath, T.P., Guirakhoo, F., Nichols, R., Yoksan, S., Schrader, R., Murphy, C., Blum, P., Woodward, S., McCarthy, K., Mathis, D., Johnson, C., and Bedford, P. (2003). Chimeric live, attenuated vaccine against Japanese encephalitis (ChimeriVax-JE): phase 2 clinical trials for safety and immunogenicity, effect of vaccine dose and schedule, and memory response to challenge with inactivated Japanese encephalitis antigen. J. Infect. Dis. *188*, 1213–1230.

Monath, T.P., Myers, G.A., Beck, R.A., Knauber, M., Scappaticci, K., Pullano, T., Archambault, W.T., Catalan, J., Miller, C., Zhang, Z.X., Shin, S., Pugachev, K., Draper, K., Levenbook, I.S., and Guirakhoo, F. (2005). Safety testing for neurovirulence of novel live, attenuated flavivirus vaccines: infant mice provide an accurate surrogate for the test in monkeys. Biologicals 33, 131–44.

Monath, T.P., Liu, J., Kanesa-Thasan, N., Myers, G.A., Nichols, R., Deary, A., McCarthy, K., Johnson, C., Ermak, T., Shin, S., Arroyo, J., Guirakhoo, F., Kennedy, J.S., Ennis, F.A., Green, S., and Bedford, P. (2006). A live, attenuated recombinant West Nile virus vaccine. Proc. Natl. Acad. Sci. U.S.A. *103*, 6694–6699.

Monath, T.P., Cetron, M.S., and Teuwen, D.E. (2008). Yellow fever vaccine, pp. 959–1055. In Plotkin, S.A., Orenstein, W.A., and Offit, P.A. (eds.), Vaccines, 5th edn. Saunders Elsevier, Philadelphia.

Montagnon, B.J., and Vincent-Falquet, J.C. (1998). Experience with the Vero cell line. Dev. Biol. Stand. *93*, 119–123.

Mor, G., Singla, M., Steinberg, A.D., Hoffman, S.L., Okuda, K., and Klinman, D.M. (1997). Do DNA vacines induce anutoimmune disease? Hum. Gene Ther. *8*, 293–300.

Morrison, D., Legg, T.J., Billings, C.W., Forrat, R., Yoksan, S., and Lang, J. (2010). A novel tetravalent dengue vaccine is well tolerated and immunogenic against all 4 serotypes in flavivirus-naive adults. J. Infect. Dis. (*201*, 370–377.

Mutebi, J.P., Gianella, A., and Travassos da Rosa, A. (2004). Yellow fever virus infectivity for Bolivian *Aedes aegypti* mosquitoes. Emerg. Infect. Dis. *10*, 1657–1660.

Mutebi, J.P., Wang, H., Li, L., Bryant, J.E., and Barrett, A.D. (2001). Phylogenetic and evolutionary relationships among yellow fever virus isolates in Africa. J. Virol. *75*, 6999–7008.

Ni, H., Watowich, S.J., and Barrett, A.D. (1997). Molecular basis of attenuation and virulence of Japanese encephalitis virus, pp. (203–211. In Saluzzo, J.F. and Dodet, B. (ed.), Factors in the Emergence of Arbovirus Diseases. Elsevier, Paris.

Nichols, W.W., Ledwith, B.J., Manam, S.V., and Troilo, P.J. (1995). Potential DNA vaccine integratoin into host cell genome. Ann N.Y. Acad Sci *772*, 30–9.

Nimmannitya, S. (1987). Clinical spectrum and management of dengue haemorrhagic fever. Southeast Asian J. Trop. Med. Publ. Hlth *18*, 392–397.

Nimmannitya, S., Halstead, S.B., Cohen, S.N., and Margiotta, M.R. (1969). Dengue and chikungunya virus infection in man in Thailand, 1962–1964. I. Observations on hospitalized patients with hemorrhagic fever. Am. J. Trop. Med. Hyg. *18*, 954–71.

Nogueira, R.M., Zagner, S.M., Martins, I.S., Lampe, E., Miagostovich, M.P., and Schatzmayr, H.G. (1991). Dengue haemorrhagic fever/dengue shock syndrome (DHF/DSS) caused by serotype 2 in Brazil. Mem. Inst. Oswaldo Cruz *86*, 269.

Ohrr, H.C., Tandan, J.B., Sohn, Y.M., Shin, S.H., Pradhan, D.P., and Halstead, S.B. (2005). Effect of a single dose of SA 14-14-2 vaccine 1 year after immunization in Nepalese children with Japanese encephalitis: a cose-control study. Lancet *366*, 1375–8.

Oya, A. (1988). Japanese encephalitis vaccine. Acta Paediatr. Jpn. *30*, 175–184.

Ozherelkov, S.V., Kalinina, E.S., Kozhevnikova, T.N., Sanin, A.V., Timofeeva, T., Timofeev, A.V., and Stivenson, D.R. (2008). [Experimental study of the phenomenon of antibody dependent tick-borne encephalitis virus infectivity enhancement in vitro]. Zh. Mikrobiol. Epidemiol. Immunobiol., Nov–Dec, 39–43.

Parker, S.E., Borellini, F., Wenk, M.L., Hobart, P., Hoffman, S.L., Hedstrom, R., Le, T., and Norman, J.A. (1999). Plasmid DNA malaria vaccine: tissue distribution and safety studies in mice and rabbits. Hum. Gene Ther. *10*, 741–758.

Pen'evskaia, N.A., and Rudakov, N.V. (2010). [Efficiency of use of immunoglobulin preparations for the postexposure prevention of tick-borne encephalitis in Russia (a review of semi-centennial experience)]. Med. Parazitol. (Mosk), Jan–Mar, 53–59.

Pierson, T.C., Fremont, D.H., Kuhn, R.J., and Diamond, M.S. (2008). Structural insights into the mechanisms of antibody-mediated neutralization of flavivirus infection: implications for vaccine development. Cell Host Microbe *4*, 229–238.

Poland, J.D., Calisher, C.H., Monath, T.P., Downs, W.G., and Murphy, K. (1981). Persistence of neutralizing antibody 30–35 years after immunization with 17D yellow fever vaccine. Bull World Health Org. *59*, 895–900.

Porter, K., Kochel, T.J., Wu, S.J., Raviprakash, K., Phillips, I., and Hayes, C.G. (1998). Protective efficacy of a dengue 2 DNA vaccine in mice and the effect of CpG immuno-stimulatory motifs on antibody responses. Arch. Virol. *143*, 997–1003.

Prud'homme, G.J., Lawson, B.R., Chang, Y., and Theophilopoulos, A.N. (2001). Immunotherapeutic gene transfer into muscle. Trends Immunol. *22*, 149–55.

Pugachev, K.V., Ocran, S.W., Guirakhoo, F., Furby, D., and Monath, T.P. (2002). Heterogeneous nature of the genome of the ARILVAX yellow fever 17D vaccine revealed by consensus sequencing. Vaccine *20*, 996–9.

Pulendran, B. (2009). Learning immunology from the yellow fever vaccine: innate immunity to systems vaccinology. Nat. Rev. Immunol. *9*, 741–7.

Putnak, R., Barvir, D.A., Burrous, J.M., Dubois, D.R., V.M. D'Andrea, Hoke, C.H., Jr., Sadoff, J.C., and Eckels, K.H. (1996). Development of a purified, inactivated, dengue-2 virus vaccine prototype in Vero cells: immunogenicity and protection in mice and rhesus monkeys. J. Infect. Dis. *174*, 1176–1184.

Putnak, R., Cassidy, K., Conforti, N., Lee, R., Sollazzo, D., Truong, T., Ing, E., Dubois, D., Sparkuhl, J., Gastle, W., and Hoke, C.H., Jr. (1996). Immunogenic and protective response in mice immunized with a purified, inactivated, Dengue-2 virus vaccine prototype made in fetal rhesus lung cells. Am. J. Trop. Med. Hyg. *55*, 504–510.

Putnak, R.J., Coller, B.A., Voss, G., Vaughn, D.W., Clements, D., Peters, I., Bignami, G., Houng, H.S., Chen, R.C., Barvir, D.A., *et al.* (2005). An evaluation of dengue type-2 inactivated, recombinant subunit, and live-attenuated vaccine candidates in the rhesus macaque model. Vaccine *23*, 4442–4452.

Querec, T.D., Akondy, R.S., Lee, E.K., Cao, W., Nakaya, H.I., Teuwen, D., Pirani, A., Gernert, K., Deng, J., Marzolf, B., Kennedy, K., Wu, H., Bennouna, S., Oluoch, H., Miller, J., Vencio, R.Z., Mulligan, M., Aderem, A., Ahmed, R., and Pulendran, B. (2009). Systems biology approach predicts immunogenicity of the yellow fever vaccine in humans. Nat. Immunol. *10*, 116–125.

Quintos, F.N., Lim, L.E., Juliano, L., Reyes, A., and Lacson, P. (1954). Hemorrhagic fever observed among children in the Philippines. Philipp. J. Pediatr. *3*, 1–19.

Ramirez-Ronda, C.H., and Garcia, C.D. (1994). Dengue in the Western hemisphere. Infect. Dis. Clin. N. Am. *8*, 107–128.

Raviprakash, K., Apt, D., Brinkman, A., Skinner, C., Yang, S., Dawes, G., Ewing, D., Wu, S.J., Bass, S., Punnonen, J., and Porter, K. (2006). A chimeric tetravalent dengue DNA vaccine elicits neutralizing antibody to all four virus serotypes in rhesus macaques. Virology *353*, 166–73.

Raviprakash, K., Kochel, T.J., Ewing, D., Simmons, M., Phillips, I., Hayes, C.G., and Porter, K.R. (2000). Immunogenicity of dengue virus type 1 DNA vaccines expressing truncated and full length envelope protein. Vaccine *18*, 2426–2434.

Raviprakash, K., Porter, K.R., Kochel, T.J., Ewing, D., Simmons, M., Phillips, I., Murphy, G.S., Weiss, W.R., and Hayes, C.G. (2000). Dengue virus type 1 DNA vaccine induces protective immune responses in rhesus macaques. J. Gen. Virol. *81*, Pt 7, 1659–1667.

Raviprakash, K., Wang, D., Ewing, D., Holman, D.H., Block, K., Woraratanadharm, J., Chen, L., Hayes, C., Dong, J.Y., and Porter, K. (2008). A tetravalent dengue vaccine based on a complex adenovirus vector provides significant protection in rhesus monkeys against all four serotypes of dengue viruses. J. Virol. *82*, 6927–6934.

Reid, M., Mackenzie, D., Baron, A., Lehmann, N., Lowry, K., Aaskov, J., Guirakhoo, F., and Monath, T.P. (2006). Experimental infection of Culex annulirostris, *Culex gelidus*, and *Aedes vigilax* with a yellow fever/Japanese encephalitis virus vaccine chimera (ChimeriVax-JE). Am. J. Trop. Med. Hyg. *75*, 659–63.

Rey, F.A., Heinz, F.X., Mandl, C., Kunz, C., and Harrison, S.C. (1995). The envelope glycoprotein from tick-borne encephalitis virus at 2 angstrom resolution. Nature *375*, 291–298.

Reynes, J.M., Laurent, A., Deubel, V., Telliam, E., and Moreau, J.P. (1994). The first epidemic of dengue hemorrhagic fever in French Guiana. Am. J. Trop. Med. Hyg. *51*, 545–553.

Rice, C.M., Grakoui, A., Galler, R., and Chambers, T.J. (1989). Transcription of infectious yellow fever RNA from full-length cDNA templates produced by in vitro ligation. New Biol. *1*, 285–296.

Rice, C.M., Lenches, E.M., Eddy, S.R., Shin, S.J., Sheets, R.L., and Strauss, J.H. (1985). Nucleotide sequence of yellow fever viruses: Implications for flavivirus gene expression and evolution. Science *229*, 726–733.

Roche, J.C., Cordellier, R., Hervy, J.P., Digoutte, J.P., and Monteny, N. (1983). Isolement de 96 souches de virus dengue 2 a partir de moustiques captures en Cote d'Ivoire et Haute Volta. Ann. Virol. *143*, 233–244.

Roehrig, J.T., Johnson, A.J., Hunt, A.R., Bolin, R.A., and Chu, M.C. (1990). Antibodies to dengue 2 virus E-glycoprotein synthetic peptides identify antigenic conformation. Virology *177*, 668–675.

Roehrig, J.T., Johnson, A.J., Hunt, A.R., Beaty, B.J., and Mathews, J.H. (1992). Enhancement of the antibody response to flavivirus B-cell epitopes by using homologous or heterologous T-cell epitopes. J. Virol. *66*, 3385–3390.

Roehrig, J.T., Bolin, R.A., and Kelly, R.G. (1998). Monoclonal antibody mapping of the envelope glycoprotein of the dengue 2 virus, Jamaica. Virology *246*, 317–328.

Rosen, L. (1958). Observations on the epidemiology of dengue in Panama. Am. J. Hyg. *68*, 45–58.

Rudnick, A. (1978). Ecology of dengue virus. Asian J. Infect.Dis. *2*, 156–160.

Sabchareon, A., Lang, J., Chanthavanich, P., Yoksan, S., Forrat, R., Attanath, P., Sirivichayakul, C., Pengsaa, K., Pojjaroen-Anant, C., Chambonneau, L., Saluzzo, J. -F., and Bhamarapravati, N. (2004). Safety and immunogenicity of two tetravalent live attenuated dengue vaccines in 5–12 year old Thai children. Pediatr. Infect. Dis. J. *23*, 99–109.

Sabchareon, A., Lang, J., Chanthavanich, P., Yoksan, S., Forrat, R., Attanath, P., Sirivichayakul, C., Pengsaa, K., Pojjaroen-Anant, C., Chokejindachai, W., Jagsudee, A., Saluzzo, J.F., and Bhamarapravati, N. (2002). Safety and immunogenicity of tetravalent live-attenuated dengue vaccines in Thai adult volunteers: role of serotype concentration, ratio, and multiple doses. Am. J. Trop. Med. Hyg. *66*, 264–72.

Sabin, A.B. (1952). Research on dengue during World War II. Am. J. Trop. Med. Hyg. *1*, 30–50.

Sall, A.A., Faye, O., Diallo, M., Firth, C., Kitchen, A., and Holmes, E.C. (2010). Yellow fever virus exhibits slower evolutionary dynamics than dengue virus. J. Virol. *84*, 765–772.

Sanchez, V., Gimenez, S., Tomlinson, B., Chan, P.K., Thomas, G.N., Forrat, R., Chambonneau, L., Deauvieau, F., Lang, J., and Guy, B. (2006). Innate and adaptive cellular immunity in flavivirus-naive human recipients of a live-attenuated dengue serotype 3 vaccine produced in Vero cells (VDV3). Vaccine *24*, 4914–4926.

Schuller, E., Jilma, B., Voicu, V., Golor, G., Kollaritsch, H., Kaltenbock, A., Klade, C., and Tauber, E. (2008). Long-term immunogenicity of the new Vero cell-derived, inactivated Japanese encephalitis virus vaccine IC51 Six and 12 month results of a multicenter follow-up phase 3 study. Vaccine *26*, 4382–4386.

Schuller, E., Klade, C.S., Heinz, F.X., Kollaritsch, H., Rendi-Wagner, P., Jilma, B., and Tauber, E. (2008). Effect of pre-existing anti-tick-borne encephalitis virus immunity on neutralising antibody response to the Vero cell-derived, inactivated Japanese encephalitis virus vaccine candidate IC51. Vaccine *26*, 6151–6156.

Schuller, E., Klade, C.S., Wolfl, G., Kaltenbock, A., Dewasthaly, S., and Tauber, E. (2009). Comparison of a single, high-dose vaccination regimen to the standard regimen for the investigational Japanese encephalitis vaccine, IC51: a randomized, observer-blind, controlled Phase 3 study. Vaccine *27*, 2188–93.

Siler, J.F., Hall, M.W., and Hitchens, A.P. (1926). Dengue: Its history, epidemiology, mechanism of transmission, etiology, clinical manifestations, immunity, and prevention. Philippine J. Sci. *29*, 1–304.

Silva, M.L., Espirito-Santo, L.R., Martins, M.A., Silveira-Lemos, D., Peruhype-Magalhaes, V., Caminha, R.C., P. de Andrade Maranhao-Filho, Auxiliadora-Martins, M., R. de Menezes Martins, Galler, R., *et al.* (2010). Clinical and immunological insights on severe, adverse neurotropic and viscerotropic disease following 17D yellow fever vaccination. Clin. Vaccine Immunol. *17*, 118–126.

Simasathien, S., Thomas, S.J., Watanaveeradej, V., Nisalak, A., Barberousse, C., Innis, B.L., Sun, W., Putnak, J.R., Eckels, K.H., Hutagalung, Y., *et al.* (2008). Safety and immunogenicity of a tetravalent live-attenuated dengue vaccine in flavivirus naive children. Am. J. Trop. Med. Hyg. *78*, 426–433.

Simmons, J.S., St John, J.H., and Reynolds, F.H.K. (1931). Experimental studies of dengue. Philippine J. Sci. *44*, 1–252.

Simmons, M., Burgess, T., Lynch, J., and Putnak, R. (2010). Protection against dengue virus by non-replicating and live attenuated vaccines used together in a prime boost vaccination strategy. Virology *396*, 280–288.

Simmons, M., Nelson, W.M., Wu, S.J., and Hayes, C.G. (1998). Evaluation of the protective efficacy of a recombinant dengue envelope B domain fusion protein against dengue 2 virus infection in mice. Am. J. Trop. Med. Hyg. *58*, 655–662.

Simmons, M., Porter, K.R., Hayes, C.G., Vaughn, D.W., and Putnak, R. (2006). Characterization of antibody responses to combinations of a dengue virus type 2 DNA vaccine and two dengue virus type 2 protein vaccines in rhesus macaques. J. Virol. *80*, 9577–9585.

Simmons, M., Vaughn, D.W., Nisalak, A., Graham, R., and Hayes, C.G. (1998). Presented at the International Conference on Emerging Infectious Diseases, Atlanta, GA, 8–11 March 1998.

Smith, H.H., Penna, H.A., and Paoliello, A. (1938). Yellow fever vaccination with cultured virus (17D) without immune serum. Am. J. Trop. Med. *18*, 437–468.

Solomon, T. (2004). Flavivirus encephalitis. N. Engl. J. Med. *351*, 370–378.

Solomon, T., Dung, N.M., Kneen, R., Thao, T.T. le, Gainsborough, M., Nisalak, A., Day, N.P.J., Kirkham, F.J., Vaughn, D.W., Smith, S.M., and White, N.J. (2002). Seizures and raised intracranial pressure in Vietnamese patients with Japanese encephalitis. Brain *125*, 1084–1093.

Solomon, T., and Vaughn, D.W. (2002). Pathogenesis and clinical features of Japanese encephalitis and West Nile virus infections. Curr Top Microbiol Immunol *267*, 171–194.

Srivastava, A.K., Aira, Y., Mori, C., Kobayashi, Y., and Igarashi, A. (1987). Antigenicity of Japanese encephalitis virus envelope glycoprotein V3 (E) and its cyanogen bromide cleaved fragments examined by monoclonal antibodies and Western blotting. Arch. Virol. 96, 97–107.

Srivastava, A.K., Putnak, J.R., Lee, S.H., Hong, S.P., Moon, S.B., Barvir, D.A., Zhao, B., Olson, R.A., Kim, S.O., Yoo, W.D., Towle, A.C., Vaughn, D.W., Innis, B.L., and Eckels, K.H. (2001). A purified inactivated Japanese encephalitis virus vaccine made in Vero cells. Vaccine *19*, 4557–4565.

Srivastava, A.K., Putnak, J.R., Warren, R.L., and Hoke, C.H., Jr. (1995). Mice immunized with a dengue type 2 virus E and NS1 fusion protein made in Escherichia coli are protected against lethal dengue virus infection. Vaccine *13*, 1251–1258.

Staples, J.E., Gershman, M., and Fischer, M. (2010). Yellow fever vaccine: recommendations of the Advisory Committee on Immunization Practices (ACIP). MMWR Recomm. Rep. *59*, 1–27.

Strode, G.K. (ed.). (1951). Yellow Fever. McGraw-Hill Book Company, New York.

Stuart, G. (1956). Reactions following vaccination against yellow fever, pp. 143–156. In Smithburn, K.C. (ed.), Yellow Fever Vaccinations. WHO, Geneva.

Sugrue, R.J., Fu, J., Howe, J., and Chan, Y.C. (1997). Expression of the dengue virus structural proteins in Pichia pastoris leads to the generation of virus-like particles. J. Gen.Virol. *78*, 1861–1866.

Sukhavachana, P., Yuill, T.M., and Russell, P.K. (1969). Assay of arbovirus neutralizing antibody by micromethods. Trans. R. Soc. Trop. Med. Hyg. *63*, 446–455.

Sun, W., Cunningham, D., Wasserman, S.S., Perry, J., Putnak, J.R., Eckels, K.H., Vaughn, D.W., Thomas, S.J., Kanesa-Thasan, N., Innis, B.L., and Edelman, R. (2009). Phase 2 clinical trial of three formulations of tetravalent live-attenuated dengue vaccine in flavivirus-naive adults. Hum. Vaccine *5*, 33–40.

Sun, W., Edelman, R., Kanesa-Thasan, N., Eckels, K.H., Putnak, J.R., King, A.D., Houng, H.S., Tang, D., Scherer, J.M., Hoke, C.H., Jr., and Innis, B.L. (2003). Vaccination of human volunteers with monovalent and tetravalent live-attenuated dengue vaccine candidates. Am. J. Trop. Med. Hyg. *69*, 24–31.

Sun, W., Nisalak, A., Gettayacamin, M., Eckels, K.H., Putnak, J.R., Vaughn, D.W., Innis, B.L., Thomas, S.J., and Endy, T.P. (2006). Protection of Rhesus monkeys against dengue virus challenge after tetravalent live attenuated dengue virus vaccination. J. Infect. Di.s *193*, 1658–1665.

Suzano, C.E., Amaral, E., Sato, H.K., and Papaiordanou, P.M. (2006). The effects of yellow fever immunization (17DD) inadvertently used in early pregnancy during a mass campaign in Brazil. Vaccine *24*, 1421–1426.

Suzuki, R., Winkelmann, E.R., and Mason, P.W. (2009). Construction and characterization of a single-cycle chimeric flavivirus vaccine candidate that protects mice against lethal challenge with dengue virus type 2. J. Virol. *83*, 1870–1880.

Swaminathan, S., Batra, G., and Khanna, N. (2010). Dengue vaccines: state of the art. Exp. Opin. Ther. Pat. *20*, 819–835.

Tabachnik, W.J., Wallis, G.P., Aitkin, T.H.G., Miller, B.R., Amato, G.D., Lorenz, L., Powell, J.R., and Beatty, B.J. (1985). Oral infection of Aedes aegypti with yellow fever virus: geographic variation and genetic considerations. Am. J. Trop. Med. Hyg. *34*, 1219–1224.

Tandan, J.B., Ohrr, H., Sohn, Y.M., Yoksan, S., Ji, M., Nam, C.M., and Halstead, S.B. (2007). Single dose of SA 14-14-2 vaccine provides long-term protection against Japanese encephalitis: a case-control study in Nepalese children 5 years after immunization. Vaccine *25*, 5041–5045.

Tarr, G C., and Hammon, W.M. (1974). Cross-protection between group B arboviruses: Resistance in mice ot Japanese B encephalitis and St. Louis encephalitis viruses induce by dengue virus immunization. Infect. Immun. *9*, 909–915.

Tauber, E., Kollaritsch, H., Korinek, M., Rendi-Wagner, P., Jilma, B., Firbas, C., Schranz, S., Jong, E., Klingler, A., Dewasthaly, S., and Klade, C.S. (2007). Safety and immunogenicity of a Vero-cell-derived, inactivated Japanese encephalitis vaccine: a non-inferiority, phase III, randomised controlled trial. Lancet *370*, 1847–1853.

Tauber, E., Kollaritsch, H., F. von Sonnenburg, Lademann, M., Jilma, B., Firbas, C., Jelinek, T., Beckett, C., Knobloch, J., McBride, W.J., Schuller, E., Kaltenbock, A., Sun, W., and Lyons, A. (2008). Randomized, double-blind, placebo-controlled phase 3 trial of the safety and tolerability of IC51, an inactivated Japanese encephalitis vaccine. J. Infect. Dis. (*198*, 493–499.

Teitelbaum, P. (2009). Expert opinion on vaccination of travelers against Japanese encephalitis. J. Travel Med. *16*, 441; author reply 441–2.

Theiler, M., and Smith, H.H. (1937). The effect of prolonged cultivation in vitro upon the pathogenicity of yellow fever virus. J Exp Med *65*, 767–786.

Theiler, M., and Smith, H.H. (1937). Use of yellow fever virus modified by in vitro cultivation for human immunization. J. Exp. Med. *65*, 787–800.

Trent, D.W., Kinney, R.M., and Huang, C.Y. (1997). Recombinant dengue virus vaccines, pp. 379–404. In Gubler, D.J. and Kuno, G. (eds.), Dengue and dengue hemorrhagic fever. CAB International, New York.

Tsai, C.J., Kuo, C.H., Chen, P.C., and Changcheng, C.S. (1991). Upper gastrointestinal bleeding in dengue fever. Am. J. Gastroenterol. *86*, 33–35.

Tsai, T.F., Burchard, G.D., Jelinek, J., Jong, E.C., F. von Sonnenburg, Steffen, R., and Zuckerman, J. (2009). Response to letter. J. Travel Med. *16*, 441–442.

Tsai, T.F., Yu, Y.X., Jia, L.L., Putvatana, R., Zhang, R., Wang, S., and Halstead, S.B. (1998). Immunogenicity of live attenuated SA14-14-2 Japanese encephalitis vaccine--a comparison of 1- and 3-month immunization schedules. J. Infect. Dis. *177*, 221–223.

Tsai, T.R., and Solomon, T. (2004). Flaviviruses (yellow fever, dengue, dengue hemorrhagic fever, Japanese encephalitis, St. Louis encephalitis tick-borne encephalitis), pp. (1926–1950. In Mandell, G.L., Bennett, J.E., and Dolin, R. (eds.), Principles and practice of infectious diseases., 6th ed, vol. 2. Elsevier, Philadelphia.

Valdes, I., Hermida, L., Gil, L., Lazo, L., Castro, J., Martin, J., Bernardo, L., Lopez, C., Niebla, O., Menendez, T., Romero, Y., Sanchez, J., Guzman, M.G., and Guillen, G. (2010). Heterologous prime-boost strategy in non-human primates combining the infective dengue virus and a recombinant protein in a formulation suitable for human use. Int. J. Infect. Dis. *14*, e377–e383.

Valdes, I., Hermida, L., Martin, J., Menendez, T., Gil, L., Lazo, L., Castro, J., Niebla, O., Lopez, C., Bernardo, L., Sanchez, J., Romero, Y., Martinez, R., Guzman, M.G., and Guillen, G. (2009). Immunological evaluation in nonhuman primates of formulations based on the chimeric protein P64k-domain III of dengue 2 and two components of Neisseria meningitidis. Vaccine *27*, 995–1001.

Vasconcelos, P.F., Luna, E.J., Galler, R., Silva, L.J., Coimbra, T.L., Barros, V.L., Monath, T.P., Rodigues, S.G., Laval, C., Costa, Z.G., *et al.* (2001). Serious adverse events associated with yellow fever 17DD vaccine in Brazil: a report of two cases. Lancet *358*, 91–97.

Vasilakis, N., Cardosa, J., Diallo, M., Sall, A.A., Holmes, E.C., Hanley, K.A., Weaver, S.C., Mota, J., and Rico-Hesse, R. (2010). Sylvatic dengue viruses share the pathogenic potential of urban/endemic dengue viruses. J. Virol. *84*, 3726–7; author reply 3727–3728.

Vaughn, D.W. (2000). Invited commentary: Dengue lessons from Cuba. Am. J. Epidemiol. *152*, 800–803.

Vaughn, D.W., Green, S., Kalayanarooj, S., Innis, B.L., Nimmannitya, S., Suntayakorn, S., Endy, T.P., Raengsakulrach, B., Rothman, A.L., Ennis, F.A., and Nisalak, A. (2000). Dengue viremia titer, antibody response pattern, and virus serotype correlate with disease severity. J Infect. Dis. *181*, 2–9.

Vaughn, D.W., Hoke, C.H., Jr., Yoksan, S., LaChance, R., Innis, B.L., Rice, R.M., and Bhamarapravati, N. (1996). Testing of a dengue 2 live-attenuated vaccine (strain 16681 PDK 53) in ten American volunteers. Vaccine *14*, 329–336.

Velzing, J., Groen, J., Drouet, M.T., G. van Amerongen, Copra, C., Osterhaus, A.D., and Deubel, V. (1999). Induction of protective immunity against Dengue virus type 2: comparison of candidate live attenuated and recombinant vaccines. Vaccine *17*, 1312–1320.

von Lindern, J.J., Aroner, S., Barrett, N.D., Wicker, J.A., Davis, C.T., and Barrett, A.D. (2006). Genome analysis and phylogenetic relationships between east, central and west African isolates of Yellow fever virus. J. Gen. Virol. *87*, 895–907.

Waddell, M.B., and Taylor, R.M. (1945). Studies on cyclic passage of yellow fever virus in South American mammals and mosquitoes. I Marmosets (*Calithrix autira*) and Cebus monkeys (*Cebus versutus*) in combination with *Aedes aegypti* an *Haemagogus equinus*. Am. J. Trop. Med. *25*, 225–30.

Waddell, M.B., and Taylor, R.M. (1947). Studies on the cyclic passage of yellow fever virus in South American mammals and mosquitoes. III. Further observations on *Haemagogus equinus* as a vector of the virus. Am. J. Trop. Med. *27*, 471–476.

Waddell, M.B., and Taylor, R.M. (1948). Studies on the cyclic passage of yellow fever virus in South American mammals and mosquitoes. IV. Marsupials (Metachirus nudicaudatus and Marmosa) in combination with Aedes aegypti as vector. Am. J. Trop. Med. *28*, 87–100.

Waddell, M.B., and Taylor, R.M. (1946). Studies on the cyclic passage of yellow fever virus to South American mammals and mosquitoes. II. Marmosets (*Callithrix penicilliata* and *Leontocebus crysomela*) in combination with *Aedes aegypti*. Am. J. Trop. Med. *26*, 455–465.

Waldvogel, K., Bossart, W., Huisman, T., Boltshauser, E., and Nadal, D. (1996). Severe tick-borne encephalitis following passive immunization. Eur. J. Pediatr. *155*, 775–9.

Wang, E., Ni, H., Xu, R., Barrett, A.D., Watowich, S.J., Gubler, D.J., and Weaver, S.C. (2000). Evolutionary relationships of endemic/epidemic and sylvatic dengue viruses. J. Virol. *74*, 3227–34.

Webb, H.E., Wight, D.G.D., Platt, G.S., and Smith, C.E.G. (1967). Langat virus encephalitis in mice. 1. The effect of administration of specific antiserum. J. Hyg. *66*, 343–354.

Webster, D.P., Farrar, J., and Rowland-Jones, S. (2009). Progress towards a dengue vaccine. Lancet Infect. Dis. *9*, 678–687.

Whalen, R.G. (1996). DNA vaccines for emerging infectious diseases: what if ? Emerg. Infect. Dis. *2*, 168–175.

Widman, D.G., Frolov, I., and Mason, P.W. (2008). Third-generation flavivirus vaccines based on single-cycle, encapsidation-defective viruses. Adv Virus Res *72*, 77–126.

Wisseman, C.L., Jr., and Sweet, B.H. (1962). Immunological studies with group B arthropod-borne viruses: I. Broadened neutralizing antibody spectrum induced by strain 17D yellow fever vaccine in human subjects previously infected with Japanese encephalitis virus. Am.J. Trop.Med.Hyg., 550–561.

World Health Organization. (1997). Dengue Haemorrhagic Fever: Diagnosis, Treatment, Prevention And Control. 2nd edn. World Health Organization.

World Health Organization. (1988). Requirements for a yellow fever vaccine. Addendum 1987. World Health Organization.

World Health Organization. (2005). Revision of the International Health Regulations. WHO.

Wright, P.F., Durbin, A.P., Whitehead, S.S., Ikizler, M.R., Henderson, S., Blaney, J.E., Thumar, B., Ankrah, S., Rock, M.T., McKinney, B.A., *et al.* (2009). Phase 1 trial of the dengue virus type 4 vaccine candidate rDEN4Δ30-4995 in healthy adult volunteers. Am. J. Trop. Med. Hyg. *81*, 834–841.

Wu, S.J., Grouard-Vogel, G., Sun, W., Mascola, J.R., Brachtel, E., Putvatana, R., Louder, M.K., Filgueira, L., Marovich, M.A., Wong, H.K., *et al.* (2000). Human skin Langerhans cells are targets of dengue virus infection. Nat. Med. *6*, 816–820.

Yaich, M. (2009). Investing in vaccines for developing countries: How public-private partnerships can confront neglected diseases. Hum Vaccin *5*, 368–9.

Yu, Y. (2010). Phenotypic and genotypic characteristics of Japanese encephalitis attenuated live vaccine virus SA14-14-2 and their stabilities. Vaccine *28*, 3635–3641.

Zeng, L., Falgout, B., and Markoff, L. (1998). Identification of specific nucleotide sequences within the conserved 3′- SL in the dengue type 2 virus genome required for replication. J. Virol. *72*, 7510–7522.

Zhao, B.T., Prince, G., Horswood, R., Eckels, K., Summers, P., Chanock, R., and Lai, C.J. (1987). Expression of dengue virus structural proteins and nonstructural protein NS1 by a recombinant vaccinia virus. J. Virol. *61*, 4019–4022.

Antibody Therapeutics Against Flaviviruses

11

Michael S. Diamond, Theodore C. Pierson and John T. Roehrig

Abstract

Flaviviruses are a group of small RNA enveloped viruses that cause severe disease in humans worldwide. Recent advances in the structural biology of the flavivirus envelope proteins and virion have catalysed rapid progress towards understanding how the most potently inhibitory antibodies neutralize infection. These insights have identified factors that modulate the potency of neutralizing antibodies and provided insight into the design of novel antibody-based therapeutics against several members of the flavivirus genus. This chapter will discuss recent advances in the understanding of the mechanisms of antibody neutralization of flaviviruses, and review the progress towards development of antibody-based therapeutics against several different flaviviruses of global concern.

Introduction

Flaviviruses comprise a genus of greater than 70 enveloped, positive-sense RNA viruses of the *Flaviviridae* family (Lindenbach and Rice, 2001). Many flavivirus infections are transmitted through the bite of an infected mosquito or tick, and have the potential to cause severe illness ranging from encephalitis to haemorrhagic disease. Several members of this group, such as dengue virus (DENV) and West Nile virus (WNV), are considered emerging or re-emerging pathogens because in the past decade the incidence of human disease has increased at an alarming rate (Mackenzie *et al.*, 2004). Other flaviviruses of significant global importance include yellow fever virus (YFV), tick borne-encephalitis virus (TBEV), and Japanese encephalitis virus (JEV).

The ~11 kb flavivirus RNA genome is transcribed as a single polyprotein and is cleaved by host and viral proteases into three structural and seven non-structural proteins. The structural proteins include a capsid protein (C) that binds viral RNA, a pre-membrane (prM) protein that blocks premature viral fusion, and an envelope (E) protein that mediates viral attachment, membrane fusion, and virion assembly. The non-structural proteins (NS1, NS2A, NS2B, NS3, NS4A NS4B, and NS5) regulate viral translation, transcription, and replication and also attenuate host antiviral responses. NS1 has co-factor activity for the viral replicase (Khromykh *et al.*, 1999; Lindenbach and Rice, 1997), is secreted from infected cells (Flamand *et al.*, 1992; Flamand *et al.*, 1999), and antagonizes complement activation (Avirutnan *et al.*, 2010; Chung *et al.*, 2006a). NS3 has protease, NTPase, and helicase activities (Murthy *et al.*, 2000; Xu *et al.*, 2005) with NS2B serving as a required co-factor for NS3 protease activity (Yusof *et al.*, 2000). NS4A and NS4B are small hydrophobic proteins that lack conserved sequence motifs of known enzymes. NS4A induces membrane rearrangements that are observed in flavivirus-infected cells (Miller *et al.*, 2007; Roosendaal *et al.*, 2006) whereas NS4B, along with NS2A, co-localizes with replication complexes (Mackenzie *et al.*, 1998; Miller *et al.*, 2006). NS5 encodes the RNA-dependent RNA polymerase and an N-7 and 2′-O-methyltransferase (Dong *et al.*, 2008; Egloff *et al.*, 2002; Malet *et al.*, 2007; Yap *et al.*, 2007).

Flavivirus structural biology

Flaviviruses are small ~500 Å in diameter spherical virions composed of a single copy of genomic RNA, the capsid protein (C), a lipid envelope derived from the endoplasmic reticulum, and two structural glycoproteins: envelope (E) and premembrane/membrane (prM/M). The atomic structure of the ectodomain of the E protein reveals an organization of three distinct domains (Kanai *et al.*, 2006; Modis *et al.*, 2003, 2004, 2005; Nybakken *et al.*, 2006; Rey *et al.*, 1995; Zhang *et al.*, 2004). Domain III (DIII) is an immunoglobulin-like domain that has been suggested to mediate attachment interactions between the virus and the host cell (Bhardwaj *et al.*, 2001; Chin *et al.*, 2007; Chu *et al.*, 2005; Hung *et al.*, 2004; Lee *et al.*, 2006; Rey *et al.*, 1995). Domain II (DII) is an elongated finger-like domain that contains the highly conserved hydrophobic fusion loop that interacts with the membranes of the target cell during fusion (Allison *et al.*, 2001; Bressanelli *et al.*, 2004; Cherrier *et al.*, 2009; Modis *et al.*, 2004). Domain I (DI) is a β-barrel structure that connects DII and DIII via flexible hinges that participate in the conformational changes that drive the fusion process (Bressanelli *et al.*, 2004; Modis *et al.*, 2004; Mukhopadhyay *et al.*, 2005). An amphipathic stretch of residues referred to as the stem anchor connects the E protein ectodomain to two transmembrane domains that anchor the E protein within the viral membrane. The stem anchor is thought to lie flat against the viral membrane of mature flaviviruses and be involved in the rearrangements occurring within and between E proteins during the fusion process. In support of this, addition of exogenous stem anchor peptides can inhibit flavivirus infection (Hrobowski *et al.*, 2005) by blocking viral fusion (Schmidt *et al.*, 2010).

Recent progress has been made on the structure of the prM protein of DENV (Li *et al.*, 2008; Yu *et al.*, 2009). The intact DENV prM glycoprotein consists of 166 amino acids and cleavage by furin releases the N-terminal 91 'pr' residues during maturation (see below), leaving the ectodomain (residues 92 to 130) and C-terminal transmembrane region (residues 131 to 166) of 'M' in the virion. Structurally, the pr peptide consists of seven β strands that are predominantly anti-parallel with three stabilizing disulfide bonds and a single N-linked glycan. The pr peptide is positioned in the prM–E crystal structure over the fusion loop at the distal end of DII (Li *et al.*, 2008) consistent with its proposed function of preventing adventitious membrane fusion (Guirakhoo *et al.*, 1991).

Flaviviruses assemble at the endoplasmic reticulum (ER) and bud into the lumen as immature virions (Mackenzie and Westaway, 2001). Cryoelectron microscopy reconstructions of WNV and DENV immature virions reveal an icosahedral arrangement of 60 trimeric spikes, each composed of prM–E heterodimers in which the prM protein is positioned at the tip of each E protein of the trimer (Li *et al.*, 2008; Zhang *et al.*, 2003; Zhang *et al.*, 2007). Transit through the mildly acidic trans-Golgi network results in a re-organization of the E proteins into anti–parallel homodimers (E-E) that lie flat against the surface of the virion. The acidic pH of this environment also exposes a cleavage site on prM for the cellular serine endoprotease furin (Stadler *et al.*, 1997). Cleavage of the prM protein by furin is a required step in the maturation of the virion (Elshuber *et al.*, 2003). However, within the secretory pathway, pr is retained on the virion (Yu *et al.*, 2008) and blocks membrane fusion that could occur during egress (Guirakhoo *et al.*, 1991; Heinz *et al.*, 1994; Yu *et al.*, 2009). Dissociation of the pr peptide occurs once the virus is released into the neutral pH of the extracellular environment. In contrast to the spikes present on the immature precursor, mature flavivirus virions are relatively smooth and composed of 90 anti-parallel dimers arranged in rafts with T = 3 pseudo-icosahedral symmetry (Kuhn *et al.*, 2002; Mukhopadhyay *et al.*, 2003). In addition to mature and immature virions, cryoelectron microscopy studies show that individual flaviviruses may also contain a mixture of smooth mature, spiky immature and 'partially mature' particles, most probably due to incomplete processing by furin during virus maturation (Cherrier *et al.*, 2009; Zhang *et al.*, 2007).

Flavivirus fusion is orchestrated by a series of conformational changes within and between E proteins on the surface of the virion. These involve the dissociation of the E protein homodimers on the mature virion (a reversible step) and the formation of E protein trimers (an irreversible step)

(Allison *et al.*, 1995; Modis *et al.*, 2004; Stiasny *et al.*, 2001; Stiasny *et al.*, 1996). This process, which may or may not be catalysed by protonation of key histidine residues (Fritz *et al.*, 2008; Kampmann *et al.*, 2006; Nelson *et al.*, 2009) involves rotation between the three domains of the E protein, radial expansion of the virion, and projection of DII away from the surface of the virion while positioning the fusion loop to interact with the target cell membrane (Bressanelli *et al.*, 2004; Kaufmann *et al.*, 2009; Modis *et al.*, 2004; Zhang *et al.*, 2004). Following insertion of the fusion loop into the target cell membrane, viral and cellular membranes are brought into close apposition as the E protein folds back upon itself with the stem anchor region fitting into grooves on the exterior of the trimer (Harrison, 2008).

Antibodies protect against flavivirus infection

Humoral immunity is an essential component of the host response to flaviviruses to control dissemination of infection, and for encephalitic flaviviruses, to arrest viral replication before entry into the central nervous system. A majority of the flavivirus-specific protective antibody response is directed against the E protein, although antibodies against other structural and non-structural proteins are detected. Studies that have characterized the antibody profile in DENV-infected humans showed a significant anti-prM antibody response (Cardosa *et al.*, 2002; Oceguera *et al.*, 2007; Se-Thoe *et al.*, 1999) with poor neutralizing capacity that may contribute to immunopathogenesis (Dejnirattisai *et al.*, 2010). Passive transfer of polyclonal or monoclonal antibodies (MAbs) against flavivirus proteins can protect mice, hamsters, and non-human primates against lethal challenge of several different flaviviruses (see Table 11.1). Consistent with this, mice lacking B cells or secretory IgM were more vulnerable to flavivirus infection (Chambers *et al.*, 2008; Diamond *et al.*, 2003a; Diamond *et al.*, 2003b; Halevy *et al.*, 1994). Antibodies are believed to protect against virus infection though several mechanisms including direct neutralization of receptor binding, inhibition of viral fusion, Fc-γ receptor-dependent viral clearance, complement-mediated lysis of virus or infected cells, and antibody-dependent cytotoxicity of infected cells. Neutralization by antibodies against flaviviruses is a function of several related parameters including epitope location and accessibility, the mechanism of inhibition, and the strength of antibody binding.

The antigenic structure of flaviviruses

Epitopes recognized by neutralizing antibodies against flaviviruses have been identified in all three domains of the E protein (Fig. 11.1). Many of the most potent MAbs are virus type specific and bind epitopes within DIII (Beasley and Barrett, 2002; Crill and Roehrig, 2001; Gromowski and Barrett, 2007; Oliphant *et al.*, 2005; Roehrig *et al.*, 1998; Sanchez *et al.*, 2005; Shrestha *et al.*, 2010). One of the most well-characterized neutralizing MAbs, the anti-WNV E16 antibody, binds four discontinuous loops on the lateral ridge of WNV DIII (DIII-LR) (Nybakken *et al.*, 2005; Oliphant *et al.*, 2005). E16 blocks WNV infection at picomolar concentrations *in vitro* (Pierson *et al.*, 2007) and can protect animals against lethal challenge even when administered as a single dose up to 6 days after infection (Morrey *et al.*, 2007). Structural and biochemical studies identified four amino acids (306, 307, 330 and 332) on the amino-terminal and *B-C* loops as the core residues of the interface of DIII and the E16 Fab fragment; these residues are not conserved among flaviviruses (Nybakken *et al.*, 2005), which explains the type specificity of this and other mouse and human MAbs that bind this region and inhibit WNV infection strongly (Beasley and Barrett, 2002; Oliphant *et al.*, 2007; Sanchez *et al.*, 2005). Molecular docking and cryoelectron microscopy studies reveal that E16 cannot bind all 180 E proteins on the virion (Kaufmann *et al.*, 2006; Nybakken *et al.*, 2005); epitopes on the tightly clustered E proteins along the 5-fold symmetry axis remain unoccupied owing to steric conflicts. Thus, at a maximum, 120 DIII-LR epitopes on the mature virion can simultaneously bind E16. This MAb potently inhibits infection because it blocks viral fusion (Kaufmann *et al.*, 2009; Thompson *et al.*, 2009) at a low binding occupancy requirement

Table 11.1 Prophylactic and therapeutic efficacy of E protein-specific MAbs

Virus	MAb	Source	Specificity	Domain	Amino acids	Prophylaxis	Therapy	References
DENV-1	DENV1-E100	Mouse	Type	DIII	307, 329, 330, 361, 362	+	–	Shrestha *et al*. (2010)
	DENV1-E103	Mouse	Type	DIII	303, 328, 329, 330, 332	+	–	Shrestha *et al*. (2010)
	DENV1-E105	Mouse	Type	DIII	329, 332, 362, 362	+	+	Shrestha *et al*. (2010)
	DENV1-E99	Mouse	Subcomplex	DIII	309, 310, 311	+	–	Shrestha *et al*. (2010)
	DENV1-E106	Mouse	Subcomplex	DIII	310, 328, 329, 330, 361, 362, 364, 385	+	+	Shrestha *et al*. (2010)
DENV-2	9A3D-8	Mouse	Type	DIII	304, 307, 327	+	ND	Johnson and Roehrig (1999), Sukupolvi-Petty *et al*. (2007)
	3H5–1	Mouse	Type	DIII	304, 383, 384	+	ND	Gromowski and Barrett (2007); Kaufman *et al*. (1987); Sukupolvi-Petty *et al*. (2007)
	DV2–44	Mouse	Type	DII	88, 233	+	+	Sukupolvi-Petty *et al*. (2010)
	DV2–46	Mouse	Type	DII	88, 233	+	+	Sukupolvi-Petty *et al*. (2010)
	DV2–48	Mouse	Type	DI	177	+	ND	Sukupolvi-Petty *et al*. (2010)
	DV2–58	Mouse	Type	DII	88, 233	+	+	Sukupolvi-Petty *et al*. (2010)
	DV2–87	Mouse	Type	DIII	336, 340, 346	+	+	Sukupolvi-Petty *et al*. (2010)
	DV2–104	Mouse	Type	DIII	336, 340, 346	+	+	Sukupolvi-Petty *et al*. (2010)
	DV2–96	Mouse	Type	DIII	305, 307, 309, 327, 389	+	+	Sukupolvi-Petty *et al*. (2010)
	DV2–106	Mouse	Type	DIII	305, 307, 327, 330, 389	+	+	Sukupolvi-Petty *et al*. (2010)
	DV2–76	Mouse	Subcomplex	DIII	305, 307, 309	+	+	Sukupolvi-Petty *et al*. (2010)
	DV2–77	Mouse	Complex	DIII	336, 340, 346	+	+	Sukupolvi-Petty *et al*. (2010)
	DV2–30	Mouse	Complex	DII	55, 71, 107, 244	+	+	Sukupolvi-Petty *et al*. (2010)
	4G2	Mouse	Group	DII	104, 106, 107, 231	+	ND	Crill and Chang (2004); Johnson and Roehrig (1999); Kaufman *et al*. (1987)
	DV2–52	Mouse	Group	DII	101	+	+	Sukupolvi-Petty *et al*. (2010)
	WNV-E60 (N297Q)	Mouse–human	Group	DII	101, 104, 106	+	+	Balsitis *et al*. (2010)
	11	Human	Group	DII	101, 104, 106	+	ND	Sultana *et al*. (2009)
	DV82.11	Human	Complex	DII	ND	+	+	Beltramello *et al.* (2010)
	DV87.1	Human	Subcomplex	DIII	ND	+	+	Beltramello *et al.* (2010)

DENV-4	5H2	Primate	Type	DI	174	+	ND	Lai *et al*., (2007)
JEV	Hs-1	Mouse	Subtype	ND	ND	+	ND	Cecilia *et al*. (1988), Gupta *et al*. (2003)
	Hs-4	Mouse	Subtype	ND	ND	+	ND	Cecilia *et al*. (1988), Gupta *et al*. (2003)
	503	Mouse	Type	DII	52, 126, 136, 275	+	+	Kimura-Kuroda and Yasui (1988), Morita *et al*. (2001)
	E3.3	Mouse	Type	DIII	337, 360, 387	ND	ND	Wu *et al*. (2003)
	112	Mouse	Type	ND	ND	+	ND	Kimura-Kuroda and Yasui (1988)
	2H4, 2F2, mC3[a]	Mouse	Type	ND	ND	+	+	Zhang *et al*. (1989, 1992)
	7A6C-5	Mouse	Subcomplex	ND	ND	+	ND	Roehrig *et al*. (2001)
	6B4A-10	Mouse	Complex	DII-DIII	200–327	+	ND	Guirakhoo *et al*. (1992), Roehrig *et al*. (2001)
	N.04	Mouse	Subgroup	ND	ND	+	ND	Kimura-Kuroda and Yasui (1988)
	A3	Primate	ND	DI	179	+	ND	Goncalvez *et al*. (2008)
	B2	Primate	ND	DII	126	+	ND	Goncalvez *et al*. (2008)
	B3	Primate	ND	DIII	302	+	ND	Goncalvez *et al*. (2008)
MVEV	4B6C-2	Mouse	Type	DII	126, 128, 274–277	+	ND	Hawkes *et al*. (1988), McMinn *et al*. (1995)
	4B6A-2	Mouse	Type	ND	ND	+	ND	Hawkes *et al*. (1988)
	4B6B-10	Mouse	Subgroup	ND	ND	+	ND	Hawkes *et al*. (1988), McMinn *et al*. (1995), Roehrig *et al*. (2001)
SLEV	6B5A-2	Mouse	Type	DIII	ND	+	+	Mathews and Roehrig (1984)
TBEV	4D9 (A3)	Mouse	Type	DII	71	+	ND	Guirakhoo *et al*. (1989), Heinz *et al*. (1983), Mandl *et al*. (1989)
	2E7 (B2)	Mouse	Complex	DIII	ND	+	ND	Guirakhoo *et al*. (1989), Heinz *et al*. (1983)
	5D6 (B4)	Mouse	Complex	DIII	384	+	ND	Guirakhoo *et al*. (1989), Heinz *et al* (1983), Mandl *et al*., (1989)
	T9	Mouse	ND	ND	ND	+	ND	Phillpotts *et al*. (1987)
WNV	E16[b]	Mouse, humanized	Type	DIII	306, 307, 330, 332	+	+	Nybakken *et al*. (2005), Oliphant *et al*. (2005)
	E24	Mouse	Type	DIII	307, 330, 332	+	+	Oliphant *et al*. (2005)
	E34	Mouse	Type	DIII	307, 330	+	+	Oliphant *et al*. (2005)

Table 11.1 continued

Virus	MAb	Source	Specificity	Domain	Amino acids	Prophylaxis	Therapy	References
	CR4374	Human	Type	DIII	307	+	+	Oliphant *et al.* (2007), Throsby *et al.* (2006)
	CR4348	Human	Type	DII	208, 246	+	+	Throsby *et al.* (2006), Vogt *et al.* (2009)
	CR4354	Human	Type	DI-DII	136	+	+	Throsby *et al.* (2006), Vogt *et al.* (2009)
	E113	Mouse	Type	DI-DII	49, 280	+	+	Oliphant *et al.* (2006)
	E121	Mouse	Type	DI	175, 191, 193, 194	+	+	Oliphant *et al.* (2006)
	E48	Mouse	Type	DII	217, 222	+	ND	Oliphant *et al.* (2006)
	E60	Mouse	Group	DII	101, 106, 107	+	+	Oliphant *et al.* (2006)
	E53	Mouse	Group	DII	75, 76, 77, 99, 106, 107	+	+	Cherrier *et al.* (2009), Oliphant *et al.* (2006)
	E31	Mouse	Group	DII	101	+	+	Oliphant *et al.* (2006)
	E18	Mouse	Group	DII	101, 104, 106	+	–	Oliphant *et al.* (2006)
	11	Human	Group	DII	101, 104, 106	+	+	Gould *et al.* (2005), Sultana *et al.* (2009)
	15	Human	Group	DII	101, 104, 106	+	+	Gould *et al.* (2005), Sultana *et al.* (2009)
YFV	A1[c]	Mouse	Subtype	ND	ND	+	ND	Barrett *et al.* (1989)
	864	Mouse	Subtype	ND	ND	+	ND	Cammack and Gould (1986)
	2C9	Mouse	Type	DI-DII	71, 72, 125, 155, 158	+	+	Brandriss *et al.* (1986), Lobigs *et al.* (1987)
	2E10	Mouse	Type	DI-DII	71, 72, 125, 155, 158	+	+	Brandriss *et al.* (1986), Lobigs *et al.* (1987)
	4E8	Mouse	Type	DI-DII	71, 72, 125, 155, 158	+	+	Brandriss *et al.* (1986), Lobigs *et al.* (1987)
	3E9	Mouse	Type	ND	ND	+	+	Brandriss *et al.* (1986)
	825	Mouse	Type	ND	ND	+	ND	Cammack and Gould (1986)
	5H3	Mouse	Group	ND	ND	+	ND	Brandriss *et al.* (1986)
	868	Mouse	Group	ND	ND	+	ND	Cammack and Gould (1986)

[a]Mixtures of 2H4, 2F2, and mC3 were therapeutic in goats and rhesus monkeys.

[b]The humanized version of E16 is currently in phase II human clinical trials for the treatment of WNV infection.

[c]A1 represents a larger group of MAbs whose protective capacities were analysed only with vaccine strains of YFV only.

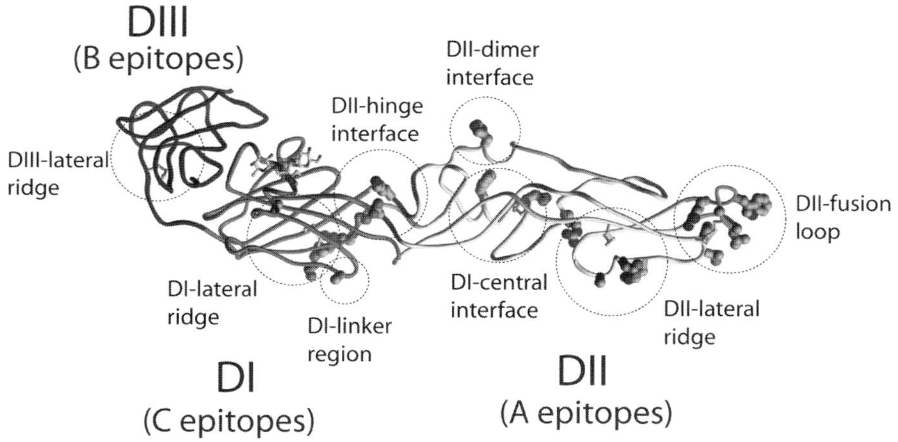

Figure 11.1 Antigenic complexity of the flavivirus E protein. Ribbon diagram of the flavivirus E protein monomer as seen from the top. Individual domains of the E protein are indicated, as is the fusion loop at the distal end of E-DII. All three domains of the E protein contain epitopes recognized by neutralizing antibodies. Residues that impact WNV antibody binding (shown in magenta) were identified using a yeast-display mapping approach for a large panel of antibodies and are labelled using a previously described nomenclature (Oliphant *et al.*, 2006). The figure is adapted from a previously published paper (Pierson *et al.*, 2008).

of as few as 30 antibody molecules (Mehlhop *et al.*, 2009; Pierson *et al.*, 2007).

Several strongly neutralizing type-specific MAbs that neutralize infection by DENV serotypes 1, 2, and 3 also map to the DIII-LR epitope (Gromowski and Barrett, 2007; Shrestha *et al.*, 2010; Sukupolvi-Petty *et al.*, 2007, 2010; Wahala *et al.*, 2010). On DENV-2 DIII, the epitope recognized by these MAbs is located at the top of the *A*-strand, and at the *B-C*, *C-C′* or *F-G* loops. Similarly, type-specific neutralizing MAbs against TBEV and JEV have been mapped to analogous regions (Stiasny *et al.*, 2007; Wu *et al.*, 2003). However, not all antibodies that bind DIII exhibit type-specific neutralizing activity. The subcomplex- and complex-specific MAbs 1A1D-2, 9F12, and 4D11 recognize an epitope centred on the *A*-strand of DIII and can neutralize infection by several DENV serotypes (Lok *et al.*, 2008; Rajamanonmani *et al.*, 2009; Sukupolvi-Petty *et al.*, 2007; Thullier *et al.*, 1999, 2001). Finally, DIII epitopes recognized by flavivirus MAbs with modest neutralizing activity also have been characterized. Typically, these map to residues that are less accessible on mature virions (Kiermayr *et al.*, 2009; Oliphant *et al.*, 2005; Pierson *et al.*, 2007; Stiasny *et al.*, 2007; Sukupolvi-Petty *et al.*, 2007). Structural and biochemical approaches identify DIII of the E protein on flaviviruses as a complex immunogen that elicits antibodies with varying specificity and functional potency.

Although it appears less common, DI is also a target for strongly neutralizing flavivirus MAbs. The i2 and IC3 MAbs against TBEV were mapped by neutralization escape selection to residues K171 and D181 in DI, respectively (Holzmann *et al.*, 1997; Mandl *et al.*, 1989). The IC3 MAb completely inhibited low pH–induced fusion of TBEV with liposomes *in vitro*, suggesting it may function analogously to DIII-LR MAbs by inhibiting structural rearrangements required for viral fusion (Stiasny *et al.*, 2007). Similarly, neutralizing MAbs that localize to a lateral ridge epitope in DI also have been described. A neutralizing MAb A3, which mapped to I179 in DI, protected against lethal JEV infection in mice (Goncalvez *et al.*, 2008) and a neutralizing DENV-4 specific chimpanzee MAb (5H2) mapped to residue K174 (Lai *et al.*, 2007). Our group recently identified a type-specific neutralizing MAb against DENV-2 in DI (DV2–48 at residue G177) that is highly protective in mice (Sukupolvi-Petty *et al.*, 2010).

Analysis of the B cell and antibody repertoire suggest that many anti-flavivirus antibodies produced in humans against the E protein recognize the fusion loop in DII (DII-FL) (Crill *et al.*, 2009; Oliphant *et al.*, 2007; Throsby *et al.*, 2006). Fusion loop-specific MAbs exhibit modest neutralization potency *in vitro* against flaviviruses, protect animal models from infection at relatively high concentration, and are usually cross-reactive (Crill and Chang, 2004; Goncalvez *et al.*, 2004; Nelson *et al.*, 2008; Oliphant *et al.*, 2006; Stiasny *et al.*, 2006). For reasons that remain uncertain, DII-FL MAbs have greater neutralizing capacity against DENV infection *in vitro* and *in vivo* (Balsitis *et al.*, 2010; Crill and Chang, 2004) compared with WNV (Oliphant *et al.*, 2006). Only one high-resolution structure of a DII-FL MAb with an E protein has been defined: the interaction of the flavivirus cross-reactive MAb E53 with WNV E protein and immature WNV and DENV-2 virions was solved by X-ray crystallography and cryo-electron microscopy, respectively (Cherrier *et al.*, 2009). The E53 MAb engages the fusion loop of the WNV E protein and binds preferentially to spikes on non-infectious, immature virions. However, it is unable to bind to mature virions, consistent with the limited solvent accessibility of the epitope and an inability to neutralize populations of mature virions (Nelson *et al.*, 2008). Thus, the neutralizing impact of E53 and likely similar fusion-loop-specific MAbs may depend on binding to the frequently observed immature, or partially mature, component of flavivirus particles (Cherrier *et al.*, 2009; Guirakhoo *et al.*, 1991; Junjhon *et al.*, 2010; Nelson *et al.*, 2008). Consistent with this, treatment of TBEV with mild non-ionic detergent increased DII-FL MAb binding to intact virions, suggesting limited accessibility of this epitope on mature viruses (Stiasny *et al.*, 2006).

Neutralizing flavivirus MAbs with complex epitopes that span multiple proteins of the oligomeric array on the virion or subviral particle also have been identified. The human MAb CR4354, which was derived from a WNV-infected patient (Throsby *et al.*, 2006), maps to the DI–DII hinge interface at residue K136 by neutralization escape selection (Vogt *et al.*, 2009). The inherent flexibility of the DI–DII hinge (Bressanelli *et al.*, 2004; Modis *et al.*, 2004; Zhang *et al.*, 2004) could explain the lack of CR4354 binding to soluble and yeast surface displayed E, as these recombinant proteins do not display the native hinge that is found on viral particles. More recent cryo-electron microscopic analysis has confirmed that CR4354 binds to a discontinuous epitope formed by protein segments from neighbouring E molecules (Kaufmann *et al.*, 2010). CR4354 effectively cross-links the six E monomers within one raft, suggesting that the antibody neutralizes WNV by blocking the pH-induced rearrangement of the E protein required for virus fusion with the endosomal membrane. *in vitro* experiments support this, as CR4354 inhibits infection at a post-attachment step and blocks WNV fusion with liposomes (Vogt *et al.*, 2009).

Although the characteristics of CR4354 appear relatively unique among anti-WNV MAbs, other anti-flavivirus neutralizing MAbs have been described that localize to this region. The anti-WNV MAb that maps closest to this region is the mouse MAb WNV E113, which binds a determinant along the DI-DII hinge interface at residues E49 and K208. E113 is protective *in vitro* and *in vivo*, but unlike CR4354, it binds recombinant and yeast-displayed E proteins efficiently (Oliphant *et al.*, 2006). Notably, MAbs (503, NARMA3, and B2) that bind this hinge region also are strongly inhibitory against the closely related JEV (Goncalvez *et al.*, 2008; Hasegawa *et al.*, 1992; Kimura-Kuroda and Yasui, 1988; Morita *et al.*, 2001).

Neutralizing antibodies against flaviviruses also target complex epitopes along the DII dimer interface. Mutation of two residues along the DII dimer interface promoted neutralization escape by the human anti-WNV MAb CR4348 (Vogt *et al.*, 2009). These amino acids (T208 and H246) lie spatially far apart (~43 Å) on the E monomer, but reside significantly closer (~19 Å) across the dimeric interface of E proteins on the virion, within the expected limits of an antibody paratope. CR4348 probably recognizes an epitope that is sensitive to the oligomeric state of the E protein, as it did not bind soluble WNV E at 4°C. However, when binding assays were performed at 37°C limited binding was detected. Consistent with this, CR4348 binds poorly to subviral particles that have been exposed to mildly acidic solutions and undergone structural rearrangement in flavivirus E proteins (Bressanelli *et al.*, 2004; Modis *et al.*, 2004).

The CR4348 MAb epitope is structurally and functionally unique among characterized anti-WNV MAbs. However, a DII–dimer interface MAb with similar properties was described for TBEV. The A5 MAb (Guirakhoo *et al.*, 1989) maps to residue E207 along the dimer interface (Mandl *et al.*, 1989), is strongly neutralizing in cell culture (Guirakhoo *et al.*, 1989), and partially blocks TBEV fusion in the liposome fusion assay (Stiasny *et al.*, 2007). Moreover, binding of at least some cross-reactive neutralizing flavivirus MAbs (e.g. 4G2 and 6B6C-1) that map to the fusion peptide in DII are also affected by mutation of residues (E231) along the dimer interface (Crill and Chang, 2004).

Stoichiometry of antibody neutralization

Two models have been proposed to for defining the number of antibodies required to bind a virion and neutralize infectivity (reviewed in (Pierson *et al.*, 2008)). 'Single hit' models describe neutralization after engagement of the virion with a single antibody (Dulbecco, 1956). These models propose that engagement of the virion at 'critical sites' causes irreversible conformational changes that render the virion non-infectious. In contrast, a 'multiple-hit' model of neutralization postulates that virus inactivation occurs as a function of the number of antibodies bound to the virion (Burnet 1937; Della-Porta, 1978). In this case, neutralization is reversible, and occurs when an individual virion is engaged with a stoichiometry that exceeds a particular threshold. While the factors that determine the stoichiometric threshold for a given virus are not clear, the requirements for neutralization may be a reflection of the number of antibodies required to 'coat' the virion surface (Burton *et al.*, 2001).

The existence of structural data and the availability of large numbers of well-characterized MAbs have made flaviviruses a useful model for investigating mechanisms of antibody-mediated neutralization. Several lines of evidence suggest that flavivirus neutralization by antibodies is a multiple-hit phenomena including (a) the demonstration that not all antibodies can bind the virion with a stoichiometry sufficient for neutralization (Pierson *et al.*, 2007); (b) the ability of C1q to augment neutralization (Mehlhop *et al.*, 2009); and (c) the ability of subneutralizing concentrations of antibodies to enhance flavivirus infection of Fc-γ-receptor bearing cells (Della-Porta and Westaway, 1978; Pierson *et al.*, 2007). Analysis of the relationship between the strength of binding and neutralization potential of several DIII-LR specific antibodies suggest that neutralization requires occupancy of only a small fraction of the accessible epitopes on the virion (Gromowski and Barrett, 2007; Gromowski *et al.*, 2008; Pierson *et al.*, 2007). In agreement, genetic approaches establish that neutralization occurs when ~25% of the 120 accessible DIII-LR epitopes on WNV are engaged by antibody, corresponding to roughly 30 MAbs (Pierson *et al.*, 2007); this number is reduced when neutralization assays are performed in the presence of C1q, which can cross-link the Fc moieties of individual antibodies (Mehlhop *et al.*, 2009). Thus, the potent inhibitory activity of DIII-LR and other MAbs is probably explained by high affinity and low occupancy requirements for neutralization.

However, not all high-affinity antibodies neutralize flavivirus infection at low concentrations or by binding a small fraction of the accessible epitopes on the virion. Many MAbs neutralize infection at concentrations that correspond to an occupancy of >99% of the accessible epitopes on the virion (Nelson *et al.*, 2008; Pierson *et al.*, 2007). Differences in the occupancy requirements for MAbs that bind distinct epitopes reflect complexities that arise from the arrangement of E proteins on the mature virion. Indeed, DENV complex-specific MAbs that recognize the *A*-strand epitope on DIII required higher occupancy levels than DIII-LR MAbs to neutralize virus infection (Gromowski *et al.*, 2008). Many MAbs with high occupancy requirements for neutralization recognize epitopes that are poorly accessible on the mature virions. Presumably when few epitopes are available, antibodies must bind a larger fraction to exceed the threshold required for neutralization. Additionally, not all antibodies have the ability to neutralize all virions in a heterogeneous population (composed of immature, mature, and partially mature virions in unknown proportions). This 'resistant' fraction reflects the existence of virions that lack a sufficient number of epitopes to exceed the threshold even when fully occupied by a given antibody (Nelson *et al.*, 2008). Thus, epitope accessibility independently governs the occupancy requirements for neutralization.

Insight into the stoichiometry of neutralization has informed our understanding of antibody-mediated enhancement of infection (ADE). ADE describes a phenomenon by which antibody increases the efficiency or duration of the interaction between virus and target cells by virtue of interactions with Fc-γ receptors (Gollins and Porterfield, 1984; Halstead, 1994). Our analysis of hundreds of MAbs raised against WNV and DENV suggest that in the absence of exogenous complement, virtually all antibodies that bind the virion and neutralize infection have the potential to promote ADE *in vitro*, irrespective of their epitope specificity (Mehlhop *et al.*, 2007; Nelson *et al.*, 2008; Oliphant *et al.*, 2006; Pierson *et al.*, 2007). The concentration of antibody that promotes maximal enhancement of infection *in vitro* is similar to those that neutralize half the virions in a given experiment (Pierson *et al.*, 2007). Neutralization and ADE are directly related by the number of antibodies bound to the virion. When half of the virions is engaged by antibody with a stoichiometry sufficient to inactivate virus infection, the other half is not, and remain infectious. Thus, at the upper limit, the number of antibodies that can promote ADE is defined by the stoichiometric threshold of neutralization. The minimum number of antibodies required for ADE, based on experimental data, appears to be approximately half the number required for neutralization (Pierson *et al.*, 2007). The addition of C1q can reduce the stoichiometric threshold for neutralization (Mehlhop *et al.*, 2009), and thus, for some MAbs and isotypes, shift the stoichiometric requirements of neutralization below that necessary for enhancement, essentially preventing ADE from occurring at any concentration of antibody (Mehlhop *et al.*, 2007).

Mechanisms of antibody protection

Steps in the flavivirus virus entry pathway may be blocked by neutralizing antibody include cell attachment, interaction with host factors required for internalization or fusion, and E protein-mediated conformational changes that drive membrane fusion. While several molecules that promote flavivirus attachment (e.g. DC-SIGN, DC-SIGNR, heparin sulfate, or αvβ3 integrin) have been characterized or reported, cellular receptors required for flavivirus entry have not been confirmed. Although the relevant interactions remain unclear, inhibition of attachment by some anti-flavivirus antibodies contributes to virus neutralization (Crill and Roehrig, 2001; He *et al.*, 1995; Hung *et al.*, 1999; Nybakken *et al.*, 2005; Rajamanonmani *et al.*, 2009), including MAbs that recognize epitopes on the *A*-strand of DIII and the DII-FL. Additionally, antibodies that inhibit flavivirus fusion or the structural changes required for a fusogenic intermediate have been described (Gollins and Porterfield, 1986; Kaufmann *et al.*, 2009; Roehrig *et al.*, 1998; Stiasny *et al.*, 2007; Thompson *et al.*, 2009; Vogt *et al.*, 2009). One caveat to these studies is that many of the fusion inhibition experiments were performed with labelled virions and synthetic liposomes; although these experiments suggest that antibodies have the capacity to inhibit fusion in cells, it remains to be established whether this correlation is perfect. The mechanism by which a given antibody neutralizes infection may depend upon circumstance, as individual antibodies may block more than one step in the viral entry pathway. For example, the MAb E16 is capable of blocking both viral fusion and attachment of WNV to cells (Nybakken *et al.*, 2005; Thompson *et al.*, 2009), depending on virion occupancy. Whether an ability to neutralize infection via multiple mechanisms is a common feature of the most potently inhibitory antibodies remains unclear.

Beyond direct virus neutralization, antibody binding to virions or virus-infected cells can trigger protective Fc-dependent antiviral activities through complement activation or Fc-γ receptor-mediated immune complex clearance mechanisms. Fc-γ receptors can activate or inhibit immune responses depending on their cytoplasmic domain and association with specific signalling molecules (Nimmerjahn and Ravetch, 2008). Recent studies indicate that interaction of the Fc-region of antibodies with activating Fc-γ receptors can augment antibody protection against flavivirus infection *in vivo*. Mice lacking all activating Fc-γ receptors required higher doses of a neutralizing E protein-specific MAb to maintain equivalent levels of protection against lethal WNV infection (Oliphant *et al.*, 2005, 2006), and Fab$'_2$ fragments of neutralizing MAbs against WNV, DENV, and YFV were less protective in mice (Balsitis *et al.*, 2010; Gould *et al.*, 2005; Schlesinger and Chapman, 1995; Zellweger *et al.*, 2010). For the specific IgG subclasses (mouse IgG2a and 2b; human IgG1 and (3) that bind the complement opsonin C1q avidly, augmented neutralization activity against WNV and DENV was observed in cell culture and in mice (Mehlhop *et al.*, 2009).

A requirement for Fc effector function is also established for protective anti-NS1 MAbs. NS1, which is expressed on the cell surface or secreted into the extracellular space, antagonizes complement control of flavivirus infection by binding the negative regulator factor H or by promoting C4 degradation (Avirutnan *et al.*, 2010; Chung *et al.*, 2006a). Passive transfer of MAbs against NS1 can protect mice against lethal infection by WNV and YFV (Chung *et al.*, 2006b; Schlesinger *et al.*, 1985) and this requires an intact Fc moiety (Schlesinger *et al.*, 1993). Mechanistic studies using immunodeficient mice demonstrate that protective anti-NS1 MAbs recognize cell surface-associated NS1 and trigger Fc-γ receptor-dependent phagocytosis and clearance of WNV-infected cells (Chung *et al.*, 2007).

Pre- and post-exposure antibody protection

Although antibody has been utilized as a therapeutic agent against several viral infections (Sawyer, 2000; Zeitlin *et al.*, 1999), it has not been used extensively against flavivirus infections in humans. A detailed list of the specificity, epitope location, and protective activity *in vivo* of MAbs against different flaviviruses is presented in Table 11.1. Most protective antibodies *in vivo* recognize the structural E protein although a subset also have been described against prM (Colombage *et al.*, 1998; Falconar, 1999; Pincus *et al.*, 1992; Vazquez *et al.*, 2002). Additionally, several groups also have described non-neutralizing, yet protective MAbs against NS1 (Chung *et al.*, 2006b; Chung *et al.*, 2007; Despres *et al.*, 1991; Falgout *et al.*, 1990; Henchal *et al.*, 1988; Putnak and Schlesinger, 1990; Schlesinger *et al.*, 1986; Schlesinger *et al.*, 1990; Schlesinger *et al.*, 1987; Schlesinger and Chapman, 1995), a protein that is absent from the virion. Thus, antibody protection against flavivirus infections *in vivo* does not always correlate with neutralizing activity *in vitro* (Brandriss *et al.*, 1986; Roehrig *et al.*, 1983; Schlesinger *et al.*, 1985).

The ability to cure rodents of flavivirus infection with immune serum or MAbs depends on the dosage and time of administration (Balsitis *et al.*, 2010; Camenga *et al.*, 1974; Chiba *et al.*, 1999; Kimura-Kuroda and Yasui, 1988; Oliphant *et al.*, 2005; Phillpotts *et al.*, 1987; Roehrig *et al.*, 2001; Shrestha *et al.*, 2010; Sukupolvi-Petty *et al.*, 2010), and polyclonal antibodies that prevent infection against one flavivirus do not provide durable cross-protection against heterologous flaviviruses (Broom *et al.*, 2000; Roehrig *et al.*, 2001). Although antibodies could have a potential therapeutic role against flaviviruses, there are at least theoretical concerns that treatment could exacerbate disease. Subneutralizing concentrations of antibody enhance flavivirus replication in myeloid cells *in vitro* (Cardosa *et al.*, 1983, 1986; Gollins and Porterfield, 1984, 1985; Peiris and Porterfield, 1979; Peiris *et al.*, 1981, 1982; Pierson *et al.*, 2007) and *in vivo* (Balsitis *et al.*, 2010; Goncalvez *et al.*, 2007; Mehlhop *et al.*, 2007; Zellweger *et al.*, 2010), and thus could complicate therapy. ADE may cause the pathological cytokine cascade that occurs during secondary dengue virus infection (Halstead, 1989; Halstead *et al.*, 1980; Kurane and Ennis, 1992; Morens, 1994); despite its extensive characterization *in vitro*, the significance of ADE *in vivo* with other flaviviruses remains uncertain. Apart from or perhaps related to ADE, an 'early death' phenomenon (Morens, 1994) has been reported that could also limit the utility of antibody therapy. According to this model, animals that have pre-existing humoral immunity but do not respond well to viral challenge may succumb to infection more rapidly than animals without pre-existing immunity. Although it has been described after passive acquisition of antibodies against yellow fever and Langat encephalitis viruses (Barrett and Gould, 1986; Gould *et al.*, 1987; Gould and Buckley, 1989; Webb *et al.*, 1968), this phenomenon has been inconsistently observed with JEV (Kimura-Kuroda and Yasui, 1988; Gupta *et al.*, 2009) but not with TBEV (Kreil and Eibl, 1997).

WNV

Passive administration of anti-WNV antibodies is both protective and therapeutic in rodents and does not appear to cause adverse effects related to immune enhancement. Transfer of immune serum prior to WNV infection protected wild type, B cell-deficient (*μMT*), and T and B-cell deficient (*RAG1*) mice from infection (Diamond *et al.*, 2003a), and no increased mortality was observed even when subneutralizing concentrations of antibodies were used. Similarly, passive administration of immune serum (Tesh *et al.*, 2002) or antiserum that recognized WNV E protein (Wang *et al.*, 2001) protected hamsters and mice against lethal WNV infection. In therapeutic trials, immune human γ-globulin protected mice against WNV-induced mortality (Ben-Nathan *et al.*, 2009; Ben-Nathan *et al.*, 2003; Engle and Diamond, 2003; Julander *et al.*, 2005). Therapeutic intervention with immune human γ-globulin even five days after infection reduced mortality in rodents; this time point is significant because WNV spreads to the brain and spinal cord by day 4. Thus, passive transfer of immune polyclonal antibody improved clinical outcome even after WNV had disseminated into the CNS.

Small numbers of human patients have received immunotherapy against WNV infection. Prophylaxis and/or therapy with neutralizing anti-WNV antibodies may be a possible intervention in the elderly and immunocompromised. Case reports (Haley *et al.*, 2003; Hamdan *et al.*, 2002; Saquib *et al.*, 2008; Shimoni *et al.*, 2001) have documented improvement in humans with neuroinvasive WNV infection after receiving immune γ-globulin from Israeli donors. Given the endemic nature of WNV in the Middle East, pooled human immunoglobulin from Israeli donors was shown to contain neutralizing titres of antibodies against WNV (Ben-Nathan *et al.*, 2003, 2009; Engle and Diamond, 2003). Although promising, γ-globulin immunotherapy against WNV infection in humans has some limitations: (a) batch variability may affect the quantitative titre, functional activity, and therapeutic efficacy of specific antibody preparations; (b) it is purified from human blood plasma, and has an inherent risk of transmitting known and unknown infectious agents; and (c) it requires a large volume of administration, which can increase adverse events in patients with cardiac or renal co-morbidities.

To overcome these limitations, humanized or human monoclonal antibodies or antibody fragments with therapeutic activity against WNV infection (Gould *et al.*, 2005; Oliphant *et al.*, 2005; Sultana *et al.*, 2009; Throsby *et al.*, 2006; Vogt *et al.*, 2009) have been developed. These human or humanized antibody fragments have high neutralizing activity *in vitro* and provide excellent protection *in vivo* in mice. If MAbs are to be an effective therapy for WNV encephalitis they should function after the onset of symptoms and ideally, after infection in the central nervous system. When murine or humanized MAbs were given as a single dose five or six days after infection 90% of mice or hamsters were protected (Morrey *et al.*, 2006, 2007; Oliphant *et al.*, 2005). Acute flaccid paralysis in hamsters also was blocked by treatment with one neutralizing mAb, E16 several days after infection (Samuel *et al.*, 2007). A phase II clinical trial has been initiated to the efficacy of the E16 antibody (also termed MGAWN1) against severe WNV infection (http://clinicaltrials.gov/ct2/show/NCT00515385). Thus, neutralizing antibody therapeutics show promise as they directly inhibit transneuronal spread of WNV infection *in vivo*. A plant-derived version of E16 MAb was recently generated in *Nicotiana benthamiana* and shown to be equally effective in mice relative to mammalian cell-generated MAb, providing a platform for cost-efficient and scalable production technologies that can transfer anti-flavivirus antibody therapeutics into the clinical setting, even in resource-poor countries (Lai *et al.*, 2010). Future use of a combination of MAbs that bind distinct epitopes and neutralize by independent mechanisms could diminish the potential risk of selecting escape variants *in vivo*

(Zhang *et al.*, 2009), especially in individuals with high-grade viraemia and tissue viral burden.

SLEV

Prior to the discovery of WNV in the United States in 1999, SLEV was the most important vector-borne, encephalitic flavivirus in the Western hemisphere. Remarkably, only one group has isolated murine MAbs specific for the E protein of SLEV and assessed their relative efficacy as prophylaxis and therapy in SLEV-infected mice (Mathews and Roehrig, 1984; Roehrig *et al.*, 1983; Roehrig *et al.*, 2001). SLEV E protein epitopes were defined originally by competitive binding studies, identifying antibodies with type, subcomplex, subgroup, and flavivirus group reactivity. Subsequent analysis of haemagglutination-inhibition and virus neutralization potency subdivided the cross-reactivity groups into eight epitopes (E-1^a, E-1^b, E-1^c, E-1^d, E-2, E-3, E-4^a, and E-4^b) (Roehrig *et al.*, 1983). The *in vitro* $PRNT_{70}$ endpoints ranged from 0.015 to 20 μg/ml using purified MAbs, with the anti-E-1^c-reacting MAb, 6B5A-2, as the most potent neutralizing antibody. As little as 5 μg of this antibody protected 100% of outbred Swiss mice from challenge with 100 i.p. LD_{50} of the SLEV-MSI-7 strain when MAb was administered 24 hours prior to infection. Two other MAbs with virus neutralizing activity (4A4C-4 (type-specific anti-E1^d; $PRNT_{70}$ titre = 1.25 μg/ml and 6B6C-1 (group-reactive anti-E-4^b; $PRNT_{70}$ titre = 5 μg/ml)) were partially protective at similar antibody doses. The E-4^b epitope was subsequently mapped to the E protein fusion loop using mutant virus-like particles (Trainor *et al.*, 2007).

The protective capacity of 6B5A-2 was studied in further detail. 6B5A-2 was highly protective, requiring less than 10 ng of antibody to protect mice from a 100 LD_{50} challenge with the MSI-7 strain of SLEV. Indeed, as little as 250 ng of antibody could protect 70% of animals from a challenge with 1,000,000 LD_{50} with the same SLEV strain. The 6B5A-2 exhibited equal $PRNT_{70}$ titres with 21 strains of SLEV isolated over a period of 45 years from both North and South America. Because of this, MAb 6B5A-2 was also examined for its therapeutic value. Administration of as little as 1 μg of MAb at 1 day post infection cured 90% of challenged mice. When mice were given 100 μg of 6B5A-2, 100% of animals were cured at 2 days p.i., and 30% of animals were cured at 4 days p.i.

MVEV

Although MVEV is the most medically important mosquito-borne, encephalitic flavivirus in Australia, it accounts for only a small number of human cases each decade. Two reports have described the isolation and characterization of murine MAbs specific for the E protein of MVEV. Only one of these studies investigated their use in prophylaxis and therapy of MVEV-infected mice (Hall *et al.*, 1990; Hawkes *et al.*, 1988). Analogous to SLEV, eight E protein epitopes were defined, and the MAbs localizing to these epitopes displayed type, subcomplex, and flavivirus group serological reactivity. The $PRNT_{70}$ titres for MVEV type-specific MAbs ranged from 0.02 μg/ml (E-1^c, MAb 4B6C-2) to > 10 μg/ml (E-1^a, MAb 4B6A-2) of purified MAb. An E-4^b-reactive MAb, 4A1B-9, demonstrated a $PRNT_{70}$ titre of 0.3 μg/ml, similar to that of the SLEV E-4^b-reactive MAb, 6B6C-1. In addition, another epitope, E-5 (MAb 4B3B-6) shared between MVEV and JEV was identified. This MAb demonstrated a $PRNT_{70}$ titre of 0.02 μg/ml with MVEV. In prophylaxis studies, the 4B6C-2 was highly protective, requiring only 500 ng of MAb to protect 80% of mice from a challenge with 100 i.p. LD_{50} of MVEV. In comparison, 5 μg of MAb 4B3B-6 was required to protect 30% of mice from virus challenge. Interestingly, 5 μg of the non-neutralizing E-1^a MAb, 4B6A-2, also protected 100% of mice from MVEV challenge. A second non-neutralizing MAb, 4B6B-10, that defined the E-6 epitope protected 40% if animals when given in a 5-μg dose. These studies were the first demonstration that E protein-specific MAbs with no measurable neutralization activity *in vitro* could protect animals from flavivirus infection.

Antibody competition assays defined three distinct regions on the MVEV E protein. The first region, comprised of epitopes E-1^a, E-1^c, E-1^d, E-5, and E-8, elicited MAbs that were type-specific, neutralizing, and protective in mice. MAbs that were flavivirus cross-reactive, had high haemagglutination inhibition activity, and relatively low neutralization activity defined the second region, comprised of epitopes E-3, E-4^{b}, and E-6. The third region was defined by a single epitope (E-7), and elicited a MAb with both neutralizing and protective capacity (McMinn *et al.*, 1995). The immunoreactivity of all epitopes was reduced at pH 6.0 (Guirakhoo *et al.*, 1992). Sequencing of the E genes of viruses resistant to neutralization with MAb 4B6C-2 identified individual amino acid changes at residues E-126, 128, 274, 276, and 277 (McMinn *et al.*, 1995). Even though E-126/128 and E-274–277 are linearly distant, they are spatially close on the upper surfaces of the E protein dimer (Rey *et al.*, 1995). Utilizing a MAb derived from JEV inoculated mice that was cross-reactive with MVEV and SLEV (MAb 6B4A-10) an additional MVEV E protein epitope, E-8, was identified and localized to E-200–327 by partial proteolysis of the E protein (Guirakhoo *et al.*, 1992).

Because of the cross-reactivity of MAb 6B4A-10, additional studies defined the ability of this and other cross-reactive MAbs to protect mice from infection with SLEV, MVEV, or JEV (Roehrig *et al.*, 2001). Cross-reactive MAbs with virtually identical binding titres for each virus demonstrated variable neutralization capacity. 6B4A-10 inhibited MVEV to a far greater degree than either JEV or SLEV. Although the reason for this was not determined, it may be related to the lower virulence in mice of MVEV as compared with JEV and SLEV, differences in maturation state of the viral particles (Nelson *et al.*, 2008) or relative antigen load.

JEV

Globally, JEV is medically the most important mosquito-borne encephalitis caused by flaviviruses. Two vaccines are effective and have been used throughout Asia: a killed vaccine derived from the Nakayama strain of JEV (JE-VAX), and a live attenuated Chinese vaccine, SA-14–14–2. More recently, a new killed JEV vaccine, Ixiaro®, has been approved by the US Food and Drug Administration for use in humans.

Several laboratories have produced murine anti-JEV MAbs and used them to dissect the antigenic structure of the JEV E protein. The Japanese JEV strain, JaGAr-01 has been studied in greatest detail (Kimura-Kuroda and Yasui, 1983, 1986, 1988). Nine epitopes were mapped on the JEV E protein, ranging in reactivity from JEV-specific to flavivirus cross-reactive. JEV type-specific MAbs had the highest virus neutralizing activity and were most protective from challenge experiments, requiring a little as 2.5 µg to completely protect mice (Kimura-Kuroda and Yasui, 1988), whereas flavivirus cross-reactive MAbs showed less protection. In subsequent studies, one of these highly protective MAbs, 503, blocked low-pH-triggered fusion of JEV-infected Vero cells (Butrapet *et al.*, 1998). This study is one of the few that demonstrates flavivirus-mediated fusion from within mammalian cells. MAb 503 neutralization-resistant viruses have been characterized (Morita *et al.*, 2001), and localized the epitope to amino acids E-52, 126, 136, and 275.

Neutralizing murine MAbs have also been identified for other strains of JEV, including viruses from Japan (Hasegawa *et al.*, 1995; Kobayashi *et al.*, 1985; Pothipunya *et al.*, 1993), India (Cecilia *et al.*, 1988; Gore *et al.*, 1990; Gupta *et al.*, 2003, 2008; Kedarnath *et al.*, 1986), China (Hasegawa *et al.*, 1995; Wu *et al.*, 1998; Zhang *et al.*, 1989) and Taiwan (Ma *et al.*, 1995). Of these reports, antibody protection from infection against the JEV Indian strain 733913 was studied in some detail (Gupta *et al.*, 2003, 2008). Both mean survival time and numbers of surviving mice increased significantly if animals were given MAb one-day prior to JEV challenge. Although protection never exceeded a 70% survival rate, MAbs were administered as ascites fluid, so the actual amount of antibody transferred was not determined. Some of these MAbs also demonstrated low-level cross-protection from challenge with WNV and DENV-2 (Gupta *et al.*, 2008).

Mixtures of three murine MAbs against JEV, with protective capacity in mice, were tested in goats and monkeys (Zhang *et al.*, 1989). A 200 µg dose of the MAb mixture (2H4, 2F2, and mC3) protected 100% of mice if given 24 hours before JEV SA14 challenge; in comparison, this cocktail cured 100% of mice if given as late as 3 days p.i., and 73% and 47% of mice if given 4 or 5 days p.i, respectively (Zhang *et al.*, 1989). A similar therapeutic effect was observed with mice following challenge with either the P3 or

JI Chinese strains of JEV. Goats receiving a 5 mg dose of the MAb mixture i.p. and intrathecally 2 days p.i. also recovered from JEV-SA14 infection. However, if the MAbs were given to goats 2 days p.i. by a single route alone (i.p. or intrathecally) no animals survived. The reason for the added protection using two routes of antibody administration was never determined, although it may be due to the increased dosage or penetration into compartments in the central nervous system. Remarkably, 100% of rhesus macaques survived SA14 challenge if given 5 mg of the MAb mixture at 8 days p.i. In this study, untreated control monkeys showed a regular onset of symptoms: at day 4 fever developed, and by days 6–8 reached 41.5°C, continuing at this level until death. At day 12, spastic limb paralysis was observed, with death occurring at day 20. Five days after MAb treatment, temperatures fell to normal and remained so for 5 months.

Because of the preliminary success with murine MAbs, three approaches have been used to prepare human or humanized anti-JEV MAbs for clinical use (Arakawa *et al.*, 2007; Goncalvez *et al.*, 2008), and a new technology using TransChromo mice is under evaluation for the production of fully human anti-JEV MAbs (Ishida *et al.*, 2002). Construction of a human Fab library using DNA from JEV hyperimmune volunteers permitted the isolation of a human monoclonal Fab (Arakawa *et al.*, 2007) with strong neutralizing capacity. Antibody repertoire cloning of a chimpanzee vaccinated with JE-VAX facilitated the isolation of 11 primate Fabs (Goncalvez *et al.*, 2008). Among these, neutralizing Fabs mapped to epitopes in all three structural domains of the JEV E protein: Fab A3 (E-179, DI), Fab B2 (E-126, DII), and Fab E3 (E-302, DIII). In mice, the MAb B2 was the most protective ($ED_{50} = 0.84\,\mu g$/mouse), followed by MAb A3 ($ED_{50} = 5.8\,\mu g$/mouse), and MAb E3 ($ED_{50} = 24.7\,\mu g$/mouse). Administration of 200 µg of MAb to mice 1 day after JEV challenge increased the mean survival time of these animals.

TBEV

Although protection against TBEV is usually provided by active immunization with a highly effective inactivated TBEV vaccine (Heinz *et al.*, 2007), passive immunization with γ-globulin enriched in antibodies against TBEV has been used in non-immunized persons who have perhaps been exposed to the virus or in immunodeficient individuals. Pharmacokinetic studies revealed a well-tolerated product that had long half-life (26 days) with sustained yet low neutralizing antibody levels (~1/15) (Adner *et al.*, 2001).

Transfer of MAbs specific for the TBEV E protein (Niedrig *et al.*, 1994; Phillpotts *et al.*, 1987)} or polyclonal immunoglobulin preparations with the same specificities (Hofmann *et al.*, 1978; Kreil and Eibl, 1997) protected mice from TBEV challenge. Although several of these formulations were highly neutralizing, limited virus replication occurs in mice passively protected by neutralizing antibodies and, as a consequence, these animals develop long-lasting immunity to TBEV challenge (Kreil *et al.*, 1998).

DENV

Several groups have described antibodies with prophylactic activity against the different serotypes of DENV. Passive prophylaxis of polyclonal sera or monoclonal MAbs against DENV E, prM, or NS1 proteins protected AG129 or BALB/c mice from lethal infection (Brandriss *et al.*, 1986; Chen *et al.*, 2009; Henchal *et al.*, 1988; Johnson and Roehrig, 1999; Kaufman *et al.*, 1987, 1989; Lai *et al.*, 2007; Sultana *et al.*, 2009). More recent studies have described anti-DENV antibodies with the therapeutic activity, which can be administered several days after infection (Balsitis *et al.*, 2010; Shrestha *et al.*, 2010; Sukupolvi-Petty *et al.*, 2010). In theory, MAb-based therapy against DENV could be complicated by ADE, and as such the administration of DENV-specific MAbs could adversely impact the outcome of infection. In non-human primates, passive transfer of subneutralizing concentrations of monoclonal or polyclonal antibody increased viraemia although no change in disease status was observed (Goncalvez *et al.*, 2007; Halstead, 1979). Because on the possibility of ADE, MAb therapeutics against DENV in humans would appear to have considerable hurdles. However, recent experiments with amino acid substitutions or deletions in the Fc region of recombinant anti-flavivirus antibodies prevented ADE *in vitro* and *in vivo* (Balsitis *et al.*,

2010; Beltramello *et al.*, 2010; Goncalvez *et al.*, 2007; Mehlhop *et al.*, 2007).

YFV

There have been effective live-attenuated YFV vaccines available for human use for several decades. These vaccines (e.g. YF 17D-204) are routinely used to interdict yellow fever outbreaks and epidemics throughout the world. Recently, however, there has been an increase in incidence of severe adverse events in otherwise healthy individuals associated with YF vaccination campaigns (Martin *et al.*, 2001a,b; Whittembury *et al.*, 2009). While YF vaccine-related adverse events remains low, inactivated YFV vaccines are under development in an attempt to prevent this complication entirely. There is also a potential use for protective human or humanized anti-YFV MAbs as alternatives to vaccination for individuals that are travelling to endemic areas for short periods of time who are deemed at high risk for vaccine-related adverse events. As there are no effective therapeutic agents for YFV, such MAbs also could be used in that capacity (Monath, 2008).

Murine MAbs have been generated after immunization with wild type or vaccine strains of YFV and used as prophylaxis and therapy in mouse models of yellow fever (Barrett *et al.*, 1986, 1989; Gould *et al.*, 1986; Schlesinger *et al.*, 1984). In all cases, MAbs specific for the YFV E protein were both protective and therapeutic. As little as 4 μg per mouse of the 'Group B' MAbs (2C9, 4E8 and 2E10) protected animals from YFV infection (Brandriss *et al.*, 1986). These same MAbs were therapeutic as late as 5 days p.i., but required 10-fold more antibody than that required for prophylaxis. An uncharacterized effector function was associated with protection by MAb 2E10 as the F(ab')2 fragment of the antibody lost activity *in vivo* (Schlesinger and Chapman, 1995). In subsequent studies, protective epitopes of other MAbs were localized using neutralization escape variants to amino acids E-71, 72, 125, 155, and 158 (Lobigs *et al.*, 1987; Ryman *et al.*, 1997). The reactivity of human-derived YFV-neutralizing Fabs generated from antibody repertoire cloning also mapped to amino acids E153–155 (Daffis *et al.*, 2005). Finally, passive transfer of anti-YFV NS1-specific MAbs also conferred protection (Schlesinger *et al.*, 1985), and required an intact Fc moiety (Schlesinger *et al.*, 1993), presumably through the effector functions of the antibody.

Conclusions

Given the continued global spread of flaviviruses, the lack of existing therapies and vaccines for many medically important flaviviruses, and the vast experience with inhibitory MAbs *in vitro* and *in vivo* in small animals, the development of antibody-based therapeutics appears a feasible direction for generating effective antiviral agents in a relatively short time period. Indeed, the recent development of plant-based production platforms for MAbs (Lai *et al.*, 2010) may allow cost-efficient transfer of these reagents to resource-poor settings, where much of the flavivirus disease burden occurs. Ultimately, a combination antibody strategy that limits the development of resistant variants, whether as monoclonal or polyclonal therapy, will probably be more effective than single agents against flaviviruses that cycle directly between humans and insect vectors.

References

Adner, N., Leibl, H., Enzersberger, O., Kirgios, M., and Wahlberg, T. (2001). Pharmacokinetics of human tick-borne encephalitis virus antibody levels after injection with human tick-borne encephalitis immunoglobulin, solvent/detergent treated, FSME-BULIN S/D in healthy volunteers. Scandinavian journal of infectious diseases 33, 843–847.

Allison, S.L., Schalich, J., Stiasny, K., Mandl, C.W., and Heinz, F.X. (2001). Mutational evidence for an internal fusion peptide in flavivirus envelope protein E.J. Virol. 75, 4268–4275.

Allison, S.L., Schalich, J., Stiasny, K., Mandl, C.W., Kunz, C., and Heinz, F.X. (1995). Oligomeric rearrangement of tick-borne encephalitis virus envelope proteins induced by an acidic pH. J. Virol. 69, 695–700.

Arakawa, M., Yamashiro, T., Uechi, G., Tadano, M., and Nishizono, A. (2007). Construction of human Fab (gamma1/kappa) library and identification of human monoclonal Fab possessing neutralizing potency against Japanese encephalitis virus. Microbiol. Immunol. 51, 617–625.

Avirutnan, P., Fuchs, A., Hauhart, R.E., Somnuke, P., Youn, S., Diamond, M.S., and Atkinson, J.P. (2010). Antagonism of the complement component C4 by flavivirus non-Structural Protein NS1. J Exp Med. 207, 793–806.

Balsitis, S.J., Williams, K.L., Lachica, R., Flores, D., Kyle, J.L., Mehlhop, E., Johnson, S., Diamond, M.S.,

Beatty, P.R., and Harris, E. (2010). Lethal antibody enhancement of dengue disease in mice is prevented by Fc modification. PLoS Pathog, e1000790.

Barrett, A.D., and Gould, E.A. (1986). Antibody-mediated early death *in vivo* after infection with yellow fever virus. J. Gen. Virol. *67*, 2539–2542.

Barrett, A.D., Pryde, A., Medlen, A.R., Ledger, T.N., Whitby, J.E., Gibson, C.A., DeSilva, M., Groves, D.J., Langley, D.J., and Minor, P.D. (1989). Examination of the envelope glycoprotein of yellow fever vaccine viruses with monoclonal antibodies. Vaccine *7*, 333–336.

Beasley, D.W., and Barrett, A.D. (2002). Identification of neutralizing epitopes within structural domain III of the West Nile virus envelope protein. J. Virol. *76*, 13097–13100.

Beltramello, M., Williams, K.L., Simmons, C.P., Macagno, A., Simonelli, L., Quyen, N.T., Sukupolvi-Petty, S., Navarro-Sanchez, E., Young, P.R., de Silva, A.M., *et al.* (2010). The human immune response to Dengue virus is dominated by highly cross-reactive antibodies endowed with neutralizing and enhancing activity. Cell Host Microbe *16*, 271–283.

Ben-Nathan, D., Gershoni-Yahalom, O., Samina, I., Khinich, Y., Nur, I., Laub, O., Gottreich, A., Simanov, M., Porgador, A., Rager-Zisman, B., *et al.* (2009). Using high titer West Nile intravenous immunoglobulin from selected Israeli donors for treatment of West Nile virus infection. BMC Infect. Dis. *9*, 18.

Ben-Nathan, D., Lustig, S., Tam, G., Robinzon, S., Segal, S., and Rager-Zisman, B. (2003). Prophylactic and therapeutic efficacy of human intravenous immunoglobulin in treating west nile virus infection in mice. J. Infect. Dis. *188*, 5–12.

Bhardwaj, S., Holbrook, M., Shope, R.E., Barrett, A.D., and Watowich, S.J. (2001). Biophysical characterization and vector-specific antagonist activity of domain III of the tick-borne flavivirus envelope protein. J. Virol. *75*, 4002–4007.

Brandriss, M.W., Schlesinger, J.J., Walsh, E.E., and Briselli, M. (1986). Lethal 17D yellow fever encephalitis in mice. I. Passive protection by monoclonal antibodies to the envelope proteins of 17D yellow fever and dengue 2 viruses. J. Gen. Virol. *67*, 229–234.

Bressanelli, S., Stiasny, K., Allison, S.L., Stura, E.A., Duquerroy, S., Lescar, J., Heinz, F.X., and Rey, F.A. (2004). Structure of a flavivirus envelope glycoprotein in its low-pH-induced membrane fusion conformation. EMBO J. *23*, 728–738.

Broom, A.K., Wallace, M.J., Mackenzie, J.S., Smith, D.W., and Hall, R.A. (2000). Immunisation with gamma globulin to Murray Valley encephalitis virus and with an inactivated Japanese encephalitis virus vaccine as prophylaxis against Australian encephalitis: evaluation in a mouse model. J. Med. Virol. *61*, 259–265.

Burton, D.R., Saphire, E.O., and Parren, P.W. (2001). A model for neutralization of viruses based on antibody coating of the virion surface. Curr. Top. Microbiol. Immunol. *260*, 109–143.

Butrapet, S., Kimura-Kuroda, J., Zhou, D.S., and Yasui, K. (1998). Neutralizing mechanism of a monoclonal antibody against Japanese encephalitis virus glycoprotein E. Am. J. Trop. Med. Hyg. *58*, 389–398.

Camenga, D.L., Nathanson, N., and Cole, G.A. (1974). Cyclophosphamide-potentiated West Nile viral encephalitis: relative influence of cellular and humoral factors. J. Infect. Dis. *130*, 634–641.

Cammack, N., and Gould, E.A. (1986). Topographical analysis of epitope relationships on the envelope glycoprotein of yellow fever 17D vaccine and the wild type Asibi parent virus. Virology *150*, 333–341.

Cardosa, M.J., Gordon, S., Hirsch, S., Springer, T.A., and Porterfield, J.S. (1986). Interaction of West Nile virus with primary murine macrophages: role of cell activation and receptors for antibody and complement. J. Virol. *57*, 952–959.

Cardosa, M.J., Porterfield, J.S., and Gordon, S. (1983). Complement receptor mediates enhanced flavivirus replication in macrophages. J. Exp. Med. *158*, 258–263.

Cardosa, M.J., Wang, S.M., Sum, M.S., and Tio, P.H. (2002). Antibodies against prM protein distinguish between previous infection with dengue and Japanese encephalitis viruses. BMC Microbiol 2, 9.

Cecilia, D., Gadkari, D.A., Kedarnath, N., and Ghosh, S.N. (1988). Epitope mapping of Japanese encephalitis virus envelope protein using monoclonal antibodies against an Indian strain. J. Gen. Virol. *69 (Pt 11)*, 2741–2747.

Chambers, T.J., Droll, D.A., Walton, A.H., Schwartz, J., Wold, W.S., and Nickells, J. (2008). West Nile 25A virus infection of B-cell-deficient ((micro)MT) mice: characterization of neuroinvasiveness and pseudoreversion of the viral envelope protein. J. Gen. Virol. *89*, 627–635.

Chen, Z., Liu, L.M., Gao, N., Xu, X.F., Zhang, J.L., Wang, J.L., and An, J. (2009). Passive protection assay of monoclonal antibodies against dengue virus in suckling mice. Curr. Microbiol. *58*, 326–331.

Cherrier, M.V., Kaufmann, B., Nybakken, G.E., Lok, S.M., Warren, J.T., Chen, B.R., Nelson, C.A., Kostyuchenko, V.A., Holdaway, H.A., Chipman, P.R., *et al.* (2009). Structural basis for the preferential binding of immature flaviviruses by a fusion-loop specific antibody. EMBO *28*, 3269–3276.

Chiba, N., Osada, M., Komoro, K., Mizutani, T., Kariwa, H., and Takashima, I. (1999). Protection against tick-borne encephalitis virus isolated in Japan by active and passive immunization. Vaccine *17*, 1532–1539.

Chin, J.F., Chu, J.J., and Ng, M.L. (2007). The envelope glycoprotein domain III of dengue virus serotypes 1 and 2 inhibit virus entry. Microbes Infect *9*, 1–6.

Chu, J.J., Rajamanonmani, R., Li, J., Bhuvanakantham, R., Lescar, J., and Ng, M.L. (2005). Inhibition of West Nile virus entry by using a recombinant domain III from the envelope glycoprotein. J. Gen. Virol. *86*, 405–412.

Chung, K.M., Liszewski, M.K., Nybakken, G., Davis, A.E., Townsend, R.R., Fremont, D.H., Atkinson, J.P., and Diamond, M.S. (2006a). West Nile virus non-structural protein NS1 inhibits complement activation by binding the regulatory protein factor H. Proc. Natl. Acad. Sci. U.S.A. *103*, 19111–19116.

Chung, K.M., Nybakken, G.E., Thompson, B.S., Engle, M.J., Marri, A., Fremont, D.H., and Diamond, M.S. (2006b). Antibodies against West Nile virus non-structural (NS)-1 protein prevent lethal infection through Fc gamma receptor-dependent and independent mechanisms. J. Virol. *80*, 1340–1351.

Chung, K.M., Thompson, B.S., Fremont, D.H., and Diamond, M.S. (2007). Antibody recognition of cell surface-associated NS1 triggers Fc-g receptor mediated phagocytosis and clearance of WNV infected cells. J. Virol. *81*, 9551–9555.

Colombage, G., Hall, R., Pavy, M., and Lobigs, M. (1998). DNA-based and alphavirus-vectored immunisation with prM and E proteins elicits long-lived and protective immunity against the flavivirus, Murray Valley encephalitis virus. Virology *250*, 151–163.

Crill, W.D., and Chang, G.J. (2004). Localization and characterization of flavivirus envelope glycoprotein cross-reactive epitopes. J. Virol. *78*, 13975–13986.

Crill, W.D., Hughes, H.R., Delorey, M.J., and Chang, G.J. (2009). Humoral immune responses of dengue fever patients using epitope-specific serotype-2 virus-like particle antigens. PloS one *4*, e4991.

Crill, W.D., and Roehrig, J.T. (2001). Monoclonal antibodies that bind to domain III of dengue virus E glycoprotein are the most efficient blockers of virus adsorption to Vero cells. J. Virol. *75*, 7769–7773.

Daffis, S., Kontermann, R.E., Korimbocus, J., Zeller, H., Klenk, H.D., and Ter Meulen, J. (2005). Antibody responses against wild-type yellow fever virus and the 17D vaccine strain: characterization with human monoclonal antibody fragments and neutralization escape variants. Virology *337*, 262–272.

Dejnirattisai, W., Jumnainsong, A., Onsirisakul, N., Fitton, P., Vasanawathana, S., Limpitikul, W., Puttikhunt, C., Edwards, C., Duangchinda, T., Supasa, S., *et al.* (2010). Cross-reacting antibodies enhance dengue virus infection in humans. Science *328*, 745–748.

Della-Porta, A.J., and Westaway, E.G. (1978). A multihit model for the neutralization of animal viruses. J. Gen. Virol. *38*, 1–19.

Despres, P., Dietrich, J., Girard, M., and Bouloy, M. (1991). Recombinant baculoviruses expressing yellow fever virus E and NS1 proteins elicit protective immunity in mice. J. Gen. Virol. *72 (Pt 11)*, 2811–2816.

Diamond, M.S., Shrestha, B., Marri, A., Mahan, D., and Engle, M. (2003a). B cells and antibody play critical roles in the immediate defense of disseminated infection by West Nile encephalitis virus. J. Virol. *77*, 2578–2586.

Diamond, M.S., Sitati, E., Friend, L., Shrestha, B., Higgs, S., and Engle, M. (2003b). Induced IgM protects against lethal West Nile Virus infection. J. Exp. Med. *198*, 1–11.

Dong, H., Zhang, B., and Shi, P.Y. (2008). Flavivirus methyltransferase: a novel antiviral target. Antiviral Res. *80*, 1–10.

Egloff, M.P., Benarroch, D., Selisko, B., Romette, J.L., and Canard, B. (2002). An RNA cap (nucleoside-2′-O-)-methyltransferase in the flavivirus RNA polymerase NS5: crystal structure and functional characterization. EMBO J. *21*, 2757–2768.

Elshuber, S., Allison, S.L., Heinz, F.X., and Mandl, C.W. (2003). Cleavage of protein prM is necessary for infection of BHK-21 cells by tick-borne encephalitis virus. J. Gen. Virol. *84*, 183–191.

Engle, M., and Diamond, M.S. (2003). Antibody prophylaxis and therapy against West Nile Virus infection in wild type and immunodeficient mice. J. Virol. *77*, 12941–12949.

Falconar, A.K. (1999). Identification of an epitope on the dengue virus membrane (M) protein defined by cross-protective monoclonal antibodies: design of an improved epitope sequence based on common determinants present in both envelope (E and M) proteins. Arch. Virol. *144*, 2313–2330.

Falgout, B., Bray, M., Schlesinger, J.J., and Lai, C.J. (1990). Immunization of mice with recombinant vaccinia virus expressing authentic dengue virus nonstructural protein NS1 protects against lethal dengue virus encephalitis. J. Virol. *64*, 4356–4363.

Flamand, M., Deubel, V., and Girard, M. (1992). Expression and secretion of Japanese encephalitis virus nonstructural protein NS1 by insect cells using a recombinant baculovirus. Virology *191*, 826–836.

Flamand, M., Megret, F., Mathieu, M., Lepault, J., Rey, F.A., and Deubel, V. (1999). Dengue virus type 1 nonstructural glycoprotein NS1 is secreted from mammalian cells as a soluble hexamer in a glycosylation-dependent fashion. J. Virol. *73*, 6104–6110.

Fritz, R., Stiasny, K., and Heinz, F.X. (2008). Identification of specific histidines as pH sensors in flavivirus membrane fusion. J Cell Biol *183*, 353–361.

Gollins, S., and Porterfield, J. (1984). Flavivirus infection enhancement in macrophages: radioactive and biological studies on the effect of antibody and viral fate. J. Gen. Virol. *65*, 1261–1272.

Gollins, S.W., and Porterfield, J.S. (1985). Flavivirus infection enhancement in macrophages: an electron microscopic study of viral entry. J. Gen. Virol. *66*, 1969–1982.

Gollins, S.W., and Porterfield, J.S. (1986). A new mechanism for the neutralization of enveloped viruses by antiviral antibody. Nature *321*, 244–246.

Goncalvez, A.P., Chien, C.H., Tubthong, K., Gorshkova, I., Roll, C., Donau, O., Schuck, P., Yoksan, S., Wang, S.D., Purcell, R.H., *et al.* (2008). Humanized monoclonal antibodies derived from chimpanzee Fabs protect against Japanese encephalitis virus *in vitro* and *in vivo*. J. Virol. *82*, 7009–7021.

Goncalvez, A.P., Engle, R.E., St Claire, M., Purcell, R.H., and Lai, C.J. (2007). Monoclonal antibody-mediated enhancement of dengue virus infection *in vitro* and *in vivo* and strategies for prevention. Proc. Natl. Acad. Sci. U.S.A. *104*, 9422–9427.

Goncalvez, A.P., Purcell, R.H., and Lai, C.J. (2004). Epitope determinants of a chimpanzee Fab antibody that efficiently cross-neutralizes dengue type 1 and type 2 viruses map to inside and in close proximity to fusion loop of the dengue type 2 virus envelope glycoprotein. J. Virol. *78*, 12919–12928.

Gore, M.M., Gupta, A.K., Ayachit, V.M., Athawale, S.S., Ghosh, S.N., and Banerjee, K. (1990). Selection of a neutralization-escape variant strain of Japanese encephalitis virus using monoclonal antibody. Indian J. Med. Res. *91*, 231–233.

Gould, E.A., A., B., B.K., G., Cane, P.A., and Doenhoff, M. (1987). Immune enhancement of yellow fever virus neurovirulence for mice: studies of mechanisms involved. J. Gen. Virol. *68*, 3105–3112.

Gould, E.A., and Buckley, A. (1989). Antibody-dependent enhancement of yellow fever and Japanese encephalitis virus neurovirulence. J. Gen. Virol. *70*, 1605–1608.

Gould, E.A., Buckley, A., Barrett, A.D., and Cammack, N. (1986). Neutralizing (54K) and non-neutralizing (54K and 48K) monoclonal antibodies against structural and non-structural yellow fever virus proteins confer immunity in mice. J. Gen. Virol. *67 (Pt 3)*, 591–595.

Gould, L.H., Sui, J., Foellmer, H., Oliphant, T., Wang, T., Ledizet, M., Murakami, A., Noonan, K., Lambeth, C., Kar, K., *et al.* (2005). Protective and therapeutic capacity of human single chain Fv–Fc fusion proteins against West Nile virus. J. Virol. *79*, 14606–14613.

Gromowski, G.D., and Barrett, A.D. (2007). Characterization of an antigenic site that contains a dominant, type-specific neutralization determinant on the envelope protein domain III (ED3) of dengue 2 virus. Virology *366*, 349–360.

Gromowski, G.D., Barrett, N.D., and Barrett, A.D. (2008). Characterization of dengue virus complex-specific neutralizing epitopes on envelope protein domain III of dengue 2 virus. J. Virol. *82*, 8828–8837.

Guirakhoo, F., Bolin, R.A., and Roehrig, J.T. (1992). The Murray Valley encephalitis virus prM protein confers acid resistance to virus particles and alters the expression of epitopes within the R2 domain of E glycoprotein. Virology *191*, 921–931.

Guirakhoo, F., Heinz, F.X., and Kunz, C. (1989). Epitope model of tick-borne encephalitis virus envelope glycoprotein E: analysis of structural properties, role of carbohydrate side chain, and conformational changes occurring at acidic pH. Virology *169*, 90–99.

Guirakhoo, F., Heinz, F.X., Mandl, C.W., Holzmann, H., and Kunz, C. (1991). Fusion activity of flaviviruses: comparison of mature and immature (prM-containing) tick-borne encephalitis virions. J. Gen. Virol. *72 (Pt 6)*, 1323–1329.

Gupta, A.K., Lad, V.J., and Koshy, A.A. (2003). Protection of mice against experimental Japanese encephalitis virus infections by neutralizing anti-glycoprotein E monoclonal antibodies. Acta Virol *47*, 141–145.

Gupta, A.K., Lad, V.J., and Koshy, A.A. (2008). Survival of mice immunized with monoclonal antibodies against glycoprotein e of Japanese encephalitis virus before or after infection with Japanese encephalitis, west nile, and dengue viruses. Acta Virol *52*, 219–224.

Halevy, M., Akov, Y., Ben-Nathan, D., Kobiler, D., Lachmi, B., and Lustig, S. (1994). Loss of active neuroinvasiveness in attenuated strains of West Nile virus: pathogenicity in immunocompetent and SCID mice. Arch. Virol. *137*, 355–370.

Haley, M., Retter, A.S., Fowler, D., Gea-Banacloche, J., and O'Grady, N.P. (2003). The role for intravenous immunoglobulin in the treatment of West Nile virus encephalitis. Clin Infect. Dis. *37*, e88–90.

Hall, R.A., Kay, B.H., Burgess, G.W., Clancy, P., and Fanning, I.D. (1990). Epitope analysis of the envelope and non-structural glycoproteins of Murray Valley encephalitis virus. J. Gen. Virol. *71 (Pt 12)*, 2923–2930.

Halstead, S.B. (1979). *in vivo* enhancement of dengue virus infection in rhesus monkeys by passively transferred antibody. J. Infect. Dis. *140*, 527–533.

Halstead, S.B. (1989). Antibody, macrophages, dengue virus infection, shock, and hemorrhage: a pathogenetic cascade. Rev. Infect. Dis. *11* Suppl. *4*, S830–839.

Halstead, S.B. (1994). Antibody-dependent enhancement of infection: a mechanism for indirect virus entry into cells. In Cellular Receptors for Animal Viruses (Cold Spring Harbor, Cold Spring Harbor Laboratory Press), pp. 493–516.

Halstead, S.B., Porterfield, J.S., and O'Rourke, E.J. (1980). Enhancement of dengue virus infection in monocytes by flavivirus antisera. Am. J. Trop. Med. Hyg. *29*, 638–642.

Hamdan, A., Green, P., Mendelson, E., Kramer, M.R., Pitlik, S., and Weinberger, M. (2002). Possible benefit of intravenous immunoglobulin therapy in a lung transplant recipient with West Nile virus encephalitis. Transpl Infect. Dis. *4*, 160–162.

Harrison, S.C. (2008). Viral membrane fusion. Nat. Struct. Mol. Biol. *15*, 690–698.

Hasegawa, H., Yoshida, M., Kobayashi, Y., and Fujita, S. (1995). Antigenic analysis of Japanese encephalitis viruses in Asia by using monoclonal antibodies. Vaccine *13*, 1713–1721.

Hasegawa, H., Yoshida, M., Shiosaka, T., Fujita, S., and Kobayashi, Y. (1992). Mutations in the envelope protein of Japanese encephalitis virus affect entry into cultured cells and virulence in mice. Virology *191*, 158–165.

Hawkes, R.A., Roehrig, J.T., Hunt, A.R., and Moore, G.A. (1988). Antigenic structure of the Murray Valley encephalitis virus E glycoprotein. J. Gen. Virol. *69 (Pt 5)*, 1105–1109.

He, R.T., Innis, B.L., Nisalak, A., Usawattanakul, W., Wang, S., Kalayanarooj, S., and Anderson, R. (1995). Antibodies that block virus attachment to Vero cells are a major component of the human neutralizing antibody response against dengue virus type 2. J. Med. Virol. *45*, 451–461.

Heinz, F., Auer, G., Stiasny, K., Holzmann, H., Mandl, C., Guirakhoo, F., and Kunz, C. (1994). The interactions of the flavivirus envelope proteins: implications for virus entry and release. Arch.Virol. *9(S)*, 339–348.

Heinz, F.X., Berger, R., Tuma, W., and Kunz, C. (1983). A topological and functional model of epitopes on the structural glycoprotein of tick-borne encephalitis virus defined by monoclonal antibodies. Virology *126*, 525–537.

Heinz, F.X., Holzmann, H., Essl, A., and Kundi, M. (2007). Field effectiveness of vaccination against tick-borne encephalitis. Vaccine *25*, 7559–7567.

Henchal, E.A., Henchal, L.S., and Schlesinger, J.J. (1988). Synergistic interactions of anti-NS1 monoclonal antibodies protect passively immunized mice from lethal challenge with dengue 2 virus. J. Gen. Virol. *69 (Pt 8)*, 2101–2107.

Hofmann, H., Frisch-Niggemeyer, W., and Kunz, C. (1978). Protection of mice against tick-borne encephalitis by different classes of immunoglobulins. Infection *6*, 154–157.

Holzmann, H., Stiasny, K., Ecker, M., Kunz, C., and Heinz, F.X. (1997). Characterization of monoclonal antibody-escape mutants of tick-borne encephalitis virus with reduced neuroinvasiveness in mice. J. Gen. Virol. *78 (Pt 1)*, 31–37.

Hrobowski, Y.M., Garry, R.F., and Michael, S.F. (2005). Peptide inhibitors of dengue virus and West Nile virus infectivity. Virol. J. *2*, 49.

Hung, J.J., Hsieh, M.T., Young, M.J., Kao, C.L., King, C.C., and Chang, W. (2004). An external loop region of domain III of dengue virus type 2 envelope protein is involved in serotype-specific binding to mosquito but not mammalian cells. J. Virol. *78*, 378–388.

Hung, S.L., Lee, P.L., Chen, H.W., Chen, L.K., Kao, C.L., and King, C.C. (1999). Analysis of the steps involved in dengue virus entry into host cells. Virology *257*, 156–167.

Ishida, I., Tomizuka, K., Yoshida, H., Tahara, T., Takahashi, N., Ohguma, A., Tanaka, S., Umehashi, M., Maeda, H., Nozaki, C., *et al.* (2002). Production of human monoclonal and polyclonal antibodies in TransChromo animals. Cloning and stem cells *4*, 91–102.

Johnson, A.J., and Roehrig, J.T. (1999). New mouse model for dengue virus vaccine testing. J. Virol. *73*, 783–786.

Julander, J.G., Winger, Q.A., Olsen, A.L., Day, C.W., Sidwell, R.W., and Morrey, J.D. (2005). Treatment of West Nile virus-infected mice with reactive immunoglobulin reduces fetal titers and increases dam survival. Antiviral Res. *65*, 79–85.

Junjhon, J., Edwards, T.J., Utaipat, U., Bowman, V.D., Holdaway, H.A., Zhang, W., Keelapang, P., Puttikhunt, C., Perera, R., Chipman, P.R., *et al.* (2010). Influence of pr–M cleavage on the heterogeneity of extracellular dengue virus particles. J. Virol. *84*, 8353–8358.

Kampmann, T., Mueller, D.S., Mark, A.E., Young, P.R., and Kobe, B. (2006). The Role of histidine residues in low-pH-mediated viral membrane fusion. Structure *14*, 1481–1487.

Kanai, R., Kar, K., Anthony, K., Gould, L.H., Ledizet, M., Fikrig, E., Marasco, W.A., Koski, R.A., and Modis, Y. (2006). Crystal structure of west nile virus envelope glycoprotein reveals viral surface epitopes. J. Virol. *80*, 11000–11008.

Kaufman, B.M., Summers, P.L., Dubois, D.R., Cohen, W.H., Gentry, M.K., Timchak, R.L., Burke, D.S., and Eckels, K.H. (1989). Monoclonal antibodies for dengue virus prM glycoprotein protect mice against lethal dengue infection. Am. J. Trop. Med. Hyg. *41*, 576–580.

Kaufman, B.M., Summers, P.L., Dubois, D.R., and Eckels, K.H. (1987). Monoclonal antibodies against dengue 2 virus E-glycoprotein protect mice against lethal dengue infection. Am. J. Trop. Med. Hyg. *36*, 427–434.

Kaufmann, B., Nybakken, G., Chipman, P.R., Zhang, W., Fremont, D.H., Diamond, M.S., Kuhn, R.J., and Rossmann, M.G. (2006). West Nile virus in complex with a neutralizing monoclonal antibody. Proc. Natl. Acad. Sci. U.S.A. *103*, 12400–12404.

Kaufmann, B., Chipman, P.R., Holdaway, H.A., Johnson, S., Fremont, D.H., Kuhn, R.J., Diamond, M.S., and Rossmann, M.G. (2009). Capturing a flavivirus pre-fusion intermediate. PLoS Pathog. *5*, e1000672.

Kaufmann, B., Vogt, M.R., Goudsmit, J., Aksyuk, A.A., Chipman, P.R., Kuhn, R.J., Diamond, M.S., and Rossman, M.G. (2010). Neutralization of West Nile virus by cross-linking of its surface proteins with Fab fragments of the human monoclonal antibody CR4354. Proc. Natl. Acad. Sci. U.S.A. *107*, 18950–18955.

Kedarnath, N., Dayaraj, C., Sathe, P.S., Gadkari, D.A., Dandawate, C.N., Goverdhan, M.K., and Ghosh, S.N. (1986). Monoclonal antibodies against Japanese encephalitis virus. Indian J Med Res *84*, 125–133.

Khromykh, A.A., Sedlak, P.L., Guyatt, K.J., Hall, R.A., and Westaway, E.G. (1999). Efficient trans-complementation of the flavivirus kunjin NS5 protein but not of the NS1 protein requires its coexpression with other components of the viral replicase. J. Virol. *73*, 10272–10280.

Kiermayr, S., Stiasny, K., and Heinz, F.X. (2009). Impact of quaternary organization on the antigenic structure of the tick-borne encephalitis virus envelope glycoprotein E.J. Virol. *83*, 8482–8491.

Kimura-Kuroda, J., and Yasui, K. (1983). Topographical analysis of antigenic determinants on envelope glycoprotein V3 (E) of Japanese encephalitis virus, using monoclonal antibodies. J. Virol. *45*, 124–132.

Kimura-Kuroda, J., and Yasui, K. (1986). Antigenic comparison of envelope protein E between Japanese encephalitis virus and some other flaviviruses using monoclonal antibodies. J. Gen. Virol. *67 (Pt 12)*, 2663–2672.

Kimura-Kuroda, J., and Yasui, K. (1988). Protection of mice against Japanese encephalitis virus by passive administration with monoclonal antibodies. J. Immunol. *141*, 3606–3610.

Kobayashi, Y., Hasegawa, H., and Yamauchi, T. (1985). Studies on the antigenic structure of Japanese encephalitis virus using monoclonal antibodies. Microbiol. Immunol. *29*, 1069–1082.

Kreil, T.R., and Eibl, M.M. (1997). Pre- and postexposure protection by passive immunoglobulin but no enhancement of infection with a flavivirus in a mouse model. J. Virol. *71*, 2921–2927.

Kreil, T.R., Maier, E., Fraiss, S., and Eibl, M.M. (1998). Neutralizing antibodies protect against lethal flavivirus challenge but allow for the development of active humoral immunity to a nonstructural virus protein. J. Virol. *72*, 3076–3081.

Kuhn, R.J., Zhang, W., Rossmann, M.G., Pletnev, S.V., Corver, J., Lenches, E., Jones, C.T., Mukhopadhyay, S., Chipman, P.R., Strauss, E.G. *et al.* (2002). Structure of dengue virus: implications for flavivirus organization, maturation, and fusion. Cell *108*, 717–725.

Kurane, I., and Ennis, F.E. (1992). Immunity and immunopathology in dengue virus infections. Semin Immunol *4*, 121–127.

Lai, C.J., Goncalvez, A.P., Men, R., Wernly, C., Donau, O., Engle, R.E., and Purcell, R.H. (2007). Epitope determinants of a chimpanzee dengue virus type 4 (DENV-4)-neutralizing antibody and protection against DENV-4 challenge in mice and rhesus monkeys by passively transferred humanized antibody. J. Virol. *81*, 12766–12774.

Lai, H., Engle, M., Fuchs, A., Keller, T., Johnson, S., Gorlatov, S., Diamond, M.S., and Chen, Q. (2010). Monoclonal antibody produced in plants efficiently treats West Nile virus infection in mice. Proc. Natl. Acad. Sci. U.S.A. *107*, 2419–2424.

Lee, J.W., Chu, J.J., and Ng, M.L. (2006). Quantifying the specific binding between West Nile virus envelope domain III protein and the cellular receptor alphaVbeta3 integrin. J. Biol. Chem. *281*, 1352–1360.

Li, L., Lok, S.M., Yu, I.M., Zhang, Y., Kuhn, R.J., Chen, J., and Rossmann, M.G. (2008). The flavivirus precursor membrane-envelope protein complex: structure and maturation. Science *319*, 1830–1834.

Lindenbach, B.D., and Rice, C.M. (1997). trans-Complementation of yellow fever virus NS1 reveals a role in early RNA replication. J. Virol. *71*, 9608–9617.

Lindenbach, B.D., and Rice, C.M. (2001). Flaviviridae: The viruses and their replication. In Fields Virology, Knipe, D.M., and Howley, P.M., eds. (Philadelphia, Lippincott Williams and Wilkins), pp. 991–1041.

Lobigs, M., Dalgarno, L., Schlesinger, J.J., and Weir, R.C. (1987). Location of a neutralization determinant in the E protein of yellow fever virus (17D vaccine strain). Virology *161*, 474–478.

Lok, S.M., Kostyuchenko, V., Nybakken, G.E., Holdaway, H.A., Battisti, A.J., Sukupolvi-Petty, S., Sedlak, D., Fremont, D.H., Chipman, P.R., Roehrig, J.T., *et al.* (2008). Binding of a neutralizing antibody to dengue virus alters the arrangement of surface glycoproteins. Nat. Struct. Mol. Biol. *15*, 312–317.

Ma, S.H., Lin, Y.L., Huang, Y.Y., Liu, C.I., Chen, S.S., Chiang, H.Y., and Chen, L.K. (1995). Generation and characterization of Japanese encephalitis virus specific monoclonal antibodies. Zhonghua Minguo wei sheng wu ji mian yi xue za zhi = Chinese journal of microbiology and immunology *28*, 128–138.

Mackenzie, J.M., Khromykh, A.A., Jones, M.K., and Westaway, E.G. (1998). Subcellular localization and some biochemical properties of the flavivirus Kunjin nonstructural proteins NS2A and NS4A. Virology *245*, 203–215.

Mackenzie, J.M., and Westaway, E.G. (2001). Assembly and maturation of the flavivirus Kunjin virus appear to occur in the rough endoplasmic reticulum and along the secretory pathway, respectively. J. Virol. *75*, 10787–10799.

Mackenzie, J.S., Gubler, D.J., and Petersen, L.R. (2004). Emerging flaviviruses: the spread and resurgence of Japanese encephalitis, West Nile and dengue viruses. Nat. Med. *10*, S98–109.

Malet, H., Egloff, M.P., Selisko, B., Butcher, R.E., Wright, P.J., Roberts, M., Gruez, A., Sulzenbacher, G., Vonrhein, C., Bricogne, G., *et al.* (2007). Crystal structure of the RNA polymerase domain of the West Nile virus non-structural protein 5. J. Biol. Chem. *282*, 10678–10689.

Mandl, C.W., Guirakhoo, F., Holzmann, H., Heinz, F.X., and Kunz, C. (1989). Antigenic structure of the flavivirus envelope protein E at the molecular level, using tick-borne encephalitis virus as a model. J. Virol. *63*, 564–571.

Martin, M., Tsai, T.F., Cropp, B., Chang, G.J., Holmes, D.A., Tseng, J., Shieh, W., Zaki, S.R., Al-Sanouri, I., Cutrona, A.F., *et al.* (2001a). Fever and multisystem organ failure associated with 17D-204 yellow fever vaccination: a report of four cases. Lancet *358*, 98–104.

Martin, M., Weld, L.H., Tsai, T.F., Mootrey, G.T., Chen, R.T., Niu, M., and Cetron, M.S. (2001b). Advanced age a risk factor for illness temporally associated with yellow fever vaccination. Emerg. Infect. Dis. *7*, 945–951.

Mathews, J.H., and Roehrig, J.T. (1984). Elucidation of the topography and determination of the protective epitopes on the E glycoprotein of Saint Louis encephalitis virus by passive transfer with monoclonal antibodies. J. Immunol. *132*, 1533–1537.

McMinn, P.C., Lee, E., Hartley, S., Roehrig, J.T., Dalgarno, L., and Weir, R.C. (1995). Murray valley encephalitis virus envelope protein antigenic variants with altered hemagglutination properties and reduced neuroinvasiveness in mice. Virology *211*, 10–20.

Mehlhop, E., Ansarah-Sobrinho, C., Johnson, S., Engle, M., Fremont, D.H., Pierson, T.C., and Diamond, M.S. (2007). Complement protein C1q inhibits antibody-dependent enhancement of flavivirus infection in an IgG subclass-specific manner. Cell Host and Microbe *2*, 417–426.

Mehlhop, E., Nelson, S., Jost, C.A., Gorlatov, S., Johnson, S., Fremont, D.H., Diamond, M.S., and Pierson, T.C. (2009). Complement protein C1q reduces the stoichiometric threshold for antibody-mediated neutralization of West Nile virus. Cell Host Microbe *6*, 381–391.

Miller, S., Kastner, S., Krijnse-Locker, J., Buhler, S., and Bartenschlager, R. (2007). The non-structural protein 4A of dengue virus is an integral membrane protein inducing membrane alterations in a 2K-regulated manner. J. Biol. Chem. *282*, 8873–8882.

Miller, S., Sparacio, S., and Bartenschlager, R. (2006). Subcellular localization and membrane topology of the dengue virus type 2 Non-structural protein 4B. J. Biol. Chem. *281*, 8854–8863.

Modis, Y., Ogata, S., Clements, D., and Harrison, S.C. (2003). A ligand-binding pocket in the dengue virus envelope glycoprotein. Proc. Natl. Acad. Sci. U.S.A. *100*, 6986–6991.

Modis, Y., Ogata, S., Clements, D., and Harrison, S.C. (2004). Structure of the dengue virus envelope protein after membrane fusion. Nature *427*, 313–319.

Modis, Y., Ogata, S., Clements, D., and Harrison, S.C. (2005). Variable surface epitopes in the crystal structure of dengue virus type 3 envelope glycoprotein. J. Virol. *79*, 1223–1231.

Monath, T.P. (2008). Treatment of yellow fever. Antiviral Res. *78*, 116–124.

Morens, D.M. (1994). Antibody-dependent of enhancement of infection and the pathogenesis of viral disease. Clin Inf Dis *19*, 500–512.

Morita, K., Tadano, M., Nakaji, S., Kosai, K., Mathenge, E.G., Pandey, B.D., Hasebe, F., Inoue, S., and Igarashi, A. (2001). Locus of a virus neutralization epitope on the Japanese encephalitis virus envelope protein determined by use of long PCR-based region-specific random mutagenesis. Virology *287*, 417–426.

Morrey, J.D., Siddharthan, V., Olsen, A.L., Roper, G.Y., Wang, H., Baldwin, T.J., Koenig, S., Johnson, S., Nordstrom, J.L., and Diamond, M.S. (2006). Humanized monoclonal antibody against West Nile virus E protein administered after neuronal infection protects against lethal encephalitis in hamsters. J. Infect. Dis. *194*, 1300–1308.

Morrey, J.D., Siddharthan, V., Olsen, A.L., Wang, H., Julander, J.G., Hall, J.O., Li, H., Nordstrom, J.L., Koenig, S., Johnson, S., *et al.* (2007). Defining limits of humanized neutralizing monoclonal antibody treatment for West Nile virus neurological infection in a hamster model. Antimicrob Agents Chemother *51*, 2396–2402.

Mukhopadhyay, S., Kim, B.S., Chipman, P.R., Rossmann, M.G., and Kuhn, R.J. (2003). Structure of West Nile virus. Science *302*, 248.

Mukhopadhyay, S., Kuhn, R.J., and Rossmann, M.G. (2005). A structural perspective of the flavivirus life cycle. Nat. Rev. Microbiol. *3*, 13–22.

Murthy, H.M., Judge, K., DeLucas, L., and Padmanabhan, R. (2000). Crystal structure of dengue virus NS3 protease in complex with a Bowman-Birk inhibitor: implications for flaviviral polyprotein processing and drug design. J. Mol. Biol. *301*, 759–767.

Nelson, S., Jost, C.A., Xu, Q., Ess, J., Martin, J.E., Oliphant, T., Whitehead, S.S., Durbin, A.P., Graham, B.S., Diamond, M.S., *et al.* (2008). Maturation of West Nile virus modulates sensitivity to antibody-mediated neutralization. PLoS Pathog. *4*, e1000060.

Nelson, S., Poddar, S., Lin, T.Y., and Pierson, T.C. (2009). Protonation of individual histidine residues is not required for the pH-dependent entry of west nile virus: evaluation of the 'histidine switch' hypothesis. J. Virol. *83*, 12631–12635.

Niedrig, M., Klockmann, U., Lang, W., Roeder, J., Burk, S., Modrow, S., and Pauli, G. (1994). Monoclonal antibodies directed against tick-borne encephalitis virus with neutralizing activity *in vivo*. Acta Virol *38*, 141–149.

Nimmerjahn, F., and Ravetch, J.V. (2008). Analyzing antibody–Fc-receptor interactions. Methods Mol Biol *415*, 151–162.

Nybakken, G., Oliphant, T., Johnson, S., Burke, S., Diamond, M.S., and Fremont, D.H. (2005). Structural basis for neutralization of a therapeutic antibody against West Nile virus. Nature *437*, 764–769.

Nybakken, G.E., Nelson, C.A., Chen, B.R., Diamond, M.S., and Fremont, D.H. (2006). Crystal structure of the West Nile virus envelope glycoprotein. J. Virol. *80*, 11467–11474.

Oceguera, L.F. 3rd, Patiris, P.J., Chiles, R.E., Busch, M.P., Tobler, L.H., and Hanson, C.V. (2007). Flavivirus serology by Western blot analysis. Am. J. Trop. Med. Hyg. *77*, 159–163.

Oliphant, T., Engle, M., Nybakken, G., Doane, C., Johnson, S., Huang, L., Gorlatov, S., Mehlhop, E., Marri, A., Chung, K.M., *et al.* (2005). Development of a humanized monoclonal antibody with therapeutic potential against West Nile virus. Nat. Med. *11*, 522–530.

Oliphant, T., Nybakken, G.E., Austin, S.K., Xu, Q., Bramson, J., Loeb, M., Throsby, M., Fremont, D.H., Pierson, T.C., and Diamond, M.S. (2007). The Induction of Epitope-Specific Neutralizing Antibodies against West Nile virus. J. Virol. *81*, 11828–11839.

Oliphant, T., Nybakken, G.E., Engle, M., Xu, Q., Nelson, C.A., Sukupolvi-Petty, S., Marri, A., Lachmi, B.E., Olshevsky, U., Fremont, D.H., *et al.* (2006). Antibody recognition and neutralization determinants on domains I and II of West Nile Virus envelope protein. J. Virol. *80*, 12149–12159.

Peiris, J.S., and Porterfield, J.S. (1979). Antibody-mediated enhancement of Flavivirus replication in macrophage-like cell lines. Nature *282*, 509–511.

Peiris, J.S.M., Gordon, S., Unkeless, J.C., and Porterfield, J.S. (1981). Monoclonal anti-Fc receptor IgG blocks antibody-dependent enhancement of viral replication in macrophages. Nature *289*, 189–191.

Peiris, J.S.M., Porterfield, J.S., and Roehrig, J.T. (1982). Monoclonal antibodies against the flavivirus West Nile. J. Gen. Virol. *58*, 283–289.

Phillpotts, R.J., Stephenson, J.R., and Porterfield, J.S. (1987). Passive immunization of mice with monoclonal antibodies raised against tick-borne encephalitis virus. Brief report. Arch. Virol. *93*, 295–301.

Pierson, T.C., Fremont, D.H., Kuhn, R.J., and Diamond, M.S. (2008). Structural insights into the mechanisms of antibody-mediated neutralization of flavivirus infection: implications for vaccine development. Cell Host Microbe *4*, 229–238.

Pierson, T.C., Xu, Q., Nelson, S., Oliphant, T., Nybakken, G.E., Fremont, D.H., and Diamond, M.S. (2007). The stoichiometry of antibody-mediated neutralization and enhancement of West Nile virus infection. Cell Host Microbe *1*, 135–145.

Pincus, S., Mason, P.W., Konishi, E., Fonseca, B.A., Shope, R.E., Rice, C.M., and Paoletti, E. (1992). Recombinant vaccinia virus producing the prM and E proteins of yellow fever virus protects mice from lethal yellow fever encephalitis. Virology *187*, 290–297.

Pothipunya, S., Patarapotikul, J., Rojanasuphot, S., and Tharavanij, S. (1993). Mapping of functional epitopes of Japanese encephalitis virus using monoclonal antibodies. Southeast Asian J Trop Med Public Health *24*, 277–283.

Putnak, J.R., and Schlesinger, J.J. (1990). Protection of mice against yellow fever virus encephalitis by immunization with a vaccinia virus recombinant encoding the yellow fever virus non-structural proteins, NS1, NS2a and NS2b. J. Gen. Virol. *71 (Pt 8)*, 1697–1702.

Rajamanonmani, R., Nkenfou, C., Clancy, P., Yau, Y.H., Shochat, S.G., Sukupolvi-Petty, S., Schul, W., Diamond, M.S., Vasudevan, S.G., and Lescar, J. (2009). On a mouse monoclonal antibody that neutralizes all four dengue virus serotypes. J. Gen. Virol. *90*, 799–809.

Rey, F.A., Heinz, F.X., Mandl, C., Kunz, C., and Harrison, S.C. (1995). The envelope glycoprotein from tick-borne encephalitis virus at 2 Angstrom resolution. Nature *375*, 291–298.

Roehrig, J.T., Bolin, R.A., and Kelly, R.G. (1998). Monoclonal antibody mapping of the envelope glycoprotein of the dengue 2 virus, Jamaica. Virology *246*, 317–328.

Roehrig, J.T., Mathews, J.H., and Trent, D.W. (1983). Identification of epitopes on the E glycoprotein of Saint Louis encephalitis virus using monoclonal antibodies. Virology *128*, 118–126.

Roehrig, J.T., Staudinger, L.A., Hunt, A.R., Mathews, J.H., and Blair, C.D. (2001). Antibody prophylaxis and therapy for flaviviral encephalitis infections. Ann NY Acad Sci, 286–297.

Roosendaal, J., Westaway, E.G., Khromykh, A., and Mackenzie, J.M. (2006). Regulated cleavages at the West Nile virus NS4A-2K-NS4B junctions play a major role in rearranging cytoplasmic membranes and Golgi trafficking of the NS4A protein. J. Virol. *80*, 4623–4632.

Ryman, K.D., Ledger, T.N., Weir, R.C., Schlesinger, J.J., and Barrett, A.D. (1997). Yellow fever virus envelope protein has two discrete type-specific neutralizing epitopes. J. Gen. Virol. *78 (Pt 6)*, 1353–1356.

Samuel, M.A., Wang, H., Siddharthan, V., Morrey, J.D., and Diamond, M.S. (2007). Axonal transport mediates West Nile virus entry into the central nervous system and induces acute flaccid paralysis. Proc. Natl. Acad. Sci. U.S.A. *104*, 17140–17145.

Sanchez, M.D., Pierson, T.C., McAllister, D., Hanna, S.L., Puffer, B.A., Valentine, L.E., Murtadha, M.M., Hoxie, J.A., and Doms, R.W. (2005). Characterization of neutralizing antibodies to West Nile virus. Virology *336*, 70–82.

Saquib, R., Randall, H., Chandrakantan, A., Spak, C.W., and Barri, Y.M. (2008). West Nile virus encephalitis in a renal transplant recipient: the role of intravenous immunoglobulin. Am J Kidney Dis *52*, e19–21.

Sawyer, L.A. (2000). Antibodies for the prevention and treatment of viral diseases. Antiviral Res. *47*, 57–77.

Schlesinger, J.J., Brandriss, M.W., Cropp, C.B., and Monath, T.P. (1986). Protection against yellow fever in monkeys by immunization with yellow fever virus nonstructural protein NS1. J. Virol. *60*, 1153–1155.

Schlesinger, J.J., Brandriss, M.W., Putnak, J.R., and Walsh, E.E. (1990). Cell surface expression of yellow fever virus non-structural glycoprotein NS1: consequences of interaction with antibody. J. Gen. Virol. *71 (Pt 3)*, 593–599.

Schlesinger, J.J., Brandriss, M.W., and Walsh, E.E. (1985). Protection against 17D yellow fever encephalitis in mice by passive transfer of monoclonal antibodies to the nonstructural glycoprotein gp48 and by active immunization with gp48. J. Immunol. *135*, 2805–2809.

Schlesinger, J.J., Brandriss, M.W., and Walsh, E.E. (1987). Protection of mice against dengue 2 virus encephalitis by immunization with the dengue 2 virus non-structural glycoprotein NS1. J. Gen. Virol. *68 (Pt 3)*, 853–857.

Schlesinger, J.J., and Chapman, S. (1995). Neutralizing F(ab')2 fragments of protective monoclonal antibodies to yellow fever virus (YF) envelope protein fail to protect mice against lethal YF encephalitis. J. Gen. Virol. *76 (Pt 1)*, 217–220.

Schlesinger, J.J., Foltzer, M., and Chapman, S. (1993). The Fc portion of antibody to yellow fever virus NS1 is a determinant of protection against YF encephalitis in mice. Virology *192*, 132–141.

Schlesinger, J.J., Walsh, E.E., and Brandriss, M.W. (1984). Analysis of 17D yellow fever virus envelope protein epitopes using monoclonal antibodies. J. Gen. Virol. *65 (Pt 10)*, 1637–1644.

Schmidt, A.G., Yang, P.L., and Harrison, S.C. (2010). Peptide inhibitors of dengue-virus entry target a late-stage fusion intermediate. PLoS Pathog. *6*, e1000851.

Se-Thoe, S.Y., Ng, M.M., and Ling, A.E. (1999). Retrospective study of Western blot profiles in immune sera of natural dengue virus infections. J. Med. Virol. *57*, 322–330.

Shimoni, Z., Niven, M.J., Pitlick, S., and Bulvik, S. (2001). Treatment of West Nile virus encephalitis with intravenous immunoglobulin. Emerg. Infect. Dis. *7*, 759.

Shrestha, B., Brien, J.D., Sukupolvi-Petty, S., Austin, S.K., Edeling, M.A., Kim, T., O'Brien, K.M., Nelson, C.A., Johnson, S., Fremont, D.H., *et al.* (2010). The Development of Therapeutic Antibodies that Neutralize Homologous and Heterologous Genotypes of dengue Virus Type 1. PLoS Pathog. *6*, e1000823.

Stadler, K., Allison, S.L., Schalich, J., and Heinz, F.X. (1997). Proteolytic activation of tick-borne encephalitis virus by furin. J. Virol. *71*, 8475–8481.

Stiasny, K., Allison, S.L., Mandl, C.W., and Heinz, F.X. (2001). Role of metastability and acidic pH in membrane fusion by tick-borne encephalitis virus. J. Virol. *75*, 7392–7398.

Stiasny, K., Allison, S.L., Marchler-Bauer, A., Kunz, C., and Heinz, F.X. (1996). Structural requirements for low-pH-induced rearrangements in the envelope glycoprotein of tick-borne encephalitis virus. J. Virol. *70*, 8142–8147.

Stiasny, K., Brandler, S., Kossl, C., and Heinz, F.X. (2007). Probing the flavivirus membrane fusion mechanism by using monoclonal antibodies. J. Virol. *81*, 11526–11531.

Stiasny, K., Kiermayr, S., Holzmann, H., and Heinz, F.X. (2006). Cryptic properties of a cluster of dominant flavivirus cross-reactive antigenic sites. J. Virol. *80*, 9557–9568.

Sukupolvi-Petty, S., Austin, S.K., Engle, M., Brien, J.D., Dowd, K.A., Williams, K.L., Johnson, S., Rico-Hesse, R., Harris, E., Pierson, T.C., *et al.* (2010). Structure and function analysis of therapeutic monoclonal antibodies against dengue virus type 2. J. Virol. *84*, 9227–9239.

Sukupolvi-Petty, S., Austin, S.K., Purtha, W.E., Oliphant, T., Nybakken, G., Schlesinger, J.J., Roehrig, J.T., Gromowski, G.D., Barrett, A.D., Fremont, D.H., *et al.* (2007). Type- and Sub-Complex-Specific Neutralizing Antibodies Against Domain III of dengue Virus Type-2 Envelope Protein Recognize Adjacent Epitopes. J. Virol. *81*, 12816–12826.

Sultana, H., Foellmer, H.G., Neelakanta, G., Oliphant, T., Engle, M., Ledizet, M., Krishnan, M.N., Bonafe, N., Anthony, K.G., Marasco, W.A., *et al.* (2009). Fusion loop peptide of the West Nile virus envelope protein is essential for pathogenesis and is recognized by a therapeutic cross-reactive human monoclonal antibody. J. Immunol. *183*, 650–660.

Tesh, R.B., Arroyo, J., Travassos Da Rosa, A.P., Guzman, H., Xiao, S.Y., and Monath, T.P. (2002). Efficacy of killed virus vaccine, live attenuated chimeric virus vaccine, and passive immunization for prevention of West Nile virus encephalitis in hamster model. Emerg. Infect. Dis. *8*, 1392–1397.

Thompson, B.S., Moesker, B., Smit, J.M., Wilschut, J., Diamond, M.S., and Fremont, D.H. (2009). A therapeutic antibody against west nile virus neutralizes infection by blocking fusion within endosomes. PLoS Pathog. *5*, e1000453.

Throsby, M., Geuijen, C., Goudsmit, J., Bakker, A.Q., Korimbocus, J., Kramer, R.A., Clijsters-van der Horst, M., de Jong, M., Jongeneelen, M., Thijsse, S., *et al.* (2006). Isolation and characterization of human monoclonal antibodies from individuals infected with West Nile Virus. J. Virol. *80*, 6982–6992.

Thullier, P., Demangel, C., Bedouelle, H., Megret, F., Jouan, A., Deubel, V., Mazie, J.C., and Lafaye, P. (2001). Mapping of a dengue virus neutralizing epitope critical for the infectivity of all serotypes: insight into the neutralization mechanism. J. Gen. Virol. *82*, 1885–1892.

Thullier, P., Lafaye, P., Megret, F., Deubel, V., Jouan, A., and Mazie, J.C. (1999). A recombinant Fab neutralizes dengue virus *in vitro*. J Biotechnol *69*, 183–190.

Trainor, N.B., Crill, W.D., Roberson, J.A., and Chang, G.J. (2007). Mutation analysis of the fusion domain region of St. Louis encephalitis virus envelope protein. Virology *360*, 398–406.

Vazquez, S., Guzman, M.G., Guillen, G., Chinea, G., Perez, A.B., Pupo, M., Rodriguez, R., Reyes, O., Garay, H.E., Delgado, I., *et al.* (2002). Immune response to synthetic peptides of dengue prM protein. Vaccine *20*, 1823–1830.

Vogt, M.R., Moesker, B., Goudsmit, J., Jongeneelen, M., Austin, S.K., Oliphant, T., Nelson, S., Pierson, T.C., Wilschut, J., Throsby, M., *et al.* (2009). Human monoclonal antibodies induced by natural infection against West Nile virus neutralize at a post-attachment step. J. Virol. *83*, 6494–6507.

Wahala, W.M., Donaldson, E.F., de Alwis, R., Accavitti-Loper, M.A., Baric, R.S., and de Silva, A.M. (2010). Natural strain variation and antibody neutralization of dengue serotype 3 viruses. PLoS Pathog. *6*, e1000821.

Wang, T., Anderson, J.F., Magnarelli, L.A., Wong, S.J., Koski, R.A., and Fikrig, E. (2001). Immunization of mice against West Nile virus with recombinant envelope protein. J. Immunol. *167*, 5273–5277.

Webb, H.E., Wight, D.G., Platt, G.S., and Smith, C.E.G. (1968). Langat virus encephalitis in mice. I. The effect of the administration of specific antiserum. J Hyg *66*, 343–354.

Whittembury, A., Ramirez, G., Hernandez, H., Ropero, A.M., Waterman, S., Ticona, M., Brinton, M., Uchuya, J., Gershman, M., Toledo, W., *et al.* (2009). Viscerotropic disease following yellow fever vaccination in Peru. Vaccine *27*, 5974–5981.

Wu, K.P., Wu, C.W., Tsao, Y.P., Kuo, T.W., Lou, Y.C., Lin, C.W., Wu, S.C., and Cheng, J.W. (2003). Structural basis of a Flavivirus recognized by its neutralizing antibody: Solution structure of the domain III of the Japanese Encephalitis virus envelope protein. J. Biol. Chem. *278*, 46007–46013.

Wu, S.C., Lian, W.C., Hsu, L.C., Wu, Y.C., and Liau, M.Y. (1998). Antigenic characterization of nine wild-type Taiwanese isolates of Japanese encephalitis virus as compared with two vaccine strains. Virus Res. *55*, 83–91.

Xu, T., Sampath, A., Chao, A., Wen, D., Nanao, M., Chene, P., Vasudevan, S.G., and Lescar, J. (2005). Structure of the dengue virus helicase/nucleoside triphosphatase catalytic domain at a resolution of 2.4 A.J. Virol. *79*, 10278–10288.

Yap, T.L., Xu, T., Chen, Y.L., Malet, H., Egloff, M.P., Canard, B., Vasudevan, S.G., and Lescar, J. (2007). Crystal structure of the dengue virus RNA-dependent RNA polymerase catalytic domain at 1.85-angstrom resolution. J. Virol. *81*, 4753–4765.

Yu, I.M., Holdaway, H.A., Chipman, P.R., Kuhn, R.J., Rossmann, M.G., and Chen, J. (2009). Association of the pr peptides with dengue virus at acidic pH blocks membrane fusion. J. Virol. *83*, 12101–12107.

Yu, I.M., Zhang, W., Holdaway, H.A., Li, L., Kostyuchenko, V.A., Chipman, P.R., Kuhn, R.J., Rossmann, M.G., and Chen, J. (2008). Structure of the immature dengue virus at low pH primes proteolytic maturation. Science *319*, 1834–1837.

Yusof, R., Clum, S., Wetzel, M., Murthy, H.M., and Padmanabhan, R. (2000). Purified NS2B/NS3 serine protease of dengue virus type 2 exhibits co-factor NS2B dependence for cleavage of substrates with dibasic amino acids *in vitro*. J. Biol. Chem. *275*, 9963–9969.

Zeitlin, L., Cone, R.A., and Whaley, K.J. (1999). Using monoclonal antibodies to prevent mucosal transmission of epidemic infectious diseases. Emerg. Infect. Dis. *5*, 54–64.

Zellweger, R.M., Prestwood, T.R., and Shresta, S. (2010). Enhanced infection of liver sinusoidal endothelial cells in a mouse model of antibody-induced severe dengue disease. Cell Host Microbe *7*, 128–139.

Zhang, M.J., Wang, M.J., Jiang, S.Z., and Ma, W.Y. (1989). Passive protection of mice, goats, and monkeys against Japanese encephalitis with monoclonal antibodies. J. Med. Virol. *29*, 133–138.

Zhang, M.J., Wang, M.X., Jiang, S.Z., Xiu, Z.Z., and Ma, W.Y. (1992). Preparation and characterization of the monoclonal antibodies against Japanese encephalitis virus. Acta Virol *36*, 533–540.

Zhang, S., Vogt, M.R., Oliphant, T., Engle, M., Bovshik, E.I., Diamond, M.S., and Beasley, D.W. (2009). The development of resistance to passive therapy by a potently neutralizing humanized West Nile virus monoclonal antibody. J. Infect. Dis. *200*, 202–205.

Zhang, Y., Corver, J., Chipman, P.R., Zhang, W., Pletnev, S.V., Sedlak, D., Baker, T.S., Strauss, J.H., Kuhn, R.J., and Rossmann, M.G. (2003). Structures of immature flavivirus particles. EMBO J. *22*, 2604–2613.

Zhang, Y., Kaufmann, B., Chipman, P.R., Kuhn, R.J., and Rossmann, M.G. (2007). Structure of immature West Nile virus. J. Virol. *81*, 6141–6145.

Zhang, Y., Zhang, W., Ogata, S., Clements, D., Strauss, J.H., Baker, T.S., Kuhn, R.J., and Rossmann, M.G. (2004). Conformational changes of the flavivirus E glycoprotein. Structure (Camb) *12*, 1607–1618.

Flavivirus Antiviral Development

12

Qing-Yin Wang, Yen-Liang Chen, Siew Pheng Lim and Pei-Yong Shi

Abstract

Many flaviviruses are human pathogens of global importance, but no clinically approved antiviral therapy is currently available to manage these diseases. Both pharmaceutical industry and academia have invested considerable efforts over the past decade on finding the flavivirus antivirals using modern drug discovery. Various high-throughput compatible target-based and cell-based assays have been developed and implemented. In this chapter, we describe in details the methodologies developed for screening inhibitors against dengue virus, and the lessons learned from our screening campaigns. Based on our experience on dengue virus and the status of hepatitis C virus drug discovery, we propose that a combined target-based approach (e.g. viral polymerase, protease, and envelope) and a cell-based approach (e.g. virus infection and replicon assays) should be persistently pursued to develop flavivirus antiviral therapy.

Introduction

The *Flaviviridae* consists of three genera: *Flavivirus, Pestivirus,* and *Hepacivirus.* The genus *Flavivirus* contains more than 53 members, among which dengue virus (DENV), West Nile virus (WNV), yellow fever virus (YFV), Japanese encephalitis virus (JEV), and tick-borne encephalitis virus (TBEV) pose global threat to public health (Gubler *et al.*, 2007). Although licensed human vaccines are available for YFV, JEV and TBEV, none has been developed for DENV and WNV. Vaccine development for dengue is complicated by the existence of four antigenically distinct serotypes and the phenomenon of antibody-dependent enhancement (ADE). Currently, there is no clinically approved antiviral therapy for flavivirus infections. The development of flavivirus treatment is urgently needed. This chapter focuses on methodologies for screening inhibitors against flavivirus infection, using DENV as an example.

Targeting virus entry

Flaviviruses enter cells by fusing their membranes with the host membrane. The fusion process is mediated by the envelope protein, E, in a pH-dependent manner (Heinz and Allison, 2000). The crystal structures of the E proteins from TBEV, WNV, and DENV show that the proteins consist of three distinct domains: the central β-barrel domain I, the elongated fingerlike domain II, and the immunoglobin-like domain III (Modis *et al.*, 2003; Nybakken *et al.*, 2006; Rey *et al.*, 1995; Kanai *et al.*, 2006; Modis *et al.*, 2005). The junctions between the three domains are flexible which allows hinge-like motions important for membrane fusion and virus maturation (Bressanelli *et al.*, 2004; Modis *et al.*, 2004; Zhang *et al.*, 2004). In addition, a hydrophobic pocket between domains I and II occupied by *n*-octyl-β-D-glucoside (β-OG) was identified in the DENV E protein structure (Fig. 12.1). Similar three-domain organization can be found in the E1 protein of alphavirus. The flavivirus and alphavirus represent class II fusion, which is different from class I fusion represented by orthomyxo-, paramyxo-, retro-, filo-, corona-, and retroviruses (Harrison, 2005; Stiasny and Heinz, 2006). Inhibition of enveloped viruses at the stage of viral

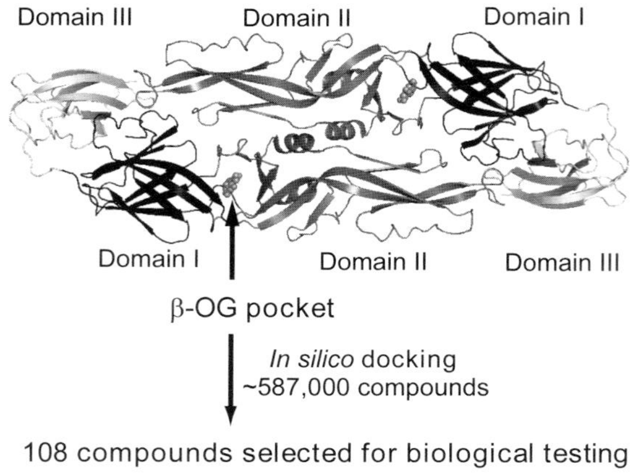

Figure 12.1 High-throughput docking of E protein (PDB entry 1OKE). Domains I, II, and III of the E protein are indicated. The *n*-octyl-β-d-glucoside (β-OG) is displayed as grey spheres. We performed an *in silico* docking of 587,000 compounds and selected 108 compounds for further biological testing, leading to the identification of an entry inhibitor reported in (Wang *et al.*, 2009).

entry provides a route for therapeutic intervention. Proof-of-concept was achieved through the clinical usage of a couple of HIV entry inhibitors, for example, Maraviroc, a small-molecule CCR5 antagonist, and enfuvirtide, an oligopeptide fusion inhibitor (Dorr *et al.*, 2005; Kilby *et al.*, 1998).

Current efforts to develop antivirals against flavivirus entry are focused on the E protein as no primary cellular receptor(s) for virus attachment and internalization has been identified. In the absence of a target-based functional assay, computational HTS of chemical libraries against dengue E protein β-OG pocket has been applied (Kampmann *et al.*, 2009; Poh *et al.*, 2009; Wang *et al.*, 2009; Li *et al.*, 2008). A few sets of low-molecular-weight compounds were selected through virtual screens. Biological testing using various cell-based assays further identified molecules that have antiviral and fusion-inhibitory activities. However, the compounds disclosed so far are of only weak potency, or have unfavourable physicochemical properties. New structural information on the mechanism by which flavivirus bind to their receptors will provide new opportunities for identifying novel classes of entry inhibitors.

Targeting viral enzymes

Methyltransferase (MTase)

The flavivirus genome contains a type 1 cap structure ($^{Me7}GpppA_{2'OMe}$) at its 5′ end, which is critical for mRNA stability and efficient translation (Cleaves and Dubin, 1979; Wengler and Wengler, 1981). The N-terminal MTase domain (1–296 aa) of NS5 protein has both N-7 and 2′-O methylation activities that catalyse two cap methylation reactions with N-7 methylation preceding 2′-O methylation (Ray *et al.*, 2006; Dong *et al.*, 2008). The N-7 MTase activity, but not the 2′-O MTase activity, has been reported to be critical for viral replication of WNV and DENV, therefore representing a potential antiviral target (Dong *et al.*, 2010). To identify suitable candidate compounds that can inhibit NS5 MTase, two HTS compatible assay systems have been developed, a streptavidin bead-based scintillation proximity assay (SPA), and a non-radioactive fluorescence polarization immuno competition assay (FPIA). One important question pertinent to the use of MTase as an antiviral target is viral specificity, because host MTase-mediated methylations of mRNA cap and protein are essential for cellular functions. The crystal structures of flavivirus MTases and the residues identified for N-7 methylation will undoubtedly aid in the rational design of antiviral agents.

SPA assay

An HTS scintillation proximity assay (SPA) has been developed using bacterial expressed dengue NS5 MTase domain (Kroschewski *et al.*, 2008; Lim *et al.*, 2008). The principle of the assay is illustrated in Fig. 12.2A. The capped biotinylated RNA template is mixed with NS5 MTase protein and [^{3}H-methyl]-AdoMet (S-Adenosyl-L-methionine) as the methyl donor. Incorporation of radiolabelled ^{3}H-methyl group into the capped biotinylated RNA substrate is catalysed by MTase. The reaction is terminated by adding 100-fold molar excess cold AdoMet in the stop buffer containing streptavidin SPA beads. Binding of the biotinylated RNA template to the streptavidin SPA beads brings the radiolabel into close proximity to the scintillant embedded in the beads, which triggers a signal that is measurable with a scintillation counter. The assay has a very low background as tritium quickly loses its energy when travelling through the liquid medium and any unbound tritium would have little effect on the final reading.

In our analysis of N-7 methylation, 100 nM enzyme, and 0.56 μM GTP capped RNA substrate (representing the first 110 nt of DENV genome) is pre-incubated with test compounds at room temperature for 20 min in a 96-well microplate. The assay buffer contains 50 mM Tris/HCl pH 7.5 and 20 mM NaCl. Similarly, for measuring the inhibition of 2′-*O* methylation, m7GTP capped 110-nt RNA is used as the RNA substrate, and the assay buffer contains 50 mM Tris/HCl pH 7.5, 10 mM KCl, and 2 mM $MgCl_2$. Enzymatic reaction is initiated by the addition of 0.64 μM [^{3}H-methyl]-AdoMet and allowed to proceed for 15 or 30 min for N-7 and 2′-*O* methylation, respectively. To terminate the reaction, 2X stop solution (100 mM Tris pH 7.0, 50 mM EDTA, 300 mM NaCl, 100-fold excess cold AdoMet, and 100 μg SPA beads) is added and the plate is read in a scintillation counter. The dissociation constants (K_i) are determined by fitting the calculated initial velocities (average counts/min) to a non-linear regression curve fit using GraphPad Prism® software. A typical inhibition curve using S-adenosyl homocysteine (SAHC), the enzymatic product, is shown in Fig. 12.2B.

FPIA assay

A non-radioactive assay based on competitive fluorescence polarization (FP) that is suitable for large scale compound testing has been developed, which can complement the aforementioned SPA assay. FP takes advantage of the inverse relationship between rotational speed of fluorescent molecules in solution and the size of the labelled molecule or complex (Jameson and Croney, 2003). Briefly, a fluorophore reaches the excited state upon absorption of a photon. The duration of the excited state of a fluorophore depends on the identity of the particular fluorophore. During the excited state, the molecule rotates or spins in solution, depending on its size. Small molecules rotate quickly and release light at a different angle than the absorbed light. Larger molecules rotate much slower, and the released light becomes polarized. This polarization is quantified using the ratio of fluorescence measured in the parallel and perpendicular planes relative to the excited plane. The value of this ratio is unitless and is referred to as polarization value (P) or commonly, is multiplied by 1000 to give millipolarization (mP) values. In general high mP values indicate that the fluorophore has a big apparent molecular weight, whereas low mP values indicate a small apparent molecular mass. Since FP is a radiometric fluorescence technique, it is less subject to variability than other non-ratiometric assays.

Competitive FP immunoassay (FPIA) was first described by Dandliker and co-workers that incorporates the use of an antibody to increase the apparent size of a small analyte analogue (Dandliker *et al.*, 1973). In principle, the concentration of unlabelled analytes (such as SAHC, the product of the methyltransferase reaction), can be determined by their ability to compete with fluorescently labelled analogues, commonly called a tracer, for antibody. We use either fluorescein- or Alexa633-conjugated SAHC as the tracer. In the absence of the enzymatic product (SAHC), all tracers are bound to the antibody, dramatically increasing its apparent mass and decelerating its rotational motion. This reduced molecular rotation results in an increase in FP indicated by

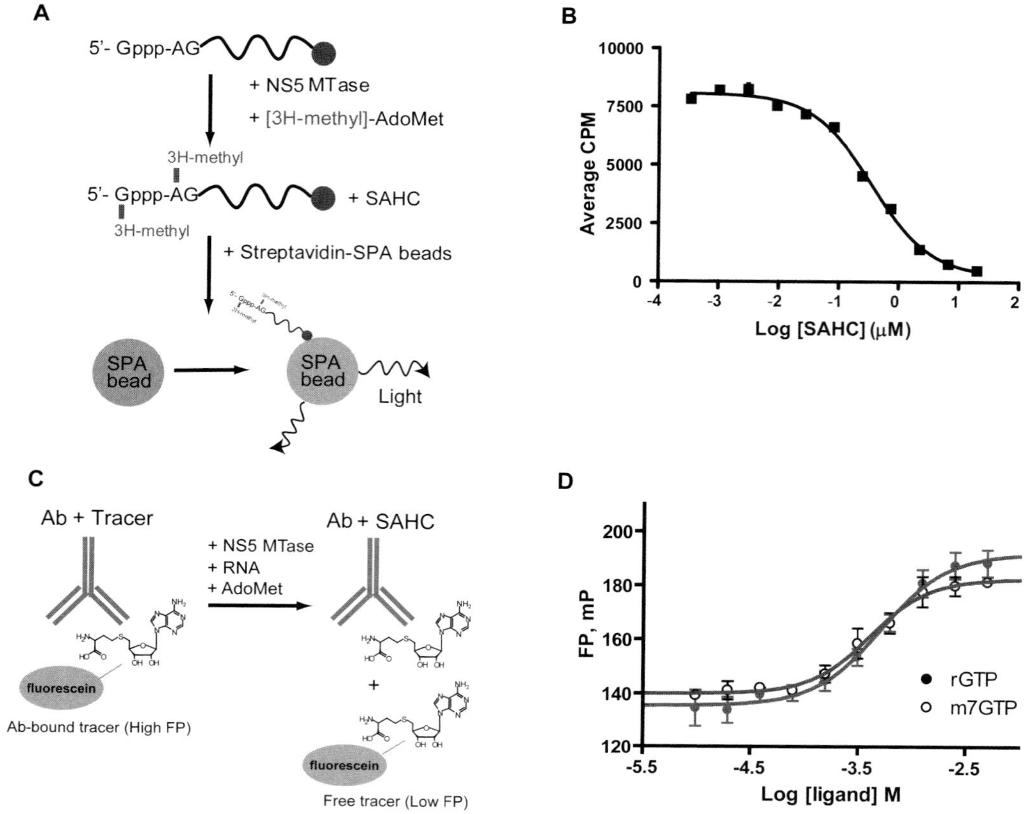

Figure 12.2 (A) Scintillation proximity assay (SPA) for screening of NS5 methyltransferase (MTase) inhibitors. The capped biotinylated RNA template is mixed with NS5 MTase protein and [^{3}H-methyl]-AdoMet (*S*-adenosyl-L-methionine) as the methyl donor. Incorporation of radiolabeled ^{3}H-methyl group into the capped biotinylated RNA substrate by the enzyme is measured using streptavidin SPA beads. The SPA beads are induced to produce light because of the binding of the [^{3}H]-biotinylated RNA template. (B) *S*-adenosyl homocysteine (SAHC) dose–response curve with SPA assay. (C) Competitive fluorescence polarization immunoassay (FPIA) for screening of NS5 methyltransferase (MTase) inhibitors. The capped RNA template is incubated with NS5 MTase protein and AdoMet. SAHC is generated as the by-product of the enzyme reaction. The SAHC competes with fluorescein-conjugated SAHC for antibody, resulting in a decrease in fluorescence polarization. (D) rGTP and m7GTP dose–response curve with FPIA assay.

high mP values. When SAHC is produced by the MTase reaction, it competes with the tracer for antibody binding, thereby decreasing polarization of the tracer. In this way, the activity of DENV MTase can be determined (Fig. 12.2C).

In our analysis of 2′-*O* methylation, 50 nM enzyme is first mixed with 40 nM RNA in 50 mM Tris/HCl, pH 7, 10 mM KCl, and 0.05% CHAPS. Reaction is then initiated by adding AdoMet and $MgCl_2$ to final concentrations of 0.64 μM and 2 mM, respectively, to give a total volume of 25 μl. Methylation is allowed to proceed for 10 min and stopped with 6 μl of 5× stop buffer containing 16× diluted fluorescein-SAHC tracer and 16× anti-SAHC antibody. The mixture is then shaken at room temperature for 1 h and read on a plate reader.

One drawback of using the fluorescein-SAHC tracer for screening is that some classes of compounds have similar excitation and emission wavelengths, thereby increasing the likelihood of false positive hits. Fluorescence spectroscopic profiling of the 71,391-member library of the National Institutes of Health (NIH) has shown that library compounds with overlapping spectra with fluorophores having red shifts beyond 550 nM are rare (Simeonov *et al.*, 2008). On the basis of this finding, we synthesized an Alexa633-conjugated SAHC tracer. All parameters remained the same as described above except that the final concentration of AdoMet used is 0.25 or 0.5 μM and the reaction time is increased to 20 min. The 5× stop buffer contains 5 nM Alexa633-SAHC tracer and 16× diluted antibody. The MTase FPIA assay is validated using GTP analogues, rGTP, and m7-GTP as inhibitors (Fig. 12.2D).

Protease

The flavivirus genome is translated into a single polyprotein that is cleaved by host proteases and a two-component viral protease, NS2B/NS3. Proper processing of the polyprotein precursor is essential for virus replication and maturation. Two main approaches have been taken to identify protease inhibitors, i.e. by HTS of chemical libraries, and by rationally designing peptide inhibitors.

A number of groups have reported of using internally quenched fluorogenic peptide substrates for characterization of protease activity (Lescar *et al.*, 2008). Cleavage of the fluorophore such as 7-amino-methyl-coumarin (AMC) by the protease results in an increase in fluorescence that can be measured on a plate reader (Fig. 12.3A). The *in vitro* assay (Yusof *et al.*, 2009) was adapted and implemented for screening of combinatorial peptide substrate libraries containing more than 130,000 substrates each (Li *et al.*, 2005). The tetrapeptide benzoyl-norleucine-lysine-arginine-arginine-AMC (Bz-nle-Lys-Arg-Arg-AMC) was identified. It served as the basis for the peptidomimetic approach by appending a range of well-known serine protease-inhibiting warheads for finding more potent peptide inhibitors. Electrophilic warheads, such as boronic acids, trifluoroketone, and aldehydes, are preferred, with boronic acids having the highest affinity, exhibiting a K_i of 43 nM (Yin *et al.*, 2006). In addition, this tetrapeptide represents the optimal substrate for HTS.

To measure the inhibition of protease activity, 50 nM recombinant protein is mixed with 20 μM Bz-nle-Lys-Arg-Arg-AMC in 50 mM Tris-HCl, pH 8.5, and 2.5 mM CHAPS in the presence of test compounds. The reaction mixture is then incubated at 37°C for 30 min followed by kinetic measurement on a plate reader. A representative inhibition curve for a classic serine protease inhibitor, aprotinin, is shown in Fig. 12.3B.

Despite the structural, biophysical, and biochemical tools available today, the search for flavivirus proteases inhibitors has not been successful. This is because the active site of flavivirus protease has a more open and less pronounced pocket in comparison with that of the HIV and HCV proteases. Nevertheless, given the importance of the target for flavivirus drug discovery, efforts have to be continued to pursue all avenues of lead discovery.

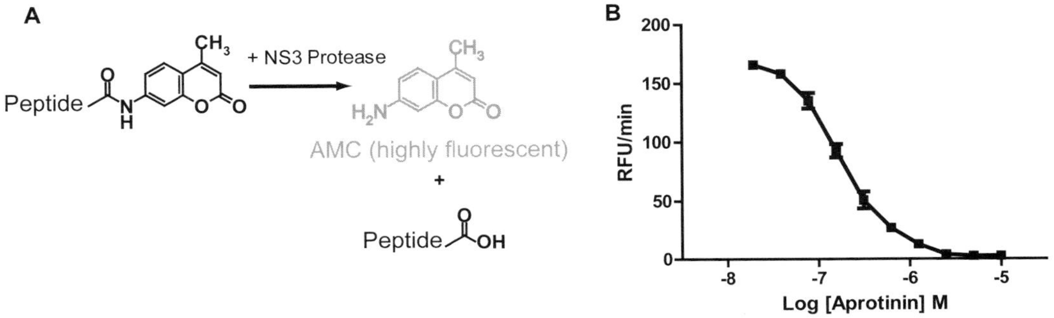

Figure 12.3 (A) Fluorescent assay for screening of NS3 protease inhibitors. An internally quenched fluorogenic peptide substrate is incubated with NS3 protease. Cleavage of the fluorophore (7-amino-methly coumarin, AMC) by the enzyme is measured by an increase in fluorescence. (B) Aprotinin dose–response curve.

Helicase/NTPase

The RNA helicase/NTPase domain of flavivirus NS3 protein is located within the C-terminal region, and it belongs to the superfamily 2 (SF2) RNA helicase (Caruthers and McKay, 2002; Gorbalenya *et al.*, 1989). NS3 helicase catalyses strand separation in an ATP-dependent manner. This unwinding of dsRNA duplex is essential for an efficient viral genomic RNA synthesis by the NS5 RNA-dependent RNA polymerase. Mutational analyses of residues within and outside the helicase/NTPase motifs show that the absence of helicase activity is lethal for virus replication (Matusan *et al.*, 2001). In this context, NS3 helicase activity appears to be an attractive target for the inhibition of viral replication. Both a structure based approach using fragment-based screen, and a screening based approach have been pursued for identifying helicase inhibitors. Similar to MTases, host helicases are involved in a variety of cellular processes involving nucleic acid metabolism. Because of the conserved nature of helicase motor function, inhibitors targeting helicases can be inhibitors of host protein, resulting in toxicity. To effectively target viral helicase, it is important to select an inhibitor that is virus specific.

ATPase assay

During the reaction catalysed by NS3 helicase, the β-γ phosphate bond of an ATP molecule is hydrolysed, releasing one inorganic phosphate (P_i). Therefore, the simplest helicase assay involves measuring the ATPase activity of NS3 helicase by monitoring inorganic phosphate production in the enzymatic reaction. To estimate the amount of inorganic phosphate released during catalysis, we used the Malachite Green reagent. In the absence of phosphate this compound has a yellow colour and absorbs only slightly at 630 nm. In the presence of inorganic phosphate, its colour changes to green and its absorbance at 630 nm increases significantly (Lanzetta *et al.*, 1979). Therefore the addition of Malachite Green reagent to a solution containing the NS3 helicase protein and ATP allows the measurement of the catalytic activity of this enzyme (Fig. 12.4A).

In our optimized assay carrying out in a 96-well microplate, 170 ng DENV NS3 helicase in 50 mM Tris-HCl pH 7.5, 2 mM $MgCl_2$, 1.5 mM dithiothreitol, 0.05% Tween 20, and 0.25 ng/μl BSA is incubated with test compounds at 37°C for 5 min. ATP (75 μM) is then added to the reaction mixture and incubated for another 10

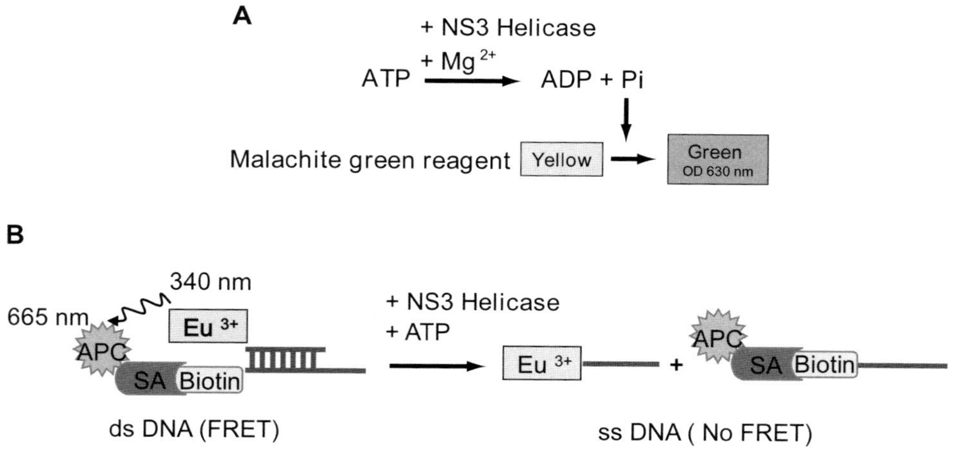

Figure 12.4 (A) ATPase assay for screening of NS3 helicase inhibitors. ATPase activity of NS3 helicase is measured by the production of inorganic phosphate using Malachite Green reagent. (B) Time-resolved fluorescence resonance energy transfer (TR-FRET) assay for screening of NS3 helicase inhibitors. The TruPoint substrate is incubated with NS3 helicase. Unwinding activity of the enzyme is measured by a decrease in FRET.

min at 37°C. At the end of incubation, 200 μl of Malachite Green reagent is added, and the plate is immediately read at the optical density at 630 nM (S. Dubaele and A. Sampath, unpublished work).

Unwinding assay (TR-FRET assay)

We have developed a time-resolved fluorescence resonance energy transfer (TR-FRET)-based helicase unwinding assay, adapted from the commercially available homogeneous TruPoint platform from Perkin Elmer (Boettcher *et al.*, unpublished work). The TruPoint helicase substrate contains two complementary oligonucleotide strands, one labelled with the energy donor, europium (Eu^{3+}) chelate, and the other labelled with biotin which after addition of streptavidin–allophycocyanin (SA-APC) forms the FRET acceptor. In the absence of helicase or inhibitors, the donor strand is in close proximity with the acceptor strand resulting in FRET that can be measured with a plate reader. While in the presence of helicase, unwinding activity of the enzyme causes separation of the pre-annealed donor/acceptor strand duplex, thereby decreasing FRET (Fig. 12.4B).

The TR-FRET unwinding assay was chosen for an HTS of the Novartis archive. The result of the HTS campaign was unfortunately not encouraging. We were only able to identify a number of promiscuous inhibitors which either formed aggregates or did not show useful structure–activity relationships (SAR). This outcome was not a complete surprise, as helicases are difficult targets for drug discovery. Helicases are complex proteins and much more work needs to be done to understand the mechanism of the reaction catalysed by the enzymes. Any resulting knowledge will undoubtedly help lead discovery for helicase inhibitors.

RNA-dependent RNA polymerase

The RNA-dependent RNA polymerase (RdRp) domain of flavivirus NS5 protein is located within its C-terminal region (Koonin, 1991; Poch *et al.*, 1989; Rice *et al.*, 1985; Sumiyoshi *et al.*, 1987). The NS5 polymerase is the most conserved flaviviral protein, with more than 75% sequence identity across dengue virus serotypes. It is essential for viral replication in all *Flaviviridae*. Mutations within the GDD motif of this protein result in the formation of non-viable viruses. Unlike helicases and MTases, human host cells are devoid of RdRp making it the most useful and favourable target class for antiviral drug discovery. Numerous compounds that are currently in clinical use for antiviral therapy are polymerase inhibitors, which can be categorized into nucleoside (NI) and non-nucleoside (NNI) inhibitors. NIs are prodrugs that have to be phosphorylated to their active 5′-triphosphate forms to act as competitive inhibitors with the natural substrates (Carroll and Olsen, 2006). While NNIs are allosteric inhibitors binding to a non-substrate binding site of the polymerase. The identification of NNIs is usually achieved through HTS of compound libraries.

SPA assay

The HTS scintillation proximity assay (SPA) was developed using purified recombinant dengue NS5 protein containing the RdRp domain (Yap *et al.*, 2007). The principle of the assay is illustrated in Fig. 12.5A. A homomeric template is annealed with the corresponding primer pair conjugated with biotin. After the polymerase reaction, the incorporated radioactive nucleotide is brought into proximity with the scintillating beads via the attached biotin. The rate of incorporation can then be easily measured by a scintillation counter. In this format, no washing step is necessary and the process is highly automated.

In our analysis, oligo $(G)_{20}$ primer containing 5′-biotin is prepared by dissolving in DEPC-treated buffer containing 50 mM Tris pH 8.0 and 150 mM NaCl. Poly(C) is dissolved in the same buffer and desalted in PD10 columns to remove excess salt. The protein is then pre-incubated with the annealed RNA template at room temperature for 1 h before adding the solution containing radioactive GTP to initiate the reaction. The final reaction mixture contains 50 mM Tris-HCl, pH 7.0, 10 mM KCl, 2 mM $MgCl_2$, 2 mM $MnCl_2$, 1 mM β-mercaptoethanol, 0.05% CHAPS, 0.25 μg poly(C)/oligo$(G)_{20}$, 0.4 μM GTP, 0.5 μCi $[^3H]$GTP (6.5 Ci/mmol), and 50 nM of the purified NS5 protein in a final volume of 50 μl. The reaction is further incubated

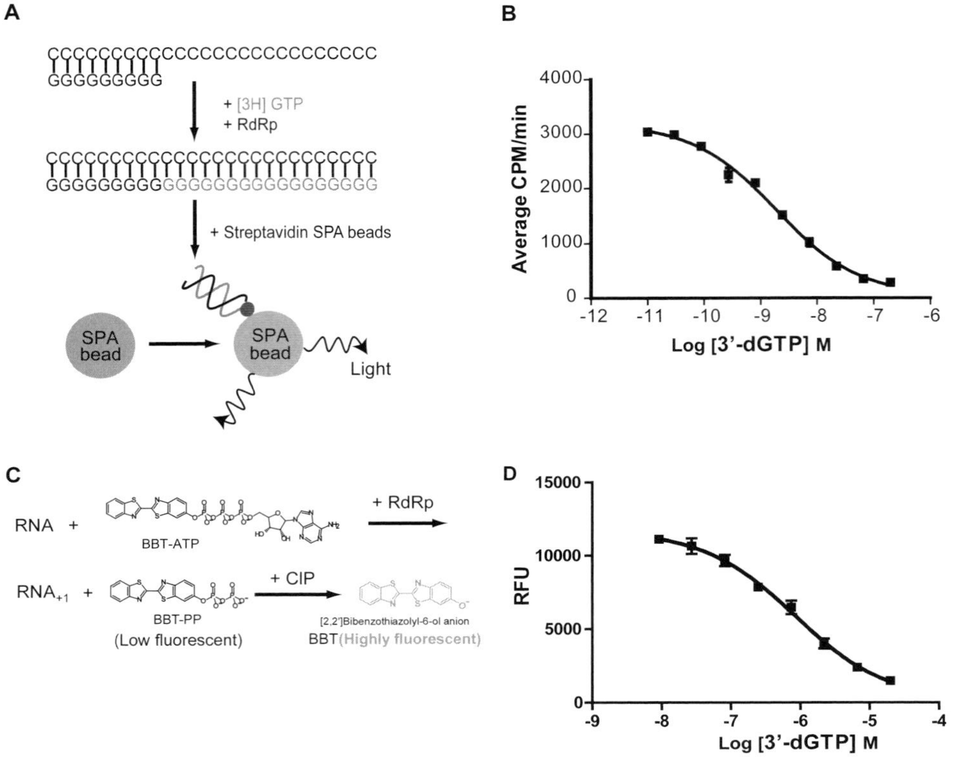

Figure 12.5 (A) Scintillation proximity assay (SPA) for screening of NS5 polymerase (RdRp) inhibitors. A poly-C template together with a biotinylated oligo G_{20} primer is incubated with RdRp in the presence of [^{3}H] GTP. Incorporation of radioactivity into the primer strand by the enzyme is measured using streptavidin SPA beads. The SPA beads are induced to produce light because of the binding of the [^{3}H]biotinylated primer. (B) 3′-dGTP dose–response curve with SPA assay. (C) Coupled fluorescent assay for screening of NS5 polymerase (RdRp) inhibitors. RNA template is incubated with RdRp in the presence of BBT-conjugated ATP. BBT pyrophosphate is generated as the by-product of the enzyme reaction. The BBT pyrophosphate concentration is determined through treatment with calf intestinal phosphatase (CIP), yielding the highly fluorescent BBT.

at room temperature for 1 h before terminating with equal volume of stop solution containing 25 mM Tris-HCl, pH 7.5, 75 mM NaCl, 20 mM EDTA, and 0.15 mg of streptavidin-coated beads. The plates are further agitated for 1 h and the level of the radioactive GTP incorporation is quantified using MicroBeta counter (Perkin Elmer). For compound screening, the compound is added at the pre-incubation step together with NS5 and RNA before initiation of reaction. IC_{50} values are determined by incubating the compounds at various concentrations with the reaction mixture followed by calculation of dose response. A representative inhibition curve for dGTP is shown in Fig. 12.5B.

Coupled fluorescent assay

The dengue RdRp activity can be measured in a coupled assay using a modified nucleotide triphosphate as the substrate in a non-radioactive format (Niyomrattanakit *et al.*, 2011). In this case, a fluorescent tag is conjugated to the gamma phosphate of the substrate nucleotide triphosphate. One example of such fluorescent tag is BBT (2′-[2-benzthiazoyl]-6′-hydroxy-benthiazole). The modified nucleotide triphosphate has very low fluorescence when intact. However, when used as a substrate for RdRp reaction, the nucleotide monophosphate is incorporated into the growing RNA chain while BBT pyrophosphate is generated as a by-product of the reaction. Under phosphatase treatment such as calf intestinal phosphatase (CIP) in alkaline condition, the fluorescent molecule is released (Fig. 12.5C).

A typical assay using the above mentioned substrate is carried out in two separate steps, one for RdRp reaction and the other for generation of fluorescent molecule. To set up the RdRp reaction, we premixed 20 nM recombinant NS5 with 50 nM RNA template in a solution containing 50 mM Tris-Cl, pH 7.0, 1 mM $MnCl_2$, and 0.01% v/v Triton X-100. The fluorescent nucleotide triphosphate substrate (2 mM) is then added to initiate the reaction. After the RdRp reaction is completed, the alkaline phosphate buffer is added to terminate the polymerase reaction and at the same time to initiate the phosphatase reaction. The final concentration for the reaction contains 10 nM CIP in 80 mM NaCl, 10 mM $MgCl_2$ and 600 mM deoxyethanolamine (DEA) at pH 10.0. The reaction mixture is incubated further until all the BBT-pyrophosphate is converted to BBT quantified at 422/566 nM. For measuring compound inhibition, the compounds are 3-fold serially diluted and incubated with the protein and substrate prior to initiation of reaction. The inhibition curve and IC_{50} is calculated by GraphPad Prism. A representative inhibition curve for 3′-dGTP is shown in Fig. 12.5D.

Cell-based viral replication assay

A number of cell-based assays have been developed and implemented for screening antiviral compounds. Compared with the target-based approach described above, viral replication-based assays cover multiple steps and targets (cellular and viral factors) involved in a virus life cycle.

Viral replicon assay system

We and others developed a luciferase-based DENV subgenomic replicon system (Ng *et al.*, 2007; Puig-Basagoiti *et al.*, 2006). The system encodes a cDNA for renilla luciferase in place of the viral structural proteins and is driven by the viral 5′ UTR (Fig. 12.6A). The DENV replicon contains all the viral non-structural (NS) proteins and RNA elements necessary for viral RNA replication. Following the transfection into cells, the viral RNA undergoes translation to generate polyprotein encompassing NS1-NS5. Cleavage by NS2B/NS3 protease leads to production of individual proteins and formation of replication complex. Antiviral agents that target any steps involved in viral RNA translation and replication will decrease the luciferase signal. This loss-of-signal end point usually results in a high number of false-positive hits, due to molecules inhibiting the activity of luciferase reporter or due to compound-mediated cytotoxicity. Therefore, appropriate counter screens, such as a cytotoxicity assay, and a cell-based luciferase screen, have to be run in parallel with the replicon assay to identify non-toxic antiviral inhibitors.

In our assay system, A549 DENV-2 replicon cells are seeded at a density of 10,000 cells per well in a 96-well microplate. After incubating at 37°C with 5% CO_2 overnight, cells are treated with compounds. Luciferase activities are measured after 48 h incubation using Promega's EndurRen live cell substrate. Following luciferase activity measurement, Promega's CellTiter-Glo reagent is added to each well to determine the cytotoxicity of the compounds. The ability of compounds to inhibit viral RNA replication is calculated as: % inhibition = $[1-(RLU_{cpd\ well}-RLU_{A549\ control})/(RLU_{replicon\ control}-RLU_{A549\ control})]\times 100$, where $RLU_{cpd\ well}$ is the relative luminescence unit measured for the compound-treated replicon cells, $RLU_{A549\ control}$ is the relative luminescence unit measured for the A549 cells and $RLU_{replicon\ control}$ is the relative luminescence unit measured for the untreated replicon cells. Cell viability is calculated as: % cell viability = $(RLU_{cpd\ well}/RLU_{replicon\ control})\times 100$. This homogeneous assay was miniaturized and implemented for an HTS of the Novartis archive. Fifty per cent inhibition was chosen arbitrarily as the cut-off for selecting antagonists to generate a hit rate of 1.4%, among which about 7000 hits were selected for dose–response testing. Using a key and fingerprint algorithm, eight scaffolds were chosen for further characterization in a secondary assay, the CFI assay, as described below (MedChem Studio™, Simulations Plus, Inc., Lancaster, CA, USA).

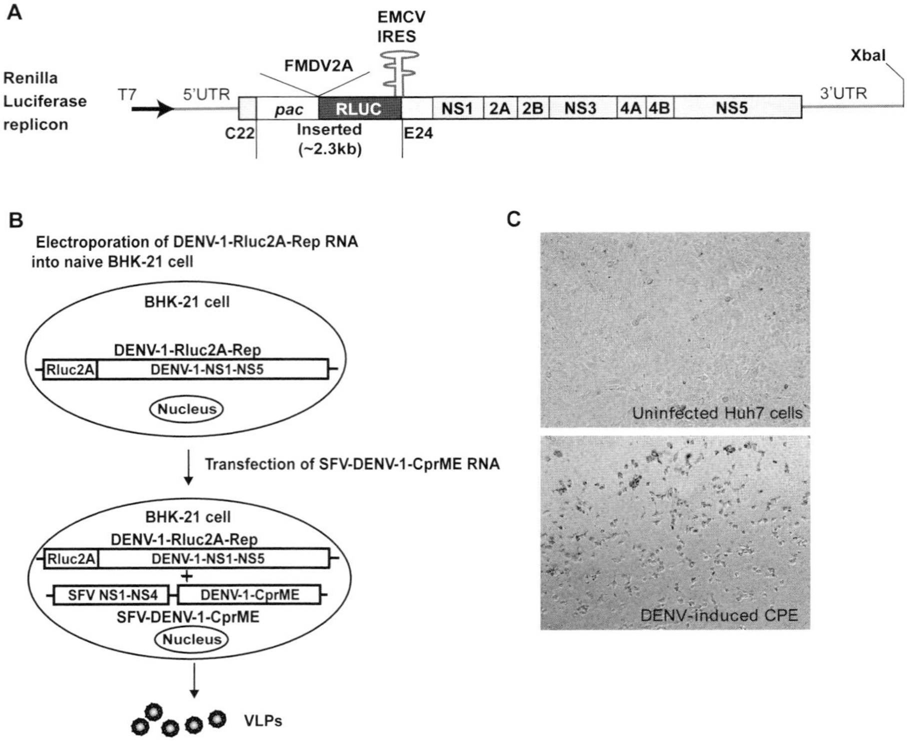

Figure 12.6 (A) The schematic diagram of DENV-2 Rluc replicon (Ng *et al.*, 2007). (B) Production of dengue virus-like particles (VLP) (Qing *et al.*, 2010). BHK-21 cells were sequentially electroporated with DENV replicon RNA, and RNA of DENV structural genes expressed by a Semliki Forest virus (SFV) replicon. (C) DENV-induced cytopathic effects (CPE) on Huh7 cells. Phase-contrast photomicrographs are shown.

Virus-like particle assay system

Virus-like particles (VLP) are formed by packaging subgenomic replicons with the structural proteins. VLP is therefore capable of entry, translation and replication (Ansarah-Sobrinho *et al.*, 2008; Harvey *et al.*, 2004; Puig-Basagoiti *et al.*, 2006). VLP of several different flaviviruses have been constructed and employed in the studies of different aspects of the flavivirus life cycle, including viral entry, replication, assembly, and secretion. To facilitate the dengue drug discovery, we have developed a DENV VLP infection assay system for screening antiviral compounds (Qing *et al.*, 2010). VLPs are generated by trans-supply of viral structural proteins (CprME) to package a dengue subgenomic replicon RNA containing a luciferase reporter described above (Fig. 12.6B). Infection of target cells (e.g. Vero cells) with VLPs leads to expression of the luciferase when the replicon replicates inside cells. Because there is no replication and expression of structural proteins, VLPs initiate only a single round of infection in target cells that reduced the biohazard level.

In our analysis, Vero cells are plated at 40,000 cells per well in a 96-well microplate. The next day, 50 μl of VLP [3×10^5 foci-forming unit (FFU)/ml] is added to the cells together with test compounds. At 48 h post infection, luciferase activity is measured with EnduRen (30 μM). The ability of compounds to inhibit VLP infection is calculated as: % inhibition = $[1-(RLU_{cpd\,well}-RLU_{cell\,control})/(RLU_{VLP\ control}-RLU_{cell\ control})]\times 100$, where $RLU_{cpd\,well}$ is the relative luminescence unit measured for the compound treated and VLP-infected cells; $RLU_{cell\ control}$ is the relative luminescence unit measured for the uninfected Vero cells; and $RLU_{VLP\ control}$ is the relative luminescence unit measured for the untreated and VLP-infected cells.

Cytopathic effects inhibition assay

Many viral infections cause cytopathic effect (CPE) on target cells (Fig. 12.6C). Inhibition of CPE represents a conventional approach for antiviral screening (Green *et al.*, 2008). The CPE assay for DENV involves infection of susceptible cells with virus in the presence of test compounds. Antiviral compounds will prevent target cells from cell death caused by DENV-induced CPE, and thereby increase cell viability. This gain-of-signal endpoint reduces false positive hits that are cytotoxic or interfere with the assay readout, but it is not a direct measure of virus replication.

In our analysis, Huh-7 cells are seeded at a density of 5000 cells per well in a 96-well microplate one day prior to infection and addition of test compounds. The cells are infected with DENV2 New Guinea C strain, at a multiplicity of infection (MOI) of 1, on the following day. After 3 days of incubation at 37°C with 5% CO_2, Promega's CellTiter-Glo® reagent is added to the infected cells to measure cell viability. The ability of compounds to inhibit DENV-induced cell killing is calculated as % inhibition = $(RLU_{cpd\ well} - RLU_{virus\ control})/(RLU_{cell\ control} - RLU_{virus\ control}) \times 100$, where $RLU_{cpd\ well}$ is the relative luminescence unit measured for the compound treated virus-infected cells; $RLU_{virus\ control}$ is the relative luminescence unit measured for the untreated virus-infected cells; and $RLU_{cell\ control}$ is the relative luminescence unit measured for the untreated and uninfected cells.

The CPE assay was chosen for an HTS of the Novartis archive. From the primary screen, about 2500 compounds were selected for dose–response testing. After hits triaging and cheminformatic analysis, seven scaffolds were identified for further characterization.

Cell-based flavivirus immunodetection (CFI) assay

All the cell-based assays described above are of homogenous format and amenable to HTS, however, none of them directly measures virus replication. To overcome this drawback, we developed a cell-based flavivirus immunodetection (CFI) assay (Wang *et al.*, 2009). The assay is based on quantitative immunodetection of the production of DENV E protein in virus-infected cells. A similar in-situ cellular ELISA approach has been reported for testing antiviral agents in many other viruses (Berkowitz and Levin, 1985; Tatarowicz *et al.*, 1991). As an immunodetection assay, the CFI assay is not a simple 'add–mix–measure' format. On the contrary, several washing steps are required that limits its adaptation to an HTS format. Nevertheless, CFI assay is a good secondary assay that has many advantages over the traditional plaque reduction assay (PRA). It is less labour intensive, requires a much shorter assay turnaround time (3 days from the time of virus inoculation versus 6–7 days in PRA), and the data collection is performed on a plate reader avoiding subjective input by manual plaque counting.

In our assay, A549 cells are seeded at a density of 20,000 cells per well in a 96-well microplate, and allow attaching overnight at 37°C with 5% CO_2. On the following day, test compounds are mixed with viruses (NGC strain; MOI=0.3) and incubated with the cells for 1 h at 37°C in 5% CO_2. After replacing the virus inoculum with fresh medium containing test compounds, the cells are incubate at 37°C in 5% CO_2 for 48 h. At the end of incubation, the cells are fixed with methanol for immunostaining. The mAb 4G2 is used as the primary antibody, horseradish peroxidase (HRP)-conjugated anti-mouse IgG is used as the secondary antibody, and 3,3′,5,5′-tetramethylbenzidine (TMB) is used as the substrate. The ability of compounds to inhibit DENV infection is calculated as: % inhibition = $[1 - (A450_{cpd\ well} - A450_{cell\ control})/(A450_{virus\ control} - A450_{cell\ control})] \times 100$, where $A450_{cpd\ well}$ is the absorbance measured at 450 nm for the compound-treated, virus-infected cells, $A450_{cell\ control}$ is the absorbance for the uninfected A549 cells and $A450_{virus\ control}$ is the absorbance for the untreated and virus-infected cells.

Conclusions

There is a high medical need for treatments for flavivirus infection. A number of approaches have been taken to develop anti-flavivirus therapy. HTS compatible target-based assays have been developed and implemented during the past few years. The atomic structures of NS3 (protease domain, helicase domain, and full-length

NS3), MTase, and RdRp proteins, together with computational advances in docking algorithms have provided solid foundation for rational drug design. In spite of having all these tools in hand, there is little success in identifying drug-like leads. The hits identified from the screens were either promiscuous inhibitors or cell impermeable. On the contrary, phenotypic cell-based screening can result in a rich source of lead-like chemical starting points. However, the targets and mechanism-of-action of the hits are not known, which makes establishing clear structure–activity relationship very challenging.

Based on the knowledge and experience gained from over two decades of HCV drug discovery efforts, protease and polymerase enzymes appear to be the most promising and druggable targets. The most advanced two compounds, telapavir and boceprevir, are NS3/4A protease inhibitors. They are currently under phase III evaluation and are expected to be approved in 2011/2012 (Lange, 2010). Furthermore, phenotypic screen using a HCV replicon assay has delivered an NS5A inhibitor, the first-in-class compound that showed proof-of-concept in the clinic for tackling HCV infection (Gao *et al.*, 2010). Comparing with the drug discovery history of HCV and HIV, drug development for flavivirus is still in its infancy. With patients and physician waiting for therapies, efforts have to be continued to pursue all avenues to achieve success.

References

Ansarah-Sobrinho, C., Nelson, S., Jost, C.A., Whitehead, S.S., and Pierson, T.C. (2008). Temperature-dependent production of pseudoinfectious dengue reporter virus particles by complementation. Virology *381*, 67–74.

Berkowitz, F.E. and Levin, M.J. (1985). Use of an enzyme-linked immunosorbent assay performed directly on fixed infected cell monolayers for evaluating drugs against varicella-zoster virus. Antimicrob. Agents Chemother. *28*, 207–210.

Bressanelli, S., Stiasny, K., Allison, S.L., Stura, E.A., Duquerroy, S., Lescar, J., Heinz, F.X., and Rey, F.A. (2004). Structure of a flavivirus envelope glycoprotein in its low-pH-induced membrane fusion conformation. EMBO J. *23*, 728–738.

Carroll, S.S. and Olsen, D.B. (2006). Nucleoside analog inhibitors of hepatitis C virus replication. Infect. Disord. Drug Targets. *6*, 17–29.

Caruthers, J.M. and McKay, D.B. (2002). Helicase structure and mechanism. Curr. Opin. Struct. Biol *12*, 123–133.

Cleaves, G.R. and Dubin, D.T. (1979). Methylation status of intracellular dengue type 2 40 S RNA. Virology *96*, 159–165.

Dandliker, W.B., Kelly, R.J., Dandliker, J., Farquahar, J., and Levin, J. (1973). Fluorescence polarization immunoassay. Theory and experimental method. Immunochemistry. *10*, 219–227.

Dong, H., Chang, D.C., Xie, X., Toh, Y.X., Chung, K.Y., Zou, G., Lescar J., Lim, S.P., and Shi P.-Y. (2010). Biochemical and genetic characterization of dengue virus methyltransferase. Virology *408*, 138–145.

Dong, H., Zhang, B., and Shi, P.Y. (2008). Flavivirus methyltransferase: A novel antiviral target. Antiviral Research *80*, 1–10.

Dorr, P., Westby, M., Dobbs, S., Griffin, P., Irvine, B., Macartney, M., Mori, J., Rickett, G., Smith-Burchnell, C., Napier, C., Webster, R., Armour, D., Price, D., Stammen, B., Wood, A., and Perros, M. (2005). Maraviroc (UK-427, 857), a potent, orally bioavailable, and selective small-molecule inhibitor of chemokine receptor CCR5 with broad-spectrum anti-human immunodeficiency virus type 1 activity. Antimicrob. Agents Chemother. *49*, 4721–4732.

Gao, M., Nettles, R.E., Belema, M., Snyder, L.B., Nguyen, V.N., Fridell, R.A., Serrano-Wu, M.H., Langley, D.R., Sun, J.H., O'Boyle II, D.R., Lemm, J.A., Wang, C., Knipe, J.O., Chien, C., Colonno, R.J., Grasela, D.M., Meanwell, N.A., and Hamann, L.G. (2010). Chemical genetics strategy identifies an HCV NS5A inhibitor with a potent clinical effect. Nature *465*, 96–100.

Gorbalenya, A.E., Koonin, E.V., Donchenko, A.P., and Blinov, V.M. (1989). Two related superfamilies of putative helicases involved in replication, recombination, repair and expression of DNA and RNA genomes. Nucleic Acids Res. *17*, 4713–4730.

Green, N., Ott, D., Isaacs, J., and Fang, H. (2008). Cell-based assays to identify inhibitors of viral disease. Exp. Opin. Drug Discov. *3*, 671–676.

Gubler, D., Kuno, G., and Markoff, L. (2007). Flaviviruses. In Fields Virology, Knipe, D.M. and Howley, P.M., eds. (Philadelphia, USA: Lippincott Williams and Wilkins), pp. 1153–1253.

Harrison, S.C. (2005). Mechanism of membrane fusion by viral envelope proteins. Adv. Virus Res. *64*, 231–261.

Harvey, T.J., Liu, W.J., Wang, X.J., Linedale, R., Jacobs, M., Davidson, A., Le, T.T.T., Anraku, I., Suhrbier, A., Shi, P.Y., and Khromykh, A.A. (2004). Tetracycline-inducible packaging cell line for production of flavivirus replicon particles. J. Virol. *78*, 531–538.

Heinz, F.X. and Allison, S.L. (2000). Structures and mechanisms in flavivirus fusion. Adv. Virus Res. *55*, 231–269.

Jameson, D.M. and Croney, J.C. (2003). Fluorescence polarization: past, present and future. Comb. Chem High Throughput. Screen. *6*, 167–173.

Kampmann, T., Yennamalli, R., Campbell, P., Stoermer, M.J., Fairlie, D.P., Kobe, B., and Young, P.R. (2009). *In silico* screening of small molecule libraries using the dengue virus envelope E protein has identified compounds with antiviral activity against multiple flaviviruses. Antiviral Research *84*, 234–241.

Kanai, R., Kar, K., Anthony, K., Gould, L.H., Ledizet, M., Fikrig, E., Marasco, W.A., Koski, R.A., and Modis, Y. (2006). Crystal Structure of West Nile Virus Envelope Glycoprotein Reveals Viral Surface Epitopes. J. Virol. *80,* 11000–11008.

Kilby, J.M., Hopkins, S., Venetta, T.M., DiMassimo, B., Cloud, G.A., Lee, J.Y., Alldredge, L., Hunter, E., Lambert, D., Bolognesi, D., Matthews, T., Johnson, M.R., Nowak, M.A., Shaw, G.M., and Saag, M.S. (1998). Potent suppression of HIV-1 replication in humans by T-20, a peptide inhibitor of gp41-mediated virus entry. Nat. Med. *4,* 1302–1307.

Koonin, E.V. (1991). The phylogeny of RNA-dependent RNA polymerases of positive-strand RNA viruses. J. Gen. Virol. *72,* 2197–2206.

Kroschewski, H., Lim, S.P., Butcher, R.E., Yap, T.L., Lescar, J., Wright, P.J., Vasudevan, S.G., and Davidson, A.D. (2008). Mutagenesis of the dengue virus type 2 NS5 methyltransferase domain. J. Biol. Chem. *283,* 19410–19421.

Lange C.M., Sarrazin C., and Zeuzem S. (2010). Review article: specifically targeted antiviral therapy for hepatitis C – a new era in therapy. Aliment. Pharmacol. Ther. *32,* 14–28.

Lanzetta, P.A., Alvarez, L.J., Reinach, P.S., and Candia, O.A. (1979). An improved assay for nanomole amounts of inorganic phosphate. Anal. Biochem. *100,* 95–97.

Lescar, J., Luo, D., Xu, T., Sampath, A., Lim, S.P., Canard, B., and Vasudevan, S.G. (2008). Towards the design of antiviral inhibitors against flaviviruses: The case for the multifunctional NS3 protein from dengue virus as a target. Antiviral Res. *80,* 94–101.

Li, J., Lim, S.P., Beer, D., Patel, V., Wen, D., Tumanut, C., Tully, D.C., Williams, J.A., Jiricek, J., Priestle, J.P., Harris, J.L., and Vasudevan, S.G. (2005). Functional profiling of recombinant NS3 proteases from all four serotypes of dengue virus using tetrapeptide and octapeptide substrate libraries. J. Biol. Chem. *280,* 28766–28774.

Li, Z., Khaliq, M., Zhou, Z., Post, C.B., Kuhn, R.J., and Cushman, M. (2008). Design, synthesis, and biological evaluation of antiviral agents targeting flavivirus envelope proteins. J. Med. Chem. *51,* 4660–4671.

Lim, S.P., Wen, D., Yap, T.L., Yan, C.K., Lescar, J., and Vasudevan, S.G. (2008). A scintillation proximity assay for dengue virus NS5 2′-O-methyltransferase – kinetic and inhibition analyses. Antiviral Res. *80,* 360–369.

Matusan, A.E., Pryor, M.J., Davidson, A.D., and Wright, P.J. (2001). Mutagenesis of the dengue virus type 2 NS3 protein within and outside helicase motifs: effects on enzyme activity and virus replication. J. Virol. *75,* 9633–9643.

Modis, Y., Ogata, S., Clements, D., and Harrison, S.C. (2004). Structure of the dengue virus envelope protein after membrane fusion. Nature *427,* 313–319.

Modis, Y., Ogata, S., Clements, D., and Harrison, S.C. (2003). A ligand-binding pocket in the dengue virus envelope glycoprotein. Proc. Natl. Acad. Sci. U.S.A. *100,* 6986–6991.

Modis, Y., Ogata, S., Clements, D., and Harrison, S.C. (2005). Variable surface epitopes in the crystal structure of dengue virus type 3 envelope glycoprotein. J. Virol. *79,* 1223–1231.

Ng, C.Y., Gu, F., Phong, W.Y., Chen, Y.L., Lim, S.P., Davidson, A., and Vasudevan, S.G. (2007). Construction and characterization of a stable subgenomic dengue virus type 2 replicon system for antiviral compound and siRNA testing. Antiviral Res. *76,* 222–231.

Niyomrattanakit, P., Abas, S.N., Lim, C.C., Beer, D., Shi, P.-Y., and Chen, Y.-L. (2010). Fluorescent-based alkaline phosphatase-coupled polymerase assay for identification of inhibitor of dengue NS5 RNA-dependent RNA polymerase. J. Biomol. Screen. *16,* 201–210.

Nybakken, G.E., Nelson, C.A., Chen, B.R., Diamond, M.S., and Fremont, D.H. (2006). Crystal Structure of the West Nile Virus Envelope Glycoprotein. J. Virol. *80,* 11467–11474.

Poch, O., Sauvaget, I., Delarue, M., and Tordo, N. (1989). Identification of four conserved motifs among the RNA-dependent polymerase encoding elements. EMBO J. *8,* 3867–3874.

Poh, M.K., Yip, A., Zhang, S., Priestle, J.P., Ma, N.L., Smit, J.M., Wilschut, J., Shi, P.Y., Wenk, M.R., and Schul, W. (2009). A small molecule fusion inhibitor of dengue virus. Antiviral Research *84,* 260–266.

Puig-Basagoiti, F., Tilgner, M., Forshey, B.M., Philpott, S.M., Espina, N.G., Wentworth, D.E., Goebel, S.J., Masters, P.S., Falgout, B., Ren, P., Ferguson, D.M., and Shi, P.Y. (2006). Triaryl pyrazoline compound inhibits flavivirus RNA replication. Antimicrob. Agents Chemother. *50,* 1320–1329.

Qing, M., Liu, W., Yuan, Z., Gu, F., and Shi, P.Y. (2010). A high-throughput assay using dengue-1 virus-like particles for drug discovery. Antiviral Research *86,* 163–171.

Ray, D., Shah, A., Tilgner, M., Guo, Y., Zhao, Y., Dong, H., Deas, T.S., Zhou, Y., Li, H., and Shi, P.Y. (2006). West Nile virus 5′-cap structure is formed by sequential guanine N-7 and ribose 2′-O methylations by nonstructural protein 5. J. Virol. *80,* 8362–8370.

Rey, F.A., Heinz, F.X., Mandl, C., Kunz, C., and Harrison, S.C. (1995). The envelope glycoprotein from tick-borne encephalitis virus at 2 A resolution. Nature *375,* 291–298.

Rice, C.M., Lenches, E.M., Eddy, S.R., Shin, S.J., Sheets, R.L., and Strauss, J.H. (1985). Nucleotide sequence of yellow fever virus: implications for flavivirus gene expression and evolution. Science *229,* 726–733.

Simeonov, A., Jadhav, A., Thomas, C.J., Wang, Y., Huang, R., Southall, N.T., Shinn, P., Smith, J., Austin, C.P., Auld, D.S., and Inglese, J. (2008). Fluorescence Spectroscopic Profiling of Compound Libraries. J. Med. Chem. *51,* 2363–2371.

Stiasny, K. and Heinz, F.X. (2006). Flavivirus membrane fusion. J. Gen. Virol. *87,* 2755–2766.

Sumiyoshi, H., Mori, C., Fuke, I., Morita, K., Kuhara, S., Kondou, J., Kikuchi, Y., Nagamatu, H., and Igarashi, A. (1987). Complete nucleotide sequence of the

Japanese encephalitis virus genome RNA. Virology *161*, 497–510.

Tatarowicz, W.A., Lurain, N.S., and Thompson, K.D. (1991). *In situ* ELISA for the evaluation of antiviral compounds effective against human cytomegalovirus. J. Virol. Methods *35*, 207–215.

Wang, Q.Y., Patel, S.J., Vangrevelinghe, E., Xu, H.Y., Rao, R., Jaber, D., Schul, W., Gu, F., Heudi, O., Ma, N.L., Poh, M.K., Phong, W.Y., Keller, T.H., Jacoby, E., and Vasudevan, S.G. (2009). A Small-Molecule dengue Virus Entry Inhibitor. Antimicrob. Agents Chemother. *53*, 1823–1831.

Wengler, G. and Wengler, G. (1981). Terminal sequences of the genome and replicative-from RNA of the flavivirus West Nile virus: absence of poly(A) and possible role in RNA replication. Virology *113*, 544–555.

Yap, T.L., Xu, T., Chen, Y.L., Malet, H., Egloff, M.P., Canard, B., Vasudevan, S.G., and Lescar, J. (2007). Crystal Structure of the dengue Virus RNA-Dependent RNA Polymerase Catalytic Domain at 1.85-Angstrom Resolution. J. Virol. *81*, 4753–4765.

Yin, Z., Patel, S.J., Wang, W.L., Wang, G., Chan, W.L., Rao, K.R.R., Alam, J., Jeyaraj, D.A., Ngew, X., Patel, V., Beer, D., Lim, S.P., Vasudevan, S.G., and Keller, T.H. (2006). Peptide inhibitors of dengue virus NS3 protease. Part 1: Warhead. Bioorganic and Medicinal Chemistry Letters *16*, 36–39.

Yusof, R., Clum, S., Wetzel, M., Murthy, H.M.K., and Padmanabhan, R. (2000). Purified NS2B/NS3 Serine Protease of dengue Virus Type 2 Exhibits co-factor NS2B Dependence for Cleavage of Substrates with Dibasic Amino Acids *in Vitro*. J. Biol. Chem. *275*, 9963–9969.

Zhang, Y., Zhang, W., Ogata, S., Clements, D., Strauss, J.H., Baker, T.S., Kuhn, R.J., and Rossmann, M.G. (2004). Conformational Changes of the Flavivirus E Glycoprotein. Structure *12*, 1607–1618.

Flavivirus Diagnostics

13

Elizabeth Hunsperger

Abstract

Within the family of *Flaviviridae* there are many medically important viruses that cause human disease worldwide. These viruses were originally categorized based on phenotype due to their antigenic relatedness and placed within groups, subgroups and types and later confirmed with nucleic acid sequence analysis. Diagnosis of disease caused by flaviviruses has been primarily based on serological identification of antiviral antibodies and virus isolation. Some of the classic serological techniques of haemagglutination inhibition assay and complement fixation were replaced with the enzyme linked immunosorbent assay (ELISA) for the detection of IgM, IgG and IgA antibodies primarily due to ease-of-use. The plaque reduction neutralization test (PRNT) provided the specificity needed for virus identification following a positive serological test by ELISA. The development of polymerase chain reaction (PCR) assays improved the ability to detect virus nucleic acid sequence when viral isolates were not obtained. Because reverse transcriptase PCR (RT-PCR) assays are easy to perform, have increased sensitivity and provide virus identification in a short period of time, RT-PCR has essentially replaced isolation techniques for rapid diagnosis. However, virus isolation is still essential for genetic analysis. The future of flaviviral disease diagnosis is new platforms for antibody and nucleic acid detection as well as the development of point-of-care diagnostics for clinical management.

Introduction

The *Flavivirus* genus belongs to the family *Flaviviridae* and encompasses 70 known viruses. Before the establishment of the defined *Flaviviridae* family, Casals and Brown classified two major antigenic groups of viruses: group A (alphaviruses) and group B (flaviviruses) (Casals and Brown, 1954). Because of the extensive serological cross-reactivity observed among flaviviruses, the haemagglutination inhibition (HI) assay was used to further characterize viruses within the group. Experiments based on the HI assay provided the differentiation between the group, complex and type-specific determinants (Edelman 1973; Westaway 1974; Trent 1977). This original classification of these viruses was based on antigenicity and was done in accordance with three basic 'rules': (1) no virus can belong to two antigenic groups; (2) if two different viruses cross-react antigenically, they are related; and (3) if viruses of different groups cross-react, they do not belong to different groups.' These three rules were used to categorized flavivirus taxonomy on the basis of phenotype (Calisher and Gould, 2003).

Monoclonal antibodies were then used to identify type-specific, subcomplex, group determinants primarily on the basis of the envelope protein (Heinz *et al.*, 1983a,b; Roehrig *et al.*, 1983). Results of further studies at the molecular level of the virus identified that the envelope glycoprotein E was responsible for the antigenic groups of the flaviviruses classified by Casals and Brown in 1954. Hence the E protein is important for both viral attachment to susceptible cells and elicits virus neutralizing and HI antibodies.

The order in which members of this family of antigenically similar viruses could be placed was determined by cross-neutralization tests with polyclonal serum (Calisher *et al.*, 1989). The family was originally divided into eight serocomplexes. The most medically relevant antigenic groups or complexes are the tick-borne encephalitis virus (TBEV) complex, the Japanese encephalitis virus (JEV) complex, the dengue viruses (DENV) complex and the yellow fever virus (YFV) complex. Because these groups were established based on antigenicity, there is cross-reactivity within these complex groups (Table 13.1). Those viruses most distantly related often are less cross-reactive and those in the same serocomplex group often are more cross-reactive in serology assays (Fig. 13.1). The cross-reactivity of these closely related viruses in the diagnostic antibody tests often complicate efforts to interpret test results especially where multiple flaviviruses co-circulate in nature.

The genomic sequence analyses of these viruses allowed for their taxonomy or grouping to be based on their nucleic acid sequence rather than phenotype. However, serological techniques are still an important component of diagnosis of flavivirus for which virus isolation is more difficult. This chapter will discuss the current nucleic acid and serological techniques used for flavivirus identification and diagnosis, confounding factors associated with the use of these techniques and other issues of flavivirus diagnosis including 'original antigenic sin' and antibody cross-reactivity. Additionally, classic molecular and serological assays used for diagnostics of flaviviruses as well as the confirmatory testing and the future direction of flavivirus diagnostics will be presented.

Viraemia and antibody response during the course of infection

Humans are first exposed to flavivirus infection when they are bitten by an infected mosquito or tick and the virus is inoculated into their skin. The next 4–14 days constitute the 'intrinsic incubation period' during which the virus replicates and moves to draining lymph nodes where viral replication continues in lymphatic tissues including dendritic and other immune cells. At this stage, an infected patient becomes viraemic and during the first 3 days the virus can be detected in or isolated from the patient's serum or cerebral spinal fluid (CSF). Following the viraemic phase, the patient develops antibodies against viral proteins with early antibodies consisting primarily of immunoglobulin M (IgM). As in other viral infections, the first line of defence to clear the virus from the body is the innate immune response followed by the acquired immune response. IgM can be detected as early as 5 days after the onset of illness and can persist as long as three months. In unusual cases such as WNV and JEV infections, IgM persists for over a year following infection (Edelman *et al.*, 1976; Kapoor *et al.*, 2004; Prince *et al.*, 2005). For most flaviviruses, however IgM is detectable during the first 14 days after the onset of symptoms. Following the IgM response is the development of long term immunity conferred by immunoglobulin G (IgG) which for most primary flavivirus infections can be detected within 9–14 days after the onset of illness. Because IgG antibodies can persist for life, those infected will have lifelong immunity to the particular strain of virus with which they were infected (Fig. 13.2).

The antibody kinetics of IgM and IgG of a patient is dependent on the infecting virus and whether the infection is a primary or a secondary flavivirus infection although they also can vary dependent on the sequence of flaviviral infections. Some basic interpretations of serological test results and testing algorithm are summarized in Table 13.2 and Fig. 13.3, respectively. The clinical manifestation of a flavivirus infection can range from unapparent to severe symptoms and these clinical outcomes are also associated with differing antibody responses. Because of this broad range of clinical symptoms, diagnosis using only clinical information is not always accurate or reliable. In addition, many flaviviruses present with very similar symptoms that are difficult to clinically differentiate. Consequently laboratory confirmation in combination with clinical presentation of the infecting virus is the preferred method for surveillance and clinical management.

Table 13.1 *Flavivirus* genus antigenic complex groups from mosquito-borne and tick-borne viruses summarized which cause disease in humans from the 7th report of the International Committee on Taxonomy of Viruses (ICTV) and cell lines used for virus isolation

Antigenic Complex Group	Species*	Subtypes/strain/ serotypes	Geographic range	Cell line for virus recovery**
Aroa virus complex	Aroa virus	Bussuquara virus	South and Central America	BHK-21
Dengue virus complex	Dengue virus	Dengue virus type 1	Tropical and subtropical regions	C6/36
		Dengue virus type 2	Tropical and subtropical regions	C6/36
		Dengue virus type 3	Tropical and subtropical regions	C6/36
		Dengue virus type 4	Tropical and subtropical regions	C6/36
Japanese encephalitis virus complex	Japanese encephalitis virus		East, South-East, South Asia, West Pacific	Vero
	Murray Valley encephalitis virus	Alfuy virus	Australasia/Indonesia	LLCK-MK2
	St. Louis encephalitis virus		North and South America	Vero
	West Nile virus		Africa, North and South America, Europe, Middle East, Asia	Vero, AP61
	Usutu virus		Europe, Africa	BHK-21
	Kunjin virus		Australasia/Malaysia	BHK-21
	Koutango virus		West and Central Africa	
Kokobera virus complex	Kokobera virus	Stratford virus	Australasia	Vero, LLCK-MK2, BHK-21
Ntaya virus complex	Bagaza virus		West and Central Africa	Vero, LLC-MK2
	Ilheus virus		South and Central America	LLC-MK2, BHK-21
		Rocio virus	South America	
	Ntaya virus		West, East and Central Africa	LLC-MK2, BHK-21
Spondweni virus complex	Zika virus		Polynesian Islands, Asia, Africa	LLC-MK2, Vero
		Spondweni virus		
Tick-borne encephalitis virus complex	Omsk haemorrhagic fever		Russia	HeLa, BHK-21
	Louping ill		British Isles	LLC-MK2, Vero
	Kyasanur Forest disease		India	HeLa

Table 13.1 continued

Antigenic Complex Group	Species*	Subtypes/strain/ serotypes	Geographic range	Cell line for virus recovery**
	Powassan		Canada and USA	LLC-MK2, Vero
	Royal Farm			HeLa
	Karshi		Uzbekistan, Kazakhstan	LLC-MK2, Vero
Yellow fever virus complex	Banzi virus		South and East Africa	Vero, BHK-21
	Edge Hill virus		Australasia	LLC-MK2, Vero
	Uganda S		Africa	Vero
	Sepik virus		Australasia	
	Wesselsbron virus		Africa, Thailand	Vero
	Yellow fever virus		Africa and South America	Vero, C6/36

*Not all viruses are listed for each complex.
**Not all cell lines listed will amplify every virus from each group. Refer to International Catalogue of Arbovirus for a complete list for each virus.

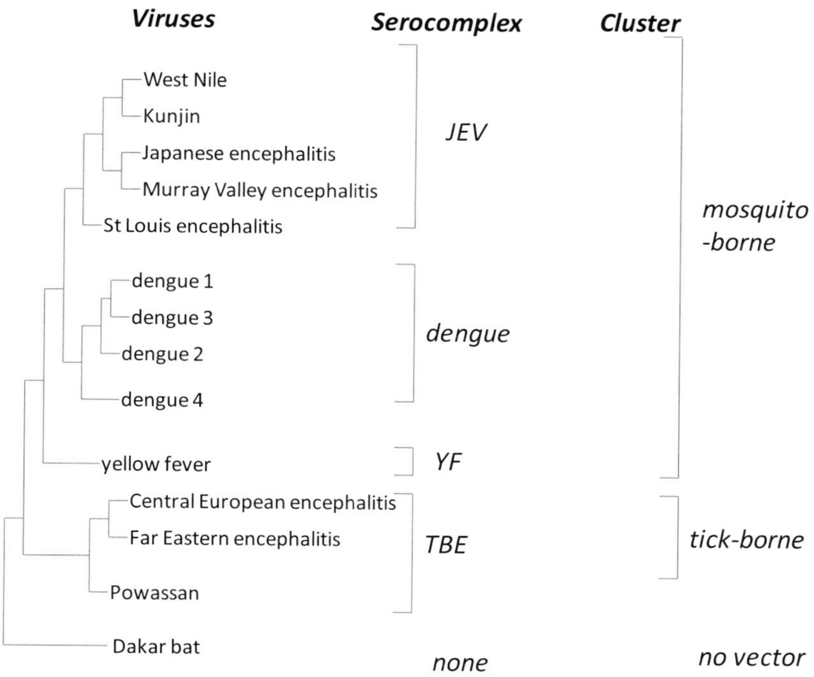

Figure 13.1 Dendogram demonstrating the organization of the *Flavivirus* genus based on phylogenetic analysis with the serocomplex group and the cluster. The evolutionary distance is not accurately represented (Kuno *et al.*, 1998).

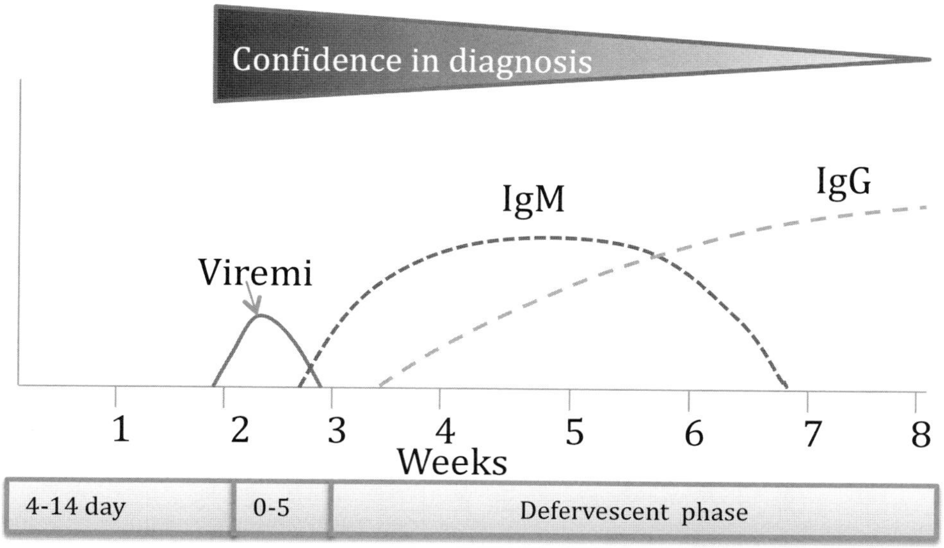

Figure 13.2 Schematic of the approximate time-line for virus and antibody detection in humans following infection.

Table 13.2 General interpretation of results from serology testing for flaviviruses. Light-grey areas are unique serology observations and not as common as those highlighted in dark grey

Acute		Convalescent			
IgM	IgG	IgM	IgG	Neutralization	Possible interpretation
Negative	Negative	Negative	Negative	No	Negative
Negative	Positive	Negative	Positive	No	Previous flavivirus exposure; no recent infection (no rise in IgG titre)
Positive	Negative	Positive	Positive	Yes (1 virus)	Recent primary flavivirus infection (IgG seroconversion)
Negative	Negative	Positive	Positive	Yes (1 virus)	Recent primary flavivirus infection
Negative	Positive	Positive	Positive	Yes (≥2 virus)	Recent secondary flavivirus infection (IgM seroconversion)
Positive	Positive	Positive	Positive	Yes ((≥2 virus)	Recent primary or secondary flavivirus infection
Negative	Negative	Positive	Negative	Yes (1 virus)	Recent primary flavivirus infection with delayed IgG
Positive	Negative	Negative	Positive	Yes (1 virus)	Recent primary flavivirus infection with rapid IgM loss
Positive	Negative	Positive	Negative	Yes (1 virus)	Recent primary flavivirus infection with delayed IgG
Positive	Positive	Negative	Positive	Yes (≥1 virus)	Recent primary or secondary flavivirus infection with rapid IgM loss

Acute, 0–5 days post onset of symptoms; convalescent, 6–14 days post onset of symptoms.

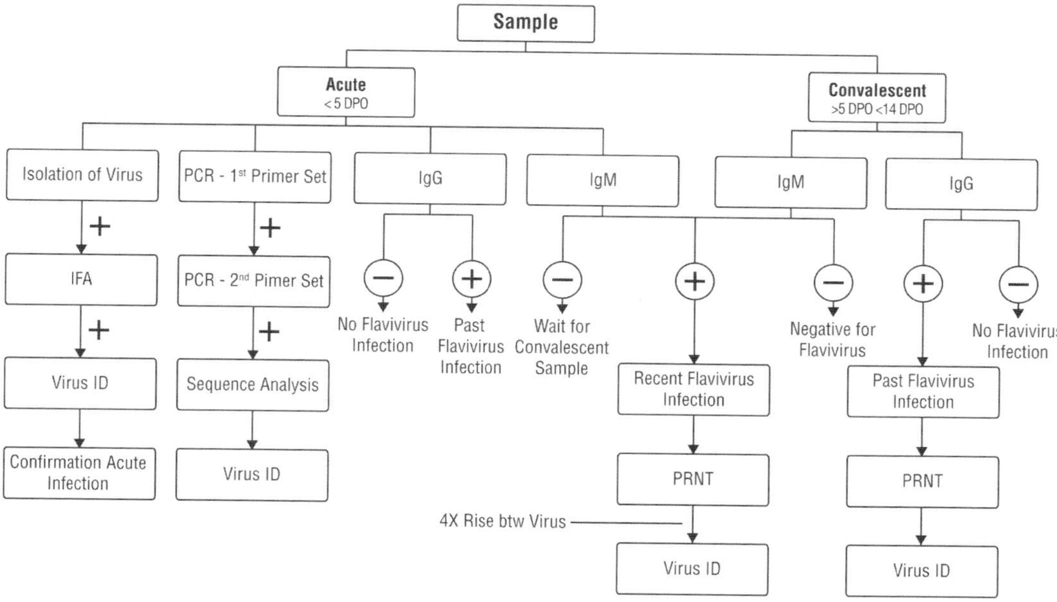

Figure 13.3 Testing algorithm and interpretation of test results based on the classification of the sample received.

Viral isolation, identification and nucleic acid detection

Viral isolation and identification by growth in cell culture and inoculated mosquitoes

The classic method of confirming a flavivirus infection is virus isolation, the amplification of the virus in a cell line that is known to propagate the virus. The success of this technique is dependent on both the virus and the sample from which the isolation is attempted. For example, the likelihood of isolating a virus from a human serum sample is dependent on the presence of antibodies and complement in the sample. Because flaviviruses are positive-sense RNA viruses with error-prone replication cycles, every replication cycle produces a mixed population. Continued passage of the virus in culture adapts the virus to the cell line in which it is being amplified. This adaptation process can introduce nucleic acid changes that are significantly different from the original viral isolate dependent on the number of times in which the virus is passaged in cell culture. Hence phylogenetic sequence analysis should be conducted from viruses which are no more than one or two passages from the original isolate.

Classic viral identification techniques include cell morphology observations and descriptions of the cytopathic effects (CPE) caused by the infection and the rate of CPE. CPE may include cell lysis, swelling of the nucleus, syncytia formation or other changes that are visually identifiable using light microscopy. Unique morphological changes can be associated with either family of viruses or specific genus. Because of the diversity within this family of viruses, CPE is dependent on the virus and the cell line used for viral growth.

In the early or acute phase of infection, isolation and identification of infecting virus is the only method to confirm the diagnosis of a current flavivirus infection. The most sensitive viral isolation technique for mosquito-borne flaviviruses is growth in mosquito cell lines. The mosquito cell line C6/36 (clone obtained from *Aedes albopictus*) has become the host cell of choice for routine isolation of most mosquito-borne flaviviruses, although the *Aedes pseudoscutellaris* cell line AP61 has also been used successfully (Singh *et al.*, 1968; Race *et al.*, 1979). Mammalian cell cultures such as Vero or LLCMK2 cells are more effective for isolating members of the JEV serocomplex viruses. Viral identification is achieved using the infected cells from the isolation and probing with monoclonal antibodies specific to a particular flavivirus group and type (Table 13.3).

A more sensitive technique for mosquito-borne virus isolation is intrathoracic inoculation of mosquitoes (*Ae. aegypti, Ae. albopictus, Toxorhynchites splendens, Tx. amboinensis*). The procedure has been described by many groups, but the most detailed description was in an article by Rosen and Gubler in 1974. In this article the authors describe the minimum infectious dose required to amplify DENV and the recovery rates for human serum inoculation in *Aedes albopictus*. They determined that female mosquitoes could tolerate larger doses of human serum inoculum and were less sensitive to the toxic effects of the serum compared with male mosquitoes. However, they also noted that using male mosquitoes was safer because males do not feed on blood and thus would pose no infectious threat if they were accidentally released. Rosen and Gubler further determined that the mosquito inoculation was always more sensitive than tissue culture isolation (Rosen and Gubler, 1974). Although intrathoracic inoculation of a mosquito is a more sensitive viral isolation method, cell culture is preferable for routine viral isolation because of the technical skill, special containment equipment required for direct inoculation of mosquitoes and access to a mosquito colony. Sensitivity of the cell culture amplification technique varies between flaviviruses. Table 13.1 provides some helpful information concerning cell lines used successfully for virus isolation.

Viral identification by nucleic acid amplification tests

Nucleic acid amplification tests (NAATs) are rapid methods of flavivirus identification compared with virus isolation. Some of the NAAT techniques commonly used include: reverse transcriptase polymerase chain reaction (RT-PCR), real-time RT-PCR, nested PCR, nucleic acid sequence based amplification (NASBA), melting curves and reverse transcriptase loop mediated isothermal amplification (RT-LAMP). The sensitivity of these NAAT is based primarily on the primer and/or probe sequences used for amplification and the method of RNA extraction. The most successful NAATs (those with greatest sensitivity and specificity) generally use the conserved regions of the genome in the primer design. Table 13.4 lists the primers and probes used for amplification and indicates their sensitivity in human serum and CSF samples (Bae *et al.*, 2003; Brown *et al.*, 1994; Callahan *et al.*, 2001; Conceicao *et al.*, 2010; Houng *et al.*, 2000; Houng *et al.*, 2001; Huang *et al.*, 2004; Ito *et al.*, 2004; Kong *et al.*, 2006; Lai *et al.*, 2007; Lanciotti and Kerst, 2001; Lanciotti *et al.*, 2000; Laue *et al.*, 1999; Liu *et al.*, 2008; Liu *et al.*, 2009; Paranjpe and Banerjee, 1998; Parida *et al.*, 2006; Santhosh *et al.*, 2007; Schwaiger and Cassinotti, 2003; Tang *et al.*, 2006; Wicki *et al.*, 2000; Yang *et al.*, 2004; Zaayman *et al.*, 2009).

RNA extraction and purification techniques

The sensitivity of any NAAT is dependent on the efficiency of the extraction and purification of the target RNA template. Commercial assays for extraction are an improvement over previously used in-house phenol/chloroform extraction techniques which produced inconsistent results because of RNA degradation during the purification process. These newer commercial assays are often faster and more reliable. The basis of most commercial assay kits is a solubilization step in which chaotropic lysis buffer is used to provide RNA stabilization to reduce RNA degradation. The RNA is bound to a silica column and after a series of stringent wash steps the RNA is purified using a low-salt eluate. Another advantage of these commercial assays is their ability to extract and purify RNA from many types of samples including serum, blood, CSF, homogenized mosquitoes and ticks and whole tissue samples. The sensitivity of any NAAT can be improved simply by extracting more starting material for RNA amplification. This technique has been used successfully to screen the blood bank specimens for WNV in addition to other modifications (Busch *et al.*, 2006). Because results from conventional RT-PCR tests of samples from people infected with viruses in the JEV and TBEV serocomplex group may have low to undetectable level of virus,

Table 13.3 Flavivirus antibodies used for serology and virology identification assays

Antibody	Virus (seed)	Reactivity	Common usage
6B6C-1	SLE virus (MSI-7)	Flavivirus group	As HRP conjugate in antigen capture and IgM capture ELISAs
4G2	Dengue 2 (New Guinea C)	Flavivirus group	Viral identification by IFA or ELISA; ELISA capture antibody
6B4A-10	JE virus (Nakayama)	JE–SLE-WN-MVE complex	Capture antibody in antigen capture ELISA
JE314H52	JE virus (Nakayama)	JE virus specific	Viral identification by IFA or ELISA
6B5A-2	SLE virus (MSI-7)	SLE virus specific	Viral identification by IFA and neutralization
4A4C-4	SLE virus (MSI-7)	SLE virus specific	Capture antibody in antigen capture ELISA
4B6C-2	MVE virus (Original)	MVE virus specific	Viral identification by IFA or ELISA
D2–1F-3	Dengue 1 (Hawaii)	Dengue 1 virus specific	Viral identification by IFA or ELISA
3H5–1–21	Dengue 2 (New Guinea C)	Dengue 2 virus specific	Viral identification by IFA or ELISA
D6–8A1–12	Dengue 3 (H87)	Dengue 3 virus specific	Viral identification by IFA or ELISA
1H10–6–7	Dengue 4 (H-241)	Dengue 4 virus specific	Viral identification by IFA or ELISA
5.00E-03	Yellow fever (17D)	Yellow fever virus specific	Capture antibody in antigen capture ELISA
117	Yellow fever (Asibi)	Yellow fever wild-type virus specific	Viral identification by IFA
864	Yellow fever (17D)	Yellow fever vaccine virus specific	Identification of vaccine strains by IFA
4D9	TBE virus	Tick-borne flavivirus complex	Viral identification by IFA or ELISA
H5.46	WN	WN virus specific	Viral identification by IFA or ELISA

Adapted from Roehrig (1986).

MVE, Murray Valley encephalitis; WN, West Nile; JE, Japanese encephalitis; TBE, tick-borne encephalitis; HRP, horseradish peroxidase; IFA, immunofluorescence assay; ELISA, enzyme-linked immunosorbent assay.

Table 13.4 RT-PCR assays, sensitivity, primer design and detection limits for flavivirus diagnosis.

Flavivirus	Type of Test	Reference	Primer Location	Primer Format	Detection Limit	Percent positivity in Human Clinical Serum (S) or CSF (No. samples)
DENV	nested RT-PCR	*Lanciotti et al., 1992*	C, PrM	2 consensus primers/4 type specific primers	10 copies	100%(93)S
	real time RT-PCR	*Laue et al., 1999*	NS5	4 type-specific primer/probe sets	2 RNA copies	80% (25)S
	real time RT-PCR	*Houng et al., 2000*	3'NC	DEN-2 type-specific primer/probe set	0.1 pfu	100% (4)S
	real time RT-PCR	*Callahan et. al. 2001*	C, NS5	4 type-specific primer/probe sets	0.1 to 1 pfu	93%(59)S
	real time RT-PCR	*Houng et al., 2001*	3'NC	6 primers + 2 consensus probes	0.04 pfu	89% (134)S
	real time RT-PCR	*Ito et. al. 2004*	E	4 type-specific primer/probe sets	0.1 to 1 pfu	100 %(35)S
	real time RT-PCR	*Johnson et. al., 2005*	PrM, C	4 type-specific primer/probe sets	.002 to .5 pfu	100%(40)S
	real time RT-PCR	*Kong et. al., 2006*	NS5	dengue consensus primer pair/4 probes	500 copies	90% (375)S
	real time RT-PCR	*Chien et. al. 2006*	NS5	flavivirus consensus primer pair/4 probes	1-10 pfu	91% (77)S
	real time RT-PCR	*Lai et. al., 2007*	3'NC	dengue consensus primer pair/ 4 probes	100 copies plasmid	ND
	qreal time RT-PCR	*Conceicao et al., 2010*	5'UTR	2 primer/1 probe	100 copies	100% (6)S
JEV	RT-PCR	*Paranjpe & Banerjee, 1998*	E	primer sets from JEV strain P 20778	4 pfu	ND
	real time RT-PCR	*Yang et al., 2004*	3'NC	primer/probe sets from JEV strain KV1899	11.2 TCID50/ml	ND
	real time RT-PCR	*Huang et al., 2004*	NS3	JEV consensus primers for all strains/probe	40 pfu/ml	ND
	RT-LAMP	*Parida et al., 2006*	E	6 primers to E of JEV/ magnesium pyrophosphate	0.1 pfu	22% (50) CSF
	SYBR Green I real time RT-PCR	*Santhosh et al., 2007*	NS3	primer sets from strain JaOArS982	20 copies	56% (32) CSF
	in situ RT-LAMP	*Liu et al., 2009*	PrM	6 primer sets from strain T1P1	ND	ND
SLE	NASBA	*Lanciotti & Kerst, 2001*	prM/M, E	ECL and Molecular beacons	0.15 pfu	ND
	real time RT-PCR	*Lanciotti & Kerst, 2001*	prM/M, E	primer probe set from MSI.7 SLE strain	0.1pfu	ND
YFV	RT-PCR	*Brown et al., 1994*	E	primer sets	30 pfu	ND
	real time RT-PCR	*Bae et al., 2003*	3'NC and NS3	primer/probe sets to YF-17D	ND	ND
	qRT-PCR	*Mantel et al., 2008*	E/NS1 and NS5	primer/probe sets to YF-17D	12 copies/reaction	ND
TBEV	real time RT-PCR	*Wicki et al., 2000*	NS5	primer/probe sets to	10 copies	ND
	real time RT-PCR	*Schwaiger & Cassinotti, 2003*	3'NC	Consensus sequence for 3 different TBEV strains	10 copies	0% (14)S; 5% (21) CSF
WNV	real time RT-PCR	*Lanciotti et al., 2000*	3'NC, E	primer/probe sets from WNV NY99	0.1 pfu	14% (28)S and 57%(28)CSF
	real time RT-PCR	*Tang et al., 2006*	3'NC	primer/probe sets from WNV NY385-99,EG-101,B-956	10 copies/ml	100%(80)S
	nested RT-PCR/FRET	*Zaayman et al., 2009*	NS5	primer/FRET probe sets from WNV lineage 1 and 2	.07 copies/ml	ND

the greater sensitivity afforded by these improved methods of RNA extraction is important (Table 13.4).

Other factors that can negatively affect the sensitivity of RT-PCR include lack of consistency between samples and inter-operator variability. These factors are most important when processing large number of samples. Many laboratories that process a large number of samples have overcome these variables by the use of automated extraction systems. The automated systems are more reproducible and extraction yields tend to be superior to manual extraction methods (Mantel *et al.*, 2008).

RT-PCR

The NAAT used most routinely by diagnostic laboratories for the identification of flaviviruses are real-time RT-PCR and nested PCR assays. The DENV nested PCR assay is the most sensitive assay for detection owing to its two-step amplification reaction consisting of an initial reverse transcription and amplification step in which universal DENV primers target a region of the virus genome (C-prM) followed by a second DENV serotype specific amplification step. The product of the reaction is identified by gel electrophoresis and the molecular mass is determine by the use of standard markers to identify the infecting serotype (Lanciotti *et al.*, 1992).

Although the real-time RT-PCR assay is not as sensitive as the nested PCR assay, it has the advantage of being a one-step assay system. In this assay, fluorescent probes enables the detection of the reaction products in real-time without the need for gel electrophoresis. Real-time RT-PCR assays are either 'singleplex' (able to detect only one flavivirus at a time) or 'multiplex' (able to detect multiple flaviviruses from a single sample) in their format. The DENV multiplex assays allow users to identify multiple DENV serotypes without the risk of introducing contamination during manipulation of the sample (Chien *et al.*, 2006; Johnson *et al.*, 2005b). In addition, real-time RT-PCR assays can be adapted to a quantitative real-time RT-PCR (qRT-PCR) in which genome copies in the patient sample are determined by using a standard curve of *in vitro* synthesized RNA quantified by spectrophotometer. The use of RNA standard curves to calculate genome copies allows for fair comparisons between techniques as well as standardization between assays. Quantifying the genome copy of a DENV specimen to determine the viral load in patients may provide important information useful for the prediction of disease severity (Vaughn *et al.*, 1997).

As previously discussed, the method of RNA extraction has a substantial effect on the sensitivity of the RT-PCR reaction. However, the method of RNA extraction is not usually accounted for in comparisons of NAATs; instead assays are usually compared on the basis of the primer design. Another factor that affects assay sensitivity is the number of cycles used in the PCR amplification step. In short, there are four main parameters that must be held constant in order to compare results between RT-PCR assays: (1) the RNA extraction method used; (2) the amount of RNA used in the reaction; (3) the primer/probe design; and (4) the number of amplification cycles.

Because of limited sample volumes, many investigators attempt to discern the infecting virus using multiplex assays with universal primers. Universal or consensus primers have been designed for flaviviruses which are primarily used for surveillance purposes. These assays identify either group reactive primers which amplify consensus sequences for a group within a family of viruses or degenerative primers designed from conserved regions of the flavivirus genome to generate a complimentary DNA (cDNA) product that can be used to target a specific virus within this genus to narrow the target within a specific family of viruses (Chao *et al.*, 2007; Dyer *et al.*, 2007; Kuno, 1998; Maher-Sturgess *et al.*, 2008; Scaramozzino *et al.*, 2001). Although these assays are very helpful for pathogen discovery, they are generally less sensitive than singleplex RT-PCR assays.

Laboratories that use NAAT assays should adhere to the following quality control and assurance measures: (1) handle the sample in a clean room to prevent laboratory contamination; (2) test samples in duplicate; (3) include a positive, negative and no template control in every reaction; (4) validate results by conducting a second NAAT targeting a different region of the genome; (5) validate NAAT positive samples with virus isolation, antigen capture ELISA or a immunofluorescence assay (IFA); and (6) add an internal RNA control to the reaction. Adherence to these quality control and assurance measures will promote confidence in a laboratories reported results.

Viral identification by *in situ* hybridization or *ex vivo* tissue amplification methods

Postmortem tissue is used for diagnosis in fatal cases of flavivirus infections. Methods of detection include immunohistochemistry using paraffin embedded sections to localize viral antigens within the tissue sample and *in situ* hybridization for viral RNA detection within tissue samples and RT-PCR. The types of tissues collected from the patient are dependent on the infecting virus. For viscerotropic viruses such as DENV, virus and viral antigens can be detected in spleen, lymph nodes, lung, thymus and liver (Bhoopat *et al.*, 1996; Fresh *et al.*, 1969; Hall *et al.*, 1991). For viruses in the JEV and TBEV serocomplex group central nervous system (CNS) tissue is harvested for analysis (Armah *et al.*, 2007; Gelpi *et al.*, 2006; Seth *et al.*, 1974).

Antibody detection

IgM antibody detection

The acquired immune response following a flavivirus infection is the synthesis of IgM antibodies. IgM circulating in the blood is generally detectable during the convalescent phase of the infection and is an indication of an acute or recent infection. During the acute phase of the infection (the first 5 days post onset of illness) a percentage of patients have detectable IgM antibodies within those first 5 days of illness however this is variable. The ideal time to measure IgM antibody response to a flavivirus is during the convalescent phase of the infection (6–14 days post onset of illness) because measurement during this period yields the most reliable results. During this period, the IgM response is measurable by most enzyme-linked immunosorbent assays (ELISA) (Fig. 13.2 and Table 13.2).

IgM antibody capture ELISA (MAC-ELISA) format is most commonly used assay in diagnostic laboratories and commercially available diagnostic kits. There are two main reasons for this widespread use. First, because the assay is based on capturing IgM antibodies on a microtitre plate using an anti-human-IgM antibody thus minimizing the interference of the high avidity IgG antibodies from binding to the antigen. Second, because many of the sources of antigen are not purified homogenous protein products and often are composed of a mixture of many different viral antigens, using the MAC-ELISA method removes non-specific antibody binding hence many different sources of antigen can be used in the assay. The antigen can be used in excess in this assay with the use of the broadly cross-reactive monoclonal antibodies that are flavivirus group specific. Another advantage of the MAC-ELISA format is that antigens can be interchanged to compare antibody response to many different flaviviruses within the same format mainly due to the use of broadly cross-reactive monoclonal antibodies conjugated to a detector molecule (Table 13.3).

Many factors can affect the sensitivity and specificity of the MAC-ELISA including the antigen, the detector, serum sample condition (unacceptable conditions include: haemolysed, lipaemic or contaminated with bacteria) and the dilution of the serum used for the assay. The sources of the antigen include suckling mouse brain antigen (SMB), tissue culture derived antigen and recombinant antigens. The most sensitive antigen is SMB because it is composed of high-titre mixture of all viral antigens. However, owing to its impurity, this antigen often has non–specific cross-reactivity. Conversely, recombinant antigens are often less sensitive but in general more specific. Most of the recombinant antigens are developed on the premise that the most immunogenic viral proteins produced are against the structural proteins. These purified sources of antigens are generally more specific but also less sensitive than SMB antigens.

The IgM response during secondary flavivirus infections often differ from that during primary infections (Table 13.2). For example as many as 20% of patients with secondary DENV infections do not have sufficient levels of IgM for detection during the convalescent phase of the infection (Innis *et al.*, 1989). In these unique cases serological confirmation is based on IgG seroconversion or 4-fold increase in IgG titres between the acute and convalescent sample (Miagostovich *et al.*, 1999). In other atypical cases, the IgM response may persist for up to three months following infection and, up to 6 months or a year as with WNV and JEV infections (Edelman *et al.*, 1976; Kapoor *et al.*, 2004; Prince *et al.*, 2005). Because this persistence of IgM can confound efforts to diagnose an acute infection, other assays were developed to assess the status of infection (acute, recent or past) on the basis of the avidity of the IgG antibody response (Levett *et al.*, 2005).

With viruses that infect the CNS (i.e. those in the JEV serocomplex group) the detection of IgM antibodies in cerebral spinal fluid (CSF) differs from the detection of IgM antibodies in the blood. For example, in JEV infection, IgM detection in the blood reaches 100% detection approximately 9–13 days following onset of symptoms; however, in the CSF it is as early as 1 day after onset of symptoms (Chanama *et al.*, 2005). These results indicate that the most efficient method of detecting IgM antibodies early in a JEV infection is CSF testing. Interestingly, the use of the HI assay was found to be inadequate to confirm JEV infection (Burke *et al.*, 1985). In addition, the IgM antibodies to JEV can persistence for as long as a year in the CSF (a phenomenon discussed further in the 'Viral persistence' section of this chapter).

Most diagnostic laboratories use in-house IgM assays however many commercial IgM assays and rapid tests have been developed for some of the medically important flaviviruses including WNV, JEV, YFV and DENV (Blacksell *et al.*, 2007; Hunsperger *et al.*, 2009; Ravi *et al.*, 2009). Some of these commercial assays have very good sensitivity and specificity. Prior to implementing a commercial assay each diagnostic laboratory should perform a comprehensive independent validation (WHO, 2009).

IgG antibody detection

The presence of IgG antibodies is an indication of a long term acquired immunity of a past flavivirus infection. In some cases this long term IgG immunity can be detected up to 60 years after the initial infection as with DENV (Imrie *et al.*, 2007). There are two methods used to measure IgG; direct and indirect ELISA. The direct IgG ELISA method is generally less sensitive than the indirect ELISA and is less often used because it requires purified viral antigen. The indirect IgG ELISA is more sensitive and uses a capture antibody to immobilize the viral antigen to a solid surface. This step is especially important when SMB antigen is used since it is less immunogenic unless it has been captured and concentrated (Johnson *et al.*, 2000). The patient's serum sample is added to the captured viral antigen and then detected with a commercially available secondary anti-human IgG-conjugated antibody. The indirect IgG ELISA has essentially replaced the HI assay because of its ease of use. This ELISA is either used quantitatively or with end-point dilutions (Johnson *et al.*, 2000; Miagostovich *et al.*, 1999). However, the IgG ELISA is often not very specific and a secondary assay such as the neutralization test is necessary to confirm the infecting virus.

Some investigators have explored the use of different antigens to increase the specificity of the IgG ELISA. Cardosa *et al.* (2002) demonstrated that the IgG response to pre-membrane (prM) protein has greater specificity when differentiating between DENV and JEV (Cardosa *et al.*, 2002). Others researchers have increased the specificity in the IgG ELISA assay by using a recombinant polypeptide located in the N-terminal portion of the envelope protein (Dos Santos *et al.*, 2004). Additionally, IgG avidity ELISAs have proven useful in differentiating primary versus secondary flavivirus infection (Cordeiro *et al.*, 2009; Matheus *et al.*, 2005). The avidity assays, which uses a stringent wash with 3–6 M urea buffer in order to remove non-specific IgG binding, have proven useful in differentiating acute from past flavivirus infections during unique circumstances in which the IgM response persists (De Souza *et al.*, 2004; Levett *et al.*, 2005).

In addition to measuring total IgG, investigators have measured different IgG subclasses to determine which is most abundant, most important for neutralization and strongly associated with disease severity (Koraka *et al.*, 2001; Thakare *et al.*, 1991; Thein *et al.*, 1993). For example in JEV, the IgG1 subclass is the most abundant antibody in the CSF following confirmed cases of clinical encephalitis. However, Thakare *et al.*, 1991 found no IgG2 in the serum or CSF of JEV patients and postulated that IgG1 antibodies are more cytophilic than other subclasses of antibodies and may be more effective in clearing the infection (Thakare *et al.*, 1991). And among DENV patients, IgG1 and IgG4 subclasses are risk markers for the development of dengue haemorrhagic fever and dengue shock syndrome (the more severe presentation of the disease) (Koraka *et al.*, 2001; Thein *et al.*, 1993).

IgA antibody detection

Immunoglobulin A (IgA) antibody detection has also been used to test for recent flaviviruses infection for DENV and JEV. Most tests for IgA use saliva samples rather than serum samples although both types of samples have been used. IgA is an attractive alternative to blood collection and it has been used successfully as a diagnostic tool for other viral diseases such as hepatitis A and B, rubella, measles, mumps and human immunodeficiency virus (Mortimer and Parry, 1988). There exist two isotypes (IgA1 and IgA2) which are important in mucosal immunity. IgA1 is found primarily in serum and IgA2 is found primarily in the saliva, tears and colostrum. The IgA2 is often referred to as secreted IgA2 because it is normally an oligomer of approximately 2–4 IgA monomers and it is resistant to proteolytic enzyme degradation to protect the body from microbes in bodily secretions.

Several studies have compared the sensitivity of IgA with that of IgM tests in both serum and saliva. The results showed that IgA in saliva had a 69% concordance compared with a 74% concordance in serum for DENV (Balmaseda *et al.*, 2003). The results from another study showed that IgA sensitivity during the first 29 days of illness was 48% (Talarmin *et al.*, 1998). In two studies among patients infected with JEV, IgA antibodies were measured in the serum and/or saliva. The percentage of patients found to have IgA antibodies in their serum compared with saliva differed dramatically. Prasad *et al.* (1994) found that 25% of patients with confirmed JEV had detectable IgA in their serum and none had detectable IgA in their saliva. A later study by Han *et al.* (1998) found that 37% of patients with JEV in the acute samples had detectable IgA in their serum and 40% had detectable IgA in their CSF. They also found that levels of IgA in the CSF of these patients was significantly lower than levels of IgM and that therefore the IgA test was less sensitive than the IgM test. Because IgA detection is in general less sensitive than IgM, IgA has not been incorporated into routine flaviviruses diagnostics.

Haemagglutination inhibition assay

The haemagglutination inhibition assay (HI) was one of the first tests used for flavivirus diagnostics and was the basis of flavivirus classification as discussed in the Introduction section of this chapter. The test was first described in 1958 by Clarke and Casals and later adapted to a microtitre plate format in 1980 (Clarke and Casals, 1958). HI assay measures total antibodies including IgM and IgG and does not differentiate between the two types of antibodies. The basis of the HI test is antibody induced inhibition of goose erythrocyte agglutination caused by the virus. The test measures the level of agglutination inhibition caused by antibodies produced against flavivirus antigens. The presence of anti-flavivirus antibodies in the serum sample inhibits the agglutination of the goose erythrocyte as observed by a button-like appearance of the erythrocyte in the 96-well micro-titre plate. The final results are read visually and determined by the experience of the laboratory technician performing the assay.

One of the main advantages of the HI test is that it is not species specific therefore it can be used to test any serum sample which is helpful for surveillance that involves multiple hosts. Since this assay is useful for surveillance there are some laboratories that still routinely use it for diagnostics however the development of the ELISA assay has essentially replaced the HI test mostly because of the difficulty in standardizing many reagents and buffers used as well as the availability of goose erythrocytes. For example, the gander was the preferred blood for the assay owing to the goose oestrus cycle yielding false positive results in the test. Many laboratories required two to three ganders to bleed every 6 weeks in order to obtain sufficient materials necessary for the assay. The quality control of the reagents used for this assay was difficult and could contribute to the difficulties of standardization. In addition, HI assay results are difficult to interpret because of cross-reactivity between viruses within the same group or subgroup. Today the HI test is used mainly for surveillance, particularly for surveillance of diseases in species other than humans when species-specific antibodies are not available.

Complement fixation

The complement fixation (CF) assay is not in widespread use but merits review since this technique was routinely used for flavivirus diagnostics prior to newly developed techniques. The CF assay is based on anti-flavivirus antibodies ability to fix complement in the presence of the viral antigen. This technique also has been replaced owing to the standardization issues similar to the HI assay. CF appears late in the infection and has a shorter half-life than IgG antibodies. CF is mainly used to distinguish between recent versus past infections. Because of the ease-of-use format of the IgM ELISA assay, it has been the preferred assay over the CF assay for recent flavivirus infection diagnosis. Today the CF assay is likely to be used when the IgM response is no longer detectable in a recent infection and paired serum samples are not available.

Antigen detection methods

Non-structural protein 1 (NS1)

The NS1 protein is a glycoprotein produced by all flaviviruses and is essential for viral replication and viability. During viral replication NS1 is secreted as a hexamer composed of dimer subunits. It produces a very strong humoral response and is known as a complement-fixing antigen. Because Ns1 is found in the blood of infected individuals, many studies have examined the usefulness of NS1 as a tool for flavivirus diagnosis especially for DENV. These studies have focused on antigen-capture ELISA and NS1-specific IgG responses.

NS1 studies have primarily been focused on DENV as a diagnostic tool. The sensitivity of the first NS1 antigen-capture ELISA tests ranged from 1 to 4 ng/ml (Alcon *et al.*, 2002; Libraty *et al.*, 2002; Young *et al.*, 2000). Results from earlier studies suggested that NS1 levels in blood correlated with disease severity among DENV patients although these results were not corroborated by those from later studies (Alcon *et al.*, 2002; Libraty *et al.*, 2002). This ELISA-based assay was also found to distinguish between DENV serotypes (Xu *et al.*, 2006). Moreover, the NS1 serotype-specific IgG indirect ELISA was found to differentiate primary from secondary DENV infections and results of this test showed good correlation with the plaque reduction neutralization test (PRNT) results (Shu *et al.*, 2003).The NS1 serotype-specific IgG ELISA worked reliably of typing DENV in convalescent-phase sera from patients with primary infection and in acute-phase sera from patients with secondary infection (thus eliminating the complicating effects of 'original antigenic sin'), but it was not reliable in typing convalescent-phase sera from patients with secondary infections.

The NS1 ELISA based antigen assay is commercially available for DENV and many investigators have evaluated the sensitivity and specificity this assay (Bessoff *et al.*, 2008; Blacksell *et al.*, 2008; Dussart *et al.*, 2006; Kumarasamy *et al.*, 2007). The commercialization of this assay and its broad usefulness during the course of infection for DENV makes this test an ideal low-tech acute-phase diagnostic test. Because of the specificity of the NS1 assay, it may also be useful in differentiating between flaviviruses. Unfortunately, many of the viruses in the JEV serocomplex group and other flaviviruses do not reach viral titres as those observed in DENV infections; therefore it is unlikely that NS1 antigen detection for other flaviviruses will be a useful diagnostic tool.

Neutralization tests

Plaque reduction neutralization test (PRNT)

The plaque reduction neutralization test (PRNT) is the most specific serological tool for the determination of type-specific antibodies to an infecting virus and is typically used to confirm the infecting flavivirus in a convalescent sera sample (Calisher *et al.*, 1989). This biological assay is based on the specific interaction of virus and antibody *in vitro* which ultimately results in the inactivation of the virus such that it is no longer able to infect and replicate in cell culture. The PRNT result is presented as the end-point titre of neutralizing antibodies from the serum to a specific virus and may be suggestive of the level of immune protection against the infecting virus.

The neutralization of flaviviruses by antibodies has been studied extensively in animal models, natural infections in humans and *in vitro* cell culture experiments. The neutralization epitopes are located throughout the E glycoprotein including Domain I, Domain II and the surface of Domain III (Crill and Roehrig, 2001; Modis *et al.*, 2005; Roehrig *et al.*, 1998). Viral neutralization by antibodies requires that multiple antibodies occupy the surface of the virion, referred to as the multihit mechanism (Gard, 1957; Westaway, 1965, 1968; Hierholzer and Dowdle, 1970; Wallis 1971). Neutralization of a virus requires groups of antigen antibody complexes on the virus surface which occupy a critical area to prevent cell penetration. Given the concept of the completion of critical areas for complete neutralization, a homologous antibody response would be more effective in neutralizing a virus than a heterologous response. However, a combination of neutralizing, non-neutralizing and subneutralizing antibodies has a synergistic neutralization effect. In other instances, heterologous antibodies could even potentiate or enhance viral entry as observed with DENV and WNV (Ferenczi *et al.*, 2008; Halstead and O'Rourke, 1977; Pierson *et al.*, 2007).

Flaviviruses have both group specific and type-specific antibody neutralization determinants. The type-specific determinants are more cross-reactive when tested in heterologous reactions, which cause the cross-reactivity observed in serological tests for IgM and IgG. The titre of neutralizing antibodies required to protect against a particular flavivirus is still not well understood. Vaccine studies have provided some information on antibody titres necessary for protection against a flavivirus infection however these studies were based on primary infections with attenuated viral strains.

The PRNT has been used to differentiate the infecting virus when cross-reactivity is observed in the standard antibody binding assays (IgM and IgG ELISA) as a confirmatory assay. Because the PRNT requires 7–10 days for a result, it is not a very useful diagnostic tool for clinical intervention and use. Because it is expensive and requires a tissue culture facility and highly skilled technical staff, it is not routinely performed in most laboratories. To obtain reliable and reproducible PRNT results, laboratories must standardize the fine details of this assay including the viral seeds and the passage of the cells used for plaquing. Furthermore, there is no consensus among laboratories on which plaque reduction percentages should be used. Some laboratories use a reduction of 70% of the total input plaquing virus and others consider a 90% reduction to be more specific. Although a 90% reduction in plaques is more specific, it is much less sensitive when using this parameter. In most vaccine trials, 50% reduction in plaques is considered significant because of the relatively low neutralization antibodies produced following vaccination with attenuated strains of viruses in naive population. The challenge of this assay is the standardization between all end-users. As stated previously, variables that account for inconsistent results between laboratories include the general interpretation of the results (PRNT 90–70–50%), the cell lines used for plaquing the virus and the viral seeds used in the assay. In an attempt to standardize the PRNT worldwide, subject matter experts from around the world recently published guidelines for PRNT standardization (Roehrig *et al.*, 2008).

Microneutralization assay

The microneutralization assays are newer formats of the neutralization assay that are quicker and easier to use. These ELISA-based assays are referred as microneutralization assay because of their 96-well microtitre plate format (Shanaka *et al.*, 2009; Taketa-Graham *et al.*, 2010; Vorndam and Beltran, 2002). They are based on the same principle as the PRNT's however the tedious task of counting the number of plaques per well is replaced with a colorimetric measurement of virus infection to determine the end-point dilution of the sample. Briefly, in the microneutralization assay susceptible cells are plated onto a 96-well microtitre format plate and infected with a previously incubated mixture containing both virus and serum sample. The cells are fixed and the virus replication is measured by using an

ELISA assay with either a group or type specific flavivirus monoclonal antibody (Table 13.3). In other formats neutral red dye has been used as the measured chromogen. The neutralization is measured by either optical density (for a colorimetric reading) or fluorescence. The per cent inhibition is determined based on the virus positive control and a graph is produced of the optical density and serum dilution. The end-point dilution is either calculated by using logistic regression curve for an absolute value or it is the serum dilution which is within two standard deviations below the virus control determined cut-off value (Putnak *et al.*, 2008; Vorndam and Beltran, 2002).

The major advantage of the microneutralization assay is that it requires less serum and fewer reagents and thus allows high-throughput testing. In addition, the 96-microtitre plate format allows for more serum dilutions thus increasing the probability of reaching an end-point dilution. Limitations of the assay include: a low correlation with PRNT especially in secondary flavivirus infections (Putnak *et al.*, 2008).

Challenges in flavivirus diagnostics

Viral persistence

In rare cases, flavivirus infection can become persistent which present unique diagnostic challenges. JEV was the first flavivirus to show evidence of viral persistence. This viral persistence has been documented primarily in cell culture and mouse models however approximately a third of all JEV cases have neurological sequelae (Ravi *et al.*, 1993) and disease recurrence has been reported (Sharma *et al.*, 1991). JEV establishes a latent infection in the peripheral blood mononuclear cells in children (Sharma *et al.*, 1991). The persistence of JEV has been described both as immunological response as well as virus in the nervous system (Ravi *et al.*, 1993). Approximately 5% of the patients either had the presence of either IgM antibodies or virus in the CSF for more than 110 days following the initial infection. These findings suggested that the virus persisted in the nervous system and in peripheral blood mononuclear cells.

WNV has also been demonstrated to persist in humans, animal models and cell culture (Appler *et al.*, 2010; Busch *et al.*, 2005; Hunsperger and Roehrig, 2005; Jarman *et al.*, 1968). Studies performed with human blood banks followed patients for 365 days in which both IgM antibodies persisted as well as virus (Busch *et al.*, 2005). These studies suggest that viral persistence is possible for flaviviruses and that the nervous system is an important reservoir for viral persistence. The persistence of both antibodies and virus complicates diagnosis of a recent or acute infection. Alternative methods of diagnosis become important to differentiate an acute, recent and past infection such as the use of antibody avidity assays (Levett *et al.*, 2005).

Original antigenic sin

The term 'original antigenic sin' was first used by an epidemiologist Thomas Francis, Jr. to describe unexplainable serological data that obtained from a 1946 influenza vaccine trial (Francis *et al.*, 1947). In that trial, Francis observed that students that had been previously infected with influenza A strain and later vaccinated with a new influenza A virus developed low serological neutralizing titres to vaccination and higher serological titres to the old virus with which they had been previously infected. The immunological basis of original antigenic sin is the expansion of lymphocyte clones that have increasing antibody avidity to cross-reactive epitopes from the viral structural proteins. This can occur when (a) sequential virus infections are antigenically related, (b) the virus is antigenically complex or (c) there is long sequential exposure intervals which allows for the selection and expansion of lymphocyte clones (Virelizier *et al.*, 1974). Original antigenic sin phenomenon makes the PRNT results for a person with a history of infection by multiple flaviviruses from the same antigenic complex uninterpretable unless previous infections are known (Halstead *et al.*, 1983). Examples of this cross-reactivity with sequential DENV infections are shown in Table 13.5 (Kuno *et al.*, 1993). In approximately 75% of the cases shown in Table 13.5 observed original antigenic sin in the second DENV infection and their PRNT results had the highest titre to the previous DENV infection and not the current infection.

Table 13.5 Original antigenic sin from sequential DENV infections taken from Kuno *et al.* (1993)

	Primary infection			Secondary infection PRNT50 titre						
Patient	Serotype	Method of confirmation	Day after onset	DENV1	DENV2	DENV3	DENV4	Serotype	Method of confirmation	With 'antigenic sin'
1	DENV1	Serological	1	10	<10	<10	<10	DENV2	Virological	Yes
			14	>10,240	2560	1280	ND			
2	DENV2	Virological	15	2560	>10240	<10	20	DENV1	Virological	Yes
			21	320	>10,240	640	80			
3	DENV4	Serological	2	<10	<10	<10	10	DENV1	Virological	No
			8	2560	640	640	160			
4	DENV1	Virological	7	1280	320	80	160	DENV4	Virological	Yes
			14	1280	320	640	640			
5	DENV2	Virological	3	ND	ND	ND	ND	DENV1 or 3	Serological	No
			20	2560	640	>10,240	320			
6	DENV4	Serological	0	<10	<10	<10	20	DENV2	Virological	Yes
			39	1600	640	1880	2480			
7	DENV4	Serological	3	<40	<40	<40	40	DENV2	Virological	Yes
			17	640	1280	320	2560			
8	DENV2	Serological	4	20	80	<10	<10	DENV 1 or 4	Serological	Yes
			38	320	2560	40	20			
9	DENV1	Serological	1	40	<10	20	<10	DENV3	Virological	Yes or No
			8	2215	80	1790	20			
			21	1080	160	1000	20			
			41	2210	160	4000	320			
			83	160	40	640	10			

Although the PRNT is considered to be the most specific test for determining the infecting virus especially following a single flavivirus exposure, multiple flavivirus infections complicate PRNT results in such a way that the results of neutralization cannot reliably confirm the infecting virus. Secondary flavivirus infections involving DENV and a non-dengue flavivirus, such as WNV, have been hypothesized to result in less common epitopes in the secondary infection and a less potent anamnestic response to the secondary infection compared with sequential dengue virus infections. Study results from Asia showed that exposure to JEV followed by a natural DENV infection caused a positive serological response to both viruses, although neutralization titres for DENV were measurably higher than those for JEV (Fukunaga *et al.*, 1983). However, sequential flavivirus infections in which exposure to DENV on one or more occasions was followed by exposure to a non-dengue flavivirus resulted in a different immune response. In a study in which animals were subjected to sequential multiple DENV and JEV infections, Edelman in 1973 found that although the animals had high neutralizing titres to both JEV and DENV, 50% of the animals displayed the effects of 'original antigenic sin,' in which the highest neutralizing titre was to DENV even though the most recent infection was to JEV (Edelman *et al.*, 1973). The incidence of 'original antigenic sin' varies and is dependent on the first acquired flavivirus infection. Interestingly, Kuno *et al.*, (1993) demonstrated that the neutralization assay was more accurate in detecting the infecting virus with acute samples than with convalescent samples (Kuno *et al.*, 1993). This was also noted by Westaway 1968 when comparing closely related flaviviruses in rabbit antisera and suggests that the IgM response is more specific and less cross-reactive than the response of higher avidity IgG antibodies (Westaway, 1968b).

Cross-protective immunity

An interesting aspect of flavivirus serology is the cross-reactivity among the different viruses that may provide cross-protective immunity to related viruses. The concept that previous flavivirus infections provide protection against heterologous flavivirus infections can be traced back to 1815 when William Pym, a medical officer in the 10th Infantry Regiment of Napoleon's army, observed that among the men who contracted yellow fever virus (YFV), all of those who had been originally stationed in India and moved to East Indies survived (100% survival). However, there were seven officers that had not been previously stationed in India, of which whom had died from YFV (30% survival). Pym had concluded from these results that the troops should be 'seasoned' in India before they are sent to East or West Indies to decrease mortality rate due to YFV (Pym 1815 (Ashcroft, 1979). 'Seasoning' referred to the process of Europeans adapting to a hot tropical climate and believed to decrease susceptibility to tropical diseases. Although it is not clear which flavivirus was protective among the soldiers, Pym observed perhaps a previous DENV infection providing cross-protective immunity to YFV as also observed by Izurieta *et al.* (2009) in a similar recent study in Colombia.

Antigenic cross-reactivity exists amongst viruses that group together genetically. However, whether this cross-reactivity results in immunity against these related viruses has been a topic of debate for many years. Edelman *et al.* (1975) found that JEV-infected patients with prior exposure to DENV had fewer symptoms during the acute phase of illness and fewer neurological residues a year after their JEV infection than did patients with no prior DENV exposure. Results of studies with the JEV vaccine (SA14–14) have indicated the presence of WNV cross-reactive antibodies in the serum of vaccinated subjects (Johnson *et al.*, 2005a; Tang *et al.*, 2008; Yamshchikov *et al.*, 2005). In another study, however, the cross-reactive antibodies from previous vaccination with JEV or YFV appeared to have no protective effect on the patients who were naturally infected with WNV (Johnson *et al.*, 2005a). In that study the neutralizing titres to YFV or JEV among patients vaccinated against these viruses ranged from undetectable levels to low titres of 1:20 prior to the patients natural WNV infection. However, when they were infected with WNV, neutralizing titres to the vaccine strains increased in YFV vaccinated patients but not to as high of levels as WNV neutralization titres. One patient who had been vaccinated against both JEV and YFV developed a higher titre to JEV than to the infecting virus thus demonstrating original antigenic sin in the JEV serocomplex group (Johnson *et al.*, 2005a). In another study of the effects of JEV and YFV vaccination, titres among vaccinated subjects ranged from 1:10 to 1:80 although titres to the vaccine strain of either JEV or YFV were 40-fold higher than titres to WNV (Yamshchikov *et al.*, 2005). These two studies both showed that the neutralization titres from vaccines of heterologous flaviviruses do not protect against natural flavivirus infections even within the same serocomplex group. Although results from many animal studies have shown previous flavivirus infections to be protective against infection with other flaviviruses, the results from these two human vaccination studies suggest that any such protective effect may be much smaller among humans (Goverdhan *et al.*, 1992; Halstead *et al.*, 1973; Sather and Hammon, 1970; Tarr and Hammon, 1974).

Cross-reactivity in flavivirus diagnosis

The process of cross-neutralization after sequential flavivirus infections is unknown, and studies using animal models yielded conflicting data. In animal models, genetically related flaviviruses have shown greater specificity to the homologous virus by IgM than by IgG (Westaway, 1968a,b). However, in studies of human infections, results have suggested that non-neutralizing antibodies could actually potentiate neutralization by inducing a conformational change in the viral antigen, resulting in increased availability of related antibodies from previous flavivirus infections capable of binding to cross-reactive epitopes of the infecting virus (Kimura-Kuroda and Yasui, 1983). In most serological assays the identification of a sequential flavivirus infection is difficult unless all previous flavivirus infections are known.

In the IgM ELISA assays, cross-reactivity occurs in the early convalescent phase of the infection. If the infection is a primary flavivirus infection, cross-reactivity is reduced in the late convalescent phase and the infecting virus can be differentiated from other flaviviruses on the basis of a fourfold increase in titre of the antibody to the viral antigen (Martin *et al.*, 2002). In secondary flavivirus infections the antibody response is dependent on the sequence of infections. There are two different methods of determining the specificity of the antibody response. One method involves generating an IgM dose–response curve for each of the viruses in question. If the antibody follows a first order kinetics from the dilution series for one viral antigen but not the other then the viral antigen with the first order kinetics is the infecting virus. This method is effective in identifying primary flavivirus infections but its effectiveness in identifying secondary flavivirus infections varies (Martin *et al.*, 2000). The second method involves testing both viral antigens in parallel and determining whether the positive to negative (P/N) values for the infecting viral antigen is at least twofold higher for the other viral antigen in question. Results obtained with both methods should always be confirmed with a PRNT (Fig. 13.3). However, as discussed in the previous section on the PRNT, the effects of 'original antigenic sin' complicate the interpretation of PRNT results for secondary flavivirus infections.

Future and conclusions

The future of flavivirus diagnostics will probably be determined by the interest in the diseases in first-world countries and non-profit organizations with the resources necessary to develop improved diagnostic technology. Because most medically relevant flavivirus diseases occur primarily in impoverished countries, the diagnostic technology should be simple and easy to implement. Diagnostic tests that require expensive specialized equipment would not be practical in these countries even though they may be considered the best diagnostic tests available on the basis of sensitivity and specificity. For example, although standard real time RT-PCR assays are considered to be the best basic diagnostic test currently available, they have not been adopted by all laboratories where flaviviruses circulate and are an important public health problem because their use requires the capacity for reagent storage and the specialized equipment.

To improve the clinical management of patients with a flavivirus infection as well as the increase the surveillance capacity for outbreak response and disaster management, health officials where flaviviruses are prevalent should focus on point-of care (POC) tests. In addition, the field of flavivirus diagnostics must develop better methods of differentiating among flaviviruses that co-circulate. The current antigens used for the IgM and IgG ELISAs are not able to provide this level of refined diagnosis. In the past 30 years there has been minimal improvement in serological assays, which begs the question of whether improved antigen development will provide better tests. Other markers in addition to antibodies could be further explored by microarrays or proteomics to discover host responses. These host response markers or biomarkers may lead to better new methods to diagnosing a flavivirus infection beyond what is currently available and provide a better alternative than the current focus in diagnostics. Again these biomarkers could be adapted to POC test for a simple, easy to use diagnostic tool for clinical intervention and surveillance purposes.

References

Alcon, S., Talarmin, A., Debruyne, M., Falconar, A., Deubel, V., and Flamand, M. (2002). Enzyme-linked immunosorbent assay specific to dengue virus type 1 nonstructural protein NS1 reveals circulation of the antigen in the blood during the acute phase of disease in patients experiencing primary or secondary infections. Journal of Clinical Microbiology *40*, 376–381.

Appler, K.K., Brown, A.N., Stewart, B.S., Behr, M.J., Demarest, V.L., Wong, S.J., and Bernard, K.A. (2010). Persistence of West Nile virus in the central nervous system and periphery of mice. PLoS ONE *5*, e10649.

Armah, H.B., Wang, G., Omalu, B.I., Tesh, R.B., Gyure, K.A., Chute, D.J., Smith, R.D., Dulai, P., Vinters, H.V., Kleinschmidt-DeMasters, B.K., *et al.* (2007). Systemic distribution of West Nile virus infection: postmortem immunohistochemical study of six cases. Brain Pathol. *17*, 354–362.

Ashcroft, M.T. (1979). Historical evidence of resistance to yellow fever acquired by residence in India. Trans R Soc Trop Med Hyg 73, 247–248.

Bae, H.-G., Nitsche, A., Teichmann, A., Biel, S.S., and Niedrig, M. (2003). Detection of yellow fever virus: a comparison of quantitative real-time PCR and plaque assay. J. Virol. Methods *110*, 185–191.

Balmaseda, A., Guzman, M.G., Hammond, S., Robleto, G., Flores, C., Tellez, Y., Videa, E., Saborio, S., Perez, L., Sandoval, E., *et al.* (2003). Diagnosis of dengue virus infection by detection of specific immunoglobulin M (IgM) and IgA antibodies in serum and saliva. Clin. Diagn. Lab.Immunol. *10*, 317–322.

Bessoff, K., Delorey, M., Sun, W., and Hunsperger, E. (2008). Comparison of two commercially available dengue virus (DENV) NS1 capture enzyme-linked immunosorbent assays using a single clinical sample for diagnosis of acute DENV infection. Clinical and Vaccine Immunology *15*, 1513–1518.

Bhoopat, L., Bhamarapravati, N., Attasiri, C., Yoksarn, S., Chaiwun, B., Khunamornpong, S., and Sirisanthana, V. (1996). Immunohistochemical characterization of a new monoclonal antibody reactive with dengue virus-infected cells in frozen tissue using immunoperoxidase technique. Asian Pacific Journal of Allergy and Immunology *14*, 107–113.

Blacksell, S.D., Bell, D., Kelley, J., Mammen Jr, M.P., Gibbons, R.V., Jarman, R.G., Vaughn, D.W., Jenjaroen, K., Nisalak, A., Thongpaseuth, S., *et al.* (2007). Prospective study to determine accuracy of rapid serological assays for diagnosis of acute dengue virus infection in Laos. Clin. Vaccine Immunol. *14*, 1458–1464.

Blacksell, S.D., Mammen Jr, M.P., Thongpaseuth, S., Gibbons, R.V., Jarman, R.G., Jenjaroen, K., Nisalak, A., Phetsouvanh, R., Newton, P.N., and Day, N.P.J. (2008). Evaluation of the Panbio dengue virus nonstructural 1 antigen detection and immunoglobulin M antibody enzyme-linked immunosorbent assays for the diagnosis of acute dengue infections in Laos. Diagn. Microbiol. Infect. Dis. *60*, 43–49.

Brown, T.M., Chang, G.J., Cropp, C.B., Robbins, K.E., and Tsai, T.F. (1994). Detection of yellow fever virus by polymerase chain reaction. Clin. Diagn. Virol. *2*, 41–51.

Burke, D.S., Nisalak, A., Ussery, M.A., Laorakpongse, T., and Chantavibul, S. (1985). Kinetics of IgM and IgG responses to Japanese encephalitis virus in human serum and cerebrospinal fluid. J. Infect.s Dis. *151*, 1093–1099.

Busch, M.P., Tobler, L.H., Saldanha, J., Caglioti, S., Shyamala, V., Linnen, J.M., Gallarda, J., Phelps, B., Smith, R.I.F., Drebot, M., *et al.* (2005). Analytical and clinical sensitivity of West Nile virus RNA screening and supplemental assays available in 2003. Transfusion *45*, 492–499.

Busch, M.P., Wright, D.J., Custer, B., Tobler, L.H., Stramer, S.L., Kleinman, S.H., Prince, H.E., Bianco, C., Foster, G., Petersen, L.R., *et al.* (2006). West nile virus infections projected from blood donor screening data, United States, (2003). Emerg. Infect. Dis. *12*, 395–402.

Calisher, C.H., and Gould, E.A. (2003). Advances in Virus Research, Vol 59 (London, Elsevier Academic Press).

Calisher, C.H., Karabatsos, N., Dalrymple, J.M., Shope, R.E., Porterfield, J.S., Westaway, E.G. and Brandt, W.E. (1989). Antigenic relationships between flaviviruses as determined by cross-neutralization tests with polyclonal antisera. J. Gen. Virol. *70 (Pt 1)*, 37–43.

Callahan, J.D., Wu, S.J.L., Dion-Schultz, A., Mangold, B.E., Peruski, L.F., Watts, D.M., Porter, K.R., Murphy, G.R., Suharyono, W., King, C.C., *et al.* (2001). Development and evaluation of serotype- and group-specific fluorogenic reverse transcriptase PCR (TaqMan) assays for dengue virus. J. Clin. Microbiol. *39*, 4119–4124.

Cardosa, M.J., Wang, S.M., Sum, M.S.H., and Tio, P.H. (2002). Antibodies against prM protein distinguish between previous infection with dengue and Japanese encephalitis viruses. BMC Microbiology *2*, 9.

Casals, J., and Brown, L.V. (1954). Hemagglutination with arthropod-borne viruses. J. Exp. Med. *99*, 429–449.

Chanama, S., Sukprasert, W., Sa-ngasang, A., A., A.*n*., Sangkitporn, S., Kurane, I., and Anantapreecha, S. (2005). Detection of Japanese Encephalitis (JE) virus-specific IgM in cerebrospinal fluid and serum samples from JE patients. Jap. J. Infect. Dis. *58*, 294–296.

Chao, D.-Y., Davis, B.S., and Chang, G.-J.J. (2007). Development of multiplex real-time reverse transcriptase PCR assays for detecting eight medically important flaviviruses in mosquitoes. J. Clin. Microbiol. *45*, 584–589.

Chien, L.J., Liao, T.L., Shu, P.Y., Huang, J.H., Gubler, D.J., and Chang, G.J.J. (2006). Development of real-time reverse transcriptase PCR assays to detect and serotype dengue viruses. J. Clin. Microbiol. *44*, 1295–1304.

Clarke, D.H., and Casals, J. (1958). Techniques for hemagglutination and hemagglutination-inhibition with arthropod-borne viruses. Am. J. Trop. Med. Hyg. *7*, 561–573.

Conceicao, T.M., Da Poian, A.T., and Sorgine, M.H.F. (2010). A real-time PCR procedure for detection of dengue virus serotypes *1, 2,* and *3,* and their quantitation in clinical and laboratory samples. J. Virol. Methods *163*, 1–9.

Cordeiro, M.T., Braga-Neto, U., Nogueira, R.M.R., and Marques Jr, E.T.A. (2009). Reliable classifier to differentiate primary and secondary acute dengue infection based on IgG ELISA. PLoS ONE *4(4)*.

Crill, W.D., and Roehrig, J.T. (2001). Monoclonal antibodies that bind to domain III of dengue virus E glycoprotein are the most efficient blockers of virus adsorption to Vero cells. J. Virol. *75*, 7769–7773.

De Souza, V.A.U.F., Fernandes, S., Araujo, E.S., Tateno, A.F., Oliveira, O.M.N.P.F., Dos Reis Oliveira, R., and Pannuti, C.S. (2004). Use of an Immunoglobulin G avidity test to discriminate between primary and secondary dengue virus infections. J. Clin. Microbiol. *42*, 1782–1784.

Dos Santos, F.B., Miagostovich, M.P., Nogueira, R.M.R., Schatzmayr, H.G., Riley, L.W., and Harris, E. (2004). Analysis of recombinant dengue virus polypeptides for dengue diagnosis and evaluation of the humoral immune response. Am. J. Trop. Med. Hyg. *71*, 144–152.

Dussart, P., Labeau, B., Lagathu, G., Louis, P., Nunes, M.R.T., Rodrigues, S.G., Storck-Herrmann, C., Cesaire, R., Morvan, J., Flamand, M., *et al.* (2006). Evaluation of an enzyme immunoassay for detection of dengue virus NS1 antigen in human serum. Clin. Vaccine Immunol. *13*, 1185–1189.

Dyer, J., Chisenhall, D.M., and Mores, C.N. (2007). A multiplexed TaqMan assay for the detection of arthropod-borne flaviviruses. J. Virol. Methods *145*, 9–13.

Edelman, R., Nisalak, A., Pariyanonda, A., Udomsakdi, S., and Johnsen, D.O. (1973). Immunoglobulin response and viremia in dengue-vaccinated gibbons repeatedly challenged with Japanese encephalitis virus. Am. J. Epidemiol. *97*, 208–218.

Edelman, R., Schneider, R.J., Chieowanich, P., Pornpibul, R., and Voodhikul, P. (1975). The effect of dengue virus infection on the clinical sequelae of Japanese encephalitis: a one year follow-up study in Thailand. Southeast Asian J. Trop. Med. Publ. Hlth *6*, 308–315.

Edelman, R., Schneider, R.J., Vejjajiva, A., Pornpibul, R., and Voodhikul, P. (1976). Persistence of virus-specific IgM and clinical recovery after Japanese encephalitis. Am. J. Trop. Med. Hyg. *25*, 733–738.

Ferenczi, E., Ban, E., Abraham, A., Kaposi, T., Petranyi, G., Berencsi, G., and Vaheri, A. (2008). Severe tick-borne encephalitis in a patient previously infected by West Nile virus. Scand.n J. Infect. Dis. *40*, 759–761.

Francis, T., Jr., Salk, J.E., and Quilligan, J.J., Jr. (1947). Experience with vaccination against influenza in the spring of 1947; a preliminary report. Am. J. Publ. Hlth *37*, 1013–1016.

Fresh, J.W., Reyes, V., Clarke, E.J., and Uylangco, C.V. (1969). Philippine hemorrhagic fever: a clinical, laboratory, and necropsy study. J. Lab. Clin. Med. *73*, 451–458.

Fukunaga, T., Okuno, Y., Tadano, M., and Fukai, K. (1983). A retrospective serological study of Japanese who contracted dengue fever in Thailand. Biken J. *26*, 67–74.

Gelpi, E., Preusser, M., Laggner, U., Garzuly, F., Holzmann, H., Heinz, F.X., and Budka, H. (2006). Inflammatory response in human tick-borne encephalitis: analysis of postmortem brain tissue. J. Neurovirol. *12*, 322–327.

Goverdhan, M.K., Kulkarni, A.B., Gupta, A.K., Tupe, C.D., and Rodrigues, J.J. (1992). Two–way cross-protection between West Nile and Japanese encephalitis viruses in bonnet macaques. Acta Virol. *36*, 277–283.

Hall, W.C., Crowell, T.P., Watts, D.M., Barros, V.L., Kruger, H., Pinheiro, F., and Peters, C.J. (1991). Demonstration of yellow fever and dengue antigens in formalin-fixed paraffin-embedded human liver by immunohistochemical analysis. Am. J. Trop. Med. Hyg. *45*, 408–417.

Halstead, S.B., and O'Rourke, E.J. (1977). Antibody-enhanced dengue virus infection in primate leukocytes. Nature *265*, 739–741.

Halstead, S.B., Rojanasuphot, S., and Sangkawibha, N. (1983). Original antigenic sin in dengue. Am. J. Trop. Med. Hyg. *32*, 154–156.

Halstead, S.B., Shotwell, H., and Casals, J. (1973). Studies on the pathogenesis of dengue infection in monkeys. II. Clinical laboratory responses to heterologous infection. J. Infect. Dis. *128*, 15–22.

Han, X.Y., Ren, Q.W., Xu, Z.Y., and Tsai, T.F. (1988). Serum and cerebrospinal fluid immunoglobulins M, A, and G in Japanese encephalitis. J. Clin. Microbiol. *26*, 976–978.

Heinz, F.X., Berger, R., Tuma, W., and Kunz, C. (1983a). Location of immunodominant antigenic determinants on fragments of the tick-borne encephalitis virus glycoprotein: evidence for two different mechanisms by which antibodies mediate neutralization and hemagglutination inhibition. Virology *130*, 485–501.

Heinz, F.X., Berger, R., Tuma, W., and Kunz, C. (1983b). A topological and functional model of epitopes on the structural glycoprotein of tick-borne encephalitis virus defined by monoclonal antibodies. Virology *126*, 525–537.

Houng, H.H., Hritz, D., and Kanesa-thasan, N. (2000). Quantitative detection of dengue 2 virus using fluorogenic RT-PCR based on 3′-noncoding sequence. J. Virol. Methods *86*, 1–11.

Houng, H.S.H., Chung-Ming Chen, R., Vaughn, D.W., and Kanesa-thasan, N. (2001). Development of a fluorogenic RT-PCR system for quantitative identification of dengue virus serotypes 1–4 using conserved and serotype-specific 3′ noncoding sequences. J. Virol. Methods *95*, 19–32.

Huang, J.L., Lin, H.T., Wang, Y.M., Weng, M.H., Ji, D.D., Kuo, M.D., Liu, H.W., and Lin, C.S. (2004). Sensitive and specific detection of strains of Japanese encephalitis virus using a one-step TaqMan RT-PCR technique. J. Med. Virol. *74*, 589–596.

Hunsperger, E., and Roehrig, J.T. (2005). Characterization of West Nile viral replication and maturation in peripheral neurons in culture. J. Neurovirol. *11*, 11–22.

Hunsperger, E.A., Yoksan, S., Buchy, P., Nguyen, V.C., Sekaran, S.D., Enria, D.A., Pelegrino, J.L., Vazquez, S., Artsob, H., Drebot, M., *et al.* (2009). Evaluation of commercially available anti-dengue virus immunoglobulin M tests. Emerg. Infect. Dis. *15*, 436–440.

Imrie, A., Meeks, J., Gurary, A., Sukhbaatar, M., Truong, T.T., Cropp, C.B., and Effler, P. (2007). Antibody to dengue 1 detected more than 60 years after infection. Viral Immunol. *20*, 672–675.

Innis, B.L., Nisalak, A., Nimmannitya, S., Kusalerdchariya, S., Chongswasdi, V., Suntayakorn, S., Puttisri, P., and Hoke, C.H. (1989). An enzyme-linked immunosorbent assay to characterize dengue infections where dengue and Japanese encephalitis co-circulate. Am. J. Trop. Med. Hyg. *40*, 418–427.

Ito, M., Takasaki, T., Yamada, K.-I., Nerome, R., Tajima, S., and Kurane, I. (2004). Development and evaluation of fluorogenic TaqMan reverse transcriptase PCR assays for detection of dengue virus types 1 to 4. J. Clin. Microbiol. *42*, 5935–5937.

Izurieta, R.O., Macaluso, M., Watts, D.M., Tesh, R.B., Guerra, B., Cruz, L.M., Galwankar, S., and Vermund, S.H. (2009). Anamnestic immune response to dengue and decreased severity of yellow fever. J. Glob. Infect. Dis. *1*, 111–116.

Jarman, R.V., Morgan, P.N., and Duffy, C.E. (1968). Persistence of West Nile virus in L-929 mouse fibroblasts. Proc. Soc. Exp. Biol. Med. *129*, 633–637.

Johnson, A.J., Martin, D.A., Karabatsos, N., and Roehrig, J.T. (2000). Detection of anti-arboviral immunoglobulin G by using a monoclonal antibody-based capture enzyme-linked immunosorbent assay. J. Clin. Microbiol. *38*, 1827–1831.

Johnson, B.W., Kosoy, O., Martin, D.A., Noga, A.J., Russell, B.J., Johnson, A.A., and Petersen, L.R. (2005a). West Nile virus infection and serologic response among persons previously vaccinated against yellow fever and Japanese encephalitis viruses. Vector-Borne Zoonotic Dis. *5*, 137–145.

Johnson, B.W., Russell, B.J., and Lanciotti, R.S. (2005b). Serotype-specific detection of dengue viruses in a fourplex real-time reverse transcriptase PCR assay. J. Clin. Microbiol. *43*, 4977–4983.

Kapoor, H., Signs, K., Somsel, P., Downes, F.P., Clark, P.A., and Massey, J.P. (2004). Persistence of West Nile Virus (WNV) IgM antibodies in cerebrospinal fluid from patients with CNS disease. J. Clin. Microbiol. *31*, 289–291.

Kimura-Kuroda, J., and Yasui, K. (1983). Topographical analysis of antigenic determinants on envelope glycoprotein V3 (E) of Japanese encephalitis virus, using monoclonal antibodies. J. Virol. *45*, 124–132.

Kong, Y.Y., Thay, C.H., Tin, T.C., and Devi, S. (2006). Rapid detection, serotyping and quantitation of dengue viruses by TaqMan real-time one-step RT-PCR. J. Virol. Methods *138*, 123–130.

Koraka, P., Suharti, C., Setiati, T.E., Mairuhu, A.T.A., Van Gorp, E., Hach, C.E., Juffrie, M., Sutaryo, J., Van Der Meer, G.M., Groen, J., *et al.* (2001). Kinetics of dengue virus-specific serum immunoglobulin classes and subclasses correlate with clinical outcome of infection. J. Clin. Microbiol. 39, 4332–4338.

Kumarasamy, V., Wahab, A.H.A., Chua, S.K., Hassan, Z., Chem, Y.K., Mohamad, M., and Chua, K.B. (2007). Evaluation of a commercial dengue NS1 antigen-capture ELISA for laboratory diagnosis of acute dengue virus infection. J. Virol. Methods *140*, 75–79.

Kuno, G. (1998). Universal diagnostic RT-PCR protocol for arboviruses. J. Virol. Methods 72, 27–41.

Kuno, G., Gubler, D.J., and Oliver, A. (1993). Use of 'original antigenic sin' theory to determine the serotypes of previous dengue infections. Trans. Roy. Soc. Trop. Med. Hyg. *87*, 103–105.

Kuno, G., Chang, G.J., Tsuchiya, K.R., Karabatsos, N., and Cropp, C.B. (1998). Phylogeny of the genus *Flavivirus*. J. Virol. *72*, 73–83.

Lai, Y.-L., Chung, Y.-K., Tan, H.-C., Yap, H.-F., Yap, G., Ooi, E.-E., and Ng, L.-C. (2007). Cost-effective real-time reverse transcriptase PCR (RT-PCR) to screen for dengue virus followed by rapid single-tube multiplex RT-PCR for serotyping of the virus. J. Clin. Microbiol. *45*, 935–941.

Lanciotti, R.S., Calisher, C.H., Gubler, D.J., Chang, G.J., and Vorndam, A.V. (1992). Rapid detection and typing of dengue viruses from clinical samples by using reverse transcriptase-polymerase chain reaction. Journal of Clinical Microbiology *30*, 545–551.

Lanciotti, R.S., and Kerst, A.J. (2001). Nucleic acid sequence-based amplification assays for rapid detection of West Nile and St. Louis encephalitis viruses. J. Clin. Microbiol. *39*, 4506–4513.

Lanciotti, R.S., Kerst, A.J., Nasci, R.S., Godsey, M.S., Mitchell, C.J., Savage, H.M., Komar, N., Panella, N.A., Allen, B.C., Volpe, K.E., *et al.* (2000). Rapid detection of west nile virus from human clinical specimens, field-collected mosquitoes, and avian samples by a TaqMan reverse transcriptase-PCR assay. J. Clin. Microbiol. *38*, 4066–4071.

Laue, T., Emmerich, P., and Schmitz, H. (1999). Detection of dengue virus RNA in patients after primary or secondary dengue infection by using the TaqMan automated amplification system. J. Clin. Microbiol. *37*, 2543–2547.

Levett, P.N., Sonnenberg, K., Sidaway, F., Shead, S., Niedrig, M., Steinhagen, K., Horsman, G.B., and Drebot, M.A. (2005). Use of immunoglobulin G avidity assays for differentiation of primary from previous infections with West Nile virus. J. Clin. Microbiol. *43*, 5873–5875.

Libraty, D.H., Young, P.R., Pickering, D., Endy, T.P., Kalayanarooj, S., Green, S., Vaughn, D.W., Nisalak, A., Ennis, F.A., and Rothman, A.L. (2002). High circulating levels of the dengue virus nonstructural protein NS1 early in dengue illness correlate with the development of dengue hemorrhagic fever. J. Infect. Dis. *186*, 1165–1168.

Liu, X.-Y., Yu, Y.-X., Li, M.-G., Xu, H.-S., Wang, H.-Y., Liang, G.-D., Jia, L.-L., and Dong, G.-M. (2008). [Study on the phenotypic characteristics of Japanese encephalitis virus strains isolated from different years]. Bingdu Xuebao *24*, 427–431.

Liu, Y., Chuang, C.K., and Chen, W.J. (2009). *in situ* reverse-transcription loop-mediated isothermal amplification (*in situ* RT-LAMP) for detection of Japanese encephalitis viral RNA in host cells. J. Clin. Virol. *46*, 49–54.

Maher-Sturgess, S.L., Forrester, N.L., Wayper, P.J., Gould, E.A., Hall, R.A., Barnard, R.T., and Gibbs, M.J. (2008). Universal primers that amplify RNA from all three flavivirus subgroups. Virol. J. *5*, 16.

Mantel, N., Aguirre, M., Gulia, S., Girerd-Chambaz, Y., Colombani, S., Moste, C., and Barban, V. (2008). Standardized quantitative RT-PCR assays for quantitation of yellow fever and chimeric yellow fever-dengue vaccines. J. Virol. Methods *151*, 40–46.

Martin, D.A., Biggerstaff, B.J., Allen, B., Johnson, A.J., Lanciotti, R.S., and Roehrig, J.T. (2002). Use of immunoglobulin M cross-reactions in differential diagnosis of human flaviviral encephalitis infections in the United States. Clin. Diagn. Lab.Immunol. *9*, 544–549.

Martin, D.A., Muth, D.A., Brown, T., Johnson, A.J., Karabatsos, N., and Roehrig, J.T. (2000). Standardization of immunoglobulin M capture enzyme-linked immunosorbent assays for routine diagnosis of arboviral infections. J. Clin. Microbiol. *38*, 1823–1826.

Matheus, S., Deparis, X., Labeau, B., Lelarge, J., Morvan, J., and Dussart, P. (2005). Discrimination between primary and secondary dengue virus infection by an immunoglobulin G avidity test using a single acute-phase serum sample. J. Clin. Microbiol. *43*, 2793–2797.

Miagostovich, M.P., Nogueira, R.M.R., Dos Santos, F.B., Schatzmayr, H.G., Araujo, E.S.M., and Vorndam, V. (1999). Evaluation of an IgG enzyme-linked immunosorbent assay for dengue diagnosis. J. Clin. Virol. *14*, 183–189.

Modis, Y., Ogata, S., Clements, D., and Harrison, S.C. (2005). Variable surface epitopes in the crystal structure of dengue virus type 3 envelope glycoprotein. J. Virol. *79*, 1223–1231.

Mortimer, P.P., and Parry, J.V. (1988). The use of saliva for viral diagnosis and screening. Epidemiol Infect *101*, 197–201.

Paranjpe, S., and Banerjee, K. (1998). Detection of Japanese encephalitis virus by reverse transcription/polymerase chain reaction. Acta Virol. *42*, 5–11.

Parida, M.M., Santhosh, S.R., Dash, P.K., Tripathi, N.K., Saxena, P., Ambuj, S., Sahni, A.K., Lakshmana Rao, P.V., and Morita, K. (2006). Development and evaluation of reverse transcription-loop-mediated isothermal amplification assay for rapid and real-time detection of Japanese encephalitis virus. J. Clin. Microbiol. *44*, 4172–4178.

Pierson, T.C., Xu, Q., Nelson, S., Oliphant, T., Nybakken, G.E., Fremont, D.H., and Diamond, M.S. (2007). The stoichiometry of antibody-mediated neutralization and enhancement of West Nile virus infection. Cell Host Microbe *1*, 135–145.

Prasad, S.R., Yergolkar, P.N., Walhekar, B.D., and Dandawate, C.N. (1994). Virus specific IgM, IgG and IgA antibodies in serum and saliva of Japanese encephalitis patients. Ind.J. Pediatr. *61*, 109–110.

Prince, H.E., Tobler, L.H., Lape-Nixon, M., Foster, G.A., Stramer, S.L., and Busch, M.P. (2005). Development and persistence of West Nile Virus-specific immunoglobulin M (IgM), IgA, and IgG in viremic blood donors. J. Clin. Microbiol. *43*, 4316–4320.

Putnak, J.R., De La Barrera, R., Burgess, T., Pardo, J., Dessy, F., Gheysen, D., Lobet, Y., Green, S., Endy, T.P., Thomas, S.J., *et al.* (2008). Comparative evaluation of three assays for measurement of dengue virus neutralizing antibodies. Am. J. Trop. Med. Hyg. *79*, 115–122.

Ravi, V., Desai, A.S., Shenoy, P.K., Satishchandra, P., Chandramuki, A., and Gourie-Devi, M. (1993). Persistence of Japanese encephalitis virus in the human nervous system. J. Med.Viro. *40*, 326–329.

Ravi, V., Robinson, J.S., Russell, B.J., Desai, A., Ramamurty, N., Featherstone, D., and Johnson, B.W. (2009). Evaluation of IgM antibody capture enzyme-linked immunosorbent assay kits for detection of IgM against Japanese encephalitis virus in cerebrospinal fluid samples. Am. J. Trop. Med. Hyg. *81*, 1144–1150.

Roehrig, J.T., Bolin, R.A., and Kelly, R.G. (1998). Monoclonal antibody mapping of the envelope glycoprotein of the dengue 2 virus, Jamaica. Virology *246*, 317–328.

Roehrig, J.T., Hombach, J., and Barrett, A.D.T. (2008). Guidelines for plaque-reduction neutralization testing of human antibodies to dengue viruses. Viral Immunol. *21*, 123–132.

Roehrig, J.T., Mathews, J.H., and Trent, D.W. (1983). Identification of epitopes on the E glycoprotein of Saint Louis encephalitis virus using monoclonal antibodies. Virology *128*, 118–126.

Rosen, L., and Gubler, D. (1974). The use of mosquitoes to detect and propagate dengue viruses. Am. J. Trop. Med. Hyg. *23*, 1153–1160.

Santhosh, S.R., Parida, M.M., Dash, P.K., Pateriya, A., Pattnaik, B., Pradhan, H.K., Tripathi, N.K., Ambuj, S., Gupta, N., Saxena, P., *et al.* (2007). Development and evaluation of SYBR Green I-based one-step real-time RT-PCR assay for detection and quantitation of Japanese encephalitis virus. J. Virol. Methods *143*, 73–80.

Sather, G.E., and Hammon, W.M. (1970). Protection against St. Louis encephalitis and West Nile arboviruses by previous dengue virus (types 1–4) infection. Proc. Soc. Exp. Biol. Med. *135*, 573–578.

Scaramozzino, N., Crance, J.M., Jouan, A., DeBriel, D.A., Stoll, F., and Garin, D. (2001). Comparison of flavivirus universal primer pairs and development of a rapid, highly sensitive heminested reverse transcription-PCR assay for detection of flaviviruses targeted to a conserved region of the NS5 gene sequences. J. Clin. Microbiol. *39*, 1922–1927.

Schwaiger, M., and Cassinotti, P. (2003). Development of a quantitative real-time RT-PCR assay with internal control for the laboratory detection of tick borne encephalitis virus (TBEV) RNA. J. Clin. Virol, *27*, 136–145.

Seth, G.P., Rodrigues, F.M., and Sarkar, J.K. (1974). The post-mortem diagnosis of Japanese encephalitis. J. Ind. Med. Assoc. *63*, 72–73.

Shanaka, W.W., Rodrigo, I., Alcena, D.C., Rose, R.C., Jin, X., and Schlesinger, J.J. (2009). An automated dengue virus microneutralization plaque assay performed in human Fcγ receptor-expressing CV-1 cells. Am. J. Trop. Med. Hyg. *80*, 61–65.

Sharma, S., Mathur, A., Prakash, V., Kulshreshtha, R., Kumar, R., and Chaturvedi, U.C. (1991). Japanese encephalitis virus latency in peripheral blood lymphocytes and recurrence of infection in children. Clin. Exp. Immunol. *85*, 85–89.

Shu, P.-Y., Chen, L.-K., Chang, S.-F., Yueh, Y.-Y., Chow, L., Chien, L.-J., Chin, C., Lin, T.-H., and Huang, J. -H. (2003). Comparison of capture immunoglobulin M (IgM) and IgG enzyme-linked immunosorbent assay (ELISA) and nonstructural protein NS1 serotype-specific IgG ELISA for differentiation of primary and secondary dengue virus infections. Clin. Diagn. Lab. Immunol. *10*, 622–630.

Taketa-Graham, M., Powell Pereira, J.L., Baylis, E., Cossen, C., Oceguera, L., Patiris, P., Chiles, R., Hanson, C.V., and Forghani, B. (2010). High throughput quantitative colorimetric microneutralization assay for the confirmation and differentiation of West Nile Virus and St. Louis encephalitis virus. Am. J. Trop. Med. Hyg. *82*, 501–504.

Talarmin, A., Labeau, B., Lelarge, J., and Sarthou, J.L. (1998). Immunoglobulin A-specific capture enzyme-linked immunosorbent assay for diagnosis of dengue fever. J. Clin. Microbiol. *36*, 1189–1192.

Tang, F., Zhang, J. -S., Liu, W., Zhao, Q.-M., Zhang, F., Wu, X.-M., Yang, H., Ly, H., and Cao, W.-C. (2008). Failure of Japanese encephalitis vaccine and infection in inducing neutralizing antibodies against West Nile virus, People's Republic of China. Am. J. Trop. Med. Hyg. *78*, 999–1001.

Tang, Y., Anne Hapip, C., Liu, B., and Fang, C.T. (2006). Highly sensitive TaqMan RT-PCR assay for detection and quantification of both lineages of West Nile virus RNA. J. Clin. Virol. *36*, 177–182.

Tarr, G.C., and Hammon, W.M. (1974). Cross-protection between group B arboviruses: resistance in mice to Japanese B encephalitis and St. Louis encephalitis viruses induced by dengue virus immunization. Infect. Immun. *9*, 909–915.

Thakare, J.P., Gore, M.M., Risbud, A.R., Banerjee, K., and Ghosh, S.N. (1991). Detection of virus specific IgG subclasses in Japanese encephalitis patients. Ind. J. Med. Res. – Section A Infect. Dis. *93*, 271–276.

Thein, S., Aaskov, J., Myint, T.T., Shwe, T.N., Saw, T.T., and Zaw, A. (1993). Changes in levels of anti-dengue virus IgG subclasses in patients with disease of varying severity. J. Med. Virol. *40*, 102–106.

Vaughn, D.W., Green, S., Kalayanarooj, S., Innis, B.L., Nimmannitya, S., Suntayakorn, S., Rothman, A.L., Ennis, F.A., and Nisalak, A. (1997). dengue in the early febrile phase: Viremia and antibody responses. J. Infect.Dis. *176*, 322–330.

Virelizier, J.L., Allison, A.C., and Schild, G.C. (1974). Antibody responses to antigenic determinants of influenza virus hemagglutinin. II. Original antigenic sin: a bone marrow-derived lymphocyte memory phenomenon modulated by thymus-derived lymphocytes. J. Exp. Med. *140*, 1571–1578.

Vorndam, V., and Beltran, M. (2002). Enzyme-linked immunosorbent assay-format microneutralization test for dengue viruses. Am. J. Trop. Med. Hyg. *66*, 208–212.

Westaway, E.G. (1968a). Antibody responses in rabbits to the group B arbovirus Kumjin: serologic activity of the fractionated immunoglobulins in homologous and heterologous reactions. J. Immunol. *100*, 569–580.

Westaway, E.G. (1968b). Greater specificity of 19S than 7S antibodies on haemagglutination-inhibition tests with closely related group B arboviruses. Nature *219*, 78–79.

WHO (2009). Evaluation of Commercially Available anti-dengue virus immunoglobulin M test. World Health Organization Evaluation Report Series No *3*, 1–52.

Wicki, R., Sauter, P., Mettler, C., Natsch, A., Enzler, T., Pusterla, N., Kuhnert, P., Egli, G., Bernasconi, M., Lienhard, R., *et al.* (2000). Swiss Army Survey in Switzerland to determine the prevalence of *Francisella tularensis*, members of the *Ehrlichia phagocytophila* genogroup, *Borrelia burgdorferi sensu lato*, and tick-borne encephalitis virus in ticks. Eur. J. Clin. Microbiol. Infect. Dis. *19*, 427–432.

Xu, H., Di, B., Pan, Y.-X., Qiu, L.-W., Wang, Y.-D., Hao, W., He, L.-J., Yuen, K.-Y., and Che, X.-Y. (2006). Serotype 1-specific monoclonal antibody-based antigen capture immunoassay for detection of circulating nonstructural protein NS1: Implications for early diagnosis and serotyping of dengue virus infections. J. Clin. Microbiol. *44*, 2872–2878.

Yamshchikov, G., Borisevich, V., Kwok, C.W., Nistler, R., Kohlmeier, J., Seregin, A., Chaporgina, E., Benedict, S., and Yamshchikov, V. (2005). The suitability of yellow fever and Japanese encephalitis vaccines for immunization against West Nile virus. Vaccine *23*, 4785–4792.

Yang, D.K., Kweon, C.H., Kim, B.H., Lim, S.I., Kim, S.H., Kwon, J.H., and Han, H.R. (2004). TaqMan reverse transcription polymerase chain reaction for the detection of Japanese encephalitis virus. J. Vet. Sci. *5*, 345–351.

Young, P.R., Hilditch, P.A., Bletchly, C., and Halloran, W. (2000). An antigen capture enzyme-linked immunosorbent assay reveals high levels of the dengue virus protein NS1 in the sera of infected patients. Journal of Clinical Microbiology *38*, 1053–1057.

Zaayman, D., Human, S., and Venter, M. (2009). A highly sensitive method for the detection and genotyping of West Nile virus by real-time PCR. J. Virol. Methods *157*, 155–160.

Flavivirus–Vector Interactions

14

Ken E. Olson and Carol D. Blair

Abstract

Flaviviruses such as dengue, yellow fever and West Nile viruses continue to cause a significant amount of disease in humans. Most flaviviruses are maintained in nature by cycling between haematophagous arthropod vectors and vertebrate hosts, and the viruses must replicate in both vectors and hosts. This review focuses on flavivirus–vector interactions to present a current understanding of events and processes that lead to vector infection, virus amplification and dissemination, and transmission. This chapter will focus mainly on dengue viruses (DENVs) and their interactions with *Aedes aegypti*, but will include interactions between other flaviviruses and their vectors where appropriate. Flavivirus–mosquito cell interactions will be discussed first to give the reader a cellular view of the infection process but this will be followed with a view of the infection process in vectors. This review will describe flavivirus interactions with the vector's recognized innate immune (Toll, Jak-STAT, apoptosis) and antiviral (RNA interference) pathways and discuss flavivirus evolution and its consequences for vector infection, DENV transmission and genotype displacement. The review will discuss how our understanding of vector genetics is enhanced by the availability of genome databases for *A. aegypti* and *Culex quinquefasciatus*, tissue-specific transcriptomes and microarrays and small RNA databases. The chapter will also discuss RNA silencing and vector transgenesis as tools for defining gene function. Finally, we will review several recently described vector-based approaches that may result in new strategies for flavivirus disease control.

Introduction

Infection of, replication and amplification in, and transmission to a vertebrate host by an arthropod vector are essential parts of the life cycle of human pathogenic flaviviruses. In addition to recent advances in knowledge and technologies for investigation of virus structure, replication and genetics that have aided research on infections in both vertebrate hosts and arthropods, major developments in mosquito genetics and genomics have allowed new insights into the molecular biology of virus–vector interactions. Complete genome sequences of two major flavivirus vectors, *A. aegypti* (Nene *et al.*, 2007) (http://aaegypti.vectorbase.org/SequenceData/Genome/) and *Cx. quinquefasciatus* (http://cquinquefasciatus.vectorbase.org/SequenceData/Genome/) are available as supercontigs, although not yet assembled; these databases allow identification of vector genes important in infection and transmission of flaviviruses.

Small laboratory animal models of disease (principally mice) are available for most of the encephalitogenic flaviviruses, such as West Nile virus (WNV), and some aspects of dengue disease can be observed after infection of IFN-deficient mice (Johnson and Roehrig, 1999; Kyle *et al.*, 2007; Shresta *et al.*, 2006) or severely immunocompromised mice engrafted with human immune system tissues (An *et al.*, 1999; Blaney *et al.*, 2002; Kuruvilla *et al.*, 2007) or stem cells (Bente *et al.*, 2005); however, no vertebrate animal that accurately models all aspects of human disease for DENVs and yellow fever virus (YFV) has been developed. For both groups of viruses, mosquitoes can provide the advantages of authentic animal models for a vital aspect of the natural flavivirus transmission and infection cycle. The observations that most flavivirus infections of both intact arthropods and their cultured cells are persistent and non-pathogenic, whereas infection of vertebrate hosts can lead to disease and death, suggest that there are fundamental differences in interactions with the two hosts at the cellular as well as the whole organism level. In this chapter, we will attempt to point out critical/key stages where these differences occur. We will also point out how knowledge of arthropod–virus interactions can aid development of strategies for flavivirus disease control.

This chapter will focus mainly on DENVs because of their importance as the major arboviral disease agent in humans and recent advances in knowledge of their interactions with their mosquito vectors, with references to WNV and other flaviviruses where additional studies provide unique information. Dengue fever is the most frequently occurring arthropod-borne viral disease of humans, with almost half the world's population at risk of infection. The high prevalence, lack of a licensed vaccine, and absence of specific treatment make dengue fever a global public health threat. *A. aegypti* is the most important vector for humans of not only DENVs (serotypes 1–4) but also YFV (Gubler, 2002; Monath, 1994). *A. albopictus* is a less common and less competent vector for DENVs (Lambrechts *et al.*, 2010).

WNV has a natural transmission cycle between birds and *Culex* spp. mosquitoes. It was known as the agent of relatively benign human disease until more virulent infections with frequent neurological involvement emerged in the middle east and southern Europe in the 1990s, then invaded North America in 1999 (Lanciotti *et al.*, 1999). In contrast to DENVs, WNV can be transmitted by a large number of mosquito genera and species, although individual species and populations vary widely in vector competence (Turell *et al.*, 2002, 2005). *Culex pipiens*, *Cx. quinquefasciatus*, and *Cx. tarsalis* are the most prominent vectors in North America.

During a natural mosquito infection, a flavivirus is taken up in a viraemic blood meal and must establish an infection in the mosquito midgut by overcoming a midgut infection barrier (MIB), replicate in the midgut epithelium, overcome a midgut escape barrier (MEB) to pass through the midgut basal lamina and disseminate to and amplify in other tissues such as fat body, haemocytes, nervous tissue and most importantly salivary glands, in order to be transmitted in saliva during host probing for a subsequent blood meal. There is little evidence of a salivary gland infection barrier to flavivirus infection (Black *et al.*, 2002).

The ability of a flavivirus to infect and escape from the midgut [vector competence, (VC)] is determined in part by mosquito genetics, and may in part be a function of mosquito antiviral immunity discussed later in this chapter. Various strains or serotypes of a particular flavivirus also exhibit genetic variability in competence to establish mosquito infection. There is phylogenetic evidence for rapid evolution of both mosquito and viral genetic determinants of infection. Lambrechts *et al.* (2009) have shown that VC for DENV is governed to a large extent by local adaptation of DENV genotypes to vectors in genetically diverse, natural populations. Arthropod vectors indeed play a vital role in flavivirus maintenance in nature, geographic expansion, and disease epidemics.

Molecular biology of flavivirus infection of mosquito cells

Fundamental molecular processes in flavivirus replication such as translation and virus-directed RNA synthesis are thought to be the same in all eukaryotic cells. Much of our basic understanding of the molecular biology of flavivirus infection and replication comes from studies in mammalian cell cultures. These are advantageous not only because almost all cells in a culture can be simultaneously infected to allow examination of synchronous steps in replication in a one-step growth experiment, but also because of the ability to establish either primary cultures or continuous lines from defined tissues that may mimic the natural target cell types of each flavivirus.

Most mosquito cell lines used currently for study of flaviviruses were established in the 1960s from larvae of *A. aegypti* and *A. albopictus* (Singh, 1967) or embryos of *A. aegypti* (Peleg, 1968). Subcultures or clonal lines have been selected and characterized from the original mosquito cell culture lines on the basis of properties such as support of high titre growth of DENV and chikungunya virus (Igarashi, 1978), presence/lack of arbovirus-induced cytopathology (Lan and Fallon, 1990; Sarver and Stollar, 1977) or culture medium requirements (Lan and Fallon, 1990). While these cell lines have been very useful for propagation of high-titre virus stocks and studies of viral interactions at the cellular level, the tissues of origin of these cells and their resemblance to adult mosquito tissues are unknown and thus interpretations of study results are limited, as will be pointed out further in this chapter.

DENVs can infect and replicate in cultured mosquito cells from a number of different species such as *A. albopictus* (C6/36 and ATC15), *A. aegypti* (Aag2 and ATC10), and *A. pseudoscutelaris* (AP61) (Thaisomboonsuk *et al.*, 2005). Flaviviruses also can be adapted to growth in cultured *Drosophila melanogaster* cell lines and flies. Advantages of using *Drosophila* cells and animals are ease of genetic manipulation of host cells and availability of molecular biological and serological reagents. For example, 116 candidate insect host factors required for DENV2 propagation were identified by carrying out a genome-wide RNA interference (RNAi) screen in cultured *D. melanogaster* cells. A human siRNA library screen showed that 42 of these genes with human homologues also are required for DENV replication in human cells (Sessions *et al.*, 2009).

Cell attachment and entry

In mammalian cells, the flaviviral envelope (E) protein mediates attachment to specific receptor(s), entry by receptor-mediated endocytosis, and fusion with the endosomal membrane, and it is assumed that these same steps are required to initiate infection of mosquito cells. Specificity of both ligands on the virion surface (E protein) and cellular receptors is involved. There has been much speculation regarding the identity of cellular receptors, and whether or not they are the same for both vertebrate and mosquito cells, or even if receptors are the same in all mosquito tissues (Black *et al.*, 2002). Characteristics of various mosquito cell receptors have been described, but none has been definitely identified.

A variety of cell-surface glycosaminoglycans (GAGs) such as heparan sulfate are expressed on many cell types and are commonly bound by many microbial agents, including DENVs, to gain access to the inside of cells (Rostand and Esko, 1997). Virus binding to GAGs involves primarily the electrostatic interaction of clusters of basic amino acids on the virion surface with negatively charged sulfate groups on the GAG polysaccharide. A number of studies have implicated cell surface GAGs in initial binding of DENVs to cultured mammalian cells (Chen *et al.*, 1997; Germi *et al.*, 2002; Hilgard and Stockert, 2000; Hung *et al.*, 1999). Hung *et al.* showed that soluble heparin blocked binding of recombinant domain III of DENV2 E-protein (EDIII) to mammalian cells (BHK21), but not to cultured *A. albopictus* (C6/36) mosquito cells (Hung *et al.*, 2004). Talarico *et al.* (2005) also showed that sulfated polysaccharides do not inhibit DENV binding to C6/36. Mosquitoes appear to have the genetic capability to synthesize GAGs (Sinnis *et al.*, 2007), but apparently they are not used in flavivirus attachment to cultured mosquito cells, and binding to specific plasma membrane protein(s) has been postulated to explain the cell tropism of DENVs *in vivo*.

Several groups have identified putative cell membrane receptors on invertebrate host cells capable of binding DENV E protein that include heat shock proteins (Reyes-Del Valle *et al.*, 2005; Salas-Benito *et al.*, 2007) and other proteins of various sizes that remain unidentified (Munoz *et al.*, 1998), suggesting that DENV may be capable of attaching to various mosquito cellular receptors, possibly via more than one E protein motif. The presence of the RGD motif in an extended loop of the E protein of encephalitogenic mosquito-borne flaviviruses suggested integrins as potential host cell attachment proteins (Chu and Ng, 2004b; van der Most *et al.*, 1999).

The crystal structures have been solved for the E protein ectodomains of the flaviviruses tick-borne encephalitis virus (TBEV) (Rey *et al.*, 1995), DENV2 (Modis *et al.*, 2003), DENV3 (Modis *et al.*, 2005), and WNV (Kanai *et al.*, 2006), revealing three prominent structural domains, DI, DII, and DIII. DIII is an immunoglobulin-like structure shown in many studies with mammalian cells to be required for binding to cellular receptors. A surface-exposed, extended loop motif on EDIII is present on all mosquito-borne flaviviruses and absent in TBEV and other tick-borne flaviviruses (Rey *et al.*, 1995; Zhang *et al.*, 2004). The absence of this structure, termed the FG loop, in the tick-borne viruses, which typically are not able to infect C6/36 cells, has led to speculation that it plays a specific role in binding to mosquito cells. In a study with DENV2 E protein FG loop mutants Erb *et al.* (2010) showed that the presence, but not the amino acid sequence, of the EDIII FG loop is critical for infection and replication in mosquito midguts and head tissues as well as cultured mammalian cells but is dispensable for infection of C6/36 mosquito cell cultures.

The flavivirus E-protein is a Class II fusion protein. When exposed to low pH, the E-protein of mature virions undergoes an irreversible conformational change in which surface dimers reassociate as trimers, resulting in a cell membrane fusion-competent form. The fusion loops in EDII at the tip of each monomer in the trimeric structure (aa 98–111 in DENV2) are exposed for insertion into a cellular membrane (Modis *et al.*, 2004). In mammalian cells, after viral attachment and entry via receptor-mediated endocytosis, acidification of the endosome causes the conformational shift that results in fusion of the virion envelope and the endocytic vesicular membrane to release the nucleocapsid into the cytoplasm. Although there is evidence that flaviviruses enter C6/36 cells by receptor mediated endocytosis (Acosta *et al.*, 2008; Chu and Ng, 2004a,b; Mosso *et al.*, 2008), electron microscopic studies have also suggested that flaviviruses can infect these cells by direct fusion with the plasma membrane (Hase *et al.*, 1989a,b; Nawa *et al.*, 2007). Certain mutations when introduced into the hydrophobic amino acids of the EDII fusion loop are lethal for infection of both cultured mammalian and mosquito cells, suggesting that virus envelope–cell membrane fusion is essential for flavivirus infection of both cell types. Interestingly, seven viable fusion loop mutants that could replicate to high titre in C6/36 cells had significantly lower infection rates in mosquito tissues after intrathoracic inoculation (Huang *et al.*, 2010). This study, like other findings cited in this chapter, indicated that DENV infection of C6/36 cells does not necessarily reflect viral infectivity for mosquitoes.

Genome expression and replication

After release from the nucleocapsid, viral genome RNA translation and replication in mosquito cell cytoplasm are assumed to use similar mechanisms to mammalian cells, although kinetics might be slower because of lower incubation temperatures of mosquito cell cultures and mosquito colonies. Most flavivirus infections of cultured mosquito cells produce little or no cytopathology. After an initial burst of RNA replication and high-titre virion production, most flavivirus-infected mosquito cell cultures shift to a persistently infected mode, in which a lower titre of infectious virus is released over extended time periods.

A number of microscopic and biochemical studies have shown that flavivirus RNA replicative complexes are found in the cytoplasm of infected mammalian cells enclosed in double-membrane structures derived from the endoplasmic reticulum (Uchil and Satchidanandam, 2003; Welsch *et al.*, 2009). Cellular organelle fractionation techniques and immunofluorescent staining for dsRNA were used to show that double-membrane vesicles that arise from the endoplasmic reticulum are associated with both positive-sense and negative-sense viral RNA in DENV2-infected mosquito cell cultures and *A. aegypti* (Poole-Smith, 2010).

Virion assembly and release

During viral assembly and maturation in low-pH exocytic vesicles in mammalian cells, the prM protein functions as a chaperone for the E-protein, maintaining its native conformation until the mature virion exits the cell. The immature virus particles assembled in the endoplasmic reticulum have 60 spikes, each composed of three prM–E heterodimers (Zhang *et al.*, 2003). The low-pH environment of the trans-Golgi network causes a rearrangement of the heterodimers to expose prM for cleavage by a cellular furin protease. The prM to M cleavage occurs late in maturation, and the pr peptide remains associated until the virion is released into a neutral pH extracellular environment (Yu *et al.*, 2009). The prM–M junction in all four DENV serotypes has a glutamic acid residue in P3 of the furin cleavage site, partially inhibiting cleavage and resulting in release of a high proportion of prM-containing virions, particularly from C6/36 cells (Junjhon *et al.*, 2010). Virions containing a high proportion of uncleaved prM protein are less infectious and not capable of efficiently initiating cell membrane fusion, and the significance of their production in mosquito cells is unknown.

In a classic electron microscopic study of St. Louis encephalitis virus (SLEV)-infected *Culex pipiens*, Whitfield *et al.* (1973) showed that maturation of a flavivirus in tissues of its natural vector occurs by an exocytic pathway in which virus particles assemble within endoplasmic reticulum-associated cytoplasmic vesicles that appear to fuse with the intact plasma membrane to release virus from cells. Electron microscopic examination of DENV2-infected C6/36 cells indicated that mature virions formed inside cytoplasmic vacuoles and were released from cells by exocytosis in a similar manner to that described for whole mosquitoes, but also suggested that budding from the plasma membrane could occur (Hase *et al.*, 1987).

The composition of virions released from mosquito cells differs from mammalian cell-produced virus in several respects. Recent studies in mammalian cell cultures and model liposome systems have indicated that flavivirus fusion activity is strongly promoted by the presence of cholesterol in the target membrane (Moesker *et al.*, 2010), that infection causes up-regulation of cholesterol biosynthesis and redistribution of cholesterol to viral replication membranes (Mackenzie *et al.*, 2007) and replication and possibly virion egress requires the presence of cholesterol and cholesterol-rich membrane microdomains (Puerta-Guardo *et al.*, 2010). Paradoxically, insects cannot synthesize cholesterol *de novo* and depend on dietary cholesterol for their physiological requirements (Krebs and Lan, 2003). Insect cells in culture contain reduced levels of cholesterol in their plasma membranes compared with mammalian cells (Mitsuhashi *et al.*, 1983). Nevertheless, structurally robust virus particles with specific infectivity equivalent to those from mammalian cells are released from cholesterol-deficient mosquito cells, as was directly demonstrated for alphaviruses (Hafer *et al.*, 2009).

The DENV E-protein contains two possible N-linked sites for glycosylation, N67 and N153; however, the DENV glycosylation site at E amino acid 67 is unique among flaviviruses. DENV1 E-glycoprotein grown in C6/36 cells is glycosylated at both N67 and N153; DENV 2 E-glycoprotein is glycosylated only at N67 (Johnson *et al.*, 1994). Hsieh *et al.* (1983) demonstrated that the composition of N-linked glycans on arbovirus glycoproteins is determined by the biosynthetic machinery of the host cell. When an alphavirus was grown in C6/36 mosquito cells, Man3GlcNAc2 glycans, which are identical to the trimannosyl core of N-linked oligosaccharides in vertebrate cells, were preferentially located at the glycosylation sites, which were previously shown to have complex glycans in virus grown in vertebrate cells. They concluded that Man3GlcNAc2 structures in *A. albopictus* are the equivalent of complex-type oligosaccharides in mammalian cells (Hsieh and Robbins, 1984). A lectin-binding assay showed that the attached carbohydrate chains for E proteins of both DENV1 and 2 grown in C6/36 cells were probably of the oligomannose type and contained no galactose or sialic acid (Johnson *et al.*, 1994).

Flavivirus genetic determinants of mosquito cell infection

Some attenuating mutations in flavivirus vaccines and vaccine candidates confer reduced replication efficiency in mosquito cells and tissues (Hanley *et al.*, 2003; Huang *et al.*, 2003). Vaccine candidate replication efficiency in mosquitoes potentially could help define molecular mechanisms comparable to those better-characterized in mammalian cells.

YFV Asibi strain infects and disseminates in a high proportion of *A. aegypti* mosquitoes. YFV Asibi was used to develop the 17D live attenuated virus vaccine (Theiler and Smith, 1937) and shortly thereafter, Whitman (1939) demonstrated that the vaccine virus could not be transmitted by *A. aegypti*. Subsequent studies have shown that YFV 17D can poorly infect mosquito midguts after oral infection (midgut infection rate of 30% as compared with 72% for Asibi); furthermore, 17D does not disseminate to the salivary glands by comparison to an 83% dissemination rate for

Asibi (McElroy *et al.*, 2006b). Genome sequence comparison of the Asibi and 17D strains shows 67 nucleotide differences encoding 33 amino acid substitutions throughout the viral genome (Hahn *et al.*, 1987). By construction of chimeric infectious cDNA clones in which genes from parental Asibi and attenuated 17D occurred in various combinations, McElroy *et al.* (2006a, b) investigated the influence of specific genes and/or protein domains on infection and dissemination in *A. aegypti*. Substitution of DIII of the envelope protein from a midgut-restricted YFV into a wild-type YFV resulted in a marked decrease in virus dissemination, suggesting an important role for EDIII in this process (McElroy *et al.*, 2006b). Substitution of the 17D NS2A or NS4B genes into Asibi also significantly attenuated YFV dissemination, demonstrating that dissemination has multiple genetic determinants (McElroy *et al.*, 2006a).

A 30 nt deletion in the 3′ untranslated region (3′Δ30) of DENV4 was shown to significantly decrease dissemination from the midgut (Troyer *et al.*, 2001). A single mutation in the NS4B protein (P101L) of 3′Δ30 caused further significant reduction of C6/36 replication and midgut infection while enhancing replication in monkey kidney (Vero) and human hepatoma cell (Huh7) cultures (Hanley *et al.*, 2003). Studies on the roles of protein NS4B in mammalian cells have shown that it acts as an interferon antagonist (Munoz-Jordan *et al.*, 2005) and might play a role in endoplasmic reticulum membrane proliferation (Miller *et al.*, 2006); however, the functions of the flaviviral non-structural proteins in mosquito cells are not known. The studies cited here emphasize the dearth of specific information on molecular biology of flavivirus replication in mosquito cells by comparison to mammalian cells and point to the need for further research in this area.

Aedes aegypti, the vector of dengue viruses

DENVs are transmitted to humans by day-biting mosquito vectors, *A. aegypti* and *A. albopictus*. From its origins in Africa, *A. aegypti*, like DENVs, has spread to tropical and subtropical regions of the world mirroring the increase of human commerce and trade and the rise of tropical urban centres (Tabachnick, 1991). During the last 30 years, *A. albopictus* also has spread from its origins in Asia and is now found in many tropical and temperate regions of the world (Benedict *et al.*, 2007; Lambrechts *et al.*, 2010). *A. aegypti* is mainly an anthropophilic vector with a limited flight range and most of this vector's daily activities occur indoors (Delatte *et al.*, 2010; Ooi *et al.*, 2006). *A. albopictus* is an opportunistic feeder seeking blood meals from a variety of mammalian hosts including humans and is mostly an outdoor mosquito (Delatte *et al.*, 2010).

A. aegypti is the primary vector of explosive urban DENV epidemics and is the vector species most often associated with the emergence of severe forms of DEN disease [DEN haemorrhagic fever (DHF) and DEN shock syndrome (DSS)] (Weaver and Reisen, 2010). All four DENV serotypes are readily maintained in an urban transmission cycle involving *A. aegypti* and humans (Mackenzie *et al.*, 2004). *A. albopictus* is an important secondary vector of DENV and in the absence of *A. aegypti* has been associated with outbreaks of DEN fever but usually not severe forms of the disease (Lambrechts *et al.*, 2010). The role of *A. albopictus* in DENV outbreaks appears to be less important than *A. aegypti* owing to its feeding preferences, behaviour, and reduced ability to disseminate DENVs in the infected vector (Lambrechts *et al.*, 2010).

At least three of the four DENV serotypes are maintained in sylvatic transmission cycles (Vasilakis *et al.*, 2008b). Here, DENVs are transmitted between canopy-dwelling *Aedes* spp. such as *A. niveus*, *A. furcifer* and *A. luteochephalus* and non-human primates (*Macaca*, *Presbytis*, *Erythrocebus patas* monkeys) (Vasilakis *et al.*, 2008b). The sylvatic DENVs are genetically distinct from DENVs usually associated with the urban transmission cycle (Vasilakis *et al.*, 2007a,b). However, sylvatic DENVs have been reported to cause disease in humans and serological and phylogenetic evidence suggests that sylvatic DENVs have caused limited outbreaks in urban environments (Vasilakis *et al.*, 2008a). Therefore, the potential remains that sylvatic DENVs can be a source for DENV re-emergence in urban cycles (Rico-Hesse, 1990; Vasilakis *et al.*, 2008b).

A. aegypti originated in Africa where two subspecies, *A. aegypti aegypti* (Aaa) and *A. aegypti formosus* (Aaf), exist. In West Africa, Aaa and Aaf subspecies are monophyletic and Aaa appears to have evolved from an ancestor of Aaf (Sylla *et al.*, 2009). The Aaa subspecies is a paler, browner form of *A. aegypti* that is highly domesticated, laying its eggs in artificial containers found in close proximity with homes. In East Africa, the Aaf subspecies (darker form) tends to breed in natural containers such as tree holes, and feeds on a variety of host species; however, in West Africa the Aaf form can be found in peridomestic environments (Mattingly, 1957; Nasidi *et al.*, 1989; Vazeille-Falcoz *et al.*, 1999). F2 progeny from crosses of Aaa and Aaf parents show chromosomal regions of recombination suppression and significant alterations in gene order from Aaa revealing multiple paracentric inversions in each of the three chromosomes of Aaf (Bernhardt *et al.*, 2009). Distinct genetic lines having high and low susceptibility for DENV2 have been generated from crosses between Aaa and Aaf in the laboratory and currently serve as genetic tools for identifying genes that play an important role in VC (Bennett *et al.*, 2005a).

Vector competence (VC) and vectorial capacity (V) for DENV

The following discussions will refer to subspecies *A. aegypti aegypti* as *A. aegypti* since this is epidemiologically the most important vector in DENV transmission.

The mathematical concept of vectorial capacity (V):

$$V = \frac{(ma^2)(p^n)(b)}{-\ln p}$$

calculates V by accounting for the influence of several important epidemiological factors that predict the likelihood of vector-borne disease (Anderson and Rico-Hesse, 2006; Black and Moore, 1996; Delatte *et al.*, 2010). Negatively impacting one or more components of this equation significantly reduces vector-borne disease. One key variable explaining V is the 'infectibility' of vectors, specifically, the proportion of vectors ingesting an infectious blood meal that successfully become infective and transmit the virus (b). This variable is also known as vector competence (VC) and is an important intrinsic measure of a mosquito's ability to become infected with virus and then transmit DENVs. The development of strategies that block VC by reducing virus entry, clearing vectors of their pathogens or eliminating pathogen transmission could bring the b value to zero, eliminating V and potentially breaking the cycle of virus transmission. While a determination of VC is an important indicator of a mosquito population's ability to transmit DENV, other components of V play a large role in determining DEN disease transmission. For instance, V is heavily influenced by (p) the daily probability of vector survival, (n) the extrinsic incubation period (EIP) or length of time from initial infection to transmission of the virus, and (ma_2) the daily probability of a mosquito feeding on a human, which combines blood feeding frequency with human biting rate. Anderson and Rico-Hesse (2006) have shown that DENV genetics also plays a role in determining V by showing that South-East Asia (SEA) genotypes of DENV2 are more efficient in infecting field strains of *A. aegypti* than the American (AM) genotype in addition to the SEA genotypes having a shorter EIP (n) than AM genotypes. Additionally, if humans are the only hosts for efficient transmission of DENV, a will be higher and so will V (Delatte *et al.*, 2010). In other words, any zoophilic behaviour by *A. albopictus* will decrease V. DEN disease control strategies that decrease (p) or increase (n) should reduce the probability that mosquitoes will survive long enough to complete the EIP for DENVs and adversely impact V.

DENV infection of *A. aegypti*

A. aegypti adult females seek a human blood meal within days after emerging from the pupal stage. Females are attracted to humans by a complex array of cues that include humidity, visual cues, CO_2, lactic acid and other chemical odorants on or near the skin surface (Bernier *et al.*, 2000; Lazzari and Stephen, 2009). The blood meal is rich in amino acids that are required for egg production (Ribeiro *et al.*, 2009). *A. aegypti* females can ingest blood meals every several days during their adult life, leading to multiple gonotrophic cycles (Scott *et al.*, 2000; Scott *et al.*, 1993). *A. aegypti* obtain blood from humans in two ways. During feeding, the mosquito mouth parts can cause skin lacerations forming haemorrhagic pools or haematomas that accumulate in the tissue; however, the female mouth parts can penetrate deeper into the skin to obtain blood from a cannulated venule or arteriole (Ribeiro *et al.*, 2009). The mosquito acquires DENV from a viraemic human during the acute febrile period of DEN disease, at which time the virus is circulating in the peripheral blood (Halstead, 2008). A viraemic individual may have up to 10^7 to 10^8 50% tissue culture infectious doses ($TCID_{50}$) of virus per ml of blood (Halstead, 2008). The blood enters the posterior midgut where virus breaches the MIB and replicates in midgut epithelial cells. About 6–7 days later the virus overcomes the MEB and escapes from midgut and disseminates to secondary tissues such as fat body, nerve tissue, haemocytes and the salivary glands (Linthicum *et al.*, 1996; Salazar *et al.*, 2007). Once in the salivary glands, virus replicates and enters saliva for DENV transmission to a new host when the female mosquito takes an additional blood meal. The EIP for DENVs in *A. aegypti* is 7–14 days under laboratory conditions (~28°C, 80% RH), although some mosquitoes can transmit as soon as 4 days post infection (dpi) (Salazar *et al.*, 2007). Mosquito infection with DENV2 is summarized in Fig. 14.1. Environmental conditions including temperature, humidity and nutritional status affect EIP (Black *et al.*, 2002). In nature, VC varies among *A. aegypti* populations for the same DENV2 strain (Bennett *et al.*, 2002; Lozano-Fuentes *et al.*, 2009) (Fig. 14.2). A significant part of the observed intraspecies variation can be attributed to genetic differences among vector populations that may involve differences in the mosquito's efficiency to support virus infection and may in part be associated with differences in the vector's innate immune responses that target the virus (Bennett *et al.*, 2002; Sánchez-Vargas *et al.*, 2009; Souza-Neto *et al.*, 2009; Waterhouse *et al.*, 2007; Xi *et al.*, 2008). The midgut appears to be the major organ that determines VC for DENV in *A. aegypti* (Black

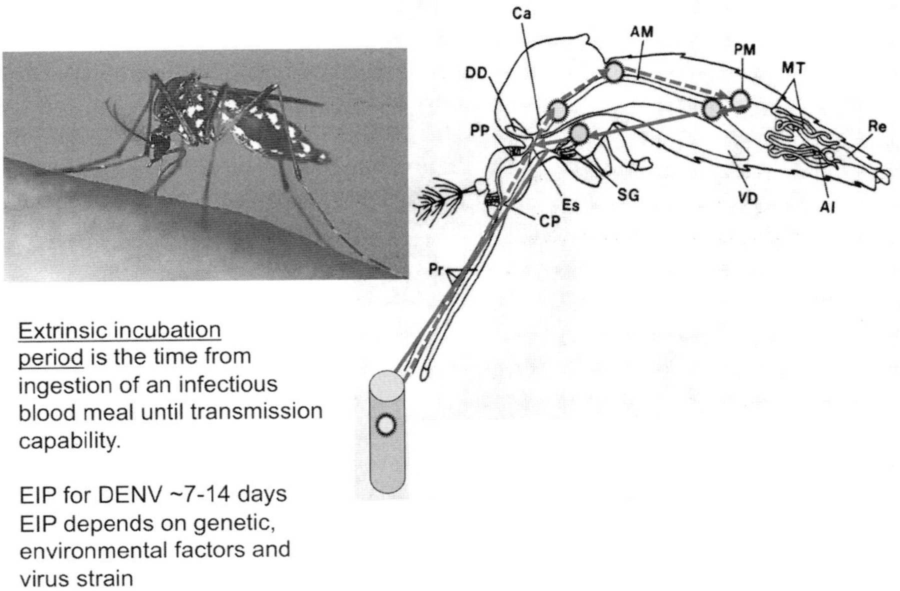

Figure 14.1 Overview of DENV infection in *Aedes aegypti*. Diagram shows cross-section of mosquito. Dashed arrows show virus movement from capillary of host to posterior midgut (PM) of the vector. Solid arrows show virus escape from PM to salivary glands (SG) and back to a susceptible host. AI, anterior intestine; AM, anterior midgut; Ca, cardia; CP, cibarial pump; DD, dorsal diverticulum; Es, oesophagus; MT, Malpighian tubule; PP, pharyngeal pump; Pr, proboscis; RE, rectum; VD, crop. Modified from Snodgrass, R.E. (1959). Smithsonian Misc. Coll. 139 No. 8, Washington DC.

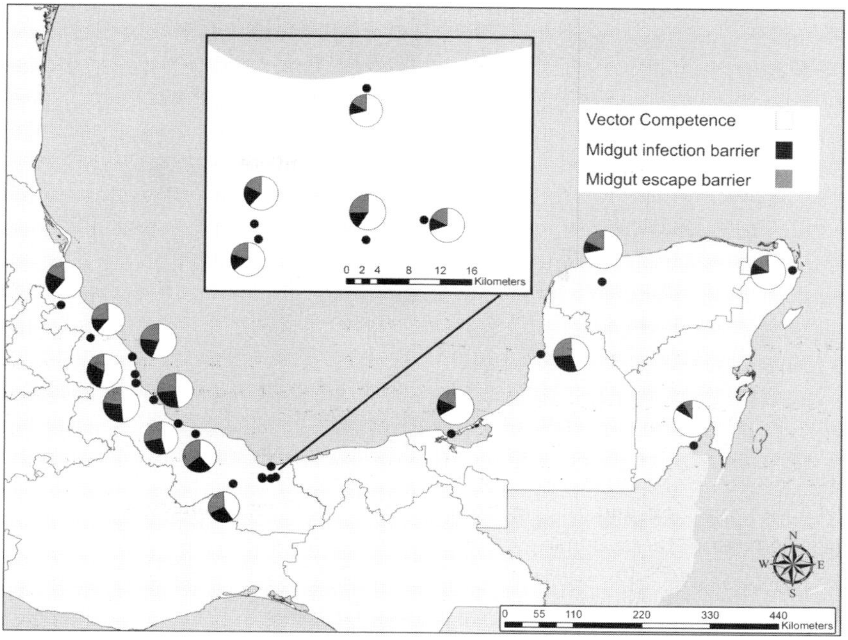

Figure 14.2 Geospatial analysis of vector competence (VC). Contributions of MIB and MEB to observed VC variation in 19 *A. aegypti* collections made in 2005 from Mexico (Lozano-Fuentes *et al.*, 2009).

et al., 2002). Depending on the particular vector strain/virus strain combination, quantitative trait loci (QTL) that determine MIB and MEB have been described for DENVs (Bennett *et al.*, 2002, 2005b; Bosio *et al.*, 2000; Lozano-Fuentes *et al.*, 2009). In presence of a MIB, the virus is not able to efficiently infect the midgut epithelium. In the presence of a MEB, the virus readily infects the midgut epithelium but its dissemination to secondary tissues is inhibited. Identification of genes controlling MIB and MEB will go a long way towards understanding VC and lead to the development of new vector control strategies (Bennett *et al.*, 2005a; Bennett *et al.*, 2005b; Black *et al.*, 2002; Bosio *et al.*, 2000; Gomez-Machorro *et al.*, 2004).

The pattern of DENV2 (Jamaica 1409) replication in various strains of *A. aegypti* has been described in several studies and is shown in Fig. 14.3 (Black *et al.*, 2002; Salazar *et al.*, 2007; Sánchez-Vargas *et al.*, 2009). After *A. aegypti* [Puerto Rico RexD/Higgs White Eye (HWE)] ingest an artificial infectious blood meal containing 10^6–10^7 plaque-forming units (pfu)/ml DENV2, an eclipse phase occurs in which virus titres decline to an undetectable level by 3 dpi. Virus titres increase in the midgut beginning at 3–5 dpi, peak at 10 dpi ($9\times10^3\pm3.6\times10^3$ pfu/mosquito), and begin to decline by 12 dpi ($7.4\times10^2\pm2.9\times10^2$ pfu/mosquito). In HWE mosquitoes, DENV2 E antigen is readily detected by IFA in the midgut at 7 and 14 dpi, virus dissemination to the fat body can be detected at about 7 dpi, and antigen is found abundantly in salivary glands at 14 dpi (Fig. 14.3). Viral E antigen and infectious virus titre decline in the midgut by 14 dpi but increase in the salivary glands from 14–21 dpi ($9.4\times10^4\pm3.6\times10^4$ pfu per mosquito in tissues outside the midgut at 16 dpi). Viral RNA can be detected from 5 to 14 dpi by Northern blot analysis of total midgut RNA, and in tissues outside the midgut from 9 to 14 dpi or longer (Sánchez-Vargas *et al.*, 2009). If infected mosquitoes are allowed to probe an artificial feeding membrane, infectious DENV2 can be detected as early as 10 dpi, and can consistently be detected in saliva of HWE mosquitoes at 14 dpi (Franz *et al.*, 2006). Both dsRNA-specific and E antigen-specific monoclonal antibodies can be used to produce an extensive immunofluorescent

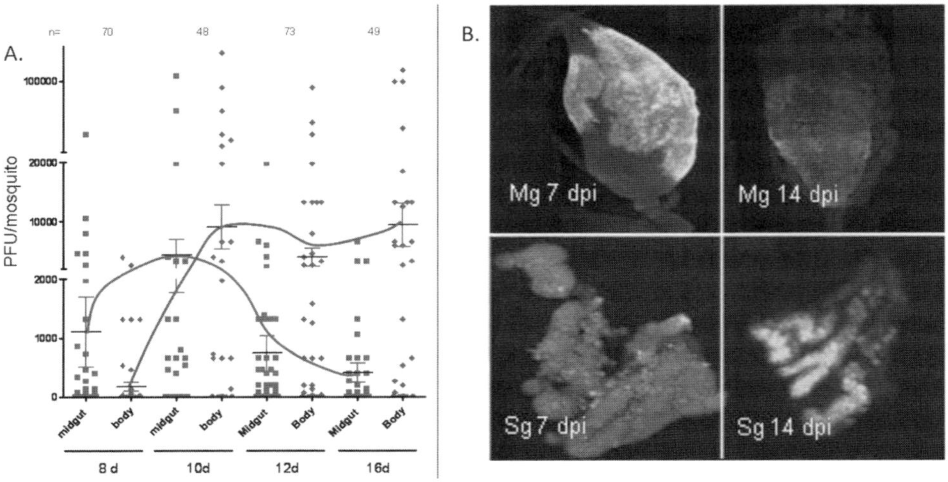

Figure 14.3 DENV2 infection of *Aedes aegypti* HWE mosquitoes. (A) DENV2 titres in midguts (red symbols/line) and remaining tissues (carcass; blue symbols/line) of individual mosquitoes at predetermined dpi. (B) Detection of DENV2 E antigen by indirect IFA in midgut (top panels) and salivary gland (bottom panels) tissues at 7 and 14 dpi. Data and IFA images from Sánchez-Vargas *et al.* (2009). A colour version of this figure is located in the plate section at the back of the book.

signal in midguts at 7 days after acquisition of a DENV2-containing blood meal, signifying active DENV replication (Fig. 14.3).

Infection of the *A. aegypti* midgut

The viraemic blood moves through the mouth parts, foregut and anterior midgut and into the posterior midgut. Within hours of entering the posterior midgut a type 1 peritrophic matrix (PM1) forms surrounding the blood, which is made of proteins, glycoproteins and chitin (Devenport and Jacobs-Lorena, 2004; Moskalyk *et al.*, 1996). Synthesis of PM1 is triggered by distension of the midgut from the blood meal (Devenport and Jacobs-Lorena, 2004). PM1 most probably is formed to isolate the blood meal and protect the vector from potentially harmful mechanical and chemical damage during digestion and from pathogens in the blood meal (Kato *et al.*, 2008). However, *A. aegypti* PM1 may not be an efficient barrier to DENV infection since disruption of PM1 by RNA silencing of *A. aegypti* chitin synthase (AeCs) expression does not increase DENV2 infection rates of midguts when compared with control mosquitoes not silenced for AeCS (Kato *et al.*, 2008). The attachment (or entry) of the virus to the epithelial cells may occur during the first 6–8 hours after ingesting the blood meal and before the PM1 matures.

The midgut is the site in adult female mosquitoes for blood meal digestion (Pennington and Wells, 2004). Virus in the blood meal encounters a highly proteolytic environment in the midgut lumen, which could lead to virus degradation. A number of trypsins, chymotrypsins, and elastins are expressed for blood meal ingestion (Brackney *et al.*, 2008, 2010; Pennington and Wells, 2004). Early expression of the serine protease early trypsin is translationally controlled; late expression of trypsin genes such as 5G1 is transcriptionally controlled (Lu *et al.*, 2006; Noriega *et al.*, 1999). Although evidence exists that arboviral infections of the midgut epithelial cells occur prior to accumulation of digestive enzymes, others have found that viable virions can be associated with the midgut for hours (Houk *et al.*, 1985; Whitfield *et al.*, 1973). For example, the flavivirus SLEV can persist in the midgut lumen of *Cx. pipiens* for at least 8 h after a blood meal before infecting the midgut epithelium (Whitfield *et al.*, 1973). In light of these data, midgut infection with certain arboviruses may be a continual process that occurs over a period of hours post ingestion. Serine proteases do not activate DENV2 as reported earlier (Molina-Cruz *et al.*, 2005), but rather proteolytic activity, specifically late trypsin activity, within the midgut actually limits virus infectivity (Brackney *et al.*, 2008). Support for this conclusion was presented by showing that RNA silencing of 5G1 (late trypsin) expression in *A. aegypti* but not early trypsin activity increases midgut infection rates, suggesting that DENV2 particles remaining in the blood meal or spreading extracellularly in the midgut are vulnerable to degradation by 5G1 trypsin. It is probable that virus sensitivity to trypsins is unique for each virus genotype/mosquito species (or genotype) interaction, and the rate at which viruses infect the midgut is dependent upon both the virus and mosquito species. However, those DENVs that attach to (or enter) epithelial cells within the first few hours after ingestion of the blood meal probably have a distinct advantage in successfully infecting the vector. QTL analysis among *A. aegypti* populations may implicate genes encoding late trypsin activity as important determinants of VC for DENV2 (Bosio *et al.*, 2000).

A major failing in our understanding of DENV midgut infection is how the DENVs enter epithelial cells. As stated earlier, to date no midgut surface proteins have been identified that are defined receptors for any of the DENV serotypes, although Muñoz and colleagues have reported that 64, 67 and 80 kDa proteins are putative receptors for the DENV (Mercado-Curiel *et al.*, 2008; Mercado-Curiel *et al.*, 2006). The genes expressing these proteins were not identified and antibody competition assays were not performed in the context of midgut infections. Identifying midgut receptors for DENV attachment and entry, rather than receptors on cultured mosquito cells, none of which are midgut derived, should be a major focus of investigation and could reveal important components of the MIB associated with DENV. Since MIB plays a large role in VC, the receptor(s) when identified could serve as an important target in developing vector control strategies.

DENV infections of midguts from different mosquito populations show similar patterns and cell-type tropisms. In *A. aegypti* mosquitoes, the infection usually starts in the posterior region of the midgut, although in some instances foci of infection are more randomly distributed throughout the midgut (Salazar *et al.*, 2007). The initial infection involves a limited number of individual cells, forming foci that spread laterally to neighbouring cells, reaching peak infection that frequently involves the entire organ, between 7 and 10 dpi, and by 14 dpi begins to decline (Salazar *et al.*, 2007). Whether the lateral spread of virus in midgut epithelial cells is important for efficient virus escape and dissemination is open to question. The epithelial cells form a single layer in the midgut (Pennington and Wells, 2004). Midgut cells are highly polarized for taking in nutrients from the midgut lumen and arbovirus antigen accumulates near the basal side of the epithelial cells within hours of infection (Myles *et al.*, 2004; Salazar-Sanchez, 2006; Salazar *et al.*, 2007) (Fig. 14.4). The few epithelial cells initially infected after the mosquito imbibes an infective blood meal may be sufficient to lead to midgut escape and dissemination.

In a study of kinetics of DENV2 RNA replication, Richardson *et al.* (2006) used strand-specific qRT-PCR to quantify viral RNA from the midguts (infection) and legs (dissemination) of individuals of a highly competent/susceptible strain (Chetumal) of *A. aegypti* at each of 14 days following an infectious artificial blood meal (Richardson *et al.*, 2006). A typical eclipse phase associated with decrease in infectious virus titres during digestion of the blood meal was seen on days 0–2. Although there was wide variation between mosquitoes, a steady increase in mean values of both positive- and negative-sense viral RNA was observed between 2 and 6 dpi, from ~10^3 to 10^6 RNA copies per mosquito. Negative-sense RNA was consistently detected, indicating formation of the template for genome synthesis by the virus-encoded RNA-dependent RNA polymerase, and the ratio of positive- to negative-sense RNA increased from ~0.5 $\log_{10}$ at 2 dpi to ~1.5 $\log_{10}$ at day 6, signifying asymmetric amplification of the genome. Midgut levels of DENV2 RNA fluctuated between 6 and 8–9 dpi, when RNA peaked, then decreased until 10 dpi, then steadily rose again until 14 dpi when the study was terminated. Dissemination out of the midgut, antiviral responses or mosquito cell metabolic changes might have caused the pronounced and rapid fluctuations in viral RNA copy numbers in the midgut between

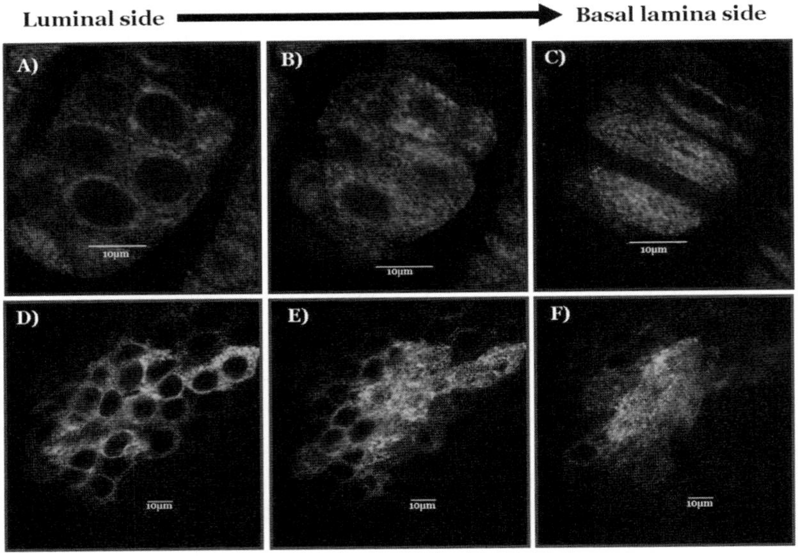

Figure 14.4 DENV2 antigen accumulates in the basal lamina side of the midgut epithelium. Confocal microscope images showing intensity gradient of DENV2 infection foci by indirect immunofluorescence assay for DENV E-protein antigen in cells of *A. aegypti* Chetumal midgut epithelial cells at 3 (A–C) and 5 dpi (D–F) reveals basolateral concentration of viral antigen (C and F) (Salazar-Sanchez, 2006).

6 and 9 dpi. Dissemination rates first reached 50% at 6 dpi and rose to 90% by 8 dpi in this highly susceptible mosquito strain. Detection of positive-sense (genome) RNA in legs preceded detection of negative-sense RNA. Infectious virus expressed as plaque-forming units (pfu) was correlated with DENV2 RNA copy number; pfu were consistently lower than RNA copy number by 2–3 $\log_{10}$.

DENV2 infection is clearly modulated in the vector's midgut. The amount of viral antigen and the number of infectious viral particles in the midgut decline over time. DENV2 replication in the mosquito midgut may be affected by antiviral responses causing either transcriptional or translational repression (Black *et al.*, 2002). For instance, impairing the innate antiviral immune RNAi pathway, a known antiviral response, in the midgut can significantly increase arbovirus infection and dissemination from that organ (Campbell *et al.*, 2008; Cirimotich *et al.*, 2009; Sánchez-Vargas *et al.*, 2009). Increasing the infectious dose and the number of infection foci may increase the likelihood that arboviruses will overcome these barriers to infection and disseminate to secondary tissues (Mahmood *et al.*, 2006; Pierro *et al.*, 2007, 2008).

In productive infections, as previously noted, DENV antigen accumulates at the basal side of midgut epithelial cells, suggesting that virus may escape by crossing the basal lamina into the haemocoel (Fig. 14.4) (Salazar-Sanchez, 2006; Salazar *et al.*, 2007). This arbovirus escape route from midgut cells is not well characterized but has been suggested by transmission electron microscopy (TEM) of *Culiseta melanura* infected with eastern equine encephalitis virus (*Alphavirus, Togaviridae*) (Weaver *et al.*, 1990; Weaver *et al.*, 1988). However, TEM analysis of SLEV infected *Cx. pipiens* shows virus particles between the basal lamina and midgut epithelia that appear to be trapped in that space (Whitfield *et al.*, 1973). Others have suggested that the mosquito tracheal system may play a role in early arbovirus dissemination. The tracheal system, which distributes 0_2 throughout the vector, acts as a route of dissemination for a number of viruses in other insects. Notable examples are nuclear polyhedrosis viruses infecting *Autographa californica* (alfalfa looper), *Trichoplusia ni* (cabbage looper), *Bombyx mori* (silk worm), and Sindbis virus in *A. albopictus* (Bowers *et al.*, 1995; Engelhard *et al.*, 1994). The tracheal system also may play a role in midgut escape for Venezuelan equine encephalitis virus (VEEV; *Alphavirus*) in *Ochlerotatus taeniorhynchus* (Romoser *et al.*, 2004). Further, WNV-like particles (VLP) localize focally in tracheal cells after intrathroracic inoculation of *Cx. quinquefasciatus* (Scholle *et al.*, 2004). DENV dissemination using trachea may have significant advantages for the virus, since this system reaches virtually every cell in the insect. Association of specific vector genes with virus entry in the midgut, sustaining virus replication in the midgut, and mechanisms of midgut escape should be important areas of research in the future.

DENV infection of other organs and tissues in *A. aegypti*

The tissue tropism over time of DENV2 strain Jamaica 1409 has been determined by examining non-midgut tissues of mosquitoes exposed to an infectious blood meal. Viral antigen is readily detected in fat body (abdomen or thorax) following dissemination (Salazar *et al.*, 2007). Fat body appears to be a major site for DENV2 replication in the vector. DENV2 dissemination to abdominal fat body or thoracic fat body can be detected as early as 3 dpi. Other infected organs and tissues include trachea, Malpighian tubules, haemocytes, oesophagus, ommatidia of the compound eye, nervous tissue, and salivary glands (Salazar *et al.*, 2007). DENV2 Jam 1409 displays a significant tropism for tissue of the central nervous system and salivary glands. Oesophagus, cardia, hindgut and especially muscle rarely contain viral antigen (Salazar *et al.*, 2007). The importance of virus amplification in secondary tissues other than salivary glands, which is critical for virus transmission, is unclear, but as new female-, adult- and tissue-specific transcriptional promoters become known, researchers should be able to express effector genes that inhibit virus replication in specific organs to assess whether virus amplification in these tissues affects virus transmission (Franz *et al.*, 2006; Franz *et al.*, 2009). For instance, the *A. aegypti* carboxypeptidase A promoter can express a DENV2-specific inverted repeat (IR) RNA in midgut epithelial cells of transgenic mosquitoes within hours of receiving an infective blood meal. The IR RNA is capable of triggering the RNAi pathway and silencing viral RNA replication in the midgut, which in turn significantly disrupts virus dissemination and transmission (Franz *et al.*, 2006). Similarly, when the IR RNA is expressed from the female adult salivary gland-specific 30 kb protein promoter in the salivary gland distal lateral lobes, DENV2 salivary gland infection and saliva transmission is significantly reduced (Mathur *et al.*, 2010). However, expression of IR RNA in a transgenic strain (Vg40) from the *vitellogenin* (*Vg*) promoter, which drives vitellogenin protein expression in fat body within 24h after ingestion of a non-infective blood meal, followed by challenge by intrathoracic inoculation of 100 pfu of DENV2 within two days, diminished virus infection of fat body but failed to significantly alter salivary gland infection or transmission rates (Franz *et al.*, 2009). In addition, challenging Vg40 mosquitoes with an infectious blood meal followed by ingestion of additional non-infectious blood meals every 2 days thereafter also had no effect on transmission.

A. aegypti females have a pair of salivary glands and each gland consists of three lobes attached to a common salivary duct (James, 2003). Each lobe comprises a secretory epithelium with a basal and apical surface. The basal ends of the epithelial cells form the outside surface of the glands and are bound to the basal lamina. How the virus crosses the basal lamina and the mechanism of virus entry into salivary gland epithelial cells is poorly understood. Female salivary glands differentiate into two lateral and one medial lobes, which produce distinctive secretions (James, 2003). The proximal region of the lateral lobes secretes enzymes involved in sugar feeding such as amylases and α–1–4 glycosidase and proteins from the medial and the distal lateral lobes participate in haematophagy and include apyrases, esterase, anticoagulants and vasodilators (Argentine and James, 1995; Beerntsen *et al.*, 1999; James, 2003; Smartt *et al.*, 1995). Virus-associated pathological changes in salivary glands and other organs occur late in infection by Sindbis virus (*Alphavirus*) in

A. albopictus and WNV in *Cx. quinquefasciatus* (Bowers *et al.*, 2003; Girard *et al.*, 2005). Analysis of DENV2 Jamaica 1409 infected salivary glands from the *A. aegypti* Chetumal strain showed E-protein antigen could be detected by IFA in one third of mosquitoes analysed as early as 4 dpi (Salazar *et al.*, 2007). However, in the majority virus antigen was detected in salivary glands after 7 dpi. In Chetumal mosquitoes, the infection often started in the distal lateral lobes, spread to the proximal section of the lateral lobes and then in the entire organ later. Virus antigen concentration in salivary glands remained high from 14 to 21 dpi, suggesting that older females in the population are excellent vectors for transmitting DENVs (Salazar *et al.*, 2007).

DENV transmission

Infection of salivary glands permits horizontal transmission of the virus to the next susceptible vertebrate host when the female feeds again after the EIP. The lack of authentic mouse models for DENV infections has hampered investigations of the role of mosquito saliva in virus transmission to vertebrates. Several studies have shown that individual mosquito vectors can inoculate 10^4 to 10^6 pfu of West Nile virus (Styer *et al.*, 2007) and 10^3 pfu of VEEV (Smith *et al.*, 2005). As stated in 'Virion assembly and release' section earlier, the envelope glycoprotein oligosaccharides on mosquito-derived DENV and other arboviruses, are mannose-rich in comparison to mammalian cell-derived E-proteins. In an initial step in infection of humans DENV attaches to dendritic cell (DC)-specific ICAM3-grabbing non-integrin (DC-SIGN) on the surface of immature DCs in the skin at the mosquito bite site (Klimstra *et al.*, 2003; Navarro-Sanchez *et al.*, 2003; Nielsen, 2009; Tassaneetrithep *et al.*, 2003). DC-SIGN is a mannose-specific, C-type lectin that facilitates virus entry and replication in DCs, suggesting that enhanced infectivity of these cells might be critical in mosquito transmission of DENV (Klimstra *et al.*, 2003). Once cells of the macrophage-DC lineage are infected with DENV, the host immune response is activated and the virus can spread to regional lymph nodes draining the site of the mosquito bite in the skin (Kelsall *et al.*, 2002; Navarro-Sanchez *et al.*, 2003).

Several investigators also have shown that protein factors in vector saliva can alter host immune responses and potentiate infection of arboviruses (Edwards *et al.*, 1998; Girard *et al.*, 2010; Styer *et al.*, 2006; Wasserman *et al.*, 2004; Zeidner *et al.*, 1999). Mosquito saliva contains a number of anti-haemostatic, anti-inflammatory and immunomodulatory molecules that are immunosuppressive in the host at high concentrations and suppress Th1 and antiviral cytokine responses at lower concentrations (Calvo *et al.*, 2006; Champagne and Ribeiro, 1994; Champagne *et al.*, 1995; Ribeiro, 1987; Ribeiro *et al.*, 2007; Schneider and Higgs, 2008; Schneider *et al.*, 2004; Titus *et al.*, 2006). As an example, D7 proteins are multifunctional and are among the most abundantly expressed (and secreted) salivary gland proteins. D7 proteins are potent vasodilators and bind to the immune response molecules biogenic amines and leukotrienes (Calvo *et al.*, 2009). Leukotrienes are pro-inflammatory members of the innate immune system that are important in recruiting leucocytes, up-regulation of phagocytosis and generation of cytokines (Peters-Golden *et al.*, 2005), suggesting that D7 salivary proteins are involved in modulating local inflammation at the bite site. Another salivary protein, SAAG-4, reduces host CD4 T cell expression of the Th1 cytokine IFN-gamma and increases expression of IL-4, giving SAAG-4 the ability to programme Th2 effector CD4 T cell differentiation (Boppana *et al.*, 2009). The *A. aegypti* sialome (mRNAs and expressed proteins of salivary glands) has been described and includes previously characterized proteins, such as amylase, apyrase, D7 family of proteins, sialokinin, 30 kDa protein as well as 31 novel proteins including two potentially related to invertebrate immunity (Ribeiro *et al.*, 2007). The sialomes for a number of vector species have now been described and will enhance our ability to understand the role of vector saliva during early events in viral pathogenesis in the host.

Mosquito innate immunity and flaviviruses

There is considerable knowledge about flavivirus interactions with vertebrate immunity; however, relatively little is known about how vectors such as mosquitoes control arbovirus infections. There is no evidence for an invertebrate adaptive immune response after virus infection, but the mosquito innate immune system has received increased attention since the publication of the complete genome sequences of several important vectors.

Drosophila model

Drosophila melanogaster is a well-established model system for understanding the molecular mechanisms of innate immunity in all metazoans, and has been particularly relevant to mosquito innate immunity because both are members of the order Diptera. In addition, work with *Drosophila* is facilitated by availability of the complete and fully annotated genome sequence, easily manipulated forward genetics systems and readily available molecular reagents such as whole-genome microarrays and RNAi knock-down libraries. The first descriptions of *Drosophila* responses to microbial infection were of two evolutionarily conserved signalling pathways, Toll and Imd, each of which is induced by recognition of particular pathogen-associated molecular patterns (PAMPs) such as lipopolysaccharides, peptidoglycans, and β-1,3-glucans of bacteria and fungi, and results in expression of antimicrobial peptides as well as cellular immune responses (Hoffmann, 2003). These signalling pathways and their regulation have largely been defined by use of DNA microarrays to detect induction of key components after injection of infectious agents and by effects of knock-out mutations in key genes on microbial load.

More recently, attention has turned to the *Drosophila* antiviral response. Studies of *Drosophila* antiviral immunity have been conducted with insect pathogenic RNA viruses from several families including Flock House virus (*Nodaviridae*), *Drosophila* C virus (DCV) and cricket paralysis virus (*Dicistroviridae*), and *Drosophila* X virus (DXV; *Birnaviridae*). *Drosophila* are not naturally infected by arboviruses, however, some investigations of the *Drosophila* antiviral response have included Sindbis (*Togaviridae*), DEN and WN (*Flaviviridae*), and vesicular stomatitis (*Rhabdoviridae*) arboviruses. These studies initially focused on pathways previously identified in anti-bacterial and anti-fungal immune responses, using transcriptional analysis and gene knock-out mutants.

By examining flies with mutations in genes with known or suggested immune activity, Zambon *et al.* (2005) showed that the Toll pathway plays a role in the *Drosophila* antiviral response to DXV. A study of the global transcriptional response of flies to infection with DCV showed that infection activates an additional antimicrobial signalling response, the Jak-STAT pathway, which interestingly is also a part of the mammalian antiviral interferon innate immune response (Dostert *et al.*, 2005). The specific viral PAMPs that triggered these pathways were not identified. Genome-wide microarray analysis of DCV-infected flies revealed that viral infection triggers expression of some 150 genes. There is little overlap between these genes and those induced by bacteria and fungi, indicating that flies respond differently to infection by DCV and cellular pathogens (Dostert *et al.*, 2005; Sabatier *et al.*, 2003).

RNA interference (RNAi) was first shown to be a potent antiviral defence mechanism in plants, and orthologues in the *Drosophila* genome indicated that it also plays an important role in the control of viral infection in insects (Ding and Voinnet, 2007). *Drosophila* research identified the key proteins and events of insect antiviral RNAi (Deddouch *et al.*, 2008; van Rij *et al.*, 2006; Wang *et al.*, 2006; Zambon *et al.*, 2005, 2006). In *Drosophila,* virus-derived long dsRNA is cleaved by the RNase III enzyme Dicer 2 (Dcr2) into small interfering (si)RNAs of 21–25 bp, often called viRNAs. Dcr2 and the dsRNA-binding protein R2D2 interact and load the viRNA into a multi-protein RNA-induced silencing complex (RISC); the viRNA is unwound and one guide strand is retained to direct the RISC to a sequence-complementary target ssRNA. Sequence–specific cleavage of viral mRNA is mediated by the 'slicer' endonuclease Argonaute 2 (Ago2) protein (Ding and Voinnet, 2007; Fragkoudis *et al.*, 2008; Kemp and Imler, 2009).

Although significant up-regulation of genes in the RNAi pathway such as *dcr2, r2d2* and *ago2* after infection has not been reported (Khoo *et al.*, 2010), several lines of evidence suggested the importance of RNAi in *Drosophila* antiviral immunity: (i) RNAi pathway mutants are hypersensitive to RNA virus infections and develop an increased viral load; (ii) many insect-pathogenic viruses, like plant viruses, encode suppressors of RNAi that counteract the immune defence of the fly; and (iii) siRNAs derived from the infecting virus genome (viRNAs) have been characterized in infected cells/flies (Kemp and Imler, 2009). The rapid evolution of RNAi genes as compared with microRNA genes in *Drosophila* suggests an on-going arms race between insect viruses and hosts, and points to their importance in antiviral defences (Obbard *et al.*, 2006).

DExD/H box helicases appear to play a crucial role in the cytosolic detection of viral RNAs in flies and mammals. Dicer-2 belongs to the same DExD/H-box helicase family as do the RIG-I–like receptors, which sense viral PAMPs and mediate interferon induction in mammals. It has been proposed that members of this gene family represent an evolutionarily conserved set of sensors that detect viral nucleic acids and direct antiviral responses (Deddouche *et al.*, 2008; Kemp and Imler, 2009).

Mosquito immunity

The *Drosophila* model has been important in understanding the innate immune response in mosquitoes, despite obvious limitations. The publication of the complete genome sequence of *A. aegypti* (Nene *et al.*, 2007) enabled a phylogenomic comparison of the known immune pathway genes in *Drosophila melanogaster* with this flavivirus vector mosquito and the malaria vector mosquito *Anopheles gambiae* (Waterhouse *et al.*, 2007). The much larger size of the *A. aegypti* genome [~1376 million base pairs (Mbp) compared with ~180 Mbp for *Drosophila*] and the evolutionary divergence (mosquitoes radiated from *Drosophila* ~250 million years ago), as well as differences in pathogen exposure owing to haematophagy have probably been responsible for rapid evolutionary expansion of many gene families in mosquitoes. In the absence of facile systems to induce mutations in mosquito genomes, further comparisons have largely relied on RNAi knock-downs and microarray gene expression analyses.

To characterize mechanisms of *A. aegypti* control of persistent DENV infection, genome-wide transcriptional analysis by microarray and RNAi-based reverse genetic analyses were carried out at 10 dpi. Up-regulation of genes putatively linked with the Toll immune signalling pathway was observed (Xi *et al.*, 2008). DENV infection in mosquitoes also activates the Jak/STAT immune signalling pathway. The mosquito's susceptibility to DENV infection increases when the Jak/STAT pathway is suppressed through RNAi silencing of pathway regulators at 3 and 7 dpbm, resulting in three- to fivefold increases in virus load (Souza-Neto *et al.*, 2009). However, Sim and Dimopoulos (2010) observed that heat-inactivated DENV induced twice as many immune-related genes as did live DENV in cultured *A. aegypti* cells, suggesting that transcriptional activation of immune pathway components might not be a specific response to virus infection.

Orthologues of *dcr2*, *r2d2* and *ago2* exist in vector mosquitoes such as *Anopheles gambiae*, *Cx. pipiens* and *A. aegypti*. Induction of mutations in these genes is not readily accomplished for mosquitoes, but RNA silencing by injection of long dsRNA cognate to specific targets demonstrated that Dcr2, R2D2, and Ago2 are crucial for limiting virus production and dissemination. This was first demonstrated by RNAi-mediated knockdown of these genes in *A. gambiae* infected with an alphavirus (Keene *et al.*, 2004) and subsequently in *A. aegypti* for both alphaviruses and flaviviruses (Cirimotich *et al.*, 2009; Sánchez-Vargas *et al.*, 2009). Knock-down of either *dcr2* or *ago2* resulted in increased viral load and shorter EIP (Keene *et al.*, 2004; Sánchez-Vargas *et al.*, 2009) (Fig. 14.5).

Flavivirus infection of cultured mosquito cells and vector mosquitoes engages the exogenous RNAi pathway, resulting in production of virus-specific small RNAs (viRNAs). The RNAi response to WNV in the midguts of orally infected *Cx. quinquefasciatus* was characterized by high-throughput, massively parallel sequencing of viRNAs (Brackney *et al.*, 2009). Deep sequencing also was used to analyse flavivirus-specific small RNAs produced during acute infection of *A. aegypti* mosquitoes (Fig. 14.6) and *A. aegypti* Aag2 cell cultures by DENV2 and persistent infection of Aag2 cells by CFAV (Scott *et al.*, 2010). In all cases the virus-specific small RNAs were predominantly 21 nt long and derived almost equally from the positive and negative strands of

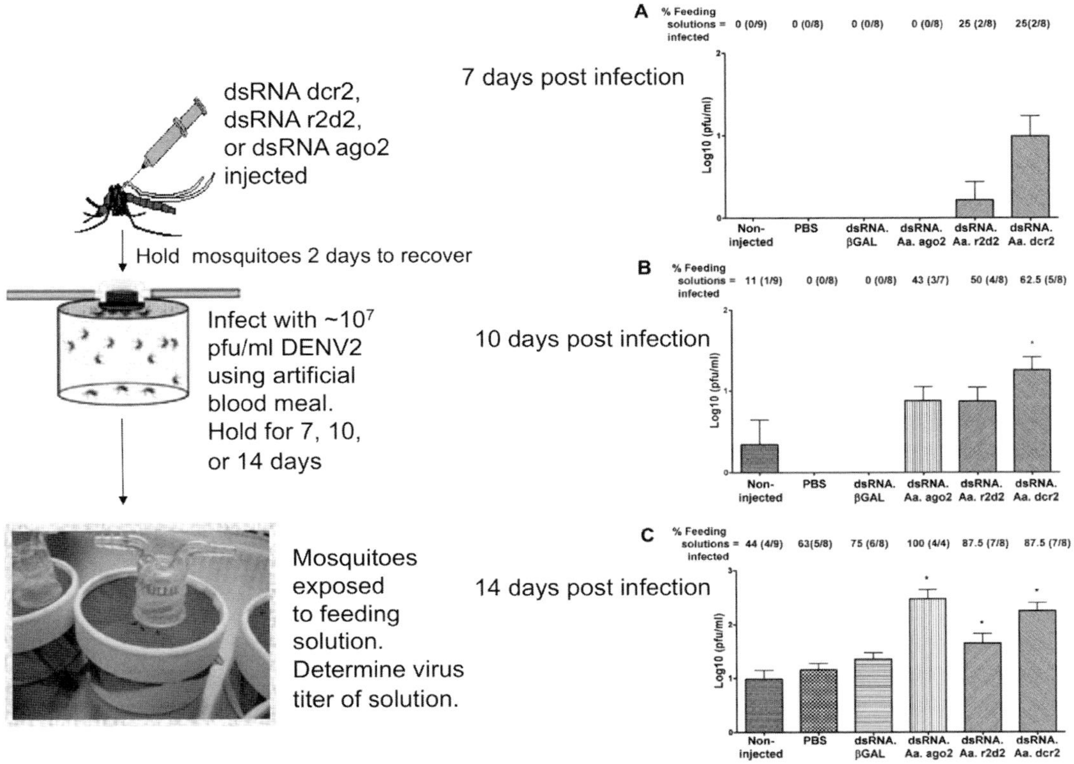

Figure 14.5 Transmission of DENV2 by *A. aegypti* at 7, 10, and 12 days post-infection after RNAi knock-down. Left: Procedure for depleting RNAi gene expression using RNAi knock-down. Groups of 200 mosquitoes were non-injected, injected with PBS, or injected with 500 ng dsRNA.βGAL, dsRNA.ago2, dsRNA.r2d2, or dsRNA.dcr2. Two days later all mosquito groups were orally infected with DENV2. Right: At 7 (A), 10 (B) and 12 (C) dpi batches of 10 mosquitoes were allowed to probe artificial feeding solutions. The feeding solutions were assayed for virus titre (*$P<0.05$). Data from Sánchez-Vargas *et al.* (2009).

the genome, indicating that they were products of Dcr2 cleavage of long dsRNA. It also appears possible that the PIWI-associated RNAi pathway, which is Dcr-independent, involves different Ago proteins and functions primarily in the control of mobile genetic elements (Gunawardane *et al.*, 2007), is involved in antiviral defence. Silencing of the PIWI family protein Ago3 affects RNAi responses to an alphavirus in *A. gambiae* (Keene *et al.*, 2004), and production of DENV-specific piRNAs was observed in C6/36 cells infected with DENV (Scott *et al.*, 2010).

Bernhardt (2010) compared rates of molecular evolution in the antiviral siRNA genes *dcr2, r2d2,* and *ago2* with their microRNA counterparts *dcr1, r3d1,* and *ago1* within and among geographically dispersed populations of *A. aegypti* by examining the ratio of replacement to silent substitutions (K_A/K_S). In *A. aegypti,* unlike *Drosophila,* both siRNA and miRNA pathway genes appear to be subject to diversifying selection, further suggesting possible overlap and redundancy in the functions of the small RNA pathways in mosquito immunity. RNA silencing may have evolved in mosquitoes, as it did in *Drosophila,* as a means to combat infection by insect pathogenic viruses but this innate immune response also acts to modulate arbovirus replication.

Other innate immune pathways used in mammalian cells may also be used by mosquitoes in antiviral defence. Midguts of a laboratory colony of *Cx. pipiens* refractory to experimental transmission of WNV were examined three days after feeding WNV-infected blood meals. Midgut

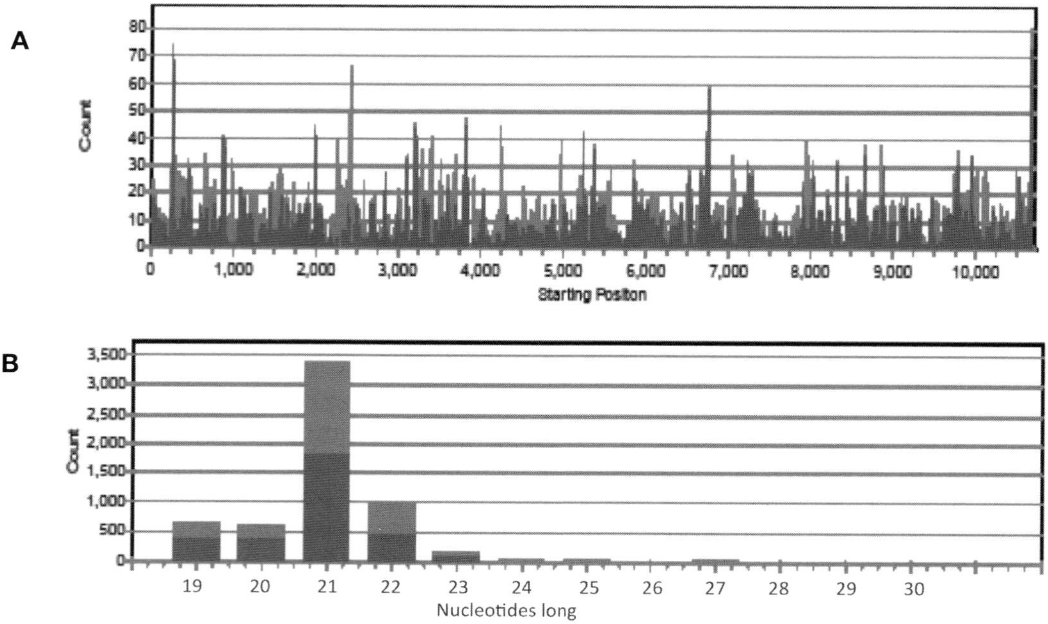

Figure 14.6 Characterization of DENV-specific small RNA (viRNA) sequences from *Aedes aegypti* infected with DENV at 9 days post infection. (A) viRNA numbers, polarity and distribution across the DENV2 genome from *A. aegypti* mosquito library. (B) viRNA size distribution in infected *A. aegypti* mosquitoes. Red bars, negative-sense viRNAs; blue bars, positive-sense viRNAs. Deep sequencing performed using SOLiD System Next Generation platform (ABI) and viRNAs analysed by NextGENe software (Softgenetics, LLC, State College, PA, USA). Modified from Scott *et al.* (2010). A colour version of this figure is located in the plate section at the back of the book.

epithelial cells of mosquitoes fed on WNV, but not those given a non-infectious blood meal, showed biochemical and ultrastructural changes consistent with apoptosis, suggesting that this may have been used as antiviral response in the refractory phenotype (Vaidyanathan and Scott, 2006). Both biochemical and ultrastructural evidence of severe cellular degeneration and apoptotic-like cell death were observed in *Cx. quinquefasciatus* salivary glands after long-term (25–28 days) WNV infection (Girard *et al.*, 2005; Girard *et al.*, 2007). Differential regulation of enzymes involved in apoptosis also was observed in DENV-infected *A. aegypti* (Xi *et al.*, 2008).

Viral counterdefence

Flaviviruses are known to inhibit or suppress a number of vertebrate host innate immune responses, and the inhibition has frequently been attributed to one or more of the virus-encoded proteins; however, no specific means of flavivirus evasion of the mosquito innate immune response has been identified. Insect-pathogenic viruses in several families encode viral suppressors of RNA silencing (VSRs) (Ding and Voinnet, 2007); for example, alphanodaviruses such as Flock House virus encode a B2 protein that has been experimentally shown to enhance virus pathogenicity in *Drosophila* cells and flies by suppressing the host RNAi response (Venter and Schneemann, 2008). In contrast to insect-pathogenic viruses, there is no evidence that arboviruses encode VSRs. Expression of nodavirus B2 proteins from the genomes of recombinant alphaviruses resulted in increased cytopathology in infected mosquito cell cultures and mortality in infected mosquitoes (Cirimotich *et al.*, 2009; Myles *et al.*, 2008). Pathogenic rather than persistent infections of mosquitoes by arboviruses, including flaviviruses, would be detrimental to virus transmission and maintenance in nature. Alphanodaviruses have been shown to be capable of causing persistent infections in *Drosophila* cell cultures (Dasgupta *et al.*, 1994; Friesen *et al.*, 1980) but the status of B2 protein expression was not examined. In a recent study, it was shown that related betanodaviruses, which are pathogens of fish, can cause persistent, vertically transmitted infections in their natural hosts in which the B2 protein is not expressed (Mézeth *et al.*, 2009). It is possible that modulation of VSR expression in potentially pathogenic viruses is required for natural persistent infections in insects as well; nevertheless, it seems probable that arboviruses have evolved mechanisms that do not involve VSRs to counteract mosquito RNAi.

Flavivirus evolution in mosquito vectors

Origins and dispersal of flaviviruses

Although cell fusing agent virus (CFAV) was isolated from *A. aegypti* cell cultures in 1975 (Stollar and Thomas, 1975), it was considered an interesting anomaly until 2003, when Kamiti River virus (KRV) was isolated from *A. macintoshi* mosquito pools collected in Kenya (Crabtree *et al.*, 2003; Sang *et al.*, 2003). Isolation and characterization of a number of insect-only flaviviruses from field collections of both *Culex* spp. (Bolling, 2010; Hoshino *et al.*, 2007; Morales-Betoulle *et al.*, 2008) and *Aedes* spp. (Hoshino *et al.*, 2009) mosquitoes followed rapidly. Phylogenetic analyses suggest that insect-specific viruses are the most divergent group within the genus *Flavivirus* and may represent the earliest forms of flaviviruses (Hoshino *et al.*, 2007). Furthermore, it was shown that the DNA genomes of *Aedes* spp. mosquitoes contain flavivirus genome-like sequences most closely related to CFAV and KRV (Crochu *et al.*, 2004), confirming previously unpublished observations by a number of vector biologists. The significance of these findings and their potential roles in flavivirus origins and evolution and arbovirus transmission are still highly speculative.

A phylogenetic tree based on 123 complete envelope gene sequences of tick- and mosquito-borne flaviviruses shows a continual and asymmetric branching process in the tick-borne flaviviruses, compared with an explosive radiation in the last 200 years in viruses transmitted by mosquitoes. The distinction between these two groups is thought to reflect differences in modes of dispersal, propagation, and changes in the size of host populations (Zanotto *et al.*, 1996).

Phylogenetic analyses of DENV E-gene sequences have led to the hypotheses that the viruses probably appeared about 1000 years ago (Twiddy *et al.*, 2003) and that the four serotypes evolved independently from sylvatic lineages that circulate in West Africa and South-East Asia (Wang *et al.*, 2000). Transfer of DENV from sylvatic transmission by canopy-dwelling *Aedes* spp. to endemic transmission by *A. aegypti* is thought to be fairly recent –about 125–300 years ago when human population densities became sufficient to sustain the transmission cycle. Evolutionary analyses have shown that the major selective pressure acting on endemic DENV genomes is purifying (negative) selection, with only sporadic occurrences of positive selection (Holmes, 2003). Recent genetic diversification within each of the serotypes correlates with human population growth, urbanization, and mass transport (Twiddy *et al.*, 2003), although Vassilakis *et al.* (2007a) showed that both the rate of evolutionary change and the pattern of selection are similar among endemic and sylvatic DENV.

Contribution of vector transmission to flavivirus evolution

RNA viruses are thought to evolve more rapidly than organisms with DNA genomes (mutation rates of the order of 10^{-4} substitutions per nucleotide per round of copying, so a genome of 10 kb accumulates approximately one mutation each time it is copied.) This has been attributed to their rapid replication rates coupled with an error-prone RNA-dependent RNA polymerase and lack of proofreading enzymes encoded by either viruses or host cells. The results are intrahost virus populations called a quasispecies, defined as heterogeneous mixtures of closely related genomes (Domingo *et al.*, 1996; Domingo and Holland, 1997).

Analyses of arbovirus genomes in individual mammalian and mosquito hosts frequently have found levels of intrahost genetic variation comparable to those seen in infections by non-arthropod-borne RNA viruses (Holmes, 2003). However, large-scale phylogenetic studies have shown that arboviruses, including flaviviruses, have significantly lower rates of non-synonymous nucleotide substitution than other RNA viruses (Jenkins and Holmes, 2003), signifying extensive purifying selection (Twiddy *et al.*, 2003). These observations have led to the hypothesis that RNA arboviruses evolve more slowly than single-host RNA viruses owing to constraints imposed by the need to replicate in alternative, phylogenetically diverse hosts, since a fitness increase in one host might result in diminished fitness in the other (Scott *et al.*, 1994; Weaver *et al.*, 1999). Corollary hypotheses are (i) repeated passages of a flavivirus in a single host facilitates adaptation to that host owing to an increase in mutation accumulation; (ii) adaptation to a single host results in fitness declines in the other host; and (iii) replication in alternate hosts selects for virus populations that are genetically conserved, with minimal adaptation to either host (Vasilakis *et al.*, 2009). An additional hypothesis is that natural transmission cycles of flaviviruses promote loss of genetic fitness owing to bottlenecks created by effective biological cloning and random selection of deleterious mutations imposed by small virus populations in both infecting and transmitted doses of mosquitoes (Weaver *et al.*, 1999). These hypotheses have been tested in laboratory studies of flavivirus evolution in both cell cultures and natural hosts.

In the absence of an authentic laboratory vertebrate animal model, Vasilakis *et al.* (2009) carried out serial passages of DENV either exclusively in a human liver cell line (Huh-7) or a commonly used *A. albopictus* mosquito cell line (C6/36), or alternating between the two cell lines. Conclusions were that viruses passed in a single host cell type exhibited fitness gains in that cell type, and fitness losses in the bypassed cell type, but that alternating passage resulted in fitness gains in both cell types. DENV that was biologically cloned before alternating passage also displayed fitness gains, possibly resulting from recovery from the genetic bottleneck. Mutations accumulated more rapidly in viruses passed in Huh-7 cells than in those passed in C6/36 cells or alternately. Mutations common to both passage series suggested positive selection for adaptations to replication in cell culture.

Jerzak *et al.* (2007, 2008) carried out similar studies; however, they measured the genetic diversity that arose in WNV genomes during alternating passage in the natural hosts, mosquitoes and birds. Intrahost genetic diversity of WNV populations that had been alternately passed or passed exclusively in mosquitoes was greater than WNV that had been passed exclusively in chickens. However, the nucleotide substitution rate in consensus WNV genomes that had undergone alternating passage in mosquitoes and birds was similar to WNV passed exclusively in either chickens or mosquitoes. They concluded that mosquitoes serve as sources for WNV genetic diversity, that birds are selective sieves, and that the genetic variation contributed to WNV populations during infection of mosquitoes and the strong purifying selection during infection of birds were maintained during alternating host switching.

To examine WNV genome diversity in nature, sequences of genome fragments from ten infected birds and ten infected mosquito pools collected in New York during the 2003 WNV transmission season were analysed. WNV sequences in mosquitoes were significantly more genetically diverse than WNV in birds. No host-dependent bias for particular types of mutations was observed, although two avian specimens had highly similar genetic signatures, suggesting that WNV genetic diversity is maintained throughout the enzootic transmission cycle, rather than arising independently during each infection. The study demonstrated that strong purifying selection has occurred in natural WNV populations (Jerzak *et al.*, 2005).

Influence of evolution in vectors on flavivirus transmission, establishment and displacement

There are now several examples of DENV strains associated with severe disease displacing less virulent strains (Messer *et al.*, 2003; Rico-Hesse, 2003; Rico-Hesse *et al.*, 1997). Epidemiological and phylogenetic evidence indicates that Southeast Asian (SEA) DENV2 genotypes are more likely to cause DHF/DSS than American (AM) or South Pacific genotypes (Leitmeyer *et al.*, 1999). After it was introduced into the Americas, the SEA DENV2 genotype began to displace the endogenous AM genotype and there is evidence that the ability to replicate in the vector has played a role in this displacement (Anderson and Rico-Hesse, 2006; Cologna *et al.*, 2005; Hanley *et al.*, 2008).

To examine the role of the vector in spread and displacement by more virulent flaviviruses, infection and growth experiments comparing DENV2 that caused dengue haemorrhagic fever epidemics (SEA genotype) with viruses that cause dengue fever only (AM genotype) were conducted. The SEA genotype viruses could outcompete AM genotype viruses in both cultured human dendritic cells and *A. aegypti* mosquitoes (Cologna *et al.*, 2005). As noted in 'Vector competence and vectorial capacity for DENV', when *A. aegypti* mosquitoes collected in South Texas were orally infected with DENV2 representing SEA or AM genotypes, viral replication in the midgut was significantly higher in SEA-infected mosquitoes, and dissemination to salivary glands occurred 7 days earlier in SEA- than AM-infected mosquitoes. This much earlier appearance of DENV in salivary glands resulted in an estimated 2- to 65-fold increase in the vectorial capacity of these mosquitoes for the viruses that can cause DHF (Anderson and Rico-Hesse, 2006).

WNV emergence in North America has presented a unique opportunity to prospectively study evolution of a flavivirus introduced into novel environmental conditions (geography, climate, etc.) in previously naive host populations. After its introduction in 1999, a new WNV genotype (WN02) emerged in New York in 2002 and became dominant in 2003. NY99 genotype viruses have EIPs in *Culex pipiens*, the main WNV vector in the north-eastern and north central United States, that are 2–4 days longer than the EIPs of WN02 genotype viruses (Ebel *et al.*, 2004). Phylogenetic analyses revealed that the dominant variant emerged across much of North America between 2002 and 2005 in a temporally and geographically dependent manner (Davis *et al.*, 2005). Moudy *et al.* (2007) examined *Cx. tarsalis*, a dominant vector in western North America, and found that after peroral infection, the EIP of the WN02 genotype also was up to 4 days shorter than the EIP of the NY99 genotype. However, after intrathoracic inoculation, there was no difference in EIP between the genotypes, suggesting that differences in genotype interaction with the mosquito midgut are likely to play a role in this phenotype. These results suggest a model for the displacement of the NY99 genotype, where earlier transmission of WN02 viruses leads to higher WN02 infection rates in avian reservoir hosts (Ebel *et al.*, 2004; Moudy *et al.*, 2007). The two examples above of emergence and displacement by flaviviruses with a growth advantage in their mosquito vector suggest the import role of the vector in virus evolution.

Tools for studying flavivirus–vector interactions

Gaining an understanding of how vector genome transcription and translation products interact with arboviruses is an essential requirement for understanding arbovirus transmission and development of new vector and disease control strategies. Since *A. aegypti* transmits DENVs, YFV, and chikungunya (*Alphavirus*) virus to humans, this vector became a high priority species for generating a complete genome sequence. In 2007, the genome sequence was published and an *A. aegypti* genome database is now available through VectorBase (http://www.vectorbase.org) for data-mining the genome. VectorBase annotates and curates vector genomes, providing an internet-accessible integrated resource for the research community. Currently VectorBase provides genome information for disease vector mosquitoes *A. aegypti*, *Anopheles gambiae* and *Cx. quinquefasciatus* (Lawson et al., 2009). VectorBase also provides other databases for genome analysis including a microarray and gene expression repository and software tools for interrogating the genomes. The VectorBase database is useful for cross-referencing with other vector genome databases found in GenBank (http://www.ncbi.nlm.nih.gov) and vector proteome databases such as UniProt protein records (Lawson *et al.*,

2009). VectorBase genome databases are also useful tools for microRNA target prediction in mirBase (http://microRNA.sanger.ac.uk). Complete vector genome sequences, RNA genome sequences from multiple DENV serotypes and genotypes, availability of microarrays and EST databases, and advances in massively parallel, deep sequencing of mosquito transcriptomes and small RNAs (sRNAs) provide the research community with an unprecedented array of tools and databases for identifying mechanisms of VC and virus–vector interactions.

The new sequencing and bioinformatics tools developed to study virus–vector interactions have revealed a complex array of protein interactions during DENV infection. Guo *et al.* (2010) have undertaken a comprehensive mapping of mosquito protein–DENV interactions that possibly affect DENV replication and persistence in mosquitoes. Microarray assay studies suggest broad responses to DENV infection in mosquitoes that involve a variety of molecules and cellular pathways. In addition, studies of physical interactions between flaviviruses and their vectors also provided insight on how viruses may overcome the host pathways during infection. With the mosquito protein interaction network constructed by Guo *et al.* (2010), researchers can now perform integrated analyses to identify vector pathways activated by DENVs.

Research tools such as high-throughput, massively parallel deep sequencing present unprecedented opportunities to probe vector–virus interactions in exquisite detail. Little is known about how small, non-translatable RNAs modulate infection in vectors. Understanding how small (s)RNAs such as exogenous and endogenous siRNAs, miRNAs, and piRNAs in the vector interact with flaviviruses will undoubtedly lead to new insights into DENV infection and transmission. The sRNA datasets will be important resources for identifying changes in the size and polarity of sRNA populations during the extrinsic incubation period for flaviviruses in their natural vectors. The datasets generated will be a valuable asset for the vector biology community and this research should identify fundamental virus–vector interactions that may be important for disease control.

RNAi-mediated silencing of mosquito genes

RNAi has revolutionized the field of functional genomics, allowing characterization of mosquito genomes that are otherwise difficult to manipulate. RNAi is an important tool for reverse genetics to characterize DENV–vector interactions. RNAi combined with bioinformatics from the *A. aegypti* genome sequence permits targeted silencing of gene expression to reveal molecules and mechanisms of vector biology that are relevant to VC and disease transmission. Several approaches for RNAi-based silencing have been described for performing functional analysis in the mosquito vector. The simplest approach is to identify a target sequence (>200 bp) from a gene of interest, design forward and reverse primers with bacteriophage T7 RNA polymerase promoter sequences at the 5′ end of each primer, amplify the sequence by PCR and allow *in vitro* transcription of each strand to form dsRNA (Garver and Dimopoulos, 2007; Luna *et al.*, 2007; Sánchez-Vargas *et al.*, 2009). The dsRNA can be transfected into cultured mosquito cells or intrathoracically injected into mosquitoes to observe an RNAi phenotype. Variations on this method have been used to determine the effect of RNAi-mediated silencing of specific vector genes involved in blood meal digestion and innate immunity during infection (Brackney *et al.*, 2010; Keene *et al.*, 2004; Sánchez-Vargas *et al.*, 2009; Souza-Neto *et al.*, 2009; Xi *et al.*, 2008). For instance, the depletion of specific RNAi genes by treating mosquitoes with dsRNA derived from endogenous *dcr2, ago2,* and *r2d2* genes prior to infection shortens EIP and allows faster virus transmission (Fig. 14.5) (Sánchez-Vargas *et al.*, 2009). Drawbacks to this method are usually partial, transient RNA silencing (or knockdown) of the gene of interest, off-target effects, and lack of tissue specificity when used in the vector.

RNAi libraries in which dsRNA are generated for each open reading frame in the genome have been developed from the *Drosophila* genome to use as a model system for rapidly screening and identifying genes that perturb DENV infection in cultured *Drosophila* cells. Sessions *et al.* (2009) identified insect host factors required for DENV2 propagation with a genome-wide RNAi screen (based on over 22,000 dsRNAs) from *Drosophila melanogaster* cells. As mentioned earlier, this screen identified candidate insect host factor genes implicated in DENV2 propagation and many had human homologues, indicating notable conservation of host factors between dipteran and human hosts. This work may suggest new gene targets and approaches to control infection in the mosquito vector and the mammalian host. Developing new genome-wide RNAi screening tools for use with new mosquito cell lines of known tissue origin may reveal additional genes that are important for VC and vector infection and transmission.

Mosquito transgenesis and applications for flavivirus–vector interactions

Genetic manipulation of the *A. aegypti* genome, while still inefficient compared with *Drosophila* transgenesis, has become a more accepted technique in the past decade for studying pathogen–vector interactions. Microinjection of preblastoderm embryos is the commonly used method to deliver DNA into the mosquito germ-line (Handler, 2001; Jasinskiene *et al.*, 1998). *A. aegypti* transformation technology has relied on the use of non-autonomous transposable elements (TEs) as gene vectors (Coates *et al.*, 1998; Jasinskiene *et al.*, 1998). Non-autonomous forms of three different TEs have been used for gene insertion for insect germ-line transformation, *piggyBac*, *mariner* (*Mos1*), and *Hermes* (Coates *et al.*, 1998; Fraser *et al.*, 1996; Jasinskiene *et al.*, 1998; Medhora *et al.*, 1991; Warren *et al.*, 1994). The TEs belong to the class II family possessing short inverted terminal repeats (ITRs) (O'Brochta *et al.*, 2003). For insect transformation, the non-autonomous TEs have the transposase open reading frame replaced by a selection marker expression cassette such as EGFP driven by the eye-specific 3xP3 promoter and a multiple cloning site allowing the insertion of a gene-of-interest for expression (Horn and Wimmer, 2000). The transposase is inserted into a separate plasmid expression vector, the helper plasmid. The helper and non-autonomous TE plasmids are co-injected into preblastoderm embryos of *A. aegypti* leading to a one-time integration event of the TE into the mosquito germ-line. Using non-autonomous TE-based systems, transformation frequencies of up to 10% can be achieved in *A. aegypti* (Adelman *et al.*, 2002). Transgenic *A. aegypti* have been developed to study innate immune pathways by tissue-specifically overexpressing dysfunctional Relish or by RNA silencing of REL1 (Bian *et al.*, 2005; Shin *et al.*, 2003). Transgenic *A. aegypti* also have been successfully developed that over-express the endogenous anti-bacterial peptide defensin (DefA) (Kokoza *et al.*, 2000) or silence DENV2 or deplete expression of Dicer2 by an RNAi-based method (Franz *et al.*, 2006; Franz *et al.*, 2009; Khoo *et al.*, 2010).

Since TE-based transgene insertion can occur randomly throughout the genome, expression levels of TE inserted transgenes can vary significantly owing to position effects. As a promising new alternative, the site-directed recombination system derived from the bacteriophage ΦC31 (*Siphoviridae*) infecting *Streptomyces ambofaciens* has recently been shown to be an ideal tool in yeast, mammalian and other insect cells as well as in *Drosophila* and *A. aegypti* (Alphey *et al.*, 2008; Franz *et al.*, 2006, 2009; Groth *et al.*, 2000, 2004; Khoo *et al.*, 2010; Nakayama *et al.*, 2006; Thomason *et al.*, 2001). The integrase of ΦC31 belongs to the family of resolvase/invertase-like enzymes that are associated with bacteriophages such as *Mu* and *Tn3* (Thorpe and Smith, 1998; Thorpe *et al.*, 2000). For this application a standard TE-based transgenic mosquito (docking) strain is made by first inserting a transgene into a region of the genome that allows strong expression, thus showing little potential for positional dysfunction. The insertion also contains a specific phage attachment site (*attP*) associated with the transgene. Embryos from the docking strain are then super-transformed with a plasmid containing a second transgene with the bacterium genome site, *attB* and *in vitro* transcribed mRNA of the ΦC31 integrase. The phage integrase binds to both DNAs that contain the *attP* and *attB* sites and facilitates homologous recombination between the sites. Upon integration the *attP* and *attB* sites are converted into *attL* and *attR* sites, the recognition sites for excision (Thorpe and Smith, 1998). What makes the ΦC31 recombination system become an ideal tool for site-directed integration of DNA is (1) the directionality of the integration mechanism; (2) the integrase does not need accessory proteins to catalyse an integration event; (3) the integrase always requires accessory protein(s) for DNA excision, because without these co-factors it cannot recognize the *attL/attR* sites. Furthermore there seems to be no limitation on the size of DNA (at least up to 40 kb, which is the size of the ΦC31 phage genome) to be inserted into the host genome. Once generated, mosquito docking strains with an anchored *attP* site can be colonized and used for future 'super-transformation' experiments involving new transgenes. Transgenic mosquitoes have been successfully engineered by using wild type strains of *A. aegypti* (strain Bangkok) and *Anopheles stephensi* and offer improved efficiency for genetically manipulating vectors and perhaps controlling vector-borne diseases such as dengue and malaria (Amenya *et al.*, 2010; Nimmo *et al.*, 2006).

Finally, new genetic tools are on the horizon to facilitate vector–pathogen interaction studies by enabling targeted gene knock-outs in genetically modified mosquitoes. No methodologies are currently available to efficiently generate precise genetic lesions in mosquitoes that would knock out expression of a specific gene. Efficient production of mosquitoes with targeted gene knock-outs would allow researchers to analyse the influence of specific genes on VC. In drosophila, null mutations (knock-outs) have been used to investigate how flies cope with entomopathic RNA virus infections in the absence of antiviral Dicer-2 or Argonaute-2 expression (Galiana-Arnoux *et al.*, 2006; van Rij *et al.*, 2006; Wang *et al.*, 2006). Null mutations are important because they set the baseline for all comparisons of expression for a given gene. Researchers have now exploited custom-designed zinc-finger nucleases (ZFNs) to cleave a gene of interest and then search for cells (or organisms) that have inaccurately repaired the lesion by non-homologous end joining (NHEJ) (Wilson, 2008). This leads to the complete abolishment of the targeted gene's activity. This approach has been used with ZFNs that induce gene-specific, NHEJ-based mutations at high frequency in *Drosophila* (Beumer *et al.*, 2008). New techniques employing ZFN's should soon be available not only to knock out genes, but also to replace a defective gene, correct gene mutations, or integrate an antiviral gene into mosquito genome at sites that minimize impacts on fitness and avoid positional dysregulation (Wilson, 2008).

Tissue-specific promoters of *Ae. aegypti* for transgene expression

Functional promoter elements are critically important components for gene-of-interest expression in transgenic *A. aegypti* and will allow tissue-specific overexpression or knockdown of genes critical to understanding VC and DENV transmission. The repertoire of DNA promoters for gene expression studies in mosquitoes is limited, but growing. Drosophila promoters such as *actin 5C* have been shown to express GOI's with some success in mosquitoes (Pinkerton *et al.*, 1999). The recently described *polyubiquitin* promoter from *A. aegypti* has been shown to constitutively express marker genes in the midguts of adult mosquitoes (Anderson *et al.*, 2010). Several inducible, tissue-specific *A. aegypti* promoters are also available for transgene expression in a predictable manner. These promoters include gut-specific *AeCPA* (*carboxypeptidase A*), fat body-specific *Vg (vitellogenin)*, and salivary gland-specific *Maltase* and *Apyrase* (Coates *et al.*, 1999; Edwards *et al.*, 2000; James *et al.*, 1991; Kokoza *et al.*, 2000; Moreira *et al.*, 2000). Studies performed with anopheline mosquitoes identified a salivary gland-specific gene encoding anopheline antiplatelet protein (AAPP) and transgenic *An. stephensi* expressed the dsRED marker abundantly in glands when driven by the *AAPP* promoter (Yoshida and Watanabe, 2006). An orthologous gene termed *aegyptin* has been described in *A. aegypti* by Mathur *et al.* (2010), who used the bidirectional 30K a/b (*aegyptin*) promoter to constitutively express both GFP and an anti-DENV effector gene in adult, female transgenic mosquitoes.

Aedes aegypti vector and DEN disease control

Vector-based control of DEN disease, which is transmitted primarily in or around the home, is difficult. The dynamics of DENV transmission has changed over the last quarter century owing to the rise of urbanization in the tropics, globalization of trade and travel, decline in vector control infrastructure, hyperendemnicity of DENVs throughout the world. The optimal means of controlling DEN disease is a combination of vector control, specific antiviral drugs, management of clinical illness and effective vaccines. Although drug and vaccine development for DEN disease have received more attention in the last decade, vector control will be critical in continuing the battle against this disease. During the last century, vector control was successful in eliminating *A. aegypti* from many parts of the western hemisphere through rigorous top-down, paramilitary style practices that significantly reduced larval habitat sources and deployed residual insecticides in and near the home (Ooi *et al.*, 2006). However, these measures proved to be unsustainable and *A. aegypti* has returned to many of the regions where it was previously eradicated. Consequently, DEN disease has resurged in the western hemisphere and elsewhere during the last 30 years, with increased incidences of DHF and DSS, and today DENV is considered to be the most medically important of the arboviruses (Mackenzie *et al.*, 2004). Unfortunately, vector control technology today is essentially the same as in the previous century. Below we will briefly describe three vector-based approaches that may offer some hope for DEN disease control in the 21st century.

Wolbachia as a vector and disease control strategy

Wolbachia are maternally inherited intracellular bacterial symbionts that infect more than half of all insect species. However, *Wolbachia* appear to be absent in natural populations of *A. aegypti*. O'Neill and colleagues have introduced a *Wolbachia* strain (wMelPop-CLA) into *A. aegypti* and have demonstrated that this strain significantly reduces the lifespan of the vector (McMeniman *et al.*, 2009). Shortening mosquito survival (p) in the vectorial capacity (V) equation will have a large impact on DENV transmission by killing the mosquito during EIP. Additionally, O'Neill and colleagues also showed that infection of *A. aegypti* with this *Wolbachia* strain directly inhibits the ability of the vector to support DENV and chikungunya virus infections owing to bacterial priming of the mosquito innate immune system and competing for limited cellular resources required for virus replication (Moreira *et al.*, 2009). Thus, this attribute of the wMelPop-CLA strain negatively impacts VC or (b) in the V equation. The combination of *Wolbachia*-mediated arbovirus interference and life-shortening may be powerful approach for the control of *A. aegypti* transmitted diseases. In theory *Wolbachia* would be vertically inherited in *A. aegypti* populations and persist over time. If this strategy is borne out in cage and field studies, this approach should be an inexpensive way to reduce vector populations and control DENV, particularly in urban settings where current vector control strategies are less effective (McMeniman *et al.*, 2009). This *Wolbachia* strain is efficiently maternally inherited and induces complete cytoplasmic incompatibility, which should facilitate spread of the bacteria within natural populations of *A. aegypti* (McMeniman *et al.*, 2009).

RIDL (release of insects carrying a dominant lethal)

A vector control strategy based on the release of transgenic *A. aegypti* carrying a conditional dominant lethal gene is an attractive approach to reduce vector populations in a targeted area (Alphey *et al.*, 2008). Transgenic *A. aegypti* have now been generated that have a female-specific flightless phenotype following expression of a tetracycline repressible transactivator (tTA) from a female-specific indirect flight muscle promoter that normally controls expression of *A. aegypti* Actin-4 (Fu *et al.*, 2010). The *Ae Act4* promoter controls effector gene expression that is highly penetrant, and adult-, tissue-, and sex-specific. Males carrying the 'flightless' trait can mate with wild-type females to pass the trait on to the offspring. In the absence of tetracycline, the effector gene in female offspring is expressed, negatively impacting the indirect flight muscle and preventing female flight. The 'flightless' transgenic strains eliminate the need for mosquito sterilization by irradiation, permit male-only release, increase larval competition for limited resources since 'flightless' RIDL affects only adult females, allow effective suppression of low-density vectors (like *Ae. aegypti*), and enable the release of eggs instead of adults giving this approach a distinct advantage over sterile insect technique (Fu *et al.*, 2010). Additionally, the 'flightless' female adults cannot produce offspring or increase biting rates in the absence of tetracycline since they cannot fly. RIDL-based strains should facilitate area-wide vector control and as a consequence impact DENV transmission as part of an integrated pest management strategy. RIDL-based control will require continual monitoring of vector populations and additional releases will be required as needed to maintain suppression of the vector population. This vector suppression strategy will be thoroughly tested and evaluated in laboratory and field cages prior to use in the field.

Vector population replacement

Another DEN control strategy is to replace vector populations competent to transmit DENVs with pathogen-incompetent vectors (James, 2007; Olson and Franz, 2009). This approach would impact VC (*b* in the *V* equation) by essentially driving (*b*) to zero. Essential features of this control strategy are to identify or introduce genes that express antiviral effector molecules in the vector, link this gene (or genes) to a gene drive system [transposable elements (TE), meiotic drive, or homing endonuclease genes] and introgress the anti-DENV effector gene(s) into field populations (Deredec *et al.*, 2008; Gould *et al.*, 2006; Sinkins and Gould, 2006). An important first step is to identify effector genes that, when expressed at the right time and place in the vector, inhibit DENV replication. Two classes of DENV2-specific effector genes have been identified that transcribe effector RNAs that target and destroy the viral RNA genome. One class is based on hammerhead ribozymes (HR) that target a highly conserved sequence in the DENV capsid coding region and cleave viral RNA, inhibiting virus production

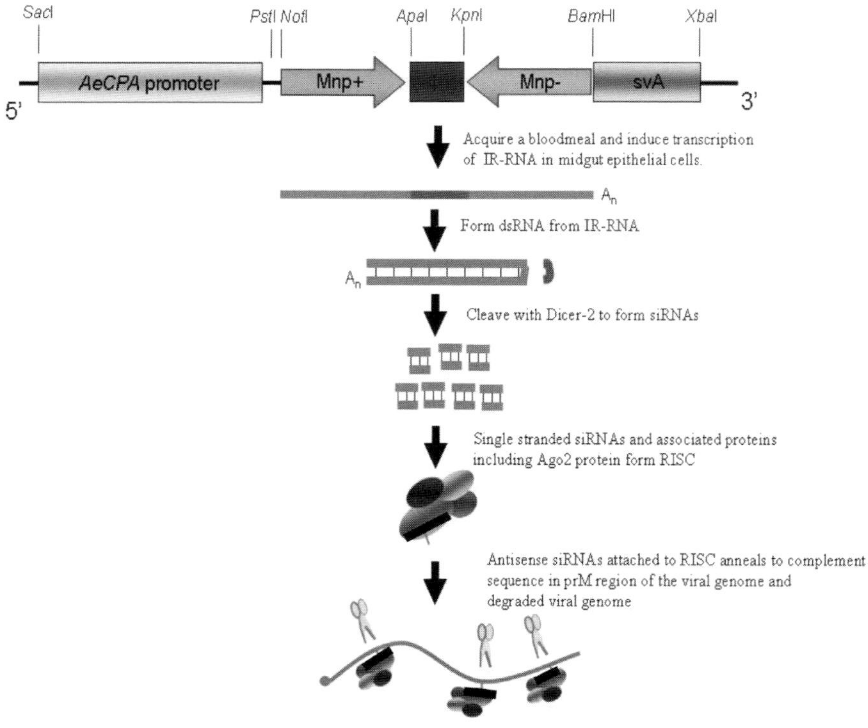

Figure 14.7 Mechanism of triggering the RNAi pathway by transcription of DENV RNA-derived IR-RNA in a transgenic line of *A. aegypti*. From Franz *et al.* (2006).

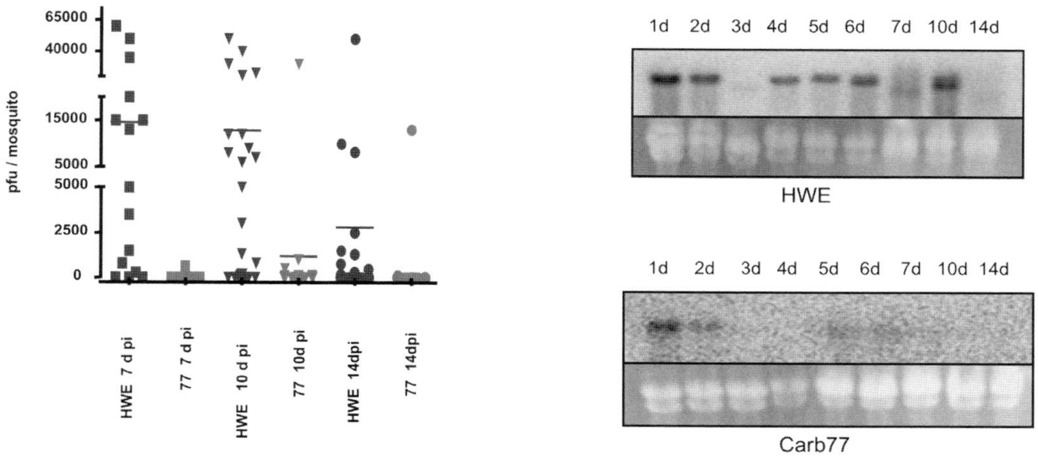

Figure 14.8 Characterization of the DENV-2 resistance phenotype in Carb77 mosquitoes expressing the IR-RNA effector. Left: Infectious DENV2 titres of single whole body Carb77 and HWE mosquitoes 7, 10, and 14 dpbm (bars indicate mean titres). Right: Northern blot analysis of DENV2 genome RNA in midguts of control HWE (top) and Carb77 (bottom) 1–14 dpbm, using a probe for sense-strand RNA from the prM gene of DENV2 RNA. Ribosomal RNA loading controls are shown. From Franz *et al.* (2009).

(Nawtaisong *et al.*, 2009). The HR class of effectors has strong potential for inhibiting all four DENV serotypes but to date has only been tested in the context of cultured mosquito cells. The second class of effector genes express a DENV2 genome-derived inverted-repeat (IR) RNA in the mosquito at the time virus enters midguts or salivary glands (Olson and Franz, 2009). The IR-RNA forms dsRNA that is detected by the mosquito's RNAi pathway, leading to destruction of the viral RNA by Ago2 slicer activity associated with RISC (Fig. 14.7). This RNAi-based effector gene strategy has been evaluated in both midguts and salivary glands of transgenic mosquitoes and shows strong RNA silencing of DENV2 infection leading to significantly reduced transmission (Fig. 14.8) (Franz *et al.*, 2006; Franz *et al.*, 2009; Mathur *et al.*, 2010). RNAi is highly sequence homology-dependent and subject to escape mutants from rapidly evolving RNA viruses. However, the IR-RNA is large (>300 bp), generating many 21 bp siRNAs that should be able to compensate for sequence variation among quasispecies. In addition, the IR effector gene has been redesigned to include target DENV sequences from each of the four DENV serotypes.

A major hurdle for realizing a vector replacement strategy is the development of a gene drive system that could introgress the effector gene in natural populations of *A. aegypti*. Gene drive systems will require considerable development to introgress effector genes in a non-mendelian manner to increase their allele frequency so it becomes fixed in the natural population. Several gene drive systems have been hypothesized and several review papers are cited here for the reader (Gould *et al.*, 2006; Sinkins and Gould, 2006). One gene drive system currently under investigation is the synthetic MEDEA (maternal effect dominant embryo arrest) engineered into *Drosophila* by Hay and colleagues (Chen *et al.*, 2007). Synthetic MEDEA is based on a naturally occurring MEDEA gene driver in *Tribolium castaneum* (Beeman *et al.*, 1992). MEDEA has two tightly linked genes, a maternally expressed toxin and a zygotically expressed antidote gene. In the *Drosophila* model, a maternal toxin gene expresses a microRNA that targets mRNA encoding Myd88, causing translational arrest of Myd88 and preventing normal embryo development (Chen *et al.*, 2007). The antidote gene, a Myd88 gene variant (Myd88v) is expressed by the zygote, compensating for the loss of normal Myd88. The Myd88v gene sequence is altered so that the microRNA toxin cannot recognize Myd88v mRNA. Only offspring that have the antidote gene can compensate for the effects of the maternally expressed microRNA toxin and survive. A cargo gene (or anti-DENV effector gene) closely linked to MEDEA then can potentially be driven into DENV-transmitting wild type populations of *A. aegypti*. The trick here is to synthesize a suitable MEDEA (toxin/antidote system) for *A. aegypti*. Population models of the MEDEA drive system suggest that the cargo gene can be introgressed into natural populations at a frequency of >99% within 50 generations (Chen *et al.*, 2007). However, the modelling of MEDEA makes a number of assumptions (limited fitness cost, infinite population size, random mating, etc.) and the approach will need considerable testing with laboratory and field cage experiments to determine its efficacy in driving cargo genes into natural populations. This approach has merit over population reduction strategies in that it will not create a vacuum in the niche that could be filled by other vectors. Additionally, population replacement strategies linked to a gene driver could be self-sustaining in DEN endemic areas for extended periods of time. Of course, ethical considerations for releasing genetically modified vectors with a gene drive system into the wild will need to be intensely discussed and the approach broadly accepted prior to any release.

Conclusion

Flaviviral diseases continue to have a significant negative impact on human health in many parts of the world. We are only now beginning to appreciate the complexity of interactions between mosquito vectors and flaviviruses. Flaviviruses, as all arboviruses, spend a significant part of their life cycle in the vector and flaviviruses must adapt to roadblocks the vector mounts to invading pathogens. Co-adaptation of flaviviruses and vectors must influence virus evolution. Advances in viral and vector genetic techniques and bioinformatic approaches over the last decade have significantly improved our understanding of these complex interactions. We are developing a much better appreciation for how the mosquito's innate immune system reacts to flavivirus infection and how genetic differences among vectors (and viruses) in a population may account for observed differences in VC and virus transmission. A new synthesis in our understanding of the molecular biology of vectors and flaviviruses is occurring and will be further advanced over the next decade. This will undoubtedly help researchers identify critical vulnerabilities in virus–vector interactions that lead to new viral and vector gene targets that can be exploited for flavivirus disease control.

Acknowledgements

We wish to thank Susan Rogers for her assistance during the preparation of this manuscript. This work was supported by NIH/NIAID grants AI034014 and AI063434 and the Grand Challenges in Global Health, Bill and Melinda Gates Foundation, through Foundation of the NIH.

References

Acosta, E.G., Castilla, V., and Damonte, E.B. (2008). Functional entry of dengue virus into *Aedes albopictus* mosquito cells is dependent on clathrin-mediated endocytosis. J. Gen. Virol. *89*, 474–484.

Adelman, Z.N., Jasinskiene, N., and James, A.A. (2002). Development and applications of transgenesis in the yellow fever mosquito, *Aedes aegypti*. Mol. Bioch. Parasitol. *121*, 1–10.

Alphey, L., Nimmo, D., O'Connell, S., and Alphey, N. (2008). Insect population suppression using engineered insects. Advanc. Exp. Med. Biol. *627*, 93–103.

Amenya, D.A., Bonizzoni, M., Isaacs, A.T., Jasinskiene, N., Chen, H., Marinotti, O., Yan, G., and James, A.A. (2010). Comparative fitness assessment of *Anopheles stephensi* transgenic lines receptive to site-specific integration. Insect Mol. Biol. *19*, 263–269.

An, J., Kimura-Kuroda, J., Hirabayashi, Y., and Yasui, K. (1999). Development of a novel mouse model for dengue virus infection. Virology *263*, 70–77.

Anderson, J.R., and Rico-Hesse, R. (2006). *Aedes aegypti* vectorial capacity is determined by the infecting genotype of dengue virus. Am. J. Trop. Med. Hyg. *75*, 886–892.

Anderson, M.A., Gross, T.L., Myles, K.M., and Adelman, Z.N. (2010). Validation of novel promoter sequences derived from two endogenous ubiquitin genes in transgenic *Aedes aegypti*. Insect Mol. Biol. *19*, 441–449.

Argentine, J.A., and James, A.A. (1995). Characterization of a salivary gland-specific esterase in the vector mosquito, *Aedes aegypti*. Insect Biochem. Mol. Biol. *25*, 621–630.

Beeman, R.W., Friesen, K.S., and Denell, R.E. (1992). Maternal-effect selfish genes in flour beetles. Science *256*, 89–92.

Beerntsen, B.T., Champagne, D.E., Coleman, J.L., Campos, Y.A., and James, A.A. (1999). Characterization of the Sialokinin I gene encoding the salivary vasodilator of the yellow fever mosquito, *Aedes aegypti*. Insect Mol. Biol. *8*, 459–467.

Benedict, M.Q., Levine, R.S., Hawley, W.A., and Lounibos, L.P. (2007). Spread of the tiger: global risk of invasion by the mosquito *Aedes albopictus*. Vector Borne Zoon. Dis. *7*, 76–85.

Bennett, K.E., Beaty, B.J., and Black, W.C. (2005a). Selection of D2S3, an *Aedes aegypti* (Diptera: Culicidae) strain with high oral susceptibility to Dengue 2 virus and D2MEB, a strain with a midgut barrier to dengue 2 escape. J. Med. Entomol. *42*, 110–119.

Bennett, K.E., Flick, D., Fleming, K.H., Jochim, R., Beaty, B.J., and Black, W.C. (2005b). Quantitative trait loci that control dengue-2 virus dissemination in the mosquito *Aedes aegypti*. Genetics *170*, 185–194.

Bennett, K.E., Olson, K.E., Munoz, M.L., Fernandez-Salas, I., Farfan-Ale, J.A., Higgs, S., Black, W.C., and Beaty, B.J. (2002). Variation in vector competence for dengue 2 virus among 24 collections of *Aedes aegypti* from Mexico and the United States. Am. J. Trop. Med. Hyg. *67*, 85–92.

Bente, D.A., Melkus, M.W., Garcia, J.V., and Rico-Hesse, R. (2005). Dengue fever in humanized NOD/SCID mice. J. Virol. *79*, 13797–13799.

Bernhardt, S.A., Blair, C., Sylla, M., Bosio, C., and Black, W.C. (2009). Evidence of multiple chromosomal inversions in *Aedes aegypti formosus* from Senegal. Insect Mol. Biol. *18*, 557–569.

Bernhardt, S.A. (2010). *Aedes aegypti* and dengue virus: Investigation of anatomic, genomic and molecular determinants of vector competence. PhD. Dissertation, Colorado State University, Fort Collins, Colorado.

Bernier, U.R., Kline, D.L., Barnard, D.R., Schreck, C.E., and Yost, R.A. (2000). Analysis of human skin emanations by gas chromatography/mass spectrometry. 2. Identification of volatile compounds that are candidate attractants for the yellow fever mosquito (*Aedes aegypti*). Anal.Chem. *72*, 747–756.

Beumer, K.J., Trautman, J.K., Bozas, A., Liu, J.L., Rutter, J., Gall, J.G., and Carroll, D. (2008). Efficient gene targeting in Drosophila by direct embryo injection with zinc-finger nucleases. Proc. Natl. Acad. Sci. U.S.A. *105*, 19821–19826.

Bian, G., Shin, S.W., Cheon, H.M., Kokoza, V., and Raikhel, A.S. (2005). Transgenic alteration of Toll immune pathway in the female mosquito *Aedes aegypti*. Proc. Natl. Acad. Sci. U.S.A. *102*, 13568–13573.

Black, W.C., and Moore, C.G. (1996). Population biology as a tool for studying vector-borne diseases. In The Biology of Disease Vectors, Beaty, B.J., and Marquardt, W.C., eds. (Niwot, CO, University Press of Colorado), pp. 393–416

Black, W.C., Bennett, K.E., Gorrochotegui-Escalante, N., Barillas-Mury, C.V., Fernandez-Salas, I., de Lourdes Munoz, M., Farfan-Ale, J.A., Olson, K.E., and Beaty, B.J. (2002). Flavivirus susceptibility in *Aedes aegypti*. Arch. Med. Res. *33*, 379–388.

Blaney, J.E., Jr., Johnson, D.H., Manipon, G.G., Firestone, C.Y., Hanson, C.T., Murphy, B.R., and Whitehead, S.S. (2002). Genetic basis of attenuation of dengue virus type 4 small plaque mutants with restricted replication in suckling mice and in SCID mice transplanted with human liver cells. Virology *300*, 125–139.

Bolling, B. (2010). Flavivirus surveillance in mosquitoes from northern Colorado, with the detection and description of two insect-specific flaviviruses. PhD Dissertation, Colorado State University, Fort Collins, Colorado.

Boppana, V.D., Thangamani, S., Adler, A.J., and Wikel, S.K. (2009). SAAG-4 is a novel mosquito salivary protein that programmes host CD4 T cells to express IL-4. Parasite Immunol. *31*, 287–295.

Bosio, C.F., Fulton, R.E., Salasek, M.L., Beaty, B.J., and Black, W.C. (2000). Quantitative trait loci that control vector competence for dengue-2 virus in the mosquito *Aedes aegypti*. Genetics *156*, 687–698.

Bowers, D.F., Abell, B.A., and Brown, D.T. (1995). Replication and tissue tropism of the alphavirus Sindbis in the mosquito *Aedes albopictus*. Virology *212*, 1–12.

Bowers, D.F., Coleman, C.G., and Brown, D.T. (2003). Sindbis virus-associated pathology in *Aedes albopictus* (Diptera: Culicidae). J. Med. Entomol. *40*, 698–705.

Brackney, D.E., Beane, J.E., and Ebel, G.D. (2009). RNAi targeting of West Nile virus in mosquito midguts promotes virus diversification. PLoS Pathogens *5*, e1000502.

Brackney, D.E., Foy, B.D., and Olson, K.E. (2008). The effects of midgut serine proteases on dengue virus type 2 infectivity of *Aedes aegypti*. Am. J. Trop. Med. Hyg. *79*, 267–274.

Brackney, D.E., Isoe, J., WBC (IV), B., Zamora, J., Foy, B.D., Miesfeld, R.L., and Olson, K.E. (2010). Expression profiling and comparative analyses of seven midgut serine proteases from the yellow fever mosquito, *Aedes aegypti*. J. Insect Physiol. *56*, 736–744.

Calvo, E., Mans, B.J., Andersen, J.F., and Ribeiro, J.M. (2006). Function and evolution of a mosquito salivary protein family. J. Biol. Chem. *281*, 1935–1942.

Calvo, E., Mans, B.J., Ribeiro, J.M., and Andersen, J.F. (2009). Multifunctionality and mechanism of ligand binding in a mosquito antiinflammatory protein. Proc. Natl. Acad. Sci. U.S.A. *106*, 3728–3733.

Campbell, C.L., Keene, K.M., Brackney, D.E., Olson, K.E., Blair, C.D., Wilusz, J., and Foy, B.D. (2008). *Aedes aegypti* uses RNA interference in defense against Sindbis virus infection. BMC Microbiol. *8*, 47.

Champagne, D.E., and Ribeiro, J.M. (1994). Sialokinin I and II: vasodilatory tachykinins from the yellow fever mosquito *Aedes aegypti*. Proc. Natl. Acad. Sci. U.S.A. *91*, 138–142.

Champagne, D.E., Smartt, C.T., Ribeiro, J.M., and James, A.A. (1995). The salivary gland-specific apyrase of the mosquito *Aedes aegypti* is a member of the 5′-nucleotidase family. Proc. Natl. Acad. Sci. U.S.A. *92*, 694–698.

Chen, C.H., Huang, H., Ward, C.M., Su, J.T., Schaeffer, L.V., Guo, M., and Hay, B.A. (2007). A synthetic maternal-effect selfish genetic element drives population replacement in *Drosophila*. Science *316*, 597–600.

Chen, Y., Maguire, T., Hileman, R.E., Fromm, J.R., Esko, J.D., Linhardt, R.J., and Marks, R.M. (1997). Dengue virus infectivity depends on envelope protein binding to target cell heparan sulfate. Nat. Med. *3*, 866–871.

Chu, J.J., and Ng, M.L. (2004a). Infectious entry of West Nile virus occurs through a clathrin-mediated endocytic pathway. J. Virol. *78*, 10543–10555.

Chu, J.J., and Ng, M.L. (2004b). Interaction of West Nile virus with alpha v beta 3 integrin mediates virus entry into cells. J. Biol. Chem. *279*, 54533–54541.

Cirimotich, C.M., Scott, J.C., Phillips, A.T., Geiss, B.J., and Olson, K.E. (2009). Suppression of RNA interference increases alphavirus replication and virus-associated mortality in *Aedes aegypti* mosquitoes. BMC Microbiol. *9*, 49.

Coates, C.J., Jasinskiene, N., Miyashiro, L., and James, A.A. (1998). Mariner transposition and transformation of the yellow fever mosquito, *Aedes aegypti*. Proc. Natl. Acad. Sci. U.S.A. *95*, 3748–3751.

Coates, C.J., Jasinskiene, N., Pott, G.B., and James, A.A. (1999). Promoter-directed expression of recombinant fire-fly luciferase in the salivary glands of Hermes-transformed *Aedes aegypti*. Gene *226*, 317–325.

Cologna, R., Armstrong, P.M., and Rico-Hesse, R. (2005). Selection for virulent dengue viruses occurs in humans and mosquitoes. J. Virol. *79*, 853–859.

Crabtree, M.B., Sang, R.C., Stollar, V., Dunster, L.M., and Miller, B.R. (2003). Genetic and phenotypic characterization of the newly described insect flavivirus, Kamiti River virus. Arch. Virol. *148*, 1095–1118.

Crochu, S., Cook, S., Attoui, H., Charrel, R.N., De Chesse, R., Belhouchet, M., Lemasson, J.J., de Micco, P., and de Lamballerie, X. (2004). Sequences of flavivirus-related RNA viruses persist in DNA form integrated in the genome of *Aedes* spp. mosquitoes. J. Gen. Virol. *85*, 1971–1980.

Dasgupta, R., Selling, B., and Rueckert, R. (1994). Flock house virus: a simple model for studying persistent infection in cultured Drosophila cells. Arch. Virol. Suppl. *9*, 121–132.

Davis, C.T., Ebel, G.D., Lanciotti, R.S., Brault, A.C., Guzman, H., Siirin, M., Lambert, A., Parsons, R.E., Beasley, D.W., Novak, R.J., *et al.* (2005). Phylogenetic analysis of North American West Nile virus isolates, 2001–2004: evidence for the emergence of a dominant genotype. Virology *342*, 252–265.

Deddouche, S., Matt, N., Budd, A., Mueller, S., Kemp, C., Galiana-Arnoux, D., Dostert, C., Antoniewski, C., Hoffmann, J.A., and Imler, J.L. (2008). The DExD/H-box helicase Dicer-2 mediates the induction of antiviral activity in drosophila. Nat. Immunol.*9*, 1425–1432.

Delatte, H., Desvars, A., Bouetard, A., Bord, S., Gimonneau, G., Vourc'h, G., and Fontenille, D. (2010). Blood-feeding behavior of *Aedes albopictus*, a vector of Chikungunya on La Reunion. Vector Borne Zoon. Dis. *10*, 249–258.

Deredec, A., Burt, A., and Godfray, H.C. (2008). The population genetics of using homing endonuclease genes in vector and pest management. Genetics *179*, 2013–2026.

Devenport, M., and Jacobs-Lorena, M. (2004). The peritrophic matrix of hematophagous insects. In Biology of Disease Vectors, Marquardt, W., ed. (Burlington, MA, Elsevier), pp. 297–310.

Ding, S.W., and Voinnet, O. (2007). Antiviral immunity directed by small RNAs. Cell *130*, 413–426.

Domingo, E., Escarmis, C., Sevilla, N., Moya, A., Elena, S.F., Quer, J., Novella, I.S., and Holland, J.J. (1996). Basic concepts in RNA virus evolution. FASEB Journal *10*, 859–864.

Domingo, E., and Holland, J.J. (1997). RNA virus mutations and fitness for survival. Annual Reviews of Microbiology *51*, 151–178.

Dostert, C., Jouanguy, E., Irving, P., Troxler, L., Galiana-Arnoux, D., Hetru, C., Hoffmann, J.A., and Imler, J.L. (2005). The Jak-STAT signaling pathway is required but not sufficient for the antiviral response of drosophila. Nat. Immunol. *6*, 946–953.

Ebel, G.D., Carricaburu, J., Young, D., Bernard, K.A., and Kramer, L.D. (2004). Genetic and phenotypic variation of West Nile virus in New York, 2000–2003. Am. J. Trop. Med. Hyg. *71*, 493–500.

Edwards, J.F., Higgs, S., and Beaty, B.J. (1998). Mosquito feeding-induced enhancement of Cache Valley Virus (Bunyaviridae) infection in mice. J. Med. Entomol. *35*, 261–265.

Edwards, M.J., Moskalyk, L.A., Donelly Doman, M., Vlaskova, M., Noriega, F.G., Walker, V.K., and Jacobs-Lorena, M. (2000). Characterization of a carboxypeptidase A gene from the mosquito, *Aedes aegypti*. Insect Mol.r Biol. *9*, 33–38.

Engelhard, E.K., Kam-Morgan, L.N., Washburn, J.O., and Volkman, L.E. (1994). The insect tracheal system: a conduit for the systemic spread of *Autographa californica* M nuclear polyhedrosis virus. Proc. Natl. Acad. Sci. U.S.A. *91*, 3224–3227.

Erb, S.M., Butrapet, S., Moss, K.J., Luy, B.E., Childers, T., Calvert, A.E., Silengo, S.J., Roehrig, J.T., Huang, C.Y., and Blair, C.D. (2010). Domain-III FG loop of the dengue virus type 2 envelope protein is important for infection of mammalian cells and *Aedes aeqypti* mosquitoes. Virology *406*, 328–335.

Fragkoudis, R., Chi, Y., Siu, R.W., Barry, G., Attarzadeh-Yazdi, G., Merits, A., Nash, A.A., Fazakerley, J.K., and Kohl, A. (2008). Semliki Forest virus strongly reduces mosquito host defence signaling. Insect Mol. Biol. *17*, 647–656.

Franz, A.W., Sánchez-Vargas, I., Adelman, Z.N., Blair, C.D., Beaty, B.J., James, A.A., and Olson, K.E. (2006). Engineering RNA interference-based resistance to dengue virus type 2 in genetically modified *Aedes aegypti*. Proc. Natl. Acad. Sci. U.S.A. *103*, 4198–4203.

Franz, A.W., Sánchez-Vargas, I., Piper, J., Smith, M.R., Khoo, C.C., James, A.A., and Olson, K.E. (2009). Stability and loss of a virus resistance phenotype over time in transgenic mosquitoes harbouring an antiviral effector gene. Insect Mol. Biol.*18*, 661–672.

Fraser, M.J., Ciszczon, T., Elick, T., and Bauser, C. (1996). Precise excision of TTAA-specific lepidopteran transposons piggyBac (IFP2) and tagalong (TFP3) from the baculovirus genome in cell lines from two species of Lepidoptera. Insect Mol. Biol. 5, 141–151.

Friesen, P., Scotti, P., Longworth, J., and Rueckert, R. (1980). Black beetle virus: propagation in Drosophila line 1 cells and an infection-resistant subline carrying endogenous black beetle virus-related particles. J. Virol. 35, 741–747.

Fu, G., Lees, R.S., Nimmo, D., Aw, D., Jin, L., Gray, P., Berendonk, T.U., White-Cooper, H., Scaife, S., Kim Phuc, H., *et al.* (2010). Female-specific flightless phenotype for mosquito control. Proc. Natl. Acad. Sci. U.S.A. *107*, 4550–4554.

Galiana-Arnoux, D., Dostert, C., Schneemann, A., Hoffmann, J.A., and Imler, J.L. (2006). Essential function *in vivo* for Dicer-2 in host defense against RNA viruses in drosophila. Nat. Immunol. 7, 590–597.

Garver, L., and Dimopoulos, G. (2007). Protocol for RNAi assays in adult mosquitoes (*A. gambiae*). J. Vis. Exp. 5, 230.

Germi, R., Crance, J.M., Garin, D., Guimet, J., Lortat-Jacob, H., Ruigrok, R.W., Zarski, J.P., and Drouet, E. (2002). Heparan sulfate-mediated binding of infectious dengue virus type 2 and yellow fever virus. Virology *292*, 162–168.

Girard, Y.A., Mayhew, G.F., Fuchs, J.F., Li, H., Schneider, B.S., McGee, C.E., Rocheleau, T.A., Helmy, H., Christensen, B.M., Higgs, S., *et al.* (2010). Transcriptome Changes in Culex quinquefasciatus (Diptera: Culicidae) Salivary Glands During West Nile Virus Infection. J. Med. Entomol. *47*, 421–435.

Girard, Y.A., Popov, V., Wen, J., Han, V., and Higgs, S. (2005). Ultrastructural study of West Nile virus pathogenesis in *Culex pipiens quinquefasciatus* (Diptera: Culicidae). J. Med. Entomol. *42*, 429–444.

Girard, Y.A., Schneider, B.S., McGee, C.E., Wen, J., Han, V.C., Popov, V., Mason, P.W., and Higgs, S. (2007). Salivary gland morphology and virus transmission during long-term cytopathologic West Nile virus infection in Culex mosquitoes. Am. J. Trop. Med. Hyg. *76*, 118–128.

Gomez-Machorro, C., Bennett, K.E., del Lourdes Munoz, M., and Black, W.C. (2004). Quantitative trait loci affecting dengue midgut infection barriers in an advanced intercross line of *Aedes aegypti*. Insect Mol. Biol. *13*, 637–648.

Gould, F., Magori, K., and Huang, Y. (2006). Genetic strategies for controlling mosquito-borne diseases. Am. Sci. *94*, 238–246.

Groth, A.C., Fish, M., Nusse, R., and Calos, M.P. (2004). Construction of transgenic Drosophila by using the site-specific integrase from phage phiC31. Genetics *166*, 1775–1782.

Groth, A.C., Olivares, E.C., Thyagarajan, B., and Calos, M.P. (2000). A phage integrase directs efficient site-specific integration in human cells. Proc. Natl. Acad. Sci. U.S.A. *97*, 5995–6000.

Gubler, D.J. (2002). The global emergence/resurgence of arboviral diseases as public health problems. Arch. Med. Res. *33*, 330–342.

Gunawardane, L.S., Saito, K., Nishida, K.M., Miyoshi, K., Kawamura, Y., Nagami, T., Siomi, H., and Siomi, M.C. (2007). A slicer-mediated mechanism for repeat-associated siRNA 5′ end formation in *Drosophila*. Science *315*, 1587–1590.

Guo, X., Xu, Y., Bian, G., Pike, A.D., Xie, Y., and Xi, Z. (2010). Response of the mosquito protein interaction network to dengue infection. BMC Genomics *11*, 380.

Hafer, A., Whittlesey, R., Brown, D.T., and Hernandez, R. (2009). Differential incorporation of cholesterol by Sindbis virus grown in mammalian or insect cells. J. Virol. *83*, 9113–9121.

Hahn, C.S., Dalrymple, J.M., Strauss, J.H., and Rice, C.M. (1987). Comparison of the virulent Asibi strain of yellow fever virus with the 17D vaccine strain derived from it. Proc. Natl. Acad. Sci. U.S.A. *84*, 2019–2023.

Halstead, S.B. (2008). Dengue virus–mosquito interactions. Annu. Rev. Entomol. *53*, 273–291.

Handler, A.M. (2001). A current perspective on insect gene transformation. Insect Biochem. Mol. Biol. *31*, 111–128.

Hanley, K.A., Manlucu, L.R., Gilmore, L.E., Blaney, J.E., Jr., Hanson, C.T., Murphy, B.R., and Whitehead, S.S. (2003). A trade-off in replication in mosquito versus mammalian systems conferred by a point mutation in the NS4B protein of dengue virus type 4. Virology *312*, 222–232.

Hanley, K.A., Nelson, J.T., Schirtzinger, E.E., Whitehead, S.S., and Hanson, C.T. (2008). Superior infectivity for mosquito vectors contributes to competitive displacement among strains of dengue virus. BMC Ecol. *8*, 1.

Hase, T., Summers, P.L., and Cohen, W.H. (1989a). A comparative study of entry modes into C6/36 cells by Semliki Forest and Japanese encephalitis viruses. Arch. Virol. *108*, 101–114.

Hase, T., Summers, P.L., and Eckels, K.H. (1989b). Flavivirus entry into cultured mosquito cells and human peripheral blood monocytes. Arch. Virol. *104*, 129–143.

Hase, T., Summers, P.L., Eckels, K.H., and Baze, W.B. (1987). An electron and immunoelectron microscopic study of dengue-2 virus infection of cultured mosquito cells: maturation events. Arch. Virol. *92*, 273–291.

Hilgard, P., and Stockert, R. (2000). Heparan sulfate proteoglycans initiate dengue virus infection of hepatocytes. Hepatology *32*, 1069–1077.

Hoffmann, J.A. (2003). The immune response of *Drosophila*. Nature *426*, 33–38.

Holmes, E.C. (2003). Patterns of intra- and interhost nonsynonymous variation reveal strong purifying selection in dengue virus. J. Virol. *77*, 11296–11298.

Horn, C., and Wimmer, E.A. (2000). A versatile vector set for animal transgenesis. Dev. Genes Evol. *210*, 630–637.

Hoshino, K., Isawa, H., Tsuda, Y., Sawabe, K., and Kobayashi, M. (2009). Isolation and characterization of a new insect flavivirus from *Aedes albopictus* and *Aedes flavopictus* mosquitoes in Japan. Virology *391*, 119–129.

Hoshino, K., Isawa, H., Tsuda, Y., Yano, K., Sasaki, T., Yuda, M., Takasaki, T., Kobayashi, M., and Sawabe, K. (2007). Genetic characterization of a new insect

flavivirus isolated from *Culex pipiens* mosquito in Japan. Virology *359*, 405–414.
Houk, E.J., Kramer, L.D., Hardy, J.L., and Chiles, R.E. (1985). Western equine encephalomyelitis virus: *in vivo* infection and morphogenesis in mosquito mesenteronal epithelial cells. Virus Res. *2*, 123–138.
Hsieh, P., and Robbins, P.W. (1984). Regulation of asparagine-linked oligosaccharide processing. Oligosaccharide processing in Aedes albopictus mosquito cells. J. Biol. Chem. *259*, 2375–2382.
Hsieh, P., Rosner, M.R., and Robbins, P.W. (1983). Host-dependent variation of asparagine-linked oligosaccharides at individual glycosylation sites of Sindbis virus glycoproteins. J. Biol. Chem. *258*, 2548–2554.
Huang, C.Y.-H., Butrapet, S., Moss, K.J., Childers, T., Erb, S.M., Calvert, A.E., Silengo, S.J., Kinney, R.M., Blair, C.D., and Roehrig, J.T. (2010). The dengue virus type 2 envelope protein fusion peptide is essential for membrane fusion. Virology *396*, 305–315.
Huang, C.Y., Butrapet, S., Tsuchiya, K.R., Bhamarapravati, N., Gubler, D.J., and Kinney, R.M. (2003). Dengue 2 PDK-53 virus as a chimeric carrier for tetravalent dengue vaccine development. J. Virol. *77*, 11436–11447.
Hung, J.J., Hsieh, M.T., Young, M.J., Kao, C.L., King, C.C., and Chang, W. (2004). An external loop region of domain III of dengue virus type 2 envelope protein is involved in serotype-specific binding to mosquito but not mammalian cells. J. Virol. *78*, 378–388.
Hung, S.L., Lee, P.L., Chen, H.W., Chen, L.K., Kao, C.L., and King, C.C. (1999). Analysis of the steps involved in Dengue virus entry into host cells. Virology *257*, 156–167.
Igarashi, A. (1978). Isolation of a Singh's *Aedes albopictus* cell clone sensitive to dengue and chikungunya viruses. J. Gen. Virol. *40*, 531–544.
James, A.A. (2003). Blocking malaria parasite invasion of mosquito salivary glands. J. Exp. Biol. *206*, 3817–3821.
James, A.A. (2007). Preventing the spread of malaria and dengue fever using genetically modified mosquitoes. J. Vis. Exp. *5*, *231*.
James, A.A., Blackmer, K., Marinotti, O., Ghosn, C.R., and Racioppi, J.V. (1991). Isolation and characterization of the gene expressing the major salivary gland protein of the female mosquito, *Aedes aegypti*. Mol. Biochem. Parasito. *44*, 245–253.
Jasinskiene, N., Coates, C.J., Benedict, M.Q., Cornel, A.J., Rafferty, C.S., James, A.A., and Collins, F.H. (1998). Stable transformation of the yellow fever mosquito, *Aedes aegypti*, with the Hermes element from the housefly. Proc. Natl. Acad. Sci. U.S.A. *95*, 3743–3747.
Jenkins, G.M., and Holmes, E.C. (2003). The extent of codon usage bias in human RNA viruses and its evolutionary origin. Virus Res. *92*, 1–7.
Jerzak, G., Bernard, K.A., Kramer, L.D., and Ebel, G.D. (2005). Genetic variation in West Nile virus from naturally infected mosquitoes and birds suggests quasispecies structure and strong purifying selection. J. Gen. Virol. *86*, 2175–2183.
Jerzak, G.V., Bernard, K., Kramer, L.D., Shi, P.Y., and Ebel, G.D. (2007). The West Nile virus mutant spectrum is host-dependant and a determinant of mortality in mice. Virology *360*, 469–476.
Jerzak, G.V., Brown, I., Shi, P.Y., Kramer, L.D., and Ebel, G.D. (2008). Genetic diversity and purifying selection in West Nile virus populations are maintained during host switching. Virology *374*, 256–260.
Johnson, A.J., Guirakhoo, F., and Roehrig, J.T. (1994). The envelope glycoproteins of dengue 1 and dengue 2 viruses grown in mosquito cells differ in their utilization of potential glycosylation sites. Virology *203*, 241–249.
Johnson, A.J., and Roehrig, J.T. (1999). New mouse model for dengue virus vaccine testing. J. Virol. *73*, 783–786.
Junjhon, J., Edwards, T.J., Utaipat, U., Bowman, V.D., Holdaway, H.A., Zhang, W., Keelapang, P., Puttikhunt, C., Perera, R., Chipman, P.R., *et al.* (2010). Influence of pr–M cleavage on the heterogeneity of extracellular dengue virus particles. J. Virol. *84*, 8353–8358.
Kanai, R., Kar, K., Anthony, K., Gould, L.H., Ledizet, M., Fikrig, E., Marasco, W.A., Koski, R.A., and Modis, Y. (2006). Crystal structure of West Nile virus envelope glycoprotein reveals viral surface epitopes. J. Virol. *80*, 11000–11008.
Kato, N., Mueller, C.R., Fuchs, J.F., McElroy, K., Wessely, V., Higgs, S., and Christensen, B.M. (2008). Evaluation of the function of a type I peritrophic matrix as a physical barrier for midgut epithelium invasion by mosquito-borne pathogens in *Aedes aegypti*. Vector Borne Zoon. Dis. *8*, 701–712.
Keene, K.M., Foy, B.D., Sánchez-Vargas, I., Beaty, B.J., Blair, C.D., and Olson, K.E. (2004). RNA interference acts as a natural antiviral response to O'nyong-nyong virus (*Alphavirus*; *Togaviridae*) infection of Anopheles gambiae. Proc. Natl. Acad. Sci. U.S.A. *101*, 17240–17245.
Kelsall, B.L., Biron, C.A., Sharma, O., and Kaye, P.M. (2002). Dendritic cells at the host–pathogen interface. Nat. Immunol. 3, 699–702.
Kemp, C., and Imler, J.L. (2009). Antiviral immunity in drosophila. Curr. Opin. Immunol. *21*, 3–9.
Khoo, C.C., Piper, J., Sánchez-Vargas, I., Olson, K.E., and Franz, A.W. (2010). The RNA interference pathway affects midgut infection- and escape barriers for Sindbis virus in *Aedes aegypti*. BMC Microbiol. *10*, 130.
Klimstra, W.B., Nangle, E.M., Smith, M.S., Yurochko, A.D., and Ryman, K.D. (2003). DC-SIGN and L-SIGN can act as attachment receptors for alphaviruses and distinguish between mosquito cell- and mammalian cell-derived viruses. J. Virol. *77*, 12022–12032.
Kokoza, V., Ahmed, A., Cho, W.L., Jasinskiene, N., James, A.A., and Raikhel, A. (2000). Engineering blood meal-activated systemic immunity in the yellow fever mosquito, *Aedes aegypti*. Proc. Natl. Acad. Sci. U.S.A. *97*, 9144–9149.
Krebs, K.C., and Lan, Q. (2003). Isolation and expression of a sterol carrier protein-2 gene from the yellow fever mosquito, *Aedes aegypti*. Insect Mol. Biol. *12*, 51–60.

Kuruvilla, J.G., Troyer, R.M., Devi, S., and Akkina, R. (2007). Dengue virus infection and immune response in humanized RAG2(–/–)gamma(c) (–/––) (RAG-hu) mice. Virology *369*, 143–152.

Kyle, J.L., Beatty, P.R., and Harris, E. (2007). Dengue virus infects macrophages and dendritic cells in a mouse model of infection. J. Infecti. Dis. *195*, 1808–1817.

Lambrechts, L., Chevillon, C., Albright, R.G., Thaisomboonsuk, B., Richardson, J.H., Jarman, R.G., and Scott, T.W. (2009). Genetic specificity and potential for local adaptation between dengue viruses and mosquito vectors. BMC Evol. Biol. *9*, 160.

Lambrechts, L., Scott, T.W., and Gubler, D.J. (2010). Consequences of the expanding global distribution of *Aedes albopictus* for dengue virus transmission. PLoS Neglected Trop. Dis. *4*, e646.

Lan, Q., and Fallon, A.M. (1990). Small heat shock proteins distinguish between two mosquito species and confirm identity of their cell lines. Am. J. Trop. Med. Hyg. *43*, 669–676.

Lanciotti, R.S., Roehrig, J.T., Deubel, V., Smith, J., Parker, M., Steele, K., Crise, B., Volpe, K.E., Crabtree, M.B., Scherret, J.H., *et al.* (1999). Origin of the West Nile virus responsible for an outbreak of encephalitis in the northeastern United States. Science *286*, 2333–2337.

Lawson, D., Arensburger, P., Atkinson, P., Besansky, N.J., Bruggner, R.V., Butler, R., Campbell, K.S., Christophides, G.K., Christley, S., Dialynas, E., *et al.* (2009). VectorBase: a data resource for invertebrate vector genomics. Nucl. Acids Res. *37*, D583–587.

Lazzari, C.R., and Stephen, J.S. (2009). Orientation towards hosts in haematophagous insects: An integrative perspective. In Advances in Insect Physiology (Academic Press), pp. 1–58.

Leitmeyer, K.C., Vaughn, D.W., Watts, D.M., Salas, R., deVillalobos, I., Ramos, C., and Rico-Hesse, R. (1999). Dengue virus structural differences that correlate with pathogenesis. J. Virol. *73*, 4738–4747.

Linthicum, K.J., Platt, K., Myint, K.S., Lerdthusnee, K., Innis, B.L., and Vaughn, D.W. (1996). Dengue 3 virus distribution in the mosquito *Aedes aegypti*: an immunocytochemical study. Med. Vet.Entomol. *10*, 87–92.

Lozano-Fuentes, S., Fernandez-Salas, I., de Lourdes Munoz, M., Garcia-Rejon, J., Olson, K.E., Beaty, B.J., and Black, W.C. (2009). The neovolcanic axis is a barrier to gene flow among *Aedes aegypti* populations in Mexico that differ in vector competence for dengue 2 virus. PLoS Neglected Trop. Dis. *3*, e468.

Lu, S.J., Pennington, J.E., Stonehouse, A.R., Mobula, M.M., and Wells, M.A. (2006). Reevaluation of the role of early trypsin activity in the transcriptional activation of the late trypsin gene in the mosquito *Aedes aegypti*. Insect Biochem. Mol. Biol. *36*, 336–343.

Luna, B.M., Juhn, J., and James, A.A. (2007). Injection of dsRNA into female *A. aegypti* mosquitos. J. Vis. Exp. *5*, 215.

Mackenzie, J.M., Khromykh, A.A., and Parton, R.G. (2007). Cholesterol manipulation by West Nile virus perturbs the cellular immune response. Cell Host Microbe *2*, 229–239.

Mackenzie, J.S., Gubler, D.J., and Petersen, L.R. (2004). Emerging flaviviruses: the spread and resurgence of Japanese encephalitis, West Nile and dengue viruses. Nat. Med. *10*, S98–109.

Mahmood, F., Chiles, R.E., Fang, Y., Green, E.N., and Reisen, W.K. (2006). Effects of time after infection, mosquito genotype, and infectious viral dose on the dynamics of *Culex tarsalis* vector competence for western equine encephalomyelitis virus. J. Am. Mosquito Control Assoc. *22*, 272–281.

Mathur, G., Sánchez-Vargas, I., Alvarez, D., Olson, K.E., Marinotti, O., and James, A.A. (2010). Transgene-mediated suppression of dengue viruses in the salivary glands of the yellow fever mosquito, *Aedes aegypti*. Insect Mol. Biol. *19*, 753–763.

Mattingly, P.F. (1957). Genetical aspects of the *Aedes aegypti* problem. I. Taxonom: and bionomics. Ann. Trop. Med. Parasitol. *51*, 392–408.

McElroy, K.L., Tsetsarkin, K.A., Vanlandingham, D.L., and Higgs, S. (2006a). Manipulation of the yellow fever virus non-structural genes 2A and 4B and the 3′non-coding region to evaluate genetic determinants of viral dissemination from the *Aedes aegypti* midgut. Am. J. Trop. Med. Hyg. *75*, 1158–1164.

McElroy, K.L., Tsetsarkin, K.A., Vanlandingham, D.L., and Higgs, S. (2006b). Role of the yellow fever virus structural protein genes in viral dissemination from the *Aedes aegypti* mosquito midgut. J. Gen. Virol. *87*, 2993–3001.

McMeniman, C.J., Lane, R.V., Cass, B.N., Fong, A.W., Sidhu, M., Wang, Y.F., and O'Neill, S.L. (2009). Stable introduction of a life-shortening Wolbachia infection into the mosquito *Aedes aegypti*. Science *323*, 141–144.

Medhora, M., Maruyama, K., and Hartl, D.L. (1991). Molecular and functional analysis of the mariner mutator element *Mos1* in Drosophila. Genetics *128*, 311–318.

Mercado-Curiel, R.F., Black, W.C., and Munoz M.de L. (2008). A dengue receptor as possible genetic marker of vector competence in *Aedes aegypti*. BMC Microbiol. *8*, 118.

Mercado-Curiel, R.F., Esquinca-Aviles, H.A., Tovar, R., Diaz-Badillo, A., Camacho-Nuez, M., and Munoz M.de L. (2006). The four serotypes of dengue recognize the same putative receptors in *Aedes aegypti* midgut and *Ae. albopictus* cells. BMC Microbiology *6*, 85.

Messer, W.B., Gubler, D.J., Harris, E., Sivananthan, K., and de Silva, A.M. (2003). Emergence and global spread of a dengue serotype 3, subtype III virus. Emerg. Infect. Dis. *9*, 800–809.

Mézeth, K.B., Patel, S., Henriksen, H., Szilvay, A.M., and Nerland, A.H. (2009). B2 protein from betanodavirus is expressed in recently infected but not in chronically infected fish. Dis. Aqua. Organ. *83*, 97–103.

Miller, S., Sparacio, S., and Bartenschlager, R. (2006). Subcellular localization and membrane topology of the Dengue virus type 2 Non-structural protein 4B. J. Biol. Chem. *281*, 8854–8863.

Mitsuhashi, J., Nakasone, S., and Horie, Y. (1983). Sterol-free eukaryotic cells from continuous cell lines of insects. Cell Biology International Reports *7*, 1057–1062.

Modis, Y., Ogata, S., Clements, D., and Harrison, S.C. (2003). A ligand-binding pocket in the dengue virus envelope glycoprotein. Proc. Natl. Acad. Sci. U.S.A. *100*, 6986–6991.

Modis, Y., Ogata, S., Clements, D., and Harrison, S.C. (2004). Structure of the dengue virus envelope protein after membrane fusion. Nature *427*, 313–319.

Modis, Y., Ogata, S., Clements, D., and Harrison, S.C. (2005). Variable surface epitopes in the crystal structure of dengue virus type 3 envelope glycoprotein. J. Virol. *79*, 1223–1231.

Moesker, B., Rodenhuis-Zybert, I.A., Meijerhof, T., Wilschut, J., and Smit, J.M. (2010). Characterization of the functional requirements of West Nile virus membrane fusion. J. Gen. Virol. *91*, 389–393.

Molina-Cruz, A., Gupta, L., Richardson, J., Bennett, K., Black, W.C., and Barillas-Mury, C. (2005). Effect of mosquito midgut trypsin activity on dengue-2 virus infection and dissemination in *Aedes aegypti*. Am. J. Trop. Med. Hyg. *72*, 631–637.

Monath, T.P. (1994). Dengue: the risk to developed and developing countries. Proc. Natl. Acad. Sci. U.S.A. *91*, 2395–2400.

Morales-Betoulle, M.E., Monzon Pineda, M.L., Sosa, S.M., Panella, N., Lopez, M.R., Cordon-Rosales, C., Komar, N., Powers, A., and Johnson, B.W. (2008). Culex flavivirus isolates from mosquitoes in Guatemala. J. Med. Entomol. *45*, 1187–1190.

Moreira, L.A., Edwards, M.J., Adhami, F., Jasinskiene, N., James, A.A., and Jacobs-Lorena, M. (2000). Robust gut-specific gene expression in transgenic *Aedes aegypti* mosquitoes. Proc. Natl. Acad. Sci. U.S.A. *97*, 10895–10898.

Moreira, L.A., Iturbe-Ormaetxe, I., Jeffery, J.A., Lu, G., Pyke, A.T., Hedges, L.M., Rocha, B.C., Hall-Mendelin, S., Day, A., Riegler, M., *et al.* (2009). A Wolbachia symbiont in *Aedes aegypti* limits infection with dengue, chikungunya, and *Plasmodium*. Cell *139*, 1268–1278.

Moskalyk, L.A., Oo, M.M., and Jacobs-Lorena, M. (1996). Peritrophic matrix proteins of *Anopheles gambiae* and *Aedes aegypti*. Insect Mol. Biol.*5*, 261–268.

Mosso, C., Galvan-Mendoza, I.J., Ludert, J.E., and del Angel, R.M. (2008). Endocytic pathway followed by dengue virus to infect the mosquito cell line C6/36 HT. Virology *378*, 193–199.

Moudy, R.M., Meola, M.A., Morin, L.L., Ebel, G.D., and Kramer, L.D. (2007). A newly emergent genotype of West Nile virus is transmitted earlier and more efficiently by Culex mosquitoes. Am. J. Trop. Med. Hyg. *77*, 365–370.

Munoz-Jordan, J.L., Laurent-Rolle, M., Ashour, J., Martinez-Sobrido, L., Ashok, M., Lipkin, W.I., and Garcia-Sastre, A. (2005). Inhibition of alpha/beta interferon signaling by the NS4B protein of flaviviruses. J. Virol. *79*, 8004–8013.

Munoz, M.L., Cisneros, A., Cruz, J., Das, P., Tovar, R., and Ortega, A. (1998). Putative dengue virus receptors from mosquito cells. FEMS Microbiol. Lett. *168*, 251–258.

Myles, K.M., Pierro, D.J., and Olson, K.E. (2004). Comparison of the transmission potential of two genetically distinct Sindbis viruses after oral infection of *Aedes aegypti* (Diptera: Culicidae). J. Med. Entomol. *41*, 95–106.

Myles, K.M., Wiley, M.R., Morazzani, E.M., and Adelman, Z.N. (2008). Alphavirus-derived small RNAs modulate pathogenesis in disease vector mosquitoes. Proc. Natl. Acad. Sci. U.S.A. *105*, 19938–19943.

Nakayama, G., Kawaguchi, Y., Koga, K., and Kusakabe, T. (2006). Site-specific gene integration in cultured silkworm cells mediated by phiC31 integrase. Mol. Genet. Genom. *275*, 1–8.

Nasidi, A., Monath, T.P., DeCock, K., Tomori, O., Cordellier, R., Olaleye, O.D., Harry, T.O., Adeniyi, J.A., Sorungbe, A.O., Ajose-Coker, A.O., *et al.* (1989). Urban yellow fever epidemic in western Nigeria, (1987). Trans. Roy. Soc. Trop. Med. Hyg. *83*, 401–406.

Navarro-Sanchez, E., Altmeyer, R., Amara, A., Schwartz, O., Fieschi, F., Virelizier, J.L., Arenzana-Seisdedos, F., and Despres, P. (2003). Dendritic-cell-specific ICAM3-grabbing non-integrin is essential for the productive infection of human dendritic cells by mosquito-cell-derived dengue viruses. EMBO Rep. *4*, 723–728.

Navarro-Sanchez, E., Despres, P., and Cedillo-Barron, L. (2005). Innate immune responses to dengue virus. Arch. Med. Res. *36*, 425–435.

Nawa, M., Machida, S., Takasaki, T., and Kurane, I. (2007). Plaque formation by Japanese encephalitis virus bound to mosquito C6/36 cells after low pH exposure on the cell surface. Japan. J. Infect. Dis. *60*, 118–120.

Nawtaisong, P., Keith, J., Fraser, T., Balaraman, V., Kolokoltsov, A., Davey, R.A., Higgs, S., Mohammed, A., Rongsriyam, Y., Komalamisra, N., *et al.* (2009). Effective suppression of Dengue fever virus in mosquito cell cultures using retroviral transduction of hammerhead ribozymes targeting the viral genome. Virol. J. *6*, 73.

Nene, V., Wortman, J.R., Lawson, D., Haas, B., Kodira, C., Tu, Z.J., Loftus, B., Xi, Z., Megy, K., Grabherr, M., *et al.* (2007). Genome sequence of *Aedes aegypti*, a major arbovirus vector. Science *316*, 1718–1723.

Nielsen, D.G. (2009). The relationship of interacting immunological components in dengue pathogenesis. Virology Journal *6*, 211.

Nimmo, D.D., Alphey, L., Meredith, J.M., and Eggleston, P. (2006). High efficiency site-specific genetic engineering of the mosquito genome. Insect Mol. Biol.*15*, 129–136.

Noriega, F.G., Colonna, A.E., and Wells, M.A. (1999). Increase in the size of the amino acid pool is sufficient to activate translation of early trypsin mRNA in *Aedes aegypti* midgut. Insect Biochem. Mol. Biol. *29*, 243–247.

O'Brochta, D.A., Sethuraman, N., Wilson, R., Hice, R.H., Pinkerton, A.C., Levesque, C.S., Bideshi, D.K., Jasinskiene, N., Coates, C.J., James, A.A., *et al.* (2003). Gene vector and transposable element behavior in mosquitoes. J. Exp. Biol. *206*, 3823–3834.

Obbard, D.J., Jiggins, F.M., Halligan, D.L., and Little, T.J. (2006). Natural selection drives extremely rapid evolution in antiviral RNAi genes. Curr. Biol. *16*, 580–585.

Olson, K., and Franz, A. (2009). Controlling dengue virus transmission in the field with genetically modified mosquitoes. In Advances in Human Vector Control, Clark, J.M., Bloomquist, J.R., and Kawada, H., eds. (Washington DC, American Chemical Society Publications), pp. 123–141.

Ooi, E.E., Goh, K.T., and Gubler, D.J. (2006). Dengue prevention and 35 years of vector control in Singapore. Emerg. Infect. Dis. *12*, 887–893.

Peleg, J. (1968). Growth of arboviruses in monolayers from subcultured mosquito embryo cells. Virology *35*, 617–619.

Pennington, J., and Wells, M. (2004). The adult midgut. In Biology of Disease Vectors, Marquardt, W., ed. (Burlington, MA, Elsevier), pp. 289–295.

Peters-Golden, M., Canetti, C., Mancuso, P., and Coffey, M.J. (2005). Leukotrienes: underappreciated mediators of innate immune responses. J. Immunol. *174*, 589–594.

Pierro, D.J., Powers, E.L., and Olson, K.E. (2007). Genetic determinants of Sindbis virus strain TR339 affecting midgut infection in the mosquito *Aedes aegypti*. J. Gen. Virol. *88*, 1545–1554.

Pierro, D.J., Powers, E.L., and Olson, K.E. (2008). Genetic determinants of Sindbis virus mosquito infection are associated with a highly conserved alphavirus and flavivirus envelope sequence. J. Virol. *82*, 2966–2974.

Pinkerton, A.C., Whyard, S., Mende, H.A., Coates, C.J., O'Brochta, D.A., and Atkinson, P.W. (1999). The Queensland fruit fly, *Bactrocera tryoni*, contains multiple members of the hAT family of transposable elements. Insect Mol. Biol. *8*, 423–434.

Poole-Smith, B.K. (2010). Isolation and Characterization of Dengue Virus Membrane Assoicated Replication Complexes from *Aedes aegypti*. Ph.D. Dissertation, Colorado State University, Fort Collins, Colorado.

Puerta-Guardo, H., Mosso, C., Medina, F., Liprandi, F., Ludert, J.E., and del Angel, R.M. (2010). Antibody-dependent enhancement of dengue virus infection in U937 cells requires cholesterol-rich membrane microdomains. J. Gen. Virol. *91*, 394–403.

Rey, F.A., Heinz, F.X., Mandl, C., Kunz, C., and Harrison, S.C. (1995). The envelope glycoprotein from tick-borne encephalitis virus at 2 Å resolution. Nature *375*, 291–298.

Reyes-Del Valle, J., Chavez-Salinas, S., Medina, F., and del Angel, R.M. (2005). Heat shock protein 90 and heat shock protein 70 are components of dengue virus receptor complex in human cells. J. Virol. *79*, 4557–4567.

Ribeiro, J.M. (1987). Role of saliva in blood-feeding by arthropods. Annu. Rev. Entomol. *32*, 463–478.

Ribeiro, J.M., Arca, B., Lombardo, F., Calvo, E., Phan, V.M., Chandra, P.K., and Wikel, S.K. (2007). An annotated catalogue of salivary gland transcripts in the adult female mosquito, *Aedes aegypti*. BMC Genomics *8*, 6.

Ribeiro, J.M.C., Arcà, B., and Stephen, J.S. (2009). From sialomes to the sialoverse: An insight into salivary potion of blood-feeding insects. In Advances in Insect Physiology (Academic Press), pp. 59–118.

Richardson, J., Molina-Cruz, A., Salazar, M.I., and Black, W.C. (2006). Quantitative analysis of dengue-2 virus RNA during the extrinsic incubation period in individual *Aedes aegypti*. Am. J. Trop. Med. Hyg. *74*, 132–141.

Rico-Hesse, R. (1990). Molecular evolution and distribution of dengue viruses type 1 and 2 in nature. Virology *174*, 479–493.

Rico-Hesse, R. (2003). Microevolution and virulence of dengue viruses. Adv. Virus Res. *59*, 315–341.

Rico-Hesse, R., Harrison, L.M., Salas, R.A., Tovar, D., Nisalak, A., Ramos, C., Boshell, J., de Mesa, M.T., Nogueira, R.M., and da Rosa, A.T. (1997). Origins of dengue type 2 viruses associated with increased pathogenicity in the Americas. Virology *230*, 244–251.

Romoser, W.S., Wasieloski, L.P., Jr., Pushko, P., Kondig, J.P., Lerdthusnee, K., Neira, M., and Ludwig, G.V. (2004). Evidence for arbovirus dissemination conduits from the mosquito (Diptera: Culicidae) midgut. J. Med. Entomol. *41*, 467–475.

Rostand, K.S., and Esko, J.D. (1997). Microbial adherence to and invasion through proteoglycans. Infect. Immun. *65*, 1–8.

Sabatier, L., Jouanguy, E., Dostert, C., Zachary, D., Dimarcq, J.L., Bulet, P., and Imler, J.L. (2003). Pherokine-2 and -3. Two *Drosophila* molecules related to pheromone/odour-binding proteins induced by viral and bacterial infections Eur. J. Biochem. *270*, 3398–3407.

Salas-Benito, J., Reyes-del Valle, J., Salas-Benito, M., Ceballos-Olvera, I., Mosso, C., and del Angel, R.M. (2007). Evidence that the 45-kD glycoprotein, part of a putative dengue virus receptor complex in the mosquito cell line C6/36, is a heat-shock related protein. Am. J. Trop. Med. Hyg. *77*, 283–290.

Salazar-Sanchez, I. (2006). Determinants of dengue Type 2 Virus Infection in the Mosquito *Aedes aegypti*. PhD Dissertation, Colorado State University, Fort Collins, Colorado.

Salazar, M.I., Richardson, J.H., Sánchez-Vargas, I., Olson, K.E., and Beaty, B.J. (2007). Dengue virus type 2: replication and tropisms in orally infected *Aedes aegypti* mosquitoes. BMC Microbiol. *7*, 9.

Sánchez-Vargas, I., Scott, J.C., Poole-Smith, B.K., Franz, A.W., Barbosa-Solomieu, V., Wilusz, J., Olson, K.E., and Blair, C.D. (2009). Dengue virus type 2 infections of *Aedes aegypti* are modulated by the mosquito's RNA interference pathway. PLoS Pathog. *5*, e1000299.

Sang, R.C., Gichogo, A., Gachoya, J., Dunster, M.D., Ofula, V., Hunt, A.R., Crabtree, M.B., Miller, B.R., and Dunster, L.M. (2003). Isolation of a new flavivirus related to cell fusing agent virus (CFAV) from field-collected flood-water Aedes mosquitoes sampled from a dambo in central Kenya. Arch. Virol. *148*, 1085–1093.

Sarver, N., and Stollar, V. (1977). Sindbis virus-induced cytopathic effect in clones of *Aedes albopictus* (Singh) cells. Virology *80*, 390–400.

Schneider, B.S., and Higgs, S. (2008). The enhancement of arbovirus transmission and disease by mosquito saliva is associated with modulation of the host immune response. Transactions of the Royal Society of Tropical Medicine and Hygiene *102*, 400–408.

Schneider, B.S., Soong, L., Zeidner, N.S., and Higgs, S. (2004). *Aedes aegypti* salivary gland extracts modulate antiviral and TH1/TH2 cytokine responses to Sindbis virus infection. Viral Immunol. *17*, 565–573.

Scholle, F., Li, K., Bodola, F., Ikeda, M., Luxon, B.A., and Lemon, S.M. (2004). Virus–host cell interactions during hepatitis C virus RNA replication: impact of polyprotein expression on the cellular transcriptome and cell cycle association with viral RNA synthesis. J. Virol. *78*, 1513–1524.

Scott, J.C., Brackney, D.E., Campbell, C.L., Bondu-Hawkins, V., Hjelle, B., Ebel, G.D., Olson, K.E., and Blair, C.D. (2010). Comparison of dengue virus type 2-specific small RNAs from RNA interference-competent and -incompetent mosquito cells. PLoS Neglected Trop. Dis. *4*, e848.

Scott, T.W., Amerasinghe, P.H., Morrison, A.C., Lorenz, L.H., Clark, G.G., Strickman, D., Kittayapong, P., and Edman, J.D. (2000). Longitudinal studies of *Aedes aegypti* (Diptera: Culicidae) in Thailand and Puerto Rico: blood feeding frequency. J. Med. Entomol. *37*, 89–101.

Scott, T.W., Chow, E., Strickman, D., Kittayapong, P., Wirtz, R.A., Lorenz, L.H., and Edman, J.D. (1993). Blood-feeding patterns of *Aedes aegypti* (Diptera: Culicidae) collected in a rural Thai village. J. Med. Entomol. *30*, 922–927.

Scott, T.W., Weaver, S.C., and Mallampalli, V.L. (1994). Evolution of mosquito-borne viruses. In Evolutionary Biology of Viruses, Morse, S.S., ed. (New York, NY, Raven Press), pp. 293–324.

Sessions, O.M., Barrows, N.J., Souza-Neto, J.A., Robinson, T.J., Hershey, C.L., Rodgers, M.A., Ramirez, J.L., Dimopoulos, G., Yang, P.L., Pearson, J.L., *et al.* (2009). Discovery of insect and human dengue virus host factors. Nature *458*, 1047–1050.

Shin, S.W., Kokoza, V.A., and Raikhel, A.S. (2003). Transgenesis and reverse genetics of mosquito innate immunity. J. Exp. Biol. *206*, 3835–3843.

Shresta, S., Sharar, K.L., Prigozhin, D.M., Beatty, P.R., and Harris, E. (2006). Murine model for dengue virus-induced lethal disease with increased vascular permeability. J. Virol. *80*, 10208–10217.

Sim, S., and Dimopoulos, G. (2010). Dengue virus inhibits immune responses in *Aedes aegypti* cells. PLoS ONE *5*, e10678.

Singh, K.R.P. (1967). Cell cultures derived from larvae of *Aedes albopictus* (Skuse) and *Aedes aegypti* (L.). Curr. Sci. India 36, 506–508.

Sinkins, S.P., and Gould, F. (2006). Gene drive systems for insect disease vectors. Nature Reviews. Genetics 7, 427–435.

Sinnis, P., Coppi, A., Toida, T., Toyoda, H., Kinoshita-Toyoda, A., Xie, J., Kemp, M.M., and Linhardt, R.J. (2007). Mosquito heparan sulfate and its potential role in malaria infection and transmission. J. Biol. Chem. *282*, 25376–25384.

Smartt, C.T., Kim, A.P., Grossman, G.L., and James, A.A. (1995). The Apyrase gene of the vector mosquito, *Aedes aegypti*, is expressed specifically in the adult female salivary glands. Experimental Parasitology *81*, 239–248.

Smith, D.R., Carrara, A.S., Aguilar, P.V., and Weaver, S.C. (2005). Evaluation of methods to assess transmission potential of Venezuelan equine encephalitis virus by mosquitoes and estimation of mosquito saliva titers. Am. J. Trop. Med. Hyg. *73*, 33–39.

Souza-Neto, J.A., Sim, S., and Dimopoulos, G. (2009). An evolutionary conserved function of the JAK-STAT pathway in anti-dengue defense. Proc. Natl. Acad. Sci. U.S.A. *106*, 17841–17846.

Stollar, V., and Thomas, V.L. (1975). An agent in the *Aedes aegypti* cell line (Peleg) which causes fusion of *Aedes albopictus* cells. Virology *64*, 367–377.

Styer, L.M., Bernard, K.A., and Kramer, L.D. (2006). Enhanced early West Nile virus infection in young chickens infected by mosquito bite: effect of viral dose. Am. J. Trop. Med. Hyg. *75*, 337–345.

Styer, L.M., Kent, K.A., Albright, R.G., Bennett, C.J., Kramer, L.D., and Bernard, K.A. (2007). Mosquitoes inoculate high doses of West Nile virus as they probe and feed on live hosts. PLoS Pathogens *3*, 1262–1270.

Sylla, M., Bosio, C., Urdaneta-Marquez, L., Ndiaye, M., and Black, W.C. (2009). Gene flow, subspecies composition, and dengue virus-2 susceptibility among *Aedes aegypti* collections in Senegal. PLoS Neglected Trop. Dis. *3*, e408.

Tabachnick, W.J. (1991). Evolutionary genetics and arthropod-borne disease: the yellow fever mosquito. Ame. Entomol.14–24.

Talarico, L.B., Pujol, C.A., Zibetti, R.G., Faria, P.C., Noseda, M.D., Duarte, M.E., and Damonte, E.B. (2005). The antiviral activity of sulfated polysaccharides against dengue virus is dependent on virus serotype and host cell. Antiviral Res. *66*, 103–110.

Tassaneetrithep, B., Burgess, T.H., Granelli-Piperno, A., Trumpfheller, C., Finke, J., Sun, W., Eller, M.A., Pattanapanyasat, K., Sarasombath, S., Birx, D.L., *et al.* (2003). DC-SIGN (CD209) mediates dengue virus infection of human dendritic cells. J. Exp. Med. *197*, 823–829.

Thaisomboonsuk, B.K., Clayson, E.T., Pantuwatana, S., Vaughn, D.W., and Endy, T.P. (2005). Characterization of dengue-2 virus binding to surfaces of mammalian and insect cells. Am. J. Trop. Med. Hyg. *72*, 375–383.

Theiler, M., and Smith, H.H. (1937). The use of yellow fever virus modified by *in vitro* cultivation for human immunizationJ. Exp. Med. *65*, 787–800.

Thomason, L.C., Calendar, R., and Ow, D.W. (2001). Gene insertion and replacement in *Schizosaccharomyces pombe* mediated by the *Streptomyces* bacteriophage phiC31 site-specific recombination system. Mol. Genet. Genom. *265*, 1031–1038.

Thorpe, H.M., and Smith, M.C. (1998). *in vitro* site-specific integration of bacteriophage DNA catalyzed by a recombinase of the resolvase/invertase family. Proc. Natl. Acad. Sci. U.S.A. *95*, 5505–5510.

Thorpe, H.M., Wilson, S.E., and Smith, M.C. (2000). Control of directionality in the site-specific recombination system of the Streptomyces phage phiC31. Mol. Microbiol. *38*, 232–241.

Titus, R.G., Bishop, J.V., and Mejia, J.S. (2006). The immunomodulatory factors of arthropod saliva and the potential for these factors to serve as vaccine targets to prevent pathogen transmission. Parasite Immunol. *28*, 131–141.

Troyer, J.M., Hanley, K.A., Whitehead, S.S., Strickman, D., Karron, R.A., Durbin, A.P., and Murphy, B.R. (2001). A live attenuated recombinant dengue-4 virus vaccine candidate with restricted capacity for dissemination in mosquitoes and lack of transmission from vaccinees to mosquitoes. Am. J. Trop. Med. Hyg. *65*, 414–419.

Turell, M.J., Dohm, D.J., Sardelis, M.R., Oguinn, M.L., Andreadis, T.G., and Blow, J.A. (2005). An update on the potential of North American mosquitoes (Diptera: Culicidae) to transmit West Nile Virus. J. Med. Entomol. *42*, 57–62.

Turell, M.J., Sardelis, M.R., O'Guinn, M.L., and Dohm, D.J. (2002). Potential vectors of West Nile virus in North America. Curr. Topics Microbiol. Immunol. *267*, 241–252.

Twiddy, S.S., Holmes, E.C., and Rambaut, A. (2003). Inferring the rate and time-scale of dengue virus evolution. Mol. Biol. Evol. *20*, 122–129.

Uchil, P.D., and Satchidanandam, V. (2003). Architecture of the flaviviral replication complex. Protease, nuclease, and detergents reveal encasement within double-layered membrane compartments. J. Biol. Chem. *278*, 24388–24398.

Vaidyanathan, R., and Scott, T.W. (2006). Apoptosis in mosquito midgut epithelia associated with West Nile virus infection. Apoptosis *11*, 1643–1651.

van der Most, R.G., Corver, J., and Strauss, J.H. (1999). Mutagenesis of the RGD motif in the yellow fever virus 17D envelope protein. Virology *265*, 83–95.

van Rij, R.P., Saleh, M.C., Berry, B., Foo, C., Houk, A., Antoniewski, C., and Andino, R. (2006). The RNA silencing endonuclease Argonaute 2 mediates specific antiviral immunity in *Drosophila melanogaster*. Genes Dev. *20*, 2985–2995.

Vasilakis, N., Deardorff, E.R., Kenney, J.L., Rossi, S.L., Hanley, K.A., and Weaver, S.C. (2009). Mosquitoes put the brake on arbovirus evolution: experimental evolution reveals slower mutation accumulation in mosquito than vertebrate cells. PLoS Pathogens *5*, e1000467.

Vasilakis, N., Durbin, A.P., da Rosa, A.P., Munoz-Jordan, J.L., Tesh, R.B., and Weaver, S.C. (2008a). Antigenic relationships between sylvatic and endemic dengue viruses. Am. J. Trop. Med. Hyg. *79*, 128–132.

Vasilakis, N., Fokam, E.B., Hanson, C.T., Weinberg, E., Sall, A.A., Whitehead, S.S., Hanley, K.A., and Weaver, S.C. (2008b). Genetic and phenotypic characterization of sylvatic dengue virus type 2 strains. Virology *377*, 296–307.

Vasilakis, N., Holmes, E.C., Fokam, E.B., Faye, O., Diallo, M., Sall, A.A., and Weaver, S.C. (2007a). Evolutionary processes among sylvatic dengue type 2 viruses. J. Virol. *81*, 9591–9595.

Vasilakis, N., Shell, E.J., Fokam, E.B., Mason, P.W., Hanley, K.A., Estes, D.M., and Weaver, S.C. (2007b). Potential of ancestral sylvatic dengue-2 viruses to re-emerge. Virology *358*, 402–412.

Vazeille-Falcoz, M., Failloux, A.B., Mousson, L., Elissa, N., and Rodhain, F. (1999). Oral receptivity of *Aedes aegypti formosus* from Franceville (Gabon, central Africa) for type 2 dengue virus. Bulletin de la Societe de Pathologie Exotique *92*, 341–342.

Venter, P.A., and Schneemann, A. (2008). Recent insights into the biology and biomedical applications of Flock House virus. Cell.Mol. Life Sci. *65*, 2675–2687.

Wang, E., Ni, H., Xu, R., Barrett, A.D., Watowich, S.J., Gubler, D.J., and Weaver, S.C. (2000). Evolutionary relationships of endemic/epidemic and sylvatic dengue viruses. J. Virol. *74*, 3227–3234.

Wang, X.H., Aliyari, R., Li, W.X., Li, H.W., Kim, K., Carthew, R., Atkinson, P., and Ding, S.W. (2006). RNA interference directs innate immunity against viruses in adult *Drosophila*. Science *312*, 452–454.

Warren, W.D., Atkinson, P.W., and O'Brochta, D.A. (1994). The Hermes transposable element from the house fly, *Musca domestica*, is a short inverted repeat-type element of the hobo, Ac, and Tam3 (hAT) element family. Genet. Res. *64*, 87–97.

Wasserman, H.A., Singh, S., and Champagne, D.E. (2004). Saliva of the yellow fever mosquito, *Aedes aegypti*, modulates murine lymphocyte function. Parasite Immunology *26*, 295–306.

Waterhouse, R.M., Kriventseva, E.V., Meister, S., Xi, Z., Alvarez, K.S., Bartholomay, L.C., Barillas-Mury, C., Bian, G., Blandin, S., Christensen, B.M., *et al.* (2007). Evolutionary dynamics of immune-related genes and pathways in disease-vector mosquitoes. Science *316*, 1738–1743.

Weaver, S.C., Brault, A.C., Kang, W., and Holland, J.J. (1999). Genetic and fitness changes accompanying adaptation of an arbovirus to vertebrate and invertebrate cells. J. Virol. 73, 4316–4326.

Weaver, S.C., and Reisen, W.K. (2010). Present and future arboviral threats. Antiviral Res. *85*, 328–345.

Weaver, S.C., Scott, T.W., and Lorenz, L.H. (1990). Patterns of eastern equine encephalomyelitis virus infection in *Culiseta melanura* (Diptera: Culicidae). J. Med. Entomol. *27*, 878–891.

Weaver, S.C., Scott, T.W., Lorenz, L.H., Lerdthusnee, K., and Romoser, W.S. (1988). Togavirus-associated pathologic changes in the midgut of a natural mosquito vector. J. Virol. *62*, 2083–2090.

Welsch, S., Miller, S., Romero-Brey, I., Merz, A., Bleck, C.K., Walther, P., Fuller, S.D., Antony, C., Krijnse-Locker, J., and Bartenschlager, R. (2009). Composition and three-dimensional architecture of the dengue virus replication and assembly sites. Cell Host Microbe *5*, 365–375.

Whitfield, S.G., Murphy, F.A., and Sudia, W.D. (1973). St. Louis encephalitis virus: an ultrastructural study of infection in a mosquito vector. Virology *56*, 70–87.

Whitman, L. (1939). Failure of *Aedes aegypti* to transmit yellow fever cultured virus (17D). Am J. Trop. Med. *s1-19*, 19–26.

Wilson, J.H. (2008). Knockout punches with a fistful of zinc fingers. Proc. Natl. Acad. Sci. U.S.A. *105*, 5653–5654.

Xi, Z., Ramirez, J.L., and Dimopoulos, G. (2008). The *Aedes aegypti t*oll pathway controls dengue virus infection. PLoS Pathogens *4*, e1000098.

Yoshida, S., and Watanabe, H. (2006). Robust salivary gland-specific transgene expression in *Anopheles stephensi* mosquito. Insect Mol. Biol.*15*, 403–410.

Yu, I.M., Holdaway, H.A., Chipman, P.R., Kuhn, R.J., Rossmann, M.G., and Chen, J. (2009). Association of the pr peptides with dengue virus at acidic pH blocks membrane fusion. J. Virol. *83*, 12101–12107.

Zambon, R.A., Nandakumar, M., Vakharia, V.N., and Wu, L.P. (2005). The Toll pathway is important for an antiviral response in Drosophila . Proc. Natl. Acad. Sci. U.S.A. *102*, 7257–7262.

Zambon, R.A., Vakharia, V.N., and Wu, L.P. (2006). RNAi is an antiviral immune response against a dsRNA virus in *Drosophila melanogaster.* Cellular Microbiology *8*, 880–889.

Zanotto, P.M., Gould, E.A., Gao, G.F., Harvey, P.H., and Holmes, E.C. (1996). Population dynamics of flaviviruses revealed by molecular phylogenies. Proc. Natl. Acad. Sci. U.S.A. *93*, 548–553.

Zeidner, N.S., Higgs, S., Happ, C.M., Beaty, B.J., and Miller, B.R. (1999). Mosquito feeding modulates Th1 and Th2 cytokines in flavivirus susceptible mice: an effect mimicked by injection of sialokinins, but not demonstrated in flavivirus resistant mice. Parasite Immunol.*21*, 35–44.

Zhang, Y., Corver, J., Chipman, P.R., Zhang, W., Pletnev, S.V., Sedlak, D., Baker, T.S., Strauss, J.H., Kuhn, R.J., and Rossmann, M.G. (2003). Structures of immature flavivirus particles. EMBO Journal *22*, 2604–2613.

Zhang, Y., Zhang, W., Ogata, S., Clements, D., Strauss, J.H., Baker, T.S., Kuhn, R.J., and Rossmann, M.G. (2004). Conformational changes of the flavivirus E glycoprotein. Structure *12*, 1607–1618.

Vectors of Flaviviruses and Strategies for Control

15

Lee-Ching Ng and Indra Vythilingam

Abstract

Recent worsening of global dengue situation, geographical spread of the West Nile virus to the United States and the unexpected emergence of the Zika virus on the Yap island in the Pacific, have placed mosquito-borne flaviviruses in the limelight. Vector control remains as a key measure for prevention and control of these diseases. Mosquito borne flaviviruses are vectored by an array of mosquito species, with different behaviour and habitats. This chapter discusses the principles of vector control using Singapore's dengue control programme to illustrate a strategy for urban peri-domestic *Aedes* mosquitoes; and control of rural WNV and JEV to demonstrate strategies for rural *Culex* mosquitoes. In both situations, the incorporation of measures that consider the complex interplay of factors, including disease ecology, vector bionomics, land use, human activities and other social economic development, is essential. Despite the different strategies demanded by different vectors of diverse ecology and bionomics, the organizational framework that guide vector control remains consistent. The management system demands cost-effectiveness, which seeks synergies among the various tools used, and among various stakeholders. Sustainability, insecticide resistance and negative impact on the environment remain as some of the challenges faced by vector control programmes.

Introduction

The Flavivirus family comprises many medically important pathogens vectored by an array of different arthropods. Such zoonotic viruses are typically transmitted from a viraemic vertebrate host to a susceptible host via the bites of arthropod vectors. The pathogen–vector partnerships include dengue virus (DENV) with *Aedes* mosquitoes; West Nile virus (WNV) and Japanese Encephalitis virus (JEV) with *Culex* mosquitoes; and tick-borne encephalitis viruses with ticks. This chapter will focus on mosquito borne flaviviruses, and Table 15.1 lists some of these pathogenic viruses, their vectors and reservoirs. The major hosts of many flaviviruses, such as JEV and WNV, are animals which serve as reservoirs of the pathogens; and humans are incidental hosts. Viraemic levels of these pathogens in human do not reach sufficiently high level for infecting a vector, and humans end the transmission cycle as a dead end host. On the other hand, other flaviviruses, such as the DENV and Yellow Fever virus (YFV), are transmissible from human to human through a vector. Together with the presence of urban vectors, the high viraemic levels of these viruses in the human host have resulted in urban cycles, which have posed a daunting challenge to the increasingly urbanised and densely populated world.

To mitigate the transmission of vector-borne flaviviruses, vector control is a key option. An effective vector control programme with the intention of preventing and controlling the diseases caused by these viruses requires a sound understanding of the epidemiology of the diseases and the bionomics of the arthropods. It demands a species-specific strategy that deprives the vector of breeding habitats or exploits the vector bionomics, such as oviposition characteristics and

Table 15.1 Flaviviruses and their vectors

	Distribution	Disease	Vector	Reservoir
Dengue virus	Global, tropical and subtropical, with threats to temperate region	Fever, haemorrhagic fever, shock syndrome	*Ae. aegypti*, *Ae. albopictus* (urban cycle) *Ae. niveus*, *Ae. furcifer*, *Ae. vitattus*, *Ae. taylori*, and *Ae. luteocephalus* (sylvatic cycle)	Human (major) and non-human primates
Yellow fever virus	Tropical and subtropical region of the Americas, Africa	Fever, haemorrhagic fever	*Ae. aegypti* (urban cycle), *Ae. africanus*, *Haemagogus* and *Sabethes* spp. mosquitoes (sylvatic cycle)	Non-human primates
Japanese encephalitis virus	Tropical areas of Asia	Fever, encephalitis	*Culex* species	Pigs, birds
Koutango virus	Africa	Fever	Unknown	Rodent suspected
Murray Valley encephalitis virus	Australia/Papua New Guinea	Fever, encephalitis	*Aedes normanensis*, *Culex annulostris*, *Culex bitaeniorhynchus*	Birds
St. Louis encephalitis virus	Americas	Fever, encephalitis	*Culex* species including *Culex quinquefasciatus*	Wild birds, monkeys, armadillos, sloths and marsupials
West Nile virus (including Kunjin virus)	Global	Fever, encephalitis	*Culex* spp. including *Culex pipiens*, *Culex tarsalis* and *Culex quinquefasciatus*	Birds
Ilheus virus	Central and South America	fever	Unknown, but isolated from *Psorophora ferox*. *Ae. serratus* suspected	Birds
Zika virus including Spondweni virus	Asia, Pacific Islands and Africa	fever	*Aedes* spp.	unknown

feeding behaviour. However, regardless of vectors and strategies, three main components are required for an effective control programme:

1 cost-effective vector control tools;
2 vector control strategies;
3 effective management system.

This chapter will discuss the principles of vector control using the Singapore's dengue control programme to illustrate a strategy for urban peridomestic *Aedes* mosquitoes; and control of rural WNV and JEV to demonstrate strategies for rural *Culex* mosquitoes. In both situations, the incorporation of measures and strategies that consider the complex interplay of factors, including disease ecology, vector bionomics, land use, human activities and other social economic development, is essential.

Flaviviruses and their vectors

Flavivirus transmission in humans is a result of a spillover from a zoonotic transmission, or an escape from the zoonotic cycle to an urban cycle. Zoonotic and urban cycles are facilitated by different vectors and the species involved are influenced by the hosts and their ecologies.

Dengue virus

DENV circulates along the tropical belt. More recently, temperate countries such as Nepal and France have reported autochthonous transmissions (French Ministry of Health and Sport, 2010; Pandey *et al.*, 2008). Urban endemic and epidemic strains of DENV, transmitted by peridomestic *Ae. aegypti* and *Ae. albopictus*, originated from sylvatic ancestors that circulate in forests of Southeast Asia and West Africa, between non-human primates and arboreal *Aedes* mosquitoes. In Asia, sylvatic DENV-1, -2, and -4, which are genetically distinct from their urban counterparts, have been isolated from *Macaca* spp. and *Presbytis* spp. monkeys. Zoophagic arboreal *Ae. niveus* has been suggested to be the principal vector of the sylvatic cycle (Rudnick and Lim, 1986). Sylvatic circulation of DENV-3 is implicated through seroconversion of sentinel monkeys. In Asia, the sylvatic cycle is limited to the forest, with no evidence of involvement of sylvatic DENV in outbreaks of human dengue. Only occasion isolated human cases of sylvatic dengue have been detected in Malaysia (Cardosa *et al.*, 2009). This is probably due to the incompetence of peridomestic vectors *Ae. aegypti* and *Ae. albopictus* in transmitting sylvatic DENV strains, and is consistent with a study that demonstrated that *Ae. aegypti and Ae. albopictus* from various geographical regions were not susceptible to a sylvatic DENV-2 (Moncayo *et al.*, 2004). Several escapes of the virus to the urban cycle, some hundreds of years ago, had involved adaptation to peridomestic *Aedes* mosquitoes, possibly via *Ae. albopictus* as a bridge vector (Rudnick and Lim, 1986).

In West Africa, sylvatic circulation was previously believed to be dominant and only DENV-2 had been uncovered. There was evidence of it being circulated in Senegal and Nigeria among *Erythrocebus patas* monkeys, involving sylvatic mosquitoes, including *Ae. furcifer*, *Ae. vitattus*, *Ae. taylori*, and *Ae. luteocephalus* (Diallo *et al.*, 2003; Vasilakis *et al.*, 2008). In contrast to the Asian epidemiology, human clusters in Africa had been associated with sylvatic lineages (Vasilakis *et al.*, 2008). In Senegal, it was postulated to be facilitated by the forest–dwelling *Ae. furcifer*, vector of sylvatic DENV, which has been shown to disperse from the forest into villages (Diallo *et al.*, 2003). However, a changed epidemiology of dengue is evident in Africa. Outbreaks in Africa are, more recently, caused by imported epidemic urban dengue strains with possible involvement of peridomestic *Ae. aegypti* and *Ae. albopictus*. These peridomestic vectors appear to have gained a foothold in Africa recently (Johnson *et al.*, 1982; Leroy *et al.*, 2009; Ng and Hapuarachchi, 2010; Ninove *et al.*, 2009).

Yellow fever virus

Yellow fever virus is maintained in the forest of South American monkeys by forest canopy mosquitoes of *Haemogogus* and *Sabethes* genera. In Africa, the sylvatic cycle is maintained by *Ae. africanus*, and the monkey-human cycle in villages by *Ae. simpsoni* (Brès, 1986; Gubler, 2004). In both regions, the virus has been transferred by infected people to urban settings, where human to human transmission is vectored by peridomestic *Ae. aegypti*. There have been repeated events of *Ae. aegypti*-driven yellow fever outbreaks, that were attributed to the spread of sylvatic cycle to human settlements. Such events have previously been reported in Bolivia, Nigeria and Trinidad (Brès, 1986; Monath, 1999; Nasidi *et al.*, 1989; Van der Stuyft *et al.*, 1999). It thus suggests that at least some strains of sylvatic YFV readily infect peri-domestic *Ae. aegypti*, unlike sylvatic DENV which seemed to need rather rare events of adaptation to the urban vector to enter the urban cycle. The competency of peridomestic *Ae. aegypti* in transmitting sylvatic yellow fever have been demonstrated by multiple studies (Johnson *et al.*, 2002; Lourenco-de-Oliveira *et al.*, 2002; Lourenco de Oliveira *et al.*, 2003; Mutebi *et al.*, 2004; Tabachnick *et al.*, 1985). Interestingly, as in the case of dengue, sylvatic *Ae. aegypti formosus* plays no role in the transmission of sylvatic yellow fever (Brès, 1986).

Japanese encephalitis virus

JEV is naturally circulating in Asia among water birds, mainly egrets and herons; and vectored primarily by *Culex tritaeniorhynchus* which feed on water birds and larger mammals such as pigs and humans. *Culex pipiens*, *Culex gelidus*, *Culex vishnui* and *Culex bitaeniorhynchus* have also been shown to play a possible role in transmission (Arunachalam *et al.*, 2009; Hasegawa *et al.*, 2008; Vythilingam *et al.*, 1997). JEV is endemic in the Australasia region, with Pakistan as the most western country to have reported circulation of JEV. While migratory birds may play a role in spreading the virus over long distances and causing more sporadic cases, pigs are amplifying vertebrate hosts that may cause large local outbreaks. In central Java, cattle were also suspected to be effective hosts involved in JEV transmission (Weaver and Barrett, 2004).

West Nile virus

The geographical spread of WNV is wide – Australia, South Asia, Middle East, Europe and more recently the United States. The five viral genetic lineages recognized are maintained in a bird–mosquito–bird cycle. Belonging to the Japanese encephalitis serogroup, it is also believed to be dispersed by migratory birds and mainly vectored by *Culex* mosquitoes (Hubalek and Halouzka, 1999). Though the virus has also been found in other mosquito genera and species such as *Aedes, Anopheles, Aedomyia, Coquilletidia, Mansonia,* the vector role of these mosquitoes has yet to be validated. In Europe and North America, WNV is maintained in the rural cycle among wetland birds and ornithophilic mosquitoes such as *Culex tarsalis*; and the urban cycle among domestic birds and mosquitoes like *Culex pipiens* and *Cx. quinquefasciatus,* which bite both human and birds (Hubalek and Halouzka, 1999). As in JEV transmission, human is a dead end host of WNV, with no evidence of human–mosquito–human transmission. In India, the virus has been isolated from human beings (George *et al.*, 1984; Paul, 1970) frugivorous bats (Paul, 1970), domestic pigs (Paramasivan, 2003) and mosquitoes (*Cx. vishnui* and *Cx. quinquefasciatus*) (Rodrigues, 1980). Experimental studies have shown that *Cx. tritaeniorhynchus, Cx. vishnui, Cx. bitaeniorhynchus* and *Cx. quinquefasciatus* could act as vectors (Iikal, 1997). WNV has been isolated from both hard and soft ticks (Komar, 2000). The role of non-Culicine arthropods has also been considered in the maintenance of WNV during inter-enzootic periods.

Vector bionomics

The bionomics of peridomestic *Ae. aegypti* (vector of DENV and YFV) and the rural *Culex* mosquitoes (vector of JEV and WNV) are described below to illustrate mosquito behaviours that could contribute to the challenges we face in controlling their populations and the diseases they carry.

Aedes aegypti

There are two main forms of *Ae. aegypti* – sylvatic *formosus* strain from African forest, which is not susceptible to DENV and YFV; and the peri-domestic *aegypti* strains, which are the primary vector of the two flaviviruses. The susceptible female *Ae. aegypti* gets infected through picking up blood from a viraemic patient. It may become infectious in 6–14 days of extrinsic incubation. Its capacity to cause a global dengue problem and a potential yellow fever spread is contributed by its ability to adapt to our built environment, its feeding and skip-oviposition behaviours.

Female *Ae. aegypti* seeks and feeds mainly on human hosts. This anthropophagic tendency results in frequent contacts between the vector and human. A female *Ae. aegypti* takes multiple blood meals during each egg-laying cycle (Scott *et al.*, 1993a,b). The frequent feeding increases the opportunities for the species to acquire and transmit the viral pathogen as the female survives 3–5 gonotrophic cycles. The species feeds during the day, when her hosts are usually active. This leads to interrupted feedings that could further contribute to the number of human hosts in contact with each mosquito. Female *Ae. aegypti* breeds in artificial containers, typically in and around built structures and homes; and each would distribute her eggs in numerous containers (skip-oviposition) to ensure the survival of at least some of her progenies in the harsh environment, and to reduce competition among the progenies (Reiter, 2007). The eggs can withstand desiccation up to about six months, and are thus able to survive long drought periods. This characteristic has contributed to the global spread of the vector. The initial successful spread of the species was believed to be facilitated by sailing vessels (Tabachnick, 1991). Today, the species continues to spread, covering higher altitude, latitude, isolated islands and rural areas. In 1988, Guerrero State, Mexico, first reported a dengue outbreak at 1700 metres above sea level and the authors discussed the capability of *Ae. aegypti* to adapt to new environments (Herrera-Basto *et al.*, 1992). In 2008, Nepal reported its first dengue outbreak, caused by recently introduced *Ae. aegypti*. The vector has evidently extended its reach into temperate regions in the last decade, as far north as Nepal (Pandey *et al.*, 2004, 2008; WHO, 2007) and as south as Buenos Aires in Argentina (Vezzani and Carbajo, 2008; Vezzani *et al.*, 2004). Though classically known as an urban vector, the vector has also spread to and established itself in rural areas, for example in Indonesia and Cambodia (Jumali *et al.*, 1979; Seng *et al.*, 2008b). Furthermore, its infiltration into relatively isolated islands in the Arabian Sea, where only 52,000 people reside among 36 islands of 32 km^2, has been reported (Sharma and Hamzakoya, 2001).

Ae. aegypti's association with our dwelling place offers us an opportunity to suppress the mosquito population through environmental management and source reduction. The feasibility of suppressing its population has been demonstrated by the success of the Americas and Singapore, in the 1950–1980s.

Culex species

Several species of *Culex* are vectors of JEV and WNV. Unlike *Ae. aegypti*, *Culex* mosquitoes are more associated with rural areas and outdoor breeding. They are not particularly fastidious, breeding in all sorts of water bodies, including rice fields, drains and pools of standing water (Gould *et al.*, 1962; Vythilingam *et al.*, 2001). In most regions, *Cx. tritaeniorhynchus* mosquitoes are present in enormous numbers for a short period each year, following periods of heavy rains (Vaughn and Hoke, 1992). They particularly thrive well in regions affected by monsoon, which allows for rice cultivation in watered rice fields, and provides favourable breeding sites. Studies in India have shown that the use of urea such as fertilizer in rice fields increased the number of *Cx. tritaeniorhynchus* and *Cx. vishnui* egg rafts in the rice fields (Sunish *et al.*, 2003). In pig farming areas, *Cx. tritaeniorhynchus* also preferred to breed in stagnant water where fresh organic matter was constantly added.

Culex quinquefascatus and Cx. *pipiens* (WNV vectors) were found to feed predominantly on birds (Vythilingam *et al.*, 1996), and human. *Cx. tritaeniorhynchus* (primary JEV vector) tends to bite pigs and cattle more readily than it bites human (Vythilingam *et al.*, 1996). Dogs and birds are also attractive to the *Culex* mosquitoes.

A *Culex* mosquito gets infected with JEV through feeding on viraemic pigs or birds such as heron and egret (Draffan *et al.*, 1893; Soman *et al.*, 1977). It becomes infectious in 9–12 days of extrinsic incubation. Although many studies have demonstrated that pigs are the main source of the JEV, and have recommended locating pigs at least 5 km from human habitats (Solomon, 2006), the epidemiology in Malaysia suggests other significant sources of the virus. A study has shown that no cases were reported from a pig farming area where several species of mosquitoes were found to harbour JEV (Vythilingam *et al.*, 1997). The *Culex's* preference for pigs over human is probably the cause of the low risk for humans. On the other hand, JEV cases have been reported from areas with a predominantly Muslim population, where there were no pigs (Fang *et al.*, 1980; Vythilingam *et al.*, 1995). It is possible that birds may play a significant role, as in the case of WNV transmission.

Most *Culex* mosquitoes are nocturnal feeders, peak biting period generally lies between 1830 to 2130 hours. Though they are typically exophagic, they have been found to enter houses to seek for blood meals (endophagic) (Hasegawa *et al.*, 2008; Vythilingam *et al.*, 1995). While the eggs cannot withstand desiccation, the adult mosquitoes are able to overwinter in hibernation (Bailey *et al.*, 1982). Vertical transmission of JEV and WNV has been demonstrated in *Cx. tritaeniorhynchus and Cx pipiens* respectively and might account for the persistence of the virus in nature.

Vector control tools

Targeting the larvae

Environmental management and physical control

The most effective ways to control *Aedes* mosquito populations is to reduce the number and types of possible mosquito breeding habitats in and around the community. Since *Ae. aegypti* larvae thrive in natural and artificial containers, such as earthen jars, rainwater drums and ornamental plant containers, container management is a key tool. Non-essential containers such as discarded tyres, abandoned domestic containers must be destroyed, a practice that is consistent with proper waste management. Frequent change of water (every few days) is required for containers that cannot be depleted of water. In places where water storage is necessary, the use of long-lasting insecticide treated netting as jar covers have been successful in controlling the breeding of *Aedes*. In Cambodia, resting density of adult *Aedes* decreased by threefold when such nets are used (Seng *et al.*, 2008a).

For control of JEV, *Culex* control has been effective through changes in pig and rice farming practices (Keiser *et al.*, 2005; Solomon, 2006). The use of floating layers of expanded polystyrene beads/balls to suffocate mosquito immatures and to inhibit oviposition, has also been demonstrated to be effective in controlling *Culex* breedings in wet pit latrines and water storage cisterns (Curtis *et al.*, 2002; Reiter, 1978, 1985; Sivagnaname *et al.*, 2005). While *Culex* breeding in an urban setting or in farms is manageable by putting in place proper infrastructure and practices, managing breeding of the rural mosquitoes in nature settings through environmental management can be daunting.

Chemical control

A common complementary approach is the use of chemicals to control mosquito larvae. Chemical larviciding including organic synthetic insecticides have been effective against container breeding *Aedes* mosquitoes. Temephos is one of the few organophosphates registered to control mosquito larvae and has proven to be an effective control especially in porous earthen jars. Insect growth regulators (IGRs) such as methoprene and pyriproxyfen specifically interfere with the development of the mosquitoes and is believed to provide environmentally long-term residual effects (Itoh *et al.*, 1994; Seng *et al.*, 2006, 2008c; Vythilingam *et al.*, 2005). In addition studies have also shown that blood-fed *Ae. aegypti* exposed to pyriproxyfen residue can transfer the product from one container of water to another, when ovipositing and the residual amount transferred is sufficient to suppress adult emergence (Dell Chism and Apperson, 2003; Sihuincha *et al.*, 2005).

Biological control

Biological control exploits natural enemies of mosquitoes to manage mosquito populations. Mosquito pathogens, parasites and predators can be introduced directly to target larvae. Mosquito pathogens such as bacteria (*Bacillus thuringiensis israelenis* – *Bti* sero type H14) (Armengol *et al.*, 2006) protzoans (ciliates of genera *Lambornella* and *Tetrahymena*) (Washburn *et al.*, 1988) and fungi (*Culicinomyces*) (Chapman, 1974) can cause larval mortality. In recent years, *Bti*, has become a popular option. When sprayed into larval pools, it is ingested by feeding larvae and its toxin kills the larvae. It is believed to be relatively safe and environmentally friendly as it selectively kills only mosquito larvae. The residual effect of Bti is dependent on the formulation. Though some studies have shown that it has a shorter residual life than chemical insecticides like temephos (Chen *et al.*, 2009; Lee and Zairi, 2006; Setha *et al.*, 2007), in-house studies showed that a slow release 'doughnut' could have up to 3 months of residual activity (Liew, unpublished data). Biological control also includes the use of larvivorous fish such as *Poecilia*, *Apochelius* and *Panchax*, which have been particularly useful for large and permanent water storage containers (Seng *et al.*, 2008b). Fish farming in rice fields has also gained popularity in Asia. Besides increasing food production, it is an important approach to integrated pest management (Ahmad and Garnett, 2011). Some other biological control agents that have claimed success include the predator crustaceans (copepods – *Mesocyclops longisetus*) and nematodes. Studies in Vietnam have shown that within a year, *Ae. aegypti* was reduced by 90% and dengue cases reduced by 76% in communes where mesocyclops were being used, compared with control areas (Kay and Vu, 2005; Vu *et al.*, 2005). There is no documented study to show that predators such as birds, bats, dragonflies, and frogs consume enough adult mosquitoes to be effective control agents.

Targeting the adult mosquitoes

Chemical adulticiding using space spray applications

This method is generally employed in outbreak situations to break the chain of vector-borne disease transmission. A thermal fogger produces a pesticide fog or smoke by heating an oil-based solution of the chemical with a coil inside of the unit. Applications can be made from a back pack unit, a vehicle or an aircraft. Unfortunately, non-targeted and sometimes beneficial organisms are affected in the process.

Chemicals can also be mechanically generated as fine aerosols less than 50 μm by an ultra-low volume (ULV) applicator. Generally water based, the application is generally more acceptable to home owners, and can also be used indoors. Though many countries employ vehicle mounted thermal fogging or ULV machines to control *Ae. aegypti* driven dengue outbreaks, many studies have shown that such application does not provide a high percentage kill of the mosquitoes (Perich *et al.*, 2000; Reiter, 1997; Vythilingam and Panart, 1991). However, a recent study has demonstrated the effectiveness of outdoor thermal fogging, when combined with indoor ULV misting and source reduction, in controlling *Ae. albopictus* vectored chikungunya outbreak in Singapore (Tan *et al.*, 2011).

The use of insecticides to control JEV and WNV vectors has generally been effective only in limited areas for a limited amount of time and at great cost. Ultra-low-volume fenitrothion delivered from fixed-wing aircraft was effective in reducing 80% *Cx. tritaeniorhynchus* adult mosquito population for 4 days in Korea (Self, 1973). In the California, while ground ULV application of Pyrenone 25–5 was found to be ineffective, aerial ULV with the same chemical resulted in apparent reduction in the population of *Cx. tarsalis* and Cx. *quinquefascatus* (Lothrop *et al.*, 2008).

In recent years, innovative design of canister with a continuous release mechanism has emerged in the market. Field test has revealed its efficacy in targeting indoor mosquitoes in an apartment. It thus has the potential to complement indoor ULV applications (Pang *et al.*, 2009).

Residual spray application of chemicals on wall surfaces was initially developed for Malaria control. It has been adopted for the control of dengue and JE vectors. Systematic laboratory test using actellic (OP) showed that the residual activity of actellic was observed up to Week 11, when applied on wall surfaces (Pang SC and Ng LC, unpublished data). Insecticide treated curtains, tested in Mexico and Venezuela, has been shown to be effective in reducing the vector density if used by most people in the community (Kroeger *et al.*, 2006). Residual spraying of pigsties have also been carried out in countries like Japan and Thailand to control JE vectors (Wada, 1989) which would rest on the walls after taking a blood meal.

Active ingredients for adulticiding include organophosphates and the less toxic synthetic pyrethroids (SP). However, many countries, including Singapore, have in the last two decades observed development of resistance to SP among local mosquito populations (Jirakanjanakit *et al.*, 2007; Pethuan *et al.*, 2007; Ping *et al.*, 2001). More effort in identifying and developing insecticides is required from the research community.

Lethal ovitraps

Various designs of lethal ovitraps have been developed and used to attract and kill female *Aedes* adult mosquitoes seeking oviposition. They are generally designed as a pot that contains hay infusion to attract gravid female *Aedes* seeking oviposition; and equipped with paddles or surfaces that are treated with insecticides or glue that kill female mosquitoes seeking oviposition (Gama *et al.*, 2007; Rapley *et al.*, 2009). Though they have been shown to be effective surveillance tools, their effectiveness in *Aedes* mosquito control has yet to be clearly demonstrated. At EHI, a Gravitrap with sticky walls (Fig. 15.1) has been designed to be simple, economical and breeding-proof. It is currently deployed to manage dengue clusters to trap out adult mosquitoes, including the infective ones. This approach aims to exploit the skip-oviposition behaviour of *Ae. aegypti*, by positioning traps in the path of gravid females that are known to scatter their eggs in numerous containers.

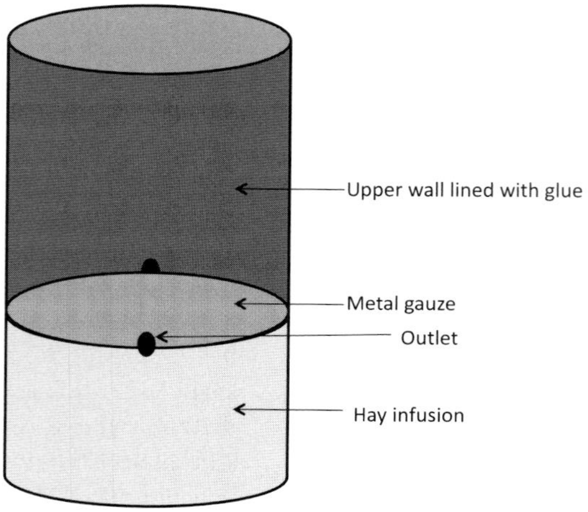

Figure 15.1 Design of EHI Gravitrap, a sticky trap that attracts and traps gravid *Aedes* females. Hay infusion attracts gravid female *Aedes*, while the metal gauze prevents emergence of adult mosquitoes in the event that eggs are laid in the pot. Water outlet at the side maintains the water level below the metal gauze.

In general, vector control tools are archaic, with very limited technological advances. There is a general over-reliance on chemicals, which is not sustainable. It is thus encouraging to note in recent years, the emergence of some potential technologies, which could provide breakthroughs in vector control after decades of stagnation.

Emerging mosquito control tools

Oxitec's RIDL technology (released insects with a dominant lethal)

Sterile insect technique (SIT), involving the release of irradiation-sterilized males, has been used in controlling insect pests in the agriculture sector (Endrichs *et al.*, 2002). Through futile mating of sterile males with wild females, the approach is species specific and environmentally friendly. Previous successes include the eradication of screwworm from North America, tsetse fly from Zanzibar, Queensland fruit fly from Western Australia, and melon fly from the Okinawa islands (Heinrich and Scott, 2000). Unfortunately, SIT on mosquitoes has faced various obstacles. Firstly, irradiated male mosquitoes had shortened lives and reduced fitness to compete in the wild. Secondly, public acceptance is limited as it involves release of 'vector species' in often densely inhabited areas. Recent advancement in molecular technology has the potential to circumvent the technical challenges. Developed in recent years is a strain of *Ae. aegypti* (RIDL) with a dominant lethal gene, which cause the progenies to die at the larval stage (Phuc *et al.*, 2007; Thomas *et al.*, 2000). Under the control of a tetracycline promoter, the lethal gene is repressed in the presence of tetracycline, allowing colonization of the strain; and expressed in the absence of tetracycline while in the wild. Using the same promoter, a second strain produces flightless females, when bred in the absence of tetracycline (Fu *et al.*, 2010). While the flightless females are designed to rapidly perish in the field with no progenies, the male will propagate the gene in the wild population for a few generations, before being diluted by the wild-type genotype. As these strains were not designed to propagate the manipulated gene in the field, the application is reversible and could gain wider public acceptance. Its aim of population suppression is also consistent with current global strategy and public messaging on source reduction, rendering it compatible with current strategies. However, it would require regular release of the genetically modified mosquito. The cost of producing and releasing large numbers of the mosquitoes could be high, and the approach is expected to be successful only in places with relatively low *Ae. aegypti* population. Nonetheless, before any release, a comprehensive risk assessment is required to install public confidence in the technology. At the point of writing, preparation for field trials is on-going in the USA (L. Alphey, personal communication) and Malaysia (Malaysia).

Use of *Wolbachia* infection

Another promising approach in controlling disease is vector population replacement, where natural competent vector population is replaced by another with reduced transmission capacity. Several lines of *Ae. aegypti* with shortened lifespan has been established through stable infections of endosymbiotic *Wolbachia* bacteria. Several physiological changes acquired from the infection are expected to significantly reduce the vector capacity of the strain. Most significantly, the strains were reported to be refractory to dengue and chikungunya infection (Bian *et al.*, 2010). The shortened lifespan also decreases the likelihood of the female adults surviving beyond the extrinsic incubation period of viruses. The infected females mate successfully and could result in the successful spread of the infections through natural population. Infected males were also found to be sterile, with a potential to reduce vector population (McMeniman *et al.*, 2009; Xi *et al.*, 2005). As the approach is based on the need to drive the infection into the natural population, a balance between the mortality/population reduction rate and the force of infection must be maintained. The theoretical feasibility of achieving 100% reduction in disease transmission was previously illustrated (Rasgon *et al.*, 2003). However, considering the fluctuations of natural *Aedes* populations owing to weather and vector control measures, striking the balance may be intricate in the field. The potential environmental impact is currently being addressed and a trial release is being planned in Cairns during the 2010–2011 wet season (Eliminate Dengue).

In more recent studies, a *Wolbachia*-infected strain was found to have less successful blood feeds with more probing activities. At the same time, saliva production was also reduced and less blood meals were taken as the mosquitoes aged (Moreira *et al.*, 2009; Turley *et al.*, 2009). These physiological changes, which are expected to significantly reduce the vectorial capacity of the strains, render the strains potentially appropriate for population reduction.

RNA interference (RNAi)

Another innovative approach seeks to exploit the defence mechanism of mosquitoes against viral materials. RNA interference (RNAi) is an innate defence pathway of plants and insects. Triggered by dsRNA formed in virus-infected cells, the small RNA molecules are products of the cleaved dsRNA, which silence homologous transcripts and hinder viral replication. Success in viral infection (e.g. DENV infection) of mosquitoes demonstrates that the virus is able to circumvent the defence mechanism. However, the mechanism offers an opportunity to develop vectors with reduced vectorial capacity (Campbell *et al.*, 2008; Cirimotich *et al.*, 2009; Myles *et al.*, 2008; Sanchez-Vargas *et al.*, 2009). A DENV-resistant strain of *Ae. aegypti* has been engineered by expressing an inverted-repeat (IR) RNA derived from the premembrane protein coding region of the DENV-2 RNA genome (Franz *et al.*, 2006). The approach is technically innovative. However, concurrent expression of an array of RNAi would be required to enable this virus-specific defence mechanism to tackle other viruses vectored by the same mosquito, and to address the possibility of an evasion of the defence system through adaptation of viruses. Indeed a study has shown that diversification of the WNV is promoted by mosquito midgets RNAi which targets WNV. Regions of the genome that are highly targeted by the host RNAi displayed a positive selection (Brackney *et al.*, 2009).

The scheme of propagating modified genetic traits in the field mosquito population may be less acceptable to the public, and may also be at odds with current initiatives on source reduction. Such a scheme may be more appropriate for controlling rural or forest mosquitoes that are often impossible to control through source reduction and environmental management.

Vector control strategies

Understanding the epidemiology of the target disease, disease ecology and bionomics of its vector is pivotal to an effective use of vector control tools and a well-designed control strategy. In the rest of the chapter, the principles of vector control will be illustrated with Singapore dengue control programme which largely targets *Ae. aegypti*, an urban peridomestic mosquito that prefers to breed and rest indoors; and general control strategies of rural *Culex* mosquitoes that breeds outdoor.

Historical success in control of urban *Ae. aegypti*, vector of DENV and YFV

The urban and peridomestic habitats of *Ae. aegypti*, offers an opportunity to suppress the vector population through careful environment management and urban planning. Practised by Singapore and many countries, the key strategy is to deprive the *Aedes* mosquitoes of stagnant water for breeding. The success of this strategy has been demonstrated in the Americas and in Singapore in the 1950s and 1960s, when *Ae. aegypti* was either eradicated or suppressed to a population that eliminate dengue or moderate dengue transmission to low endemicity.

In the Americas, *Ae. aegypti* eradication programme was initiated in the 1940s by the Pan American Health Organization to combat yellow fever. Using larvicides and insecticides in addition to breeding site elimination, the vector was eliminated from 19 countries representing 73% of the area originally infested (Gubler, 2004; Guzman and Kouri, 2003). It appeared that by the middle of the 20th century, only the American genotype DENV-2 was circulating in the continent (Halstead, 2006). In Singapore, the programme aimed to suppress the *Aedes* population. Supported by a firm legislation that penalise households and commercial entities that were found breeding mosquito vectors, and a team of trained health officers that performed regular house-to-house checks, the Singapore's premise index was brought down from about 50% (in 1960s) to less than 5% by mid 1970s. Along with the plummeting premise index, the incidence rate of dengue decreased significantly (Fig. 15.2). However, since the 1980s, both the Americas and Singapore have experienced resurgence in dengue transmission. The recent epidemiology of dengue in Singapore is characterized by a 5–6 year cycle, with increasing incidence rate within each cycle. An unprecedented outbreak in 2005 involved a total of 14,006 cases, including 27 deaths (Koh *et al.*, 2008). The American region has evolved from non-endemicity or hypoendemicity with intermittent outbreaks to one of hyperendemicity with annual outbreaks. The resurgence of dengue in the Americas has been reported to be due to the

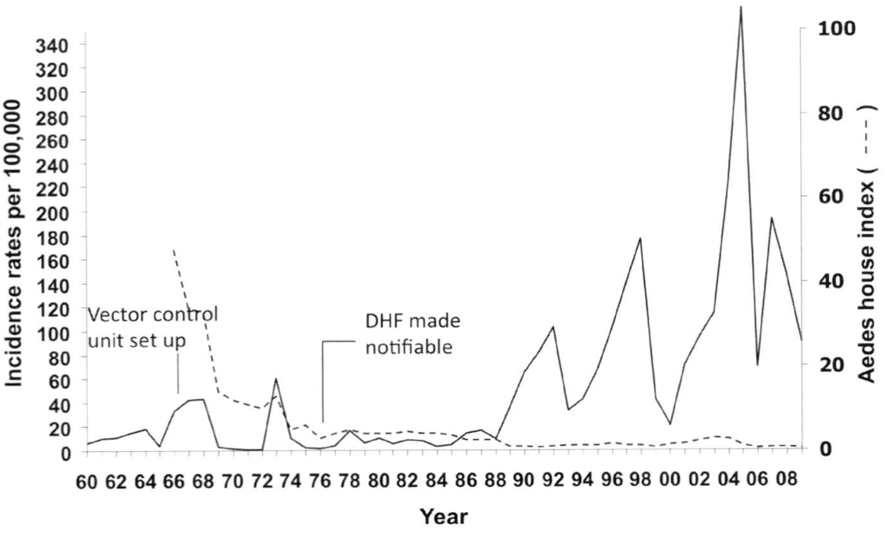

Figure 15.2 Singapore dengue trend, illustrating the emergence of dengue in the late 1980s and the reversing of the trend after 2007.

'relaxation of intensive vector control' (Halstead, 2006) and shift from a centrally administrated vertical control programme to decentralisation of health services and community based programme. However, in Singapore, the programme has been sustained through the decades. It has been postulated that the resurgence was due to factors such as an increase in human population and density, increase in travel, and low herd immunity resulting from low transmission for more than a decade. In response to the 2005 unprecedented outbreak, a thorough review of the system has revealed its failure to evolve with the changing environment and society. Population in Singapore grew from 2.1 million in 1970 to 5.0 million by 2009 (Department of Statistics, 2009) making the arduous house-to-house check unsustainable. In 2009, 18 million passengers arrived by air, a dramatic increase from the 1.7 million recorded in 1970 (Cornelius-Takahama, 1998). Locally, the construction of the first expressway was initiated in 1966 and the first operation of the mass rapid transport (MRT) system in 1987. Singapore today boasts of 160 km of expressway and 150 km of MRT lines criss-crossing the 720 km^2 island (LTA) (Department of Statistics, 2010). These developments have undoubtedly favoured the propagation of *Aedes* mosquitoes, the frequent importation of viruses and the rapid dispersal of the virus within the country – all of which contribute to the increased contact rate among *Ae. aegypti*, human and virus. It is clear from the global landscape that the Singapore's urban development is not unique but is reflecting current situation in many emerging economies, including the Americas. These global changes challenged a system and strategy that had previously been successful and called for an innovative approach for dengue control.

Evolution of Singapore's *Aedes* control programme

Integrated mosquito control

Singapore's mosquito control programme has been maintained as an integrated programme that includes environmental management, source reduction of breeding areas, regular mosquito surveillance, routine larviciding in breeding areas that cannot be eliminated, and adulticiding only when necessary. A preventative mosquito control programme is complemented with outbreak control measures. To address the increasing challenge of dengue, Singapore's vector control programme has since 2005, been enhanced. A key novel feature is the incorporation of a decision support system that is built on 4 cornerstones – case, virus and entomological surveillance, and ecological information (Fig. 15.3). Surveillance and ecological data are used for temporal and spatial risk stratification, which forms the core of the decision support system and facilitates optimal deployment of resources in time and space.

Temporal and spatial risk stratification

Analysis of retrospective data has shown that the incidence rate of dengue cases in Singapore displayed a distinct seasonal pattern, with correlation to ambient temperature with a 12-week lag (Low *et al.*, 2006). Previous data has also found an association between a shift in predominant virus serotype and pending dengue outbreak. In the enhanced system, ambient temperature and circulating serotypes are monitored weekly for early warning of outbreaks or increased transmission. More intensive control measures are triggered by early warning signs (temperature and serotype switch). In 2007, an early warning provided by these surveillance data, 6 months ahead of an outbreak, had probably contributed to the moderate size of the outbreak (Lee *et al.*, 2010).

Risk of transmission is also spatially stratified using geographical information system. Factors that are taken into consideration include the predominant virus serotype in circulation, previous exposures of the human population to the different serotypes and *Aedes* population. Through analysis of remotely sensed satellite images and population data, we have established that vegetation index (negatively correlated), age of buildings and population density are also relevant parameters for predicting risk of dengue (Tan SSY and Ng LC, unpublished data).

The improved system has been enabled by a better understanding of the epidemiology of dengue and by technologies such as the geographical information system. Case surveillance, through mandatory notification of cases, has been

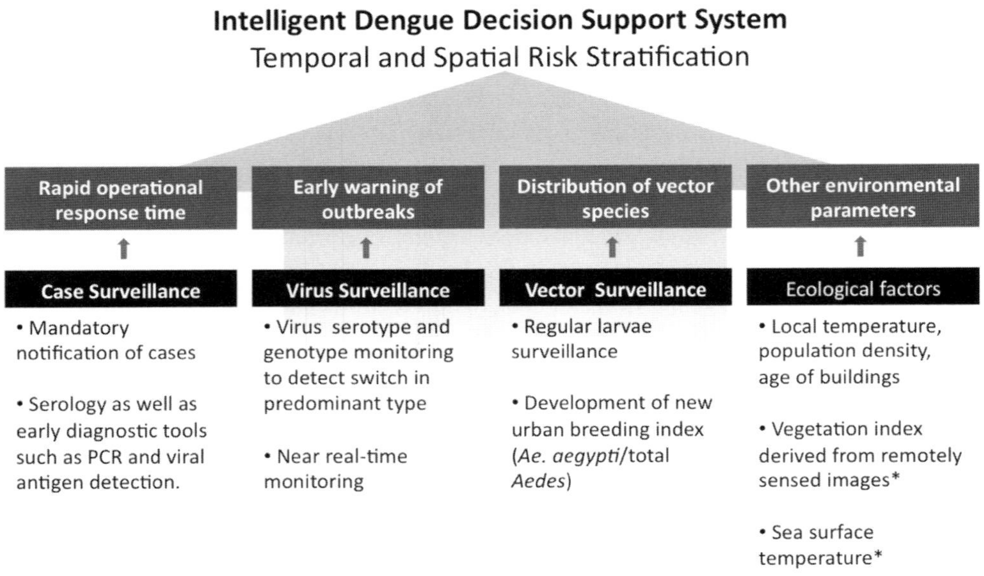

Figure 15.3 Dengue Singapore dengue decision support system, built on four pillars, which continuously feed information for temporal and spatial risk stratification. *Parameters that are currently under evaluation.

enhanced with early diagnosis technologies like PCR and viral antigen detection (Lai *et al.*, 2007; Pok *et al.*, 2010). The effort has contributed to a reduction in average operational response time from 7.3 days in 2004 to 4.3 days in 2009. Virus surveillance has been made possible by rapid serotype and genotype determination of circulating dengue viruses by PCR and sequencing technologies (Lee *et al.*, 2010). For entomological surveillance, a new index, breeding ratio (BR) which considers the ratio of *Ae. aegypti* and *Ae. albopictus* breeding, has been found to be more sensitive than the classical house index (L.C. Ng and S.S.Y. Tan, unpublished data).

Inter-epidemic effort

Considering that vector control during an outbreak could have very limited impact on transmission outcome, Singapore has focused much attention on vector control during annual inter-epidemic period, to reduce the *Aedes* population and the number of human reservoir (cases) before the traditional dengue season. The spatial and temporal risk assessment has allowed enhanced preventative vector control measures to be targeted before any onset of outbreaks.

Outcome of new effort

The new strategies have effected a reduction in annual dengue cases; particularly noticeable is the suppression of the traditional high peaks which usually occur during warmer months (June to August) (Fig. 15.4). The Singapore situation clearly showed that a strategy can become obsolete, superseded by changes in the environment and the demographics. It also demonstrates the need to be constantly updated on disease epidemiology and technologies for an effective evidence-based control programme. However, the challenge remains vexing despite the recent successes. Every successful aversion of an outbreak has heightened the susceptibility of the population. This underscores the irony of battling an endemic disease in an environment that is highly conducive for transmission.

Control of rural Culex, vectors of JEV and WNV

In contrast to urban peridomestic mosquitoes, the association of many *Culex* spp. with the wilderness and rural areas has rendered the control of their population daunting. Fortunately, the generally low human density in such rural areas reduces risk of disease; and a relatively higher threshold

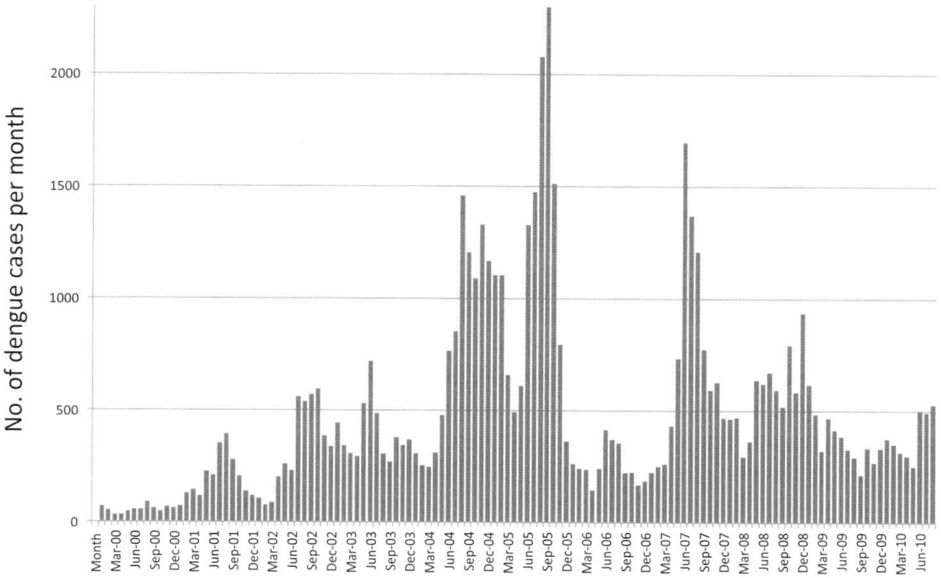

Figure 15.4 Singapore recent dengue epidemiology, with peak transmission during the warmer months between July and September. Moderation of the peak since 2008 has been effected by the revised dengue control programme.

is required for tangential transmission from the enzootic mosquito–animal amplification cycle to humans. Owing to the stark contrast in vector and reservoir host ecology, the strategy needed for control of rural *Culex* is markedly different from *Ae. aegypti*. Though chemicals are often used, especially for emergency response to break transmission of disease, ecologically based approach has proven to be more sustainable and effective.

Control of JEV

Several Asian countries, such as Japan, Korea, China, Taiwan and Thailand, had carried out routine immunization of school children. However, the efficacy of JE vaccines remains questionable. While vector control complemented vaccination in some regions in early years, it forms the mainstay of JEV prevention in current time. Besides thermal fogging during outbreaks, residual sprays of pigsties are performed as a preventative measure. Locating pigs farm away (at least 5 km) from human habitation has also reduced the incidence of JEV. In areas with rice fields, which are just as attractive to birds that are potentially harbouring the virus as to the mosquito vectors, changes in farming practices have been shown to yield results. These include the use of rice cultivars that require less water and the use of intermittent irrigation. During the annual approximately 100 days of rice planting period, intermittent flooding (20 times) reduced the breeding of the *Culex* mosquitoes compared with conventional irrigation where the fields are flooded throughout the rice growing period (Lu, 1987). In Japan, a decrease in areas of irrigation fields and changes in rice cultivation methods wherein water is drained repeatedly from the fields during mosquito season has helped to decrease the population of *Cx. tritaeniorhynchus* (Kamimura, 1998). In India it was shown that while nitrogenous fertilizers with blue-green algae (BGA) increased grain yield and immature mosquito population, BGA alone increased paddy yield without enhancing mosquito production. Thus use of BGA with less nitrogenous fertilizers was recommended to control JE vectors in paddy fields (Victor, 2000). Complementing these practices is the use of biological control agents such as larvivorous fish, which concurrently provides farmers an additional source of income. Other environmentally friendly agents are neem cakes, mermithid nematode (*Romanomermis culicivorax*), a fungus (*Lagenidium giganteum*) and bacteria (*Bacillus thuringiensis* var. *israelensis* and *Bacillus sphaericus*), many of which had little or no adverse effects on non-target populations of vertebrate and invertebrates (Lacey and Lacey, 1990; Solomon, 2006). Vector–human contact can be further reduced by putting a distance or a dry belt between farming and the communities.

Control of WNV

WNV transmission in the USA can be largely categorized into *Cx. tarsalis*-driven rural transmission among birds and urban transmission, where *Culex* breedings (e.g. *Cx. quinquefasciatus* and *Cx. pipiens*) can be found around homes and in urban parks (Reisen and Brault, 2007). Though source reduction can be applied in the latter scenario, control in rural areas mainly relies on the use of chemicals. To facilitate cost effective preventative control, multiple spatial risk models have been generated to guide vector control. In the USA, remotely sensed vegetation indices and distance from water bodies were used to discriminate among mosquito habitats within a densely populated urban environment and to identify areas with WNV vectors (Brown *et al.*, 2008; Brownstein *et al.*, 2002). In Canada, a MultiAgent GeoSimulation prototype system was developed to model population dynamics of mosquitoes and birds, and the spread and transmissions of WNV in time and space. The system aimed to facilitate short term decision making in vector control measures, according to climatic parameters (Bouden *et al.*, 2008). Pre-epidemic risk stratification allows for proactive vector control, which has been reported to be effective in California. Early intensive aerial ULV applications of chemicals as soon as the virus was detected in the spring, resulted in an apparent reduction of *Cx. tarsalis* population and consequently a significant decline in transmission by *Cx. quinquefasciatus* populations and low human incidence (Lothrop *et al.*, 2008). Nevertheless, timely diagnosis and reporting of human cases is important for prompt reactive or emergency control by adulticides to remove infected mosquitoes. To succeed in interrupting transmission, at least 10 days of adulticiding was recommended to surpass the typical length of the viraemia period in bird reservoir hosts (Reisen and Brault, 2007).

Interestingly, a study showed that WNV infection rates among *Culex* mosquitoes declined with increasing wetland cover. Wetland area was not significantly associated with either vector density or amplification host abundance, but was strongly correlated with non-passerine and passerine species ratio. Non-passerine birds are non-reservoirs and could play a role in diluting disease transmission. Contrary to common belief and understanding, the study suggested that preserving large wetland areas may represent a feasible ecosystem-based approach for controlling WNV outbreaks (Ezenwa *et al.*, 2007). The result resonates with an observation in Malaysia, where no cases were reported from a pig farming area where several species of mosquitoes were found to harbour JEV, but cases were detected in areas with no pigs (Vythilingam *et al.*, 1997). This underscores the importance of going beyond investigating disease reservoirs and vectors in our pursuit of disease control strategies. The overall ecology, including players that may not be immediately apparent, plays a significant role in disease transmission and could be exploited.

Vector management system

Despite the different strategies demanded by different vectors of diverse ecology and bionomics, the organizational framework that guide vector control remains consistent among the vectors. Evidence-based integrated vector management (IVM) is a key to vector control. Promoted by the World Health Organization (WHO) and many experts, IVM is a 'rational decision-making process for the optimal use of resources for vector control.' The management system should demand cost-effectiveness, which seeks synergies among the various tools used, and among various stakeholders. Therefore intersectoral cooperation is critical to ensure that the activities of other sectors like urban development, agriculture or water resources do not compromise any vector control programme. Instead, vector control should be included as an agenda of each of these sectors.

In Singapore, vector control function resides with the environment, rather than the health sector. The close coordination between the two sectors is therefore imperative to her ability to tackle vector-borne diseases. Surveillance of vector diseases rides on national disease surveillance programme at the Ministry of Health. Daily communication of surveillance data enables prompt vector control response from the National Environment Agency. Through a coordinated inter-agency task force, vector control has also become an important agenda of each government agency. Together with community involvement and a pest control industry that support the national vector control effort, IVM in Singapore is aligned with the whole government effort in establishing public, private and people (3P) partnership to develop innovative and sustainable environmental initiatives that promote environmental ownership amongst the local community. Through coordination between ground vector control and laboratory case and virus surveillance, the ground effort is guided by scientific evidences. The stakeholders are many in the control of vectors and the associated diseases. An effective IVM thus must comprise effective communication among various stakeholders, with elements of feedbacks and data sharing.

IVM also seeks to target other vectors using the same system. Singapore's success in controlling chikungunya and malaria, which was largely vectored by *Ae. albopictus* and *Anopheles* respectively, demonstrated how a small modification of the same system according to the bionomics of various vectors, can lead to effectiveness against other mosquito-borne diseases (Lee *et al.*, 2009; Leo *et al.*, 2009; Ministry of Health, 2008; Ng *et al.*, 2009). Regardless of the target mosquitoes, an integrated programme should include a consortium of tools such as environmental management, source reduction of breeding areas, routine larviciding in those breeding areas that cannot be eliminated, and reduce the reliance of chemicals.

Challenges

Vulnerability increases with successful vector control

Despite many successes, the challenge of vector control remains vexing. In an environment that is conducive for high endemicity of a disease, every successful aversion of an outbreak could heighten the susceptibility of the population, and predispose the population to an outbreak. As demonstrated by a mathematical model for dengue – the estimated number of *A. aegypti* pupae per person required to result in a 10% or greater rise in seroprevalence of antibody correlates with the level of herd immunity of the population (Focks *et al.*, 2000). For instance, at 26°C, 3.41 pupae person is required for herd immunity of 67%, while 1.55 is required for 33%. It is clear that a lowered herd immunity requires less mosquitoes per person to mount an outbreak. This paradox of vector control is clearly experienced in Singapore, where decades of successful control had led to a population with low herd immunity, which has contributed to the rapid re-emergence of dengue outbreaks in the last two decades. It underscores the need for sustained and perhaps increasingly effective vector control efforts, particularly for the control of flaviviruses, such as DENV, that exploit human as a vehicle for spread. Success would hinge on a global or regional concerted effort, to reduce the level of transmission in the region, which in turn could reduce the exchanges of viruses and further spread of the vectors.

Sustainability

Unfortunately, today's vector control approaches are labour intensive, monotonous and dependent on individuals' competence. Deliberate efforts are required for sustainability. Vector control requires constant concerted efforts from many stakeholders, various government agencies, private sectors and the community. Globally, it remains a challenge to ensure that every stakeholder plays a sustainable effective role. As the strength of a system is not unlike a chain, which is only as strong as its weakest link, constant advocacy and educational efforts that reach out to all stakeholders are essential for a fruitful programme.

Insecticide resistance

Currently, three main families of insecticides are recommended by WHO for mosquito control – organophosphate (OPs), synthetic pyrethroids (SPs) and carbamate. Unfortunately, resistance to these insecticides, is prevalent and poses a global concern. Cross-resistance within the SP family, and between SP and OP are well known (Brooke *et al.*, 1999), e.g. resistance to deltamethrin also confers resistance to permethrin and other SPs or OP. Resistance of *Aedes* mosquitoes to SPs has been reported in China, Singapore, Thailand and Vietnam, Brazil, Cuba, Venezuela, other Latin American countries (Cui *et al.*, 2006; Jirakanjanakit *et al.*, 2007; Kawada *et al.*, 2009a; Kawada *et al.*, 2009b; Paeporn *et al.*, 2004; Ping *et al.*, 2001; Ponlawat *et al.*, 2005; Rodriguez *et al.*, 2001; Rodriguez *et al.*, 2007; Yaicharoen *et al.*, 2005). While OPs may remain effective against most populations of *Aedes* adult population, the Aedes larvae have displayed various levels of resistance to temephos (OP) (Braga *et al.*, 2004; Lima *et al.*, 2003; Ponlawat *et al.*, 2005). For *Culex* species of mosquitoes, *Cx. quinquefasciatus* has shown resistance to SPs, OPs and carbamate insecticides in many parts of the world, including Cuba, Brazil, Africa, Florida and Malaysia (Bisset *et al.*, 1991; Bracco *et al.*, 1997; Chandre *et al.*, 1998; Liu, 2004; Nazni *et al.*, 2005). Though the impact of resistance on control efforts is not well documented, vector control programmes are facing an impending serious public health challenge.

The causes of resistance development include the widespread use of insecticides through fogging and vehicle mounted ULV; and the use of SPs in household aerosol insecticides (Brogdon and McAllister, 1998). Owing to the cross-knockdown resistance between DDT and SPs (Soderlund, 2008), it is speculated that the rapid emergence of resistance to SPs were most probably due to cross-resistance developed by the previous widespread use of DDT.

To overcome resistance of insecticide in mosquitoes, synergists are often being used. Synergists are compounds which work by inhibiting the activities of detoxifying enzymes, such as DEF (*S,S,S*-tributyl enzymes and PBO (piperonyl butoxide) that inhibits MFO (mixed function oxidases) (Scott, 1990). The exposure of *Ae. albopictus* and *Ae. aegypti* to PBO followed by exposure to permethrin showed that the LT50 values were lower compared with exposure to permethrin alone (Wan-Norafikah, 2008, 2010). This shows that PBO can increase the susceptibility of pyrethroid resistance mosquitoes. However, the multiple findings on the widespread involvement of *kdr* mutations in resistance development, suggest that PBO may not be useful in many situations.

Environmental impact

The impact of vector control on environment and ecology has invited much debate. Besides the well-known direct impact of chemicals on non-target animals and the ecology, a recent study has demonstrated the negative impact of mosquito reduction on insectivorous birds (Gilbert, 2010; Poulin *et al.*, 2010). The loss of mosquitoes as a result of extensive Bti-treatment in Camargue, France, had resulted in a marked reduction in the breeding success of house martins (*Delichon urbicum*). Though elimination or low population of mosquitoes is appreciated in an urban setting, its potential impact on the ecology in nature could pose an additional hindrance to vector control for zoonotic flaviviruses such as WNV, transmission of which is driven by hosts residing in natural habitats.

Adaptation of viruses to vectors

High transmission of viruses leads to genetic diversity. It offers the virus ample opportunities to improve its fitness and to adapt, resulting in viruses with high epidemic potential. The 2007 dengue outbreak in Singapore was found to be associated with a clade replacement within the cosmopolitan genotype of DENV-2, with merely nine amino acid substitutions in the whole viral genome (Lee *et al.*, 2010). A report from Cuba suggested the emergence of a variant of a DENV2 with improved fitness, during the 1997 outbreak (Rodriguez-Roche *et al.*, 2011). Similarly, the high genetic diversity and adaptability of WNV have facilitated the adaptation of the virus to North American ecology. The emergence of a virulent strain had contributed to the rapid spread of the virus across the continent (Weaver and Barrett, 2004).

Though not demonstrated in dengue, it is clear from the evolution of the newly emerged chikungunya virus (an alphavirus), that a virus can quickly adapt to a new vector, *Ae. albopictus*, when high transmission in classical vector, *Ae. aegypti* coincides with presence of *Ae. albopictus* (Ng and Hapuarachchi, 2010). Unfortunately, the diversity of dengue virus, which is further complexed by the four serotypes, has made such studies in dengue virus more difficult.

At this instance, viruses and vectors continue to evolve, occasionally with formations of new partnerships and with increased fitness. Vector control will thus remain as a pursuit of solutions to the evolving epidemiology of flaviviruses and other vector-borne pathogens. As we address and highlight vector control for pathogenic flaviviruses that we are aware of, more flaviviruses may be emerging. This is evidently demonstrated by the recent epidemic of Zika virus in the Yap State, Micronesia (Duffy *et al.*, 2009; Lanciotti *et al.*, 2008).

References

Ahmad, N., and Garnett, S.T. (2011). Integrated rice–fish farming in Bangladesh: meeting the challenges of food security. Food Sec. *3*, 81–92.

Armengol, G., Hernandez, J., Velez, J.G., and Orduz, S. (2006). Long-lasting effects of a *Bacillus thuringiensis* serovar *israelensis* experimental tablet formulation for *Aedes aegypti* (Diptera: Culicidae) control. J. Econ. Entomol. *99*, 1590–1595.

Arunachalam, N., Murty, U.S., Narahari, D., Balasubramanian, A., Samuel, P.P., Thenmozhi, V., Paramasivan, R., Rajendran, R., and Tyagi, B.K. (2009). Longitudinal studies of Japanese encephalitis virus infection in vector mosquitoes in Kurnool district, Andhra Pradesh, South India. J. Med. Entomol. *46*, 633–639.

Bailey, C.L., Faran, M.E., Gargan, T.P. 2nd, and Hayes, D.E. (1982). Winter survival of blood-fed and nonblood-fed *Culex pipiens* L. Am. J. Trop. Med. Hyg. *31*, 1054–1061.

Bian, G., Xu, Y., Lu, P., Xie, Y., and Xi, Z. (2010). The endosymbiotic bacterium *Wolbachia* induces resistance to dengue virus in *Aedes aegypti*. PLoS Pathog. *6*, e1000833.

Bisset, J.A., Rodriguez, M.M., Hemingway, J., Diaz, C., Small, G.J., and Ortiz, E. (1991). Malathion and pyrethroid resistance in *Culex quinquefasciatus* from Cuba: efficacy of pirimiphos-methyl in the presence of at least three resistance mechanisms. Med. Vet. Entomol. *5*, 223–228.

Bouden, M., Moulin, B., and Gosselin, P. (2008). The geosimulation of West Nile virus propagation: a multiagent and climate sensitive tool for risk management in public health. Int. J. Health Geogr. *7*, 35.

Bracco, J.E., Dalbon, M., Marinotti, O., and Barata, J.M. (1997). [Resistance to organophosphorous and carbamates insecticides in a population of *Culex quinquefasciatus*]. Rev Saude Publica *31*, 182–183.

Brackney, D.E., Beane, J.E., and Ebel, G.D. (2009). RNAi targeting of West Nile virus in mosquito midguts promotes virus diversification. PLoS Pathog. *5*, e1000502.

Braga, I.A., Lima, J.B., Soares Sda, S., and Valle, D. (2004). *Aedes aegypti* resistance to temephos during 2001 in several municipalities in the states of Rio de Janeiro, Sergipe, and Alagoas, Brazil. Mem Inst Oswaldo Cruz *99*, 199–203.

Brès, P.L.J. (1986). A century of progress in combating yellow fever. Bull. WHO *64*, 775–786.

Brogdon, W.G., and McAllister, J.C. (1998). Insecticide resistance and vector control. Emerg. Infect. Dis. *4*, 605–613.

Brooke, B.D., Hunt, R.H., Koekemoer, L.L., Dossou-Yovo, J., and Coetzee, M. (1999). Evaluation of a polymerase chain reaction assay for detection of pyrethroid insecticide resistance in the malaria vector species of the *Anopheles gambiae* complex. J. Am. Mosq. Control Assoc. *15*, 565–568.

Brown, H., Duik-Wasser, M., Andreadis, T., and Fish, D. (2008). Remotely sensed vegetation indices identify mosquito clusters of West Nile virus vectors in an urban landscape in the northeastern United States. Vector Borne Zoonotic Dis. *8*, 197–206.

Brownstein, J.S., Rosen, H., Purdy, D., Miller, J.R., Merlino, M., Mostashari, F., and Fish, D. (2002). Spatial analysis of West Nile virus: rapid risk assessment of an introduced vector-borne zoonosis. Vector Borne Zoonotic Dis. *2*, 157–164.

Campbell, C.L., Keene, K.M., Brackney, D.E., Olson, K.E., Blair, C.D., Wilusz, J., and Foy, B.D. (2008). *Aedes aegypti* uses RNA interference in defense against Sindbis virus infection. BMC Microbiol. *8*, 47.

Cardosa, J., Ooi, M.H., Tio, P.H., Perera, D., Holmes, E.C., Bibi, K., and Abdul Manap, Z. (2009). Dengue virus serotype 2 from a sylvatic lineage isolated from a patient with dengue hemorrhagic fever. PLoS Negl. Trop. Dis. *3*, e423.

Chandre, F., Darriet, F., Darder, M., Cuany, A., Doannio, J.M., Pasteur, N., and Guillet, P. (1998). Pyrethroid resistance in *Culex quinquefasciatus* from west Africa. Med. Vet. Entomol. *12*, 359–366.

Chapman, H.C. (1974). Biological control of mosquito larvae. Annu. Rev. Entomol. *19*, 33–59.

Chen, C.D., Lee, H.L., Nazni, W.A., Seleena, B., Lau, K.W., Daliza, A.R., Ella Syafinas, S., and Mohd Sofian, A. (2009). Field effectiveness of *Bacillus thuringiensis israelensis* (Bti) against *Aedes* (*Stegomyia*) *aegypti* (Linnaeus) in ornamental ceramic containers with common aquatic plants. Trop Biomed *26*, 100–105.

Cirimotich, C.M., Scott, J.C., Phillips, A.T., Geiss, B.J., and Olson, K.E. (2009). Suppression of RNA interference increases alphavirus replication and virus-associated mortality in *Aedes aegypti* mosquitoes. BMC Microbiol *9*, 49.

Cornelius-Takahama, V. (1998). Paya Lebar Airport (National Library Singapore. Available at http://infopedia.nl.sg/articles/SIP_130_2005-01-22.html).

Cui, F., Raymond, M., and Qiao, C.L. (2006). Insecticide resistance in vector mosquitoes in China. Pest. Manag. Sci. *62*, 1013–1022.

Curtis, C.F., Malecela-Lazaro, M., Reuben, R., and Maxwell, C.A. (2002). Use of floating layers of polystyrene beads to control populations of the filaria vector *Culex quinquefasciatus*. Ann. Trop. Med. Parasitol. *96* Suppl. *2*, S97–104.

Dell Chism, B., and Apperson, C.S. (2003). Horizontal transfer of the insect growth regulator pyriproxyfen to larval microcosms by gravid *Aedes albopictus* and *Ochlerotatus triseriatus* mosquitoes in the laboratory. Med. Vet. Entomol. *17*, 211–220.

Department of Statistics, M.o.T.I., Republic of Singapore. (2009). Time Series on Population (Mid-Year Estimates). (Available at http://www.singstat.gov.sg/stats/themes/people/hist/popn.html).

Department of Statistics, M.o.T.I., Republic of Singapore. (2010). Yearbook of Statistics Singapore. Available at http://wwwsingstatgovsg/pubn/reference/yos10/yos2010pdf.

Diallo, M., Ba, Y., Sall, A.A., Diop, O.M., Ndione, J.A., Mondo, M., Girault, L., and Mathiot, C. (2003). Amplification of the sylvatic cycle of dengue virus type *2*, Senegal, 1999–2000: entomologic findings and epidemiologic considerations. Emerg. Infect. Dis. *9*, 362–367.

Draffan, R.D.W., Garnett, S.T., and Malone, G.J. (1893). Birds of the Torres Straits: an annotated list and biogeographical analysis. Emu *83*, 207–234.

Duffy, M.R., Chen, T.H., Hancock, W.T., Powers, A.M., Kool, J.L., Lanciotti, R.S., Pretrick, M., Marfel, M., Holzbauer, S., Dubray, C., *et al.* (2009). Zika virus outbreak on Yap Island, Federated States of Micronesia. N. Engl. J. Med. *360*, 2536–2543.

Eliminate Dengue. Project e-newsletter. Available at http://www.eliminatedengue.com/Portals/58/enews/eNews1_2010.htm.

Ezenwa, V.O., Milheim, L.E., Coffey, M.F., Godsey, M.S., King, R.J., and Guptill, S.C. (2007). Land cover variation and West Nile virus prevalence: patterns, processes, and implications for disease control. Vector Borne Zoonotic Dis. *7*, 173–180.

Fang, R., Hsu, D.R., and Lim, T.W. (1980). Investigation of a suspected outbreak of Japanese encephalitis in Pulau Langkawi. Malays J. Pathol. *3*, 23–30.

Focks, D.A., Brenner, R.J., Hayes, J., and Daniels, E. (2000). Transmission thresholds for dengue in terms of *Aedes aegypti* pupae per person with discussion of their utility in source reduction efforts. Am. J. Trop. Med. Hyg. *62*, 11–18.

Franz, A.W., Sanchez-Vargas, I., Adelman, Z.N., Blair, C.D., Beaty, B.J., James, A.A., and Olson, K.E. (2006). Engineering RNA interference-based resistance to dengue virus type 2 in genetically modified *Aedes aegypti*. Proc. Natl. Acad. Sci. U.S.A. *103*, 4198–4203.

French Ministry of Health and Sport, F. (2010). ProMED mail. Available at http://www.promedmail.org/pls/otn/f?p=2400:1001:57555::NO::F2400_P1001_BACK_PAGE,F2400_P1001_PUB_MAIL_ID:*1010*, 84835.

Fu, G., Lees, R.S., Nimmo, D., Aw, D., Jin, L., Gray, P., Berendonk, T.U., White-Cooper, H., Scaife, S., Kim Phuc, H., *et al.* (2010). Female-specific flightless phenotype for mosquito control. Proc. Natl. Acad. Sci. U.S.A. *107*, 4550–4554.

Gama, R.A., Silva, E.M., Silva, I.M., Resende, M.C., and Eiras, A.E. (2007). Evaluation of the sticky MosquiTRAP for detecting *Aedes* (Stegomyia) aegypti (L.) (Diptera: Culicidae) during the dry season in Belo Horizonte, Minas Gerais, Brazil. Neotrop. Entomol. *36*, 294–302.

Gilbert, N. (2010). Mosquito spray affects bird reproduction. NatureNews. Available at http://www.nature.com/news/2010/100615/full/news.2010.296.html#B1.

Gould, D.J., Barnett, H.C., and Suyemoto, W. (1962). Transmission of Japanese encephalitis virus by *Culex gelidus* Theobald. Trans. R. Soc. Trop. Med. Hyg. *56*, 429–435.

Gubler, D.J. (2004). The changing epidemiology of yellow fever and dengue, 1900 to 2003: full circle? Comp. Immunol. Microbiol. Infect. Dis. *27*, 319–330.

Guzman, M.G., and Kouri, G. (2003). Dengue and dengue hemorrhagic fever in the Americas: lessons and challenges. J. Clin. Virol. *27*, 1–13.

Halstead, S.B. (2006). Dengue in the Americas and Southeast Asia: do they differ? Rev. Panam Salud Publica *20*, 407–415.

Hasegawa, M., Tuno, N., Yen, N.T., Nam, V.S., and Takagi, M. (2008). Influence of the distribution of host species on adult abundance of Japanese encephalitis vectors *Culex vishnui* subgroup and *Culex gelidus* in a rice-cultivating village in northern Vietnam. Am. J. Trop. Med. Hyg. *78*, 159–168.

Heinrich, J.C., and Scott, M.J. (2000). A repressible female-specific lethal genetic system for making transgenic insect strains suitable for a sterile-release program. Proc. Natl. Acad. Sci. U.S.A. *97*, 8229–8232.

Hendrichs, J., Robinson, A.S., Cayol, J.P., and Enkerlin, W. (2002). Medfly area wide sterile insect technique programmes for prevention, suppression or eradication: the importance of mating behaviour studies. Florida Entomologist *85*, 1–13.

Herrera-Basto, E., Prevots, D.R., Zarate, M.L., Silva, J.L., and Sepulveda-Amor, J. (1992). First reported outbreak of classical dengue fever at *1*, 700 meters above sea level in Guerrero State, Mexico, June (1988). Am. J. Trop. Med. Hyg. *46*, 649–653.

Hubalek, Z., and Halouzka, J. (1999). West Nile fever – a reemerging mosquito-borne viral disease in Europe. Emerg. Infect. Dis. *5*, 643–650.

Iikal, M.A., Mavale, M.S., Prasanna, Y., Jacob, P.G., Geevarghese, G., Banerjee (1997). Experimental studies on the vector potential of certain *Culex* species to West Nile virus. Ind. J. Med. Res. *106*, 225–228.

Itoh, T., Kawada, H., Abe, A., Eshita, Y., Rongsriyam, Y., and Igarashi, A. (1994). Utilization of bloodfed females of *Aedes aegypti* as a vehicle for the transfer of the insect growth regulator pyriproxyfen to larval habitats. J. Am. Mosq. Control Assoc. *10*, 344–347.

Jirakanjanakit, N., Rongnoparut, P., Saengtharatip, S., Chareonviriyaphap, T., Duchon, S., Bellec, C., and Yoksan, S. (2007). Insecticide susceptible/resistance status in *Aedes* (*Stegomyia*) *aegypti* and

Aedes (Stegomyia) albopictus (Diptera: Culicidae) in Thailand during 2003–2005. J. Econ. Entomol. *100*, 545–550.

Johnson, B.K., Ocheng, D., Gichogo, A., Okiro, M., Libondo, D., Kinyanjui, P., and Tukei, P.M. (1982). Epidemic dengue fever caused by dengue type 2 virus in Kenya: preliminary results of human virological and serological studies. East Afr. Med. J. *59*, 781–784.

Johnson, B.W., Chambers, T.V., Crabtree, M.B., Filippis, A.M., Vilarinhos, P.T., Resende, M.C., Macoris Mde, L., and Miller, B.R. (2002). Vector competence of Brazilian *Aedes aegypti* and *Ae. albopictus* for a Brazilian yellow fever virus isolate. Trans. R. Soc. Trop. Med. Hyg. *96*, 611–613.

Jumali, Sunarto, Gubler, D.J., Nalim, S., Eram, S., and Sulianti Saroso, J. (1979). Epidemic dengue hemorrhagic fever in rural Indonesia. III. Entomological studies. Am. J. Trop. Med. Hyg. *28*, 717–724.

Kamimura, K. (1998). Studies on population dynamics of the principal vector mosquito of Japanese encephalitis. Med. Entomol. Zool. *49*, 181–185.

Kawada, H., Higa, Y., Komagata, O., Kasai, S., Tomita, T., Thi Yen, N., Loan, L.L., Sanchez, R.A., and Takagi, M. (2009a). Widespread distribution of a newly found point mutation in voltage-gated sodium channel in pyrethroid-resistant *Aedes aegypti* populations in Vietnam. PLoS Negl. Trop. Dis. *3*, e527.

Kawada, H., Higa, Y., Nguyen, Y.T., Tran, S.H., Nguyen, H.T., and Takagi, M. (2009b). Nationwide investigation of the pyrethroid susceptibility of mosquito larvae collected from used tires in Vietnam. PLoS Negl. Trop. Dis. *3*, e391.

Kay, B., and Vu, S.N. (2005). New strategy against *Aedes aegypti* in Vietnam. Lancet *365*, 613–617.

Keiser, J., Maltese, M.F., Erlanger, T.E., Bos, R., Tanner, M., Singer, B.H., and Utzinger, J. (2005). Effect of irrigated rice agriculture on Japanese encephalitis, including challenges and opportunities for integrated vector management. Acta Trop. *95*, 40–57.

Koh, B.K., Ng, L.C., Kita, Y., Tang, C.S., Ang, L.W., Wong, K.Y., James, L., and Goh, K.T. (2008). The 2005 dengue epidemic in Singapore: epidemiology, prevention and control. Ann. Acad. Med. Singapore *37*, 538–545.

Komar, N. (2000). West Nile Viral Encephalitis. Rev. Sci. Tech. *19*, 166–176.

Kroeger, A., Lenhart, A., Ochoa, M., Villegas, E., Levy, M., Alexander, N., and McCall, P.J. (2006). Effective control of dengue vectors with curtains and water container covers treated with insecticide in Mexico and Venezuela: cluster randomised trials. BMJ *332*, 1247–1252.

Lacey, L.A., and Lacey, C.M. (1990). The medical importance of riceland mosquitoes and their control using alternatives to chemical insecticides. J. Am. Mosq Control Assoc. Suppl. *2*, 1–93.

Lai, Y.L., Chung, Y.K., Tan, H.C., Yap, H.F., Yap, G., Ooi, E.E., and Ng, L.C. (2007). Cost-effective real-time reverse transcriptase PCR (RT-PCR) to screen for Dengue virus followed by rapid single-tube multiplex RT-PCR for serotyping of the virus. J. Clin. Microbiol. *45*, 935–941.

Lanciotti, R.S., Kosoy, O.L., Laven, J.J., Velez, J.O., Lambert, A.J., Johnson, A.J., Stanfield, S.M., and Duffy, M.R. (2008). Genetic and serologic properties of Zika virus associated with an epidemic, Yap State, Micronesia, (2007). Emerg. Infect. Dis. *14*, 1232–1239.

Lee, K.S., Lai, Y.L., Lo, S., Barkham, T., Aw, P., Ooi, P.L., Tai, J.C., Hibberd, M., Johansson, P., Khoo, S.P., *et al.* (2010). Dengue virus surveillance for early warning, Singapore. Emerg. Infect. Dis. *16*, 847–849.

Lee, Y.C., Tang, C.S., Ang, L.W., Han, H.K., James, L., and Goh, K.T. (2009). Epidemiological characteristics of imported and locally acquired malaria in Singapore. Ann Acad Med Singapore *38*, 840–849.

Lee, Y.W., and Zairi, J. (2006). Field evaluation of *Bacillus thuringiensis* H-14 against Aedes mosquitoes. Trop Biomed *23*, 37–44.

Leo, Y.S., Chow, A.L., Tan, L.K., Lye, D.C., Lin, L., and Ng, L.C. (2009). Chikungunya outbreak, Singapore, (2008). Emerg. Infect. Dis. *15*, 836–837.

Leroy, E.M., Nkoghe, D., Ollomo, B., Nze-Nkogue, C., Becquart, P., Grard, G., Pourrut, X., Charrel, R., Moureau, G., Ndjoyi-Mbiguino, A., *et al.* (2009). Concurrent chikungunya and dengue virus infections during simultaneous outbreaks, Gabon, (2007). Emerg. Infect. Dis. *15*, 591–593.

Lima, J.B., Da-Cunha, M.P., Da Silva, R.C., Galardo, A.K., Soares Sda, S., Braga, I.A., Ramos, R.P., and Valle, D. (2003). Resistance of *Aedes aegypti* to organophosphates in several municipalities in the State of Rio de Janeiro and Espirito Santo, Brazil. Am. J. Trop. Med. Hyg. *68*, 329–333.

Liu, H., Cupp, E.W., Micher, K.M., Guo, A., and Liu, N. (2004). Insecticide resistance and cross resistance in Alabama and Florida strain of *Culex quinquefasciatus*. J. Med. Entomol. *41*, 408–413.

Lothrop, H.D., Lothrop, B.B., Gomsi, D.E., and Reisen, W.K. (2008). Intensive early season adulticide applications decrease arbovirus transmission throughout the Coachella Valley, Riverside County, California. Vector Borne Zoonotic Dis. *8*, 475–489.

Lourenco-de-Oliveira, R., Vazeille, M., Bispo de Filippis, A.M., and Failloux, A.B. (2002). Oral susceptibility to yellow fever virus of *Aedes aegypti* from Brazil. Mem Inst Oswaldo Cruz *97*, 437–439.

Lourenco de Oliveira, R., Vazeille, M., de Filippis, A.M., and Failloux, A.B. (2003). Large genetic differentiation and low variation in vector competence for dengue and yellow fever viruses of *Aedes albopictus* from Brazil, the United States, and the Cayman Islands. Am. J. Trop. Med. Hyg. *69*, 105–114.

Low, J.G., Ooi, E.E., Tolfvenstam, T., Leo, Y.S., Hibberd, M.L., Ng, L.C., Lai, Y.L., Yap, G.S., Li, C.S., Vasudevan, S.G., *et al.* (2006). Early Dengue infection and outcome study (EDEN) – study design and preliminary findings. Ann. Acad. Med. Singapore *35*, 783–789.

Lu, B., ed. (1987). Environmental management for the control of ricefield breeding mosquitoes in China (Philippines, International Rice Research Institute).

Malaysia, Application, B.C.H. for the Limited Release of Male *Aedes aegypti* Mosquito and to Conduct Mark-release-recapture Experiment Involving a Wild

Type Strain and a Living Modified Strain (OX513A) Available at http://wwwbiosafetynregovmy/ongoing_consult/aedesaegyptishtml.

McMeniman, C.J., Lane, R.V., Cass, B.N., Fong, A.W., Sidhu, M., Wang, Y.F., and O'Neill, S.L. (2009). Stable introduction of a life-shortening *Wolbachia* infection into the mosquito *Aedes aegypti*. Science *323*, 141–144.

Ministry of Health, S. (2008). Singapore's first chikungunya outbreak – surveillance and response. Epidemiological News Bulletin 35, 25–28. Available at http://www.moh.gov.sg/mohcorp/publicationsnewsbulletins.aspx?id=19542.

Monath, T.P. (1999). Facing up to re-emergence of urban yellow fever. Lancet *353*, (1541).

Moncayo, A.C., Fernandez, Z., Ortiz, D., Diallo, M., Sall, A., Hartman, S., Davis, C.T., Coffey, L., Mathiot, C.C., Tesh, R.B., *et al.* (2004). Dengue emergence and adaptation to peridomestic mosquitoes. Emerg. Infect. Dis. *10*, 1790–1796.

Moreira, L.A., Saig, E., Turley, A.P., Ribeiro, J.M., O'Neill, S.L., and McGraw, E.A. (2009). Human probing behavior of *Aedes aegypti* when infected with a life-shortening strain of *Wolbachia*. PLoS Negl. Trop. Dis. *3*, e568.

Mutebi, J.P., Gianella, A., Travassos da Rosa, A., Tesh, R.B., Barrett, A.D., and Higgs, S. (2004). Yellow fever virus infectivity for Bolivian *Aedes aegypti* mosquitoes. Emerg. Infect. Dis. *10*, 1657–1660.

Myles, K.M., Wiley, M.R., Morazzani, E.M., and Adelman, Z.N. (2008). Alphavirus-derived small RNAs modulate pathogenesis in disease vector mosquitoes. Proc. Natl. Acad. Sci. U.S.A. *105*, 19938–19943.

Nasidi, A., Monath, T.P., DeCock, K., Tomori, O., Cordellier, R., Olaleye, O.D., Harry, T.O., Adeniyi, J.A., Sorungbe, A.O., Ajose-Coker, A.O., *et al.* (1989). Urban yellow fever epidemic in western Nigeria, (1987). Trans. R. Soc. Trop. Med. Hyg. *83*, 401–406.

Nazni, W.A., Lee, H.L., and Azahari, A.H. (2005). Adult and larval insecticide susceptibility status of *Culex quinquefasciatus* (Say) mosquitoes in Kuala Lumpur Malaysia. Trop. Biomed. *22*, 63–68.

Ng, L.C., and Hapuarachchi, H.C. (2010). Tracing the path of Chikungunya virus – Evolution and adaptation. Infect. Genet. Evol. *10*, 876–885.

Ng, L.C., Tan, L.K., Tan, C.H., Tan, S.S.Y., Hapuarachchi, H.C., Pok, K.Y., Lai, Y.L., Lam-Phua, S.G., Bucht, G., Lin, R.T.P., *et al.* (2009). Entomologic and virologic investigation of Chikungunya, Singapore. Emerg. Infect. Dis. *15*, 1243 (1249).

Ninove, L., Parola, P., Baronti, C., De Lamballerie, X., Gautret, P., Doudier, B., and Charrel, R.N. (2009). Dengue virus type 3 infection in traveler returning from west Africa. Emerg. Infect. Dis. *15*, 1871–1872.

Paeporn, P., Ya-umphan, P., Supaphathom, K., Savanpanyalert, P., Wattanachai, P., and Patimaprakorn, R. (2004). Insecticide susceptibility and selection for resistance in a population of *Aedes aegypti* from Ratchaburi province, Thailand. Trop. Biomed. *21*, 1–6.

Pandey, B.D., Morita, K., Khanal, S.R., Takasaki, T., Miyazaki, I., Ogawa, T., Inoue, S., and Kurane, I. (2008). Dengue virus, Nepal. Emerg. Infect. Dis. *14*, 514–515.

Pandey, B.D., Rai, S.K., Morita, K., and Kurane, I. (2004). First case of eengue virus infection in Nepal. Nepal Med. Coll. J. *6*, 157–159.

Pang, S.C., Foo, S.Y., Png, A.B., Deng, L., Lam-Phua, S.G., Tang, C.S., and Ng, L.C. (2009). Evaluation of a 'fogging' canister for indoor elimination of adult *Aedes aegypti*. Dengue Bull. *33*, 203–208.

Paramasivan, R., Mishra, Mourya, A.C., D.T. (2003). West nile virus:the Indian scenario. Indian J. Med. Res. *118*, 101–108.

Paul, S.D., Rajagopalan, P.K., and Sreenivasan, M.A. (1970). Isolation of West Nile virus from the frugivorous bat, *Rousettus leschenaulti*. Ind. J. Med. Res. *58*, 1169–1171.

Perich, M.J., Davila, G., Turner, A., Garcia, A., and Nelson, M. (2000). Behavior of resting *Aedes aegypti* (Culicidae: Diptera) and its relation to ultra-low volume adulticide efficacy in Panama City, Panama. J. Med. Entomol. *37*, 541–546.

Pethuan, S., Jirakanjanakit, N., Saengtharatip, S., Chareonviriyaphap, T., Kaewpa, D., and Rongnoparut, P. (2007). Biochemical studies of insecticide resistance in *Aedes* (*Stegomyia*) *aegypti* and *Aedes* (*Stegomyia*) *albopictus* (Diptera: Culicidae) in Thailand. Trop. Biomed. *24*, 7–15.

Phuc, H.K., Andreasen, M.H., Burton, R.S., Vass, C., Epton, M.J., Pape, G., Fu, G., Condon, K.C., Scaife, S., Donnelly, C.A., *et al.* (2007). Late-acting dominant lethal genetic systems and mosquito control. BMC Biol. *5*, 11.

Ping, L.T., Yatiman, R., and Gek, L.P. (2001). Susceptibility of adult field strains of *Aedes aegypti* and *Aedes albopictus* in Singapore to pirimiphos-methyl and permethrin. J. Am. Mosq Control Assoc. *17*, 144–146.

Pok, K.Y., Lai, Y.L., Sng, J., and Ng, L.C. (2010). Evaluation of nonstructural 1 antigen assays for the diagnosis and surveillance of dengue in Singapore. Vector Borne Zoonotic Dis. *10*, 1009–1016.

Ponlawat, A., Scott, J.G., and Harrington, L.C. (2005). Insecticide susceptibility of *Aedes aegypti* and *Aedes albopictus* across Thailand. J. Med. Entomol. *42*, 821–825.

Poulin, B., Lefebvre, G., and Paz, L. (2010). Red flag for green spray: adverse trophic effects of *Bti* on breeding birds. J. Apl. Ecol. *47*, 884–889.

Rapley, L.P., Johnson, P.H., Williams, C.R., Silcock, R.M., Larkman, M., Long, S.A., Russell, R.C., and Ritchie, S.A. (2009). A lethal ovitrap-based mass trapping scheme for dengue control in Australia: II. Impact on populations of the mosquito *Aedes aegypti*. Med. Vet. Entomol. *23*, 303–316.

Rasgon, J.L., Styer, L.M., and Scott, T.W. (2003). *Wolbachia*-induced mortality as a mechanism to modulate pathogen transmission by vector arthropods. J. Med. Entomol. *40*, 125–132.

Reisen, W., and Brault, A.C. (2007). West Nile virus in North America: perspectives on epidemiology and intervention. Pest Manag. Sci. *63*, 641–646.

Reiter, P. (1978). Expanded polystyrene balls: an idea for mosquito control. Ann. Trop. Med. Parasitol. *72*, 595–596.

Reiter, P. (1985). A field trial of expanded polystyrene balls for the control of *Culex* mosquitoes breeding in pit latrines. J. Am. Mosq. Control Assoc. *1*, 519–521.

Reiter, P., and Gubler, D.J., eds. (1997). Surveillance and Control of Urban Dengue Vectors (Oxford, CAB International).

Rodrigues, f.M., Bright Singh, P., dandawate, C.N., Soman, R.S., Guntikar, S.N., and Kaul, H.N. (1980). Isolation of Japanese encephalitis and west Nile viruses from mosquitoes collected in Andhra Pradesh. Indian Journal Parasitology *4*, 149–153.

Rodriguez-Roche, R., Sanchez, L., Burgher, Y., Rosario, D., Alvarez, M., Kouri, G., Halstead, S.B., Gould, E.A., and Guzman, M.G. (2011). Virus Role During Intraepidemic Increase in Dengue Disease Severity. Vector Borne Zoonotic Dis.

Rodriguez, M.M., Bisset, J., de Fernandez, D.M., Lauzan, L., and Soca, A. (2001). Detection of insecticide resistance in *Aedes aegypti* (Diptera: Culicidae) from Cuba and Venezuela. J. Med. Entomol. *38*, 623–628.

Rodriguez, M.M., Bisset, J.A., and Fernandez, D. (2007). Levels of insecticide resistance and resistance mechanisms in *Aedes aegypti* from some Latin American countries. J. Am. Mosq. Control Assoc. *23*, 420–429.

Rudnick, A., and Lim, T.W. (1986). Dengue fever studies in Malaysia. Inst. Med. Res. Malaysia Bull. *23*, 51–152.

Sanchez-Vargas, I., Scott, J.C., Poole-Smith, B.K., Franz, A.W., Barbosa-Solomieu, V., Wilusz, J., Olson, K.E., and Blair, C.D. (2009). Dengue virus type 2 infections of *Aedes aegypti* are modulated by the mosquito's RNA interference pathway. PLoS Pathog. *5*, e1000299.

Scott, J.G., ed. (1990). Investigating Mechanisms of Insecticide Resistance: Methods, Strategies and Pitfalls (New York, Chapman & Hall).

Scott, T.W., Chow, E., Strickman, D., Kittayapong, P., Wirtz, R.A., Lorenz, L.H., and Edman, J.D. (1993a). Blood-feeding patterns of *Aedes aegypti* (Diptera: Culicidae) collected in a rural Thai village. J. Med. Entomol. *30*, 922–927.

Scott, T.W., Clark, G.G., Lorenz, L.H., Amerasinghe, P.H., Reiter, P., and Edman, J.D. (1993b). Detection of multiple blood feeding in *Aedes aegypti* (Diptera: Culicidae) during a single gonotrophic cycle using a histologic technique. J. Med. Entomol. *30*, 94–99.

Self, L.S.R., H.I., and Lofgren, C.S. (1973). Aerial applications of ultra-low-volume insecticides to control the vector of Japanese encephalitis in Korea. Bull. WHO *49*, 353–357.

Seng, C.M., Setha, T., Chanta, N., Socheat, D., Guillet, P., and Nathan, M.B. (2006). Inhibition of adult emergence of *Aedes aegypti* in simulated domestic water-storage containers by using a controlled-release formulation of pyriproxyfen. J. Am. Mosq. Control Assoc. *22*, 152–154.

Seng, C.M., Setha, T., Nealon, J., Chantha, N., Socheat, D., and Nathan, M.B. (2008a). The effect of long-lasting insecticidal water container covers on field populations of *Aedes aegypti* (L.) mosquitoes in Cambodia. J. Vector Ecol. *33*, 333–341.

Seng, C.M., Setha, T., Nealon, J., Socheat, D., Chantha, N., and Nathan, M.B. (2008b). Community-based use of the larvivorous fish *Poecilia reticulata* to control the dengue vector *Aedes aegypti* in domestic water storage containers in rural Cambodia. J. Vector Ecol. *33*, 139–144.

Seng, C.M., Setha, T., Nealon, J., Socheat, D., and Nathan, M.B. (2008c). Six months of *Aedes aegypti* control with a novel controlled-release formulation of pyriproxyfen in domestic water storage containers in Cambodia. Southeast Asian J. Trop. Med. Publ. Hlth *39*, 822–826.

Setha, T., Chantha, N., and Socheat, D. (2007). Efficacy of *Bacillus thuringiensis israelensis*, VectoBac WG and DT, formulations against dengue mosquito vectors in cement potable water jars in Cambodia. Southeast Asian J. Trop. Med. Publ. Hlth *38*, 261–268.

Sharma, S.K., and Hamzakoya, K.K. (2001). Geographical Spread of *Anopheles stephensi*, Vector of Urban Malaria, and *Aedes aegypti*, Vector of Dengue/DHF, in the Arabian Sea Islands of Lakshadweep, India. Dengue Bull. *25*, 88–91.

Sihuincha, M., Zamora-Perea, E., Orellana-Rios, W., Stancil, J.D., Lopez-Sifuentes, V., Vidal-Ore, C., and Devine, G.J. (2005). Potential use of pyriproxyfen for control of *Aedes aegypti* (Diptera: Culicidae) in Iquitos, Peru.J Med. Entomol. *42*, 620–630.

Sivagnaname, N., Amalraj, D.D., and Mariappan, T. (2005). Utility of expanded polystyrene (EPS) beads in the control of vector-borne diseases. Ind. J. Med. Res. *122*, 291–296.

Soderlund, D.M. (2008). Pyrethroids, knockdown resistance and sodium channels. Pest Manag. Sci. *64*, 610–616.

Solomon, T. (2006). Control of Japanese encephalitis – within our grasp? N. Engl. J. Med. *355*, 869–871.

Soman, R.S., Rodrigues, F.M., Guttikar, S.N., and Guru, P.Y. (1977). Experimental viraemia and transmission of Japanese encephalitis virus by mosquitoes in ardeid birds. Indian Journal of Medical Research *66*, 709–718.

Sunish, I.P., Rajendran, R., and Reuben, R. (2003). The role of urea in the oviposition behaviour of Japanese encephalitis vectors in rice fields of South India. Mem Inst Oswaldo Cruz *98*, 789–791.

Tabachnick, W.J. (1991). The evolutionary relationships among arboviruses and the evolutionary relationships of their vectors provides a method for understanding vector–host interactions. J Med Entomol *28*, 297–298.

Tabachnick, W.J., Wallis, G.P., Aitken, T.H., Miller, B.R., Amato, G.D., Lorenz, L., Powell, J.R., and Beaty, B.J. (1985). Oral infection of *Aedes aegypti* with yellow fever virus: geographic variation and genetic considerations. Am. J. Trop. Med. Hyg. *34*, 1219–1224.

Tan, C.H., Wong, P.S.J., Li, M.Z.I., Tan, S.Y.S., Lee, T.K.C., Pang, S.C., Lam-Phua, S.G., Nasir, M., Png, A.B., Koou, S.Y., *et al.* (2011). Entomological investigation and control of a chikungunya cluster in Singapore. Vector Borne Zoonotic Dis. *11*, 383–390.

Thomas, D.D., Donnelly, C.A., Wood, R.J., and Alphey, L.S. (2000). Insect population control using a dominant, repressible, lethal genetic system. Science *287*, 2474–2476.

Turley, A.P., Moreira, L.A., O'Neill, S.L., and McGraw, E.A. (2009). *Wolbachia* infection reduces blood-feeding success in the dengue fever mosquito, *Aedes aegypti*. PLoS Negl. Trop. Dis. *3*, e516.

Van der Stuyft, P., Gianella, A., Pirard, M., Cespedes, J., Lora, J., Peredo, C., Pelegrino, J.L., Vorndam, V., and Boelaert, M. (1999). Urbanisation of yellow fever in Santa Cruz, Bolivia. Lancet *353*, 1558–1562.

Vasilakis, N., Tesh, R.B., and Weaver, S.C. (2008). Sylvatic dengue virus type 2 activity in humans, Nigeria, (1966). Emerg. Infect. Dis. *14*, 502–504.

Vaughn, D.W., and Hoke, C.H., Jr. (1992). The epidemiology of Japanese encephalitis: prospects for prevention. Epidemiol Rev. *14*, 197–221.

Vezzani, D., and Carbajo, A.E. (2008). *Aedes aegypti, Aedes albopictus*, and dengue in Argentina: current knowledge and future directions. Mem Inst Oswaldo Cruz *103*, 66–74.

Vezzani, D., Velazquez, S.M., and Schweigmann, N. (2004). Seasonal pattern of abundance of *Aedes aegypti* (Diptera: Culicidae) in Buenos Aires City, Argentina. Mem Inst Oswaldo Cruz *99*, 351–356.

Victor, T.J. a.R., R. (2000). Effects of organic and inorganic fertilisers on mosquito populations in rice fields of southern India. Med Vet Entomol. *14*, 361–368.

Vu, S.N., Nguyen, T.Y., Tran, V.P., Truong, U.N., Le, Q.M., Le, V.L., Le, T.N., Bektas, A., Briscombe, A., Aaskov, J.G., *et al.* (2005). Elimination of dengue by community programs using *Mesocyclops* (Copepoda) against *Aedes aegypti* in central Vietnam. Am. J. Trop. Med. Hyg. *72*, 67–73.

Vythilingam, I., Jeffery, J., and Mahadevan, S. (2001). Japanese encephalitis is Sepang: studies on juvenile mosquito population. Trop.Biomed. *18*, 89–95.

Vythilingam, I., Luz, B.M., Hanni, R., Beng, T.S., and Huat, T.C. (2005). Laboratory and field evaluation of the insect growth regulator pyriproxyfen (Sumilarv 0.5G) against dengue vectors. J. Am. Mosq Control Assoc. *21*, 296–300.

Vythilingam, I., Mahadevan, S., Tan, S.K., Abdullah, G., and Jeffery, J. (1996). Host feeding pattern and resting habits of Japanese encephalitis vectors found in Sepang, Selangor, Malaysia. Trop. Biomed. *13*, 45–50.

Vythilingam, I., Oda, K., Chew, T.K., Mahadevan, S., Vijayamalar, B., Morita, K., Tsuchie, H., and Igarashi, A. (1995). Isolation of Japanese encephalitis virus from mosquitoes collected in Sabak Bernam, Selangor, Malaysia in (1992). J. Am. Mosq. Control Assoc. *11*, 94–98.

Vythilingam, I., Oda, K., Mahadevan, S., Abdullah, G., Thim, C.S., Hong, C.C., Vijayamalar, B., Sinniah, M., and Igarashi, A. (1997). Abundance, parity, and Japanese encephalitis virus infection of mosquitoes (Diptera:Culicidae) in Sepang District, Malaysia. J. Med. Entomol. *34*, 257–262.

Vythilingam, I., and Panart, P. (1991). A field trial on the comparative effectiveness of malathion and Resigen by ULV application on *Aedes aegypti*. Southeast Asian J. Trop. Med. Publ. Hlth *22*, 102–107.

Wada, Y. (1989). Control of Japanese Encephalitis vectors. Southeast Asian J. Trop. Med. Publ. Hlth *20*, 623–626.

Wan-Norafikah, O., Nazni, W.A., Lee, H.L., Zainol-Ariffin, P., and Sofian-Azirun, M. (2010). Permethrin resistance in *Aedes aegypti* (Linnaeus) collected from Kuala Lumpur, Malaysia. J. Asia-Pacific Entomol. *13*, 175–182.

Wan-Norafikah, O., Nazni, W.A., Lee, H.L., Chen, C.D., Wan-norjuliana, W.M., Azahari, A.H., and Sofian, M.A. (2008). detrection of permethrin resistance in *Aedes albopictus* Skuse, collected from Titiwangsa zone, Kuala Lumpur, Malaysia. Proc. ASEAN Congr. Trop. Med. Parasitol. *3*, 69–77.

Washburn, J.O., Egerter, D.E., Anderson, J.R., and Saunders, G.A. (1988). Density reduction in larval mosquito (Diptera: Culicidae) populations by interactions between a parasitic ciliate (Ciliophora: Tetrahymenidae) and an opportunistic fungal (Oomycetes: Pythiaceae) parasite. J. Med. Entomol. *25*, 307–314.

Weaver, S.C., and Barrett, A.D. (2004). Transmission cycles, host range, evolution and emergence of arboviral disease. Nat. Rev. Microbiol. *2*, 789–801.

WHO Trend of dengue case and CFR in SEAR Countries. (2007).

Xi, Z., Khoo, C.C., and Dobson, S.L. (2005). *Wolbachia* establishment and invasion in an *Aedes aegypti* laboratory population. Science *310*, 326–328.

Yaicharoen, R., Kiatfuengfoo, R., Chareonviriyaphap, T., and Rongnoparut, P. (2005). Characterization of deltamethrin resistance in field populations of *Aedes aegypti* in Thailand. J. Vector Ecol. *30*, 144–150.

Index

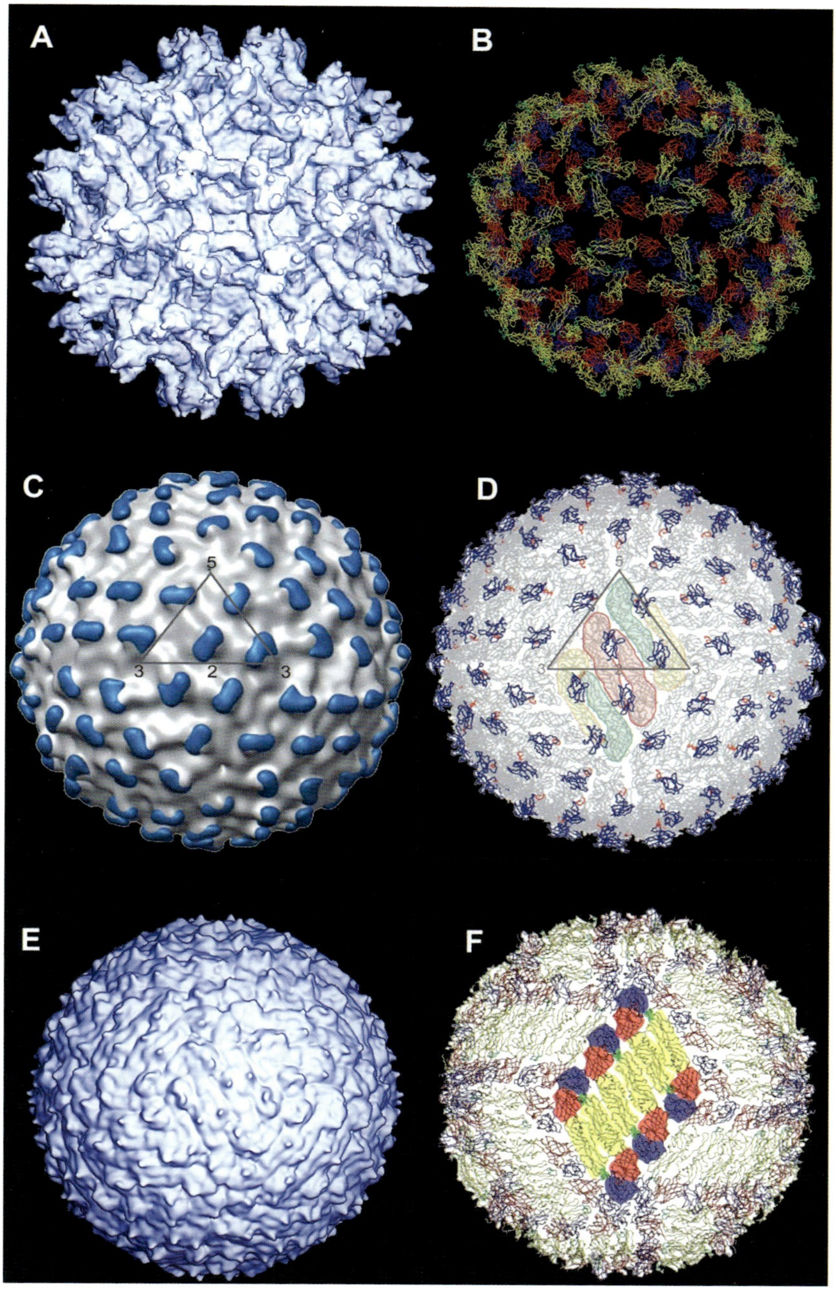

Plate 2.1 Cryo-electron microscopy (EM) reconstructions of dengue virus. Surface shaded views (A, C, E) or fitted E protein (B, D, F) of immature virus at neutral pH (A, B), immature virus at pH 6 (C, D), or mature virus at neutral pH (E, F). In B and F, the C-α backbone of the E protein is presented in standard colours with domain I in red, domain II in yellow, and domain III in blue. In D, the E protein is drawn in grey and the C-α backbone of the pr protein is shown in blue. The density corresponding to pr is also shown in blue in C, along with the symmetry axis, which is the same in all panels. The particles are not drawn to scale.

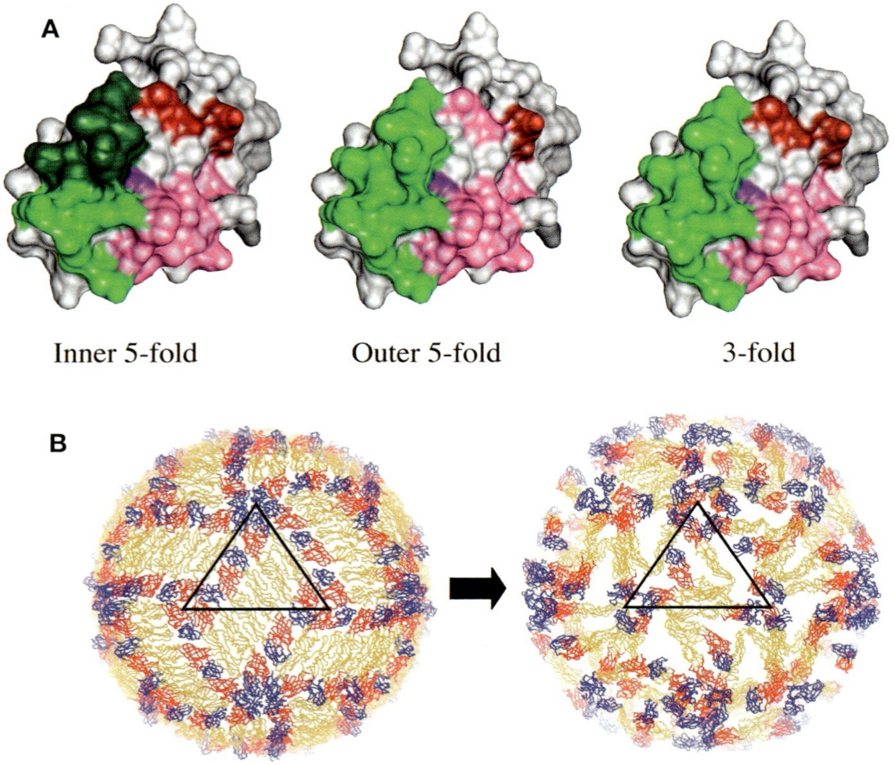

Plate 2.2 Neutralizing monoclonal binding to flaviviruses. (A) Surface shaded view of E protein domain III and the buried surface area of the epitopes recognized by Fab E16 and Fab1A1D-2 on each of the three unique E-DIIIs in the icosahedral asymmetric unit of the mature virus. Epitopes of Fab E16 and 1A1D-2 (green and pink, respectively) are partially buried under neighbouring E subunits in the mature flavivirus. Buried regions are coloured dark green and dark red. Note that 1A1D-2 has buried surface in each of the E-DIIIs of the virion, whereas E16 is only partially buried at the true 5-fold (inner 5-fold). (B) The radical change in E protein position captured by monoclonal 1A1D-2. The panel shows the Cα chains of E proteins from uncomplexed mature DENV (left), and the Fab complex structure (right). The black triangle represents the asymmetric unit of the virus. The E domains are coloured red for domain I, yellow for domain II, and blue for domain III.

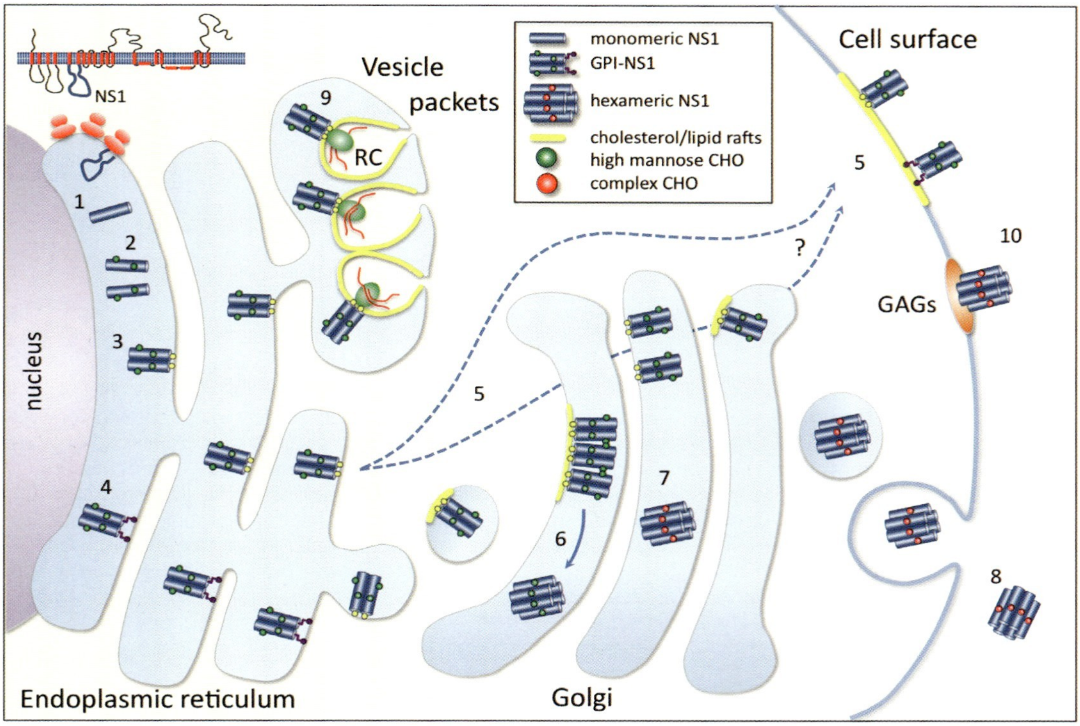

Plate 4.1 Schematic summary of NS1 trafficking in mammalian cells. NS1 is initially expressed in association with the endoplasmic reticulum (ER) as part of a long polyprotein that is encoded by a single open reading frame of the viral RNA genome. A signal sequence at the C-terminus of the virion glycoprotein E targets NS1 to the lumen of the ER and co-translational cleavage within the ER lumen at both its N- and C-terminus generates a hydrophilic monomeric subunit (1). This monomer is modified by the addition of high mannose carbohydrate moieties at multiple sites (2) and rapidly forms a dimeric species leading to the acquisition of a hydrophobic character (mNS1), resulting in membrane association (3). A subset of dengue virus NS1 acquires a glycosyl-phosphatidylinositol (GPI) anchor in the ER as a consequence of the recognition of a GPI addition signal at the N-terminus of NS2A (4). Both mNS1 and GPI-anchored NS1 are trafficked to the cell surface via an unknown pathway where they have been shown to associate with lipid rafts (yellow membrane highlights) (5). A proportion of NS1 traffics from the ER through the Golgi where dimeric units associate to form soluble hexamers (6), although it is also possible that these hexamers are formed earlier in the ER. Passage of this soluble hexameric species (sNS1) through the Golgi results in one of the high mannose carbohydrate moieties on each monomer being trimmed and processed to a complex carbohydrate form (7) and is then secreted from the cell (8). An alternative pathway for mNS1 from the ER sees a subset of the high mannose form becoming associated with vesicle packets (VP) where it co-localises with other non-structural viral proteins that comprise the viral replication complex (RC) (9). Some of the cell-surface-associated NS1 may be previously secreted NS1 that has bound directly to cell surface glycosaminoglycans (GAGs) (10) The oligomeric nature of the various cell-bound forms of NS1 remains largely unknown.

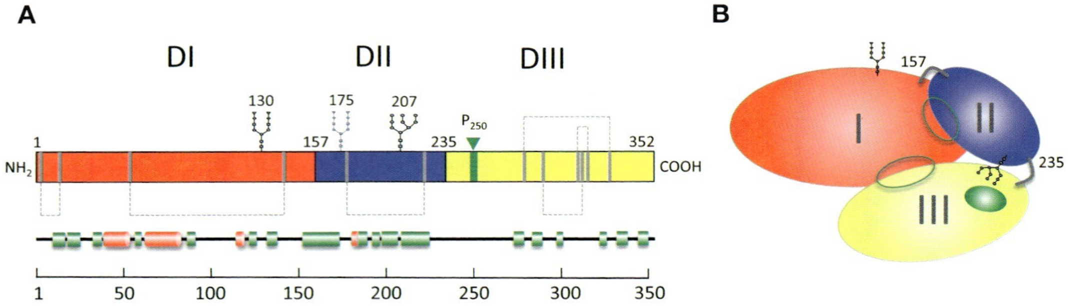

Plate 4.2 Antigenic and secondary structure models of NS1. (A) Linear representation of NS1 highlighting the three structural fragments, domain I (red), domain II (blue) and domain III (yellow) identified for WNV NS1. Disulfide linkages for all 12 conserved cysteines are shown in grey and the two conserved N-linked glycosylation sites are shown in black (complex CHO at Asn130 and high mannose CHO at Asn207). A third carbohydrate addition site at position 175 and found in the NS1 species of the JE serocomplex of flaviviruses is shown in grey. Destabilization of dimers by a naturally occurring mutation of residue 250 (green arrowhead) in WNV and MVE suggests that the dimerization domain is located in this region. A consensus secondary structure prediction based on an alignment of more than 40 flavivirus sequences and showing putative β sheets (green cylinders) and α helices (red cylinders) is shown below. (B) Proposed tertiary arrangement of the three structural domains of NS1 based on overlapping epitope reactivity by monoclonal antibodies (schematically represented by the green ovals).

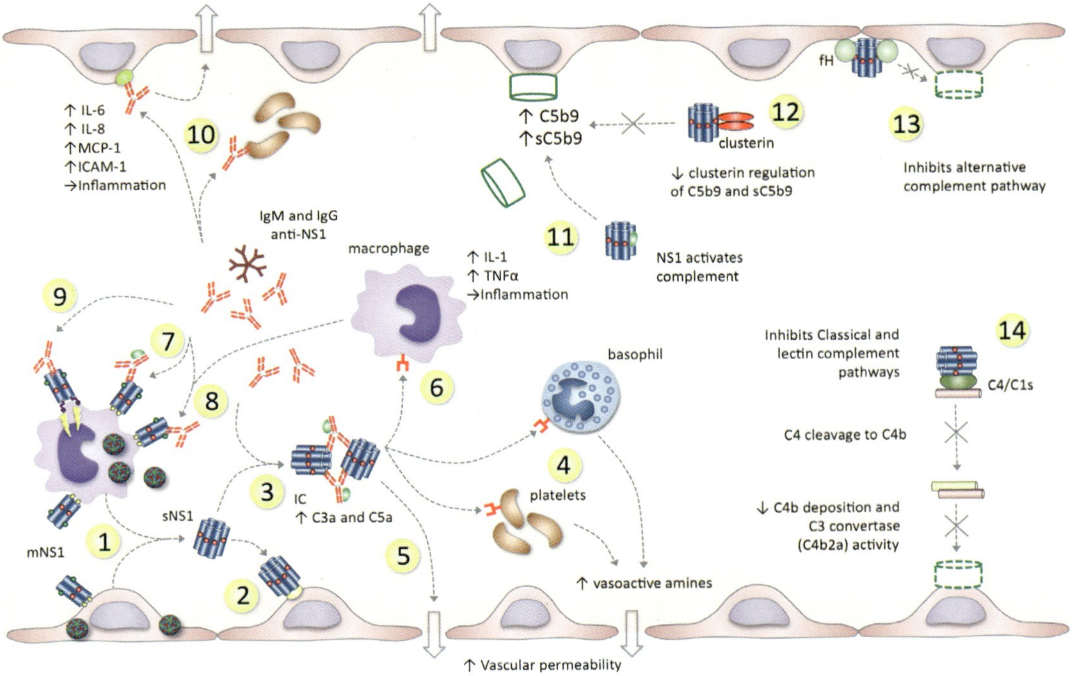

Plate 4.3 Schematic of flavivirus NS1 engagement with host cell components. The production of sNS1 and mNS1 from flavivirus infected cells and their interaction with selected host cell components is depicted. NS1 is expressed on the surface of, and secreted from, flavivirus infected cells (1). Circulating sNS1 can subsequently bind to the surface of both infected and uninfected cells via charged interactions with GAGs, heparin sulphate and chondroitin sulphate E (2). The consequences of that binding are yet to be fully determined. In dengue virus secondary infections, an anamnestic IgM and IgG antibody response to NS1 during the acute phase of disease can lead to the formation of immune complexes (3) that are capable of triggering a range of inflammatory processes including the activation of complement (green) to generate the anaphylatoxins C3a and C5a. ICs also act on basophils and platelets via Fc receptor engagement to release vasoactive amines (4) that can cause endothelial cell retraction and increase vascular permeability. This in turn may lead to IC deposition (5) inducing platelet aggregation and further complement activation. Binding of ICs to macrophages leads to their activation and the release of cytokines, further increasing the inflammatory response (6). NS1 specific antibody binding to cell surface-exposed NS1 targets infected cells for complement mediated lysis (7) and/or complement-independent phagocytosis (8). NS1-specific antibody binding of cell surface-exposed GPI anchored NS1 may also mediate activation of infected cells (9). A role for auto-immune, cross-reactive anti-NS1 antibodies in pathogenesis has also been proposed. These auto-antibodies have been shown to bind to host determinants on the surface of both platelets and endothelial cells resulting in the release of inflammatory cytokines and nitric oxide leading to inflammation and/or apoptosis. In the absence of antibody, circulating sNS1 has been shown to modulate/inhibit complement pathways through its interaction with various complement components (11–14). sNS1 has been shown to activate complement directly resulting in increased formation of the membrane attack complex C5b9 and sC5b9 (11) while its interaction with the complement inhibitory factor, clusterin is thought to result in the increased formation of sC5b9 in the serum of infected patients (12). sNS1 has also been shown to display an immune evasion function through its binding to the alternative complement pathway regulatory protein, fH (13) and its interaction with the classical complement pathway components C4 and C1s (14) with the resulting decrease in the deposition of C4b and C3 convertase leading to a decrease in the terminal membrane attack complex. The diverse array of ways in which NS1 and NS1-specific antibodies engage with the host as depicted in this schematic is by no means exhaustive.

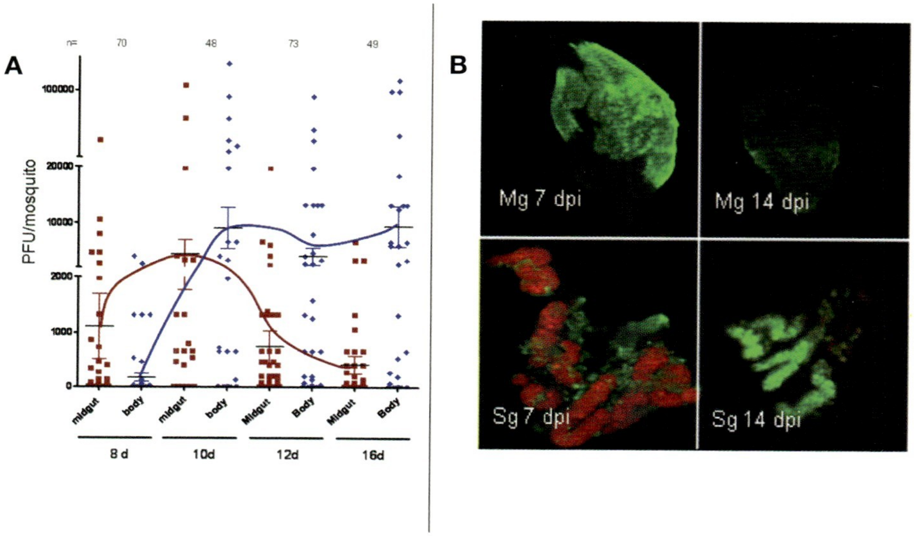

Plate 14.3 DENV2 infection of *Aedes aegypti* HWE mosquitoes. (A) DENV2 titres in midguts (red symbols/line) and remaining tissues (carcass; blue symbols/line) of individual mosquitoes at predetermined dpi. (B) Detection of DENV2 E antigen by indirect IFA in midgut (top panels) and salivary gland (bottom panels) tissues at 7 and 14 dpi. Data and IFA images from Sánchez-Vargas *et al.* (2009).

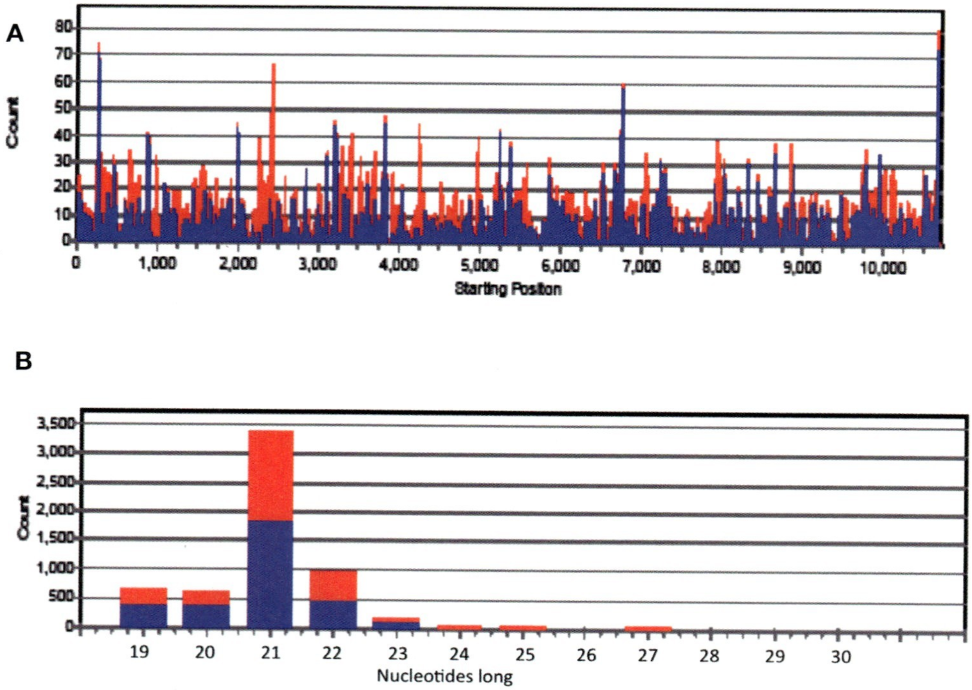

Plate 14.6 Characterization of DENV-specific small RNA (viRNA) sequences from *Aedes aegypti* infected with DENV at 9 days post infection. (A) viRNA numbers, polarity and distribution across the DENV2 genome from *A. aegypti* mosquito library. (B) viRNA size distribution in infected *A. aegypti* mosquitoes. Red bars, negative-sense viRNAs; blue bars, positive-sense viRNAs. Deep sequencing performed using SOLiD System Next Generation platform (ABI) and viRNAs analysed by NextGENe software (Softgenetics, LLC, State College, PA, USA). Modified from Scott *et al.* (2010).